EVOLUTION
and the
DIVERSITY of LIFE

EVOLUTION
and the
DIVERSITY of LIFE

SELECTED ESSAYS

ERNST MAYR

The Belknap Press of Harvard University Press
Cambridge, Massachusetts
and
London, England

Library of Congress Cataloging-in-Publication Data

Mayr, Ernst, 1904–
Evolution and the diversity of life.
Bibliography: p.
Includes index.
1. Evolution—Addresses, essays, lectures.
2. Species—Addresses, essays, lectures.   I. Title.
QH366.2.M393   575'.08   75-42131.
ISBN 0-674-27104-1 (cloth)
ISBN 0-674-27105-X (pbk.)

To
Wendy, Christoph, Niki,
Michael, and Becky

CONTENTS

IV. PHILOSOPHY OF BIOLOGY

V. THEORY OF SYSTEMATICS

VI. THE SPECIES

EVOLUTION
and the
DIVERSITY of LIFE

General Introduction

When I recently looked over my publications, covering more than 40 years of research, I realized that all of them, directly or indirectly, deal with evolution. This is true even though some focus on birds, others on biogeography, speciation, the theory of systematics, the history and philosophy of biology, population biology, and behavior. If we define *evolution* as changes in the diversity and adaptation of populations of living organisms—and this is the least vulnerable definition of organic evolution known to me—then indeed the process of evolution is the one subject that binds all my work together.

The theme of evolution is not at the center of interest for all of biology. As I pointed out in 1961, there are two biologies, functional and evolutionary biology. Functional biologists deal with proximate causes and ask "how" questions; they study (mostly physiological) processes, favor the reductionist approach, and have no particular interest in the history or meaning of the genetic programs of the organisms investigated. Conclusions are, in the main, reached by means of experimentation. That part of biology has never been my special field of interest.

As a lifelong naturalist, I have instead been interested in the well-nigh inconceivable diversity of the living world, its origin and its meaning. To study the life of the tropics, as had Humboldt, Wallace, and Darwin, as well as my teachers Stresemann and Rensch, was the greatest ambition of my youth. I fulfilled it when I lived as a naturalist and collector for two years and a half (1928–1930) in the interior of New Guinea and the Solomon Islands; this experience had an impact on my thinking that cannot be exaggerated.

At the American Museum of Natural History in New York, whose staff I joined in January 1931, evolution was everybody's interest, but its interpretation was highly diverse and the source of great controversies. I was unbelievably fortunate in my experiences over the ensuing 40 years: my association with the zoologists, particularly Dobzhansky, at Columbia University; summer research at Cold Spring Harbor (1943–1952); and my friendship with evolutionary biologists all over the world. Appointment (in 1953) as Alexander Agassiz Professor in Harvard's Museum of Comparative Zoology gave me the opportunity to have my own students and to develop an interest in the history and philosophy of biology.

Curiously, from an interest in all sorts of specialized topics, again and again I came back to the fundamentals laid down in Darwin's *Origin of Species*. Indeed, when I first read the *Origin* I was rather amazed at how close Darwin's interests were to my own. A physiologist might wonder why, so long after Darwin, all mysteries of evolution have not already been solved. A legitimate question, but one not easily answered. In part, the delay is due to the nature of the Darwinian explanation, which was, and in many ways still is, alien to the Western tradition. Shaped by Plato's essentialism, the Western mind has tended to think in terms of unchanging types and essences and to regard variation as fleeting and unimportant. The thinking of the dominant schools of philosophy has throughout been incompatible with an adequate consideration of the importance of variation.

To an equal extent the delay in the acceptance of Darwinism is due to the enormous complexity of the evolutionary process, which seems to defy explanation. Every evolutionary phenomenon must be painstakingly analyzed and reanalyzed, and the findings synthesized with new discoveries in other areas of evolutionary biology.

What has been most needed is a clarification of evolutionary concepts and processes. Species, speciation, isolating mechanisms, the population, mutation, mutation pressure, the target of selection, evolutionary trends, evolutionary inertia, and many other aspects of evolution had to be clarified, new terms had to be coined and old ones redefined. So complex a phenomenon as evolution can be brought closer to understanding only by such repeated factual and conceptual analyses. It is to this objective that I have dedicated my scientific career.

The essays collected here represent those of my findings that I think are of more than purely ephemeral or specialized interest. They are grouped into eight broad areas: evolution, speciation, history of biology, philosophy of biology, theory of systematics, the species, man, and biogeography. For each of these sections I have written an explanatory introduction. When I have included several essays on the same subject, as on the species concept, I have done so in order to illuminate the maturation in my thinking. Many essays, particularly those on biogeography, have been drastically shortened, in order to eliminate specialized detail, repetition, or peripheral material. Others, particularly recent ones, have been reproduced in their entirety. In many places I have updated references to the literature, and in a few places I have added in brackets new interpretations or facts that have come to light since original publication. Unless otherwise indicated, all translations are my own.

Not every essay in this collection will appeal equally to every reader. The ornithologist and the zoogeographer will be more interested in certain essays, the historian and philosopher in others. Some essays are addressed to specialists, others to a more general audience. Essays 1 and 5 were first presented to continental European audiences and were addressed particularly to people who are still somewhat unfamiliar with population thinking and thus uncomfortable with the concept of an all-powerful natural selection. In spite of their superficial heterogeneity, these essays nevertheless share a pervasive basic theme: They all deal with the diversity of living nature, the product of the evolutionary process.

I
EVOLUTION

Introduction:
Darwin Vindicated

When H. J. Muller exclaimed in 1959, "One hundred years without Darwin is enough!" he reflected the disappointment of the evolutionist, frustrated by the seeming inability of nonbiologists to accept Darwinian evolutionism. In spite of the worldwide Darwin jubilees in that year and an unprecedented level of evolutionary research in the past 25 years, the situation has not greatly improved since Muller's outcry. Perhaps it would have been too much to expect fundamentalists to abandon their dogmas and accept the scientific evidence, but what is rather discouraging is that in considerable areas of the European continent there is still very active support for a number of non-Darwinian interpretations that are not supported by any evidence. The situation is made worse when physical scientists and mathematicians attempt to "prove" by calculations and computer simulations that the Darwinian model will not work (Moorhead 1967). Many philosophers, perhaps the majority, including such a distinguished person as Karl Popper, are still holding out for an alternative to Darwinism (Popper 1972). Well-read, well-educated lay people like the jurist Macbeth (1971) show in their writings how little they understand the Darwinian theory.

All the evolutionist can do in the face of such resistance is to attempt again and again to explain the nature of the Darwinian argument, analyzing it down to its basic components and citing the evidence supporting the argument. This task has been performed extremely well in the past, as for instance by G. G. Simpson in a series of books (1949, 1964), by Dobzhansky (1970), Julian Huxley, and others. I myself have devoted a series of essays to this task, based on lectures to various audiences in different countries. They deal either with the basic concepts of evolutionary biology or with such more specific problems as directional evolution, rates of evolution, kinds of selection, and the origin of evolutionary novelties. These are brought together in this first section of the collection. The problem of the multiplication of species (speciation), being one of my major fields of research, is dealt with in a separate section following this first group of nine essays.

An attempt was made during the editing to eliminate repetition ex-

cept *where I considered it important for didactic reasons to state the
same principles or facts repeatedly in different contexts. Considering
how frequently certain links in the chain of the evolutionary argument
are misunderstood, it seemed advisable to reiterate some of the major
arguments concerning the meaning of natural selection, the invalidity of
essentialist thinking, and the unity of the genotype.*

REFERENCES

Dobzhansky, T. 1970. *Genetics of the evolutionary process.* Columbia University
    Press, New York.
Macbeth, N. 1971. *Darwin retried.* Delta Books, New York.
Moorhead, P. S., and M. M. Kaplan, eds. 1967. *Mathematical challenges to the
    neo-Darwinian interpretation of evolution.* Wistar Inst. Symp. Mongr., no. 5.
    Wistar Institute Press, Philadelphia.
Muller, H. J. 1959. One hundred years without Darwinism are enough. *School Sci.
    Math.,* 59:304–316.
Popper, K. 1972. *Objective knowledge: an evolutionary approach.* Clarendon Press,
    Oxford.
Simpson, G. G. 1949. *The meaning of evolution.* Yale University Press, New Haven.
—— 1964. *This view of life.* Harcourt, Brace and World, New York.

1
Basic Concepts of
Evolutionary Biology

That evolution is a fact and that the astonishing diversity of animals and plants evolved gradually was accepted quite universally soon after 1859. But how this evolution proceeded, particularly the nature of its moving force, has been a source of controversy from the very beginning.

Among specialists, almost complete agreement has been reached in recent decades. Whether they are botanists or zoologists, paleontologists or geneticists, all of them interpret the results of the evolutionary process in the same manner and find the same causal connections. With the nonspecialists the situation is different; whether biologists or not, they often remain unconvinced. Again and again some colleague has told me: "The story you present sounds quite logical and irrefutable, but I still can't get rid of the feeling that something isn't quite right." When I insist on being told what it is that is not right, it turns out that the doubter either has an altogether insufficient knowledge of basic facts or suffers from certain conceptual misunderstandings. In order to preclude this possibility, I shall begin by (1) stating the essential aspects of the modern interpretation of the causality of evolution in a few simple sentences; and then (2) attempt to explain the conceptual, indeed the philosophical, foundations of the evolutionary theory as it is now generally accepted.

THE MODERN THEORY OF EVOLUTION

How does the modern biologist see the process of evolution? Most of the earlier theories of evolution based their explanation on a single factor, such as mutation, environment, or isolation; it was Darwin's genius to have proposed a two-factor explanation. The first factor, genetic variability, is entirely a matter of chance, whether it is produced by mutation, recombination, or by whatever other mechanism. Pre-

Translated and abridged from "Grundgedanken der Evolutionsbiologie," *Naturwissenschaften,* 56, no. 8 (1969):14–25.

cisely the opposite is true of the second factor, natural selection, which is decidedly an "anti-chance" factor. Among the millions of individuals that are produced in every generation, selection always favors certain ones, whose advantageous attributes are due to specific genetic combinations. It must be emphasized once more that the most important component of Darwin's achievement was pointing out the duality of the evolutionary process. It is precisely this combination of chance and anti-chance that gives evolution both its great flexibility and its goal-directedness (Mayr 1963).

Darwin's explanation of evolution, in a sense, is dualistic. Dualism, however, is a word that has a bad reputation among biologists because for centuries they have suffered under the dualism of body and mind that Descartes brought back into philosophy. Biologists reject this dualism quite emphatically for reasons Bernhard Rensch has perceptively presented (Rensch 1968). Admittedly, there are quite a number of dualisms in biology, but they usually are not "either-or" dualisms but rather "first-then" dualisms, which we perhaps could designate as tandem dualisms. Mutation–selection is such a tandem dualism. Another dualism equally important for the understanding of evolution is the dualism of genotype–phenotype. We must fully understand this particular dualism before we can hope to understand the process of evolution.

The genotype is the totality of the genetic endowment that an individual received from his parents at conception (formation of the original zygote). The phenotype is the totality of the characteristics (the appearance) of an individual resulting from the interaction of the genotype (genetic program) with the environment during ontogeny.

But why is this duality of genotype and phenotype so important for the evolutionist? Embryologists have always silently assumed that the fertilized egg cell, including its nucleus, participates completely and directly in the development of the embryo. The only one who, for theoretical reasons, did not agree with this assumption was August Weismann, who was ahead of his time by several decades in this as in so many other ways. Weismann's solution, a separation of soma and germ line, was not validated. But his basic idea was nevertheless right. The ultimate solution of Weismann's problem was provided only recently by molecular genetics: the genetic material (DNA) does not participate itself in the development of the embryo but functions only as a blueprint. The instructions of the DNA are translated (with the help of RNA) into polypeptides and proteins, and it is only these which participate directly in the development of the embryo. The genetic material itself, the DNA, remains unchanged during this entire process.

I will not discuss further the highly interesting consequences for embryonic development that result from this functional separation of the DNA and the proteins. Instead I want to point out the importance of this separation for certain evolutionary problems. First of all, it is now perfectly evident why a direct influence of the environment on the genetic material is impossible, an influence postulated by the majority of the Lamarckians. The way from the DNA (via the RNA) to the proteins is a one-way street. The environment can influence the developmental process but it cannot affect the blueprint that controls it. Changes in the proteins cannot be translated back into nucleic acids.

The second consequence is perhaps equally important. The complete separation of the DNA genotype from the protein phenotype has the result that much of the potential of the genotype of a given individual is not translated into the phenotype at all and thus is not exposed to selection. This is shown by the great number of recessive genes in diploid organisms and by the suppressor genes in epistatic systems. Such potentials can be mobilized in later generations through recombination. This method of reacting to fluctuations in the environment is clearly superior to a process of direct environmental induction as postulated by many Lamarckians. For instance, suppose we submit a cold-adapted experimental stock to increasingly higher temperatures over five generations and then suddenly expose it to cold temperatures. If this stock had been uniformly induced to great heat tolerance during the preceding five generations, it would surely be exterminated by the sudden cold shock. On the other hand, if the phenotype is not the direct product of the environment, then recombination can develop a whole population of new reaction norms in every generation, some of which—with great probability—will be preadapted to highly aberrant environmental conditions. The rapidity with which insects have developed DDT-resistant populations is evidence for the type of preadaptation that is stored in the blueprint of the DNA. In the ensuing discussion I shall return to the enormous importance of this difference in the roles of the genotype and the phenotype.

PHILOSOPHICAL CONSIDERATIONS

Darwinism has a well-defined philosophical basis, an understanding of which is a prerequisite for the understanding of the evolutionary process. It has long been a puzzle for the historian of biology why the key to the solution of the problem of evolution was found in England rather than on the European continent. No other country in the world had such a shining galaxy of famous biologists in the middle of the last century as the Germany of Rudolphi, Ehrenberg, Karl E. von Baer, Schleiden, Leuckart, Siebold, Koelliker, Johannes Müller, Virchow, and Leydig, and yet the solution to the problem of evolution was found by two English amateurs, Darwin and Wallace, neither of whom had had thorough zoological training. How can one explain this? My answer is that philosophical thinking on the continent was dominated at the time by essentialism. This philosophy, as was shown by Reiser (1958), is quite incompatible with the assumption of gradual evolution. Essentialism had its roots in Plato's concept of the *eidos*. We all know Plato's famous allegory, according to which we see reality only indirectly like shadows on a cave wall, while the real nature of things, the *essence* of the scholastics, can be inferred only indirectly. Owing to the central importance of *essence* for this school of philosophy, it has been designated essentialism by Karl Popper (1950). By contrast, a very different kind of thinking, strongly supported by empiricism, had developed in England: the so-called population thinking, for which gradual evolution poses no difficulties. Population thinking is based on assumptions opposite to those of essentialism. It claims that only individual phenomena have reality and that every endeavor to infer from them an essence is a process of abstraction. Population thinking thus turns the dogma of essentialism upside down. The replacement of typological (essentialistic) thinking by population thinking was perhaps the most important conceptual revolution in the history of biology (See Essay 3 for further discussion of typological and population thinking).

It is possible to make the philosophical problems even clearer by discussing some of the objections that are frequently raised against the Darwinian interpretation. I have heard, for instance, the statement "I cannot believe that so perfect an organ as the eye could have been produced by accidental mutations." The evolutionist entirely agrees with this. Accidental mutations alone, quite obviously, could not have done this. All a mutation does is to enrich the genetic variability of the gene pool. Mutation has nothing to do with adaptation. Selection is what achieves that. It is therefore also a misrepresentation of the situation to say that this or that product of evolution was the result of mutation pressure. There is no such thing as mutation pressure. Unfortunately, there are still some authors who have not yet abandoned the entirely misleading ideas on mutation advanced by de Vries and other mutationists of the first decade of this century. These early geneticists saw in mutation a process that could produce in a single step a new type, be it a new species or even a higher taxon. These mutationists' essentialism forced them to assume that all evolutionary change had to proceed by saltations. We now know that this idea conflicts with the factual evidence. Each error in the replication of DNA is a mutation. Since every higher animal has enough DNA in its genome for about 5 million genes, and since each gene has on the average a thousand mutable base pairs, it is quite possible that each individual differs from every other individual by at least one new mutation of the 5 billion base pairs. If a mutation does not lead to the immediate death of the cell in which it arises, it may change the cell's physiology so slightly that extremely sensitive methods are required to determine that a mutation has occurred. The essential characteristics of the process of mutation are therefore: (1) mutation enriches the gene pool with genetic variation, (2) mutation is a relatively frequent process, and (3) most mutations in higher organisms are cryptic. Each of these three conclusions conflicts with the beliefs of the early mutationists.

The misunderstandings concerning the nature of selection were equally large. There is, for instance, the ever-repeated claim that Darwin's explanation of evolution is a tautology. The Darwinian argument is misrepresented in the following manner: First question: "Who survives?"; answer: "He who is fittest." Second question: "Who is fittest?"; answer: "He who survives." This formulation totally misrepresents Darwin's position. Darwin says: "Any variation . . . if it be in any degree profitable to an individual of any species, in its infinitely complex relations to other organic beings and to external nature, will tend to the preservation of that individual." What Darwin says, and I agree, is that it is the possession of certain characteristics which determines evolutionary success and that such characteristics have, at least in part, a genetic basis. An individual that has these genetic properties will survive and reproduce with a much greater probability than another that lacks them. It is obvious that this correct formulation is not at all tautological.

Another matter is often misunderstood and thus requires clarification. Selection does not deal with single genes because its target is the phenotype of the entire individual. To assume that a given gene has a fixed selective value is an error because the contribution a gene makes to the fitness of an individual depends to a considerable extent on the composition of the genotype, that is, of the interaction of this gene with other genes. At best, one can calculate a statistical mean value for the fitness of a gene. Population geneticists were rather surprised when they found that certain genes with above-average fitness in some combinations became lethal in other combinations.

One should never ignore the difference between genotype and phenotype. In particular, one must always remember that a given component of the phenotype is rarely or never directly determined by a single gene. The influence of a gene is usually rather indirect. Neglect of this consideration has led to such assertions as "Natural selection cannot explain why all mammals have seven cervical vertebrae." A characteristic like the number of cervical vertebrae is the product of a large part of the genotype, and the greater the phylogenetic age of a character, the more extensively it is built into the blueprint of the DNA, and the more resistant it becomes to change. What is decisive is the selective value of the total genotype.

Molecular genetics has demonstrated how complicated the realization of a character is, even in such relatively simple organisms as bacteria. Each gene has its repressors and inducers and a complicated regulatory system. A great deal of imagination is not required to understand how many billions or trillions of mutual interactions of genes are possible in multicellular organisms and how few of these are actually realized. It would be easy to cite from the literature hundreds of objections to selection, all of which rest on the erroneous assumption that genotype and phenotype are one and the same phenomenon and that each component of the phenotype is directly and immediately controlled by a specific gene. The typical objection is well represented in this statement by a well-known biologist: "The appearance of teeth in the embryos of animals which, like the baleen whales, are toothless as adults, or of the splint bones in the foot of the horse, cannot in the slightest be explained in terms of natural selection." To be sure, we still lack an understanding of the details of the process of differentiation responsible for the long retention of these features in embryonic development, but no difficulty in principle exists. We now know that each gene participates, or can participate, in many developmental processes, and that the manifold interactions among genes impart, so to speak, a holistic stamp to the totality of embryonic development. Provided the development of a young baleen whale as a whole is harmonious, selection will erode the genetic basis for the tooth anlage only very gradually. To eliminate it too rapidly could possibly lead to drastic disturbances in the developmental process and would not be tolerated by selection.

Behavior plays an important role in evolution. It does not, of course, have a direct influence on the genotype, as was believed by Lamarck, but it sets up new selection pressures and may facilitate the invasion of new adaptive zones. (This is discussed in Essay 47.)

UNSOLVED PROBLEMS

In conclusion, I would like to raise the question of the future of evolutionary biology. If the current theoretical framework of this field, as it was developed in the 1930s and 1940s, is truly able to counter all objections successfully, does this not mean that evolutionary biology has become a dead science? If evolutionary biology had the theoretical framework of physics, the answer to this question might indeed be yes. The goal of the physicist is to establish general laws and to reduce all phenomena to a minimal number of such laws. General laws, however, play a much smaller role in biology. Just about everything in biology is unique: every animal and plant community, fauna or flora, species or individual. The strategy of research in biology must, for this reason, be quite different from the strategy of the physicist. In this respect, biology may be nearer to such sciences as archeology and linguistics. As interesting and important as the general laws of linguistics are, they by no means diminish the importance of studying individual languages. However important the laws derived from a comparison of cultures, they by no means constitute a reason for abandoning the study of individual cultures. Indeed, a study of specific phenomena is an indispensable prerequisite for all comparative studies. It is the comparison of different languages, different cultures, different faunas, or different groups of animals and plants that leads to the most interesting generalizations.

In this respect we are still very much at the beginning in evolutionary biology. We are still utterly unable to answer questions such as these: Why have certain blue-green algae (*Cyanophycea*) hardly changed in 1½ billion years? Why have the faunas of certain geological periods been exposed to great catastrophes and why have these catastrophes occurred sometimes simultaneously in the ocean and on land and at other times in one or the other? Or: what were the special conditions that permitted only a single phyletic line to become man among probably more than a billion phyletic lines that existed on earth? Such questions indicate some of the problems that currently preoccupy the evolutionary biologist.

Molecular biology has posed some highly interesting new questions. I shall mention merely the riddle posed for us by the amount of DNA in each nucleus. We have been wondering for a good many years why there is evidence for only 5 to 50 thousand genes among higher animals, even though there is enough DNA for 5 million genes. More recently, it has been found that DNA can be rather heterogeneous, but that only some of it, perhaps not more than 25 percent, is clearly repetitive (Britten and Kohne 1968). As yet, we have no idea what the physiological and evolutionary significance of this heterogeneity is. I suspect that we are in for some major surprises.

The structure of the chromosomes in the eucaryotes likewise poses many puzzles. Why is the number of chromosomes relatively constant in many groups of animals and variable in others? Why is the number of chromosomes high in certain groups of animals and plants and low in others? Why are nearly all chromosomes approximately the same size in some groups of animals whereas in others there is an enormous range of sizes? Even though we are convinced that all these phenomena are controlled by natural selection, we must nevertheless admit that we still lack clues to the reasons for all these difficulties.

The problem of the rate of evolution is another package of unknowns. I have mentioned already the evolutionary stability of certain blue-green algae that have not changed visibly in more than a billion years, as compared with the enormous rapidity of evolution in freshwater fishes, which seem capable of generating new species in less than 5 thousand years. Is this great difference entirely a matter of recombination and selection, or are there actually differences in mutability? And if the latter, what chemistry regulates rates of mutation?

Unsolved puzzles exist at every level of integration. The mutation process itself is not yet entirely explained. The phylogenetic rate of change in macromolecules is a highly controversial subject and uncertainties exist at every succeeding level of organization, up to animal and plant communities. It is now possible even to take a new look at the problems of phylogeny. The new findings of chemical taxonomy provide clues to relationship that were hitherto unavailable. Concurrently, functional morphology has provided insights that were not accessible to the typologist in his *Bauplan* studies. It is evident that this is an area in which we can expect all sorts of new findings in the course of coming years. Curiously, none of this has weakened the basic Darwinian thesis. To the contrary, the new findings again and again have fully confirmed Darwin's brilliant vision.

Every biological phenomenon, every structure, every function, indeed everything in biology, has a history. And the study of this history, the reconstruction of the selection pressures that have been responsible for the biological world of today, is as much a part of the causal explanation of the world of organisms as physiological or embryological explanations. A purely physiological-ontogenetic explanation that omits the historical side is only half an explanation. It is not sufficient to know how the DNA program is translated into the phenotype if one wants to acquire a correct view of the phenomena of life: one must also attempt to explain the historical origin of the DNA program. Only evolutionary biology can make that contribution; thus the study of evolution remains an important branch of biology.

REFERENCES

Mayr, E. 1963. *Animal species and evolution.* Belknap Press of Harvard University Press, Cambridge, Mass.

Popper, K. R. 1950. The open society and its enemies. In *The Spell of Plato,* vol. 1. Routledge and Kegan Paul, London.

Reiser, O. L. 1958. In *A book that shook the world,* ed. R. Buchsbaum, p. 68. University of Pittsburgh Press, Pittsburgh.

Rensch, B. 1968. *Biophilosophie auf erkenntnistheoretischer Grundlage.* G. Fischer, Stuttgart.

2
The Evolution of
Living Systems

The number, kind, and diversity of living systems is overwhelmingly great, and each system, in its particular way, is unique. So different indeed are the kinds of organisms that it would be futile to try to understand evolution as a whole by describing the evolution of viruses and fungi, whales and sequoias, or elephants and humming birds. Perhaps we can arrive at valid generalizations by approaching the task in a rather unorthodox way. Living systems evolve in order to meet the challenge of the environment. We can ask, therefore, what *are* the particular demands that organisms have to meet?

The first challenge is to cope with a continuously changing and immensely diversified environment, the resources of which, however, are not inexhaustible. Mutation, the production of genetic variation, is the recognized means of coping with the diversity of the environment in space and time. Let us go back to the beginning of life. A primeval organism in need of a particular complex molecule in the primordial "soup" in which it lived gained a special advantage by mutating in such a way that, after having exhausted this resource in its environment, it was able to synthesize the needed molecule from simpler molecules that were abundantly available. Simple organisms such as bacteria or viruses, with a new generation every 10 or 20 minutes and with enormous populations consisting of millions and billions of individuals, may well be able to adjust to the diversity and to the changes of the environment by mutation alone. Indeed, a capacity for mutation is perhaps the most important evolutionary characteristic of the simplest organisms. Furthermore, their system of phenotypic adaptation is remarkably flexible, permitting a rapid adjustment to changes of the environment.

More complex organisms, those with much longer generation times, much smaller population size, and particularly with a delicately balanced coadapted genotype, would find it hazardous to rely so greatly on mutation to cope with changes in the environment. The chances that

Reprinted from "The evolution of living systems," *Proceedings of the National Academy of Sciences,* 51, no. 5 (1964):934–941.

the appropriate mutation would occur at the right time so that mutation alone could supply appropriate genetic variability for sudden
changes in the environment of such organisms are virtually nil. What,
then, is the prerequisite for the development of more complex living
systems? It is the ability of different organisms to exchange "genetic
information" with each other, the process the geneticist calls recombination, more popularly known as sex. The selective advantage of sex is
so direct and so great that we can assume it arose at a very early stage in
the history of life. Let us illustrate this advantage by a single example.
A primitive organism able to synthesize amino acid A, but dependent
on the primordial soup for amino acid B, and another organism able to
synthesize amino acid B, but dependent on the primordial soup for
amino acid A, would be able by genetic recombination to produce
offspring with the ability to synthesize both amino acids and thus the
ability to live in an environment deficient in both of them. Genetic
recombination can speed up evolutionary change enormously and assist
in emancipation from the environment.

Numerous mechanisms evolved in due time to make recombination
increasingly precise in every respect. The result was the evolution of
elaborately constructed chromosomes; of diploidy through two
homologous chromosome sets, one derived from the father, the other
from the mother; of an elaborate process of meiosis during which
homologous chromosomes exchange pieces so that the chromosomes of
father and mother are transmitted to the grandchildren not intact, but
as newly reconstituted chromosomes with a novel assortment of genes.
These mechanisms regulate genetic recombination among individuals,
by far the major source of genotypic variability in higher organisms.

The amount of genetic diversity within a single interbreeding population is regulated by a balance between mechanisms that favor inbreeding and those that favor outbreeding. The extremes in this respect are
much greater among plants and lower animals than among higher animals. Extreme inbreeding (self-fertilization) and extreme outbreeding
(regular hybridization with other species) are rare in higher animals.
Outbreeders and inbreeders are drastically different living systems in
which numerous adaptations are correlated in a harmonious manner.

The result of sexuality is that ever-new combinations of genes can be
tested by the environment in every generation. The enormous power of
the process of genetic recombination by sexual reproduction becomes
evident if we remember that in sexually reproducing species no two
individuals are genetically identical. We must admit, sex is wonderful!

However, even sex has its drawbacks. To make this clear, let me set
up for you the model of a universe consisting entirely of genetically
different individuals that are *not* organized into species. Any individual
may engage in genetic recombination with any other individual in this
model. Occasionally, as a result of chance, new gene complexes will be
built up that have unique adaptive advantages. Yet, because in this

particular evolutionary system there is no guarantee that such an exceptional individual will engage in genetic recombination *only* with individuals having a similarly adaptive genotype, it is inevitable that this exceptionally favorable genotype will eventually be destroyed by recombination during reproduction.

How can such a calamity be avoided? There are two possible means, and nature has adopted both. One method is to abandon sexual reproduction. Indeed we find all through the animal kingdom, and even more often among plants, a tendency to give up sexuality temporarily or permanently in order to give a successful genotype the opportunity to replicate itself unchanged, generation after generation, taking advantage of its unique superiority. The history of the organic world makes it clear, however, that such an evolutionary opportunist reaches the end of its rope sooner or later. Any sudden change of the environment will convert its genetic advantage into a handicap and, not having the ability to generate new genetic variability through recombination, it will inevitably become extinct.

The other solution is the "invention," if I may be pardoned for using this anthropomorphic term, of the biological species. The species is a protective system guaranteeing that only such individuals interbreed and exchange genes as have largely the same genotypes. In this system there is no danger that breakdown of genotypes will result from genetic recombination, because all the genes present in the gene pool of a species have been previously tested, through many generations, for their ability to recombine harmoniously. This does not preclude considerable variability within a species. Indeed, all our studies make us realize increasingly how vast is the genetic variability within even comparatively uniform species. Nevertheless, the basic developmental and homeostatic systems are the same, in principle, in all members of a species.

By simply explaining the biological meaning of species, I have deliberately avoided the tedious question of how to define a species. Let me add that the species can fulfill its function of protecting well-integrated, harmonious genotypes only by having some mechanisms (called "isolating mechanisms") by which interbreeding with individuals of other species is prevented.

In our design of a perfect living system, we have now arrived at a system that can cope with the diversity of its environment and that has the means to protect its coadapted, harmonious genotype. As described, this well-balanced system seems so conservative as to offer no opportunity for the origin of additional new systems. This conclusion, if true, would bring us into a real conflict with the evolutionary history of the world. The paleontologists tell us that new species have originated continuously during geological time and that the multiplication of species, in order to compensate for the extinction of species, must occur at a prodigious rate. If the species is as well balanced, well protected, and as delicate as we have described it, how can one species be divided into two? This serious problem puzzled Darwin greatly, and evolutionists have argued about it for more than a hundred years.

Eventually it was shown that there are two possible solutions, or perhaps I should say two normally occurring solutions. The first mode occurs very frequently in plants, but is rare in the animal kingdom. It consists in the doubling of the chromosome set so that the new individual is no longer a diploid with two sets of homologous chromosomes, but, let us say, a tetraploid with four sets of chromosomes or, if the process continues, a higher polyploid with an even higher chromosome number. The production of a polyploid constitutes instantaneous speciation; it produces an incompatibility between the parental and the daughter species in a single step.

The other mode of speciation is simplicity itself. Up to now, we have spoken of the species as something rigid, uniform, and monolithic. Actually, natural species, particularly those that are widespread, consist like the human species of numerous local populations and races, all of them differing more or less from each other in their genetic composition. Some of these populations, particularly those at the periphery of the species range, are completely isolated from each other and from the main body of the species. Let us assume that one of these populations is prevented for a long time from exchanging genes with the rest of the species, because the isolating barrier—be it a mountain range, a desert, or a waterway—is impassable. Through the normal processes of mutation, recombination, and selection, the gene pool of the isolated population becomes more and more different from that of the rest of the species, finally reaching a level of distinctness that normally characterizes a different species. This process, called "geographic speciation," is by far the most widespread mode of speciation in the animal kingdom and quite likely the major pathway of speciation also in plants.

Before such an incipient species can qualify as a genuine new species, it must have acquired two properties during its genetic rebuilding. First, it must have acquired isolating mechanisms that prevent it from interbreeding with the parental species when the two come again into contact. Second, it must also have changed sufficiently in its demands on the environment, in its niche utilization (as the ecologist would say), so that it can live side by side with mother and sister species without succumbing to competition.

KINDS OF LIVING SYSTEMS

In our discussion of the evolution of living systems, I have concentrated, up to now, on major unit processes or phenomena, such as the role of mutation, of genetic recombination and sex, of the biological species, and of the process of speciation. These processes give us the mechanisms that make diversification of the living world possible, but they do not explain why there should be such an enormous variety of life on earth. There are surely more than 3 million species of animals and plants living on this earth, perhaps more than 5 million. What principle permits the coexistence of such a wealth of different kinds? This question troubled Darwin, and he found an answer for it that has stood the test of time. Two species, in order to coexist, must differ in their utilization of the resources of the environment in a way that reduces competition. During speciation there is a strong selective premium on becoming different from preexisting species by trying out new ecological niches. This experimentation in new adaptations and new specializations is the principal evolutionary significance of the process of speciation. Once in a long while one of these new species finds the door to a whole new adaptive kingdom. Such a species, for instance, was the original ancestor of the most successful of all groups of organisms, the insects, now numbering more than a million species. The birds, the bony fishes, the flowering plants, and all other kinds of animals and plants all originated ultimately from a single ancestral species. Once a species discovers an empty adaptive zone, it can speciate and radiate until this zone is filled by its descendants.

To avoid competition, organisms can diverge in numerous ways, for instance in size. Even though there is a general trend toward large size in evolution, some species and genera, often in the same lines as large species and genera, have evolved toward decreased size. Small size is by no means always a primitive trait.

Specialization for a very narrow niche is perhaps the most common evolutionary trend. This is the characteristic approach of the parasites. Literally thousands of parasites are restricted to a single host, indeed restricted to a small part of the body of the host. There are, for instance, three species of mites that live on different parts of the honeybee. Such extreme specialization is rare if not absent in the higher plants, but is characteristic in insects and explains their prodigious rate of speciation. The deep sea, lightless caves, and the interstices between sand grains along the seashore are habitats leading to specialization.

The counterpart of the specialist is the generalist. Individuals of such species have a broad tolerance to all sorts of variations of climate, habitat, and food. It seems difficult to become a successful generalist, but the very few species that can be thus classified are widespread and abundant. Man is the generalist par excellence, with his ability to live in all latitudes and altitudes, in deserts and in forest, and to subsist on the pure meat diet of the Eskimos or on an almost pure vegetable diet. There are indications that generalists have unusually diversified gene pools and, as a result, produce rather high numbers of inferior genotypes by genetic recombination. Widespread and successful species of *Drosophila* seem to have more lethals than rare or restricted species. It is not certain that this observation can be applied to man, but this much is certain, that populations of man display much genetic variation. In man we do not have the sharply contrasting types ("morphs") that occur in many polymorphic populations of animals and plants. Instead we find rather complete intergradation of mental, artistic, manual, and physical capacities (and their absence). Yet, whether continuous or discontinuous, genetic variation has long been recognized as a useful device by which a species can broaden its tolerance and enlarge its niche. That the same is true for man is frequently forgotten. Our educators, for instance, have for far too long tended to ignore man's genetic diversity and have tried to force identical educational schedules on highly diverse talents. Only within recent years have we begun to realize that equal opportunity calls for differences in education. Genetically different individuals do not have equal opportunities unless the environment is diversified.

Every increase in the diversity of the environment during the history of the world has resulted in a veritable burst of speciation. This is particularly easily demonstrated for changes in the biotic environment. The rise of the vertebrates was followed by a spectacular development of trematodes, cestodes, and other vertebrate parasites. The insects, whose history goes back to the Paleozoic nearly 400 million years ago, did not really become a great success until the flowering plants (angiosperms) evolved some 150 million years ago. These plants provided such an abundance of new adaptive zones and niches that the insects entered a truly explosive stage in their evolution. By now three quarters of the known species of animals are insects, and their total number (including undiscovered species) is estimated to be as high as 2 or 3 million.

PARENTAL CARE

Let me discuss just one additional aspect of the diversity of living systems, care of the offspring. At one extreme we have the oysters, which do nothing whatsoever for their offspring. They cast literally millions of eggs and male gametes into the sea, providing the opportunity for the eggs to be fertilized. Some of the fertilized eggs will settle in a favorable place and produce new oysters. The statistical probability that this will happen is small, owing to the adversity of the environment, and although a single full-grown oyster may produce more than 100 million eggs per breeding season, it will have on the average only two descendants. That numerous species of marine organisms practice this type of reproduction, many of them enormously abundant and many of them with an evolutionary history going back several hundred million years, indicates that this shotgun method of thrusting offspring into the world is surprisingly successful.

How different is reproduction in species with parental care! This always requires a drastic reduction in the number of offspring, and it usually means greatly enlarged yolk-rich eggs, it means the development of brood pouches, nests, or even internal placentae, and it often means the formation of a pair bond to secure the participation of the male in the raising of the young. The ultimate development along this line of specialization is unquestionably man, with his enormous prolongation of childhood.

Behavioral characteristics are an important component of parental care, and our treatment of the evolution of living systems would be incomplete if we were to omit reference to behavior and to the central nervous system. The germ plasm of a fertilized egg contains in its DNA a coded genetic program that guides the development of the young organism and its reactions to the environment. However, there are drastic differences among species concerning the precision of the inherited information and the extent to which the individual can benefit from experience. The young in some species appear to be born with a genetic program containing an almost complete set of ready-made, predictable responses to the stimuli of the environment. We say of such an organism that its behavior is unlearned, innate, instinctive, that its behavior *program* is *closed*. The other extreme is provided by organisms that have a great capacity to benefit from experience, to learn how to react to the environment, to continue adding "information" to their behavior program, which consequently is an *open program*.

Let us look a little more closely at open and closed programs and their evolutionary potential. We are all familiar with the famous story of how Konrad Lorenz imprinted goslings on himself. Young geese or ducklings just hatched from the egg will adopt as parent any moving object (but preferably one producing appropriate sounds). If hatched in an incubator, they will follow their human caretaker and not only consider him their parent but consider themselves as belonging to the human species. For instance, upon reaching sexual maturity they may tend to display to and court a human individual rather than another goose or duck. The reason for this seemingly absurd behavior is that the hatching bird does not have an inborn knowledge of the Gestalt of its parent; all it has is a readiness to *fill in* this Gestalt. Its genetically coded program is open; it provides for a readiness to adopt as parent the first moving object seen after hatching. In nature, of course, this is invariably the parent.

Let us contrast this open program with the completely closed one of another bird, the parasitic cowbird. The mother cowbird, like the European cuckoo, lays her eggs in the nests of various kinds of songbirds, such as yellow warblers, vireos, or song sparrows, and then abandons them completely. The young cowbird is raised by its foster parents, and yet, as soon as it is fledged, it seeks other young cowbirds and gathers into large flocks with them. For the rest of its life, it associates with members of its own species. The Gestalt of its own species is firmly imbedded in the genetic program with which the cowbird is endowed from the very beginning. It is—at least in respect to species recognition—a completely closed program. In other respects, much of the behavioral program of the cowbird is open, that is, ready to incorporate experiences by learning. Indeed, there is probably no species of animals, not even among the protozoans, that does not, at least to some extent, derive benefit from learning processes. On the whole, and certainly among the higher vertebrates, there has been a tendency to replace rigidly closed programs by open ones or, as the student of animal behavior would say, to replace rigidly instinctive behavior by learned behavior. This change is not a change in an isolated character. It is part of a whole chain reaction of biological changes. Since man is the culmination of this particular evolutionary trend, we naturally have a special interest in it. Capacity for learning can best be utilized if the young are associated with someone from whom to learn, most conveniently their parents. Consequently, there is strong selection pressure in favor of extending the period of childhood. And since parents can take care of only a limited number of young, there is selection in favor of reducing the number of offspring. We have here the paradoxical situation that parents with a smaller number of young may nevertheless have a greater number of grandchildren, because the mortality among well-cared-for and well-prepared young may be reduced even more drastically than the birth rate. (For further discussion, see Essay 47.)

The sequence of events I have just outlined describes one of the dominating evolutionary trends in the primates, a trend that reaches its extreme in man. A broad capacity for learning is an indispensable prerequisite for the development of culture, of ethics, of religion. But the oyster proves that there are avenues to biological success other than parental care and the ability to learn.

One final point: how can we explain the harmony of living systems? Attributes of an organism are not independent variables but interdependent components of a single system. Large brain size, the ability to learn, long childhood, and many other attributes of man all belong together; they are parts of a single harmoniously functioning system. And so it is with all animals and plants. The modern population geneticist stresses the same point. The genes of a gene pool have been brought together for harmonious cooperation; they are coadapted. This harmony and perfection of nature (to which the Greeks referred in the word *kosmos*) has impressed philosophers from the very beginning. Yet there seems to be an unresolved conflict between this harmony of nature and the apparent randomness of evolutionary processes, beginning with mutation and including also much of reproduction and mortality. Opponents of the Darwinian theory of evolution have claimed that the conflict between the harmony of nature and the apparent haphazardness of evolutionary processes can *not* be resolved.

The evolutionist, however, points out that this objection is valid only if evolution is a one-step process. In reality, every evolutionary change involves two steps. The first is the production of new genetic diversity through mutation, recombination, and related processes. On this level randomness is indeed predominant. The second step, however—selection of those individuals that are to make up the breeding population of the next generation—is largely determined by genetically controlled adaptive properties. This is what natural selection means; only that which maintains or increases the harmony of the system will be selected for.

The concept of natural selection, the heart of the evolutionary theory, is still widely misunderstood. Natural selection says no more and no less than that certain genotypes have a greater than average statistical chance to survive and reproduce under given conditions. Two aspects of this concept need emphasis. The first is that selection is not a theory but a straightforward fact. Thousands of experiments have proved that the probability that an individual will survive and reproduce is not a matter of accident, but a consequence of its genetic endowment. The second point is that selective superiority gives only a statistical advantage. It increases the probability of survival and reproduction, other things being equal.

Natural selection is measured in terms of the contribution a genotype makes to the genetic composition of the next generation. Reproductive success of a wild organism is controlled by the sum of the adaptive properties possessed by the individual, including its resistance to weather and its ability to escape enemies and find food. General superiority in these and other properties permits an individual to reach the age of reproduction.

In civilized man these two components of selective value, adaptive superiority and reproductive success, no longer coincide. The individuals with above-average genetic endowment do not necessarily make an above-average contribution to the gene pool of the next generation. Indeed, shiftless, improvident individuals who have a child every year are certain to add more genes to the gene pool of the next generation than those who carefully plan the size of their families. Natural selection has no answer to this predicament. The separation in the modern human society of mere reproductive success from genuine adaptedness poses a serious problem for man's future.

3
Typological versus
Population Thinking

Rather imperceptibly a new way of thinking began to spread through biology soon after the beginning of the nineteenth century. It is now most often referred to as population thinking. What its roots were is not at all clear, but the emphasis of animal and plant breeders on the distinct properties of individuals was clearly influential. The other major influence seems to have come from systematics. Naturalists and collectors realized increasingly often that there are individual differences in collected series of animals, corresponding to the kind of differences one would find in a group of human beings. Population thinking, despite its immense importance, spread rather slowly, except in those branches of biology that deal with natural populations.

In systematics it became a way of life in the second half of the nineteenth century, particularly in the systematics of the better-known groups of animals, such as birds, mammals, fishes, butterflies, carabid beetles, and land snails. Collectors were urged to gather large samples at many localities, and the variation within populations was studied as assiduously as differences between localities. From systematics, population thinking spread, through the Russian school, to population genetics and to evolutionary biology. By and large it was an empirical approach with little explicit recognition of the rather revolutionary change in conceptualization on which it rested. So far as I know, the following essay, excerpted from a paper originally published in 1959, was the first presentation of the contrast between essentialist and population thinking, the first full articulation of this revolutionary change in the philosophy of biology.

The year of publication of Darwin's *Origin of Species*, 1859, is rightly considered the year in which the modern science of evolution was born. It must not be forgotten, however, that preceding this zero year of history there was a long prehistory. Yet, despite the existence in

Reprinted from pp. 409–412 of "Darwin and the evolutionary theory in biology," in *Evolution and anthropology: a centennial appraisal* (Washington, D.C.: The Anthropological Society of Washington, 1959).

1859 of a widespread belief in evolution, much published evidence on its course, and numerous speculations on its causation, the impact of Darwin's publication was so immense that it ushered in a completely new era.

It seems to me that the significance of the scientific contribution made by Darwin is threefold:

1. He presented an overwhelming mass of evidence demonstrating the occurrence of evolution.

2. He proposed a logical and biologically well-substantiated mechanism that might account for evolutionary change, namely, natural selection. Muller (1949:459) has characterized this contribution as follows:

> Darwin's theory of evolution through natural selection was undoubtedly the most revolutionary theory of all time. It surpassed even the astronomical revolution ushered in by Copernicus in the significance of its implications for our understanding of the nature of the universe and of our place and role in it. . . . Darwin's masterly marshalling of the evidence for this [the ordering effect of natural selection], and his keen-sighted development of many of its myriad facets, remains to this day an intellectual monument that is unsurpassed in the history of human thought.

3. He replaced typological thinking by population thinking.

The first two contributions of Darwin's are generally known and sufficiently stressed in the scientific literature. Equally important but almost consistently overlooked is the fact that Darwin introduced into the scientific literature a new way of thinking, "population thinking." What is this population thinking and how does it differ from typological thinking, the then prevailing mode of thinking? Typological thinking no doubt had its roots in the earliest efforts of primitive man to classify the bewildering diversity of nature into categories. The *eidos* of Plato is the formal philosophical codification of this form of thinking. According to it, there are a limited number of fixed, unchangeable "ideas" underlying the observed variability, with the *eidos* (idea) being the only thing that is fixed and real, while the observed variability has no more reality than the shadows of an object on a cave wall, as it is stated in Plato's allegory. The discontinuities between these natural "ideas" (types), it was believed, account for the frequency of gaps in nature. Most of the great philosophers of the seventeenth, eighteenth, and nineteenth centuries were influenced by the idealistic philosophy of Plato, and the thinking of this school dominated the thinking of the period. Since there is no gradation between types, gradual evolution is basically a logical impossibility for the typologist. Evolution, if it occurs at all, has to proceed in steps or jumps.

The assumptions of population thinking are diametrically opposed to those of the typologist. The populationist stresses the uniqueness of everything in the organic world. What is true for the human species—that no two individuals are alike—is equally true for all other species of animals and plants. Indeed, even the same individual changes continuously throughout its lifetime and when placed into different environments. All organisms and organic phenomena are composed of unique features and can be described collectively only in statistical terms. Individuals, or any kind of organic entities, form populations of which we can determine only the arithmetic mean and the statistics of variation. Averages are merely statistical abstractions; only the individuals of which the populations are composed have reality. The ultimate conclusions of the population thinker and of the typologist are precisely the opposite. For the typologist, the type (*eidos*) is real and the variation an illusion, while for the populationist the type (average) is an abstraction and only the variation is real. No two ways of looking at nature could be more different.

The importance of clearly differentiating these two basic philosophies and concepts of nature cannot be overemphasized. Virtually every controversy in the field of evolutionary theory, and there are few fields of science with as many controversies, was a controversy between a typologist and a populationist. Let me take two topics, race and natural selection, to illustrate the great difference in interpretation that results when the two philosophies are applied to the same data.

RACE

The typologist stresses that every representative of a race has the typical characteristics of that race and differs from all representatives of all other races by the characteristics "typical" for the given race. All racist theories are built on this foundation. Essentially, it asserts that every representative of a race conforms to the type and is separated from the representatives of any other race by a distinct gap. The populationist also recognizes races but in totally different terms. Race for him is based on the simple fact that no two individuals are the same in sexually reproducing organisms and that consequently no two aggregates of individuals can be the same. If the average difference between two groups of individuals is sufficiently great to be recognizable on sight, we refer to such groups of individuals as different races. Race, thus described, is a universal phenomenon of nature occurring not only in man but in two thirds of all species of animals and plants.

Two points are especially important as far as the views of the populationist on race are concerned. First, he regards races as potentially overlapping population curves. For instance, the smallest individual of a large-sized race is usually smaller than the largest individual of a small-sized race. In a comparison of races the same overlap will be found for nearly all examined characters. Second, nearly every character varies to a greater or lesser extent independently of the others. Every individual will score in some traits above, in others below the average for the population. An individual that will show in all of its characters the precise mean value for the population as a whole does not exist. In other words, the ideal type does not exist.

NATURAL SELECTION

A full comprehension of the difference between population and typological thinking is even more necessary as a basis for a meaningful discussion of the most important and most controversial evolutionary theory, namely, Darwin's theory of evolution through natural selection. For the typologist everything in nature is either "good" or "bad," "useful" or "detrimental." Natural selection is an all-or-none phenomenon. It either selects or rejects, with rejection being by far more obvious and conspicuous. Evolution to him consists of the testing of newly arisen "types." Every new type is put through a screening test and is either kept or, more probably, rejected. Evolution is defined as the preservation of superior types and the rejection of inferior ones, "survival of the fittest" as Spencer put it. Since it can be shown rather easily in any thorough analysis that natural selection does not operate in this described fashion, the typologist comes by necessity to the conclusions: (1) that natural selection does not work, and (2) that some other forces must be in operation to account for evolutionary progress.

The populationist, on the other hand, does not interpret natural selection as an all-or-none phenomenon. Every individual has thousands or tens of thousands of traits in which it may be under a given set of conditions selectively superior or inferior in comparison with the mean of the population. The greater the number of superior traits an individual has, the greater the probability that it will not only survive but also reproduce. But this is merely a probability, because under certain environmental conditions and temporary circumstances, even a "superior" individual may fail to survive or reproduce. This statistical view of natural selection permits an operational definition of "selective superiority" in terms of the contribution to the gene pool of the next generation.

REFERENCES

Muller, H. J. 1949. The Darwinian and modern conceptions of natural selection. *Proc. Amer. Phil. Soc.,* 93:459–470.

4
Accident or Design: The Paradox of Evolution

No consequence of Darwin's theory of natural selection was a source of greater dismay to his opponents than the elimination of design from nature. Those who studied the countless superb adaptations of animals and plants had been most gratified by the explanation that such perfection was clearly the result of design by the maker of this world. To explain the perfection of adaptation mechanistically, as a result of random genetic variation and selection, seemed to demand more from these natural processes than they could satisfy. Natural philosophers, in particular, continued to insist that the evolutionary theory is unable to explain "plan and purpose in nature." Such doubt goes back to the beginning of the evolutionary theory. Darwin was fully aware of this difficulty, and in a letter to Asa Gray (26 Nov. 1860) he expressed the paradox as follows: "I am conscious that I am in an utterly hopeless muddle. I cannot think that the world, as we see it, is the result of chance; and yet I cannot look at each separate thing as the result of design."

Expressed in terms of the modern evolutionary theory, the paradox of evolution is the apparent contradiction between, on the one hand, the seeming purposefulness of organic nature, and, on the other hand, the haphazardness of evolutionary processes. In order to understand the magnitude of this contradiction, we must look more closely at the two contradictory phenomena.

DESIGN

Let us begin with a contemplation of design in nature. Any biologist could quote literally hundreds of examples of the most incredible and miraculous adaptations. Let me mention only a few.

Adapted from "Accident or design: the paradox of evolution," pp. 1–14 in *The evolution of living organisms* (proceedings of the Darwin Centenary Symposium of the Royal Society of Victoria) (Melbourne: Melbourne University Press, 1962).

The Yucca moth is specially adapted to the Yucca plant and depends on it throughout its life cycle (Rau 1945). The Yucca plant in turn is adapted to be fertilized by this insect and by no other. The female moth collects a ball of pollen from several flowers, then finds a flower suitable for ovipositing. After depositing her egg in the soft tissue of the ovary, by means of a lancelike ovipositor, she pollinates the flower by pushing the pollen to the bottom of the funnel-shaped opening of the pistil. This permits the larva to feed on some of the developing seeds of the fertilized flower, and yet guarantees the development of enough seeds in the nonparasitized sectors of the fruit to permit the Yucca plant abundant reproduction. This perfection of the mutual adaptation of flower and moth is indeed admirable. And in addition to this pollination and egg-laying relationship, there are numerous other adaptations, such as the emergence of the pupating larvae only after rain and the emergence of the moths in early summer some 10 months after pupation, precisely at the time when the Yucca plants are in flower. "Could blind chance have achieved such perfection?" ask the skeptics.

Or take the field of intracellular symbionts studied by Buchner and his students. Here we have innumerable structures which permit insects and other metazoans to house the bacteria, yeasts, or other microorganisms that supply enzymes for the better utilization of the food of the host. Most of the host organisms have developed various mechanisms in order to supply their offspring with the right kind of symbiotic microorganisms. The number of the sometimes almost incredible adaptations is legion (Buchner 1953).

Or take the division of labor and the harmonious collaboration of individuals of various castes in colonies of social insects, on which so many of our outstanding naturalists have reported in detail (Wilson 1971). Or consider the orientation of nocturnally migrating birds. Even individuals that have been raised in complete isolation in order to eliminate the possibility of any form of learning are able to undertake a normal migration, because they have not only a perfect clock mechanism telling them the time of the day and the season of the year, but also an innate knowledge of the stellar constellations that gives them the compass and the map by which to guide their wanderings (Sauer 1957). As a matter of fact, we need not resort to such spectacular examples. There is a similar perfection in the various functions of the intestinal tract when it is digesting various types of food, or in the growing tissues of a developing organism. Whole books have been devoted to the discussion of such perfect adaptations and to what appears to be the extraordinary foresight of much of organic function and behavior (e.g., Russell 1945). Given all this, the conclusion is inevitable: we find in all organisms a fitting together of inborn actions or structures so perfect that one can hardly avoid such terms as "design" or "purposefulness."

CHANCE

But when we ask how this perfection is brought about, we seem to find only arbitrariness, planlessness, randomness, and accident. Again let us study this in some detail. Any evolutionary change is brought about by a series of steps. The first of these is the change of a genetic factor, a mutation. The question whether there are any laws controlling the direction of mutation must be answered negatively. Mutations never seem to happen in response to a *need* for a particular genetic change, nor is a particular mutation released by a particular constellation of environmental factors. We still have a great deal to learn about the process of mutation; yet there is little reason to doubt that most mutations are merely copying mistakes during the replication of the genetic material. Hence, it is correct to state that mutations are strictly accidental occurrences.

The next process of interest to us is the combining of genes (recently mutated or not) into genotypes. In sexually reproducing higher organisms these genes are organized in chromosomes. The reassortment of genes within the chromosomes, which normally happens once in every generation, is governed largely by chance. Each individual in sexually reproducing organisms has two homologous sets of chromosomes, one from the father and one from the mother. At some time prior to the formation of the gametes, the two homologous chromosomes exchange equivalent pieces with each other by a process called "crossing over." By and large (there are many exceptions) no laws seem to determine where the chromosomes will break or how large the pieces will be that are exchanged. Which particular combination of pieces of maternal and paternal chromosomes making up the new chromosome will enter a given egg or spermatozoon is largely a matter of chance, at least in most chromosomes and most species. Likewise, it is largely a matter of accident *which* chromosomes will go into which germ cell, provided only that each cell receives its full set of chromosomes.

Chance is of overwhelming importance in the next step, the fate of the gametes. Each male produces millions or billions of spermatozoa. Only a few will fertilize eggs and contribute to the maintenance of the species; of the eggs likewise, not all will be fertilized. Since by far the majority of the spermatozoa and eggs are fully viable, it is again largely a matter of accident which will participate in the genetic contribution to the next generation.

Chance plays a far smaller role in the next step, the fate of the fertilized egg, or zygote, as the biologist calls it. Here we find a great difference between organisms like man, on the one hand, which produce only a few zygotes, or, on the other hand, certain parasites and marine organisms that produce millions. The greater the number of zygotes, the greater will be the mortality caused by pure chance; a whale feeding on planktonic crustaceans does not choose among them on the basis of their selective inferiority!

The series of partially or largely accidental steps leading to the production of a new potential parent can be listed as follows:

Mutation at one or several loci
Crossing over
Distribution of chromosomes during reduction division
Success of gametes, involving choice of partner and choice of gamete
Success of zygote

The aforementioned evolutionary events recurring during each generation are only a few of many such events completely or in part controlled by accident.

By now you will appreciate the true magnitude of Darwin's predicament. Nearly all steps leading to evolutionary change seem to be controlled entirely or largely by accident; yet the final product of evolution is perfection in adaptation. How can this seemingly hopeless contradiction be resolved? Many philosophers and even some biologists adopted a defeatist attitude at this point and abandoned all endeavor to find a causal explanation. Instead they introduced vitalistic or finalistic principles into their considerations, such as Bergson's *élan vital*, Driesch's *Entelechie*, the "inherent improvement drive" of some Lamarckians and similar euphemisms for the unknown. Others have appealed to saltation as the source of the sudden origin of new perfections. Such capitulations to the unknown have had a paralyzing effect on the spirit of scientific inquiry. They have proven themselves utterly sterile pseudo-solutions and are unanimously rejected by those who have a grasp of modern evolutionary theory and of modern genetics. Yet a purely negative rejection of these pseudo-solutions is not enough. The evolutionist must come up with a constructive solution to the great paradoxical contradiction of evolution. What, then, is the solution?

Darwin emphasized again and again—and on this point he was in complete agreement with his opponents—that design cannot be the result of "blind chance." The choice of the expression "blind chance" implies an explanation of evolutionary change by a single step, an all-or-none phenomenon. Such an explanation is in direct conflict with the Darwinian theory of evolution, which regards any evolutionary change, including the perfecting of adaptations, as a two-step process. The first step is the production of genetic variation; the second step is the sorting of the many phenotypes into successful and unsuccessful (or less successful) ones. In the first step, indeed, blindness reigns, whether in the process of mutation itself or in recombination in its several aspects, from crossing over to mate selection. It is here that the multiple blind accidents described above occur. Yet no damage is done by accident at this level, because the ultimate effect of all the accidents is to maintain high genetic variability. The less determinacy at this level, the less prejudging of probable fitness, the greater the ultimate choice for natural selection. The function of this first step is to keep the material of evolution variable—to keep pliable, so to speak, the clay that natural selection shapes into the finished sculpture.

Precisely the opposite is true for the second step, the evaluation of phenotypes. Here, where survival and differential reproduction are concerned, anything but blindness prevails. We have a proverb that is applicable here, "Nothing succeeds like success," and this is the secret of natural selection. Success, in this case, means leaving offspring. But what is it that determines this success? If success were determined by blind chance, as are most processes that lead to genetic variation, we would not be justified in speaking of natural selection, for selection implies discrimination. But, and this is the cornerstone of the evolutionary theory since Darwin, it is justifiable to refer to differential reproduction as natural selection because individuals differ from each other in their genetic endowment, and it is, at least in part, the nature of this genetic endowment that determines reproductive success.

Merely asserting this claim dogmatically will not convince those who, until now, have been disbelievers. It is more important to point out that most of their objections are directed against obsolete views. No one can fully appreciate the strong position of the Darwinian theory of evolution who is not fully aware of some of the recent improvements in our understanding of genetics and evolution. New findings in these areas have given us a much better insight into the working of natural selection. Let me demonstrate this through a discussion of (1) the nature of mutation, (2) the relation of gene and character, and (3) the nature and object of selection.

MUTATION

The formerly held view that mutations are always drastic, and nearly always deleterious, has undergone considerable revision in recent years. This erroneous view was due chiefly to the choice of material by the early Mendelians, particularly De Vries, but was unconsciously favored also by the working method of the classical geneticists, like T. H. Morgan, who quite naturally selected conspicuous and clear-cut mutations to study the laws of inheritance. Newer evidence indicates that these mutations of classical genetics are by no means typical; indeed, they may well be in the minority. It is now believed that many, if not most, mutations have only slight effects or are entirely invisible because they affect only nonmorphological characters. It would lead us too far afield to present here the detailed evidence for this assertion. It is based, in part, on a new concept of the gene, and he who would acquire a better understanding of the nature of mutation must investigate the nature of the mutating structure, the gene.

Within recent years, it has become firmly established that DNA is the essential carrier of the genetic properties. Furthermore, it appears that the double helix of the DNA molecule serves as a complicated program of information that regulates the formation of the species-specific proteins and controls all development. Genes can be considered the carriers of "bits of information," to use the happy term of information theory. Such bits of information are handed down from generation to generation, unless a mistake is made during their replication; such a mistake is called a "gene mutation." The first mutation for which the exact chemical change has been analyzed is that from normal hemoglobin to sickle-cell hemoglobin (Ingram 1956). The genetic information is contained in the precise sequence of the nucleotide pairs, and mutation consists of a change in this sequence.

Because all mutations segregate in a typically Mendelian manner, we treat mutations as a unit phenomenon. This is not necessarily correct. Since different kinds of DNA perform different functions, such as programing messenger RNA or regulating the activity of a gene locus, it is likely that, functionally, we should recognize different mutations. Multiplicity of types of mutations is even more firmly established for the evolutionary significance of mutations. There is a broad spectrum extending from lethal and other drastic mutations to rather inconsequential changes in the DNA. It is particularly important that many of these slight changes are almost equivalent in their selective significance and that they result in no visible change of the phenotype. These are the genes called "isoalleles" by Stern and others (see Stern and Schaeffer 1943). An isoallele is the product of a mutation that represents not a breakdown of the physiological machinery, as mutations are so often visualized, but merely a slight "variation on a theme." This may lead to an alternate metabolic pathway. The particular environment and the particular genetic background of such a mutation may determine whether or not it is superior to the gene from which it has mutated. The fact that an occasional mutation produces a superior gene must be emphasized in refutation of the widely held belief that mutation is always deleterious and destructive. The smaller the effect of a mutation, the greater the probability that it will be advantageous. Microbiology in particular has taught us how often such mutations can be immediately constructive, for instance, in producing resistance to antibiotics or toxic substances. The recent work of Dobzhansky and his school, and of other population geneticists, has shown that even in higher organisms an appreciable percentage of mutations enhances viability. The picture we now have of mutation is certainly very different from that prevailing in the older evolutionary literature.

GENE AND CHARACTER

A second major revolution has occurred in our thinking about the relation of gene and character. In the early days of genetics there was the naive assumption of a one-to-one relation between gene and character: that each character is controlled by a gene and each gene controls a character. Accordingly, one spoke of the blue eye-color gene, the red hair gene, the pink flower gene, etc. Such a concept of gene action led to the further assumption that each mutation was either favorable or unfavorable and, indeed, that mutation itself was the major evolutionary force. These assumptions permitted only one interpretation of evolutionary change; that evolutionary progress depends on the opportune occurrence of favorable mutations and, thus, that it proceeds at the mercy of blind chance.

The modern picture is altogether different: a gene elaborates a gene product, added during development to the stream of differentiation. The products of innumerable genes collaborate to produce a terminal organ or character. Indeed, extremists have suggested, perhaps not quite seriously, that every gene affects all characters and that every character is affected by all genes. Clearly, it has become necessary to study gene *interaction* rather than gene action. Gene A may be deleterious when combined with gene B, but may give superiority when combined with gene C. It is evident that this changed interpretation of mutation and of the genetic basis of the phenotype leads to an evaluation of natural selection entirely different from that of the earlier geneticists.

NATURAL SELECTION

An understanding of the working of natural selection is the key to the Darwinian theory of evolution. I know of no other scientific theory that has been misunderstood and misrepresented as greatly as the theory of natural selection. First of all, it is usually represented as strictly negative, as a force that eliminates, a force that kills and destroys. Yet Darwin, by his choice of the name "selection," clearly emphasized the positive aspects of this force. Indeed, we now know that one can go even further and call natural selection a creative force. Second, natural selection is not an all-or-none phenomenon. The typologist, the follower of Plato, seems to think that alternatives are always either good or bad, black or white, worthy of preservation or doomed to rejection. This viewpoint is represented in two statements by well-known contemporary philosophers, chosen at random from the recent literature: "Natural selection requires life and death utility before it can come into play"; and "Unsuccessful types will be weeded out by the survival of the fittest but it cannot produce successful types."

Actually, types in the sense of these statements do not exist; only variable populations exist. No one will ever understand natural selection until he realizes that it is a statistical phenomenon. In order to appreciate this fully one must think in terms of populations rather than in terms of types. There may be a thousand or several thousand variable gene loci in any species. Some individuals have more genes, some have fewer genes that are favorable under particular conditions of the contemporary environment. The more favorable genes an individual has, the greater the probability that it will survive, and what is more important, that it will reproduce successfully. This probability of reproducing, of contributing to the gene pool of the next generation, describes the true nature of natural selection. And this is why natural selection is now often referred to as "differential reproduction."

I must emphasize here Darwin's genius in having recognized this point quite clearly. When speaking of the struggle for existence, he states in the first edition (1859:62) of the *Origin of Species* that he uses this term *"in a large and metaphorical sense, including dependence of one being on another, and including (which is more important) not only the life of the individual, but success in leaving progeny."* Since natural selection is a statistical phenomenon we must visualize it in terms of a population curve with a minus tail and a plus tail. Natural selection tends to clip off the tail of the curve with the minus variants in favor of the plus half, automatically resulting in a shift of the mean toward the plus end of the curve.

A further consideration will help to make the role of natural selection even clearer. Not the "naked gene" but the total phenotype is exposed to selection. A gene occurring in a population will contribute toward very many phenotypes. In some cases these phenotypes will be successful, in others they will not. The success of the phenotypes will depend on the fitness of the particular gene, within the framework of the gene pool of this population. And this again will be an essentially statistical phenomenon.

Let us also remember that recombination, not mutation as such, is the primary source of the phenotypic variation encountered by natural selection. The usual argument of the anti-Darwinian is: "How can an organism rely on the opportune occurrence of a favorable mutation whenever one is needed, considering that most mutations are deleterious? Surely all organisms would be doomed to extinction if in times of need they had to rely on such rare events?" Those who ask such questions confuse genetic variability and phenotypic variability. To be sure, mutation is ultimately the source of all genetic variation. But natural selection operates not at the level of the gene but at the level of the phenotype. Further, the main source of phenotypic variation is recombination rather than mutation, and this source of variation is ever present. With every individual differing genetically from every other one, every phenotypic character is variable, showing deviations of varying intensities and directions around the mean. Under normal conditions, selection will favor the mean (stabilizing selection), but if a deviation in any direction should be required by a newly arising selective force, the material is instantaneously available to respond to this force (directive selection).

Natural selection in this modern nontypological interpretation is an exceedingly sensitive instrument. The phenotype in nearly every case is actually a compromise between a number of conflicting selective forces. Let us take, for instance, egg number in birds. On the one hand, there is a selective force to increase it because the larger the number of eggs, the more young will be produced. On the other hand, with young birds requiring parental care, there is an optimum number beyond which parental care deteriorates, so that the broods from the largest clutches actually have a lower survival than those from optimal clutches.

The almost unbelievable power of selection is demonstrated by much experimental work published during the last three decades. But even those unfamiliar with the results of experimental genetics need only to think of the products of animal and plant breeding to appreciate the power of selection. For instance, one might consider such physiological monstrosities as the modern dairy cow or some extreme breeds of dogs—all produced by selection!

I hope that this discussion has made clear how unfortunate such terms as "struggle for existence" or "survival of the fittest" are, because they tend to distract our attention from the central aspect of the phenomenon of natural selection, its purely statistical nature. Anything adding to the probability of survival and reproductive success will automatically be selected for.

Evolutionary Accidents and Genetic Information

We are now ready to take a second look at the various levels of accidents discussed earlier. When we take into consideration (1) that most mutations have minor effects, (2) that selection deals with phenotypes and only very indirectly with genes, and (3) that all organisms live in a variable environment, it becomes quite apparent that there is nothing negative or deleterious about these accidents. All they do is increase phenotypic variability or at least maintain it. They do not determine the course of evolution, they only supply the material with which natural selection works. With every species containing ten thousands, millions, or even hundreds of millions of individuals, genetic losses are not losses of evolutionary ground. They obey the rules of statistics, and gene frequencies will remain unchanged, as stated in the Hardy-Weinberg formula. On the contrary, these accidents lead to an unbiased method of testing whatever novel gene recombinations may produce superior phenotypes, particularly if there have been unprecedented or otherwise irregular changes of the environment.

This role of evolutionary accidents will become even clearer if we once more think of genes as carriers of information. Every genotype is a unique program of information that directs the development and the behavior of an individual organism. Some programs lead to the production of phenotypes that, in the particular environments in which they are placed, are "better adapted" than others. These we may call successful programs, and there is every probability that they will make a greater contribution to the gene pool of the next generation than will programs that result in less well-adapted individuals. The programs of information of the next generation will be formed from the gene pool, that is, from a recombination of the successful programs of the previous generation. These programs and the phenotypes they produce are statistical populations. Admittedly, recombination will result in the destruction of some perfectly satisfactory programs of the previous generation, yet some of the new programs may be in the plus tail of the curve and may be superior to anything that existed in the parental generation. The possible combinations of genes being infinite for all practical purposes, and with the fitness of the phenotypes to a large extent unpredictable (on the basis of parental genotypes), the undeniable role of accident in maintaining variation in populations is on the whole beneficial.

Objections to a Selectionist Interpretation of Adaptation

Numerous cases of adaptation are cited in the anti-Darwinian literature as defying a selectionist interpretation. The new understanding of the nature of genetic material and of the working of natural selection permits us to look at these objections in a new light and to test their validity. Let me select, among the many conventional objections, some that are typical of the arguments of anti-selectionists.

The Origin of Excess Structures

This objection may be phrased as follows: "How can natural selection be all-powerful if it permits the development of excessive structures that are either useless or definitely deleterious, like the antlers of the giant Irish elk?"

The answer to this objection is at least threefold: (1) the structure may actually have selective value at a certain stage in the life cycle, for instance, in the case of the young elk, or in certain environments; (2) sexual selection, as already pointed out by Darwin, leads to reproductive success, and is therefore favored unless counteracted by other components of natural selection (and most of the "excessive" structures cited in the literature are secondary sexual characters); and (3) what we see evolve is only the visible phenotype. The genes producing it may have been selected for other cryptic functions that contributed positively to the survival of its bearer. The widespread occurrence of the giant elk and its relative abundance indicate that in its particular environment, and at the time it lived, it very definitely had superior survival ability and was not handicapped by its giant antlers (Gould 1974).

The possibility of cryptic contributions of genes to the phenotype bears also on the objection that many of the differences between related species seem to have no adaptive significance. "How could natural selection have brought this about?" it is asked. Again, we must be certain that these characters have no adaptive significance, and if this should be proven, we must next determine what other functions the genes have that produce these so-called neutral characters.

Chance Mutations and Parallel Evolution

This objection contends that chance mutations cannot be involved in adaptations that have been acquired repeatedly by separate evolutionary lines. How, it is asked, can unrelated organisms have had the same mutation in response to the same need?

This question confuses mutation and phenotype. It is indeed highly unlikely that parallel evolutionary developments have an identical genetic basis. Yet evidence is accumulating that similar phenotypes may be built up on a very different genotypic basis. Selection has an extraordinary power to steer variation in the right direction.

Evolution is strictly opportunistic and whenever a change of phenotype is advantageous, whatever aspect of the phenotype is variable at the time will be utilized by selection. Let us look, for instance, at the various structures that facilitate floating in pelagic marine animals. An illustration of such devices in various types of organisms shows that almost any part of the body may be used. Depending on the original variation of the particular genus or species that is shifting from a benthonic to a pelagic mode of life, various parts will be elaborated for the purpose of floating, that is, for the purpose of an enlargement of the body surface. The same opportunism is true for the plumes of the birds of paradise. The selective premium in this case is on conspicuousness, but it depends on genus or species whether the feathers of the crown, neck, flanks, wings, or tail are utilized for this purpose. A study of the eye in various types of organisms is a further illustration. The essential components of the eye are a light-sensitive tissue, a lens, a focusing device, and a pigment that shields the undesirable light penetration. Eyes have evolved in the animal kingdom at least a dozen times independently and the stated basic needs have been answered quite differently in each case.

Selection and Incipient Structures

Another objection is based on the claim that natural selection cannot act on a newly developing organ until it has reached an elaboration that permits perfect functioning.

This claim overlooks several facts. Even a very rudimentary organ, like the first beginnings of a gliding wing in the "pro-avis," may be of distinct selective advantage provided that none of its enemies or competitors has the same or a more perfect structure. Again, as in previous cases, the incipient new character may be a pleiotropic by-product of a gene or gene complex selected for a different reason. The new structure, although of no selective significance in the beginning, is in this case the product of the selectively advantageous total genotype. Furthermore, the acquisition of a new function may have given a pre-existing structure a new evolutionary significance. Perhaps the most decisive answer to this argument is that even a very slight change of the phenotype may have a statistically significant effect, since natural selection is a statistical process. Indeed, all selection experiments of recent years have substantiated precisely this point (see Essay 6).

Selection and the Anticipation of a Need

This objection asks, "How can natural selection explain the acquisition of an organ long before it is needed?"

The usual illustration for this objection cites animals like the wart-hog or the ostrich that as embryos develop calluses where later in life their bare skin is exposed to much friction. To me this argument seems based on a deplorable confusion of ontogeny and phylogeny. It is just as silly as asking, "Why should the embryo have a brain and extremities long before it uses them?" Why would it not be of advantage to the young wart-hog to have calluses when it first starts to kneel down to feed? Nor is the question "Why were genes ever selected that would provide for calluses?" too difficult to answer. If 10 people who are unused to physical labor were to work strenuously with an axe or a spade, some would form calluses, others blisters that might be very painful or might even become infected. It would be of obvious selective advantage to have genes that would facilitate callus formation without a blister stage and this would be far more crucial for a feeding wart-hog than for a working man. Considering how many different genes presumably contribute to such a simple phenotypic phenomenon as calluses, it is not difficult to understand that such genes will accumulate in a population as soon as such a modification of the phenotype becomes of selective advantage. The more of these polygenes there are present, the easier it will be to modify the phenotype in the desirable direction until the character appears in the embyros. One should never overlook, in the interpretation of selection experiments, that most characters have a highly polygenic basis.

Selection and Highly Complex Structures

This objection can be phrased in the question: "How can natural selection produce so highly complex an organ as the middle ear of mammals, or so complex a behavior pattern as that of the New Zealand cave fungus gnats?"

I have discussed this question in detail elsewhere (Essay 9). Let me summarize my findings by saying that in all these cases one particular structure or behavior is the key to the situation and once it has developed it will favor the acquisition by selection of all the subsidiary and auxiliary structures or behavior patterns. Furthermore, in at least some cases, opportunistic natural selection has taken advantage of a purely accidental neighborhood of two structures to mold them into a new, complex structure with a new function. This has been demonstrated particularly decisively by W. Bock (1959) for the secondary jaw articulation of birds.

Let me now summarize the answers to these and other objections against natural selection. I hope I have shown in every case that the phenomena which form the basis of the objections can be interpreted without difficulty in terms of the modern evolutionary theory, that is, in terms of genetic variation sorted out by natural selection. Indeed I would like to go further, and say that such an interpretation produces far fewer obstacles to our understanding than any alternative interpretation so far advanced.

ADAPTATION AND PURPOSE

The final argument of the anti-selectionists usually is: "But all these adaptations are so obviously purposive that there must be some internal purposive force!" One must object vigorously to this line of argument. An individual can have purpose but an evolutionary line cannot. There is no need whatsoever, indeed there is no excuse whatsoever, for considering adaptation as evidence of purpose. (See Essay 26 for a detailed discussion of teleology).

The use of the word "purpose" is very ambiguous. It implies a conscious, or at least instinctive, striving for a definite goal. The evolution of the whales from the Eocene zeuglodonts, or that of the birds from the pseudosuchians via *Archaeopteryx,* seems indeed as decisively directed toward perfect adaptation in the new medium, water in one case, air in the other, as though someone had directed the course of evolution. Yet, the study of those few evolutionary lines, for instance the horses, for which enough fossil material is available to permit detailed analysis shows that evolution is only rarely smoothly rectilinear. Progress, instead, is by trial and error. One organ may run far ahead, the others lag behind; periods of stagnation may alternate with periods of explosive advance. There is a continued trend toward improved adaptation to the shifting environment, but to call this purposive only clouds the issue.

It is far more illuminating to interpret the purposive nature of all adaptations in terms of the genetic information. Every generation of organisms consists of innumerable slight variations in the DNA program of information that is characteristic for the particular species. Each individual phenotype is the product of one such program, and adaptation of the phenotype to its surroundings is based on the information derived from this genetic program. The carriers of successful information will be the progenitors of the next generation. The individual is enabled to act purposefully because it is endowed with the proper information. Like an electronic computer, it is "programed," permitting it to cope with the vicissitudes of development and of life. No such program exists, however, for the "life" of an evolutionary line. Whatever purposiveness the organic world seems to have is thus not a finalistic one but, if I may say so, an *a posteriori* one, or, in other words, the result of past natural selection. A steady selection pressure, generation after generation, can push an evolutionary line steadily closer to an ultimate goal without the species population of a given generation at any time showing any purposive behavior whatsoever.

We are now ready to sum up the verdict on Darwin's great paradox. To be sure, the individual events contributing to the inexhaustible variability of phenotypes in natural species are random events and are thus, so to speak, accidents. But this is the extent of the role played by chance. The molding of the highly variable gene pools, as the geneticist calls them, is in the hands of natural selection. The product of selection is adaptation, and the adaptedness of organisms and their utilization of the environment is improved from generation to generation until it appears as perfect as if it were the product of design. In short, the solution of Darwin's paradox is that natural selection itself turns accident into design.

REFERENCES

Bock, W. J. 1959. Preadaptation and multiple evolutionary pathways. *Evolution,* 13:194–211.

Buchner, P. 1953. *Endosymbiose der Tiere mit Pflanzlichen Mikroorganismen.* Birkhäuser, Basel and Stuttgart.

Darwin, C. 1859. *On the origin of species by means of natural selection, or the preservation of favored races in the struggle for life.* J. Murray, London.

Gould, S. J. 1974. The origin and function of "bizarre" structures: antler size and skull size in the "Irish Elk," *Megaloceros giganteus. Evolution,* 28(2):191–221.

Ingram, V. M. 1956. A specific chemical difference between the globins of normal human and sickle-cell anaemia haemoglobin. *Nature,* 178:792–794.

Rau, P. 1945. The Yucca plant, *Yucca filamentosa,* and the Yucca moth, *Tegeticula (pronuba) yuccasella* Riley: an ecologico-behavior study. *An. Mo. Bot. Gd.,* 32:373–394.

Russell, E. S. 1945. *The directiveness of organic activities.* Cambridge University Press, Cambridge.

Sauer, F. 1957. Die Sternenorientierung nächtlich ziehender Grasmücken. *Z. Tierpsychol.,* 14:29–70.

Stern, C., and E. W. Schaeffer. 1943. On wild-type iso-alleles in *Drosophila melanogaster. Proc. Nat. Acad. Sci.,* 29:361–367.

Wilson, E. O. 1971. *The insect societies.* Belknap Press of Harvard University Press, Cambridge, Mass.

5
Selection and Directional Evolution

Considering the present agreement among evolutionary biologists on the explanation of the processes and mechanisms of evolution, it is rather puzzling that so many nonbiologists, and even some physiologists and molecular biologists, still refuse to accept the Darwinian explanation. One has the feeling that there must be fundamental misunderstandings. It is the major objective of this essay to remove some of these misunderstandings. The claim is made particularly often that the variation(mutation)-selection model is unable to explain evolutionary trends, that is, evolutionary changes that seemingly continue in the same direction over long periods of time. The doubts are often expressed as follows: "How can regularities that extend over millions of years be explained with the help of random mutations?" Admittedly, attempts to explain such directional evolution were only partially successful in the beginning. But new facts and new concepts have changed this state of affairs quite drastically in recent decades. What was most important for this clarification was the removal of certain erroneous conceptions.

Let us begin with *mutation*. Mutation was conceived as a thoroughly drastic phenomenon by De Vries and other mutationists at the beginning of the century. Mutations, for them, had the capacity in a single stroke to create entirely new characters, indeed, entirely new species. This concept of the drastic nature of mutation has been thoroughly invalidated by subsequent genetic research. In the course of the last 50 years it has been confirmed again and again that most mutations have hardly visible or entirely invisible effects. This has made it necessary to reformulate the concept of mutation: there is no genetic difference between drastic mutations and those that are components of so-called continuous variation. The genetics of both types of variation is the same. The modern evolutionist no longer sees in mutation a *deus ex machina* suddenly engendering new species or other drastic evolution-

Translated and adapted from "Selektion und die gerichtete Evolution," *Naturwissenschaften*, 52, no. 8 (1965):173–180.

ary inventions; mutation is simply the source of inexhaustible genetic variability.

A second widespread misunderstanding concerns the material on which natural selection acts. Selection does not deal with genes, or even with genotypes, but only with individuals, that is, phenotypes. All that is being evaluated by evolution is the selective value, the "fitness," of individuals. The individual organism is not a conglomerate of accidentally, and rather chaotically, assembled genes, as has often been claimed by the opponents of the Darwinian theory of evolution. Instead, it is a remarkably harmonious system, the individual components of which, even though they owe their origin to the process of random mutation, were fitted together into coadapted systems under the constant supervision of natural selection.

To repeat the most important point: selection evaluates not single genes, but rather gene combinations, that is, whole systems. From this fact it is clear why the harmonious interaction of genes is of such great importance in evolution. It creates an internal environment, a gene environment, that is as important for evolution as the external environment of climatic, geological, and biotic factors. Whenever we deal with evolutionary processes that extend over many generations, sometimes over hundreds of thousands, or millions, of years, it is evident that a reconstruction of the gene environment is involved. Such a rebuilding of the system of genes can proceed only slowly because a system can be modified only gradually, for all parts must continue to remain in step with the other parts of the system.

There are analogies for this in human technology. The first motor car was nothing but a horse carriage with a gasoline motor. It required tens of years of reconstruction until it became a genuine motor car. Even in the modern auto there are components which can be explained only as remnants of this history, for instance the "running board" of the original Volkswagen. Evolution proceeds in a similar way; it consists not in the replacement of independent genes, but rather in the gradual modification of harmonious gene systems.

A third misunderstanding concerns the nature of natural selection. An essentialist (see Essay 3) seems to be utterly unable to understand that a population can be gradually transformed by the favoring of certain individuals. Each individual has a unique combination of genetic characteristics and consequently differs from all other members of its population in the probability of successful reproduction under the conditions of the current environment.

When natural selection acts, step by step, to improve such a complex system as the genotype, it does not operate as a purely negative force, as many opponents of Darwinism maintained. It does not confine itself to the elimination of inferior gene combinations; rather, its most important contribution is to bring superior gene combinations together. It acts as a positive force that pays a premium for any contribution toward an improvement, however small. For this reason some of the most profound thinkers about evolution, such as Theodosius Dobzhansky, Julian Huxley, and G. G. Simpson, have called selection "creative." Such a characterization is particularly apt for diploid organisms in which variation of the genotype is largely the result of recombination. A character that in its most perfect form is determined, let us say, by 50 genes can, through natural selection, be brought steadily closer to an optimal combination. It is evident from this consideration that natural selection can effect gradual, simultaneous improvement in dozens of characteristics independent of one another.

These preliminary considerations place us in a position to look anew at the problem of so-called directional, or rectilinear, evolution.

DIRECTIONAL EVOLUTION

What evolutionary trends are claimed to be in conflict with the Darwinian theory of evolution? All of them have been provided by paleontology. Anyone who is familiar with the paleontological literature is acquainted with unidirectional evolutionary series among ammonites, graptolites and foraminiferans, to mention only a few of the best-known examples (Rensch 1960; Schindewolf 1950a, b). Rectilinear evolutionary series have also been described for fishes, reptiles, and mammals. At a time when Lamarckian ideas were still prevailing, such series were interpreted as proof of an intrinsic tendency toward perfection. Now, we question not only this interpretation, but even the fact of undeviating, straight-line evolution. With far more material available than at the beginnings of paleontology, it has become apparent that most of the so-called orthogenetic series in the textbooks were grossly oversimplified. Simpson (1953) and Romer (1949) have shown for the horses, which have always been cited as the paradigm of unidirectional evolution, that such trends exist only in very broad outlines. As soon as one analyzes individual phyletic lines in detail, one finds that two daughter lines can proceed in rather different directions from that of their ancestors and that evolutionary changes can come to a complete standstill in certain lines and occasionally even regress. In other words, there is no evidence at all for an obligatory intrinsic drive toward perfection. The phylogeny of the horses is by no means a uniform series, but rather a branching bundle of separate evolutionary lines that are independent of each other and may show very different trends. This conclusion is valid not only for the evolution of horses as a whole but for individual characters that display evolutionary trends, such as body size, brain size, specializations of the teeth, proportions of the extremities, and reduction of the toes (Simpson 1953).

Two aspects of this evolution, characteristic not only of horses but of all carefully studied evolutionary lines, must be stressed. First, a relatively rapid evolution can come almost or entirely to a halt in some lines. Second, when a line divides into separate phyletic lines, the daughter lines invariably differ in direction and rate of evolution.

Detailed modern research has rather thoroughly refuted the regularity and strict goal-directness previously claimed for the so-called orthogenetic series. However, one must not go too far. No one, of course, will deny that certain tendencies are widespread in evolution.

EXPLANATION OF DIRECTIONAL EVOLUTIONARY TRENDS

We have now arrived at the most fundamental question of our investigation. How can evolutionary trends that seem to continue in the same general direction for millions of years be explained with the help of random mutations?

The answer to this question has two parts. First, selection itself often shows a directional trend, as has long been emphasized by those evolutionists who have designated this phenomenon as orthoselection. If, for example, the climate becomes steadily drier through a period of millions of years, as was the case in the second half of the Tertiary, and if in consequence grasses become harsher and richer in silicates, this trend exerts a continuing selection pressure on grazing mammals favoring increased resistance of their teeth to wear. This is a typical case of orthoselection. Any major ecological shift, as to life in caves, the deep sea, or the air, results in a lengthy period of such orthoselection.

Orthoselection alone, of course, cannot explain the balanced harmony in the restructuring of organisms during evolution. For this we must consider the relationship between gene and character. The idea that one gene controls one character has long been abandoned. This idea itself goes back to preformism: the concept that all characters are fully preformed in the original germ cell. Actually, the relationship between gene and character is by no means so narrow and direct. We now know (1) that particularly in higher organisms, most genes are pleiotropic, that is, they affect many different components of the phenotype; and (2) that at the same time most characters are polygenic, that is, they are simultaneously controlled by many genes.

Discoveries in molecular biology indicate that harmony in the interaction of genes is perhaps even more important than was imagined only a few years ago (although conclusions in this area still rest almost entirely upon inference and must really be called speculation). The majority of metabolically important enzymes are very widely distributed in the organic world, from the simplest bacteria to the most complex plants and animals. Although the function and often the chemical structure of the active site remain quite constant, each enzyme molecule has been considerably modified in the course of phylogeny. As a result of mutation, certain amino-acid residues have been replaced by different ones, and these changes have become fixed, presumably because of a higher selective value. The functional importance of at least some of these changes results from modifications in the three-dimensional conformation of the folded proteins. A selective advantage of such changes undoubtedly is to facilitate or prevent interactions with other physiologically active molecules. Since other macromolecules likewise are being modified in the course of evolution, natural selection would, of necessity, favor eventual modification in all of them so as to maintain harmonious interaction among them (see Essay 7).

In recent years a great deal of evidence has indicated that there are not only enzyme-forming genes, but also various kinds of regulatory genes controlling the activity of one or several enzyme genes. Like the enzyme genes, the regulator genes consist of DNA, are organized in systems, and are involved in the regulation of protein synthesis. It is still entirely uncertain how many kinds of such regulator genes exist, how they are organized, and how they function (Jacob and Monod 1961a, b; Britten and Davidson 1969).

This problem is particularly puzzling among higher animals. A *Drosophila* or a mammal has enough DNA for 5 or 6 million genes (cistrons), and yet there is much evidence from genetic analysis and the study of enzymes that a higher organism has perhaps only 5,000, or at most 50,000, enzyme genes. Are the millions of other genes, the presence of which one can infer from the DNA content of the chromosomes, all regulatory genes? Or are there multiple cistrons for certain enzymes, all of which control the production of an identical enzyme? One must admit that our ignorance in this area is still virtually complete. As uncertain as all these figures are, the facts nevertheless indicate that regulator genes presumably play a much greater role among the higher organisms than among the bacteria. This, of course, would not be surprising.

Growth, development, and differentiation are processes that are the product of an extraordinarily precise regulation of biochemical processes. To control these is the function at least in part of regulator genes. They determine at what time and for how long enzyme genes produce their specific enzymes. If regulator genes, owing to their DNA structure, mutate like other genes, then this system of genes would have enormous plasticity. The organism is, so to speak, a gigantic clockwork in which innumerable regulatory mechanisms control the optimal harmony of the system. It is possible, indeed probable, that many of the phenomena ascribed in the past to pleiotropy and polygeny should now be ascribed to the effects of regulatory genes. The details of these mechanisms are not important. What is important is to realize how wrong were the original ideas of the mutationists, who held that a single mutation would produce a new character, indeed even a new species. Now we know that each character is the product of a successful collaboration of dozens, if not hundreds, of genes, some of which control enzymes and some of which may be regulatory. This is why we speak of the unity of the genotype, and this is what we have to keep in mind when we attempt to interpret the phenomena of directional evolution.

The pathway from the genotype to the phenotype is extraordinarily complex and characterized by all sorts of feedbacks. It is no longer possible to regard the phenotype as a mosaic in which each part can be replaced without any effect on the neighboring components. On the contrary, the genotype is an organic whole with an internal harmony that, when exposed to a new selection pressure, will change almost automatically in a harmonious manner.

The phenotype, as we find it in nature, is a compromise between all the selection pressures to which an organism is exposed. Thus, under domestication or in the laboratory it is possible through selection to generate almost any specialization of an organism, but always at the expense of some other characteristics. A highly selected domestic animal would not long survive in the wild. In *Drosophila* it is possible by appropriate selection to increase the number of bristles on various segments of the body quite drastically, sometimes almost to double them. However, in these extremely selected stocks one encounters many difficulties, particularly sterility. Extreme selection has clearly destroyed the harmony of the genotype. In nature, where the premium is always on overall performance, such disturbances are only rarely encountered. However, the fact that more than 99 percent of all phyletic lines that existed in the past are now extinct proves that species in nature are by no means always capable of achieving feasible adaptations to new selection pressures.

CONSEQUENCES OF THESE FINDINGS

Many of the most controversial phenomena of evolution appear in a new light as a consequence of the new concepts of genetics.

Evolutionary Tendencies

When the distribution of characters in various groups of plants and animals is studied—as naturalists have done for hundreds of years—very definite tendencies may be observed. Among the artiodactyls (deer, bovids, antelopes, sheep, goats, and others), for instance, there is a tendency to develop horns. This tendency is completely absent in other groups of mammals, such as rodents, insectivores, or lagomorphs. There is a tendency for spots to appear in certain groups of beetles, such as the coccinellids, while in other families there is a tendency to form stripes. Formation of bands is widespread among land and marine snails, but among the bivalves such a tendency seems to be absent. Certain groups of birds tend to form crests, whereas other groups tend to form ornamental plumes on the flanks, on the wings, or on the tail.

Similar tendencies are found not only by systematists and comparative anatomists, but also by biochemists, comparative physiologists, and cytologists. The cytologist M. J. D. White (1962) has spoken of the "principle of karyotypic orthoselection":

> In each particular lineage the structural changes that manage to establish themselves tend to be of the same type. In one lineage pericentric inversions convert rod-shaped chromosomes into V-shaped ones; in another fusions progressively diminish the number of chromosomes; in yet another heterochromatic arms are added to what were originally effectively one-armed chromosomes. Thus one chromosome after another tends to undergo the same type of structural change, thus keeping the members of the karyotype far more uniform in size, shape and other features than would be the case if structural changes were even approximately random.

One who sees the driving force in evolution in single mutations is unable to explain tendencies that are confined to certain families and orders of animals. But one who interprets such characters as the product of a harmonious gene complex considers it only natural that the entire gene complex of a higher taxon has certain proclivities. Once the ontogenetic pathway has been organized in such a manner that it tends to form horns on the skull, it will be relatively easy for the formation of horns to be actually realized in various related groups. The potential is presumably still present even in lines in which the formation of horns is suppressed and therefore can again be expressed at a later evolutionary stage. However, a group that lacks such ontogenetic tendencies will form horns only with a very low probability.

Periods of Stagnation and Florescence

The characteristics of harmonious gene complexes can also explain two well-known evolutionary phenomena that were exceedingly puzzling to the old mutationists: periods of evolutionary stagnation and flowering. Paleontologists have described many cases in which a certain morphological type experienced no essential evolutionary change for hundreds of millions of years. The horseshoe crab *Limulus,* the lamp shell (brachiopod) *Lingula,* and the fairy shrimp (phyllopod) *Triops* are among the best-known examples of this phenomenon. Each of these three genera has remained essentially unchanged during the last 2 to 4 hundred million years. Since all other interpretations have been a failure, one is forced to postulate that the physiology of the entire genotype of these species is so harmoniously balanced that selection eliminates all new mutations as disturbing elements. The importance of this type of selection, called stabilizing selection, has become ever more apparent in the course of recent years. Selectionists have found that individuals near the mean value of the population in their morphological characters often show higher fitness than individuals with greater deviations from the mean (Falconer 1960; Lerner 1954).

The correct explanation for phylogenetic stagnation in morphological characters seems to be that all gene-controlled processes during ontogeny are so precisely adjusted to each other that even modest phenotypic deviations have disturbing effects. Lerner has designated as genetic homeostasis the process through which each more pronounced deviation from the mean value is correlated through a feedback mechanism with a reduced reproductive success.

Phyletic lines that have become too rigid usually become extinct sooner or later. Occasionally, however, they succeed in escaping their homeostatic straightjacket and in loosening up their morphotype. How this is done is still a puzzle. Hybridization is perhaps important among plants. In animals the passage through the bottleneck of a founder population seems to be crucial. Whatever the explanation might be, it is well established that many evolutionary lines have suddenly experienced an almost explosive revitalization after long periods of stagnation. In the course of a few million years they may form a dozen or more new genera, some of which may become the founders of entirely new families. One has the impression that the gene complex becomes loosened up to such an extent that altogether novel combinations and new specializations can be experimented with. Such periods of new vitality have been described by many paleontologists (Rensch 1960; Schindewolf 1950a). Eventually, stabilization occurs and the gene complex seems again to become more rigid from generation to generation. A new period of evolutionary stagnation has begun.

I hope my analysis has made clear that the phenomena of directional evolution are by no means in conflict with the Darwinian theory of evolution. One who understands that selection deals with individual genes only quite indirectly through the phenotype will understand that genes which jointly participate in the formation of the phenotype are fitted together in a harmonious gene complex and that such a gene complex can, therefore, evolve only harmoniously.

REFERENCES

Britten, R. J., and E. H. Davidson. 1969. Gene regulation for higher cells: a theory. *Science*, 165:349–357.

Falconer, D. S. 1960. *Introduction to quantitative genetics.* Oliver and Boyd, Edinburgh.

Jacob, F., and J. Monod. 1961a. Genetic regulatory mechanisms in the synthesis of proteins. *Jour. Mol. Biol.,* 3:318–356.

—— 1961b. On the regulation of gene activity. *Cold Spring Harbor Symp. Quant. Biol.,* 26:193–211.

Lerner, I. M. 1954. *Genetic homeostasis.* Oliver and Boyd, Edinburgh.

Rensch, B. 1960. *Evolution above the species level.* Columbia University Press, New York.

Romer, A. S. 1949. Time series and trends in animal evolution. In *Genetics, paleontology, and evolution,* ed. G. L. Jepsen, E. Mayr, and G. G. Simpson. Princeton University Press, Princeton.

Schindewolf, O.H. 1950a. *Grundfragen der Paläontologie.* Schweizerbart, Stuttgart.

—— 1950b. *Der Zeitfaktor in Geologie und Paläontologie.* Schweizerbart, Stuttgart.

Simpson, G. G. 1953. *The major features of evolution.* Columbia University Press, New York.

White, M. J. D. 1962. Genetic adaptation. *Austr. Jour. Sci.,* 25:179–186.

6
Population Size and Evolutionary Parameters

In the first 100 years after 1859, no objection was raised more frequently against Darwinism than that it was too materialistic and that it did not allow for the special properties of living organisms and their design. In this, the theologians agreed with the Cartesian dualists. Although this objection is still being heard, no matter how often it has been refuted, a new opposition has developed in recent decades based on entirely different arguments. When I lectured in the mid-1950s to a small audience in Copenhagen, the great physicist Niels Bohr stated in the discussion that he could not conceive how accidental mutations could account for the immense diversity of the organic world and its remarkable adaptation. As far as he was concerned, the period of 3 billion years since life had originated was too short by several orders of magnitude to achieve all of this. In the intervening years several mathematicians have attempted to simulate the evolutionary process on the computer, and they likewise concluded that the Darwinian paradigm is incapable of accounting for the facts of evolution. This conclusion was, of course, vigorously opposed by the evolutionary biologists, whose painstaking analysis of the facts of evolution led them to confirm the original Darwinian thesis that genetic variability combined with natural selection is indeed able to account for the seeming perfection of the living world.

During these discussions it became obvious that the opposing camps talked two different languages and held quite different conceptions. It was finally agreed to hold a conference of mathematicians and evolutionary biologists in an endeavor to clarify the issues and to remove as many misunderstandings as possible. This conference was held April 25 and 26, 1966, under the auspices of the Wistar Institute of Anatomy and Biology at Philadelphia (see Moorhead and Kaplan 1967).

There probably has been no other recent meeting which has brought

Adapted from "Evolutionary challenges to the mathematical interpretation of evolution," pp. 47–58 in *Mathematical challenges to the neo-Darwinian interpretation of evolution*, ed. P. S. Moorhead and M. M. Kaplan, Wistar Institute Symposium Monograph no. 5 (Philadelphia: Wistar Institute Press, 1967).

out as clearly the differences in thinking of physical scientists and biologists. For instance, all the entities of a given class with which chemists and physicists work are identical with one another, except for movement and location. A sodium atom is identical with any other one and each pi-meson is like any other. This absolute identity of "individuals" in contrast to the unlimited diversity of individual organisms is one of the greatest differences between animate and inanimate nature. Even though physical scientists think statistically when describing the probability of the occurrence of certain processes, they tend to think typologically when the description of things is involved. They are justified in doing so, for indeed the entities they deal with are usually identical and can thus be described typologically. However, to transfer this assumption to biology may lead to serious error. The mathematicians participating in the conference subconsciously made the assumption that individuals of a species were genetically identical, necessitating a long time span before a gene locus could pass from one homozygous to a different homozygous condition. For them, evolution seemed to be "tandem evolution": the succession of genetic events produces a succession of homogeneous populations or species. They did not consider the possibility of a simultaneous genetic variability of ten thousands of gene loci because it would be impossible to calculate its effects. When they spoke of the evolutionary improvement of "the eye," they made the silent assumption that all eyes in a species are identical, although there is much reason to believe, as biologists have pointed out, that there may be a billion different eyes in a species with a billion individuals. Physical scientists find such variability quite inconceivable, and yet no one can truly understand the evolutionary process until he accepts the magnitude of genetic variation.

Another issue at the conference which prompted a great deal of discussion was that of the "constants" or "parameters" of evolutionary change. If one wants to simulate the evolutionary process on the computer, one must assign a numerical value to the frequency of mutation, selection pressure, population size, fertility, gene flow, and all other biological factors of evolutionary significance. Here again there was a great difference of opinion between mathematicians and biologists. The mathematicians quite naturally wanted clear-cut, hard figures to be inserted into their equations. The biologists, on the other hand, demonstrated that most of these factors varied by several orders of magnitude in different species under different conditions. In other words, if one wants to make computer simulations one must repeat them for the entire range of possible values of the controlling factors.

The great range of variation of another factor, the size of populations, must not be overlooked. That even the same population may greatly fluctuate in size was emphasized 50 years ago by Chetverikov and Timofeeff-Ressovsky, who further stressed that the total population of a species very often consists of numerous more or less isolated small populations, each of which up to a point can go its own way.

Methods to measure these evolutionary constants in natural popula-tions are very difficult to apply and for this reason I suggested at the 1966 Wistar conference that we approach the problem from the other end. If we look at some particular evolutionary situation and try to reconstruct the pathways by which the populations and species reached their present condition, what suggestions can we make about which factors are important and the order of magnitude of the various param-eters? I suggested that we look at three kinds of evolutionary phenom-ena—evolutionary rates, rates of speciation, and rates of extinction—and attempt to determine numerical constants.

EVOLUTIONARY RATES

Let us begin with evolutionary rates. You all know what extra-ordinarily rapid rates of evolution were discovered when pesticides were first applied in agriculture and antibiotics in medicine. Evolutionists are also familiar with the rapidity by which industrial melanism spread in a number of species of cryptically colored moths.

The house sparrow (*Passer domesticus*) has evolved a large number of genetically different local populations since its introduction into the western world in the 1850s. In spite of frequent subsequent trans-plantations and of active dispersal, some of the populations, such as those in Mexico, in British Columbia, and in Hawaii, and to a lesser degree those in localities that are far less distant, became as different as some populations in other species that ornithologists had called differ-ent subspecies. An extraordinarily rapid rate of evolution is evident for size as well as for pigmentation (Johnston and Selander 1964).

Such rapidity of evolution is one side of the coin. Let us now turn it over and contemplate slow rates of evolution. The paleontological literature records many cases like those of *Lingula* and *Limulus* and other genera of marine invertebrates that have existed since the Cambrian, or Ordovician, or Silurian for periods of between 300 and 600 million years. There is a species of freshwater fairy shrimp (*Triops cancriformis*) that exhibits extraordinarily complex structural detail in its extremities. This species is known to have existed in the Triassic, a period of something like 180 or 200 million years ago. The only differ-ence biologists have been able to discover between modern specimens and the Triassic ones is that the latter are a little smaller than the average of the modern populations.

Evolution among microorganisms seems to have proceeded even more slowly, at least as far as morphological detail is concerned. There are blue-green algae and various other primitive organisms of unknown relationship going back to 1 or 2 billion years ago that are virtually identical with living forms. Some of this material is extraordinarily well preserved, so that the gelatinous sheath and other detail is quite clearly visible. Even if some slight morphological change occurred in 1000 million years, one must admit that it has been negligible compared to the rate of evolution of the angiosperms, carnivores, or elephants during the Tertiary. The point I am trying to make is now quite obvious. On the one hand, we have visible evolutionary change in a period of 100 years, and in the other case we have hardly visible or invisible changes in a thousand million years, that is, in a period that is larger by seven orders of magnitude. If we want to make a computer simulation of evolution what kind of factors should be entered into our equations?

RATES OF SPECIATION

Let us now take a similar look at the process of speciation and again look at both sides of the coin. Lake Victoria in East Africa is a huge freshwater lake with a rich fauna of fish. The most common genus is the cichlid genus *Haplochromis* with several hundred species in the lake. On the northwest side of Lake Victoria is a small lagoon, Lake Nabugabob, separated from the big lake by a ridge of sand dunes. This lake is about 1 square kilometer in size and has a fauna of six species of *Haplochromis*. One of them is identical with one of the species in Lake Victoria; the other five are considered by the specialists to be endemic species, although each of them is related to one particular species in Lake Victoria (Greenwood 1965).

Lake Nabugabob was clearly a bay of Lake Victoria when the water level was higher and the five endemic species have evolved since the two lakes became separated. In order to determine the rate of speciation we must date the separation of the two lakes. This date has indeed been determined with the help of methods of quaternary geology and carbon dating and has been found to be a maximum of 4 or 5 thousand years. The conclusion is thus firmly established that the five endemic species evolved in the incredibly short time of less than 5000, and possibly less than 4000, years.

Let us now turn the coin over and study the effects of another geographic barrier, the Isthmus of Panama, which connects Colombia with Central and North America and which separates the Caribbean from the Pacific. Formerly there was a broad oceanic connection between the Caribbean and the Pacific that permitted the mingling of the faunas of the two oceans. Indeed, they had a single fauna. The final closing of this water gap took place about 5 to 6 million years ago when the Isthmus of Panama arose. The question we must raise is: What happened to the formerly uniform fauna that was separated into two faunas through this geological event? In many instances, particularly in the case of intertidal and shallow-water organisms, speciation has occurred, and we now find species pairs, so-called geminate species, with one representative species in the Caribbean and the other in the Pacific. What is far more interesting, however, is that there are many other species in which the Caribbean and the Pacific populations are either indistinguishable or so slightly different that they do not deserve to be considered different species. This seems to be particularly true for pelagic species. What could be the reason for this exceedingly slow rate of speciation? It seems to me that population size is involved. In the case of the African cichlid fishes, rapid speciation occurred where the total population size was very small. The rising of the Isthmus of Panama, however, separated two colossal gene pools. And when a gene pool is very large, evolutionary change occurs very slowly. The rates of speciation in the two cases differ by at least three orders of magnitude.

RATES OF EXTINCTION

The rate at which species and other taxa become extinct is very difficult to determine. Paleontologists have long known that massive extinction occurred in certain geological periods, such as at the end of the Permian and at the end of the Cretaceous. At other periods there seems to have been relatively little extinction. Recently I have found indications that on islands the rate of extinction seems to depend on the size of the island (Mayr 1965). If we plot the percentage of endemic species of birds against the size of islands, we find that this percentage rises very regularly with the increase in the size of the island. This is true not only for relatively small islands but also for such large ones as New Caledonia and Madagascar. To me this was rather unexpected. I would have predicted a rather rapid turnover on small islands, but I would not have expected island size to influence the percentage of endemic species when the islands had reached a size of 1000 to 5000 square kilometers. I was mistaken. The question that needs to be answered is the meaning of this correlation. The interpretation which best fits the observations is that the larger the island, the lower the faunal turnover, that is, the lower the frequency of extinctions. A very large island has a well-balanced fauna that is relatively immune to the impact of new colonists. Being a well-balanced universe, it has a high percentage of endemic species that have lived there a long time and have sometimes even had adaptive radiations on the island.

Conversely, the smaller the island, the smaller the gene pools of all the species and the less their storage capacity for genetic variation. The less the genetic variation, the greater the vulnerability of these species and the greater their susceptibility to extinction, particularly in the face of new immigrants. As a result, the percentage of endemic species is the smaller, the smaller the island. Indeed, the turnover of bird faunas is so rapid on many islands that they harbor no endemic species at all.

The most important message to be derived from the rates of evolution, speciation, and extinction is the importance of population size. It seems to me that this evolutionary factor has been far too much neglected in the past. The size of the gene pool has an important influence on the interaction between genes, that is, on the cohesion of the total genotype.

KINDS OF EVOLUTION

Up to this point my emphasis has been on *evolutionary change.* Indeed, as far as the nonspecialist is concerned, evolution to a considerable extent means progressive evolution, that is, evolution toward a condition of ever-greater perfection, as illustrated by the changes during the horse phylogeny in adaptation to changing climatic and vegetational conditions and such other evolutionary trends as increasing brain size in the hominid line and evolution toward an ever-greater perfection of the eye. This concept of progressive evolution is largely a remnant of the *scala naturae* idea of the old philosophers, which is only very imperfectly substantiated by the fossil record, as has been demonstrated by Simpson and others.

A second type of evolution, which we might call *maintenance evolution,* is more frequent and perhaps more important. This comprises all the pressures of stabilizing and normalizing selection that maintain balance and stability in the gene pool of a species once that species or gene pool has reached an adaptive equilibrium with its environment. Much, if not most, natural selection is concerned with maintenance evolution.

Perhaps we should recognize a third type of evolution, one that is more difficult to elucidate than the two types just discussed. It is what we might call *switch evolution,* evolution occurring when, in the simplest case, an evolutionary line invades a new niche. In an exceptional case, the new niche may turn out to be the entrance to a whole new adaptive universe, what Simpson (1953) calls a new "adaptive zone." The appearance of the first bird or pro-avis able to glide successfully from one tree to another was the beginning of the invasion of the wholly new adaptive universe of the aerial reptile. Anything that leads to the acquisition of evolutionary novelties or to entry into a new adaptive zone involves switch evolution. In order not to be misunderstood, let me emphasize that during switch evolution there is always simultaneously maintenance evolution. This is particularly true during a relatively rapid change of the environment. An organism, in order to maintain an adequate adaptation during a rapid change of the environment, is sometimes forced to adopt a new utilization of environmental resources.

The fourth kind of evolution we might distinguish is *speciation,* the splitting of an evolutionary line into several daughter lines, each of them acquiring its own isolating mechanisms and its own niche.

RELATIVE SIZE OF EVOLUTIONARY PARAMETERS

Let us now look at the relative size of various evolutionary parameters in relation to these four kinds of evolution. Let us take, for instance, genetic input into gene pools or species. Mathematical popula-

tion geneticists, for the sake of simplicity, have tended to lump mutation and gene flow together in their calculations (gene flow being the
movement of genes from one population to another). In the early
history of population genetics this was a legitimate simplification,
because ultimately all variation is, of course, due to mutation. Nevertheless, when we study concrete evolutionary situations, we find that
the role of gene flow and mutation are quite different.

When is gene flow more important and when is mutation more
important? Under what circumstances are either of them deleterious?
During progressive evolution high gene flow is desirable because it helps
the spread of favorable new gene combinations and thus speeds the
adaptive advance of the species as a whole.

In maintenance evolution gene flow is likewise important, although
for somewhat different reasons. Genes are steadily lost during reproduction through errors of sampling in temporarily isolated small
populations—a phenomenon quite rightly emphasized by Sewall Wright.
Temporarily lost genes are restored by gene flow and the population
thereby regains its level of heterozygosity, which seems necessary for
the overall genetic homeostasis of genotypes. What is so important
about gene flow, in contrast to mutation, is that it supplies "pretested"
genes and gene combinations, that is, genes and gene combinations that
have already been tested by natural selection in other populations. Even
though such pretested gene combinations may have to be adjusted, by
selection, for optimal coadaptation in the new gene pool, they nevertheless provide the type of new variability that is beneficial for a
population and that can be adjusted by natural selection. The input of
untested new mutations, on the other hand, very frequently has a
deleterious effect.

There are other kinds of evolution, however, in which the opposite
condition is important and advantageous: a cessation of gene flow. This
would seem to be true in switch evolution. When a population is in the
process of switching into a new adaptive zone, it is, so to speak, painfully building up a new gene complex that adapts it to this new situation. To have this population continuously polluted by genes from the
parental gene pool would greatly delay (or prevent altogether) the
process of genetic reconstruction. This is in part the reason that evolution can advance so rapidly in small, peripherally isolated populations.
Evidently, then, both in switch evolution and in speciation a reduction
of gene flow is advantageous. Under both circumstances mutation could
be highly advantageous because it might provide just the kind of new
genes that were not present previously. Such genes would be most
welcome to build up new isolating mechanisms in the case of speciation
or to improve adaptations in the case of switch evolution.

In these cases the evolutionary importance of small population size is
particularly evident. Gulick, Hagedoorn, and Sewall Wright have rightly
emphasized the importance of small populations for evolution. For a

long time, or so it seems to me, they emphasized too strongly the stochastic aspects of such phenomena as random fixation. My own conclusion is, as I first pointed out in 1954 (Essay 15), that the most important aspect of population size is that it affects the selective value of certain genes and gene combinations. There are reasons to believe that some genes have a higher selective value in big, populous species with a great deal of gene flow, while other genes are more advantageous in small, isolated populations with an increased amount of homozygosity due to inbreeding.

There are many reasons, some of which I have discussed in detail in Essay 15, why evolution in these small, isolated populations is different from evolution in large ones. Locally isolated populations, for instance, can respond to local selection pressures far better because they no longer have to absorb a great deal of gene flow from the remainder of the species population.

These considerations of time, of space, and of population size lead us to a picture of evolution that in some respects, particularly in emphasis, is appreciably different from the classical, neo-Darwinian picture given in the literature. When mathematicians talk about the neo-Darwinian model, they generally think of R. A. Fisher of 1930 and Sewall Wright of 1931. To be sure, what Fisher and Wright said is still largely correct, but it is only part of the answer owing to the new ideas that have to be superimposed on it.

It appears that every once in a while a new species originates that has a highly successful genotype. As a result it can spread widely, can become very populous, and can continue to improve slightly by progressive evolution. However, once it has reached its adaptive peak, it will be subject primarily to maintenance evolution. When a species has reached the status of a successful, widespread species, it seems to become rather incapable of undertaking major switches or acquiring major evolutionary novelties. Furthermore, it is unable to speciate, at least within the main body of the species.

Significantly, what the paleontologist usually finds in fossil deposits are remnants of such widespread, populous species. But it would be misleading to base our interpretation of evolution on such a nonrandom sample. In fact, it provides only part of the story.

In contrast to the widespread, populous species is the second kind of species, consisting of many small, isolated populations, most of them being peripheral, geographical isolates, more or less completely separated from the contiguous set of populations in the core area of the species. Such isolated populations, particularly those established by founders, can undergo a great deal of genetic change in a very short time. They can acquire new isolating mechanisms and experiment with new kinds of niche utilization and the invasion of new adaptive zones. The vast majority of such experiments are unsuccessful, just as the vast majority of all mutations have no lasting evolutionary effects; yet occasionally an evolutionary experiment is successful and an evolutionary advance is made. This leads sometimes to the replacement of one of the previously successful species. In other cases, the new evolutionary line finds a previously unoccupied ecological niche or adaptive zone and succeeds in spreading there without greatly disturbing the existing organic universe.

COMPUTER SIMULATION

To me, as an evolutionist, the greatest advantage of computer simulation is the fact that the computer can simulate such a wide range of possible conditions. It should be possible to develop programs that reflect the actual conditions in nature. Evolution is the change of systems, the modification of one extremely complex system into a different one under the impact of extremely complex sets of selective and random forces. I consider it very doubtful that simplistic, deterministic models will give us a realistic picture of evolutionary events.

In my *Animal Species and Evolution* (1963), I devoted an entire chapter to discussing the unity of the genotype. I stressed that one cannot expose *a gene* to selection. The organism responds to selection as a whole. Yet, in due time, it may be possible to dissect at least part of this wholeness into its components. Much light has been shed on this through the study of macromolecules. If an enzyme has 100 amino-acid residues and the active site is occupied by only 8 residues, this does not mean that the other 92 are irrelevant or "neutral," as is sometimes claimed. They are not "noise," as the information theorist might say, or "garbage," as one biochemist once said. It is now quite evident that these other 92 sites have something to do with the folding of the molecule, and this in turn affects the interactions of the molecule with others and facilitates the operations of the active site. How important the precise configuration of such molecules is, is indicated by their evolutionary constancy. For instance, so far as I know, the molecule of cytochrome C is uniform in the entire order of ungulates. Even though in this large and species-rich order of mammals there must have been millions of mutations at the 103 sites since the early Oligocene, they were all sooner or later eliminated by selection. This shows how important even the smallest detail of a molecule is and how optimal efficiency is constantly maintained by natural selection.

The interaction of genes is more and more recognized as one of the most important evolutionary factors. The longer a genotype has resisted a major reorganization, the stronger will be its developmental homeostasis, its canalizations, and its system of internal feedbacks. Such a genotype will respond to selection pressures as a whole rather than as an aggregate of independent genes. I have already mentioned evolutionary lines that have remained essentially unchanged for 50, 100, or even 900 millions of years. At a time when evolutionists did not make a clear distinction between genotype and phenotype, such evolutionary constancy was explained in terms of low mutation rates or low selection pressures. As soon, however, as one adopts the model of the internally balanced, almost totally homeostatic genotype, one can explain evolutionary constancy in the presence of normal mutation and selection rates.

One of the real puzzles of evolution is how a perfectly coadapted system can be loosened up in such a way as not to induce extinction. Such a loosening up of previously exceedingly stable evolutionary lines has been observed several times in the fossil record. Evolutionists have referred to this as "explosive evolution" or "evolutionary flowering," usually combined with strong adaptive radiation. Paleontologists have described cases in which a genus had remained virtually unchanged for 50 or 100 million years until it suddenly burst out into 12, 15, or even more descendant genera. After a relatively short time span, geologically speaking, such evolutionary lines usually experience a period of heavy extinction but then return to the previously existing stability.

I have stressed this unity of the genotype because most mathematical models of evolution are based on changes in the frequency of single genes. I do not believe that such models are realistic. Admittedly, it is an exaggeration to claim that every character of an organism is affected by all genes and that every gene affects all characters, and yet this claim is perhaps closer to the truth than the early Mendelian concept of the organism as an aggregate of independent genes.

The most important evolutionary phenomena are almost without exception changes in quality. Some critics conceded to Darwin that the animal breeder could select the "best," that is, the most productive, individuals. But how can one say that "nature," personified, exercises a similar choice, they asked? Is this not going back to invoking supernatural forces, they asked? Indeed, by using the word *select* Darwin had given the impression of a deliberate, conscious process. Such selection would be an *a priori* activity, but natural selection is actually a reward system for superior performance, that is, it is an *a posteriori* process. The difference between artificial and natural selection is, however, not as great as here suggested. The selection by an animal breeder is also a reward for superior performance, and the implied prediction of superior performance by the offspring frequently fails to come true. The reason for the unpredictability is the extremely large number of potentially interacting components. The quality of the phenotype depends on the interaction of a very large number of genetic combinations with a large number of components of the environment. Absolute prediction is therefore not possible, but probabilistic prediction is feasible, and is the basis not only of the success of animal and plant breeders, but also of all evolutionary adaptation.

What does all this mean to him who wants to simulate evolution with the help of the computer? I think it should mean one thing in particular, which is that the approach adopted should not be too simplistic. To be sure, one will have to start with a set of simplified assumptions and expand from them gradually; but in the end one would have to adopt for every set of factors a far greater range of extremes than was believed necessary or even possible only 20 years ago. Evolution, again and again, has resulted in unique phenomena and in startlingly unpredictable phenomena. If we set up our programs in too deterministic a manner, I am afraid we will never be able to arrive at a realistic interpretation of evolution.

REFERENCES

Greenwood, P. H. 1965. The cichlid fishes of Lake Nabugabo, Uganda. *Bull. Brit. Mus. (Nat. Hist.)*, 12:315–357.

Johnston, R. F., and R. K. Selander. 1964. House sparrows: rapid evolution of races in North America. *Science,* 144:548–550.

Mayr, E. 1963. *Animal species and evolution.* Belknap Press of Harvard University Press, Cambridge, Mass.

—— 1965. Avifauna: turnover on islands. *Science,* 150:1587–1588.

Moorhead, P. S., and M. M. Kaplan, eds. 1967. *Mathematical challenges to the neo-Darwinian interpretation of evolution.* Wistar Inst. Symp. Monogr., no. 5. Wistar Institute Press, Philadelphia.

Simpson, G. G. 1953. *The major features of evolution.* Columbia University Press, New York.

7
From Molecules to
Organic Diversity

The words chemistry and evolution, when used together, tend to evoke a very specific image in our minds. It includes questions about the origin of life, about the primacy of heterotrophism over autotrophism, at what stage nucleic acids entered the picture, the shift from anaerobic to aerobic pathways, and many related topics discussed in numerous recent symposia.

What fascinates me about these evolutionary steps is that they all occurred so relatively soon after the origin of life. The vast majority of the kind of molecules and macromolecules that are of primary importance in living organisms, and of the chemical pathways that lead either to more complex molecules or to their breakdown, exist already among the bacteria.

Since these questions have been abundantly discussed in many symposia, I want to concentrate on biochemical evolution in higher organisms. I am particularly concerned with what seems to me to be one of the great paradoxes of nature: the contrast between the essential chemical unity of organisms from the lowest bacteria to the highest animals and plants, with the truly staggering diversity of life on land, in the water, and in the air. To what extent can we hope to explain the evolution of this diversity in terms of chemical evolution? What role has chemical evolution played in the evolution of this diversity?

I hardly need to stress that extraordinary uniformity of life, not only in its cellular constituents but also in its chemical components and metabolic pathways. Whether we deal with information transfer through precisely replicating nucleic acids, or the role and nature of enzymes, or energy transfer through ADP and ATP, the unity of life from bacteria and viruses to the higher animals and plants is astounding. Even the simplest of all free-living organisms, the procaryotes, are metabolically complete. Energy transfer by polyphosphates, information transfer through nucleic acids, protein formation, catalysis through enzymes, all that is already present to perfection.

Reprinted and revised from "From molecules to organic diversity," *Federation Proceedings,* 23(1964):1231–1235.

In order to bring the magnitude of our problem into sharper focus, let us have a quick look at organic diversity, the extent of which only the specialist in zoology and botany can fully appreciate. We find astounding diversity even at the level of unicellular organisms, viruses, rickettsias, bacteria of all sorts, blue-green algae, innumerable kinds of fungi, regular algae, and protozoans. The diversity becomes even more overwhelming when we consider multicellular organisms, the classical kingdoms of plants and animals.

Some 16 major categories are known among the multicellular plants, ranging from algae, fungi, and mosses to ferns and seed plants. The most advanced division among these groups, the angiosperms, contains several hundred thousand species.

The diversity in the animal kingdom is, if anything, even greater. There are probably more than 3 million species of animals in existence, of which 1 million have already been described by systematists. The living animals are grouped in about 25 so-called phyla, each of which is set apart by structural characteristics that distinguish it clearly from all others. In order to bring home to you the enormous, the well-nigh inconceivable diversity of the animal kingdom, let me single out just 2 of the 25 phyla, the arthropods and the chordates. The phylum arthropods is subdivided into about 15 classes, the three best-known among which are the spiders and their relatives (including scorpions and mites), the crustaceans (including lobsters, shrimp, and countless small plankton types), and finally the insects, with about three quarters of a million described species. In the phylum chordates I need only mention mammals, birds, reptiles, amphibians, and fishes, all of them chordates, in order to call to mind the diversity in that phylum, and there are still some 23 other phyla of animals, not counting extinct types.

Let me now make my point. Can we explain this almost inconceivable amount of organic diversity in biochemical terms? What role has chemical evolution played in the evolution of organic diversity? Were the many new evolutionary developments initiated by chemical mutations, or are the chemical changes observed in various phyletic lines merely a secondary response to selection pressures set up by evolutionary shifts in behavior or ecology and by the invasion of new adaptive zones?

To be honest, we must admit that we do not yet have the facts to give a complete answer. But we can advance some tentative formulations.

Let us look at two decisive steps in the phylogeny of organisms, the step from the procaryotes to the eucaryotes and the step from unicellular eucaryotes to multicellular plants and animals.

The procaryotes, which include the bacteria and the blue-green algae, differ from the higher protists, the eucaryotes, by greater simplicity, that is, by a lower level of organization. They lack mitochondria and chloroplasts, that is, specialized organs for respiration and photosynthesis, and the nucleus is not separated from the cytoplasm by a membrane. In contrast, the most striking characteristic of the eucaryotic cell, as clearly stated by Stanier and van Niel (1962), is that its major work areas of cellular function, that is, respiration, photosynthesis, and genetic information transfer, are surrounded by individual membranes, thus creating a new level of intracellular organization. This organizational difference between procaryotes and eucaryotes is greater than any chemical difference between the two groups, except for the relative amount of DNA, which is far greater in the nucleated protists than in the bacteria. And this is part of our story.

Let us now look at the second major phylogenetic step, that from unicellular to multicellular eucaryotes. Let me first say that this step was probably taken repeatedly by independent lines, perhaps several times among plants and possibly two or three times among animals. In no case is there a striking chemical difference, so far as I know, between the unicellular ancestor and the multicellular descendant. Rather, what the multicellular organization does is to provide the organism with an enormous new potential for a division of function and a specialization of parts and components. And in this the evolutionist can discern rather clear-cut trends.

Before discussing these, let me say a few words about evolutionary mechanisms, because this is an area in which there are still many misconceptions.

The evolutionary role of mutation was badly misunderstood by the geneticists early in the century. They regarded it as a driving force and one that moved forward in distinct, clear-cut steps, indeed in the form of veritable saltations. This idea may have some validity for haploid, uniparentally reproducing microorganisms. Among higher, sexually reproducing organisms, however, the role of mutation is simply to replenish the genetic diversity of the gene pool. It is natural selection that, at any given moment, gives direction to the evolutionary trend. Yet even in haploid microorganisms with minimal genetic recombination, it seems that natural selection is far more important for evolution than mutation.

Natural selection is another often misunderstood concept. Is it the "traditional law of tooth and claw," as one biochemist has recently phrased it? Actually, nothing could give a more misleading conception of the working of natural selection than Tennyson's metaphor. Natural selection is a statistical process, referring to reproductive success as expressed in the relative frequency of descendants in the next generation. It is not a bloody struggle.

"Natural selection works on genes." This statement is also totally misleading. Natural selection describes the reproductive success of individuals, that is, of phenotypes. To be sure, genes contribute to the selective valence of phenotypes and are therefore indirectly affected by natural selection. However, the selective superiority of higher organisms is mostly due to such components of the phenotype as tolerance to heat and cold and the ability to secure adequate food, to recognize and escape enemies, to reproduce, and to reproduce abundantly. Virtually all these characteristics are due to highly complex interactions of many genes. They are the products of whole gene systems. Evolutionary advances in multicellular organisms, therefore, are far more often due to the improvement of genetic and biochemical systems rather than to singular biochemical inventions.

When we apply current concepts of the role of mutations and of natural selection and of the relation between gene and the phenotype to the process of evolution, we come to the conclusion that evolution among higher organisms must be an exceedingly gradual process. This conclusion is, indeed, entirely confirmed by the studies of biochemists, geneticists, paleontologists, and general evolutionists. Let me single out two important evolutionary processes to test this conclusion, the origin of new species and the origin of new structures or organs, both processes interpreted by the early Mendelians on a macromutational basis. Actually, the origin of new species, except for cases of polyploidy, is an exceedingly gradual process (Mayr 1963). It consists in the slow genetic reorganization of geographically isolated populations (I do not have the space here to discuss the genuine difficulties posed by haploid and uniparently reproducing organisms). The acquisition of new structures or entirely new organs is likewise a gradual process, as discussed in Essay 16. There was never a threshold where a sudden jump was made. New systems evolved very gradually, requiring undoubtedly many biochemical adjustments but not requiring any sudden biochemical revolutions.

There are usually several possible answers to a need of the phenotype. For instance, when multicellular animals increased in size during evolution, they evolved a skeleton, but it so happens that this was an internal skeleton in some phyletic lines, such as the vertebrates, but an external skeleton in others, such as the arthropods. This engineering aspect of the skeleton at once determined the limits of further evolution in these lines. An internal skeleton permits continuous and virtually unlimited growth; hence the development of elephants, whales, and gigantic dinosaurs among the vertebrates. Growth in arthropods with an external skeleton is possible only through molts, and the soft-shelled creature emerging after a molt is not only vulnerable to attacks by enemies, but also unable to support its weight properly. Horseshoe crabs, lobsters, and some long-legged crabs are the ultimate in size that can be achieved by an organism with an external skeleton. The dramatic difference between these phyla is probably due to the fortuitous accident that, in comparatively small ancestral forms, in one case an internal and in the other case an external skeleton was invented. A further limit on the size of terrestrial arthropods is set by the inefficiency of their respiratory organ, the tracheae.

Evolutionists have often commented on the haphazardness and opportunism of evolution. But one should not exaggerate this aspect of evolution. Huxley (1942) and Rensch (1960a, b) have shown how many regularities can be found in evolutionary lines and have demonstrated, furthermore, that the resulting trends are caused by natural selection rather than by any teleological, finalistic principles.

In most, if not all, evolutionary lines of multicellular organisms, there is a tendency toward a change in size, usually an increase, but under special circumstances a decrease. Among the advantages of size increase are that it permits (1) greater specialization of body parts, (2) emancipation from the environment, and (3) protection against enemies (see also Newell 1949:103). Increased size, however, generates very specific demands. For instance, since all metabolic processes are more or less associated with membrane surfaces, that is, with two-dimensional structures, while body size increases by the cube, the growth of all surfaces has to be exponential in comparison to the increase in body size.

The interior of the body is farther and farther removed from the surface with increased size, and simple osmosis is no longer sufficient to transport oxygen and nutrients to all the cells and to remove the waste products of metabolism. A premium is set up for the evolution of transport systems, that is, organs of circulation, respiration, digestion, and excretion, that bring cells in every part of the body in close and efficient contact with the outside world.

Increased size places still other demands on the organism, of which I shall mention only two: (1) a firm skin, epidermis, shell, or bark, which protects the organism against the loss of moisture and against attacks by microorganisms or small predators; and (2) a skeleton to provide stability and prevent the collapse of the organism, particularly in the case of terrestrial organisms.

Similar generalizations can be made concerning trends in metabolism, control mechanisms, reproductive biology, sensory organs, the central nervous system, behavior patterns, and so forth. All these trends and specializations are made possible by the extraordinary flexibility that a multicellular organization gives to organisms. No case is known to me in which a change in body chemistry initiated a new evolutionary trend. Invariably it was a change in habits or habitat that created a selection pressure in favor of chemical adjustments.

In principle, the evolution of new kinds of higher organisms demonstrates the same phenomenon as the shift from procaryotic to eucaryotic protists. It is the development of new biological systems consisting largely of new constellations and differing proportions of the same basic unit elements. In other words, biological evolution is dominated by the continuous emergence of new systems: and systems often display characteristics that one could not have predicted on the basis of the properties of the unit elements.

I pointed this out, in the mid-1950s, in a lecture in Copenhagen that was attended by Niels Bohr. In the ensuing discussion, he agreed with my conclusions, except for reminding me that an emergence of new characteristics in systems was not peculiar to living systems. He cited the chemical elements, which are systems that owe their highly specific properties to the quantity and pattern of their simple unit components, the nuclei and electrons. These properties, Bohr said, could not have been predicted in detail on the basis of a knowledge of isolated protons, neutrons, and electrons. On the other hand, a study of the properties of these systems has taught us a great deal about the properties of protons and electrons.

The same is true for biological systems. Much of the difference among organisms is a matter of difference in systems rather than in unit components. A giraffe, an elephant, and a rabbit differ much more from one another in the relative size of extremities and organs and in other quantitative characteristics such as density of fur, thickness of the epidermis, relative frequency of various glands, and so forth, than in the presence or absence of certain basic chemical constituents.

The systems approach, which I have espoused here, strongly contrasts with an approach usually referred to as reductionism. It is the belief that systems can be fully understood by dissecting them, by analyzing them down to the lowest basic element. As valuable as an analytical approach is, in my opinion it is only a first step and should never be considered the final goal.

This still leaves us with a number of basic questions. Why is the organic world characterized by such uniformity in its basic chemical constituents and metabolic processes? Because there is usually only one kind of molecule or macromolecule that is able to perform a specific task with optimal efficiency. Considering the high selective premium on the acquisition of optimal processes, all basic inventions were made at the bacterial level or earlier. Since macromolecules are endowed with unique properties, there is often only a single way to construct more complex macromolecules, and as a consequence, specialized macromolecules may independently originate repeatedly in different phyletic lines, formed by the same universally available constituents. The polyphyletic origin of hemoglobin is one example. Another is the independent utilization of carotenoids as visual pigments by the arthropods, mollusks, and vertebrates. The reason, of course, is that the "carotenoids alone among natural pigments have straight-chain conjugated systems, capable of readily undergoing cis-trans isomerization by light" (Wald 1963: 18).

The simple theory of a conspicuous parallelism between organic evolution and biochemical evolution is thus not supported. Nevertheless, higher organisms differ chemically from their ancestors in manifold ways. Our revised question, then, must be phrased as follows: What is the nature of the chemical changes that accompany the elaboration of ever more complex biological systems? The answer to this question is beginning to emerge from modern genetics and from developments in biochemistry, which combine the best features of an analytical and a systems approach. In biochemistry we have acquired a whole new dimension of understanding through the systematic comparative phylogenetic study of characteristic macromolecules, such as hemoglobin and certain enzymes.

Let us look, in a very general way, at some of the results. A large enzyme may have some 300-500 or more amino-acid residues, yet the active site may involve only 10 or fewer residues. What then, may we ask, is the function of all those other residues that make up 98 percent of the molecule? These are the residues that used to be ignored and discarded in the study of the active site!

Ever since I became a zoologist, I have been impressed with the functional significance of structure. It is perhaps for this reason that I am particularly intrigued by the frequently made suggestion that the tertiary folding of the polypeptide chains at the nonactive sites contributes significantly to the specific biochemical milieu of a species. As an evolutionist I am furthermore impressed by the power of natural selection, and I doubt for this reason that any particular amino-acid configuration was ever replaced by a different one unless this replacement was favored by natural selection. The reconstruction of an enzyme outside its active site has almost surely something to do with the control of enzyme activity, and precision in this control would be of high selective value.

An organism is not just a bag full of enzymes. As I have repeatedly stressed, everything in an organism is part of a system, and I agree with those biochemists who believe that much in the structure of an enzyme has significance for the interaction of the protein with other proteins. If, in the course of evolution, some of the proteins of an organism undergo evolutionary changes, it is easy to understand that this might create a selection pressure in favor of remodeling other proteins in order to improve interaction. A single amino-acid replacement in a polypeptide chain may require additional substitutions in order to improve conditions for tertiary and quaternary folding. As an evolutionist, I am firmly convinced that natural selection exerts a continuous pressure on the interaction of the various chemical constituents of an organism. An improvement in one protein is apt to lead to unbalances in its interaction with other proteins and thus favor their reconstruction.

Our ignorance about the physiology of such changes and their selective significance is still almost complete. Yet such work as that on the phylogeny of the hemoglobins, which clarifies the evolutionary changes in the alpha, beta, delta, and gamma chains, is preparing the ground for a better understanding. It has been shown by population geneticists that balanced heterozygosity, that is, the simultaneous presence of two slightly different homologous enzymes, sometimes leads to greater physiological flexibility in an organism and thus to greater tolerance for changing environmental conditions. It would reduce the genetic load of the population if the same versatility were achieved by the simultaneous possession of both enzymes even in homozygotes, owing to duplication, translocation, or other genetic mechanisms. The hemoglobins characterized by the simultaneous presence of both alpha and gamma chains may illustrate this principle.

Regulation and feedback are the most important aspects of systems that function through the interaction of components. Our ignorance is still nearly complete in this area. Concerning the genetic information by which DNA governs the construction of complex systems, we now believe that in addition to the structural genes, which control enzymes, there are operator and regulator genes, which in turn control the activities of the structural genes. Their existence introduces a whole new dimension of flexibility and diversity into the developmental pathways (Monod et al. 1963:306; Pontecorvo 1963:1). This is surely part of the reason why the amount of DNA in higher organisms is so much greater than seems necessary for the programing of their basic enzyme systems. All aspects of relative growth are presumably governed by such regulator genes.

Many, if not most, of the differences among closely related higher organisms consist of differences in body proportions or in the degree by which certain characters are expressed. Again, these quantitative differences are presumably affected, if not wholly controlled, by regulator genes. At present, we know virtually nothing about their functional pathways. I would not be surprised, however, if research on the action of operator and regulator genes would not some day converge upon the study of amino-acid substitutions in enzymes at sites other than the active sites. I rather suspect that these other sites have a good deal to do with the regulation of enzyme activity.

Let me summarize my conclusions. The evolution of organic diversity is to a large extent due to the elaboration of ever more complex, ever more diverse systems. During much of this evolution the same basic chemical constituents or classes of chemical constituents have been utilized. However, there has been a great deal of modification of macromolecules, which presumably is correlated with the interaction of various proteins and with regulation of enzyme activity. Evolution of higher organisms must be studied both as an elaboration of systems and through the analysis of the elementary units. A one-sided emphasis of one to the exclusion of the other will fail to produce true understanding.

In biology, as already acknowledged in physics, the ideal approach is the combination of an analytical and a systems approach. The two approaches complement each other beautifully. The biologist studies systems of increasing complexity from molecules, macromolecules, cellular components, cells, tissues, organs, organ systems, individuals, families, populations, and species up to species aggregates. A system at any level is composed of elementary units that are the systems of the next lower level. On each level it is equally legitimate to study either the system as a whole or the elementary units of the system, but we will not get the whole truth unless we study both. It is fortunate, both for physics and for biology, that systems at higher levels can be studied with profit long before the elementary units at the lower levels are fully understood. The past history of biology has shown that progress is equally inhibited by an anti-analytical holism and a purely atomistic reductionism. A healthy future for biology can be guaranteed only by a joint analytical and systems approach.

REFERENCES

Huxley, J. S. 1942. *Evolution: the modern synthesis.* Allen and Unwin, London.

Mayr, E. 1963. *Animal species and evolution.* Harvard University Press, Cambridge, Mass.

Monod, J., J.-P. Changeux, and F. Jacob. 1963. Allosteric proteins and cellular control systems. *Jour. Mol. Biol.,* 6:306-329.

Newell, N.D. 1949. Phyletic size increase: an important trend illustrated by fossil invertebrates. *Evolution,* 3(2):103-124.

Pontecorvo, G. 1963. Microbial genetics: retrospect and prospect. *Proc. Roy. Soc. London,* ser. B, 158(970):1-23.

Rensch, B. 1960a. *Evolution above the species level.* Columbia University Press, New York.

—— 1960b. *The evolution of life.* University of Chicago Press, Chicago.

Stanier, R. Y., and C. B. van Niel. 1962. A concept of a bacterium. *Arch. Mikrobiol.,* 42:17-35.

Wald, G. 1963. *Evolutionary biochemistry.* Pergamon, London.

8
Sexual Selection and Natural Selection

Julian Huxley (1938a) remarked quite rightly, "None of Darwin's theories has been so heavily attacked as that of sexual selection." The reasons for this widespread criticism are manifold, and some of them will be analyzed in the present essay. Even though some of the criticism was justified, it is now clear that Darwin was right in principle and that the label "sexual selection" has helped to bring together and organize a vast body of scattered observations (for a summary of Darwin's views, see *The Descent of Man*, pp. 613–617). The vitality of the principle is best documented by the large number of recent review papers or chapters in general books (for instance, Boesiger 1967, Campbell 1972, Ghiselin 1969, 1974, Huxley 1938a, b, c, and Maynard Smith 1958).

Darwin devoted two thirds of his *Descent of Man* (1871) to the presentation and substantiation of his principle of sexual selection. This induced some of his critics to claim that he had invented this principle because he was unable to explain many attributes of man as due to natural selection. This insinuation is quite unfounded, as Darwin himself points out in the preface to the second edition (1874:vi) of *The Descent of Man*. Indeed, an entire section in Chapter 4 of the first edition of *The Origin of Species* (1859:87–90) is devoted to this subject and headed "Sexual Selection." His principal ideas on the subject are clearly sketched out in this early treatment. *The Descent of Man* was, as Darwin said, his first opportunity for a full-length treatment of the subject.

Darwin introduced the concept of sexual selection to explain certain aspects of the reproductive biology of animals that he was unable to ascribe to natural selection. The sharp distinction that he thereby made between two different kinds of selection was one of the reasons for the attacks on him. The controversy was aggravated by repeated shifts in the meaning of key words such as *fitness, struggle,* and *female choice*.

In order not to run into the same difficulties, we must define our terms. We must begin by finding out how Darwin himself defined

Abridged from "Sexual selection and natural selection," pp. 87–104 in *Sexual selection and the descent of man,* ed. Bernard Campbell (Chicago: Aldine Publishing Co., 1972).

sexual selection. He writes that it "depends on the advantage which certain individuals have over others of the same sex and species solely in respect of reproduction" (p. 209). In all cases in which the males have acquired a particular structure, "not from being better fitted to survive in the struggle for existence, but from having gained an advantage over other males, and from having transmitted this advantage to their male offspring alone, sexual selection must here have come into action. It was the importance of this distinction which led me to designate this form of selection as Sexual Selection" (p. 210). Darwin was fully aware that not all differences between the sexes are the result of sexual selection. Many are either sex-associated aspects of reproduction (primary and accessory sex organs) or are correlated with niche specializations of the two sexes, and of these Darwin states expressly, "They have no doubt been modified through natural selection" (p. 209). Sexual selection, in contradistinction, deals only with those components of sexual dimorphism that were acquired as a result of mere reproductive advantage.

A separation of sexual and natural selection makes sense only if one adopts the same definition of fitness as Darwin, who employed the term in an uncomplicated, everyday sense. Fit, to him, meant well adapted, and anything that improved the chance for survival in the struggle for existence increased fitness. Fitness for Darwin was the property of a whole individual, or, as we might say, of an entire genotype. Subsequently, the mathematical geneticists (Haldane, Fisher, Wright) redefined fitness rather drastically when they introduced the concept of the fitness of single genes. This required defining fitness in terms of the "contribution to the gene pool of the next generation." Fitness under this new definition could be due either to superior fitness in the Darwinian sense or to reproductive advantage that does not add to the adaptedness of the species. Sexual selection under this new definition merely becomes one of several various forms of natural selection. To be sure, this new definition facilitates the mathematical treatment, and yet I have a feeling that something rather important was lost in the process. In questioning the usefulness of this redefinition of fitness I am not alone. Huxley, for instance, said:

> When we examine the problem more critically, we find that we must differentiate between two quite distinct modes of natural selection, leading to different types of evolutionary trend, which we may call survival selection and reproductive selection. . . . In the actual processes of biological evolution, survival selection is much the more important; selection exerts its effects mainly on individual phenotypes, and operates primarily by means of their differential survival to maturity. This will produce evolutionary effects because, as Darwin saw, (a) the majority of individuals which survive to maturity will mate and leave offspring; (b) much

of the phenotypic variance promoting survival has a genetic basis. (1963: xviii–xix)

The questions that need to be answered now begin to become apparent. There are two in particular: Is it possible to make as sharp a distinction between natural selection and sexual selection as was claimed by Darwin and Huxley? And if the existence of genuine sexual selection can be established, what different forms does it take in various groups of organisms and how widespread in the animal kingdom are biological characteristics that can be interpreted as due to sexual selection?

Sexual selection as envisioned by Darwin usually resulted in sexual dimorphism, that is, in a difference between males and females. However, not all sexual dimorphism is the result of sexual selection. Most of the differences between the sexes are clearly the result of natural selection. Among these are accessory sexual characters (like claspers) that facilitate copulation and fertilization, as well as a wide range of characters having to do with parental care, such as the pouch of female marsupials, the mammae of female mammals, and the brood pouches of male seahorses and of many invertebrates. Females often show additional adaptations that have nothing to do with sexual selection. For instance, cryptic coloration is widespread among those species of birds in which females alone perform the duties of incubation, but in butterflies also the females are often more cryptically colored than the males. There is a strong selection pressure for reduced size in the females of the hole-nesting species of ducks (only the females incubate!). Consequently, there is an increased size dimorphism in these species that, at least initially, had nothing to do with sexual selection. The same is true for the sexual size dimorphism of many other species of animals. Additional cases of sexual dimorphism resulting from natural selection pressures will be listed below. However, they do not refute the possibility of genuine sexual selection.

WHAT CHARACTERS ARE DUE TO SEXUAL SELECTION?

With so many components of sexual dimorphism explicable through natural selection, it is legitimate to ask whether any are left that cannot be explained that way. Of this, Darwin was convinced:

There are many other structures and instincts [of males] which must have been developed through sexual selection—such as the weapons of offence and the means of defence of the males for fighting with and driving away their rivals—their courage and pugnacity—their various ornaments—their contrivances for producing vocal or instrumental music—and their glands for emitting odors, most of these latter structures serving only to allure or excite the female. It is clear that these characters are the result of sexual and not of ordinary selection, since unarmed, unorna- mented, or unattractive males would succeed equally well in the battle for life and in leaving a numerous progeny, but for the presence of better endowed males. We may infer that this would be the case, because the females which are unarmed and un- ornamented, are able to survive and procreate their kind. (pp. 210–211)

Darwin's supporters as well as his critics have pointed out that Dar- win bracketed together in this statement a rather heterogeneous set of phenomena. In particular, two classes of phenomena must be dealt with separately. First, there are the weapons and other characteristics of males displayed in fighting among themselves. The role of the females in these fights is strictly passive. Wallace and others denied that such characters qualify as products of sexual selection. Second, there are all those other characteristics by which males attempt to attract females and induce them to copulate. Let us begin with the discussion of this second category.

Male Ornaments and Attractants

There are numerous species of birds in which the males are brightly colored and adorned with special plumes, such as the birds of paradise or the peacocks. Such spectacular structures could never have evolved, says Darwin, unless females exercise a choice among various eligible males and, more than that, unless females have a sense for beauty.

There were two elements in Darwin's "female choice": first, a de- liberate preference by the female for one particular male chosen from a group of available suitors; second, an esthetic sense on the part of the female that resembles very much our human appreciation of beauty. Indeed, it was this deliberate choice of the female that suggested the term "sexual selection" to Darwin, as the equivalent to the conscious selection of the animal breeder:

> Just as man can give beauty according to his standard of taste, to his male poultry, or more strictly can modify the beauty originally acquired by the parent species . . . so it appears that female birds in a state of nature, have by a long selection of the more attractive males added to their beauty or other attractive qualities. No doubt this implies powers of discrimination and taste on the part of the female which will at first appear extremely improbable; but by the facts to be adduced hereafter, I hope to be able to shew that the females actually have these powers. (p. 211)

Alas, Darwin did not convince his contemporaries; their reaction was almost wholly negative. We must remember that Darwin published his theory of sexual selection during a period when even natural selection was rejected by most of his contemporaries. Not surprisingly, sexual selection had to face even rougher sledding. A. R. Wallace (1889) found it unacceptable, but, curiously, for the only time in his entire career he did not invoke natural selection in order to explain male beauty. He ascribed the brilliance of the male's plumage to excess vitality and found the cause "for the origin of ornamental appendages of birds and other animals in a surplus of vital energy leading to abnormal growth in those parts of the integument where muscular and nervous action are greatest" (p. 293). Wallace's almost "Lamarckian" explanation has been refuted often enough in the literature not to require any further attention. Yet it must be conceded to Wallace and other critics that Darwin's analysis was incomplete.

Female Discrimination

On the basis of his own observations and of the abundant records of the naturalists, Darwin was convinced of the existence of "female choice" (however defined). Students of animal courtship have since substantiated Darwin's assertion. They are now virtually unanimous in stating that the females are far more discriminating than the males. It is easy to observe that males tend to display not only to their own females, but also to females of related species, to males of their own or related species, and, in the absence of appropriate display partners, to even less appropriate objects. If the females were lacking discrimination to a similar extent, an enormous amount of hybridization would take place. Males are very easily stimulated to engage in courtship, while females are often inactive and respond to the overtures of the males not at all or only after long-continued male displays. The reasons for this "coyness" have been provided by Richards (1927), Bateman (1948), and Trivers (1972). They point out that this difference in the behavior of the sexes is a result of the highly unequal investment of the two sexes in reproduction. The male produces millions of gametes, and his role in reproduction usually ends with copulation, which requires a negligible expenditure of physiological energy. The female, by contrast, usually produces only a limited number of yolk-rich eggs and in addition devotes a large amount of time and energy to caring for eggs and young. The male has little to lose by courting numerous females and by attempting to fertilize as many of them as possible. Anything that enhances his success in courtship will be favored by selection. The situation is quite different in the case of the female. Any failure of mating with the right kind of male may mean total reproductive failure and a total loss of her genes from the genotype of the next generation.

Failure for a female mammal may mean weeks or months of wasted time. The mechanical and nutritional burden of pregnancy may mean increased vulnerability to predators, decreased disease resistance, and other dangers for a long time. . . . Once she starts

on her reproductive role, she commits herself to a certain high
minimum of reproductive effort. Natural selection should regulate
her reproductive behavior in such a way that she will assume the
burdens of reproduction only when the probability of success is at
some peak value that is not likely to be exceeded. (Williams
1966:183)

There is thus a high selective premium on the discrimination by
females of the most appropriate mate. Assuming that there is some
correlation between the vigor and the persistence of the male's courtship with his fitness, it will be of selective advantage for the female to
have a prolonged refractory period in order to test the male's perseverance and coordination. That it is normally the female who exercises
choice in mate selection is the direct consequence of the highly unequal
energy expenditure of the two sexes.

In Darwin's time female choice was largely conjectured. Wallace
(1889) emphasized that there was no observational evidence for it, and
he, therefore, characterized Darwin's argument as speculative. This view
was countered by Poulton, one of the few naturalists who stood up for
Darwin and said that such a lack of evidence is not surprising "because
the vast majority of those interested in nature are either anatomists,
microscopists, systematists, or collectors. There are comparatively few
true naturalists, men who would devote much time and the closest
study to watching living animals amid their natural surroundings, and
who would value a fresh observation more than a beautiful dissection or
a rare specimen" (1890:287). The deficiency deplored by Poulton has
since been largely repaired. There is now abundant observational evidence (Trivers 1972) that most females are very fickle indeed and
usually remain for a long time unimpressed by the displays of large
numbers of suitors before finally accepting one of them. Furthermore,
it has been clearly shown that very specific characteristics may determine this choice. This has been demonstrated for positive and negative
imprinting, by the Petit principle (Petit 1958, Boesiger 1967), and by
observations of pair formation in birds and the sexual success of com-
munally displaying birds. Similar preferences have been described for
mammals (for example, Beach and LeBoeuf 1967). All of this recent
work is a vindication of Darwin's original assumption.

However, the question as to the nature of the criteria on which
females base their choice is still open. It is really true, as claimed by
Darwin, that female animals appreciate "beauty"? The first evolutionist
who had the courage to come out in favor of a well-developed esthetic
sense among females was Poulton (1890). To say, as had Darwin's
critics, that the male characteristics are simply species recognition
marks does not explain why they appear beautiful to the human eye,
says Poulton. "For the purposes of recognition, beauty is entirely
superfluous and indeed undesirable; strongly marked and conspicuous

differences are alone necessary" (p. 316). "The musical value of the song of birds can not be explained as a means of recognition between the sexes. The beauty of song is something more than its clearness, loudness and individuality" (p. 319). The existence of an esthetic sense in birds, he says, is clearly demonstrated by the bowerbirds, who decorate their bowers with flowers and other colorful objects. The subsequent discovery of bower painting (Marshall 1954) further strengthens this case.

Poulton finally summarizes additional evidence supporting Darwin:

> There are one or two general facts which seem to me to strongly support the theory of Sexual Selection and to oppose any theory which is not based on selective breeding.
>
> (1) Sexual colors only developed in species which court by day or twilight, or have probably done so at no distant date. [Compares butterflies and moths.]
>
> (2) Sexual colors are not developed on parts of the body which move so rapidly that they become invisible. [Wings of hummingbirds and rapidly flying moths.]
>
> (3) Colors are best seen from the direction which corresponds to the position from which the female would see them. [Examples among butterflies and pheasants.] (p. 331)

PREREQUISITES FOR THE FUNCTIONING OF SEXUAL SELECTION

A Surplus of Males

Darwin saw clearly that sexual selection, narrowly defined, can operate only if there is a considerable surplus of males. "When the sexes exist in exactly equal numbers, the worst endowed males will (except where polygamy prevails), ultimately find females, and leave as many offspring, as well fitted for their general habits of life, as the best endowed males" (1871:213). This disturbed Darwin greatly because "after investigating, as far as possible, the numerical proportion of the sexes, I do not believe that any great inequality in number commonly exists" (1871:213). Darwin's own attempts to escape this dilemma (for example, a large reservoir of unmated birds) were not successful and need not be discussed. It has now become obvious that different interpretations pertain to species with polygyny (or promiscuity) and to those with strict monogamy. In the former case there is a clear-cut competition among males, and this will be discussed below. The sexual dimorphism of monogamous species has little if anything to do with sexual selection and can be explained in terms of straight natural selection (Hamilton 1961; see below under "Epigamic Selection" and "Isolating Mechanisms").

Struggle among Males

Female choice was one half of sexual selection for Darwin, fighting among males was the other half. Here also Darwin used the term "struggle," which had already got him into a good deal of trouble with respect to natural selection. He does not inform us whether he uses it in a "metaphorical" (*Origin of Species*, p. 62) or literal sense. Actually, he seems to use the term in both senses. "It is certain that amongst almost all animals there is a struggle between the males for the possession of the female. This fact is so notorious that it would be superfluous to give instances. Hence, the females have the opportunity of selecting one out of several males, on the supposition that their mental capacity suffices for the exertion of a choice" (p. 212). The logic of this statement is not compelling. There is no evidence whatsoever that the choice of the females is influenced by the size or form of the weapons that help a particular male to be victorious over another male. It seems to me that if the outcome of the struggle between the males had been decisive, there would be little opportunity left for choice by the females. The victorious male would gather all the females of the neighborhood into a harem, while the vanquished males would be driven out of bounds. This is the classical situation in many mammal societies. At best the females would have the choice of which victorious male's harem they would join. Wallace is even more impressed than Darwin by the importance of fighting between males. He considers it "a form of natural selection which increases the vigor and fighting power of the male animal, since in every case the weaker are either killed, wounded or driven away. . . . It is evidently a real power in nature; and to it we must impute the development of the exceptional strength, size and activity of the male, together with the possession of special offensive and defensive weapons" (pp. 282–283). The modern naturalist is far less impressed. Cases of actual fighting between males for the possession of females seem to be the exception. What males usually fight for are territories that serve either as mating stations or as the place where the animal expects to raise his family. In such cases what the female chooses is not necessarily the strongest or most beautiful male, but sometimes simply the male with the most attractive territory. There is not necessarily a close correlation between strength, beauty, and acquisition of the best territory.

In monogamous species of birds virtually all males acquire a territory, and most of them also acquire a mate. Under these circumstances there is little leeway for sexual selection. The situation is quite different in species of birds and mammals in which the mating system is either polygyny or promiscuity. I refer the reader to the analyses of Verner (1964), Orians (1969), and Bartholomew (1970) for a more detailed discussion of these mating systems. What is becoming increasingly evident is that even in these polygynous mating systems the development of sexual dimorphism is far more due to natural selection than to sexual selection, at least in the species in which breeding takes place within the territory of the male. Although bitter fights sometimes occur between territory neighbors, particularly among the pinnipeds, the objective of the fights seems to be territory. The defense of such territories is largely carried out by threats and bluffing (Huxley 1938c). Warning colors, ruffs, and manes play an important role during these threat displays.

Several recent studies of birds as well as mammals show it is highly probable that the choice of the female is largely determined by the quality of the territory and not by the particular appearance of the male territory holder. If this could be confirmed and if the males have a similar ability to discriminate between good and bad territories, then the fighting ability of males would indeed be closely correlated with the well-being of their offspring. It would be a matter of natural rather than of sexual selection.

SEXUAL SELECTION OR NATURAL SELECTION?

In the preceding discussion of the fights among males for superior territories, it has become apparent that phenomena which at first had appeared to be clear-cut evidence for sexual selection ("struggle among males") were largely forms of natural selection. Darwin did not see this as clearly as later workers. Indeed, he classified not only fighting among males but also various other phenomena under sexual selection that we would now unhesitatingly designate as components of natural selection. Darwin was not unaware of these ultimate explanations (see below), but he chose to disregard them, perhaps for didactic reasons. Wallace went to the other extreme: "The term sexual selection must, therefore, be restricted to the direct result of male struggle and combat. This is really a form of natural selection, and is a matter of direct observation" (1889:296). Hence, sexual selection did not really exist for him. In the ensuing years many attempts were made to draw a sharp line between characteristics that had been acquired through sexual selection and those due to natural selection. None of these attempts was particularly successful, although Richards (1927:300) was quite correct in saying, "A character that has been acquired or preserved by the action of Sexual Selection must either be displayed to the other sex in courtship or used to drive away rivals." However, as we shall presently see, many characters that owe their existence to natural selection are likewise employed in these two contexts, and Richards himself demonstrated this in his outstanding review of the subject. It is now evident, in part owing to the analyses of Richards and of Huxley (1938a, b), that there are three major, and presumably several minor, selection pressures which favor the development or enhancement of sexual dimorphism, without requiring sexual selection.

Epigamic Selection

Copulation, in most species of animals, is only the last step in a long series of interactions between male and female. The first is the mutual finding of the sex partners. Here we generally find one of three possible situations: (1) The male is stationary (often on a well-defended territory); in this case the male makes his station known by songs and calls, as among birds, frogs, cicadas, and many orthopterans. These signals are usually highly stereotypic, and whatever difference between individual males may exist is limited to the loudness and persistence of these signals. Or (2) the female is stationary, as among many moths, and attracts the males by chemical means (scents). In these cases the female is usually ready to mate as soon as the first male arrives. Or (3) both male and female are mobile, as among many butterflies, and mating is usually preceded by rather protracted displays. It is only in the third alternative that discrimination of sex plays a role, because sex recognition is automatically given in case one (by the advertising song of the males) and case two (by the scent of the female). As Richards (1927) pointed out correctly (and so had Wallace and others before him), characteristics that facilitate the finding of one sex by the other will be favored by natural selection because their existence will reduce the length of time during which the sex partners are vulnerable to predation and other dangers during their search for each other.

Courtship consists of an exchange of stimuli between male and female until both have reached a state of physiological readiness in which successful copulation can occur. The analysis of these signals is made difficult because they serve at least three independent functions: (1) to suppress fleeing or attacking tendencies in the sex partner (this has been particularly well described by the Peckhams [1890] for spiders), (2) to advertise the presence of a potential mate, and (3) to synchronize mating activity (for their role as isolating mechanisms, see below).

All these functions are important for natural selection, but the question is still open as to what extent these signals, particularly those of the more aggressive sex, also serve sexual selection. Darwin assumed rather naively that "the best armed males" were also the strongest and that "the more attractive" males were "at the same time more vigorous" (p. 220). As stated above, there is no demonstration of an automatic correlation between the two characteristics. However, Fisher (1930:152) saw quite clearly that a plumage character of a male bird which was originally acquired through natural selection for purely physiological reasons may "proceed, by reason of the advantage gained in sexual selection, even after it has passed the point in development at which its advantage in natural selection has ceased." The fact that display characters are most pronounced in species in which there is a surplus of males or a special mating system (such as promiscuity or polygyny) seems to confirm this inference. Indirectly, it is also confirmed by the fact that in monogamous species, such as herons (egrets), in which the pair bond is continuously tested and strengthened by mutual displays, there has been a "transference" of the display characters from the males to the females with the result that both sexes have elaborate display plumes.

We can summarize the findings on epigamic characters by saying that their development was probably favored originally by natural selection, to synchronize the physiological state of the two sexes, but that sexual selection is presumably superimposed in all cases in which a male may gain reproductive advantage owing to an extreme development of an epigamic character.

Isolating Mechanisms

It is well known that the mating drive in the males of many species is so strong that they display not only to females of their own species, but also to females of related species. If the females were equally lacking in discrimination, an enormous amount of hybridization among closely related species would take place. Since hybrids are ordinarily of considerably lower fitness, natural selection will favor two developments: first, any genetic change that would make the females more discriminating, and second, any characteristics in the males that would reduce the probability that they be confused with the males of another species. Such characteristics are called isolating mechanisms. Darwin, curiously, almost entirely ignored the role of species-specific male characteristics as isolating mechanisms. Wallace was considerably more perceptive on this point. He was convinced that one of the chief meanings of sexual coloration is to enable "the sexes to recognize their kind [=species], and thus avoid the evils of infertile classes. . . . The wonderful diversity of color and of marking that prevails, especially in birds and insects, may be due to the fact that one of the first needs of a new species would be, to keep separate from its nearest allies, and this could be most readily done by some easily seen external mark of difference" (pp. 217–218). He continues: "Among insects the principle of distinctive coloration for recognition has probably been at work in the production of the wonderful diversity of color and marking we find everywhere, more especially among the butterflies and moths; and here its chief function may have been to secure the pairing together of individuals of the same species" (p. 226). Curiously, later writers on the subject, such as Poulton (1890), Richards (1927), and Huxley (1938a), although mentioning the role of species-specific male characters as isolating mechanisms, paid very little attention to them. It was not until Dobzhansky (1937) emphasized the great evolutionary importance of isolating mechanisms that their role was fully appreciated. It has now become apparent that many male characteristics which Darwin regarded as products of sexual selection actually serve as isolating mechanisms and were acquired through natural selection or at least as by-products of speciation and subsequently reinforced by natural selection. Mayr (1942) gave one of the early summaries of the new viewpoint and called attention to the fact that the conspicuous male characteristics some-times were lost in island birds when there were no other closely related species on the same island. The loss of these characters was apparently due to a relaxation of selection for the distinctive isolating mechanisms.

Behavioral isolating mechanisms and epigamic selection grade into each other imperceptibly. Considering that there will be a constant selection pressure on the females to respond as quickly and precisely as possible to the displays of males of their own species, such species-specific isolating mechanisms function simultaneously as epigamic characters to facilitate the physiological coordination of the two dis-play partners.

Different Niche Utilization

In many species, particularly of birds, males and females differ from each other in niche utilization. This was fully appreciated by Darwin (1871:208–209), who recognized that sexual dimorphism associated with a difference in ecology has nothing to do with sexual selection. "When . . . the two sexes differ in structure in relation to habits of life, they have no doubt been modified through natural selection" (p. 209). Much of the literature on such ecological sexual dimorphism has been recently summarized by Selander (1966, 1969).

NATURAL FITNESS AND REPRODUCTIVE ADVANTAGE

Darwin in several statements implied that natural selection and sexual selection were mutually exclusive phenomena. For instance: "Sexual selection depends on the success of certain individuals over others of the same sex, in relation to the propagation of the species; while natural selection depends on the success of both sexes, at all ages, in relation to the general conditions of life" (1874:614). And yet, throughout his discussions, he stressed how many of the secondary sex characters of males contribute to general fitness. Huxley (1938a) ques-tions whether or not it is desirable to mark off sexual selection sharply from natural selection and whether the male secondary sexual char-acters involved with combat and display are clearly distinguishable from other male characters. He concludes that "display characters are inex-tricably entangled with those subserving threat and also sex recognition. . . . It is clear that Darwin's original contention will not hold. Many of the characters which he considered to owe their evolution to sexual selection do have value to the species in the general struggle for exis-tence, and not merely in the struggle between males for reproduction" (pp. 33–34).

Both Darwin and Wallace, with virtually no tangible evidence, assumed that the males which won out in the struggle with other males would mate "with the most vigorous and best nourished females. . . . If such females select the more attractive, and at the same time vigorous males, they will rear a larger number of offspring" (Darwin, 1874:220). In other words, they assumed that there was a correlation between general fitness and the kind of characters that would lead to victory in the struggle among males. Wallace likewise believed that reproductive success of males was the result of their "vigor and fighting power" and that reproductive success could all be explained in terms of natural selection. Now, 80 years later, there is still little tangible evidence that there really is such a correlation. However, in one study of *Drosophila melanogaster* it was found that the males with the highest sexual drive also produced the highest number of offspring (Fulker 1966).

The preceding discussion of epigamic selection, of the importance of superior male territories, and the role of isolating mechanisms has demonstrated that male secondary sexual characters, perhaps originally acquired through sexual selection, almost invariably also contribute to general fitness. This conclusion leads to the question whether there are any forms of sexual selection which result in the evolution of characters that are useless or deleterious to the species. The excessive plumes of birds of paradise and peacocks as well as the antlers of certain cervids, such as the Irish elk (Gould 1974), have to be mentioned until somebody proves that they are not deleterious. But even if we disregard such excesses, a belief in sexual selection forces us to ask this question: If the female chooses her mate primarily on the basis of esthetic criteria, how can this truly benefit the species? Or, to use Darwin's own words, how can it contribute to fitness for males to have acquired a particular structure, "not from being better fitted to survive in the struggle for existence, but from having gained an advantage over other males" (1874:210)? It is distinctly conceivable that extreme courtship adaptations acquired as a result of sexual selection may actually reduce the ecological success, that is, the fitness, of a species.

This point raises once more the question of the definition of fitness and of modes of selection. It is quite possible, as suggested by Haldane (1932) and Huxley (1938a), that one should make a distinction between intrapopulation (intraspecies) and interspecies selection. As long as density-dependent factors regulate population size, sexual selection for ornamental characters may not have any effect on the interspecific component of selection. If, as everyone seems to agree, the females have no difficulty in getting fertilized and producing offspring, a certain amount of selection for "useless male characters" may have no effect whatsoever on the fitness of the species in interspecific selection. Natural selection will surely come into play as soon as this sexual selection leads to the production of excesses that significantly lower the fitness of the species in interspecific encounters. I feel that here is an area that has not been thought out completely.

CONCLUSION

1. Darwin's assumption that the females in many species of animals (sometimes also the males) make a definite choice of their sex partner has been confirmed in numerous recent investigations.

2. Darwin was wrong, however, in assuming that most aspects of sexual dimorphism in animals are the result of sexual selection.

3. Nevertheless, a residue of male ornaments and attractants remains that can hardly be ascribed only to natural selection. A male can achieve an increased contribution to the gene pool of the next generation through features which do not contribute to the fitness of the species but merely to the reproductive success of the possessor of those characters.

4. Whenever such sex-limited features affect interspecific selection, the pressure of natural selection will tend to eliminate deleterious excesses.

REFERENCES

Bartholomew, G. A. 1970. A model for the evolution of pinniped polygyny. *Evolution*, 24:546–559.

Bateman, A. J. 1948. Intra-sexual selection in *Drosophila. Heredity*, 2:349–368.

Beach, F. A., and B. J. LeBoeuf. 1967. Preferential mating in the bitch. *Animal Behavior*, 15:546–558.

Boesiger, E. 1967. La signification évolutive de la selection sexuelle chez les animaux. *Scientia*, 111:1–17.

Campbell, B., ed. 1972. *Sexual selection and the descent of man, 1871-1971.* Aldine Publishing Co., Chicago.

Darwin, C. 1859. *On the origin of species by means of natural selection, or the preservation of favoured races in the struggle for life.* J. Murray, London.

—— 1871. *The descent of man, and selection in relation to sex*, 1st ed. J. Murray, London.

—— 1874. *The descent of man, and selection in relation to sex*, 2nd ed. J. Murray, London.

Dobzhansky, T. 1937. *Genetics and the origin of species.* Columbia University Press, New York.

Ehrman, L., and C. Petit. 1968. Genotype frequency and mating success in the *Willistoni* species group of *Drosophila. Evolution*, 22:649–658.

Fisher, R. A. 1930. *The genetical theory of natural selection.* Clarendon Press, Oxford.

Fulker, D. W. 1966. Mating speed in male *Drosophila melanogaster*: a psycho-genetic analysis. *Science*, 153:203–205.

Ghiselin, M. T. 1969. *The triumph of the Darwinian method.* University of California Press, Berkeley.

—— 1974. *Economy of nature and the evolution of sex.* University of California Press, Berkeley.

Gould, S. J. 1974. The origin and function of "bizarre" structures: antler size and skull size in the "Irish elk"—*Megaloceros giganteus. Evolution*, 28(2):191–221.

Haldane, J. B. S. 1932. *The causes of evolution.* Longmans, Green, London.

Hamilton, T. H. 1961. On the functions and causes of sexual dimorphism in breeding plumage characters of North American species of warblers and orioles. *Amer. Nat.*, 95:121–123.

Huxley, J. 1938a. Darwin's theory of sexual selection and the data subsumed by it, in the light of recent research. *Amer. Nat.*, 72:416–433.

—— 1938b. The present standing of the theory of sexual selection. In *Evolution: essays on aspects of evolutionary biology,* ed. G. R. de Beer. Clarendon Press, Oxford.

—— 1938c. Threat and warning coloration in birds with a general discussion of the biological functions of colour. *Proc. 8th Int. Ornith. Congr.*, (1934):430–455.

—— 1963. *Evolution: the modern synthesis,* 2nd ed. Allen and Unwin, London.

Marshall, A. J. 1954. *Bower-birds: their displays and breeding cycles.* Oxford University Press, London.

Maynard Smith, J. 1958. Sexual selection. In *A century of Darwin,* ed. S. A. Barnett, Harvard University Press, Cambridge, Mass.

Mayr, E. 1942. *Systematics and the origin of species.* Columbia University Press. New York.

—— 1963. *Animal species and evolution.* Belknap Press of Harvard University Press, Cambridge, Mass.

Orians, G. H. 1969. On the evolution of mating systems in birds and mammals. *Amer. Nat.*, 103:589–603.

Peckham, G. W., and E. G. Peckham. 1890. Additional observations on sexual selection in spiders of the family Attidae, with some remarks on Mr. Wallace's theory of sexual ornamentation. *Occas. Pap. Nat. Hist. Soc.* (Wisconsin), 1:117–151.

Petit, C. 1958. Le déterminisme génétique et psycho-physiologique de la compétition sexuelle chez *Drosophila melanogaster. Bull. Biol.*, 92:248–329.

Poulton, E. D. 1890. *The colours of animals: their meaning and use, especially considered in the case of insects.* Kegan Paul, Trench, Trübner, and Co., London.

Richards, O. W. 1927. Sexual selection and allied problems in the insects. *Biol. Rev.*, 2:298–364.

Selander, R. K. 1966. Sexual dimorphism and differential niche utilization in birds. *Condor*, 68:113–151.

—— 1969. The ecological aspects of the systematics of animals. In *Systematic biology, Nat. Acad. Sci. Publ.* no. 1692:213–247.

Tinbergen, N. 1954. The origin and evolution of courtship and threat display. In *Evolution as a process*, ed. J. Huxley, A. C. Hardy and E. B. Fords. Allen and Unwin, London.

Trivers, R. L. 1972. Parental investment and sexual selection. In *Sexual selection and the descent of man, 1871–1971,* ed. B. Campbell, pp. 136–179. Aldine Publishing Co., Chicago.

Verner, J. 1964. Evolution of polygamy in the long-billed marsh wren. *Evolution,* 18:252–261.

Wallace, A. R. 1889. *Darwinism: an exposition of the theory of natural selection with some of its applications.* Macmillan, London.

Williams, G. C. 1966. *Adaptation and natural selection: a critique of some current evolutionary thought.* Princeton University Press, Princeton.

9
The Emergence of
Evolutionary Novelties

One of the favorite objections raised by Darwin's critics was the incompatibility between the theory of gradual Darwinian evolution and the seemingly sudden origin of new structures in phylogeny, such as the lungs of vertebrates or the wings of insects. How could natural selection build such new structures gradually? Darwin answered this question as satisfactorily as possible in an age that lacked any understanding of genetics. A more meaningful analysis has become possible only in the last few decades.

Prior to Darwin natural selection was seen primarily as a negative force, as something that had the power to eliminate the unfit. This was also, on the whole, how most of Darwin's critics interpreted selection. Such selection, obviously, could not account for new structures. But even if one granted natural selection the power to improve existing organs, it would still leave us with the problem of the first origin of these organs. "How can natural selection explain the origin of entirely new structures?" asked Darwin's opponents. Is not evolution characterized by the continuous production of complete novelties, such as the lungs of vertebrates, the limbs of tetrapods, the wings of insects and birds, the middle ear of mammals, and literally thousands of structures in all the phyla of animals and plants? To explain these by sudden saltations is unsatisfactory, because a major mutation would surely disturb the harmony of the type. Yet, it was asked, how can an entirely new structure originate without complete reconstruction of the entire type? And how can a new structure be gradually acquired when the incipient structure has no selective advantage until it has reached a considerable size and complexity?

These were some of the questions that bothered Darwin and that have continued to occupy the minds of evolutionists to the present day. Darwin, to some extent, fell back on Lamarckian[1] explanations, and it

Reprinted and revised from "The emergence of evolutionary novelties," pp. 349–380 in *The evolution of life,* ed. S. Tax (vol. 1 of *Evolution after Darwin*) (Chicago: University of Chicago Press, 1959); reprinted by permission of The University of Chicago Press.
1. For the sake of simplicity I shall combine under the term "Lamarckian" all theories that

is not surprising that during the remainder of the nineteenth century the majority of evolutionists ascribed the origin of evolutionary novelties to Lamarckian causes. Yet as time went on, the fallacy of Lamarckian explanations became obvious, and the mutationism of De Vries and Bateson, no matter how wrong it was, was in a way a wholesome reaction against Lamarckism. At that period it seemed somehow impossible to find an interpretation that avoided the opposing evils of Lamarckism and saltationism.

The situation has changed greatly during the past 50 years. The saltationism of the early Mendelians has been refuted in all its aspects. But the pendulum has perhaps swung too far in the opposite direction. There has been such an exclusive emphasis on the gradual nature of all evolutionary change that the problem of the origin of evolutionary novelties has largely been neglected. It is now time to redress the balance by considering this question once again. Recent advances in evolutionary theory make it possible to give an answer not in conflict with the synthetic theory of evolution and, more specifically, an answer not requiring the occurrence of macromutations.

WHAT ARE "EVOLUTIONARY NOVELTIES"?

The discussion will gain in precision if I state at the very beginning what I include in the category "evolutionary novelties." I include any newly arisen character, structural or otherwise, that differs more than quantitatively from the character that gave rise to it. Consequently, not every change of the phenotype qualifies, because change of size or of pigmentation would be a change of phenotype not necessarily qualifying as "emergence of an evolutionary novelty." What particular changes of the phenotype, then, would qualify? Certainly any change that would permit an organism to perform a new function. Tentatively, one might restrict the designation "evolutionary novelty" to any newly acquired structure or property that permits the assumption of a new function. This working definition must remain tentative until it is determined how often it is impossible to decide whether or not a given function is truly "new."

The exact definition of an "evolutionary novelty" faces the same insuperable difficulty as the definition of the species. As long as we believe in gradual evolution, we must be prepared to encounter intermediate evolutionary stages. Equivalent to the cases in which it is impossible to decide whether a population is not yet a species or already a species, will be cases of doubt as to whether a structure is already or not yet an evolutionary novelty. The study of this difficult transition from the quantitative to the qualitative is one of the objects of this

---

postulate the occurrence of "induction" of directed genetic changes by the environment, by use, or by various other finalistic and vitalistic forces, and the inheritance of the characters thus acquired.

paper. Unwillingness to face such a difficult situation is one of the reasons so many authors have adopted a saltationist interpretation.

The origin of new taxa, from species to higher categories, will be considered as lying outside the scope of this discussion. Even though, admittedly, the origin of new higher categories is often correlated with the emergence of a new structure or other character, the natures of the two problems are sufficiently different to necessitate separate treatment.

Even so, our scope is wide. In the days of classical comparative anatomy, the term "evolutionary novelty" referred unequivocally to a new structure. With the broadening of biology, attention has been directed to evolutionary novelties that are not morphological or at least not primarily morphological. New habits and behavior patterns are very often as important in evolution as are new structures. Their origins will not be dealt with, since so little is known about the evolution of behavior, even though it seems that the evolution of behavior patterns obeys the same laws as the evolution of structures. The study of cellular physiology (biochemistry of metabolic pathways) and of microorganisms has likewise demonstrated the occurrence of important evolutionary novelties that do not involve the origin of gross new structures such as lungs, extremities, or brood pouches. The uric acid and fat metabolism of the amniote egg of the terrestrial vertebrates is an example. These are chemical innovations at the cellular level. Such cellular inventions have improved the efficiency of almost every organ. Granick (1953) has investigated some of the inventions necessary for the efficient functioning of iron metabolism in organisms possessing hemoglobin: (1) the change of a portion of the intestinal tract into an acid state (by HCl secretion) to permit the reduction of the inorganic iron in the food from the insoluble ferric to the soluble ferrous state; (2) an invention regulating uptake of ferrous iron by the mucosal cells of the intestines; (3) the invention of a special protein—transferrin—which transports the iron from the capillaries of the intestinal tract to various storage places in the liver, spleen, and bone marrow; (4) the invention of still another protein—ferritin—which serves as a storage mechanism, for times of need (hemorrhage).

Smith (1953) described the numerous inventions characterizing excretion in the various classes of vertebrates. Sharks (elasmobranchs), for instance, prevent water loss to the surrounding sea water by reducing the renal excretion of urea whenever the urea concentration in the blood drops to as low as 2–2.5 percent. This involves acquisition of special properties by two separate sets of cells: (1) the respiratory epithelium of the gills must become impermeable to urea so that it does not permit the urea molecules to diffuse into the surrounding sea water, and this without seriously impairing the permeability of this epithelium to oxygen and carbon dioxide; (2) the cells of the renal tubules must be able to recover the urea from the glomerular filtrate by tubular reabsorption; about 90 percent of the urea lost from the blood by the renal glomeruli is saved from excretion by this reabsorption.

Many similar biochemical inventions are recorded in the literature. They may have played a great role in the replacement of the major classes and phyla of the animal kingdom throughout geological history, but no one knows. Our knowledge of comparative physiology is still so elementary that we do not know, for instance, whether or not the cellular biochemical pathways of the mollusks give them superiority over the brachiopods, as one might suspect from a study of the geological record of these phyla. There is a wide field for comparative physiology and comparative biochemistry.

Evolutionary novelties at the cellular level tend to differ quite drastically, in several respects, from structural novelties. First of all, the genetic basis is usually simpler—indeed, a single gene mutation may be the primary basis of the novelty. Second, the new function may not require any reconstruction of the "type." No lengthy developmental pathway is involved that would necessitate an adjustment in the harmonious interaction of numerous genes. With the individual cell being the phenotype, the pathway from gene to phenotype is short and direct. Third, such a cellular invention more often than not will lead to an improvement of "general adaptation," whether it concerns respiration, digestion, excretion, or environmental tolerance, while a new structure frequently results in an adaptation to a more specialized situation. These average differences between "cellular" novelties and "structural" novelties are recorded in full cognizance of considerable overlap. For instance, new structures may also, on occasion, rest on a single primary mutation, and they may also lead to general rather than special adaptation. Yet the two classes of novelties are, as classes, rather different from each other. And the major evolutionary problem concerns the origin of new structures, since the preservation of gene mutations that permit adaptive improvement at the cellular level is no problem for the modern geneticist. My discussion, then, will center on the origin of new structures.

THE ORIGIN OF NEW STRUCTURES

The comparative anatomist and paleontologist, when comparing related taxa, occasionally find what appears to be an entirely new structure. Examples that come readily to mind are the bird feather, the ear bones of mammals, the swim bladder of fish, the wings of insects, and the sting of aculeate Hymenoptera. In the case of most of these structures, one might argue whether or not they are "really" new, and this is even more true for numerous other structures cited in the literature. The line between a quantitative and a qualitative change is not always sharply defined; indeed, to anticipate the outcome of my analysis, this border line is always indistinct. Far more structures were labeled "entirely new" in Darwin's day, when the fossil record was less completely known and when far less was known than is today about homologies among distant relatives. Thus there was far more evidence in Darwin's day that seemed to be quite incompatible with gradual improvement of the type through natural selection. And yet gradual improvement was the interpretation that Darwin continued to advance in the face of all the attacks by anti-selectionists. It soon became evident that ultimately there were only two alternative interpretations: appearance of evolutionary novelties by sudden saltation or by gradual emergence. Dispute over these alternatives has continued to the present day.

The Origin of Novelties by Saltation

The saltationists' theories in evolution have many roots, some of them going as far back as Plato's essentialism (see Essay 21). Indeed, all saltationists have been typologists, and most typologists have been saltationists of one sort or another. Genuine variation and gradual change are, of course, incompatible with the typological viewpoint. One of the usual arguments of the typologists is that organs and structures form a harmonious whole, characterizing an entire morphological type, like the mollusks, coelenterates, or vertebrates, and that new structures could have arisen only as a by-product of the origin of these new, major types. Furthermore, the typologist argues, since the origin of these types goes back to the early Cambrian or Pre-Cambrian (antedating the fossil record), it will never be possible to explain the origin of new structures. Admittedly, it may never be possible to reconstruct the origin of the chordates or of the arthropods on the basis of their fossil record, but this is no reason for defeatism. Some of the "minor" types, such as birds or mammals, differ strikingly in many structures from the groups from which they have arisen, and yet we have a fairly good fossil record indicating the pathway of the changes. There is already sufficient material available to describe the way in which many "evolutionary novelties" have come into being.

The theory of the origin of new structures by saltation was strong in Darwin's day and was an important component in the anti-selectionist argument of his opponents. Chief among these was Mivart (1871), who devoted an entire volume, *The Genesis of Species,* to a point-by-point refutation of Darwin. The problem of the origin of new structures is one of Mivart's major concerns. This is of special interest to the student of Darwin, because most of the major revisions which Darwin made in the sixth edition of the *Origin of Species* (1872) were rebuttals of Mivart's arguments. Mivart was a saltationist who assumed, for instance, that the differences between the extinct three-toed *Hipparion* and the horse (*Equus*) had arisen suddenly. He thought it difficult to believe that the wing of a bird "was developed any other way than by a comparatively sudden modification of a marked and important kind," and he applied the same explanation to the wings of bats and pterodactyls. Darwin (1872:261) opposed this assumption quite emphatically: "This conclusion, which implies great breaks or discontinuity in the series, appears to me improbable in the highest degree." He supports his objection by arguing: "He who believes that some ancient form was transformed suddenly through an internal force or tendency, into, for instance, one furnished with wings, will be almost compelled to assume . . . that many individuals varied simultaneously." The absurdity of believing in the simultaneous appearance of numerous "hopeful monsters," as Goldschmidt (1940) has called them, was far more clearly appreciated (p. 265) by Darwin than by some recent evolutionists, and yet such a multiple origin would be a necessity in sexual organisms. That the saltationist theory produces far more difficulties than it explains was pointed out by Darwin in the following words (p. 265):

> He [the saltationist] will further be compelled to believe that many structures beautifully adapted to all the other parts of the same creature and to the surrounding conditions, have been suddenly produced; and of such complex and wonderful co-adaptations, he will not be able to assign a shadow of an explanation. He will be forced to admit that these great and sudden transformations have left no trace of their action on the embryo. To admit all this is, as it seems to me, to enter the realms of miracle and to leave those of science.

Darwin's contention is fully supported by modern genetics. If one had to rely on mutation pressure as the only evolutionary factor, one would need such a high rate of mutation that it would result in an enormous production of "hopeful monsters." All available evidence is opposed to such an assumption. Indeed, most mutations appear to have only a slight, if not an invisible, effect on the phenotype. More penetrant mutations are usually disruptive and produce disharmonious phenotypes, as correctly implied by Darwin, and will therefore be selected against. The real function of mutation is to replenish the gene pool and to provide material for recombination as a source of individual variability in populations.

It was a long time before this role of mutation was clearly appreciated. The "one character—one mutation" reasoning, implicit in much thinking of the Darwinian period, was eventually made the basis of a major evolutionary theory, the mutationism of the early Mendelians (De Vries, Bateson). According to this theory, any new character, any new species, any new higher taxon, comes into being through mutation. The genetic work of the last four decades has refuted mutationism (saltationism) so thoroughly that it is not necessary to repeat once more the evidence against it. Most important, of course, is the realization that the phenotype (in higher organisms) is the product of a long developmental pathway and that any part of it, any "character," depends on the harmonious interaction of many, if not all, of the genes of the organism. A mutation affecting one of the numerous genes contributing to the phenotype of a character will have only a minor effect, or, if it has a major one, such drastic interference with the harmony of development will almost certainly be deleterious.

There has been much confusion in the literature on the purely semantic problem of how to define a "big" mutation. It seems to me that this must be measured not in terms of visible change but in terms of adjustment to the environment. When speaking of the "bigness" of a mutation, we must specify whether we are speaking of the level of the gene (amount of reorganization of the DNA), the level of the phenotype, or the level of the resulting fitness. A mutation that affects a growth pattern, such as branching in a plant or a sessile invertebrate, may produce drastic visible changes in the phenotype without much effect on fitness. Individuals of corals with different types of septa could well coexist in a single, interbreeding population, as could graptolites with different systems of branching, or ammonites with different patterns of lobe formation. A sinistral and a dextral snail are conspicuously different from each other, and yet the slight change in the direction of the mitotic spindle causing this shift, as well as the ultimate difference in phenotype, is not likely to have drastic effects on selective values, unless the shift interferes with the interbreeding of dextral and sinistral individuals. Such changes in growth pattern may have a very small differential at the cellular level and be of negligible selective significance, regardless of the considerable phenotypic difference. It would seem to me that it is at the cellular level that a single mutation may be of the greatest effect. The emergence of evolutionary novelties due to a single mutation will occur most likely among microorganisms or, indeed, all unicellular organisms of simple structure.

This conclusion does not deny that a single mutation may add to the fitness of an organism or make it better adapted for a slightly different environmental niche. Huxley (1942:52, 118, 449) has cited numerous mutations that affect temperature tolerance, growth rate, fecundity, seasonal adjustment, and other components of fitness. The recent literature on balanced polymorphism has added many other cases. Industrial melanism (Kettlewell 1959) is a specially well-analyzed example of the fitness-enhancing property of single genes. Wherever soot darkens the bark of trees, the melanic moths gain a cryptic advantage over the normally pale-colored individuals of the species. If one is so inclined, one may call the incorporation of any such mutation into a population an emergence of an evolutionary novelty. To me it seems, however, that this would dilute beyond all usefulness a legitimate phenomenon, that of the emergence of new structures. The stated genetic changes lead usually only to what I have called (Essay 16) "ecotypic adaptation," not to a shift of phylogenetic significance.

The Gradual Acquisition of New Structures

The evidence, whether genetic, morphological, or functional, is so uniformly opposed to a saltationist origin of new structures that no choice is left but to search for explanations in terms of a gradual origin. The role of natural selection in evolution would indeed be a very inferior one if, as was believed by the saltationists, it did nothing but weed out "hopeless monsters" in favor of "hopeful monsters." Darwin was fully aware of this situation: "If it could be demonstrated that any complex organ existed which could not possibly have been formed by numerous, successive, slight modifications, my theory would absolutely break down. But I can find out no such cases" (p. 189). Yet the problem remains of how to push a structure over the threshold where it has a selective advantage. The problem of the emergence of evolutionary novelties then consists in having to explain how a sufficient number of small gene mutations can be accumulated until the new structure has become sufficiently large to have selective value. Or is there an explanation that avoids this troublesome threshold problem? This has been discussed by a number of authors, usually under the heading "the origin of adaptations." The publications of Sewertzoff (1931), Huxley (1942), Rensch (1947), and Davis (1949) might be mentioned as recent works devoting special attention to this problem. The following possibilities of the origin of new structures are apparent:

The new structure originates (1) as a pleiotropic by-product of a changing genotype, (2) as the result of an intensification of function, or (3) as the result of a change of function.

Pleiotropic By-Product

This hypothesis assumes that not all phenotypic expressions of pleiotropic genes have a definite selective value but that a "natural" character may subsequently acquire selective value under certain circumstances. Darwin suggests (p. 91) that plants may excrete some sweet liquid accidentally from the flower and that this would in time lead to a well-organized system of pollination by insects. The secretion of nectar may well have had such an accidental origin; yet Darwin seems to have forgotten that the collecting of pollen was undoubtedly the original reason that insects visited flowers. Nectar is merely an additional "bonus." Darwin also suggests that some of the variation in the structure of the spines of sea urchins may have led to the development of the pedicellarias (p. 249). Even though it is very probable that neutral genes do not exist, there is no reason for denying the possibility of neutral aspects of the phenotype, that is, "neutral characters." The many differences among phenotypes that are independent but equivalent selective responses to the same functional needs (see below under "Multiple Pathways") indicate the existence of a certain amount of morphological leeway in the selective response. To be sure, it may be impossible to prove whether or not a minor structure has selective significance.

Intensification of Function

Most evolutionary changes take place without the origin of new structures. Even when we compare birds or mammals with their strikingly different reptilian ancestors, we are astonished at how few are the truly new structures. Most differences are merely shifts in proportions, fusions, losses, secondary duplications, and similar changes that do not materially affect what the morphologist calls the "plan" of the particular type. An intensification of the running function has led to a conversion of the five-toed mammalian foot (or hand) to the two-toed foot of the artiodactyls or the one-toed foot of the perissodactyls. Many glands are the result of intensified function and local concentration of previously scattered secretory cells. The intensification of function in these cases does not lead to the emergence of anything that is basically new, and yet it may result in a reorganization of the phenotype so drastic that the first impression is that of the emergence of an entirely new organ. Of importance to the evolutionist is the fact that no essentially new selection pressure is involved but merely the intensification of a previously existing selection pressure. At no time is there a stage in which "the incipient structure is not yet of selective value," to cite a frequently heard objection of the anti-selectionists. Sewertzoff (1931:183–236) has made a special analysis of this process of intensification of function. Darwin was fully aware of it. In fact, he used this principle to explain the origin of what is, perhaps, the most complex of all structures, the eye.

"To suppose that the eye, with all its inimitable contrivances for adjusting the focus to different distances, for admitting different amounts of light, and for the correction of spherical and chromatic aberration, could have been formed by natural selection, seems, I freely confess, absurd in the highest degree" (p. 186). But then he shows, step by step, that this "difficulty" should not be considered "insuperable." The evolution of the eye ultimately hinges on one particular property of certain types of protoplasm—photosensitivity. This is the key to the whole selection process. Once one admits that the possession of such photosensitivity may have selective value, all else follows by necessity. And if one visualizes the enormous number of extinct organisms—we know only the smallest fraction of them—"the difficulty ceases to be very great in believing that natural selection may have converted the simple apparatus of an optic nerve, coated with pigment and invested by transparent membrane, into an optical instrument as perfect as is possessed by any member of the Articulate Class" (p. 188).

It is somewhat oversimplified to explain the origin of the eye in terms of an intensification of the function of a piece of optic nerve. Yet there is a correct nucleus in this claim. In other cases the situation is far more clear-cut. The improvement of a single key component of a structure may result in an "evolutionary avalanche." Schaeffer (1948), for instance, showed that an improvement in the mechanical efficiency of the tarsal joint in a group of Condylarthra, occurring during a period of about 15 million years, gave rise to the highly characteristic foot structure of the artiodactyl ungulates. No really new structure originated, only a shift of proportions and positions, accompanied by an ever-increasing efficiency of function.

A similar improvement of a structure is the conversion of an orthodox mammalian claw in one line of taeniodont mammals into an efficient digging claw. This resulted in sufficient adaptive shift and increased success to set off a whole series of correlated changes in dentition and skull structure (Patterson 1949:262). Yet this taeniodont digging claw is not an entirely new structure.

It is often difficult to say to what extent a structure is new and to what extent it is merely an improvement on an old one. Let us take, for instance, the evolution of the amniote egg, which enabled the reptiles to complete the shift, initiated by the amphibians, from water to land (Needham 1931:1132). This shift is characterized not only by the acquisition of a hard shell and of new embryonic membranes (amnion, allantois) but also by certain changes in the metabolism of the egg. The uric acid catabolism permits an easy elimination of waste products without poisoning the embryo. Likewise, the shift from a largely protein to an essentially fat metabolism has numerous obvious advantages for an amniote egg. As stated above, clear-cut shifts, such as in these metabolic processes, are most often observed on the cellular-molecular level.

The area indicated by the term "intensification of function" is large and ought to be subdivided. However, I am not entirely certain that the subdivisions proposed by Sewertzoff are the best possible ones. What is needed at the present time, more than anything else, is the collecting of many cases falling under this category from all groups of animals and plants, preparatory to a more detailed analysis.

Change of Function

By far the most important principle in the interpretation of the origin of new structures is that of the "change of function." The discovery of this principle is usually ascribed to Anton Dohrn (1875), but it was clearly recognized and sufficiently emphasized by Darwin, whom Dohrn cites in his essay, in the sixth edition of the *Origin of Species* (1872)—indeed it was mentioned in the first edition (p. 454). Two subsequent authors who have made a special analysis of this principle are Plate (1924) and Sewertzoff (1931). The latter distinguishes no less than seven subdivisions or separate forms of change of function, a scheme that does not seem to add appreciably to an understanding of the problem. Indeed, it makes it appear more complex than it is.

Darwin recognized quite clearly that the possibility for a change of function usually depends on two prerequisites. The first of these is that a structure or an organ can simultaneously perform two functions. "Numerous cases could be given amongst the lower animals of the same organ performing at the same time wholly distinct functions" (p. 190). The other one is the principle of duplication. "Again, two distinct organs, or the same organ under two different forms, may simultaneously perform in the same individual the same function, and this is an extremely important means of transition." As an example Darwin quotes the fish that were ancestral to the tetrapods and had two separate organs of respiration—gills and primitive lungs.

A change of function is easily explained on the basis of these two premises, either a simultaneous multiple functioning of a single structure or the performance of the same function in different or duplicated organs. For this there are several alternative possibilities. The second structure may secondarily acquire a new function, and this new accessory function may eventually become the primary function. Or if two structures have two simultaneous functions from the beginning, one of them may become the primary and eventually exclusive function of one of the structures, and the other function for the second structure. This second structure is in many cases a simple duplication of the first. The duplication of structures is a frequent phenomenon in segmented, as well as in radially symmetrical, organisms. Morphologists, however, know that it can also take place independently of segmentation. "It occurs quite often, that a primarily undivided organ, a part of the skeleton, a muscle, or a nerve, is divided in the course of phylogeny and that several more or less independent organs originate in this manner" (Sewertzoff 1931:232). Sewertzoff's "similation" and Gregory's (1934) "polyisomerism" are related phenomena.

How a duplicated structure arose initially is not always completely clear. The origin of the mammalian middle ear may be a case in point. The location on the prearticular bone of a tympanic ring in a South American Triassic mammallike reptile indicates that this organism had, simultaneously, two tympanic membranes. One of these was the original reptilian tympanic membrane; the other, lying in front of it, was a secondary window, the presence of which may have facilitated sound transfer in these rather heavy-boned creatures. The origin of the second window was part of a slow reconstruction of the jaw and ear region in this branch of the reptiles, one change leading, by necessity, to the next. It appears that the functional value and hence selective significance of the primary (reptilian) tympanic membrane deteriorated at a subsequent stage following the reorganization of the jaw articulation. The stage was now set for a gradual obliteration of the primary tympanic membrane and the transfer of its function to the secondary membrane. (Hopson 1966; Crompton and Jenkins 1973).

In all the cases known to us in which there is a transfer of function from one structure to a duplicate one, there is always a transitional stage during which both structures function simultaneously. This is, for instance, well established for the transfer of respiration in the fish-amphibian series from the gills to the lungs. It has recently been demonstrated for the double jaw articulation of birds (Bock 1959).

Not only was Darwin aware of the principle, but he cited several illustrations of such a change of function. Perhaps the most frequently quoted one in the evolutionary literature is the shift of function between swim bladder and lungs. Darwin, like the majority of writers since his time, assumed that the swim bladder was the original condition. "The illustration of the swimbladder in fishes is a good one, because it shows us clearly the highly important fact that an organ originally constructed for one purpose . . . may be converted into one for a widely different purpose" (p. 192). Recent discoveries among fossil fishes have shown that diverticula of the respiratory tract first functioned in them as primitive lungs and only secondarily, in some fishes, as swim bladders. This, however, does not affect the correctness of Darwin's statement that this organ is involved in a transfer of function.

Darwin continued: "In considering transitions of organs, it is so important to bear in mind the probability of conversion from one function to another, that I will give one more instance" (p. 191). He then cites the case of egg-carrying folds in one family of cirripedes that become respiratory gills in another family, owing to a change of function. One would never have been able to trace the pathway of this change if the more primitive family had become extinct. "If all pedunculated cirripedes (with the egg-bearing folds) had become extinct . . . who would ever have imagined that the branchiae (gills) in this latter family had originally existed as organs for preventing the ova from being washed out of the sack?" (p. 192).

Such cases of a change in function are legion. The cited cases are given merely as illustrations of the stated principles. To give a complete catalogue would mean listing a good portion of all animal structures. The change of the ovipositor of bees into a sting, the development of the thyroid from the endostyle, of teeth from scales, and of various parts of the angiosperm flower are other examples. The electric organs in fish, so puzzling to Darwin, also belong here. Lissman (1958) presents suggestive indirect evidence that the electric field created by the contracting muscles is utilized in orientation, gradually evolves into a regular series of pulses, and eventually results in the shock discharges that in some species have such powerful offensive and defensive effects. The muscles are converted during this evolution into electric organs, as the subsidiary function becomes the primary function.

A change of function is not always tied to a duplication of structures. Consider what happens if a structure serving locomotion undergoes a change in function or a shift to a new primary function. The anterior extremity of unspecialized mammals serves several functions. In addition to its primary function of ordinary terrestrial locomotion, it may be used for digging, swimming, or, in arboreal gliders, for gliding (with the help of a patagium). The intensification of such a secondary function has led to such greatly modified structures as the shovel arm of the moles (*Talpa,* etc.), the flipper of whales (with secondary polyisomerism), and the wing of bats.

Preadaptation

The two kinds of situations I have described differ only in minor detail. In one case the structure exists in duplicate, and a new function is acquired by the duplicated structure. In the other case a secondary function is added to the primary function without duplication of structure. Both cases have the essential feature in common: that an existing structure is preadapted to assume a new function without interference with the original function. This is preadaptation, as now understood (Bock 1959).

The term "preadaptation" has been applied to diverse concepts. It was coined by Cuénot during the heyday of mutationism. All evolutionary change at that time was believed to be due to major saltations, and the new "hopeful monster" was either preadapted for a new niche or doomed to immediate extinction. Preadaptation in the modern theory of gradual evolution is something quite different from the concept held by the mutationists.

Discussions on the significance of preadaptation will gain greatly in precision if a distinction is made between preadaptation for a functional shift and that for a habitat shift. In the first case, a single structure is involved that can assume a new function while still carrying out the primary function. Illustrations of this are the wing of a diving bird, preadapted to become a paddle; or the primitive lungs of the early fishes, preadapted to become a hydrostatic mechanism (swim bladder); or the large antennae of the cladocerans, preadapted to become paddles.

On the other hand, an organism as a whole may be preadapted to undertake a major habitat shift. The aquatic branch of the vertebrates that gave rise to the first partially terrestrial amphibians must not only have had a crawling locomotion, but must also have been partially air-breathing and have had other characteristics of skeleton, epidermis, and sense organs that preadapted them for the habitat shift. The proavis must have had a considerable number of structural characteristics, such as a light body build and partial bipedalism, along with well-developed anterior extremities, to have been preadapted for flight. Admittedly, the preadaptation of a whole organism for an entirely new adaptive zone grades rather insensibly into the limited preadaptation for a single new function; yet it may be useful to distinguish categorically between these two kinds of preadaptations.

The Selective Value of Incipient Structures

This discussion of the multiple function of structures and of the consequent preadaptation of structures for new functions has prepared us for a consideration of an old anti-selectionist objection. It claims that many structures could not possibly have had any selective value until these structures were sufficiently large and elaborate to perform the function that gives them selective advantage. If selection is not responsible for getting them through this "incipient stage," what else can be but some kind of "internal force?" This claim of an absence of selective value in an incipient structure was one of the strongest arguments in Mivart's (1871) attempt to refute Darwin. He devoted the entire second chapter of his book on the *Genesis of Species* to the "incompetence of natural selection to account for the incipient stages of useful structures." This is an eminently reasonable account, which cites many structures like the baleens of large plankton-feeding whales and the milk glands of mammals, a gradual origin of which is indeed not easily imagined. Darwin was struck by the strength of these arguments and went to great lengths to refute them in the sixth edition of the *Origin of Species*. The debate between the two authors is still of interest in our day, even though it is very apparent that both contestants were misled by their ignorance of genetics and by a lack of appreciation of the statistical nature of natural selection.

It is now easy to see that two different types of phenomena are involved. Some new structures are advantageous from the very beginning. Darwin (p. 230) counters quite effectively Mivart's claim that the lengthening of the neck in the giraffe could not have been brought about by natural selection. Darwin shows that an ability to reach higher branches would be most useful in a continent like Africa that is overrun by grazing and browsing ungulates. One might add that no doubt the detection of lions in high grass is likewise facilitated by the lengthening of the neck. That this could have come about gradually and that every increase might well have been of selective advantage can be asserted with good reason. Indeed, this example of Mivart's is not particularly well chosen because only a rather slight modification of an already existing structure is involved, and not the origin of a new structure.

An immediate selective value is, however, evident even in some cases of genuinely new organs. Let us consider this in connection with the lungs of fishes. Their earliest recorded occurrence is in the Antiarchi (Denison 1941), in which a pair of sacs with a common duct grows out from the floor of the pharynx. *Polypterus,* one of the most primitive actinopterygians, and some of the choanichthyians also have ventral "lungs." These are obviously primitive structures, preceding in time the dorsal swim bladder of the advanced actinopterygian fishes (Goodrich 1930). How were the first ventral lungs of fishes developed? It can be assumed that oxygen uptake took place in the lowest fishes through all membranes, external skin, gills, and the intestinal tract. As the outer skin became increasingly unsuitable for gas exchange (partly owing to the development of dermal armor) and, even more important, as the gills became temporarily rather useless in oxygen-poor stagnant swamps during Devonian drought periods, active air uptake by "air swallowing" became at times the most important source of oxygen. At this stage, any enlargement of the surface of the inner throat or esophagus, any formation of diverticles, etc., was favored by natural selection. It is apparent that such a ventral diverticle from the floor of the pharynx was the beginning of the respiratory system of the higher vertebrates. At this early stage, however, it was not truly a new organ, but merely an enlargement (an "intensification," as Sewertzoff would say) of an existing organ: the total internal membranaceous surface used for oxygen uptake. This rather rudimentary organ was exposed to a renewed and increased selection pressure when the tetrapods became truly terrestrial. This shift of habitat resulted in the elaborate lungs of mammals and birds.

In a similar manner it can be shown for many structures that they must have been useful from the very beginning. This is true for almost any of the improvements of the digestive apparatus and all mechanisms having to do with heat regulation. It is presumably true for the majority of improvements at the cellular level. It is, however, also true for certain aspects of the general phenotype. Experimental work on mimicry and warning coloration has shown that exceedingly slight changes may be of selective value. It would seem inconceivable that the elaborate "eyes" on the hind wing of certain moths, serving so effectively as warning patterns, could be the result of selection. Yet Blest (1957a, b) has shown that the sudden revelation of a very simple contrasting spot on the hind wings has considerable protective value.

Opposed to these evolutionary novelties that add to fitness from the very beginning are others that one cannot consider useful until they have reached a certain size or perfection. Many of the novelties discussed above under "Change of Function" belong in this category. When we allow for the various auxiliary assumptions mentioned above, it becomes apparent in one case after another how an incipient structure could have continued to evolve until it was large enough to assume a new function. Mivart's argument that natural selection is incompetent to account for the early stages of useful structures has now lost most of its force.

One of the questions of Darwin's opponents was: Why are not more "transitional stages" of new structures found in nature? Darwin was able to counter this objection rather easily (p. 183):

Animals displaying early transitional grades of the same structure will seldom have survived to the present day, for they will have been supplanted by their successors, which were gradually rendered more perfect through natural selection. Furthermore we may conclude that transitional states between structures fitted for varying habits of life will rarely have been developed at an early period in great numbers and under many subordinate forms.

We would say, nowadays, that adaptive radiation will not take place until after the evolutionary novelty has reached a certain degree of perfection. Furthermore, forms representing intermediate stages will not be common enough to be encountered in the scanty fossil record until after such adaptive radiation and increase in numbers have taken place.

THE ENVIRONMENTAL SITUATION

With the fossil record preserving only the morphology of organs, it is natural that the morphological aspects are always stressed in the discussion of the origin of new organs. However, as Sewertzoff has said so correctly, "The morphological change of structure in an organ is important for a species only to the extent that it achieves an improvement in the function of this organ" and thus adds to the fitness of the species. Yet a change in function may precede a structural reorganization, or the selective value of a structure may change, owing to a change in selection pressures caused by a change in the environment. No discussion of the emergence of evolutionary novelties can be considered exhaustive that does not include a treatment of the environmental situation. Indeed, most evolutionary changes of structures cannot be fully understood without an analysis of the accompanying environmental changes. What categories of environmental change may be important in the origin of evolutionary novelties? And what type of adaptive change would occur most frequently in response to each class of environmental change?

Changes in the Physical or Biotic Surroundings

The environment is never constant. There are always climatic changes, among which long-term climatic trends are particularly important. There are general vegetational trends, as well as specific extinctions or invasions of individual species. There is the steady coming of new sources of food, new competitors, and new enemies and the steady loss of old ones. The organism is, more or less passively, exposed to all these changes and must be prepared to cope with them. Our knowledge of historical ecology is still too slight to permit detailed description of the effects of such changes on individual species and their adaptations. As a general rule, one might suggest that broad, general adaptations will prove most useful in coping with secular changes of the environment. Darwin commented on the superior fitness of species that live on continents in the midst of ever-changing faunas and floras. This is not the place to follow up the nature of this adaptation to broad tolerance and ever-changing conditions.

The Invasion of a New Niche or Adaptive Zone

The active shift of an organism into a novel niche or entirely new adaptive zone will set up a powerful array of new selection pressures (Simpson 1953). An organism must have a special set of characteristics to cope with the demands of the new environment. It must be "pre-adapted" for the new world in which it will henceforth live. The change from water living to land living is a particularly instructive illustration of this. Indeed, the combination of properties permitting terrestrial locomotion and respiration and preventing desiccation is sufficiently improbable to have been mastered only a very few times. The number of independent invasions of land by animals (vertebrates, several arthropods, and mollusks) is incredibly small in spite of the rich opportunities of the plant-covered land, opportunities made particularly apparent by the prodigious adaptive radiation of those few animals that successfully accomplished the shift. Among all the marine animals, only benthonic ones, because they already lived a somewhat "terrestrial" life underwater, were able to emerge onto land. The peculiar pedunculated fins of the crossopterygian fishes, presumably used in part for moving along rocks and over the bottom, were ideally preadapted for locomotion along land. A similar situation is probable for the arthropods (Manton 1953). Such adaptation of the extremities for a shift from aquatic to terrestrial environments are cases in which a structure can function in two adaptive zones in an essentially similar manner.

Perhaps most astonishing is the relative slightness of reconstruction that seems to be necessary for successful adaptation to rather drastic shifts of adaptive zones. For instance, there is every reason to believe that the group of reptiles ancestral to the birds already had feathers, even though they had been acquired either for temperature control, as an epigamic character, or in some other way not connected with flight. These pro-aves were furthermore preadapted in being arboreal, bipedal, and furnished with well-developed, functional anterior extremities. They had all the necessary equipment for becoming a flying machine, and not a single major new structure has appeared in the birds since they branched off from the reptiles. This, of course, does not belittle the many modifications in the bird skeleton, musculature, central nervous system, and sense organs. All these are avian modifications of the reptilian heritage, not the origin of entirely new structures.

No niche is too aberrant or too forbidding to preclude invasion. The bathypelagic niche is one of the most specialized and in some ways most demanding habitats open to living organisms; its fauna comes from two sources, surface pelagic and deep-sea bottom (bathybenthonic). That the deep-sea-bottom fauna should produce pelagic descendants seems particularly unexpected; yet it has been clearly established as the source of some of the most extraordinary inhabitants of the oceans, such as pelagic holothurians and octopuses.

Whenever a novel type of ecological niche is explored by naturalists, a new fauna is discovered in it. The more aberrant the niche, the more extraordinary its fauna. The psammofauna of the interstitial spaces in sea-bottom sand, discovered by Remane, is a typical example. Who would have expected to find a jellyfish in such a habitat? And yet this medusa (*Halammohydra*) has become completely adapted to this niche, which would at first sight appear to be totally unsuitable for it. Any textbook of ecology will give further examples of such niches, like hot springs, alkali flats, oil puddles, shifting sand dunes, and caves, that have been successfully colonized by organisms.

Each of these major shifts of habitat is a major evolutionary experiment. Each of the successful branches of the animal kingdom, for example, the insects, the tetrapods, the birds, is a product of such a shift. However, not all such shifts are equally successful. No spectacular adaptive radiation has followed the invasion of the sand niche by a coelenterate. The shift of a carnivore to a herbivorous diet (giant panda) has not led to a new phylogenetic breakthrough (Davis 1964). The tree kangaroos, the return of a specialized line of terrestrial marsupials to arboreal life, likewise seems to have reached an evolutionary dead end. The penguins, on the other hand, in a really extraordinary conquest of the aquatic niche by birds, may be considered reasonably successful, to judge from the enormous size of the penguin populations.

There has been much speculation in the evolutionary literature as to whether a "primitive" or a "specialized" creature has a higher evolutionary potential. Cope (1896:172) proposed the "law of the unspecialized," according to which every specialization is a dead-end street and true evolutionary advance is to be expected only from amorphous, unspecialized forms. This generalization is certainly not supported by the known evolutionary facts. Most major evolutionary advances depended on a shift into a new adaptive zone, and the feasibility of this shift, in turn, depended on available preadaptations. There is certainly nothing "unspecialized" about the earliest fishes and particularly about those that gave rise to the tetrapods. And those reptiles that gave rise to the mammals and birds were certainly as specialized in their way, or more so, than branches of the reptiles that did not give rise to successful offshoots or are still surviving. On the other hand, some of the nonspecialized "primitive" groups seem to be so successful in surviving that their evolutionary potential is questionable. For example, the opossum (*Didelphis*) represents an ancient group that goes back to the Eocene or earlier and that gave rise to most of the marsupial fauna of Australia and Tertiary South America. Yet many, if not most, of these specialized derivatives have become extinct, while *Didelphis* continues to survive and to be quite successful, even though it has remained essentially unchanged. It is therefore completely correct to stress the success of many unspecialized forms, but it is wrong to claim that they are the only forms with a future. Amadon (1943) and Romer (1946) have correctly emphasized the importance of specialization for evolutionary progress. This, of course, does not mean that every specialization will preadapt to the conquest of new adaptive zones. One may state that most specializations lead into dead-end alleys; yet new conquests could not be made without incessant experimentation.

One of the reasons for the former insistence on the "unspecialized" was archetypal thinking. When reconstructing the common ancestor of several evolutionary lines, those who are addicted to archetypal thinking tend to eliminate all specializations. As a result, all their putative ancestors are in every respect generalized and unspecialized. Osborn's (1936) phylogenetic tree of the proboscidians is a typical example of this way of thinking. It never seems to have occurred to these students that the creatures they reconstructed in this manner could never have existed in nature. Nor did it ever occur to them that this method would lead to the establishment of phylogenies in which all fossils were always aberrant side lines.

To sum up the evolutionary aspects of a shift into a new niche or adaptive zone: Such a shift can occur only if the organism is preadapted for it. However, as soon as the shift has been achieved, a whole new set of selection pressures will tend to modify all those structures that are particularly concerned with life in the new environment. The more drastic the change in environment, the more rapid will be the evolutionary change and the more far-reaching, in general, the structural reorganization.

A Change in Behavior

A shift into a new niche or adaptive zone requires, almost without exception, a change in behavior. In the days of mutationism (De Vries, Bateson), there was much heated argument over the question whether structure precedes habit or vice versa. The choice was strictly between saltationism and Lamarckism. The entire argument has become meaningless in the light of our new genetic insights. It is now quite evident that every habit and behavior has some structural basis but that the evolutionary changes that result from adaptive shifts are often initiated by a change in behavior, to be followed secondarily by a change in structure (see Essay 46). The new habit often serves as the pacemaker that sets up the selection pressure that shifts the mean of the curve of structural variation. Let us assume, for instance, that a population of fish acquires the habit of eating small snails. In such a population any mutation or gene combination would be advantageous that would make the teeth stronger and flatter, facilitating the crushing of snail shells. In view of the ever-present genetic variation, it is virtually a foregone conclusion that the new selection pressures (owing to the changed habit) would soon have an effect on the facilitating structure.

Darwin was fully aware of this sequence of events. The parasitic wasp *Polynema natans,* in the family Proctotrupidae, lays its eggs under water, mostly in the eggs of dragonflies. Most of its life cycle, including copulation, takes place under water. "It often enters the water and dives about by the use not of its legs, but of its wings, and remains as long as four hours beneath the surface; yet it exhibits no modification in structure in accordance with its abnormal habits" (Darwin 1872:185). Other aquatic species of parasitic wasps have since been discovered in the families Chalcididae, Ichneumonidae, Braconidae, and Agriotypidae. As Darwin stated correctly, none of them has undergone any major structural reorganization following the shift into a new adaptive zone.

The shift from water to land, as mentioned above, was likewise made possible by a prior shift in habits, in this case, in locomotor habits. Students of vertebrates (Westoll 1958) and of arthropods (Manton 1953) agree about this. With habitat selection playing a major role in the shift into new adaptive zones and with habitat selection being a behavioral phenomenon, the importance of behavior in initiating new evolutionary events is self-evident. A study of behavior differences among related species and genera is apt to throw much light on the sequence of events that trigger the emergence of evolutionary novelties.

A Change in the Structural Environment

Many functions are performed not by simple structures, but by a combination of structures. For an articulation, for instance, a minimum of two bones is needed, as well as the muscles that move these bones and the ligaments that help to bind them. To achieve efficient vision, a highly complex organ is needed, consisting of a receptor and its nervous connections, a lens and other focusing devices, pigments, etc. It is probable that some evolutionary novelties have emerged as the result of a more or less incidental coming together of such components. This may happen either because each component has a primary function and the complex structure evolved in response to a selection pressure exerted in connection with this primary function or because the components are potentialities that are realized singly in various related species or genera but cannot perform with full efficiency until brought together in a single individual. It is probable that the improvement of a primary structure through accessory organs is usually delayed until the proper gene combination arises that permits the accessory structure to emerge. Without the primary structure, there would be no selective value in the secondary structure. This is true, for instance, of many of the accessory structures of the eye.

Let us now consider a specific case of the quasi-accidental coming together of two structures, resulting in a new character complex with a unified function of high selective advantage. This is the case of the secondary jaw articulation in birds, discovered and beautifully analyzed by Bock (1959). In certain types of birds in which the open jaw (particularly the mandible) is exposed to heavy impact during the pecking of food or catching of prey, there is a strong selection for a heavy musculature permitting the rapid raising of the mandible (closing of the bill) against considerable resistance. To permit the insertion of the increased muscle mass, a bony spur grows out at the inside of the mandible toward the skull. When this spur has grown so long that it comes in contact with the skull, the stage is set for the development of a new character complex composed of previously independent organs (part of the mandible and part of the skull). This new character complex serves as a new articulation that functions simultaneously with the primary articulation. The new articulation has considerable value in all species with feeding habits that expose the jaw to possible dislocation. The secondary jaw articulation reaches its highest perfection in the skimmer (*Rhynchops*), which skims along the surface of the water with its mandible submerged until striking an object, which is then grasped. This secondary jaw articulation is an almost ideal illustration of the formation of a new structure as a result of a coming-together of two structures formed for entirely independent reasons.

The origin of the mammalian ear may well be another example. During the development of the new jaw articulation of the mammals (Crompton and Jenkins 1973), the contact of the quadrate bone to the skull became loosened, and it acquired, at least in the South American forms, the stapes as a medial brace. This simple structural change initiated the establishment of the mammalian chain of ossicles. Natural selection utilized the accidental proximity of these ossicles and the second tympanic membrane and fused them into a vastly improved new character complex. Not all the steps of this process are yet entirely apparent, but I think that little doubt is left as to the principle involved.

Such a fusing together of individual characters into a new character complex is not restricted to structural characters. It may also play a role in the emergence of complex new behavior patterns. Let me discuss a specific case. Goldschmidt (1948) described the extraordinary behavior complex of the larvae of a New Zealand cave gnat (*Araschnocampa luminosa*) of the family of Mycetophilidae. These larvae live on the ceiling of caves in self-spun webs, and lower trapping threads covered with sticky droplets on which they catch midges (Chironomidae) as they emerge in large numbers from the cave waters. To make their "trapping system" more effective, they have evolved bioluminescence. Goldschmidt asks: How could such a combination of characters have evolved gradually by the selection of favorable genetic variants? These fungus gnats have eight adaptations, says Goldschmidt, none of which would be of selective value to them without all the others. Actually, most of the eight prerequisites cited by Goldschmidt, such as the ability to select a habitat, are fairly widespread among fungus gnats or among animals in general, and the list reduces to three essential components of this interesting habit: (1) a carnivorous instead of a fungus diet, (2) the ability to spin the sticky trapping threads, and (3) luminescence. Subsequent researches have shown Goldschmidt's belief that none of these characters could occur without the others to be mistaken. There are other carnivorous fungus gnats, luminescence is not unique in the family, and, as Goldschmidt himself mentions, there are even other species that spin slimy trapping threads. There is little difficulty in seeing how these various potentialities of the family could have been combined into a single, highly effective device. Indeed, it seems to me that the assumption that all these adaptations could have appeared simultaneously as a single, efficient, new behavior complex in a single orthodox fungus gnat would be infinitely harder to understand. We do not know what the key invention of the New Zealand fungus gnat was, but it is possible that the new behavior complex started with a species which varied its fungus diet by scavenging, that is, by eating dead insects that had become stuck to the moist cave wall. Once such an extension of food habits had occurred, a high selection pressure for all the other components of the character complex would be obvious.

The three cited cases have in common the essential feature of pre-existing building blocks, which, when pieced together, give rise to an "improbable" new character complex of high selective value. The particular organisms are preadapted to acquire the new character complex because they already possess the potentiality for it, that is, the individual building stones. The role of natural selection in these cases is apparently not the bringing together of the individual units; this is done by forces independent of the prospective new structure. Natural selection enters the scene as soon as the pieces have been combined into a new complex that can function as a unit and can respond to natural selection as a unit.

Multiple pathways

These cases of the "piecing together" of character complexes illustrate most graphically the ever-ready opportunism of evolution. Whenever the need for a new structure arises, a high premium is placed on anything that satisfies this need. If the same need arises independently in unrelated organisms, independent solutions may be found. There is perhaps no better way to learn how evolutionary novelties emerge than by carefully comparing similar structures that have evolved independently in response to similar selection pressures. The fact that so many independent answers may be found to satisfy a single need proves three points: (1) the ever-present pressure of selection, (2) the opportunism of evolution, and (3) the potential variability of any structure. Whichever structure is the first to vary in a desirable direction will be the one on which natural selection can work. That component of the variation of accessory structures will be favored by natural selection which best fits with the modification of the primary structure. The almost innumerable ways by which beetles stridulate is a good illustration. Poison organs throughout the animal kingdom are another one. Any specialist can give numerous examples from the group with which he is most familiar, whether it be web construction in spiders, plume development in birds of paradise, floating devices in pelagic animals, or whatnot. At least five families of songbirds have independently discovered the usefulness of mud in nestbuilding—the South American ovenbirds (*Furnariidae*), the *Hirundo* group among the swallows, the nuthatches (Sittidae), certain thrushes (Turdidae) (for the inside of the nest), and the Australian Grallinidae. Methods for passing cellulose-rich plant food repeatedly through the intestinal tract have been invented by herbivorous mammals independently four times—the ruminating artiodactyls, certain kangaroos (marsupials), the beavers (rodents), and some, if not all, lagomorphs.

One would imagine that social bees with their colonies full of honey and larvae would be exceedingly vulnerable to raids by various nest robbers if they were not protected by their stings. And yet the stings have been lost in one group of social bees, the Meliponinae. Lindauer (1957) has described the numerous methods by which various species of stingless bees in the genus *Trigona* defend their nests. Most of them do it by biting; thousands of bees attack the intruder and make it very uncomfortable for him. *Oxytrigona* has acquired an accessory gland to the mandibles to pour an acid and very painful secretion into the wound. *Trigona droryana* of South America immobilizes the intruder by covering him with small pellets of a very sticky resin, and the South African *Trigona braunsii* by pouring honey over him! The very generalized need—protection against intruders—is achieved by exceedingly different, yet equally efficient, methods.

Natural selection comes up with the right answer so often that one is sometimes tempted to forget its failures. Yet the history of the earth is a history of extinction, and every extinction is in part a defeat for natural selection, or at least it has been so interpreted. Natural selection does *not* always produce the needed improvements.

Darwin was fully aware of this situation, not least because Mivart used it skillfully in an argument, restated by Darwin as follows (p. 260): "It has often been asked, if natural selection be so potent, why has not this or that structure been gained by certain species to which it would apparently have been advantageous?" The answer Darwin gives still appears to be the right one: "It may often have happened that the requisite parts did not vary in the right manner or to the right degree." And this is still our interpretation. Natural selection can operate only when it has a choice between alternate phenotypes. If a gene pool of a population does not contain the right genes, that is, genes that would permit an advantageous variation of the phenotype, natural selection is helpless. It is also often asked: "Why are not all animals as intelligent as man, if intelligence and a large brain are of as great an evolutionary advantage as is claimed by students of human evolution?" In this case it is probable that the selective premium for increased brain size was not sufficient in the other groups to set up a selection pressure anywhere near as large as that which occurred in the hominid line.

Let us not forget that the phenotype is a compromise between conflicting selection pressures and that every specialization is bought at a price. In many groups of organisms an increase in brain size may not give sufficient selective advantage to compensate for the anatomical and physiological unbalancing that it inevitably causes. Brain size is correlated in many subtle ways with the whole mode of life. Among songbirds (*Oscines*), for instance, a relatively large brain seems to be found only among omnivorous groups. All specialized feeders seem to have relatively small brains. Far more comparative anatomical work is needed, however, before this suggested correlation can be considered established.

CONCLUSION

The tentative answer to our question "What controls the emergence of evolutionary novelties?" can be stated as follows: Changes of evolutionary significance are rarely, except at the cellular level, the direct results of mutation pressure. Exceptions are purely ecotypic adaptations, such as cryptic coloration. The emergence of new structures is normally due to the acquisition of a new function by an existing structure. The resulting "new" structure is merely a modification of a preceding structure. The selection pressure in favor of the structural modification is greatly increased by a shift into a new ecological niche, by the acquisition of a new habit, or by both. A shift in function exposes the fully formed "preadapted" structure to the new selection pressure. This, in most cases, explains how an incipient structure could be favored by natural selection before reaching a size and elaboration where it would be advantageous in a new role. Mutation pressure, as such, plays a negligible role in the emergence of evolutionary novelties, except possibly at the cellular level. It should also be mentioned that the structure of the gene complex is important: too great a genetic and developmental homeostasis will result in too stabilized a phenotype and will tend to prevent a response to new selection pressures. Any population phenomenon that would tend to counteract excessive stability of the phenotype may favor evolutionary changes.

REFERENCES

Amadon, D. 1943. Specialization and evolution. *Amer. Nat.*, 77:133–141.

Barghusen, H. R., and J. A. Hopson. 1970. Dentary-squamoral joint and the origin of mammals. *Science*, 168:573–575.

Blest, A. D. 1957a. The function of eyespot patterns in the Lepidoptera. *Behaviour*, 11:209–256.

—— 1957b. The evolution of protective displays in the Saturnioidea and Sphingidae. *Behaviour*, 11:257–309.

Bock, W. 1959. Preadaptation and multiple evolutionary pathways. *Evolution*, 13:194–211.

Cope, E. D. 1896. *The primary factors of organic evolution*. Open Court, Chicago.

Crompton, A. W. 1972. The evolution of the jaw articulation of cynodonts. In *Studies in vertebrate evolution*, pp. 231–251. Winchester Press, New York.

Crompton, A. W., and F. A. Jenkins, Jr. 1973. Mammals from reptiles: a review of mammalian origins. *Ann. Rev. Earth Planet. Sci.*, 1:131–155.

Darwin, C. 1872. *Origin of species*. 6th ed. Oxford University Press reprint (1956), Oxford.

Davis, D. D. 1949. Comparative anatomy and the evolution of vertebrates. In *Genetics, palaeontology, and evolution*, ed. G. L. Jepsen, G. G. Simpson, and E. Mayr, pp. 64–89. Princeton University Press, Princeton.

—— 1964. The giant panda. *Fieldiana, Zool. Mem.*, 3:337.

Denison, R. H. 1941. The soft anatomy of Bothriolepis. *Jour. Palaeontol.*, 15:553–561.

Dohrn, A. 1875. *Prinzip des Functionswechsels*. Engelmann, Leipzig.

Goldschmidt, R. 1940. *The material basis of evolution*. Yale University Press, New Haven.

—— 1948. Glow worms and evolution. *Rev. Sci.*, 86:607–612.

—— 1951. Eine weitere Bemerkung über Glühwürmer und Evolution. *Naturwiss.*, 19:437–438.

Goodrich, E. S. 1930. *Studies on the structure and development of the vertebrates*. Macmillan, London.

Granick, S. 1953. Inventions in iron metabolism. *Amer. Nat.*, 87:65–75.

Gregory, W. K. 1934. Polyisomerism and anisomerism in cranial and dental evolution among vertebrates. *Proc. Nat. Acad. Sci.*, 20:1–9.

Hopson, J. A. 1966. The origin of the mammalian middle ear. *Amer. Zool.*, 6:437–450.

Huxley, J. 1942. *Evolution: the modern synthesis*. Harper & Bros., New York.

Kettlewell, H. B. D. 1973. *The evolution of melanism: the study of a recurring necessity, with special reference to industrial melanism in the Lepidoptera*. Clarendon Press, Oxford.

Lindauer, M. 1957. Zur Biologie der stachellosen Bienen: Ihre Abwehrmethoden. *Tagungsbericht, Deutsche Akad. Landwirtschafts, Berlin*, no. 11:71–78.

Lissman, H. W. 1958. On the function and evolution of electric organs in fish. *Jour. Exper. Biol.*, 35:156–191.

Manton, S. M. 1953. Locomotory habits and the evolution of the larger arthropodan groups. In Symp. Soc. Exper. Biol., no. 7: *Evolution*, pp. 339–376.

Mayr, E. 1958. Behavior and systematics. In *Behavior and evolution*, ed. A. Roe and G. G. Simpson, pp. 341–362. Yale University Press, New Haven.

—— 1959. Darwin and the evolutionary theory in biology. In *Evolution and anthropology: a centennial appraisal*, pp. 3–12. Anthropological Society of Washington, Washington, D.C.

Mivart, St. George. 1871. *Genesis of species*. Macmillan, London.

Needham, J. 1931. *Chemical embryology*, 3 vols. Cambridge University Press, Cambridge.

Osborn, H. F. 1936–1942. *Proboscidea: a monograph of the discovery of evolution, migration of the mastodons and elephants of the world,* 2 vols. American Museum of Natural History, New York.

Patterson, B. 1949. Rates of evolution in taeniodonts. In *Genetics, palaeontology, and evolution,* ed. G. L. Jepsen, G. G. Simpson, and E. Mayr, pp. 243–278. Princeton University Press, Princeton.

Plate, L. 1924. *Allgemeine Zoologie und Abstammungslehre.* Fischer, Jena.

Remane, A. 1951. Die Besiedelung des Sandbodens im Meere und die Bedeutung der Lebensformtypen für die Ökologie. *Verh. Deutsch. Zool. Gesellsch. (Wilhelmshaven),* pp. 327–359.

Rensch, B. 1947. *Neuere Probleme der Abstammungslehre.* Enke, Stuttgart.

Romer, A. S. 1946. The early evolution of fishes. *Quart. Rev.,* 21:33–69.

Schaeffer, B. 1948. The origin of a mammalian ordinal character. *Evolution,* 2:164–175.

Sewertzoff, A. N. 1931. *Morphologische Gesetzmässigkeiten der Evolution.* Fischer, Jena.

Simpson, G. G. 1953. *The major features of evolution.* Columbia University Press, New York.

Smith, H. 1953. *From fish to philosopher.* Little, Brown, Boston.

Westoll, S. T. 1958. The lateral fin-fold theory and the pectoral fins of ostracoderms and early fishes. In *Studies on fossil vertebrates,* pp. 180–211. University of London, London.

II
SPECIATION

Introduction

Evolution has many separate facets. The nature of genetic variation, evolutionary trends, and the origin of evolutionary novelties are among the components that were discussed in the preceding essays. The particular evolutionary process that has, on the whole, been the main concern of my own research is speciation. Speciation is the multiplication of species, that is, the division of one parent species into several daughter species. It is this process that is responsible for the extraordinary diversity of the organic world.

Lamarck, the first genuine evolutionist, did not concern himself with the problem of speciation. He had become interested in evolution primarily through a comparison of fossil mollusks with recent mollusks, and his attention was therefore centered on what was later called phyletic evolution. His concept of species was vague and in his work he nowhere discusses the multiplication of species.

It was quite different with Darwin. The problem of speciation was the core of his evolutionary interest for two reasons, one historical and the other scientific. The historical reason is that it was a problem of speciation—the question of the species status of the Galápagos mockingbirds—that made Darwin an evolutionist in March 1837; it was the problem posed by the mockingbirds, plus the weight of his previous experiences on the Beagle, *that finally tipped the scale and convinced Darwin to be an evolutionist. And for years the problem of speciation was the dominant topic in Darwin's notebooks. The scientific reason is that for Darwin the origin of new species was the key problem of evolution. Pre-Darwinian authors freely conceded the origin of new varieties, and they readily believed in upward progression owing to the unfolding ("evolution" in the older sense of the word) of immanent potentialities of the type. But they considered the origin of new species inconceivable, and thus speciation became the real touchstone of evolutionary thought.*

We still lack a thorough analysis of the changes in Darwin's thoughts on speciation. It appears, however, that in the beginning speciation was for Darwin geographic speciation, as he saw it documented by the mockingbirds (Mimus) *of the Galápagos. This interpretation of speciation he seems to have maintained into the 1840s (Essay of 1844). This view of speciation was consistent with Darwin's earlier concept of species as reproductively isolated populations (although Darwin did not use this terminology). But eventually, particularly after 1852, Darwin stressed more and more the second major characteristic of species, the occupation of a unique ecological niche by each species. This new emphasis tended to erode his previous geographical emphasis, which he replaced with a stress on the ecological divergence of species without geographic separation. Darwin's uncertainties and his ambivalence are particularly well reflected in his correspondence, later in life, with M. Wagner, K. Semper, and A. Weismann.*

The situation was aggravated because Darwin was clearly confused about the meaning of "variety," a term he used sometimes for individuals, sometimes for populations. The effect that this and other confusions had on Darwin's concept of species and his interpretation of speciation are discussed in Essay 10.

From the 1860s on the proponents of ecological (sympatric) and geographical speciation drifted farther and farther apart. There continued to be confusion about most of the underlying concepts. When the word isolation *was used, it referred sometimes to extrinsic barriers (geographic isolation) and sometimes to intrinsic ones (what we now call isolating mechanisms). There was considerable uncertainty concerning the origin of reproductive isolation, resulting in a lively correspondence between Darwin and Wallace (see Essay 11).*

The naturalists, in the meantime, vigorously advanced the cause of geographic speciation, their work culminating in a superb presentation by Karl Jordan in 1905 (see Essay 12). Yet, in spite of Karl Jordan's enlightened analysis of the issues of speciation, confusion and disagreement continued into the 1940s. The geneticists almost unanimously thought they had solved the problem of speciation when they succeeded in elucidating the underlying genetic mechanisms. This is well illustrated by R. A. Fisher's choice of a statement by W. Bateson as motto for the first chapter of his Genetical Theory of Natural Selection *(1930): "As Samuel Butler so truly said: To me it seems that the 'Origin of Variation,' whatever it is, is the only true 'Origin of Species.' " In other words, mutation is the only true origin of species. This attitude completely ignored the fact that speciation in most cases is essentially a population phenomenon.*

If we look at other treatises on evolution and the variation of species published in the 1930s, like T. H. Morgan's Scientific Basis of Evolution, *or Robson and Richards'* Variation of Animals in Nature *(1936), we discover how little the process of geographic speciation was accepted in the 1930s. Although most geneticists merely ignored it, Richard Goldschmidt launched a vigorous attack against the theory in his* Material Basis of Evolution *(1940) and attempted to prove that geographic variation led only to local adaptation and never to a multi-plication of species.*

In the face of all this resistance, the naturalists renewed their efforts to demonstrate not only the feasibility of geographic speciation, but also its widespread actual occurrence. The author who marshaled the evidence most effectively was B. Rensch (1929, 1933). He cited from the literature and from his own research numerous examples of natural populations that were on the borderline between subspecific and specific rank, and he pointed out that all of these crucial populations were geographically isolated.

My own work was a continuation of the work of Rensch. The rich collections of the Whitney South Sea Expedition had provided material on geographic speciation that was quite unmatched in the history of zoological exploration. The number of instances graphically demon-strating geographic speciation added up to several hundreds.

This rich material was summarized in a major review paper (1940) and in my book Systematics and the Origin of Species *(1942). That geographic speciation is the prevailing process of speciation, at least in animals, was no longer questioned after this date. What remained con-troversial, however, was the question to what extent other processes of speciation also occur. The conditions that have to be met to permit sympatric speciation, one of these alternate processes, were analyzed by me in 1947 (see Essay 13). The rich diversity of species encountered in the tropics gave rise to suggestions that additional mechanisms of speciation were operative there. This possibility, at least as far as birds are concerned, is discussed in Essay 14. Essay 15 deals with some of the genetic aspects of speciation, particularly the possibility of an occur-rence of genetic revolutions in peripherally isolated founder popula-tions. Finally, Essay 16 discusses the relations between speciation and climatic adaptation.*

REFERENCES

Fisher, R. A. 1930. *The genetical theory of natural selection.* Clarendon Press, Oxford.

Mayr, E. 1940. Speciation phenomena in birds. *Amer. Nat.,* 74:249–278.

Rensch, B. 1929. *Das Prinzip geographischer Rassenkreise und das Problem der Artbildung.* Borntraeger, Berlin.

—— 1933. Zoologische Systematik und Artbildungsproblem. *Verh. Deut. Zool. Gesellsch.* (Cologne), pp. 19–83.

10
Darwin and Isolation

The fact that isolation plays a role in evolution and particularly in speciation was recognized long before Darwin (Buch 1825). What this role is, however, was frequently misunderstood. Most of the great pioneers of modern evolutionary thinking, Darwin, Wallace, Huxley, Weismann, Wagner, De Vries, and others, grappled with this difficult theme, and it is a fascinating task to trace their steps and to determine where they advanced our understanding and where they "missed the boat" and why. There are few subjects in biology that lend themselves as well to historical treatment as the subject of isolation.

DARWIN

Darwin's voyage on the *Beagle* gave him abundant opportunity to observe isolation at work: "barriers of any kind, or obstacles to free migration, are related in a close and important manner to the differences between the production of various regions . . . on the opposite sides of lofty and continuous mountain-ranges, of great deserts and even of large rivers, we find different productions" (Darwin 1859:347). When chided by Moritz Wagner for underestimating the role of isolation in speciation, Darwin defended himself with the words: "It would have been a strange fact if I had overlooked the importance of isolation, seeing that it was such cases as that of the Galápagos Archipelago, which chiefly led me to study the origin of species" (F. Darwin 1888: vol. 3:159, letter of October 13, 1876). Yet, there is no doubt that Wagner's criticism was justified. Darwin admitted the occurrence of speciation on islands, but he emphasized again and again that incipient species could also evolve into full species without any spatial isolation: "I can by no means agree [with Wagner] that migration and isolation are necessary elements for the formation of new species. . . . I believe

Adapted with permission from pp. 222–225 of "Isolation as an evolutionary factor," *Proceedings of the American Philosophical Society*, 103(1959):221–230.

that many perfectly defined species have been formed on strictly continuous areas" (1872:106, 175).

All the evidence that has accumulated since Darwin indicates that this assumption is unwarranted as far as higher animals are concerned. It is of more than historical interest to determine how Darwin arrived at his erroneous conclusion. An analysis of his publications and letters indicates that this conclusion stemmed from Darwin's failure to understand clearly four evolutionary phenomena or concepts.

The Meaning of the Term "Variety"

Since species to Darwin were the result of gradual evolution, they had to pass through an intermediate stage, such as the "incipient species" and the "variety." "A well-marked variety may be called an incipient species" (1859:52). With the variety occupying such a key position in the problem of the origin of species, one might expect that Darwin would have devoted a great deal of effort to a definition of the concept and to an investigation of the circumstances under which varieties are being formed, but this is not the case. Darwin simply took over term and concept as it was current among the naturalists and systematists of his period. He never seems to have realized that since Linnaeus (and even before) the term had been applied indiscriminately to two very different kinds of phenomena, deviating individuals and deviating populations. Weismann, as we shall presently see, fell into the same error. For a typologist any deviation from the type is a "variety." For a biologist, on the other hand, it is of vital importance to know whether such a deviant is merely an intrapopulation variant or whether it is a different population. If Darwin had made a clear distinction, he would not have said: "If a variety were to flourish so as to exceed in numbers the parent species, it would then rank as the species, and the species as the variety; or it might come to supplant and exterminate the parent species; or both might coexist, and both rank as independent species" (1859:52). In the first edition of the *Origin of Species* the term "variety" is used on 24 pages. In 4 cases it clearly refers to geographical populations, in 8 cases to individual variants, and in 12 cases the usage is so ambiguous that it could have meant either.

A Morphological Species Concept

In view of Darwin's great insight into biological processes and evolutionary phenomena, one is somewhat shocked to realize how thoroughly his species definition is based on degree of difference: "I look at the term species as one arbitrarily given, for the sake of convenience, to a set of individuals closely resembling each other, and that it does not essentially differ from the term variety, which is given to less distinct and more fluctuating forms . . . the amount of difference is one very important criterion in settling whether two forms should be ranked as species or varieties" (1859:52, 56). Weismann (1872:19), who argued along exactly the same lines as Darwin, expressed the matter as follows: "The species is nothing absolute, and the differences among various species are of exactly the same nature as the differences between the sexes of one and the same species." It is on the basis of views like this that De Vries eventually developed his theory of speciation by mutation. Not taking sufficiently into consideration that the species is a "reproductively isolated population," a concept which was definitely familiar to many of his contemporaries (Mayr 1957), Darwin was quite unable to focus on the essential aspects of speciation and on the role of isolation. And yet, the same Darwin had extremely sound ideas on the development of sterility in incipient species (see Essay 11).

Failure to Distinguish between
Phyletic Evolution and Multiplication of Species

Evolutionary change within a lineage will, given enough time, produce sufficient change to justify ranking the descended population as a species different from the ancestral one. This is the "formation of a new species" of which Darwin spoke most frequently in his work. But this process of evolutionary change fails to account for the steady increase in the number of species that seems to have occurred in geological history. The splitting off of one lineage from another, the "multiplication of species," includes an evolutionary factor (the acquisition of reproductive isolation) that is not implicit in mere evolutionary change. Owing to his essentially morphological species definition, Darwin failed to give sufficient emphasis to the problem of the origin of reproductive isolation. Where he was concerned with the origin of sterility among species, it was in connection with "natural selection" rather than with speciation.

A Desire for a Single-Factor Explanation

Evolutionists prior to about 1930 were singularly reluctant to consider the interaction of various factors in the causation of evolutionary phenomena. Weismann (1872) bases his entire discussion on the choice of "natural selection *or* geographical isolation" as the factor responsible for the origin of new species. Darwin states, "Although isolation is of great importance in the production of new species, on the whole I am inclined to believe that largeness of area is still more important" (1872:107). He supports this claim by arguing that in large areas there are more individuals, and thus "there will be a better chance of favorable variations." Also he says that "the conditions of life are much more complex from the large number of already existing species" (1872:107) and this will favor the origin and spread of new "varieties." Even if it were true that these factors favor phyletic evolution, this would not shed light on the origin of reproductive isolation.

In acknowledging Wagner's volume on geographic isolation (F.

Darwin 1888: vol. 3:158), Darwin admits the importance of isolation but then continues, "I must still believe that in many large areas all the individuals of the same species have been slowly modified, in the same manner, for instance, as the English race-horse has been improved, that is by the continued selection of the fleetest individuals, without any separation." Did Darwin not see that there is no stricter isolation of gene pools than exercised by the animal and plant breeders? What would have happened to the selection of race horses if they had been permitted to interbreed freely with ponies, jumping horses, draft horses, and all sorts of utility horses? This very example should have convinced Darwin completely of the indispensability of isolation, yet he used it to argue against the importance of isolation because he felt he had to make a choice between isolation and selection. And this is the same Darwin who had said elsewhere (1868:185): "On the principle which makes it necessary for man, whilst he is selecting and improving his domestic varieties, to keep them separate, it would clearly be advantageous to varieties in a state of nature, that is to incipient species, if they could be kept from blending. . . ." One would expect Darwin to continue "by spatial segregation." Instead he concludes, "either through sexual aversion, or by becoming mutually sterile." The mechanism by which coexisting varieties might acquire sexual aversion or mutual sterility is not indicated. Darwin concludes his rejection of Wagner's thesis with the emphatic statement: "My strongest objection to your theory [of geographic speciation] is that it does not explain the manifold adaptations in structure in every organic being" (F. Darwin 1888: vol. 3:158), as if speciation and adaptation were mutually exclusive phenomena.

WAGNER, HIS FOLLOWERS, AND HIS OPPONENTS

The most ardent champion of isolation in the evolutionary literature was the great naturalist Moritz Wagner (1813–1887). During his travels in Asia, Africa, and the Americas, he had observed that, almost invariably, closely related forms or species occupy adjacent ranges, separated from each other by rivers, mountain ranges, valleys (in the case of mountain species) or, indeed, by any barrier to dispersal. He first published this observation in 1841 (p. 199), elaborated it in a major essay in 1868, and added to it in a number of later essays, republished posthumously in 1889. While Darwin had readily agreed that geographic isolation favors speciation, Wagner insisted that it was a *conditio sine qua non*: "The formation of a genuine variety which Mr. Darwin considers an 'incipient species,' will succeed in nature only when a few individuals can spatially segregate themselves for a long time from the other members of the species by transgressing the confining barriers of their range" (1889:64). It is crucial for an understanding of the controversy between Wagner and Weismann to realize that Wagner, when speaking of a variety as an incipient species, referred without exception to geographic races, while Darwin and Weismann in similar arguments more often than not referred to intrapopulation variants.

Wagner's original observation (the geographic relation of incipient species) and the empirical conclusion he drew from this observation (the importance of geographic isolation for speciation) are as true today as they were when first published. Wagner, however, was not quite so successful when he tried to find reasons for the need of geographic isolation. He advances two reasons. The first is to prevent the new population from being swamped by the parental one: "The origin and continued evolution of a race will always be endangered where numerous reinvading individuals of the same species disturb this process by general mixing and, thus, usually suppress it altogether. . . . Without a long continued separation of the colonists from the other members of their species the formation of a new race cannot succeed in my opinion" (1889:65). This first conclusion not only is substantiated by observation but has been confirmed by much recent work in population genetics that has established the overriding importance of integrating factors in an isolated gene pool.

Wagner was quite wrong in the second half of his explanation. He firmly believed that a change of environment was necessary for natural selection to become active. "Organisms which never leave their ancient area of distribution will never change" (1889:82). Even though it is, of course, true that there is less probability of evolutionary change in a stable environment, natural selection (even if only normalizing selection) will always be active and phyletic evolution will frequently take place. These marginal and ill-conceived comments on the relation between isolation and natural selection served Weismann (1872) as the basis for a severe criticism of Wagner's theory of geographic speciation. The original question "Can species multiply without geographic isolation?" was changed into the questions "Is isolation itself the factor which is responsible for the changes in isolated populations?" and "Is isolation necessary for varieties to become constant?" Weismann's refutation of Wagner, like Darwin's, is based on a morphological species definition and on the assumption that a morphologically different variety is an incipient species. Weismann states that, on the basis of Wagner's theory, on the one hand all isolated populations of species should be different varieties and, on the other hand, different varieties should not be able to originate in the area occupied by the normal type of species. Both of these postulates can, of course, be easily disproved. Failure of geographic speciation, he says, is indicated by the many cosmopolitan species, particularly among the tardigrades and the fresh water crustaceans, but also among higher organisms (for instance, the butterfly *Vanessa cardui*). On the other hand, Weismann continues, there are numerous cases of polymorphism, particularly among the butterflies, that prove that varieties can originate without isolation. To what extent in his argument he equates morphological change with origin of new species may be documented by some quotations. He asks, is it possible "that newly arising characters can become constant only through isolation and subsequent colony formations and thus give cause to the origin of a new species?" (1872:6) and asserts in a statement of *non sequiturs:* "it would be a fatal mistake if one were to assume . . . that isolation were an indispensable prerequisite to the modification of species, that not selection, but isolation alone would make possible the change of a species, that is its splitting into several forms" (1902:319). One final quotation may demonstrate how completely Weismann missed the essential point in the problem of the multiplication of species: "In this it is quite unimportant how they [endemic species that originated in isolated areas] originated, whether by Amixia in a period of variation or by natural selection, which tried to adjust the immigrants to the new environmental conditions of the isolated area. The change can even have been caused by influences which had nothing to do with the isolation, as for example the direct influence of the physical environment or the process of sexual selection" (1872:107).

The main reason for Weismann's inability to evaluate properly the role of isolation was his failure to determine what an incipient species is. This was seen much more clearly by his successors, and explicitly or implicitly it forms the basis of most discussions of isolation during the 1880s and 1890s. From the very beginning to the present day there has been a clear separation of two schools of thought. On one side are those who, with Wagner, insist that geographic isolation is an indispensable prerequisite for the multiplication of species in sexually reproducing organisms (excluding the special situation of polyploidy); on the other side are those who insist with equal determination that in addition to geographic speciation there exist various modes of sympatric speciation. The sympatrists claim that varieties with all the properties of incipient species often coexist. Their argument was, for the first time, clearly stated by Darwin himself: "I can bring forward a considerable body of facts showing that within the same area, two varieties of the same animal may long remain distinct, from haunting different stations, from breeding at slightly different seasons, or from the individuals of each variety preferring to pair together" (1872:105). Wallace (1889:150) enthusiastically supports this contention and concludes "that geographical or local isolation is by no means essential to the differentiation of species, because the same result is brought about by the incipient species acquiring different habits or frequenting a different station; and also by the fact that different varieties of the same species are known to prefer to pair with their like, and thus to bring about a physiological isolation of the most effective kind."

The first author who can be credited with a full understanding of the difference between phyletic evolution and the multiplication of species and who stated the case in a thoroughly modern manner seems to have been the ornithologist Seebohm. He points out with great logical acuity that natural selection and variation alone cannot account for the multiplication of species (1887:chap. 3):

> There is no reason why evolution should not go on indefinitely modifying a species from generation to generation until a pre-glacial monkey becomes a man, and yet no second contemporary species be originated. The origin of a second species is prevented by interbreeding. So long as the area of distribution of the species is continuous and not too large, the constant intermarriage which takes place between the males of one family and the females of another distributes the inherited and transmittable modifications throughout the race or species; which may advance or retrograde according to circumstances, but is prevented by interbreeding from originating a second species. . . . In order to originate a second species, it is necessary to counteract the leveling effect of interbreeding by isolating some of the individuals comprising the species, so that there may be two colonies, which are unable to communicate with each other, and consequently unable to interbreed. . . . In every species there is a tendency to vary in definite directions: the variations are hereditary and cumulative, so that evolution goes on steadily, though slowly, from generation to generation. If a part of the species be isolated from the rest, the evolution of the two colonies does not proceed in exactly the same direction, and the rapidity with which differentiation takes place is exactly in proportion to the difference in the circumstances in which the two colonies are placed.

One might think that the increasing frequency in the taxonomic literature of the 1880s and 1890s of conclusions similar to that of Seebohm would have led rather quickly to the universal adoption of the theory of geographic speciation, but this was not to be. On the contrary, at no period was geographic speciation more neglected by the most active students of evolution than during the years from Darwin's death (1882) to the 1930s.

The opposition came from two rather different schools. One camp consisted of those who could not believe that gradual fluctuating variation could be converted by selection into new species. They therefore postulated that the discontinuity among species could only be achieved by discontinuous genetic variation. This idea, which in its simplest form had been defended by some ancient Greek philosophers, had found a warm proponent in Maupertuis. It was later expressed by Bateson as follows: "The Discontinuity of which Species is an expression, has its origin . . . in the original Discontinuity of Variation. The evidence of

Variation . . . suggests in brief that the Discontinuity of Species results from the Discontinuity of Variation" (1894:567). What Bateson had postulated seemed to be brilliantly confirmed by De Vries's discovery of the origin of new types of *Oenothera*. This led the early Mendelians to the belief that "species arise by mutation, by a sudden step in which either a single character or a whole set of characters together becomes changed" (Lock 1906:144). In the same year De Vries said, "The theory of mutation assumes that new species and varieties are produced from existing forms by certain leaps. The parent type itself remains unchanged throughout this process and may repeatedly give birth to new forms" (1906:vii). These saltationist views swayed the thinking of contemporaries so completely that D. S. Jordan (1905) was forced to complain: Wagner's theory of geographic speciation "is accepted as almost self-evident by every competent student of species or of the geographical distribution of species . . . but in the literature of evolution of the present day the principles set forth by Wagner have been almost universally ignored." The geneticists themselves were the first to demonstrate that the *Oenothera* mutations are an exceptional situation and that the majority of mutations are small and precisely what one might postulate as a possible causation of Darwin's gradual variation. The *Drosophila* work in Morgan's laboratory, Castle's work on modifying genes in rats, and the work on plants by Nilsson-Ehle, East, and Baur, all confirmed this. Nevertheless, several naturalists attempted to apply De Vries's mutationism to the speciation problem as late as the 1920s, provoking H. F. Osborn's (1927) famous outburst against mutations. And despite the overwhelming evidence, Goldschmidt (1940) revived the theory of macromutations as an important factor in evolution and was followed in this by Schindewolf and some other paleontologists.

The other camp of opponents of geographic speciation consisted of those who took up the ideas of Darwin, Wallace, and Weismann that speciation could occur sympatrically through ecological or behavioral specialization. Although there is some modern evidence that such speciation may occur occasionally, most of the examples cited in the 1880s and 1890s revealed a complete lack of understanding of the populational character of species. The person who did more than anyone else to make this point clear and to reemphasize the geographical nature of most speciation in animals was Karl Jordan (1861–1959) (see Essays 12 and 20).

REFERENCES

Bateson, W. 1894. *Materials for the study of variation*. Macmillan, New York.
Buch, L. von. 1825 Physicalische Beschreibung der Canarischen Inseln. *Kgl. Akad. Wiss.* (Berlin), pp. 132–133.
Darwin, C. 1859. *The origin of species*. J. Murray, London.

—— 1868. *The variation of animals and plants under domestication.* J. Murray, London.

—— 1872. *Origin of species,* 6th ed. Oxford University Press reprint (1956), Oxford.

Darwin, F., ed. 1888. *The life and letters of Charles Darwin.* J. Murray, London.

De Vries, H. 1906. *Species and varieties: their origin by mutation,* ed. D. T. Mac-Dougal, 2nd ed. Open Court Publishing Co., Chicago.

Goldschmidt, R. B. 1940. *The material basis of evolution.* Yale University Press, New Haven.

Jordan, D. S. 1905. The origin of species through isolation. *Science,* 22:545–562.

Lock, R. H. 1906. *Variation, heredity, and evolution.* J. Murray, London.

Mayr, E. 1955. Karl Jordan's contribution to current concepts in systematics and evolution. *Trans. Roy. Ent. Soc. London,* 107:45–66.

—— 1957. Species concepts and definitions: the species problem. *Amer. Assn. Adv. Sci. Publ.,* 50:1–22.

Osborn, H. F. 1927. The origin of species. V. Speciation and mutation. *Amer. Nat.,* 61:5–42.

Schindewolf, O. H. 1936. *Paläontologie, Entwicklungslehre und Genetik.* Borntraeger, Berlin.

Seebohm, H. 1887. *The geographical distribution of the family Charadriidae.* Henry Sotheran, London.

Wagner, M. 1841. *Reisen in der Regentschaft Algier in den Jahren 1836, 1837, und 1838.* Leopold Voss, Leipzig.

—— 1868. *Die Darwin'sche Theorie und das Migrationsgesetz der Organismen.* Duncker and Humblot, Leipzig.

—— 1889. *Die Entstehung der Arten durch räumliche Sonderung. Gesammelte Aufsätze,* Benno Schwalbe, Basel.

Wallace, A. R. 1889. *Darwinism.* Macmillan, London.

Weismann, A. 1872. *Über den Einfluss der Isolierung auf die Artbildung.* Wilhelm Engelmann, Leipzig.

—— 1902. *Vorträge über Deszendenztheorie.* 2:315–336. Fischer, Jena.

Darwin, Wallace, and the Origin of Isolating Mechanisms

Moritz Wagner's repeated insistence on the importance of isolation in speciation eventually led to the recognition that this was indeed an important factor. Yet when we read what Weismann (1872) and Romanes (1897) wrote about isolation, we realize the magnitude of the confusion that still prevailed in the minds of the followers of Darwin. And this confusion continued into the 1940s, despite the clarifying discussions of Poulton and K. Jordan.

The majority of authors applied the term "isolation" to two entirely independent biological phenomena, geographic isolation and reproductive isolation. As important as both phenomena are, their roles in the evolutionary process are very different. *Geographic isolation* refers to the division of a gene pool into two independent ones by strictly extrinsic factors. *Reproductive isolation* refers to the prevention of interbreeding of populations by intrinsic mechanisms, now called isolating mechanisms. The sterility barrier is, of course, the best known of these mechanisms, and earlier authors generally used the term "sterility" for the phenomenon for which Dobzhansky (1937) has since introduced the term *isolating mechanisms.* It is now realized that the acquisition of isolating mechanisms is the crucial step in the process of speciation, and quite rightly a great deal of attention has been paid to various features of isolating mechanisms, such as their nature and their classification. The first systematic treatment of the subject was given by Du Rietz (1930), who provided a rather complete listing of the various kinds of barriers, but it was not until Dobzhansky coined the term "isolating mechanisms" and devoted an entire chapter to them in his classic *Genetics and the Origin of Species* (1937) that the full significance of these adaptations was recognized by evolutionists. However, even then the difference between geographic isolation and isolating mechanisms was not clearly understood: Dobzhansky listed geographical barriers among isolating mechanisms, and in 1942 I referred to the

Adapted with permission from pp. 227–228 of "Isolation as an evolutionary factor," *Proceedings of the American Philosophical Society,* 103(1959):221–230.

true isolating mechanisms as "biological isolating mechanisms," as if there were others, such as geographical ones. The confusion was not cleared up until 1963 when I provided a new definition (1963:91): *isolating mechanisms are biological properties of individuals which prevent the interbreeding of populations that are actually or potentially sympatric.* This clearly excluded geographical isolation from the category of isolating mechanisms. Chapter 5 of my *Animal Species and Evolution* provides a review of the extensive literature of the subject with ample historical references.

One major problem, still not yet entirely resolved, is what role selection plays in the perfection of isolating mechanisms. Two opposing theories can be found in the contemporary literature. According to one, isolating mechanisms are strictly an incidental by-product of the genetic changes occurring in geographically isolated populations and are perfected during this isolation without any *ad hoc* selection. According to the other, the development of isolating mechanisms is started in isolation, but their perfection is achieved through natural selection after the incipient species has reestablished contact with the parental species.

It is not realized by most contemporary students that this argument goes back to Darwin and Wallace. As I said earlier, they used the term "sterility" where we would use the term "isolating mechanisms," but the essential part of the argument, the role of selection, has remained unchanged. Darwin first stated his views in *The Origin of Species* in the chapter on hybridism (1872:320ff.):

At one time it appeared to me probable, as it has to others, that the sterility . . . [between species] might have been slowly acquired through the natural selection of slightly lessened degrees of fertility, which, like any other variation, spontaneously appeared in certain individuals of one variety when crossed with those of another variety. For it would clearly be advantageous to two varieties or incipient species, if they could be kept from blending.

But, he continues, there are a number of facts known that do not fit with this hypothesis:

In the first place, it may be remarked that species inhabiting distinct regions are often sterile when crossed; now it could clearly have been of no advantage to such separated species to have been rendered mutually sterile, and consequently this could not have been effected through natural selection; but it may perhaps be argued, that, if a species was rendered sterile with some one compatriot, sterility with other species would follow as a necessary contingency.

But of course, this would not explain how it became sterile with this "compatriot." Darwin continues, saying that the drastic differences so often observed between reciprocal crosses (for instance, ♂A × ♀B fertile, ♂B × ♀A sterile) is another phenomenon not explicable by selection, "for this peculiar state of the reproductive system could hardly have been advantageous in either species." He then continues:

> In considering the probability of natural selection having come into action, in rendering species mutually sterile, the greatest difficulty will be found to lie in the existence of many graduated steps from slightly lessened fertility to absolute sterility. It may be admitted that it would profit an incipient species, if it were rendered in some slight degree sterile when crossed with its parent form or with some other variety; for thus fewer bastardised and deteriorated offspring would be produced to commingle their blood with the new species in process of formation.

But Darwin is much too critical a thinker to be satisfied with a solution that is nothing but wishful thinking, and so he comes to the conclusion that such a scheme cannot operate (1872:321–322):

> But he who will take the trouble to reflect on the steps by which this first degree of sterility could be increased through natural selection to that high degree which is common with so many species ... will find the subject extraordinarily complex. After mature reflection it seems to me that this could not have been effected through natural selection. Take the case of any two species which, when crossed produce few and sterile offspring; now, what is there which could favor the survival of those individuals which happened to be endowed in a slightly higher degree with mutual infertility, and which thus approached by one small step toward absolute sterility?

The same argument is presented, in part with identical phrases, in *The Variation of Animals and Plants under Domestication* (1868). Here he states his conclusions even more unequivocally (p. 188): "We may conclude that with animals the sterility of crossed species has not been slowly augmented through natural selection ... As species have not been rendered mutually infertile through the accumulative action of natural selection ... we must infer that it has arisen incidentally during their slow formation in connection with other and unknown changes in their organisation," a conclusion that is now quite generally accepted.

Wallace, on the other hand, was not willing to let natural selection take the back seat so completely. After having read what Darwin had stated in the 1868 work, he writes in February 1868 (F. Darwin

1903:vol. 1:288): "I do not see your objection to sterility between allied species having been aided by Natural Selection. It appears to me that, given a differentiation of a species into two forms, each of which was adapted to a special sphere of existence, every slight degree of sterility would be a positive advantage, not to the individuals who were sterile, but to each form," and continues to explain this in detail. This started an active exchange of letters (letters 209–214 in F. Darwin 1903:vol. 1:288–297) on the subject of the origin of sterility. Darwin, however, sticks to his guns:

> I feel sure that I am right about sterility and Natural Selection. ... If sterility is caused or accumulated through Natural Selection, then, as every degree exists up to complete barrenness, Natural Selection must have the power of increasing it. Now take two species A and B, and assume that they are (by any means) half-sterile, *i.e.,* produce half the full number of offspring. Now try and make (by Natural Selection) A and B absolutely sterile when crossed, and you will find how difficult it is. . . .

This was a challenge to Wallace, which he took up at once. He developed a most detailed thesis, consisting of nineteen propositions or theorems by which he thought he could do what Darwin considered impossible (letter 211, March 1, 1868). In retrospect it is clear that Wallace made so many assumptions "to avoid complication," as he says, that he started out with virtually reproductively isolated species. Darwin's reaction is sufficiently amusing to be reproduced in full (letter 212, March 17):

> I do not feel that I shall grapple with the sterility argument till my return home; I have tried once or twice, and it has made my stomach feel as if it had been placed in a vice. Your paper has driven three of my children half mad—one sat up till 12 o'clock over it. My second son, the mathematician, thinks that you have omitted one almost inevitable deduction which apparently would modify the result. He has written out what he thinks, but I have not tried fully to understand him. I suppose that you do not care enough about the subject to like to see what he has written.

In this he misjudged Wallace, for only a week later Wallace writes (letter 212A): "I return your son's notes with my notes on them," and on goes the argument! Darwin's anguish is real (letter 213):

> I have been considering the terrible problem. Let me first say that no man could have more earnestly wished for the success of Natural Selection in regard to sterility than I did, and when I considered a general statement (as in your last note) I always felt

sure it would be worked out, but always failed in detail. The cause being, as I believe, that Natural Selection can not effect what is not good for the individual, including in this term a social community.

In the end Darwin apparently felt what he had said to Huxley a year earlier (F. Darwin 1903:vol. 1:277): "Nature never made species mutually sterile by selection, nor will men." Wallace once more summarized his views in his *Darwinism* (1889:174–179). In retrospect it is clear that in spite of the brilliance of analysis on both sides, the problem at that time was insoluble, because of lack of knowledge not only about sterility, but also about the other isolating mechanisms, genetics, and population structure. It was, though, while it lasted, a noble debate.

All the recent work indicates that by far the greatest part of the genetic basis of the isolating mechanisms is an incidental by-product of the genetic divergence of isolated gene pools and is acquired during this isolation. There is no argument about this conclusion; indeed, it is obvious that there would be complete hybridization after the breakdown of geographic barriers if such an isolating mechanism had not previously developed. The frequency of hybrid belts between formerly isolated portions of species fully substantiates this conclusion. The only remaining argument concerns one point: Will the formation of an occasional hybrid lead to an improvement of the isolating mechanisms of populations that are sympatric in the zone of overlap of two newly formed species? Logically, the answer would have to be yes. Hybrids, at least in animals, are normally of decreased fitness as far as their contribution to the gene pool of the next generation is concerned. Since the fact of their occurrence indicates that the genetic isolating mechanisms of their parents were less than perfect, such hybridization should lead to an elimination of these less than perfect genes from the gene pool. Moore (1957) points out correctly that on the whole the effects of such selection will be confined to the zone of overlap. Furthermore, such a production of hybrids will strengthen the isolating mechanisms only if the hybrids are effectively sterile or inviable, because otherwise introgression between the two parental species would occur and, with it, an accelerating weakening of the reproductive isolation. Ecological and behavioral isolating mechanisms should be the ones most easily strengthened in such zones of overlap. This conclusion is supported by the available observational evidence, as cited by Dobzhansky (1951), which indicates an increased reproductive isolation between certain pairs of species in areas of geographical overlap.

REFERENCES

Darwin, C. 1868. *The variation of animals and plants under domestication.* J. Murray, London.

—— 1872. *Origin of species,* 6th ed. Oxford University Press reprint (1956), Oxford.

Darwin, F. 1903. *More letters of Charles Darwin.* D. Appleton, New York.

Dobzhansky, T. 1937. *Genetics and the origin of species,* 1st ed. Columbia University Press, New York.

Du Rietz, G. E. 1930. The fundamental units of biological taxonomy. *Svensk. Bot. Tidskrift,* 24:333–428.

Mayr, E. 1963. *Animal species and evolution.* Belknap Press of Harvard University Press, Cambridge, Mass.

Moore, J. A. 1957. An embryologist's view of the species concept: the species problem. *Amer. Assn. Adv. Sci. Publ.,* 50:325–338.

Romanes, G. J. 1897. *Darwin and after Darwin,* vol. 3. Open Court Publishing Co., Chicago.

Weismann, A. 1872. *Über den Einfluss der Isolierung auf die Artbildung.* Wilhelm Engelmann, Leipzig.

12

Karl Jordan on Speciation

The opponents of geographic speciation in the period of 1880–1940, whether they believed in instantaneous speciation by macromutations or in speciation by ecological specialization, agreed on the universal occurrence and high frequency of speciation without preceding geographic isolation. Both Karl Jordan (1905:162) and, at about the same time, D. S. Jordan reported that the theory of sympatric speciation was virtually unanimously accepted among the biologists of the period. This theory maintained that individual variants could develop into full species without geographic isolation.

In his discussion of this problem K. Jordan notes that the concept of sympatric speciation appears at first sight to be abundantly supported by various facts. One of them is that there are many places on the earth where a great number of closely related species coexist, as the lemurs on Madagascar, certain groups of insects in the Hawaiian Islands, the marsupials in Australia, and the hummingbirds in America. However, says Jordan, some authors dissent and have come to the conclusion that one species can split into several only by way of a geographic variety. (This concept had first been clearly expressed by Moritz Wagner and had since been adopted by a considerable number of taxonomists.) "My own investigations completely confirm this theory as I want to demonstrate in the subsequent discussion of the great difference between geographic and non-geographic variation" (1905:163).

The theory of evolution postulates that species evolve from incipient species. Darwin and his successors, following up this thought, came to the conclusion that varieties within species are these incipient species. No agreement, however, was reached on the question whether or not all of the many kinds of varieties within species are equally important as incipient species. Jordan distinguishes three kinds of such varieties, or, as he expresses it, three types of polymorphism within species: individual, seasonal and geographic polymorphism. Nowadays, of course,

Adapted from pp. 58–63 of "Karl Jordan's contribution to current concepts in systematics and evolution," *Transactions of the Royal Entomological Society of London,* 107(1955): 45–66.

we would apply the term "polymorphism" only to variation within a population, and not to differences between populations. We call a species that breaks up into subspecies a polytypic species, not a polymorphic species. The clear distinction between variation within a population and variation between populations that we now make was not fully appreciated by Jordan. Fortunately, it did not affect the validity of his argument.

Full species of insects, Jordan concludes, differ from each other in coloration and other external morphological characters and usually also in the structure of the genitalia. Incipient species then should differ from each other either in external morphological characters or in the structure of the genitalia or in both. A variety, in order to qualify as an incipient species, must vary in a way consistent with this postulate. Accordingly, Jordan studied the mode of variation of the three kinds of varieties (individual, seasonal, and geographical).

SPECIATION BY MACROMUTATION

De Vries had recently emphasized the evolutionary importance of those individuals that lie completely outside the normal variation of a population, his so-called mutations. Such forms, Jordan says, are not infrequent among butterflies and are eagerly sought by the amateur collector. They are generally referred to as "sports." Jordan examined the genitalia of many such sports and this is what he found (1905:167): "In not a single aberration, either caught in nature or artificially produced by temperature shocks, have I found that the aberrant colouration was correlated with any differences in the genital organs. The sports remain with respect to these organs within the normal limits of variation. They are not new 'species' which have separated themselves from the mother species by a saltation and which now could co-exist independently."

It is curious how long Jordan's clear-cut refutation of De Vries's theory of speciation by mutation was ignored by the Mendelians. De Vries denied that the species of the naturalists, as recognized from the time of Linnaeus on, have any reality in nature. "These units are not really existing entities, they have as little claim to be regarded as such as genera and family" (1906:12). Since species are simply different types, "the origin of species may be seen as easily as any other phenomenon. It is only necessary to have a plant in mutable condition" (1906:26).

Jordan came back repeatedly in his writings to this theory of speciation by mutation. In 1911 (p. 401) he pointed out that even polymorph types are not nearly as clear-cut as is usually maintained, provided one has sufficiently large collections. The modern geneticist might say that additional modifying factors may help to bridge the phenotypic gap between two polymorph alleles. And Jordan continues "and it is perhaps also not superfluous to reiterate in this connection that the geographical races, too, are nearly always found to be connected by intergradations, if a sufficiently large number of specimens is examined."

SYMPATRIC SPECIATION

The fact that many similar but ecologically slightly different species may occur in the same geographical area was, at an early time, interpreted as evidence for speciation without geographic isolation. One author after the other tried to discover the mechanism that would make such sympatric speciation possible. The more we learn about the co-adaptation of gene complexes of local populations the more unlikely such schemes of sympatric speciation appear. Jordan, however, without this knowledge of modern genetics, had come precisely to the same conclusion merely through the study of incipient speciation and by an astonishingly lucid and logical analysis of the speciation process. Although he had presented a convincing refutation of sympatric speciation as early as 1896, he came back to it whenever a new paper was published that tried to revive the dead theory. Those who postulated such a process with the help of the hypothesis of homogamy (the mating of the most similar individuals within a population) he countered by stating that the observations on which these claims were based were ambiguous, if not mistaken, and that even if homogamy occurred it could not lead to a splitting of a single population into two (1905:173). When H. M. Vernon (1897) proposed a new version of this theory called evolution by "reproductive divergence," Jordan (1897, 1898) demonstrated its fallacy so conclusively that this hypothesis never again entered the biological literature. It is historically interesting that Jordan's premises are essentially the same as those that led, subsequently, to the establishment of the Hardy-Weinberg formula. Another contemporary who had proposed a theory of sympatric speciation was Romanes ("physiological selection") (1897), but he had already been refuted so decisively by Wallace, in his book *Darwinism* (1889), that Jordan could afford to dismiss his thesis rather summarily (1896:443).

The striking morphological difference sometimes found between seasonal forms of insects had led to the establishment of still another hypothesis of sympatric speciation by some entomologists. They argued that new species might arise if these seasonal forms became more and more different. Jordan admits that seasonal variants differ from ordinary individual variants, because they frequently "separate breeding communities. . . . In the species in which the first generation of mature individuals has disappeared before the next one hatches, these generations are reproductively isolated from each other as much as is one species from another" (1905:174). Jordan then asks himself whether or not this really could lead to species formation. He concludes finally that the odds are all against it. First of all, the summer generation is normally the direct descendant of the spring generation, and, second, there is no other evidence except phenotypic difference that these are incipient species. Jordan finds the best proof for this in the variation of the genitalia. Geographical races, which are incipient species, often deviate quite markedly in the structure of the genitalia from each other, while in the case of a large number of seasonal forms investigated by him, he found only a single species (*Papilio xuthus*) in which there is even the slightest genitalic difference. There is thus no evidence that seasonal forms are incipient species (1905:174).

There are two authors in particular whose hypothesis of sympatric speciation Jordan rejected emphatically. One is Dahl (1889), who had outlined an imaginary case of sympatric speciation in the genus *Gonopteryx* by postulating differences of incipient food plant races. "At the time when selection sets in in Dahl's illustration there were already two 'species,' and hence the specific distinctness of the two is not the outcome of psychological selection" (1896:427). Dahl had failed to explain the first crucial step, which can be explained only by geographical speciation. The other opponent was Petersen (1903), who had proposed quite a fantastic hypothesis based on blending inheritance, correlated host and mating preference, and various other improbable or actually erroneous assumptions. It is a real pleasure to read Jordan's answer (1903) with its superb knowledge of the evolutionary literature of the period and its critical emphasis on the essential points, most of which he had already listed in his 1896 paper (p. 444): "The physiological selection will, therefore, in no case result in divarication of a species into two, but the outcome of the physiological variation will be either dimorphism of the species, when both the normal and the varietal form are equally favoured in respect to the circumstances of life, or extinction of that form which is the least favoured. If however the most favourable kind of variation does not lead to the origin of a new species beside the parent one, no other variation will lead to this end. Hence we must conclude that a divarication of a species into two or more species cannot come about so long as the divergent varieties live so together that a direct or indirect intercrossing is not prevented."

Jordan's rejection of sympatric speciation is decisive. He says that there is no possibility of the origin of a new species without the help of some kind of local separation. Nor is the hypothesis of sympatric speciation needed to explain the geographical coexistence of closely related forms. Those who postulate it overlook two factors: the ability of animals to make active or passive changes in their range and the time factor. In fact, says Jordan, it can be shown quite clearly in many cases that the present coexistence of two closely related species is obviously the result of a comparatively recent invasion of each other's ranges (1905:203).

GEOGRAPHIC SPECIATION

By showing that the various schemes of sympatric speciation do not work and that neither mutations nor individual variants nor seasonal races qualify as incipient species, Jordan prepared the way for a discussion of geographic speciation. The questions he poses are whether or not geographic races are incipient species and whether or not the great diversity in nature can be explained as having arisen through geographic speciation. These questions can be answered only through a study of geographic variation.

The old concept that species are fixed, invariable types was drastically undermined by the work of Wallace in the East Indian Archipelago and that of Bates in the basin of the Amazon. Bates had pointed out that a comparison of individuals of a species from different localities revealed a graded series, ranging from differences so slight that they are not worth mentioning to differences so great that it was doubtful whether the individuals were still conspecific. Jordan comments (1905: 181–182) that "these grades of variability, which Bates found on the Amazon River can be found by the systematist also in Europe. It is a universally valid fact that there are all degrees between virtual identity of individuals from two areas and morphologically constant and well defined differences. A complete identity of all the individuals of a species in two localities occurs rarely, if ever. Most often that which is normal (=most frequent) at one locality, is not normal at the other, The curve of variability has shifted at the upper or lower end or at both. It also may occur that one organ remains the same and another is different or that a species is monomorph at one locality and dimorph at another."

Jordan takes this observation of geographic variability as his starting point and asks next whether it affects only unimportant, superficial characteristics or also those that distinguish good species. It was this question in particular that induced him to undertake his historical research in the geographic variation of genitalic structures in Lepidoptera (1896, 1905). Summing up his research, he states that since many geographic races of butterflies differ from each other in the structure of their genitalia in the same manner as good species and since such geographic races vary also in all the other characters of color and structure that distinguish species, it must be concluded that geographic races are to be considered incipient species. He continues to point out that the various characters of a species may be independent in their geographic variation. Two subspecies that differ in coloration or wing venation do not necessarily differ in their genitalia, the more so since even a certain number of good species are not distinguishable on this basis. The conclusion that species originate by geographic variation can be extended by analogy to animals other than insects. Jordan says that one must apply the same method to other species characters where species do not differ in the structure of the genitalia. This had been done already by Wagner, Seebohm, Hartert, Rothschild, and others.

Jordan was not satisfied merely to point out the fact of geographic variation, but attempted to establish general laws governing this phenomenon. I have already mentioned his discovery of the independence of different organs with respect to their variation. In the African *Papilio phorcas,* for instance, specimens from the Congo agree in coloration with those from Sierra Leone, but in the structure of the genitalia they agree with the populations from East Africa. Other rules established by Jordan are the following: The degree of geographic variation is different from species to species and even among closely related species; one species may be highly constant over large areas while another is subject to a great deal of geographic differentiation. "The amount of geographic variation depends on the nature of the environment (including the living one) and on the nature of the species. But not only on these two factors. It is also obvious that two areas between which there is an active inter-change of individuals are actually only a single area for the species because the individuals of the two areas form a single inter-breeding unit as if they were co-existing in the same area. One must therefore conclude that geographic variation depends furthermore on the degree of separation of the various geographically localised inter-breeding units. The degree of separation by one and the same geographic barrier in turn will be different according to the nature of the animal. What is a barrier for a terrestrial animal like a snail or flightless insect, and therefore separates various breeding communities, may not be any barrier at all for a winged animal. And finally is is easy to comprehend that the amount of time which has passed since the separation is a factor in the origin of a geographic variety" (1905:187). There was no other biologist in 1905 who saw and expressed the situation as clearly as did Karl Jordan.

"That species may differ according to locality has been known to systematists as far back as the 18th Century" (1905:184). Jordan's major contribution to the subject has been to treat it quantitatively by attempting to determine the amount of geographic variation in every species of an entire group, such as the Papilios or the Sphingidae, and by being the first to extend the investigation from a few superficial characters to such internal structures as the genital armatures. His treatment of the geographic variation of genitalic structures in the species *Papilio dardanus* is still outstanding as an example of such a quantitative treatment. Even the percentage occurrence of certain variants is recorded for each population: a spine on the valve occurs in 100 percent of the East African specimens, in 13 percent of Uganda individuals, and in less than 3 percent of those from the west coast of Africa.

As a true biologist Jordan is not satisfied merely with describing the fact of geographic speciation, he also wants to determine its causes. Wagner, the founder of the theory of geographic speciation, had claimed that isolation in itself is an active factor of transmutation. This is rejected by Jordan (1896:445): "Isolation as such is not an active factor which produces a character but is a factor which merely preserves a character produced by some other factor; isolation has, therefore, no direct effects." In support of this conclusion he cites numerous facts. How can isolation itself account for the fact, he asks, that the butterflies and moths of Sumatra and Borneo are quite universally dark and those from Queensland pale? Since isolation itself cannot account for this, "there is only one way possible by which the divarication of a species into two or more can come about—that is, the combination of isolation and transmuting factors" (1896:446). I believe that with this statement Jordan was the first to state clearly that speciation is the joint product of mutations and isolation. Wagner had neglected one of these factors; the contemporary Mendelians and the Darwinians had ignored the other. And how, asks Jordan, do these transmuting factors act during this period of isolation: "The geographical races thus produced we must assume to be first inconstant, to become more and more constant and divergent by the incessant influence of the transmuting factors, and to develop finally into a form which is so modified that it never will fuse either with the parent form or the sister forms, and that it therefore agrees with the definition of the term 'species'" (1896:446). It is quite evident from the context that Jordan meant "transmuting factors" to include both genetic changes and natural selection. This was written in 1896 before the birth of genetics, and a certain vagueness in terminology, like "constant" or "inconstant," is excusable. The modern geneticist would speak of the development of reproductive isolating mechanisms and of the evolution of coadapted and integrated gene complexes. What is important is that Jordan at this early date described the process of the change of geographically isolated populations just as we do today. As far as the nature of the genetic changes is concerned, he was still on the fence, allowing even the possibility of Lamarckian changes. However, he pointed out repeatedly, and as we must admit in retrospect correctly, that this point did not really make any difference so far as the theory of speciation is concerned. In this he was far more penetrating than any of his contemporaries or any other author during the 50 years that followed him.

Again and again Jordan emphasized the importance of the study of geographic variation. In 1896 (p. 446) he said with respect to geographic speciation: "as this kind of divarication of species is the only possible one . . . the study of localised varieties is of the greatest importance in respect to the theory of evolution; *the study of geographical races, or subspecies, or incipient species, is a study of the origin of species.*" And in 1903 (p. 664) he summarized his views as follows: "Geographic variation is the basis of species formation and it alone gives the explanation for the mutual sterility of species for which one has searched in vain since Darwin's day. Spatial separation alone permits the gradual divergence in morphology and physiology. It alone can prevent fusion with parental and sister populations of the varieties which develop in a new environment and become gradually constant. This permits that the originally very small, unimportant and inconstant differences in genitalia and other characters increase by accumulation to such an extent that a fusion is no longer possible."

Although the taxonomist normally works with morphological characters, it is evident from this quotation and from Jordan's other writings how highly he ranked physiological characters in the speciation process. "For us systematists that kind of psychological variation is of more [=greatest] interest which is the immediate outcome of morphological differences in the organs of sense and in the organs which are destined to affect the senses" (1896:434). In view of the fact that males discriminate between the scent of females of their own species and the scent of females of other species it should be important to look for geographic variation in this character. One such case, he continues, has already been found, namely that of Standfuss on geographic variation in smell preference between Swiss and Italian individuals of *Callimorpha* [=*Panaxia*] *dominula*. This, he says, proves how potent a factor in speciation a geographic variation in scent preference would be (1896:434). The relatively greater constancy of coloration in a given population of birds, as compared with the much greater variability of most butterflies, is due to the fact that the visual sense is the primary sense in birds while in most insects the olfactory sense is primary.

Alas, Jordan's cogent arguments in favor of the importance of geographic speciation were universally ignored by the Mendelians. Both De Vries and Bateson (as late as 1922) insisted that speciation was a phenomenon involving individuals rather than populations. The result was a tragic schism between the experimental geneticists on the one hand and the naturalists on the other. The ultimate bridging of the gap was only achieved when it was realized that the visible effects of most mutations are very small (rather than cataclysmic) and that even very small selective advantages can be of evolutionary significance. When the laboratory workers finally began to adopt populational thinking, the way was cleared for a balanced treatment of the problem of geographic speciation, an important component of the evolutionary synthesis, by Rensch (1933), Dobzhansky (1937), Huxley (1942) and Mayr (1942).

REFERENCES

Dahl, Fr. 1889. Die Bedeutung der geschlechtlichen Zuchtwahl bei der Trennung der Arten. *Zool. Anz.*, 12:262–266.

De Vries, H. 1906. *Species and varieties: their origin by mutation*, ed. D. T. MacDougal, 2nd ed. Open Court Publishing Co., Chicago.

Jordan, K. 1896. On mechanical selection and other problems. *Novit. Zool.*, 3:426–525.

—— 1897. Reproductive divergence: a factor in evolution? *Nat. Sci.*, 11:317–320.

—— 1898. Reproductive divergence not a factor in the evolution of new species. *Nat. Sci.*, 12:45–47.

—— 1903. Bemerkungen zu Herrn Dr. Petersen's Aufsatz: Entstehung der Arten durch physiologische Isolierung. *Biol. Zbl.*, 23:660–664.

—— 1905. Der Gegensatz zwischen geographischer und nicht-geographischer Variation. *Z. wiss. Zool.* 83:151–210.

Petersen, W. 1903. Entstehung der Arten durch physiologische Isolierung. *Biol. Zbl.*, 23:468–477.

Romanes, G. J. 1897. *Darwin and after Darwin,* vol. 3:41–100. Open Court Publishing Co., Chicago. (Summary of earlier publications.)

Vernon, H. M. 1897. Reproductive divergence: an additional factor in evolution. *Nat. Sci.*, 11:181–189.

Wallace, A. R. 1889. *Darwinism: an exposition of the theory of natural selection.* Macmillan, London.

13

Sympatric Speciation

The majority of evolutionists and systematists, as late as the 1930s and 1940s, believed that sympatric speciation was as frequent as geographic speciation, if not more so. This belief was particularly evident in the writings of entomologists, marine zoologists, and botanists. Sibling species with different habitat or host preferences were still recorded as biological races and interpreted as incipient species. It was for this reason that I decided in 1947 to analyze the whole question of the role of ecological factors in speciation, and in particular the problem of sympatric speciation. It seemed to me that the best strategy was to state the case against sympatric speciation as uncompromisingly as possible in order to provoke refutation if such was possible. I specifically enumerated all the conditions that had to be met to permit the occurrence of sympatric speciation.

My strategy was successful. The vague claims of sympatric speciation were henceforth replaced by well worked-out, fully documented cases. Three among the numerous proposals thus elicited in the past 25 years are the best candidates for possible sympatric speciation: (1) speciation by so-called disruptive selection (Thoday and Gibson 1962), (2) allochronic speciation (Alexander and Bigelow 1960), and (3) speciation by host specialization (Bush 1969). (For further details see Mayr 1970: 261–273). Actually only the third of the three is reasonably well established. Sympatric speciation by host specialization quite likely does occur occasionally, even though host specialization occurs allopatrically in the majority of instances of speciation in host-specific organisms.

The past decades have seen the gradual formulation of the theory of geographical speciation, principally based on work in the fields of the systematics of mammals, birds, insects, and mollusks. This theory postulates that in bisexual animals a new species can develop only "if a population, which has become geographically isolated from its parental

Reprinted and revised by permission of *Evolution* from "Ecological factors in speciation," *Evolution,* 1, no. 4 (1947):263–288.

species, acquires during this period of isolation characters which promote or guarantee reproductive isolation when the external barriers break down" (Mayr 1942). A number of workers in recent years have considered this statement an oversimplification. The omission of any reference in this description of the speciation process to the ecological factors—so vitally involved in the process of speciation—has also been objected to. Even though the importance of geographical speciation is universally admitted, there are a number of authors, particularly among the ichthyologists and entomologists, who believe that in addition to geographical speciation there is another process of speciation, variously referred to as ecological or sympatric speciation. Thorpe (1945), for example, is inclined to consider that in addition to geographical speciation there is a speciation process which is characterized by the fact "that local differences of habit may be the starting point for the evolution of new species."

It appears to me that there is no real conflict between those authors who stress the ecological aspects of speciation, such as Stresemann (1943) and Thorpe (1945), and those who, like myself, have stressed the geographical aspects. This was pointed out by Timofeeff-Ressovsky (1943). The seeming differences are due partly to misunderstandings and partly to a different emphasis placed on various aspects of a single process of speciation. A fresh analysis and possibly a synthesis of the two viewpoints is most desirable. The principal obstacle blocking such a synthesis up to now has been the general acceptance of sympatric speciation as an integral part of ecological speciation by those authors who emphasize the role of ecological factors in speciation. To remove this difficulty, a special section (pp. 152–171 below) has been devoted to a discussion of sympatric speciation. It is more important, however, to determine the respective roles of geographical and ecological factors in speciation, and to find out whether they operate consecutively or concomitantly.

The field has suffered particularly from the lack of a clear-cut *Fragestellung*. The result has been that only a few investigations have ever been undertaken that would give decisive answers to the unsolved questions. Most of the published studies sadly miss the crucial points. It is one additional object of this discussion to focus the attention of investigators on those aspects that are in particular need of further study.

THE PROCESS OF SPECIATION AND THE
ROLE OF ECOLOGICAL FACTORS

The most important aspect of speciation is that it is a process involving populations rather than individuals—contrary to the views of De Vries and other early Mendelians. The gene pool of a whole population serves as "the material basis of evolution" that is so often referred to in the genetic literature. The problem of speciation then boils down to two questions: (1) how do new populations[1] within a species develop? (2) how do such populations become reproductively isolated from other populations of the parental species?

Two steps are thus involved in speciation: (1) the establishment of new populations and (2) the establishment of intrinsic reproductive isolation. The theory of geographic speciation assumes that the sequence of these steps is 1-2.

How does the theory of geographic speciation explain the origin of new populations and of reproductive isolation between them? How does it evaluate the role of ecological factors in these processes? Furthermore, what is understood when the expression "geographically isolated" is used?

What is Geographic Isolation?

When two populations are separated by such a formidable barrier as a vast desert, an ocean, or an extensive, high mountain range, everyone will agree that they are geographically isolated. However, is a corn field geographically isolated from the next corn field by an intervening wheat field? How large must the distance be between "isolated" populations? How much gene flow is permissible?

Unfortunately, no general answer can be given to these questions. It depends in each case on the circumstances. The less suitable a given habitat is for a species, the more efficient a barrier it becomes. The term "geographically isolated" is to be construed broadly and refers to any environmental factor that effectively inhibits gene flow between two neighboring populations. In addition to the macrogeographical isolation of the geographical barriers listed above, there are many cases of *microgeographical isolation* where no great distances or conspicuous barriers are involved. The terms "topographical" or "spatial" isolation have often been used for these situations, particularly in the case of species with short average cruising ranges, such as are typical for many invertebrates. However, opposing the terms geographical and spatial isolation creates the erroneous impression that two different principles are involved. This is not the case. Whether large or small distances are involved, in all these instances there is a stretch of unsuitable terrain that inhibits dispersal and thus reduces gene flow. The term "microgeographical isolation" is thus preferable to "spatial" or "topographical" isolation.

All the terminologies so far proposed break down in those cases in which two populations that differ in their ecological requirements are

---

1. A population is a group of freely interbreeding individuals of a locality. Separate sympatric populations belong, by definition, to different species. Separate conspecific populations must, also by definition, be allopatric during their breeding season (except in the rare cases of seasonal separation in zones of secondary overlap).

in contact but neither interbreed nor overlap. Are they species or sub-species? Should they be called sympatric or allopatric? In the narrowest sense, they are, of course, allopatric, but, carrying the argument to the limit of absurdity, are not any two individuals allopatric? The terms sympatric and allopatric have been very useful in focusing attention on the spatial relationships of natural populations, but they seem to become meaningless in those cases where ecologically different populations exclude each other in space. If a meadow and a forest species occur in the same general region, are they allopatric or sympatric? This problem is particularly acute where habitats occur as widespread features of the landscape, sharply delimited against each other, as in much of the tropics and subtropics or wherever the original landscape has not yet been disturbed too severely by man. Here differences in ecology are also differences in geography. The seriousness of this problem is evident to every naturalist.

In a recent letter R. E. Moreau wrote me about the following situations concerning African birds. "In each of the mountains of East Africa a mountain forest white-eye (*Zosterops*) is found which is isolated as completely from the other montane ones as if it inhabited an oceanic island. I doubt whether any one of them comes into regular contact with the non-forest species (*Z. senegalensis*) which entirely surrounds them geographically but seems to avoid the immediate neighborhood of forest patches. An even more critical example of this sort of thing is provided by the two drongos, *Dicrurus ludwigii* and *D. adsimilis*. The former, in my experience, never ventures outside the edge of the evergreen forest, the latter seems never to perch even on the outside edge of a forest tree. Yet, both may occur within 50 yards of each other and the latter species is quick to occupy a clearing made in evergreen forest. I have come across some other nice examples, such as where *Cossypha heuglini* and *C. semirufa* occur in the same locality. The former is always restricted to its normal habitat, namely, bush country outside the forest, while the latter is found only in the forest. In one or two mountain forests where *C. semirufa* is absent the other bird replaces it. Are these species sympatric or allopatric?"

A universally applicable answer cannot be given to all the above questions. They must be answered individually and the answer will inevitably depend on the particular circumstances of each case.

The Invasion of New Habitats

The process of speciation results in ecological diversification and consequently in an ever-increasing efficiency in the utilization of the environment (Mayr 1948b; Lack 1948). One aspect of gradual speciation is therefore that it is an ecological process. Every species lives on an adaptive peak and the problem of speciation is how to reach new, not previously occupied, adaptive peaks. A species might do this either (1) by becoming locally more euryoecous (ecologically tolerant) or (2) by invading new areas with different ecological conditions. As far as the first possibility is concerned, a population that was previously restricted to the forest might become adapted to live in orchards or meadows as well as in the forest. Observations show that this occurs only very rarely. A species that is adapted for life in the forest will normally not be able to survive the competition it meets in the meadow. A population adapted to live on one species of food plant will normally not be able to enlarge its niche to include a second kind of food plant, unless its genetic make-up is altered. To be sure, individuals of many species try continuously to invade new niches. Such colonizations are often temporarily successful, but a single adverse season usually reduces the ecological amplitude of the species to its normal width. Some species are, at least locally, restricted to an extremely narrow niche, others are more tolerant in their ecological requirements. Whether a population's niche will be broad or narrow is largely a function of its genetic constitution. A local gene complex can become modified by the usual genetic processes so as to increase or decrease the ecological tolerance of the population of which it is the genetic basis, but the development and simultaneous maintenance in the same area of two different gene complexes of the same species adapted for two different ecological niches will be prevented by gene flow. Some especially successful species, the euryoecous species of the ecologist, get around this difficulty by developing a gene complex that is equally satisfactory in a number of different habitats. Such species are, however, rather rare.

How then does a species colonize a new type of habitat? A glance at the distribution map of ecologically variable species gives us a clue. The ecologically (as well as genetically) most aberrant populations are nearly always found along the periphery of the range of the species. There are several separate reasons for this. One of them is that the one-directional gene flow near the border of the range increases the rate of evolutionary change. Another is that the normal habitat in the center of the range may not offer optimal living conditions near the periphery, a situation that encourages a shift in habitat. Many specialized species have been able to develop in each district a geographical race adapted for the locally most abundant or otherwise most satisfactory ecological niche. The red crossbill (*Loxia curvirostra*) offers a particularly graphic illustration of this important principle (Kirikov 1940). The mountains of central Asia (Himalayas, Altai) were probably the original home of the species. It lives there on the prevailing coniferous trees, various species of spruce (*Picea*) or larch (*Larix*), and has developed a rather thin, slender bill. From here the crossbill has spread east and west and has reached a number of areas where pines (*Pinus* spec.) are the prevailing or exclusive tree, such as the Crimea (*L. c. mariae*), Tunisia (*L. c. polyogyna*), the Balearic Islands (*L. c. balearica*), and particularly northern Europe (*L. c. pityopsittacus*): crossbills are also found in the southwestern United States and Central America. In these areas geographical races of the crossbill have developed that have large, heavy bills adapted to the opening of the tough pine cones. However, only one race has developed in the areas where there are two prevalent conifers, such as *Pinus cembra* and *Picea excelsa* in the Alps. The cross-bill of the Alps (*L. c. curvirostra*) is a spruce race. The only apparent exception occurs in parts of northern Europe where the "subspecies" *curvirostra* and *pityopsittacus* are reported to breed in certain districts side by side without mixing. The greater portion of the ranges of these two forms is, of course, still separate, revealing their geographical origin. The evidence indicates that the most distinct of these "ecological" races originated in complete geographical isolation.

At first sight the concept of a geographically and ecologically variable species seems full of contradictions. We see that at one locality a species is restricted to a very definite habitat niche, while at another locality it occurs in a different, sometimes very different, niche. We know that through selection in each of these populations a definite gene complex has developed that permits the population to survive and thrive in spite of competition, predation, and all sorts of other adversities. I asserted in 1942 (p. 196) that geographical isolation precedes the formation of ecological preferences in every case. If this were true, how could species ever invade new habitats, as they undoubtedly do? Thorpe (1945) correctly points out this contradiction.

The adaptation for life in a given habitat includes the faculty of the individual of the species to select this habitat (from a vast array of other possible ones!) during the dispersal phase. Although genetic factors must to some extent control this ability of habitat selection, it is nevertheless not entirely rigid and undeviating. There is a certain amount of ecological plasticity, greater in some species, less in others.

The establishment of individuals in a new environment will be assisted by conditioning, as emphasized by Thorpe (1945), as well as by "organic selection," that is, the gradual substitution of modifications by mutations (Baldwin 1902; Gause 1947). A new population within the species will thus come into being and will permit the species to spread into areas that were previously outside the breeding range. It is probable that most range expansions of species are caused by the origin of such new intraspecific populations. This is well illustrated by the recent spread of the mistle thrush (*Turdus viscivorus*) in northwestern Europe (Peitzmeier 1942). This species had lived for a long time in the coniferous mountains south of Westphalia, but did not invade the lowlands until 1928. By 1939 it was common in the deciduous woods and farm gardens of the area, a habitat strikingly different from the pine and spruce forest of the nearby mountains. Peitzmeier, in tracing this invasion back to its source, presents evidence that shows that it came not from the neighboring mountains but rather from the west (northern France, Belgium), where the mistle thrush had always been an inhabitant of deciduous lowland woods and gardens. From here it spread eastward through Holland and across the Rhine into Westphalia. Thus the invasion of the deciduous woods of Westphalia is due not to a shift in the tolerance of the population from the coniferous woods of the hills around Westphalia, but to a range expansion of an already existing deciduous woods population. Unfortunately, nothing is known about the relationship of the two populations, but it is quite possible that at the present time there is very little gene exchange between these two geographical-ecological races of the mistle thrush. It would be interesting to determine through banding whether there is any interbreeding in the zone of contact. It would also seem important to determine where and how the deciduous woods population of northern France originated.

Not always are the ecological differences as clearly associated with geographical features (different distributions) as in the case of the mistle thrush. The smaller the normal cruising range of individuals of a species, the more restricted is the geographical range of a local population, and the more difficult it is to discern whether the establishment of new populations is a microgeographical phenomenon or a purely ecological one. The literature of economic entomology is replete with cases of locally segregated populations showing ecological differences. The codling moth (*Carpocapsa pomonella*) is particularly apt to develop such local races. Armstrong (1946) recently described a local population of this species from the province of Ontario (Canada) that had developed in a large, 80-year old pear orchard that was "more or less isolated from other large pear or apple orchards." While moths in Ontario apple orchards have their peak of emergence between June 26 and July 17, the moths in this pear orchard emerged mainly between July 17 and August 7, two weeks later. The delayed emergence of the pear moths coincided with the period during which the pears softened and became penetrable for the larvae. Armstrong points out that only "the absence of large apple orchards in the vicinity has prevented this strain from being swamped by crossing with the generally prevailing apple strain." For, even though the peaks and extremes of the emergence curves are well separated, there is still a large area of overlap of emergence time that would result in rapid swamping if it was not for the microgeographical segregation.

Several other populations of the apple codling moth are known that have invaded other hosts and have developed partly microgeographically segregated and biologically distinct populations. In California and in other areas the codling moth attacks the walnut. The first damage to walnut crops in California was reported in 1909. The first heavy infestation was in 1931, although apples and pears had been heavily infested in nearby districts since 1880 (Boyce 1935).

This interesting case has never been analyzed fully. It is not known whether there are two separate strains of the codling moth in the walnut districts or whether the ecological requirements of the local race have become broader, permitting life on both apple and walnut trees. There is comparatively little apple growing in the chief walnut districts, and it appears possible that the codling moth population in the walnut districts has changed completely into a "walnut strain." It would be intensely interesting to follow up these possibilities. All we know up to now is that an evolutionary change has occurred in the apple codling moth of the walnut-growing districts of California—it has invaded a new host 50 years after its introduction into California. The existence of such cases is not necessarily proof for sympatric speciation since, as stated, there are several alternative possibilities (Bush 1974).

Amount of Gene Flow between Incipient Species

All geographical barriers are relative. Even the most isolated islands are frequently visited by strays from the nearest mainland. Hundreds of continental species of animals are annually recorded from Great Britain. The interruption of gene flow even between two "isolated" populations is thus always relative and incomplete (Figure 13-1B). If two such populations diverge nevertheless, it proves that the combined effects of mutation pressure, selection, and random fixation (+ recombination) outweigh the equalizing effects of gene flow.

This leads to the simple, but important, question: How strong must the gene flow be to prevent the genetic drifting apart of two populations? It is very unlikely that the purely genetic processes of mutation

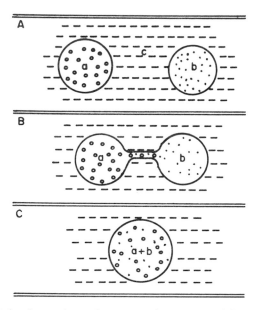

Figure 13-1. Spatial relationship of two populations *a* and *b*: (*A*) completely isolated by unsuitable terrain; (*B*) incompletely isolated—some gene exchange through partial barrier; (*C*) completely sympatric, but partially or completely segregated in different ecological niches.

pressure and random fixation cause changes of a sufficiently high order of magnitude to hold any sizable gene flow in check. It may be different with selection pressure, but unfortunately it has never been determined how much gene flow strong selection pressure can neutralize.

Let us consider, for example, two spatially segregated, contiguous populations within a species that are not separated by an extrinsic barrier. Selection pressure will tend to pull these two populations apart if they live in rather different habitats. But can selection overcome the effects of dispersal across the zone of contact? Can conditioning reduce dispersal across the line of contact to such an extent that it no longer prevents the steady divergence of this population? Sufficient facts are not available for a decisive answer to these questions. The scanty evidence that is available indicates that speciation by geographical segregation (without isolation) is rare, if it occurs at all. Ecological differences are rarely abrupt in zones of primary intergradation. There is likely to be an intermediate habitat with an intermediate population to serve as a channel for the gene flow. In the cases where two ecologically very different subspecies are found in immediate contact, it can nearly always be shown that secondary contact is involved.

The only study known to me of the effect of gene flow across a habitat border is an investigation by Blair (1947) of deer mouse (*Peromyscus maniculatus*) populations on pinkish gray and dark red soils in the Tularosa Basin. Populations that occur 18 miles apart on differently colored soils show adaptive differentiation in the frequency of buff and gray genes. However, there was no difference between the populations found in two stations on different soils only 4 miles apart. Future investigations must show to what extent a stronger habitat preference can reduce the width of the dispersal belt.

It is unfortunate that so little evidence is available on this problem. Sympatric speciation would appear more probable if speciation by geographic segregation was at all common. It is here that the chief difference of opinion seems to exist between those who believe in occasional sympatric speciation and those who do not. The adherents of sympatric speciation believe that invasions of new—geographically not isolated—habitats are sufficiently irreversible and conditioning for these new habitats sufficiently complete to prevent dispersal and permit the building up of distinct gene complexes. A thorough examination of the whole problem of sympatric speciation is therefore essential for the proper evaluation of the various possibilities.

AN ANALYSIS OF SYMPATRIC SPECIATION

The role of ecological factors in geographic speciation was discussed in the preceding section. The present section will be devoted to answering the question whether or not there is a separate process of sympatric speciation, independent of geographic speciation. What "sympatric speciation" is has never been properly defined, but it is generally characterized by one or both of the following assumptions: (1) the establishment of new populations in different ecological niches within the normal cruising range of the individuals of the parental population (Figure 13–1C); (2) the reproductive isolation of the founders of the new population from individuals of the parental population. Gene flow between daughter and parental populations is inhibited by intrinsic rather than extrinsic factors. A rapid, if not almost instantaneous, process of species formation is implied in most schemes of sympatric speciation.

The concept of sympatric speciation goes back to pre-Darwinian days. Darwin himself was uncertain about this question, as is evident from his correspondence with Wagner, Semper, and Weismann, and as indicated in numerous passages in his works, for instance: "If a variety were to flourish so as to exceed in numbers the parent species, it would then rank as the species, and the species as the variety; or both might co-exist, and both rank as independent species." "The small differences distinguishing varieties of the same species, steadily tend to increase till they come to equal the greater differences between species of the same genus, or even of distinct genera." Darwin makes no distinction between speciation through individuals (individual varieties) and speciation through populations (subspecies). While in some of his statements he seems to place due weight on the geographical element, in others he seems to ignore it altogether.

It was M. Wagner who, in a series of papers from 1868 to 1889, brought out the importance of geographical isolation of populations for the multiplication of species. At first these ideas, radical and novel as they were, found few adherents. Indeed, the last two decades of the nineteenth century witnessed the greatest flowering of the concept of sympatric speciation. Romanes devoted an entire volume of his *Darwin and after Darwin* (1897) to a theory of sympatric speciation that he called the theory of physiological selection. He and his followers insisted that geographical speciation was only of minor importance for the origin of new species. The argument of this school is based on two concepts, one unproven and very unlikely (that of homogamy), and the other definitely wrong (that of blending inheritance). Since they took blending inheritance for granted, they had to postulate the existence of homogamy (see also Pearson 1900). The concept of homogamy or associative mating states that within a population the most similar individuals will mate with each other. For "so long as there is free intercrossing, heredity cancels variability, and makes in favour of fixity of type. Only when assisted by some form of discriminate isolation [=preferential mating], which determines the exclusive breeding of like with like, can heredity make in favour of change of type, or lead to what we understand by organic evolution" (Romanes 1897). The most significant point of this argument is that new populations are formed by the nonrandom mating of individuals within populations. Romanes' theory is merely an elaboration of the theories first independently proposed by Catchpool (1884) and by Dahl (1889) (see Essay 11). All these theories have three postulates in common (which will be discussed below): homogamy, complete linkage of mate selection and habitat selection, and the absence of genetic segregation.

The confused reasoning of this school cannot be illustrated better than by the following quotation: "Suppose that on the same oceanic island the original colony has begun to segregate into secondary groups under the influence of natural selection, sexual selection, physiological selection, or any of the other forms of isolation, there will be as many lines of divergent evolution going on at the same time (and here on the same area) as there are forms of isolation affecting the oceanic colony" (Romanes 1897). Actually, of course, exactly the opposite is true. Isolated oceanic islands are exemplary for monotypic evolution, as Lack (1947) has demonstrated so convincingly in connection with the Cocos Island finch.

It would be foolish to quote these fallacies if it were not for the fact that the very same arguments are still used today to endorse sympatric speciation. The only novel feature is that nowadays the occurrence of "macromutations" is sometimes postulated, as for example by Valentine (1945): "Speciation may proceed within a population as a result of the appearance of relatively drastic changes (mutations) that are recognized by both the mutants and the norms, causing reluctance to cross. This, in brief, is the principle of associative speciation."

The Evidence for Sympatric Speciation

There must be some reasons why there are still so many authors left who believe that a process of sympatric speciation occurs in addition to allopatric (geographical) speciation. Botanists, in particular, have frequently stated that they find relatively little evidence for geographical speciation. Romanes (1897) quotes Nägeli to the effect "that in the vegetable kingdom closely allied species are most frequently found in intimate association with one another, not, that is to say, in any way isolated by means of physical barriers." In fact, *all* of the cases quoted by Romanes as proving sympatric speciation are taken from the plant kingdom. As is now known, instantaneous sympatric speciation through polyploidy is actually a common occurrence among plants. Apomixis, hybrid swarms, and the polytopic formation of ecotypes further complicate the picture. On the other hand, speciation in sexually reproducing plants seems not to differ materially from speciation among sexually reproducing animals.

Two classes of phenomena are frequently quoted as proving the existence of sympatric speciation. One consists of experiments showing that parasitic insects, as well as monophagous or oligophagous food specialists among plant feeders, can be conditioned to accept new hosts.

The other is the occurrence of "species swarms" among freshwater organisms in many rather recent lakes where, it is believed, geographic speciation could not have been operating. The significance of this evidence will be discussed below.

The main reason for postulating sympatric speciation, however, seems to be the fact that invariably there are ecological differences between closely related and morphologically similar sympatric species. This ecological specialization leads to a wonderfully efficient way in which the numerous species of a locality utilize their environment. It is argued that such perfect adaptation could not have developed in geographical segregation. Among naturalists there is a strong reluctance to admit the evolutionary role of accident. Just as the Lamarckian finds it impossible to believe that the wonderful morphological adaptations found in nature could be the result of random mutation and selection, in a parallel manner some ecologists find it hard to accept that ecological speciation should happen in the apparently roundabout way of spatial segregation rather than by the direct route of sympatric habitat choice.

Assumptions Underlying the Concept of Sympatric Speciation

Sympatric speciation cannot be discussed profitably before some of the assumptions made by the adherents of this thesis are stated more succinctly. Two recent papers contain a clear statement of the major assumptions. Thorpe (1945) visualizes the following procedure of sympatric speciation: "Imagine now an area where two types of habitat *a* and *b* (e.g. two different vegetational types) are available in mosaic distribution and a species (of bird) confined to habitat *a* within that area. In some exceptional circumstances of crisis or as a result of some slight germinal change certain individuals of the species spread into habitat *b* and the young reared there become imprinted, or otherwise specialized, to the new niche. If this niche provides room for expansion, the birds of the new habitat will rapidly come to fill it during which time they will be reproductively isolated to a considerable extent from the *a* habitat birds." In this scheme Thorpe implies thus the assumptions: (1) that young birds raised in a new habitat will "become imprinted, or otherwise specialized, to the new niche" to such a degree that they will not reenter the ancestral habitat to any appreciable extent, but rather establish at once a new population by a more or less irreversible process; and (2) that such individuals "will be reproductively isolated to a considerable extent" from the inhabitants of the ancestral habitat by being completely restricted to the new habitat. The choice of habitat and not microgeographical segregation would be the isolating factor since the two habitats are "available in mosaic distribution" (see Figure 13–1C).

Test (1946) regards sympatric speciation as a preadaptation process. "Ecological speciation is that differentiation which has taken place during isolation of extreme variants which could not survive in the parental microhabitat, but which are able to take over a microhabitat not compatible with the parental morphology and ecology. This isolation is not of a geographic type, but is due to the unsuitability for the parent stock of the microhabitat invaded by the offshoot." In connection with this hypothesis the following assumptions are made explicitly or by implication:

1. That new species may be established by extreme individuals rather than by populations.

2. That such founders and their offspring are different genetically from the parent population.

3. That such founders are preadapted for a habitat different from the parental habitat.

4. That—in order to survive—such individuals search for the habitat for which they are preadapted.

5. That during the original colonization "an intervening area" can be crossed which later no longer can "be crossed by individuals of either stock."

Each of these assumptions is arbitrary and largely unsupported by facts. Combining several of these improbable assumptions into a single hypothesis increases their improbability.

Objections to the Hypothesis of Sympatric Speciation

Every one of the various proposed hypothetical models of sympatric speciation makes certain unproven assumptions. Obviously, even if all these assumptions should turn out to be invalid, this would not necessarily prove that the basic hypothesis is wrong. In any case, a critical evaluation of the various assumptions will lead to a clearer understanding of the problem.

Homogamy

The concept of homogamy states that within a population the most similar individuals will prefer to mate with each other. The postulation of homogamy was the inevitable consequence of the theory of blending inheritance, because random mating would soon lead to a complete elimination of genetic variability in a population under the theory of blending inheritance. Dahl (1889), Romanes (1897), Pearson (1900), and many other writers of that period who were unaware of the prevalence of geographical speciation claimed that homogamy was an indispensable requisite of speciation. However, positive evidence for the occurrence of nonrandom mating within populations is exceedingly meager. Exceptions are provided by species that are very variable in size and continue to grow after reaching maturity. In the nudibranch *Chromadoris zebra*, for example, Crozier (1918) found that courtship usually broke off between individuals of too uneven size and that there was considerable correlation (about 0.6) between the sizes of the sexes in copulating pairs. However, in this and other similar cases there was complete overlapping between neighboring size classes and no evidence for the origin of any discontinuities owing to this associative mating. Furthermore, in many other less variable species (Alpatov 1925; Spett 1929) there was no correlation in size between copulating individuals. As far as color characters or other species characters are concerned, the evidence is even more completely negative. Among the numerous cases of polymorphic vertebrates that have been investigated the only known case of probable nonrandom mating is that of the snow goose–blue goose (Manning 1942). Of the 1500 blue geese that were nesting on Southampton Island among about 30,000 snow geese, 850 formed pure blue pairs instead of about 180 as would be expected on the basis of random pairing. Kosswig (1947) suggested that homogamy might explain the rapid speciation of cichlid fishes in the East African lakes. Not only does he produce no evidence to support this speculation, but he is definitely in error when he says that monogamy will necessarily result in homogamy. Gershenson (1941) has been able to prove conclusively that there is random mating among the color phases of the hamster (*Cricetus cricetus*). The same has been shown for color patterns in insects (Spett 1929). The only apparent exception is the yellow mutant in *Drosophila melanogaster* and probably in other species of the genus (Spett 1931; Diederich 1941). However, in this case there is only partial one-directional mating preference and the viability of individuals of this mutant is so inferior to the wild type that it would be rapidly eliminated in wild populations. Even where there is a slight amount of homogamy, its role in speciation can apparently be neglected, for Hogben states (1946:163) on the basis of extensive calculations: "Positive assortative mating can have very little importance as an evolutionary process unless it is exceedingly intense." This discussion can then be summarized by saying that evidence in favor of homogamy is virtually nonexistent and that homogamy, where it exists, is not of the type that would lead to the establishment of discontinuities within populations.

[More recent research has established that heterogamy is actually a far more important evolutionary factor than homogamy. Petit (1958) discovered and Ehrman (1967) confirmed that *Drosophila* females of several species show a definite preference for rare genotypes among available males. This mechanism acts to maintain genetic variability in populations and thus opposes certain potential types of sympatric speciation.]

Linkage of Mate Preference and Habitat Preference

Romanes and others seem to take it for granted that if a species is found at a given locality in several niches, mating occurs only between those individuals that live in the same ecological niche. The choice of a new niche would then automatically result in the establishment of reproductive isolation. The known facts do not support this contention.

Copulation in most butterflies and moths, for example, takes place during the dispersal phase of the life cycle (on flowers, etc.) and not on the host plant. If it does take place on or near the host plant, it is usually only the female that is sedentary while the male undertakes relatively extensive flights.

Ecological factors function simultaneously as isolating mechanisms only under two sets of circumstances. In the first case, there must be a difference in the breeding period. If one of two closely related species breeds only in sunlight, the other only in darkness, or if one breeds only in the spring, the other only in the fall, they will be effectively isolated. However, there is always a twilight and a summer to connect the separated breeding periods. Furthermore, there is no evidence or probability that such drastic differences can develop within a single local population or by single mutations (see below).

The other possibility is furnished by species in which mating always takes place on the host species and host selection is entirely rigid. Such rigid host selection has primarily a genetic basis, although it may be reinforced by conditioning. Switching over to a new host would thus be tantamount to the establishment of a new species. The difficulties in this case are several. First of all, there are the heterozygotes that serve as intermediaries between the old and the new host (see below). Second, species that are *that* rigidly adapted to one host probably have rather low survival value on a new host, unless the new host is in a different environment. In that case, the scheme would be merely a modification of spatial speciation. However, the mechanism is possible and deserves further investigation. An evolutionary analysis of monophagous families of insects is badly needed.

Conditioning without Isolation

Thorpe, Cushing, and other recent authors believe that conditioning could under certain circumstances play a considerable role in the origin of new species. This interest in conditioning goes back to a revival of Baldwin's theory of organic selection, which is most fully stated in his book of 1902. This theory states in short that if a population enters a new environment there will be a selective premium on mutations that would give a genetic basis to the previously nongenetic adaptive characters. This process of the substitution of modifications by mutations, which simulates direct action of the environment, has aroused considerable attention in Russia, as reported by Gause (1947). As a matter of fact, there is nothing strikingly new in this theory. It merely paraphrases two well-known facts, namely, that adaptive characters are favored by natural selection and that the more adaptable a species is the more easily it may enter a new environment. Both of these facts are in harmony with the concept of geographical speciation. To prove that organic selection and conditioning play an important role in speciation, it must be demonstrated that these processes reduce gene flow between the new and the parental habitat below the point where it interferes with the establishment of genetically controlled isolating mechanisms. Do the experiments on conditioning support this assumption?

An experiment by Cushing (1941) conducted in support of the hypothesis that olfactory conditioning is "one way in which an isolation may arise within insect populations" may be analyzed in more detail. A stock of the species *Drosophila guttifera,* which normally inhabits fungus but had been kept in the laboratory on the ordinary corn-molasses-agar medium, was subdivided into two strains. One continued on this laboratory medium ("controls"), the other on a mushroom extract medium ("mushroom conditioned"). Given a choice of a laboratory and a mushroom medium for egg laying, the control flies in four sets of trials laid 19.5 percent of their eggs on the laboratory medium, while the mushroom-conditioned flies laid only 8 percent of their eggs on this medium. Does this experiment indicate that conditioning can split one population into two? I believe not. To begin with, 80.5 percent of the eggs laid by flies conditioned and selected for many generations to lay on the laboratory medium chose the mushroom extract medium. This illustrates strikingly the powerful influence of the genetic constitution of these fungus flies. On the other hand, among the reconditioned flies 8 percent still laid eggs on the laboratory medium. Translated into terms of the natural environment, this means that there would be a continuous very active gene flow from the population in one food niche to the population in the other. Furthermore, contrary to Cushing's opinion, selection seems to have played a role in his and other similar experiments. Even inbred laboratory strains have much concealed variability, and the first generation or early generations in such "conditioning" experiments are often characterized by high mortality. This happened also in the experiment of Meyer and Meyer (1946), who succeeded in transferring a strain of *Chrysopa vulgaris* to a new prey (the coccid *Pseudococcus comstocki*). The success of the experiment is ascribed by the authors to natural selection by means of the survival of those larvae that were physiologically adapted to the new kind of food. There was high mortality in the first three generations. That the selection of the medium on which eggs are deposited has a genetic basis is confirmed also by the work of Masing (1946), who developed lines in a dumpy stock of *Drosophila melanogaster* in which the females preferred either a sugary or a nonsugary medium for egg deposition. The obvious correlation with the intestinal yeast flora was not examined. It may be mentioned incidentally that the attempt by Dobzhansky and Mayr (1945) to affect the degree of sexual isolation between two species of *Drosophila* by raising the larvae jointly in the same containers was not successful. In conditioning experiments by Thorpe (1939) the possible effect of selection was carefully eliminated. *Drosophila melanogaster* flies normally have a slight avoidance of air scented with essence of peppermint oil. Given a choice of pure and scented air, only 35 percent of the normal flies chose the scented air. During 21 experimental tests, 2 showed 50 percent or more preference, 5 showed 40–49 percent, 7 showed 30–39 percent, and 7 showed only 22–29 percent preference for the peppermint-scented air. Freshly hatched experimental flies that had been raised in a medium containing 0.5 percent peppermint essence showed 67.0 percent preference for scented air. These findings confirmed and extended the earlier work of Thorpe and Jones (1937) and of Thorpe (1938) on conditioning parasitic insects for new hosts. Significant as the work is, it does not prove that conditioning can prevent or even drastically reduce gene flow between the two kinds of environment if they are "available in mosaic distribution."

As stated in the previous section, the role of conditioning is that it assists in the invasion of new habitats by conspecific populations. However, extrinsic factors will have to inhibit gene flow to and from these new populations to permit them to develop into separate species.

The discussion of conditioning is not complete without reference to "locality conditioning." Thorpe (1945) is correct in giving much weight to the degree of sedentariness and the homing ability in the speciation process. As Rensch and others have pointed out (Mayr 1942), the less the degree of dispersal in a species—and homing as well as locality conditioning cuts down dispersal—the greater the extent of speciation and incipient speciation. It is well known that in social insects, as well as in adult birds and many other animals, there is an amazing degree of *Ortstreue*. However, even though this phenomenon serves to lower the magnitude of the dispersal factor *m* (Wright 1943), it by no means eliminates it completely. Dispersal operates in social insects through the nuptial flights, and in birds through the scattering of the juveniles. Least of all can such "locality conditioning" lead to the fission of a local population into several sympatric ones. The importance of *Ortstreue* is that it enhances the efficiency of external barriers. The role of this and other intrinsic factors is discussed elsewhere (Mayr 1948b).

Preadaptation

Preadaptation plays an important role in several of the hypotheses associated with sympatric speciation. It is postulated that individuals actively search for the habitat for which they are preadapted. Needless to say, there is not a shred of evidence for this claim. There is a dispersal phase in the life cycle of every species during which many individuals are carried onto unsuitable locations and perish. Others may survive during favorable seasons and form the nucleus of a new population only to be wiped out again during the first adverse season. Finally, an occasional individual may have an unusually favorable gene combination that will permit some of its descendants to flourish in the new locality. This can be observed all the time along the ecological and geographical margin of the range of a species. I know of no evidence, however, that would indicate that a dispersing individual ever actively searches for a habitat for which it is "preadapted."

Difficulties Created by the Hypothesis of Sympatric Speciation

Sympatric speciation is usually postulated to explain certain ecological aspects in the speciation pattern that some authors consider inexplicable by geographic speciation. These authors overlook that sympatric speciation creates new difficulties that are avoided by the theory of geographical speciation. These difficulties can essentially be classified in two groups.

Neglect of Dispersal

Dispersal is conspicuously neglected in all schemes of sympatric speciation, even though it is one of the basic properties of organic nature. There is a dispersal phase in the life cycle of every species. It is the adult stage in most insects with wingless larvae. In these species mating usually does not take place on the host or food plant, and hence there is no reproductive isolation between the various host races.

But even where mating takes place on the food plant, there is a dispersal period at one stage of the life cycle. If a form that usually lives on plant A develops some individuals that colonize plant B, there is no reason why the offspring of the B individuals should not recolonize plant A, or why there should not be repeated colonizations from A to B. Thus the conditions are not favorable for the establishment of a discontinuity if there is no spatial isolation between A and B. The same is true for host selection in parasites.

It is known that certain birds are extremely sedentary and live throughout the breeding season or even throughout their adult life within a restricted territory. However, even in these species there is a certain amount of dispersal before the juvenile takes up its first territory. There seems no logical connection between the first and the second part of the statement: "Birds tend to be strongly territorial animals and the claim that geographical isolation always precedes other kinds of isolation seems premature" (Thorpe 1945). Territoriality is a property of individuals, while speciation deals with populations.

Sessile marine organisms always seem to have a larval stage in which they may be dispersed by ocean currents for hundreds of miles and during which the populations within this area are thoroughly mixed. This factor is often overlooked, as, for example, by Test (1946) in her thesis of sympatric species in the limpets of the genus *Acmaea*. Ecological isolation, in order to be effective, must be able to prevent the mixing of the to-be-separated populations in spite of dispersal.

There are quite a number of cases among insects where the female ordinarily does not leave the host plant or where she selects the medium for egg laying before fertilization. Among the scolicid beetles, for example, the females bore holes under the bark. The males search them out in these tunnels and mate with them there. Here appears to be a potential mechanism for the rapid development of new monophagous races. Unfortunately, none of these cases has ever been carefully analyzed. Furthermore, even in such a scheme there are many difficulties for sympatric speciation, such as the dispersal of the males and of the young females, as well as the reversibility of all conditioning that is not genetically reinforced.

Genetic Difficulties

The most serious weakness of most schemes of sympatric speciation is that they make untenable genetic assumptions. The principal one is that mutations produce new "types," as if genetic change is an "all-or-nothing" mechanism. Actually new mutants in diploid bisexual animals always occur first in heterozygous condition. Such heterozygotes will form the bridge between individuals with the new character in homozygous condition and the individuals of the parental population. To escape this difficulty all sorts of auxiliary assumptions have to be made. A new isolating mechanism cannot appear as a dominant mutation A because the single bearer (Aa) of such a mutation would at once be effectively isolated from all other individuals of his species. Furthermore, there is much evidence, to be discussed below, that a single mutation will not produce such a drastic result as complete reproductive isolation without simultaneous deleterious effects on viability.

Hence, genetic models of sympatric speciation customarily operate with recessive mutations. Curt Stern (*in litt.*) suggested to me the following mechanism [his wording slightly modified by me]:

Let the animal AA, which is host specific on plant 1, have the mutation a which in homozygous condition produces host specificity for plant 2. Let us then make the following assumptions:

Assumption 1: Let AA live only on plant 1.
Assumption 2: Let aa live only on plant 2.
Assumption 3: Let the heterozygotes Aa be exactly like AA.
Assumption 4: Let there be little or no dispersal in the reproductive phase so that A animals do not meet aa animals.
Assumption 5: Let A be ill adapted to plant 2.
Assumption 6: Let aa be ill adapted to plant 1.
Assumption 7: Let segregated aa formed on plant 1 have difficulties in finding plant 2, even though the original aa found 2.

Then there will be little mixing between the populations on 1 and 2 and the opportunity is provided for a gradual piling up of additional genetic differences between these populations.

In this scheme seven separate assumptions are made, of which at least four are necessary to have the scheme work. The example involved a shift in the genetic basis for habitat preference. Schemes that work with the time element are dependent on a similar pyramiding of arbitrary assumptions. One such scheme may be formulated as follows: "Let there be a species AA which breeds from April 1–15, let there be in this species a mutation a which in homozygous condition shifts the breeding season to the period April 25–May 10 . . . etc., etc." Some of the objections to the assumptions made in this scheme are the following:

1. It is highly unlikely that such an important part of the life cycle of the species as the breeding time ever depends on a single gene. In fact, there will be a selective premium to have in every population a number of factors in balanced condition that affect the breeding season.

2. If a mutation occurs in one of these loci, it may shift the onset of breeding by a few days, but it is very unlikely that such a mutation would be so drastic as to eliminate overlapping of the breeding seasons.

3. If there had been available a period suitable for breeding adjoining the present breeding period, it would long since have become incorporated into the breeding season in view of the normal fluctuation of the breeding season in a large population of individuals and the ever-present mutation pressure.

4. The gene complex of an organism has been selected for thousands of generations to find optimum ecological conditions for the growth of its offspring in the period immediately following the breeding season. The offspring of individuals suddenly shifted out of their normal breeding season would be at a strong selective disadvantage.

A population can become adapted for a new breeding season only slowly by a gradual genetic reconstruction. Geographical isolation is necessary to permit this process to go on undisturbed. The shift of the breeding season is usually associated with a shift into a new breeding range where the changed season is advantageous.

The same arguments hold when a shift from diurnal to nocturnal breeding is involved or any of the other schemes that operate with time as the isolating factor.

The usual assumptions made both in the habitat preference and in the seasonal isolation schemes are: (1) that the heterozygotes are like one of the homozygotes; (2) that the two phenotypes (for habitat preference or breeding season) are discontinuous, that is, that a single mutation ($A{\rightarrow}a$) produces complete isolation; and (3) that—in the case of habitat selection—mating takes place in the newly-adopted habitat. Although all three assumptions are possible, in particular 1 and 3, the relevant evidence is only rarely produced. The most vulnerable assumption is that a harmonious, viable mutation can produce such striking effects as postulated under 2.

The whole subject is still dominated by the early De Vriesian thesis of the origin of species by single mutations. Unfortunately, no data are available on the number of gene differences between newly arisen and parental species, but some information is available on the order of magnitude involved. The weakness of such considerations is, of course, that not only the quantity but also the quality of mutations is important. A mutation on a single locus may be more important in affecting isolation between two populations than the mutations of a hundred other loci.

All individuals in sexually reproducing species (except identical twins) can be expected to be genetically different or, to put it in other words, to differ by several mutations. These are not merely minor changes, but may include a high percentage of lethals, as shown by the work in *Drosophila*. There is much evidence, on the basis of morphological, physiological, immunological, and pathological studies, that two individuals of a single interbreeding human population may differ by hundreds of genes. If isoalleles are at all common, as indicated by the work of Stern, the number of gene differences may actually be a multiple of the maximum figure now proposed. It has not been possible in most present studies to give more than a minimum estimate of gene differences, but the available data indicate that even closely related species differ in a large number of genes.

The degree of morphological difference between two species is an exceedingly unreliable measure of the number of genic differences between them. Many sibling species of *Drosophila* are virtually indistinguishable morphologically even though they may differ by scores, if not hundreds, of genes. This is indicated by their cytological, ecological, and physiological differences, as well as by their partial or complete sterility.

If the acquisition of reproductive isolation were a single gene mutation, one would expect isolating mechanisms to be simple. However, most isolating mechanisms between closely related species that have been studied thoroughly were found to be multiple. There always seem to be involved (1) differences in the ecological requirements, (2) reduction of the mutual sexual stimulation, and (3) reduction in the number and the viability of the offspring. On close analysis even these three categories turn out to be composite. For example, the mutual sexual stimulation may be reduced simultaneously by differences in scent, color patterns, rhythm, and pattern of display movements.

There is a selective advantage in this multiplicity for two reasons. One is that fewer zygotes are lost if a species can rely on the mutually reenforcing action of a number of different mechanisms rather than on a single one that is subject to an occasional breakdown. The other reason is that the isolating mechanisms must protect the species from interbreeding not only with one but with *all* related species. It is very unlikely that a single mechanism would furnish complete protection against all other species. Finally, there are no genetic difficulties in gradually building up a multiple isolating mechanism in spatially segregated populations.

Before closing the discussion on the possible genetic basis of sympatric speciation, a word needs to be said about "specific modifiers." As Muller (1942) has pointed out, there will be a selective advantage for genes that act to increase the adaptiveness of the expression of one allele as compared with others, if there is a discontinuity in the factor to which the gene reacts. However, since habitat preference and mate preference are not controlled by the same gene, it becomes necessary for two sets of specific modifiers to develop simultaneously. The probability that this will occur unassisted by a geographical segregation of populations is obviously very small. This is one of the reasons why the evolution of mimetic polymorphism in certain species of butterflies did not lead to the origin of a considerable number of new species, even though specific modifiers were involved.

The Hypothesis of Sympatric Speciation Is Unnecessary

The principal reason why a hypothesis of sympatric speciation is usually proposed is that certain taxonomic or ecological phenomena are believed to be incompatible with geographical speciation. It is the aim of this section to show that these phenomena are by no means in conflict with geographical speciation. The objections that have been raised are based on misunderstandings.

The Species as an Aggregate of Ecologically Different Populations

Ecological conditions vary throughout the range of all but the most narrowly distributed species. Thus every population of a species lives in a somewhat different environment and its genotype evolved by selection for this specific local environment. What is true for local populations is even more true for subspecies. Every geographical race shows certain ecological differences from other geographical races. It was Turesson (1922) who was the first to point this out emphatically. The taxonomist who looks at these populations looks at them from his special point of view. He investigates whether such populations show morphological differences that would be of diagnostic value. If they do, he calls such populations or groups of populations subspecies. The ecologist, on the other hand, looks for ecological differences between such populations (regardless of morphological characteristics). If he finds such differences, he calls the populations ecological races or, if he is a botanist, ecotypes. In much of the current literature these two categories, the subspecies and the ecological race, are treated as two completely distinct phenomena. The truth, however, is that they are merely two facets of a single phenomenon. Gregor (1946) has recently stated correctly that if "the term 'ecotype' [is applied] to any population differentiated in respect of any characteristic attributable to the selec-

tive action of ecological factors ... the majority of taxonomic subspecies and varieties will on experimental examination be found to bear
ecotypic characteristics." I am willing to go even further than this. I am
convinced that it will be a long time before even a single subspecies is
found that does not bear ecotypic characteristics. Not only does every
subspecies have ecotypic characteristics, but even within most subspecies there are numerous ecologically different populations. Also,
ecotypes do not live in an indiscriminate mixture; they are spatially
segregated from each other, even though they may intergrade mar-
ginally. Ecotypes are populations or groups of populations and so are
subspecies. Botanists (e.g., Gregor 1944, 1946), on the whole, have
realized this fact more clearly than have animal taxonomists.

The closest approach to geographical races in identical environments
is perhaps found on some tropical islands, but even here there are
stronger ecological differences than are usually recognized. The size and
the elevation of the islands are different and with them temperature
and precipitation. Islands in close proximity are often situated in different ocean currents. Most important of all, however, since each island
has a fauna and flora that arrived fortuitously, the composition of the
biota will be different on each island. In consequence, the biotic environment of a given species will be different even where the climatic
environment is the same. It was Darwin who discovered this principle:

> How has it happened in the several (Galapagos) islands situated
> within sight of each other, having the same geological nature, the
> same height, climate, etc., that many of the immigrants should
> have been differently modified, though only in a small degree.
> This long appeared to me a great difficulty; but it arises in chief
> part from the deeply-seated error of considering the physical con-
> ditions of a country as the most important for its inhabitants;
> whereas it cannot be disputed that the nature of the other inhabi-
> tants with which each has to compete is at least as important, and
> generally a far more important element of success. . . . When in
> former times an immigrant settled on any one or more of the
> islands, or, when it subsequently spread from one island to an-
> other, it would undoubtedly be exposed to different conditions of
> life in the different islands, for it would have to compete with
> different sets of organisms. . . . If then it varied, natural selection
> would probably favour different varieties in the different islands.
> (1859:400–401)

The converse is equally true. I do not know of a single ecological race
that is not at the same time at least a microgeographic race. This means
that random contact between individuals belonging to the two populations is drastically reduced by distance and barriers, and the gene flow
between the two populations is reduced in a parallel manner. To repeat,

all geographical races are also ecological races, and all ecological races are also geographical races.

The ecological specialization of geographical races may be far-reaching. The ornithological literature abounds with cases of ecological difference between geographical races of a single species: one in coniferous woods, the other in deciduous; one in the lowlands, the other in the high mountains; one feeding largely on ants, the other largely on beetles, etc. The most thorough analysis of adaptive features in geographical races of birds is Lack's (1947) study of Darwin's finches (Geospizinae). Most of the geographical races are shown to differ ecologically, but no evidence is found for the sympatric evolution of these ecological races. In a similar fashion Amadon (1947) has shown the importance of geographical isolation for the development of races of Drepanididae that are strikingly different in their ecologies.

The picture that emerges from all recent studies is that species show an amazing degree of local adaptability. Species adjust themselves to every occupied environment within the boundaries of their range. As long as these populations are in contiguous contact, there is no serious reduction of gene flow. The ecological variability within a species results in increased intraspecific variability and thus furthers evolutionary divergence. Without an extrinsic reduction of gene flow, however, the ecological variability cannot become a primary source of discontinuity.

The Ecological Differences between Species

In her review of speciation among the limpets of the genus *Acmaea* Test (1946) makes the following statement: "Along the west coast of North America occur nine species of one subgenus (*Collisella*) affording every indication of close relationship, yet having either completely concurrent or encompassed ranges. In as much as the obviously parental form has the greater range in every instance in which the range of one species is encompassed by that of another ... it becomes practically impossible to explain the situation in terms of geographical speciation." I have quoted this statement in full because it is quite typical of the reasoning of the authors who believe in sympatric speciation. What Test found for *Acmaea* is true for the species of nearly every polytypic genus. Among birds the woodwarblers (*Dendroica*), the buntings (*Emberiza*), the white-eyes (*Zosterops*), the whistlers (*Pachycephala*), the weavers (*Ploceus*), and the hawks (*Accipiter*)—to mention only a few genera—afford good examples of closely related sympatric species. There are many even more impressive examples among insects (*Drosophila, Anopheles, Aedes*) and in other groups of invertebrates. When these groups of closely related, partially or wholly sympatric species are studied carefully, it is found—with rare exceptions—that each of the species has its own ecological niche. It is this observation which serves as the starting point for the hypothesis of sympatric speciation by ecological segregation. It is argued that since these species are partially sympatric their ecological differences could not have developed in geographical isolation. Lack (1944, 1947) has submitted a different interpretation. He shows conclusively that on the basis of the Gause (1934) principle no two related species can persist in the same locality without possessing ecological differences. Ecological difference between related sympatric species is thus proof for the efficacy of competition and natural selection but not for any particular method of speciation. There is no conflict whatsoever between the fact that related species differ ecologically and the assumption that they have originated by geographical speciation (Mayr 1948b). It is fully consistent with the known facts to assume that the ecological differences had previously been developed in geographical segregation.

Reputed Instances of Sympatric Speciation

The most convincing proof for the importance of geographical speciation is the abundance of incipient cases. Hence if there is an additional process of sympatric-ecological speciation, it should be possible to find for it also incipient cases. Indeed, there are numerous taxonomic and ecological situations that have been interpreted in the past as constituting such incipient cases. It will be shown in this section that the evidence in these cases is misinterpreted and that the facts are completely consistent with the theory of geographical speciation. The relevant phenomena can be organized under three headings: sibling species, lake swarms, and secondary zones of intergradation.

Sibling Species

Many of the cases of so-called ecological or physiological races of the literature are nothing but sibling or cryptic species. There is no evidence that these sibling species have evolved in any other way than by geographical speciation. This has been discussed elsewhere (Mayr 1942, 1948a) in considerable detail and this point does not need to be taken up again.

Lake Swarms

It is well known that "species flocks" occur in all the larger and older freshwater lakes. This has been found independently by students of fishes, crustaceans, mollusks, and other freshwater organisms. There are, for example, 178 species of cichlid fishes in Lake Nyasa in East Africa and more than 300 species of Gammarid crustaceans in Lake Baikal. Lake Lanao on the island of Mindanao in the Philippines, the large lakes of Celebes, and other tropical and subtropical lakes in all parts of the world are also famous for their rich endemic faunas. Since it seems at first sight impossible to conceive that these sympatric faunas could have originated by geographic speciation, Woltereck, Herre, and others have postulated various processes of "explosive" speciation. None of these proposed schemes was ever worked out in detail, but they all invoke either macrogenesis (the origin of new systematic categories by a single mutation step) or various forms of homogamy (assortative mating) (e.g., Kosswig). The objections to such theories have been discussed above.

The fact that the species of these swarms are now sympatric and that they live according to the Gause principle in different ecological niches in order to minimize competition has led previous authors astray. Rensch (1933) has indicated the right solution. It is that these species have come into contact only after they had evolved and after they had acquired their ecological differences.

As I said in 1942:

> Old fresh-water lakes are, for freshwater faunas, very much what old islands are for terrestrial faunas. They permit the survival of old elements which have long since become extinct in the surrounding areas. It seems to me that students of fresh-water faunas have vastly underestimated the age of the species with which they are working. The evidence for this is quite overwhelming for Lake Baikal, Lake Tanganyika, Nyasa, and so forth. The statement by the proponents of explosive speciation that ecological speciation precedes the establishment of discontinuity is not in the least plausible, if we remember that these habitats are in continuous contact with each other and that there is no evidence for the establishment of biological isolating mechanisms as long as unrestricted interbreeding takes place between the inhabitants of the different sympatric ecological niches. On the other hand, no objections seem to exist against the assumption that species flocks originated by multiple colonizations, corresponding to the double and triple colonizations and the archipelago speciation among island animals.

None of the lake flocks has so far been analyzed from a broader point of view. Such questions need to be answered as: What species are found only in part of the lake, like *Tilapia karongae* in northern Lake Nyasa? What species have discontinuous ranges within a given lake? Are the nearest relatives of a given species found in the same lake, or in some neighboring tributary or other lake? Has the inventory of the freshwater faunas of all the present and former tributaries of these lakes been completed and how do these faunas compare with the lake fauna?

There must be situations where one might be able to study the origin of a species flock *in statu nascendi*. Hubbs and Raney (1946) have recently described such a potential case. In Lake Waccamac in North Carolina each of the genera *Notropis, Fundulus, Boleosoma,* and *Menidia* developed an endemic form during the last quarter of the Pleistocene. These streamlined lake forms may or may not be reproductively isolated from their allopatric ancestral forms. However, if repeated invasions of the respective species complexes would occur, after reproductive isolation had been established, it would lead to the formation of species flocks. The environment in the lake is very different from the environment of the tributary streams, and this undoubtedly leads to a considerable acceleration of the evolutionary rate. In this respect lakes are very much like islands, as previously pointed out (Mayr 1942). However, even though the rate of evolutionary divergence is much accelerated, there is no evidence that the origin of discontinuities differs in principle from that in terrestrial animals.

[Research on speciation in freshwater lakes has made enormous progress in the 25 years since this was written. A number of important review articles on the subject were written by Brooks (1950), Fryer (1960, 1972), Lowe-McConnell (1969), and Greenwood (1965). A monograph by Kozhov (1963) presents an up-to-date review of evolution in Lake Baikal. Most important, a number of studies have substantiated the claims of extraordinarily rapid rates of speciation in freshwater lakes. For instance, five endemic species of *Haplochromis* in Lake Nabugabob (next to Lake Victoria) in East Africa cannot be older than 5000 years (Greenwood 1965). Yet no evidence was found for sympatric speciation in any of these studies.]

Secondary Contiguity of Ecologically Different Subspecies

An ever-increasing number of cases is being described in the literature where two ecologically very distinct subspecies are found to have adjacent or even interdigitating ranges. Such cases are often quoted as proving sympatric speciation. Actually there is, of course, in these cases nearly always spatial segregation, and, furthermore, it can often be demonstrated that the present contiguity is a rather recent phenomenon. In the mayfly *Stenonema interpunctatum* there are four intergrading subspecies with only partially separated ranges (Spieth 1947). Most of the range of *S. i. heterotarsale* in the Lake Erie–Ontario region is overlapped by the subspecies *canadense* and *interpunctatum*. "Obviously there must be a definite survival value attached to each of the populations or they would quickly fuse into one variable one." *S. i. interpunctatum* and *canadense* emerge earlier in the season, *heterotarsale* later. On the basis of the present distribution pattern it appears probable that the four subspecies originated during the Pleistocene in the following districts: *interpunctatum* in the Mississippi valley, *canadense* in Canada, *frontale* on the U.S. Atlantic coast, and *heterotarsale* in the Lake Erie–Ontario district. Most of the present range overlaps are due to the post-Pleistocene range expansions of *interpunctatum* and *frontale*. The most interesting aspect of this case is that during their geographical segregation each of these forms acquired a number of physiological characteristics or adaptations to special, narrow ecological niches, so that now several of these "subspecies" can live in various parts of the same stream without losing their identity.

Similar cases have been described in fishes. The nominate race *nigrum* of the small fish *Boleosoma nigrum* has a very wide range in the

United States. Within its range is found the subspecies *eulepis,* which is more completely scaled on cheeks, nape, and breast. The nominate race occurs in lakes and streams with a firm (sandy or rocky) substratum, while *eulepis* is a fish of estuaries, rather extensive and quiet or slow-moving lowland waters, characterized by moderate or dense growths of aquatic vegetation, and on bottoms composed at least in part of mud or silt. The interesting aspect of this case is that the range of *eulepis* consists of more or less discontinuous pockets or foci of abundance, surrounded by peripheral areas of intergradation with the nominate race, but more or less restricted to the Great Lakes region. The morphological differences are not phenotypic modifications through life in the specific habitat niche, but have a genetic basis, as was shown by Lagler and Bailey (1947). Physiological differences (oxygen and temperature tolerance, etc.) are presumably associated with the morphological ones, but they have not yet been investigated. There are two alternative explanations of this case. Either one presumes that in each stream and lake a quiet-water race developed in spatial segregation, or one believes that the present proximity of the races is a secondary phenomenon. Although both solutions are possible, I tend to the second for two reasons. In many "*eulepis* habitats" in the range of *nigrum,* this subspecies appears to be absent, particularly in that part of the range of *nigrum* that is outside the Great Lakes region. The second reason is the comparative homogeneity of *eulepis* in its now disjunct range. The available evidence suggests that *eulepis* once had a continuous range during one of the later stages of the Pleistocene?) but that the somewhat superior *nigrum* invaded its range and forced it to retreat into the quiet water pockets where *eulepis* is superior. It would be very interesting to know more about the genetic mechanisms that prevent the breaking up of the *nigrum*- and *eulepis*-gene complexes that are superior in their respective habitats.

This evidence can be summarized in the statement that it appears that ecologically contrasting subspecies normally originate in spatial segregation and usually even in complete geographical isolation. Contact zones between such subspecies are usually zones of secondary intergradation (Mayr 1942).

THE FACTORS OF SPECIATION

A balanced evaluation of the respective roles in speciation of ecological and geographical factors is still missing. Both work closely together and "there is no geographic speciation that is not at the same time ecological and genetic speciation" (Mayr 1942). The first step in the speciation process is the founding of a new intraspecific population. This is very often possible only through a shift in ecological tolerance, and in such cases it is clear that an ecological rather than a geographical event may be the first step in speciation, or—since the new population will be not only ecologically different but also spatially segregated from the parental one—that the ecological event (adaptation to a new ecological niche) is at least simultaneous with the first geographical event. This first step leads to the establishment of a spa͘ally segregated population that is exposed to different selection pressures owing to the more or less differing ecological conditions under which it lives. The ecological factors lead to evolutionary divergence. These populations will drift apart genetically (probably at an accelerating rate!) until a discontinuity through reproductive isolation develops, provided extrinsic barriers reduce dispersal (= gene flow) to such an extent that it can no longer neutralize the effects of the different selection pressures in the two populations. It is still unknown whether this can happen between contiguous populations without the help of extrinsic factors that reduce gene flow.

SUMMARY

All degrees of geographical isolation are known, resulting in everything from complete interruption to only slight reduction of gene flow between the isolated populations. The term "microgeographical" isolation may be used where only short distances are involved.

The establishment of a new intraspecific population is usually associated with a shift in the ecological characteristics of such a population. All subspecies show ecological differences, and no "ecological races" are known that are not also at least "microgeographical."

It is unproven and unlikely that reproductive isolation can develop between contiguous populations.

Many of the assumptions made in the various hypotheses of sympatric speciation are erroneous. The evidence usually cited as proving sympatric speciation is fully consistent with the theory of geographic speciation. The concept of sympatric speciation creates many difficulties avoided by the concept of geographic speciation.

Sympatric speciation, if it occurs at all, must be an exceptional process. The normal process of speciation in obligatorily sexual and cross-fertilizing organisms is that of geographical speciation.

REFERENCES

Alexander, R. D., and R. S. Bigelow. 1960. Allochronic speciation in field crickets (Orthoptera: Gryllidae). *Behavior,* 17:130–223.

Amadon, D. 1947. Ecology and the evolution of some Hawaiian birds. *Evolution,* 1:63–68.

Alpatov, B. W. 1925. Über die homogame und pangame Paarung im Tierreiche. *Zool. Anz.,* 62:329–331.

Armstrong, T. 1946. Differences in the life history of the codling moth, *Carpocapsa pomonella* (L.), attacking pear and apple. *Canadian Ent.,* 77:231–233.

Baldwin, J. M. 1902. Development and evolution. Macmillan, London.

Blair, W. F. 1947. Estimated frequencies of the buff and gray genes (G, g) in adjacent populations of deermice (*Peromyscus maniculatus blandus*) living on soils of different colors. *Contr. Lab. Vert. Biol.*, 36:1-16.

Boyce, A. M. 1935. The codling moth in Persian Walnuts. *Jour. Econ. Ent.*, 28: 864-873.

Brooks, J. L. 1950. Speciation in ancient lakes. *Quart. Rev.*, 25(1):30-60.

Bush, G. L. 1974. The mechanism of sympatric host race formation in the true fruit flies (*Tephritidae*). In *Genetic mechanisms of speciation in insects*, pp. 3-23. Australian and New Zealand Book Co., Sydney.

Catchpool, E. 1884. An unnoticed factor in evolution. *Nature*, 31:4.

Crozier, W. 1918. Assortative mating in a nudibranch, *Chromodoris zebra*. *Jour. Exper. Zool.*, 27:247-292.

Cushing, J. E. 1941. An experiment on olfactory conditioning in *Drosophila guttifera*. *Proc. Nat. Acad. Sci.*, 27:496-499.

Dahl, E. 1889. Die Bedeutung der geschlechtlichen Zuchtwahl bei der Trennung der Arten. *Zool. Anz.*, 12:262-266.

Darwin, C. 1859. *On the origin of species by means of natural selection, or the preservation of favoured races in the struggle for life.* J. Murray, London.

Diederich, G. W. 1941. Non-random mating between yellow-white and wild types of *Drosophila melanogaster*. *Genetics*, 26:148.

Dobzhansky, T., and E. Mayr. 1945. Experiments on sexual isolation in *Drosophila*. IV. Modification of the degree of isolation between *Drosophila pseudoobscura* and *Drosophila persimilis* and of sexual preferences in *Drosophila prosaltans*. *Proc. Nat. Acad. Sci.*, 31:75-82.

Ehrman, L. 1967. Further studies on genotype frequency and mating success in *Drosophila*. *Amer. Nat.*, 101:415-424.

Fryer, G. 1960. Some controversial aspects of speciation of African cichlid fishes. *Proc. Zool. Soc. London*, 135:569-578.

Fryer, G. and T. D. Iles. 1972. *The cichlid fishes of the great lakes of Africa: their biology and evolution.* Oliver and Boyd, Edinburgh.

Gause, G. F. 1934. *Struggle for existence.* Williams and Wilkins, Baltimore.

—— 1947. Problems of evolution. *Trans. Conn. Acad. Arts and Sci.*, 37:17-68.

Gerhenson, S. 1941. Additional data concerning the mating system in wild populations of common hamster (*Cricetus cricetus*). *Compt. Rend. Acad. Sci. URSS*, 31:155-156.

Greenwood, P. H. 1965. The cichlid fishes of Lake Nabugabo, Uganda. *Bull. Brit. Mus. (Nat. Hist.)*, 12:315-357.

Gregor, J. W. 1944. The ecotype. *Biol. Rev.*, 19:20-30.

—— 1946. Ecotypic differentiation. *New Phytologist*, 45:254-270.

Hogben, L. 1946. An introduction to mathematical genetics. W. W. Norton, New York.

Hubbs, C. L., and E. C. Raney. 1946. Endemic fish fauna of Lake Waccamac, North Carolina. *Misc. Publ. Mus. Zool., Univ. Mich.*, no. 54:1-30.

Kosswig, C. 1947. Selective mating as a factor for speciation in cichlid fish of East African lakes. *Nature*, 159:604.

Kozhov, M. 1963. Lake Baikal and its life. W. Junk, The Hague.

Lack, D. 1944a. Ecological aspects of species formation in passerine birds. *Ibis*, 86:260-286.

—— 1944b. Correlation between beak and food in the Crossbill, *Loxia curvirostra* Linnaeus. *Ibis*, 86:552-553.

—— 1947. *Darwin's finches*. Cambridge University Press, Cambridge.

—— 1948. The significance of ecological isolation. In *Genetics, paleontology, and evolution*, ed. G. L. Jepsen, G. G. Simpson and E. Mayr, pp. 299–308. Princeton University Press, Princeton.

Lagler, K. F., and R. M. Bailey. 1947. The genetic fixity of differential characters in subspecies of the percid fish, *Boleosoma nigrum*. *Copeia*, 50–59.

Lowe-McConnell, R. H. 1969. Speciation in tropical freshwater fishes. *Biol. Jour. Linn. Soc.*, 1:51–75.

Manning, T. H. 1942. Blue and lesser snow geese on Southampton and Baffin Islands. *Auk*, 59:158–175.

Masing, R. A. 1946. Experiments on selection for selective egg-deposition in *Drosophila melanogaster*. *Compt. Rend. Acad. Sci. URSS*, 51:393–396.

Mayr, E. 1942. *Systematics and the origin of species*. Columbia University Press, New York.

—— 1948a. The bearing of the new systematics on genetical problems. *Advances in Genetics*, 2:205–235.

—— 1948b. Speciation and systematics. In *Genetics, paleontology, and evolution*, ed. G. L. Jepsen, G. G. Simpson, and E. Mayr, pp. 281–298. Princeton University Press, Princeton.

—— 1970. *Populations, species, and evolution*. Belknap Press of Harvard University Press, Cambridge, Mass.

Meyer, N. F., and Z. A. Meyer. 1946. The formation of biological forms in *Chrysopa vulgaris* Schr. (Neuroptera, Chrysopidae). *Zool. Jour.*, 25:115–120.

Muller, H. J. 1942. Isolating mechanisms, evolution and temperature. *Biol. Symp.*, 6:71–125.

Pearson, K. 1900. *The grammar of science*, 2nd ed. Adam and Charles Black, London.

Peitzmeier, J. 1942. Die Bedeutung der oekologischen Beharrungtendenz für faunistische untersuchungen. *Jour. f. Ornith.*, 90:311–322.

Petit, C. 1958. Le déterminisme génétique et psycho-physiologique de la compétition sexuelle chez *Drosophila melanogaster*, *Bull. Biol.*, 92:248–329.

Rensch, B. 1933. Zoologische Systematik und Artbildungsproblem. *Verh. Deutsch. Zool. Gesellsch.*, pp. 19–83.

Romanes, G. J. 1897. *Darwin and after Darwin*, vol. 3. Open Court Publishing Co., Chicago.

Spett, G. 1929. Zur Frage der Monogamie und Pangamie bei Tieren, Untersuchungen an einigen Coleopteren. *Biol. Zbl.*, 49:385–392.

—— 1931. Gibt es eine partielle sexuelle Isolation unter den Mutationen und der Grundform von *Drosophila melanogaster* Meig.? *Z. ind. Abst. Vererb.*, 60:63–83.

Spieth, H. T. 1947. Taxonomic studies on the Ephemeroptera. IV. The genus Stenonema. *Ann. Ent. Soc. Amer.*, 40:87–122.

Stresemann, E. 1942. Oekologische Sippen-, Rassen-, und Artunterschiede bei Vögeln. *Jour. f. Ornith.*, 91:305–328.

Test, A. R. 1946. Speciation in limpets of the genus *Acmaea*. *Contr. Lab. Vert. Biol.*, 31:1–24.

Thoday, J. M., and J. B. Gibson. 1962. Isolation by disruptive selection. *Nature*, 193:1164–1166.

Thorpe, W. H. 1938. Further experiments on olfactory conditioning in a parasitic insect: the nature of the conditioning process. *Proc. Roy. Soc. London*, 126:370–397.

—— 1939. Further studies on pre-imaginal olfactory conditioning in insects. *Proc. Roy. Soc. London*, 127:424–433.

—— 1945. The evolutionary significance of habitat selection. *Jour. Animal Ecol.*, 14:67–70.

Thorpe, W. H., and F. G. W. Jones. 1937. Olfactory conditioning in a parasitic insect and its relation to the problem of host selection. *Proc. Roy. Soc. London*, 124:56–81.

Timofeeff-Ressovsky, N. W. 1943. [Erbliche und oekologische Isolation.] *Jour. f. Ornith.*, 91:326–327.

Turesson, G. 1922. The species and the variety as ecological units. *Hereditas*, 3:100–113, 210–350.

Valentine, J. M. 1943. Insect taxonomy and principles of speciation. *Jour. Wash. Acad. Sci.*, 33:353–358.

—— 1945. Speciation and raciation in Pseudo-anophthalmus (Cave beetles). *Trans. Conn. Acad. Arts and Sci.*, 36:631–659.

Wagner, M. 1889. *Die Entstehung der Arten durch räumliche Sonderung*. Benno Schwalbe, Basel.

Wright, S. 1943. Isolation by distance. *Genetics*, 28:114–138.

14

Bird Speciation in the Tropics

The richness of tropical faunas and floras is proverbial. A local temperate zone fauna of breeding land birds consists of about 100–170 species. The breeding land bird faunas of both Colombia and Venezuela considerably exceed 1300 species. The tropical island of New Guinea has a larger number of species than the nontropical part of Australia. Admittedly, a purely geographical treatment of species diversity in the tropics is somewhat misleading. Parapatric[1] distribution patterns are particularly common among tropical birds, as we shall presently see, and figures given for politically defined areas, such as Colombia or Venezuela, are thus exaggerated. The number of species actually found at a given locality in the tropics has been determined in some modern ecological surveys, and these indicate far less of a disparity in species number between tropical and temperate zone bird faunas than comparisons between politically defined areas (MacArthur 1969). However, even allowing for these corrections, tropical bird faunas contain at least three times if not four or more times as many species as comparable temperate zone bird faunas.

The big question this situation poses is the determination of the relative importance of the various factors that contribute to the richness of the tropical bird faunas. The use of birds for an analysis of species diversity has various advantages. The first is that the inventory taking of species is virtually completed and that it is therefore possible to make rather precise quantitative statements. However, a mere inventory of taxa is not enough. An understanding of their relationship is also necessary. Here even avian systematics is often deficient, particularly in tropical South America. The second advantage of birds is that all speciation is geographic. No one questions that spatial isolation is required in birds for the acquisition of the two essential properties of new species, efficient isolating mechanisms and ecological compatibility

Adapted from "Bird speciation in the tropics," *Biological Journal of the Linnean Society,* 1(1969):1–17.
1. "Parapatric" means: in nonoverlapping geographical contact, without interbreeding.

with sister species. Polyploidy, so important in the speciation of plants, is unknown in birds.

Even given the fact that all speciation is geographical, several interpretations of the richness of tropical bird faunas are possible, genetic and ecological ones. One might, for instance, postulate that certain genetic processes are active in the tropics, such as increased mutation rates or a speeding up of the number of generations, that would accelerate the process of the multiplication of species. However, this is unlikely since there is no indication that the genetics of speciation in the tropics differs from that in the temperate region. There is no reliable evidence for the occurrence of an increased rate of mutation in the tropics, and, furthermore, mutation pressure does not result in speciation. An increased rate of mutation would result only in an increased level of variability within populations, in other words in an enrichment of gene pools, but not in speciation. Nor is there any evidence for an accelerated succession of generations among tropical birds. Indeed what little evidence we have seems to indicate an increased longevity among adult tropical birds, possibly resulting in a lengthening of the average duration of generations.

We are thus forced to look for ecological factors to account for the high diversity of species in the tropics. Here we must make a strict distinction between ecological factors that are part of the process of speciation as such and others that are independent of the process of speciation and merely facilitate the subsequent sympatric coexistence of species. A discussion of these latter factors is outside the scope of this essay and the reader is referred to the rich current literature on the subject (MacArthur 1969).

In this essay I will limit myself largely to what might be called the ecology of geographic speciation. Ideally, one would like to have accurate information on the following points:

1. The nature of the isolating barriers

2. The amount of gene flow into a geographic isolate compatible with the completion of speciation, in other words, the required efficiency of the geographic barriers

3. The size of the isolated populations and the amount of genetic turnover in the gene pool of the isolated populations, for example, the possible occurrence of a genetic revolution among isolates

4. The percentage of completed speciations among isolates in relation to population size and length of isolation

5. The nature and magnitude of ecological shifts occurring during speciation

6. The consequences of a lifting of the geographic barrier prior to the completion of the process of speciation

Unfortunately, even though we can ask questions on all these points, we do not yet have sufficient information available to answer them unambiguously and quantitatively. For a more complete discussion of the ecology of speciation see Mayr 1963:chap. 18.

METHODOLOGY

Speciation, except sometimes in the case of polyploidy, cannot be observed directly. Even in the most rapid cases it requires thousands of years for completion. It can be studied only by inference, through an analysis of incipient species and of species that have evidently recently completed speciation. This, quite properly, has been the method of the student of speciation ever since the time of Wallace and Darwin. A superimposed refinement is a comparative approach. One can gain additional insight into the relative importance of different factors in speciation by comparing speciation in different climatic zones, in areas with different types of geographic barriers, and in different organisms, let us say birds or snails.

THE OCEAN AS BARRIER

Both Darwin and Wallace first studied speciation in tropical archipelagoes. Here, speciation presents itself with the simplicity of a textbook diagram. The gaps between islands are the obvious barriers providing geographical isolation. They separate incipient species from each other, as well as species that have recently completed the process of speciation. This, as I have indicated, has been known for more than 100 years. What is not precisely known is the relative frequency of such isolations and the subsequent fate of the isolates. To fill this gap in our knowledge I undertook a quantitative analysis of all resident land birds of northern Melanesia, under which name I combine the totality of the islands comprising the Admiralty Islands, the Bismarck Archipelago, and the Solomon Islands.

Table 14-1 gives some statistics concerning this avifauna. Not included in the analysis are seabirds and freshwater birds (grebes, ducks, herons, etc.), but rails are included among the analyzed species. Two different ways are chosen to tabulate the 214 species of land birds breeding in northern Melanesia. Under Category I the species are listed on the basis of whether they belong to superspecies (76) or do not belong to superspecies in northern Melanesia or New Guinea (Papuan species that form superspecies with species that occur outside the Australo-Papuan area were not included among the 39 superspecies). Under Category II the species are divided into those that are not endemic in northern Melanesia (110 species) and those that are endemic (104 species).

Two results of the analysis are noteworthy. The first is that no less than 39 (22 percent) of the 177 zoogeographic entities of northern

Table 14–1. Analysis of the land birds of northern Melanesia
(as of 1969).

I. Species and Superspecies		
Superspecies in northern Melanesia or Papua	39	
Species not belonging to superspecies	138	177
Total species in the 39 superspecies	76	214
II. Endemic and Nonendemic Species		
Nonendemic species		
Without endemic subspecies	24	
With endemic subspecies	86	110
Endemic Species		
Monotypic (24 occurring on several islands)	67	
Polytypic	37	104
Total no. of species		214

Melanesia are superspecies. Superspecies are here defined as allopatric species assemblages of the New Guinea—Melanesian area. (Maps of the distribution of such superspecies were published in the original paper.) The second and perhaps even more surprising finding for an area of such active speciation is the fact that in only 6 cases has a newly formed species been able to "break out" of its island of origin and successfully invade adjacent islands or the mainland of New Guinea. It is only in the genera *Columba (vitiensis-pallidiceps)*, *Lorius (lory-hypoinochrous-albidinucha)*, *Phylloscopus (trivirgatus-amoenus)*, *Rhipidura (rufifrons-malaitae)*, *Monarcha (alecto-hebetior)*, and *Lonchura (forbesi-hunsteini)* that range overlaps are found which would indicate recently completed speciation, (for maps see the original paper). This situation seems to be in striking contrast to the situation both on the mainland and on impoverished archipelagoes like the Galápagos or Hawaiian Islands. I will try to interpret it after an analysis of speciation on tropical mainlands. (A more detailed analysis of the ornithogeography of Northern Melanesia by Mayr and Diamond is in preparation.)

SPECIATION ON TROPICAL MAINLANDS

The analysis of speciation on tropical mainlands encounters various complications not found in archipelagoes and it has therefore not made much headway until the last 10 or 15 years. To be sure, the role of the Amazon River and its tributaries in speciation was known to Bates and other naturalists of the mid-nineteenth century, but their findings were more prophetic than definitive, since these early naturalists did not apply the polytypic species concept and usually designated every distinct local race or subspecies as a full species.

As far as bird speciation on tropical mainlands is concerned, there is a great deal of work in preparation but little has so far been published. Keast (1961) has made such an analysis for all of Australia, including tropical Australia. Stresemann, Hartert, Paludan, and Rothschild (1936) have made a preliminary analysis for New Guinea. Moreau (1966) has discussed the history of bird populations in Africa, and Hall and Moreau (1970) have mapped speciation in African Passeres. I will refrain from duplicating their work. South America, the continent with the richest bird fauna in the world, has been strangely neglected, in part because there are still many uncertainties about the systematics of South American birds. However, Dr. J. Haffer has now prepared a set of fine studies of Amazonian birds, some published, others still unpublished. I am privileged to be able to use some of Dr. Haffer's unpublished findings [now published, Haffer 1974].

Altitudinal Barriers

Distributional barriers caused by altitude are one set of barriers on the mainland that have long been known and often described. Speciation on the Venezuelan highlands (Mayr and Phelps, 1967) is a good illustration. Among the 96 highland bird species of this area, 29 are endemic. Seven of these are derived from the Andes, 5 from even greater distances, usually by long-distance colonization. Seventeen other endemic species are derived from related lowland species. Just as valleys and lowlands are a speciational barrier for mountain birds, so are mountains for lowland birds. The Andes of South America, for instance, have isolated various populations on the Pacific coast of Ecuador and Colombia from their nearest relatives in the interior, and this interruption of gene flow has permitted the evolution of endemic Pacific species. A similar process has occurred in New Guinea where the enormous barrier of the central range (Snow Mountains, etc.) has precluded any appreciable gene flow between lowland populations in the south and in the north except around the ends of mountain ranges. However, the situation here is made more complicated by the fact that there were apparently islands parallel to the mainland of New Guinea, both in the south and in the north, and the respective contribution of the isolation by mountains and islands has not yet been worked out.

Vegetational Barriers

Of far greater importance in the tropics than either water barriers or altitudinal barriers are vegetational barriers. Savannas of all types are formidable barriers for birds of the tropical rain forest and the rain forest plays the same role for savanna birds. Vegetational barriers, however, in spite of their abundance, have one great disadvantage as factors in speciation: they are far less permanent than mountain ranges or ocean straits. Their location is determined by climatic conditions and these, as we all know, have fluctuated greatly during Quaternary times.

Before the nature and the effect of vegetational barriers can be discussed profitably, a few words must be said on geological and climatic events during the Quaternary that have affected the strength of these barriers. In particular, I want to emphasize some recent findings that have resulted in a modification or complete revision of some previously held ideas:

1. The duration of the Pleistocene was much greater than has formerly been stated, amounting to 3.5–4 million years since the beginning of the Villafranchian. This provided far more time for Pleistocene speciation than was previously conceived.

2. Mountain building continued quite actively throughout that period, particularly in the Andes and in New Guinea, and presumably also in the mountains of southeast Asia. The mountains in these areas are probably higher now than they have ever been before.

3. There are indications that the lowering of temperature was greatest at the highest altitudes of the mountains and less and less toward the tropical lowlands. The reduction of temperatures in the lowlands was relatively minor, being probably of the order of 3–5°C. This unequal drop of temperatures seems to have resulted in a compression of altitudinal zones in the mountains (Haffer 1974). These altitudinal differences must be emphasized because some authors have quite improperly extrapolated from findings made at 2000 meters to the situation existing in tropical lowlands. Indeed if their conclusions were correct, the entire tropical fauna and flora would have been exterminated during the glacial periods.

The decisive factor determining the nature of vegetation is not actual rainfall but evaporation deficiency. Any lowering of the temperature will have a major impact on reducing evaporation deficiency and will lead toward a considerable spread of the forested areas even when there is only a relatively minor increase in rainfall.

Our task is to reconstruct the sequence of vegetational changes that occurred during the Quaternary period and that might have resulted in speciation. At the present time the Amazonian rain forest is enormously widespread and of virtually unbroken distribution. Yet there is good evidence that this is a temporary and rather recent condition. Throughout the Pleistocene and concomitant with the periods of glaciation, there was a succession of wet and dry periods in the tropics resulting in a sequence of expansions and contractions of the tropical rain forest and conversely of tropical savannas and semideserts. At the height of each arid period, the rain forests contracted into isolated pockets and forest refuges, separated from each other by savannas and other forms of vegetation that act as distributional barriers for forest birds. Conversely, at the height of each humid period, arid-land vegetation was contracted into isolated pockets, particularly in the rain shadows of mountain ranges, and was thus widely separated from related populations in similar habitats.

The populations of each species that were isolated in the forest refuges during arid periods, and in the savanna refuges during the humid periods, formed geographical isolates. Each such isolate is an incipient species, and according to our postulate the majority of the tropical lowland species now in existence must have originated in such refuges. In order to test the validity of this assumption, one must make two sets of reconstructions: (1) the inferred geographical location of the refuges, and (2) the species populations isolated in each refuge. Such reconstructions are bound to be speculative and provisional, but can be tested against a considerable body of available facts.

The Forest Refuges in South America

Haffer (1974) makes use of two kinds of evidence on which he bases his reconstruction of the location of nine tentatively postulated refuges for tropical rain forest birds. One consists in the mapping of current inequalities in the distribution of annual rainfall. For instances, there is at present an area of high rainfall north of the upper Amazon River and another at the Atlantic slope of the Guianas. There is a considerably drier belt, yet still largely covered by rain forest, between the two areas of rainfall maxima. If the rainfall was reduced evenly during dry periods, it would have still been sufficient to maintain rain forest refuges in the areas of present rainfall maxima, while the intervening belt would have become a savanna. Obviously, the pattern of rainfall distribution is not necessarily exactly the same during pluvial and arid periods, but the stated assumption is legitimate as a first approach. Any other factor that might have influenced distribution of rainfall during arid periods, particularly shifts in the distribution and strength of prevailing winds, will also have to be considered.

Even more important clues are provided by the current distribution of species that are likely to have originated in forest refuges during Quaternary drought periods. After completing speciation, many species still retain a strong geographical attachment to the refuge in which they originated. However, different species display different behavior in this respect. In order to be able to evaluate the relevant evidence, it is necessary to recognize that each species has its own pattern and rate of speciation.

When a population in a geographic isolate meets again the parental population after the breakdown of geographic isolation, four stages of more or less completed speciation can be distinguished.

Completed Speciation

In the first case, the isolate has acquired not only isolating mechanisms but also ecological compatibility and behaves like a good independent species toward parental or sister species and can broadly overlap their geographic ranges. Any case of broad geographic overlap of two closely related species falls potentially in this category. In such cases, however, conclusive proof of recently completed speciation and the location of the former isolate is very difficult to obtain, as I will presently discuss.

Parapatry

In the second case, the isolate has acquired isolating mechanisms but has failed to acquire ecological compatibility. The result will be a parapatric distribution pattern in which the new species establishes contact with the parent species or with one or several sister species but is unable to overlap their ranges to any extent.

Superspecies, such as the 39 superspecies which I described above for northers Melanesia, are typical illustrations of parapatric distribution patterns. Such parapatric patterns occur also in mainland situations and have been described by Keast (1961) for Australia and Hall and Moreau (1970) for Africa. They are not infrequent in South America, as shown by Haffer (1974). In many of these cases no visible barriers seem to separate the two allopatric species and yet they seem to be quite unable to invade each other's ranges. It is still unknown, in the cases that have come to my attention, what the particular requirements of the two species are that control this seemingly competitive exclusion.

Parapatric distributions represent recently completed geographical speciation almost diagrammatically. For this reason they are often illustrated and this conveys the impression of great frequency. However, they appear to be only a small minority of the mainland species. The majority of them invade each other's ranges once they have "broken out" of their isolates after completion of speciation. Indeed, many of them seem to have been able to expand into all parts of the Amazon valley and beyond.

The respective frequency of parapatric and widely overlapping species in the Amazonian bird fauna has not yet been determined. To do this would require the exact mapping of all the species, the determination of ecological differences between them, and finally—based on all this information—the reconstruction of the ancestral distribution. This is precisely what Hall and Moreau have done for African birds and what Haffer has done for many South American birds. It should now be done on a strictly quantitative basis for the entire tropical avifauna of South America and indeed of all continents.

The occurrence of parapatric distribution patterns in which two species meet without interbreeding in the zone of contact indicates that certain species pairs are still so similar ecologically that they cannot invade each other's geographic range. On one side of the dividing line between two such species one of them is superior, on the other side the other species. Such parapatric contact zones provide ecologists with a golden opportunity to discover the environmental factors that determine the placement of the dividing line at that particular location.

Formation of Hybrid Zones

The third possibility is that the isolate has acquired certain morphological (and perhaps minor ecological and behavioral) differences but that these are not sufficient to provide for effective reproductive isolation. The result will be a well-defined zone of hybridization where contact between two semispecies is established. Such a pattern of distribution suggests two things, first, that there is too much ecological incompatibility between the incipient species to permit substantial geographical overlap and, second, that the gene complexes of the two semispecies are too different for any large-scale introgression in spite of the absence of complete reproductive isolation. Narrow zones of hybridization between otherwise allopatric semispecies are indications of an interesting stage in the process of geographic speciation. They are not infrequent in tropical South America but far more taxonomic analysis is required to determine their actual frequency. They have also been described for Australia (Keast 1961) and Africa (Hall and Moreau 1970).

Absence of Speciation

The fourth possibility is that the isolation has been too short to permit the development of appreciable differences between the isolated population and the parental population, so that both populations will completely fuse with each other as soon as the geographic isolation is terminated.

It is evident that the first and fourth alternatives, that is, complete overlap or complete fusion, are useless in our endeavor to reconstruct the location of former refuges but that the second and third will give us valuable clues.

DIFFERENCES BETWEEN ARCHIPELAGOES AND MAINLANDS

Although it is as yet impossible to provide exact quantitative data, it is nevertheless evident that there is a striking difference between speciation in an archipelago, like northern Melanesia, and in a mainland area, like Amazonia. In northern Melanesia a secondary overlap occurs in only 6 cases, while in 39 others a strictly allopatric pattern is preserved. On the mainland, secondary overlaps seem to be far more common than parapatric patterns.

What is the reason for this contrast? Perhaps the explanation is as follows: in an archipelago, distributional overlap between two recently evolved sister species can be achieved only when at least two individuals of one of the species cross a water gap and invade the range of a sister species. There are formidable odds against the successful achievement of this task. Most tropical birds are highly sedentary and respect water barriers to a high degree. Colonization is usually effected by single individuals and the new arrival is unlikely to find a conspecific mate. If it pairs at all, it will mate with an individual of the sister species and its genes will either be lost through sterility or introgress into the sister species. In either case, the incipient colonization across the water barrier will be unsuccessful.

The situation is completely different on a continent. As the refuges expand in the wake of the retreating barriers, the contained populations spread as populations. When they finally encounter a sister species from an adjacent refuge, they are not forced to interbreed, because numerous suitable conspecific mates are available in either population. Furthermore, when the populations of the two sister species come into contact, and presumably into competition, they will be sufficiently variable—as populations—to provide natural selection with an opportunity to select for sufficient differences in niche utilization to permit coexistence through exclusion. A single interisland colonist is confronted by an all-or-none situation and has no opportunity for gradual improvement through natural selection.

The cases of *Penelope, Crax,* and other genera with similar parapatric patterns (Vaurie 1968) show that the invasion of each other's ranges is occasionally difficult even for mainland species. Nevertheless, overlap occurs very frequently on the mainland, while it is the exception in a mature archipelago like northern Melanesia. Cross colonization occurs more frequently in remote island groups, like the Galápagos or Hawaii, which appear to be well below the level of faunal saturation.

A SPECIAL PROBLEM: EXTINCTION

If one were to make the assumption that a representative of each rain forest species became isolated in each forest refuge during each of the drought maxima and successfully speciated in this refuge, one should find a far greater number of species than one actually does. If one takes Amazonian species, for instance, and lists all those which seem to have had their place of origin in the Guyanan refuge, one finds that this number is remarkably small. The same, incidentally, is more or less true for any other refuge. There are two reasons for this deficiency. The first one, as I have already mentioned, is that many newly evolved species, as soon as they started to expand, were able to spread so wide and far that it is no longer possible to assign them to the Guyanan refuge as their point of origin. But there is a second and more important cause. It is highly probable that many, if not most, of these isolates became extinct. Being crowded together in a refuge during deteriorating climatic conditions, subjected to an increasing reduction in the diversity of plant life and to much inbreeding owing to decreased population size, it would seem inevitable that many of these species became extinct. Yet up to the present time we do not have a single investigation that has attempted to determine the amount of extinction that has occurred in the various refuges. Here is an area wide open for further research.

THE CHRONOLOGY OF SPECIATION

Even in the temperate zone, where our knowledge of the exact time sequence of climatic events is now so highly advanced, it is difficult to determine the exact chronology of speciation events (Selander 1965). We know that there has been a whole series of consecutive pluvial and arid periods, and yet in most of our discussions we act as if there had been only one pluvial and one arid period. This simplification is unavoidable at the present stage of our knowledge. Moreau (1966) has discussed this problem perceptively in his study of the history of the African avifauna. The Würm (Wisconsin) glaciation began about 70,000 years ago, but since then there have been perhaps three moist periods (52,000, 35,000, and 20,000 years ago) and perhaps three dry periods ($\pm 60,000$, $\pm 12,000$, ± 6000 years ago). But this continuous fluctuation of the climate is only part of the trouble.

The other difficulty is that the rate of differentiation in a given refuge seems to differ from species to species. In the same period during which one isolate may have hardly subspeciated, the isolate of another species in the same refuge may have reached full species level. It seems that the process of speciation in any refuge is completed only by a minority of species, unless the isolation is very complete and very long. For instance, Moreau (1966) shows that 82 species live in the isolated upper Guinea forest. But this figure includes only 9 endemic species and only 5 (6 percent) with a representative species in the lower Guinea forest. A similar situation exists in the case of the African nonforest species along the northern edge of the tropical forest belt. This nonforest belt was at one time separated by the formerly very extensive Lake Chad barrier into an eastern and a western portion. Nevertheless, among 178 endemic northern tropical nonforest species at most about one dozen species evolved west of the barrier. In only four or, possibly, five cases is there a superspecies represented by a different species east and west of the Lake Chad line. Moreau (1966) shows that not a single Ethiopian species that became isolated in North Africa when the Sahara became increasingly arid about 6000 years ago has diverged beyond the subspecies level. Our ignorance is still complete when it comes to a discussion of rate of speciation and to an answer for the question why certain populations in an isolate participate in a speciation while others do not.

CONCLUSION

In principle, there is no difference between speciation in tropical and nontropical birds. Different vegetation zones are the most important isolating barrier for tropical mainland birds. Newly evolved species in faunally rich archipelagoes tend to retain their allopatric position owing to difficulties of mutual colonization. Vegetational refuges during Pleistocene pluvial and arid maxima served as islands in which isolates could speciate on the mainland. Present distribution patterns indicate that, indeed, many species originated in this manner. When speciation is not completed, hybrid belts develop. Where ecological compatibility has not yet evolved, parapatric distribution patterns ensue. Narrow or broad geographical overlap between sister species seems to be the most frequent result of completed speciation on tropical mainlands.

REFERENCES

Diamond, J. 1972. *Avifauna of the eastern highlands of New Guinea.* Nuttall Ornith. Club, Cambridge, Mass.

Haffer, J. 1974. *Avian speciation in tropical South America.* Publ. Nuttall Ornith. Club, no. 14.

Hall, B. P., and R. M. Moreau. 1970. *An atlas of speciation in African passerine birds.* Brit. Mus. (Nat. Hist.), London.

Hammen, T. van der, and E. Gonzalez. 1960. Upper Pleistocene and Holocene climate and vegetation of the "Sabana de Bogota" (Colombia, South America). *Lied. Geol. Meded.,* 25:261–315.

Keast, A. 1961. Bird speciation on the Australian continent. *Bull. Mus. Comp. Zool. Harv.,* 123:305–495.

MacArthur, R. H. 1969. Patterns of communities in the tropics. *Biol. Jour. Linn. Soc.,* 1:19–30.

Mayr, E. 1963. *Animal species and evolution.* Belknap Press of Harvard University Press, Cambridge, Mass.

Mayr, E., and M. H. Moynihan. 1946. *Evolution in the* Rhipidura rufifrons *group.* Amer. Mus. Novit., no. 1321.

Mayr, E., and W. H. Phelps. 1967. The origin of the bird fauna of the South Venezuelan highlands. *Bull. Amer. Mus. Nat. Hist.,* 136:269–328.

Moreau, R. E. 1966. *The bird faunas of Africa and its islands.* Academic Press, London.

Selander, R. K. 1965. Avian speciation in the Quaternary. In *The quaternary of the United States,* ed. H. E. Wright and D. G. Frey, pp. 527–542. Princeton University Press, Princeton.

Stresemann, E., E. Hartert, K. Paludan, and L. Rothschild. 1936. Die Vögel des Weyland-Gebirges und seines Vorlandes. *Mitt. Zool. Mus. Berlin,* 21:165–240.

Vaurie, C., 1968. Taxonomy of the Cracidae (Aves). *Bull. Amer. Mus. Nat. Hist.,* 138:131–260.

15
Change of Environment and Speciation

Darwin and Wallace at an early date appreciated the importance of islands in speciation. Nevertheless, owing to the fact that the majority of species of animals and plants live on continents, the distribution and variation of continental species formed the basis for most discussions of speciation. Geographic speciation was envisioned as due to the development of a barrier in a previously continuous range and the resulting reduction or total prevention of gene flow. Most textbook diagrams of geographic speciation reflect this conception. It largely dominated my own thinking for some years, but at an early date I was impressed by the fact that the most deviant derivatives of a speciating taxon invariably occurred in a peripherally isolated situation. I pointed out as early as 1940, and I greatly emphasized in 1942, that extreme peripheral isolates sometimes differ from the parental species to such an extent that they have been classified as different genera. I cited Serresius, Todirhamphus, Thyliphaps, Oedirhinus, *and* Dicranostephes *as illustrations of this principle (1942:284). Indeed, I became convinced that higher taxa most probably originated in such populations and that they were the favorite locale for the origin of evolutionary novelties.*

What remained a great puzzle to me was why such drastic evolutionary changes should be favored in peripheral isolates. Many years of pondering this problem finally led me to the proposal of a theory of genetic revolutions in founder populations. This is perhaps the most original theory I have ever proposed. In the early years after its publication (1954), this theory was almost universally ignored or was confused with the founder principle, which I had so designated in 1942. In more recent years, however, an increasing number of authors have begun to appreciate the importance of this principle and even paleontologists have begun to see that the absence of missing links from the geological record may often be due to the fact that the crucial steps occurred in

Reprinted from "Change of genetic environment and evolution," pp. 157–180 in *Evolution as a process,* ed. J. Huxley, A. C. Hardy, and E. B. Ford (London: Allen & Unwin, 1954).

small, peripherally isolated populations where the probability of leaving testimony in the recovered fossil record is exceedingly small.

Observing the extraordinary morphological uniformity of most populous, widespread species, I placed great stress in the following essay (1954) on the equalizing power of gene flow. This interpretation was repeatedly criticized in subsequent years, but the critics never supplied an alternate explanation. Even though the importance of gene flow still needs to be stressed (it is much more important than admitted by some critics), it has become evident in the last 20 years that there are numerous balancing mechanisms that give great unity and stability to the genotype. I have discussed the relative importance of these two factors in Populations, Species, and Evolution *(1970:300).*

That mutation, recombination, selection, and isolation are the four cornerstones of evolution is now generally acknowledged. The way in which these factors interact in the various evolutionary processes and the role played by diverse subsidiary factors are, however, by no means fully clarified. In particular, the role of one factor, a sudden change in the genetic environment, seems never to have been properly considered. That this factor might be exceedingly important in the evolutionary process occurred to me when studying a puzzling phenomenon, frequently encountered by the systematist, the conspicuous difference of most peripherally isolated populations of species.

Let us look, for instance, at the range of the Papuan kingfishers of the *Tanysiptera hydrocharis-galatea* group (Figure 15-1). It is typical for hundreds of similar cases. On the mainland of New Guinea three subspecies occur which are very similar to each other. But whenever we find a representative of this group on an island, it is so different that five of the six Papuan island forms were described as separate species and four are still so regarded.

EVOLUTIONARY GENE FLOW

For such a striking dissimilarity of peripherally isolated populations two reasons are usually cited: difference of physical and biotic environment or genetic drift. It seems to me that neither one of these factors alone nor a combination of the two can provide a full explanation, even though both may be involved.

Let us first look at the possible effects of the environment. New Guinea is a tropical continent. If it were projected on the map of Europe, it would reach from England to the Black Sea. Northwestern New Guinea (near the equator) is exceedingly humid and without pronounced seasons; nearly the same amount of rain falls in every month of the year. In southeast New Guinea (Port Moresby) most of the rainfall is condensed into a short rainy season, while it is dry the rest of the year. The biotic environments in the extreme areas of New Guinea

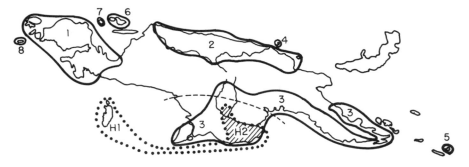

Figure 15-1. Species and subspecies of the *Tanysiptera hydrocharis-galatea* group. The subspecies *1, 2,* and *3* of *galatea* on the mainland of New Guinea are exceedingly similar to each other. The subspecies *vulcani* (*4*) and *rosseliana* (*5*) are much more distinct. The populations on Biak (*6*), Numfor (*7*), and Koffiao (*8*) have reached species level. The form on Aru Island, *hydrocharis* (H_1), has also reached species rank and now coexists in South New Guinea (H_2) (shaded area) with a subspecies of *galatea* (*3*). (From Mayr 1942.)

are as drastically different as the physical environments, only heavy rain forest in some areas, much monsoon forest in others. Still, the populations of *Tanysiptera galatea* that occur at the two ends of New Guinea are hardly distinguishable.

A similar situation is found in numerous other species; if there are subspecific differences within New Guinea they are often merely matters of degree. We can generalize and say that strong environmental differences may not lead to conspicuous morphological differentiation. (This statement does not deny the physiological adaptation of the populations to their respective ranges.)

What is the situation with respect to the islands? Numfor, Biak, Koffiao, and Rossel are less than 100 miles from the New Guinea shelf, and each island is approximately in the same climatic district as the nearest part of the mainland of New Guinea. In spite of this similarity of environment they are inhabited by populations that are phenotypically strikingly different. Thus selection alone cannot give the full answer.

Drift—in the ordinary sense—cannot be the complete answer either because in many cases very large islands are involved with populations consisting of tens of thousands, hundreds of thousands, or even millions of individuals (see below).

The phenomenon of conspicuous divergence of peripherally isolated populations, so well illustrated by the *Tanysiptera hydrocharis-galatea* group, is familiar to every taxonomist. Scores, if not hundreds, of examples can be found in every monograph or checklist. I mention merely a few more:

The kingfisher *Halcyon australasia* is virtually without geographical variation all over Australia (*sancta*), but has subspecies on New Zealand,

New Caledonia, the Loyalty Islands, and very strikingly different sub-species in the Lesser Sunda Islands.

The hawk *Accipiter novaehollandiae* shows little evidence of geographical variation in New Guinea but has endemic subspecies on many islands east and west of New Guinea, five subspecies in the Bismarck Archipelago, and five in the Solomon Islands, in a total land area considerably smaller than the area of New Guinea.

A comparison of the disrupted Mediterranean ranges of many amphibians, insects (e.g., *Bombus*), and lower invertebrates (e.g., *Dugesia gonocephala*) with the contiguous ranges of the same species in the temperate parts of the Palearctic region shows the same pattern. The lizards (*Lacerta*) of the Mediterranean area have only a few slight subspecies on the mainland, but scores on islands. The four species of *Peromyscus* most closely related to *P. maniculatus* are (or have been until recently) peripherally isolated.

It would lead too far to quote here more cases; all would merely be variations on the same theme. I have spoken in earlier publications of the "law of peripheral populations." This is not entirely accurate. Peripheral populations are not outstandingly different if they are part of a continuous series of populations. Only "peripherally isolated" populations show the pronounced deviations from the species "type" illustrated by the examples given above.

THE STRUCTURE OF SPECIES

It is evident that there are two types of geographical variation, ecotypic and typostrophic.[1]

Ecotypic Variation

This variation adapts to the local environment populations that are members of a continuous series of contiguous populations. Owing to the never-ceasing gene flow through such a system these populations are merely variations on a single theme, even though they may be sufficiently distinct to have received the attention of the taxonomist and to have been described as subspecies. Goldschmidt (1940:182) has singled out *such* subspecies to attack the concept of the subspecies as incipient species: "The differences between two subspecies are usually clinal, merging into each other. . . . While the characters of subspecies are of a gradient type, the species limit is characterized by a gap. . . . The subspecies do not merge into the species either actually or ideally. . . . Subspecies are actually, therefore, neither incipient species nor models for the origin of species. They are more or less diversified blind alleys within the species." This statement is correct so long as it is applied

---

1. "Typostrophic" is a term used by the paleontologist Schindewolf to denote the origin of a real evolutionary novelty, a new "type."

only to those subspecies that are subdivisions of a widespread array of continuous populations. Such subspecies, indeed, are not incipient species. Incipient species require isolation (see below).

The variation in such a system of contiguous populations is characteristically "ecotypical," that is, adaptive within the established "type." It is characterized by clines, and most variation within such a system obeys the various ecological rules. That the evolutionary changes due to this type of variation do not necessarily lead to species formation has long been recognized by systematists: "Clines indicate continuities, but since species formation requires discontinuities, we might formulate a rule: *The more clines are found in a region, the less active is species formation.* We can prove this if we compare regions in which clines are frequent with those in which they are rare" (Mayr 1942:97).

Nevertheless, there may be an accumulation of considerable genetic differences at the opposite ends of a cline. Nothing illustrates this better than the well-known overlapping circles of races as well as reduced fertility among geographical races. A secondary isolation of such populations may quickly lead to completion of species formation (see below).

"Typostrophic" Variation

Isolated populations such as the ones illustrated above show a type of variation so different from that of contiguous populations that we may be dealing with something entirely new. Not only do these peripherally isolated populations often have the characteristic features of incipient species, but what is more important they often are species or incipient species of an entirely new type. That is, they may have morphological or ecological features that deviate quite strikingly and unexpectedly from the "parental" pattern.

A comparison of the two kinds of geographical variation, as well as of the populations produced by them, leads inevitably to the question: What is the factor that distinguishes the isolated population from the population that is part of a large group of populations? I shall attempt to answer this in the next section.

THE EFFECT OF GENE FLOW

Of the many factors that are of importance to the evolutionist, gene flow is perhaps the most neglected one. In a given wild population genetic novelties may occur through mutation or through immigration from outside populations (with recombination making available an unlimited assortment of these factors). While there are numerous studies on all aspects of mutation, virtually nothing is known about the genetics of gene flow. In fact, the technique of the geneticist is built on the avoidance of gene flow or of immigration, hence the stoppers on

Drosophila bottles and the wire netting of mouse cages. Whenever gene flow occurs in spite of all precautions, it is called "contamination," and the contaminated cultures are carefully destroyed. Since the genetics of gene flow has not yet been studied in the laboratory, a determination of its importance depends either on field work, such as done by Timofeeff-Ressovsky, Dobzhansky, and Ford, or on guesses. In orthodox genetics the effects of gene flow are usually presented as the addition of the immigrated genes to the gene pool of the local population and their subsequent gradual elimination insofar as they are inferior to the other genes of the gene pool. Such a purely additive treatment of gene flow is not correct, as I shall presently discuss, since it neglects the fact that a gene pool is not an unconnected "pile of genes," but a well-integrated, balanced system.

For a further analysis of this problem it is of vital importance to determine how great the amount of gene flow is as compared with mutation. I venture the guess that the total amount of genetic change contributed by immigration to a given local breeding population of a prosperous widespread species is many times that contributed by mutation occurring among the members of this population.[2]

Since gene flow is due to the movement of individuals that are the carriers of genes, the phenomenon of dispersal becomes of interest to the geneticist. This was clearly realized by Timofeeff-Ressovsky, Dobzhansky, Ford, and other population geneticists. Measuring the amount of gene flow through a study of dispersal, as was attempted by naturalists and geneticists, involves various difficulties.

The chief difficulty is that at best one can determine only what percentage of individuals in a population are immigrants, but not how genetically similar or different these immigrants are as compared with the members of the "native" population. The amount of genetic difference depends largely on the distance (within the species range) from which these immigrants have come. The usual assumption is that individuals settle within a predictable and rather narrow circle around their place of birth. In studies of noncolonial bird populations it is usually found that 30–40 percent of the individuals that newly settle in a study area are born within the study area, the remainder being new arrivals. It is usually assumed that most of these come from the immediately adjacent area. This assumption is based on the further assumption that the dispersal curve is essentially a normal curve. Bateman (1950), however, has summarized numerous data, including his own and those of Dobzhansky and Wright, which indicate that the dispersal curve is not normal but leptokurtic and probably not even symmetrical but strongly skewed. In fact, so far as birds are concerned, there are some indications that the populations in many species are composed of two kinds of individuals, those with a strong locality sense and those with little or

---

2. It has no bearing on the present argument that the genetic differences of the immigrants are ultimately also due to mutation.

none. It is possible that the latter, perhaps up to 10–30 percent of the population (differing from species to species) may settle in any suitable spot up to 100 kilometers or more from the place of birth. Dobzhansky and Wright (1947) likewise suggest for *Drosophila pseudoobscura* the possibility of a composit dispersal curve. Previous calculations of the amount of gene flow have tended to ignore this minority of long-distance colonists, which are nevertheless of considerable importance in counteracting the effects of local selection pressures. Such long-distance dispersal not only has been established by bird banders, but is very characteristic for the spreading of expanding species like the serin finch (*Serinus serinus*) or the ring dove (*Streptopelia decaocto*). Individuals from far distant populations, even though few in number, are apt to contribute many new genetic factors to a population.

In previous discussions of the genetic effects of long-range dispersal attention was focused almost entirely on the fate of the alien genes in the new gene complexes. Since they are usually from regions with rather different selective factors, such alien genes are apt to be of inferior viability in the new environment and will be eliminated sooner or later. Little or no thought, however, was given to the effects of these alien genes on the relative viability of the genes of the gene complex into which they were introduced. It appears probable that the frequent introduction of such alien genes into a gene pool will lead to selection of such "native" genes as are tolerant to combination with such alien genes, that is, genes which produce viable heterozygotes with a great assortment of alien alleles or gene combinations.

A further effect of such gene flow is that it disturbs the integration of local gene complexes in response to local selection pressures. Although all populations are somewhat ecotypical, such infiltration of alien genes may prevent a complete response of the gene complex to the local selection pressure and may therefore act as a conservative ("stabilizing") element in the whole evolutionary picture.

Several proofs can be cited for the reality of this effect of gene flow. Dr. R. A. Fisher kindly called my attention to a case described by Turesson where in a specialized habitat no specific ecotype developed because the location was too small and therefore too much exposed to gene flow from adjacent localities. But even where a local ecological race develops, its great variability is evidence for the continued inflow of genes from adjacent populations (see the cases described by Clausen, Hiesey, and Keck and those summarized by Stebbins 1950).

Particularly instructive are the races of small mammals in the southwestern United States that live on lava flows. Endemic blackish races develop on small lava flows only if they are completely surrounded by sandy desert. If they are in contact (on more than 1/10 of their circumference) with areas of desert rocks there will be too much gene flow to permit the development of endemic black races (Dice and Blossom 1937; Hooper 1941). When two soils of different colors come in contact, the effects of gene flow will be noticeable for many miles on either side, or at least on one side of the zone of contact, as shown by Sumner, Dice, Blair, Hayne, and others. It is evident in all these cases that local selection pressure is partially neutralized by the effects of gene flow.

Selection will be able to work unimpeded only if the selective agent simultaneously eliminates dispersal. Contact poisons for insects, for instance DDT, are indeed such agents. In a population that is being selected for DDT resistance all nonresistant flies that enter the population are eliminated before they can counteract through their genes the genetic process of the continued improvement of the DDT resistance.

These considerations finally elucidate a problem that has been a great puzzle to all naturalists and zoogeographers, namely, the problem of the borderline of species. The species border is the line beyond which the selective factors of the environment prevent the successful reproduction of the species. However, it is well known to naturalists that through dispersal from the species range a considerable number of individuals settle down annually beyond the normal species border, where they attempt to reproduce. Some even succeed in establishing new colonies, but these are sooner or later eliminated in an adverse season. This has the result that the species border, though fluctuating, remains at a dynamic, stable line. What is puzzling is that natural selection in the belt immediately beyond the previously existing borderline has not been able to produce a population adapted to the local conditions, in the same way as the application of DDT produces a DDT-resistant strain of flies. This is particularly puzzling since conditions beyond the borderline differ from conditions within the species border usually only slightly and in degree. This puzzle can be considered solved if we assume that this process of adaptation by selection is annually disrupted by an infiltration of alien genes and gene combinations from the interior of the species range that prevents the selection of a stabilized gene complex adapted to the conditions of the border region.

THE GENETIC ENVIRONMENT

The reason for the importance of gene flow and of isolation is implied in much of the recent genetic work but has never been fully stated. Classical genetics studied the genetic changes at a given gene locus as well as the physiological and selective effects of such changes. Since—for the sake of simplifying the analysis—each locus was studied separately, the genetic factors of an organism were treated as so many beans in a large bag. That this is not so is now known to every geneticist, but "beanbag" thinking is still widespread. The fact is, of course, that genes do not exist in "splendid isolation," but are parts of an integrated system. In order to appreciate the complexity of this system it is necessary to recall some of the recent studies.

The normal model of genetic change, presented in most evolutionary studies, is that of a gene A, originally in homozygous condition, on which mutation pressure (or immigration pressure) is exerted by an allele a. A maximum of 50 percent heterozygotes may occur under these conditions (if both alleles are of equal frequency). Numerous studies indicate, however, that the situation in which there are only two competing alleles may be the exception rather than the rule. At many loci there are simultaneously three, four, five, or more alleles available.

Two alleles produce one kind of heterozygote, 3 alleles 3, 4 alleles 6, 5 alleles 10, and 6 alleles 15 kinds of heterozygotes. The series expands very rapidly: 15, 21, 28, 36, 45, 55, etc. The number of homozygotes (under the simplified assumption of equal frequency of the various alleles) is $1/n$, so that with 5 alleles present, only $1/5 = 20$ percent of the genotypes might be homozygotes. It is quite evident from these considerations that the heterozygotes are of much greater importance in such a system than the homozygotes, and the more so, the greater the number of alleles.

In view of the considerable morphological uniformity of samples from most natural species, it may be denied that multiple alleles are frequent in nature. This may be true for some loci, but it is certainly not true for others. Even alleles that produce lethal homozygotes may be indistinguishable as heterozygotes, and it is known that lethals are frequent in many wild populations. A consideration of the so-called *isoalleles* (Stern) is important in this connection. Isoalleles are alleles that are phenotypically indistinguishable in homozygous condition from the "normal" wild-type allele but have different expression when placed in heterozygous combination with tester alleles. Only a few studies have been made so far to determine the frequency of isoalleles (Stern, Timofeeff, Spencer), but, as Spencer's summary indicates (Spencer 1944), isoalleles appear to be amazingly common at some loci. The fact that there are different frequencies of hereditary diseases in different human races may in part also be due to different mutation rates in different isoalleles.

Still more important than the multiple alleles is the fact that during development all genes are members of a team. Not only has every gene that has been thoroughly studied been found to have pleiotropic effects, but it has also been found that every character is produced by the joint action of many genes. It is immaterial in this connection what particular genetic theory one adheres to: major genes and modifiers, genes and polygenes, switch genes and gene complexes, position effects, and nonlocalized genes. They all agree on the essential point, which is that the action of a given gene is strongly influenced by its genetic background, its genetic "coactors." And what is true for the function of a gene is true also for its selective value. A gene that is of high selective advantage on one genetic background may be selected against on another genetic background. *The selective value or viability of a gene is thus not an intrinsic property but is the sum total of the viabilities on all the genetic backgrounds that occur in a population.*

THE COADAPTED SYSTEM

The concept that the viability of a given allele depends on its genetic background is not new. It has been emphasized by several students of this problem. Sewall Wright (1931:155), for instance, stated: "The selection coefficient of a particular gene is really a function not only of the relative frequencies and momentary selection coefficients of its different allelomorphs, but also of the entire system of frequencies and selection coefficients of non-allelomorphs." Recently Dobzhansky has supplied experimental proof for coadapted selective changes in chromosomes or gene arrangements (Dobzhansky 1950; Dobzhansky and Levene 1951). The relativity of such viabilities is most convincingly illustrated by the findings of Wallace and King (1951) in irradiated *Drosophila* populations. They find "that while an examination of the individual chromosomes of a population may reveal that these are generally 'deleterious' when homozygous, an examination of pairs of unrelated chromosomes from the same population may reveal that these pairs are distinctly superior." Muller's (1948) work on dosage effects and much other genetic work points in the same direction.

Such a well-integrated, coadapted gene complex constitutes an evolutionary unit in spite of its intrinsic variability. Any disharmonious gene or gene combination that attempts to become incorporated in such a gene complex will be discriminated against by selection. There is much evidence for this, partly from experiments and partly from a study of natural populations. That the offspring of crosses among species or other only distantly related populations produce inferior gene combinations has long been known to students of hybridization. This is often true even in cases where there is superficial heterosis. In the cases of back-crosses of F_1 hybrids with one of the parents, it is often found that only those back-crosses are viable that are close to the parental type.

That gene complexes are well integrated units is the explanation of a phenomenon that has long been a puzzle to naturalists. There are many cases known where two incipient species come together in an allopatric zone of hybridization after a previous extrinsic barrier has been removed. Sometimes this zone is wide, but more often it is very narrow, even though all the available evidence indicates that the zone has existed for thousands of years. The border between the carrion crow and the hooded crow (*Corvus corone* and *cornix*) is an excellent example. Peus (1950) discusses such a case in a flea. In these and many similar cases it appears that the gene complexes which come together are so well balanced within themselves that combinations with alien genes lead to combinations of inferior viability and are eliminated by selection. This counter-selection reduces introgressive gene flow drastically. (See Selander, Hunt, and Yang 1969 for a recent analysis).

Even a gene mutation that leads to an improvement in the phenotype may have difficulties in such a system because it will take a long time for it to become fully fitted into the total pleiotropic-buffered gene complex. Simpson's findings that tooth elongation in fossil horses was of the order of only 1 millimeter per 1 million years is a suitable illustration of this process. To be of real value, such an improvement in the teeth has to be correlated with a strengthening of the upper and lower jaw and with numerous other readjustments of the skeleton, the muscles, and presumably even the viscera and the nervous system. All these changes require a rather thorough overhauling of the total gene complex. It is not often that selection permits a single structure to rush far ahead of the other parts of the system to which it belongs.

The better integrated such a gene complex is, the smaller the chance that a novel mutation will lead to an improvement. As Schmalhausen, Heuts, and others have pointed out, much of selection is stabilizing. A widespread species, with many local populations among which there is active gene exchange, tends to be very conservative.

Even though a species may have many local races (ecotypes), it arrives sooner or later at a geographical line, its species border, beyond which it cannot expand. As stated above, and other things being equal, this appears to be due to the fact that even the border populations are tied by gene flow to the integrated gene complex of the main body of the species. This applies to horizontal as well as to vertical (altitudinal) distributions. In most species there is a limit to ecotypic adaptation. It has long been known, for instance, that the validity of the so-called ecological rules (Bergmann's rule, Allen's rule, etc.), so far as it exists at all, is restricted to intraspecific variation. Subspecies of birds living in a cool climate tend to be of larger body size than subspecies living in the warmer parts of the range of the same species. However, a more northerly *species* is by no means always larger than its nearest more southern relative. The same is true for altitudinal variation. As Rand and others have shown, there is a steady increase of size with altitude within most sedentary species of birds with a wide altitudinal range. However, again, this does not necessarily apply to full species that replace each other altitudinally, as Dr. Rand pointed out to me. In 15 such pairs of bird species from New Guinea, the higher species was larger in 5 cases, of equal size in 3 cases, and smaller in 7 cases. It appears that in these latter cases expansion into the higher altitudes was made possible on an altogether different physiological (and hence genetic) basis. The gene flow through the lower altitude species was too stabilizing to permit range expansion into the higher altitude.

CHANGES OF THE GENETIC ENVIRONMENT

The make-up of the well-integrated gene complex, discussed in the previous section, as well as the continuous immigration of alien genes from adjacent or far distant populations, makes it evident that genes with either or both of the following two properties will be specially favored by selection: (1) genes that produce heterozygotes of high viability, preferably viability superior to the homozygotes; (2) genes that produce viable combinations on the greatest number of different genetic backgrounds. The former phenomenon leads to balanced polymorphism, first postulated by R. A. Fisher, the widespread occurrence of which has been abundantly confirmed in recent years. Dobzhansky and Levene (1951) have shown how quickly natural selection can produce such heterosis in cases where it was previously absent.

The selective advantage of genes that do well on a great variety of genetic backgrounds, "jack-of-all-trades" genes, does not seem to have received much attention in the genetic literature. The richer a population is in genetic factors (multiple allelic heterozygosity on many loci), the more important such genes are. A "good mixer," rather than a good "soloist," has a tremendous advantage in such a system.

This all may change dramatically when a few individuals are taken out of the stream of genes and placed in isolation. Let us illustrate this again with reference to the kingfishers of the *Tanysiptera galatea* complex. Let us assume, for instance, that Numfor Island, previously without *Tanysiptera*, was colonized by a couple of pairs of New Guinea birds. What changed in the conditions? The climate of Numfor is much like that of the opposite coast of New Guinea, thus the selection pressure by the physical environment will remain much as it was in the previous range. The flora is somewhat different and the fauna is somewhat impoverished but the only potentially serious predator, *Accipiter novaehollandiae*, occurs both on Numfor and New Guinea. The physical and biotic environments in these places are thus rather similar. A third environment, however, the genetic environment, is strikingly different. The Numfor population is geographically and hence also genetically completely isolated from all other populations of the species (perhaps a for New Guinea individuals may get there every 10 years), while *every New Guinea population is in the midst of a continuous stream of genes flowing back and forth across the entire island continent*. While the number of possible contacts of a given gene with other genes is exceedingly high in New Guinea, it is drastically reduced among the founders of the Numfor population.

The total sum of the relative selective values of each allele may be changed because the number of possible genetic interactions of this allele is much reduced. To illustrate the situation diagrammatically we might group the total number of possible genetic backgrounds of the

Table 15-1. Arbitrary values of the frequencies (q) of classes of
genetic backgrounds (x_1–x_{10}) in a population and of the viabilities (w)
of alleles a_1 and a_2 on these 10 backgrounds. The total viabilities of
a_1 and a_2 are about equal in this population (1.102).

	x_1		x_2		x_3		x_4		x_5	
	q	w	q	w	q	w	q	w	q	w
a_1	0.1	1.1	0.05	1.2	0.02	0.9	0.17	1.1	0.12	1.3
	0.110		0.06		0.018		0.187		0.156	
a_2	0.1	1.33	0.05	3.4	0.02	2.8	0.17	2.5	0.12	0.4
	0.133		0.17		0.056		0.425		0.048	

	x_6		x_7		x_8		x_9		x_{10}	
	q	w	q	w	q	w	q	w	q	w
a_1	0.09	0.4	0.07	1.4	0.06	1.8	0.19	1.2	0.13	0.08
	0.036		0.098		0.108		0.228		0.101	
a_2	0.09	0.1	0.07	1.0	0.06	0.5	0.19	0.3	0.13	0.8
	0.009		0.07		0.03		0.057		0.104	

parental populations in 10 classes, x_1 to x_{10}. Let us further assume that
in the population there are two alleles, of which one (a_1) is of broad,
general efficiency on many genetic backgrounds, while the other
allele (a_2) is very superior on some genetic backgrounds but inferior or
even lethal on others. We are assuming arbitrarily for the sake of illustration an extreme situation. Most genes would presumably be ranged
somewhere between the extremes of a_1 and a_2. The selective values of
the two alleles a_1 and a_2 on the 10 backgrounds might be as in Table
15-1.

The total selective value of both alleles is identical, allowing for the
relative frequencies (q) of the carriers of the genetic backgrounds
x_1–x_{10} in the population. If the founding of the Numfor population
was made by individuals with only the genetic backgrounds x_1–x_4,
each being present with equal frequency ($q = 0.25$), the total viability
of the alleles a_1 and a_2 suddenly changes to 1.075 for a_1 and to 2.5075
for a_2. Instead of retaining equal viability, the viability of a_2 is now 2½
times that of a_1. It is evident that a formidable selection pressure will
be exerted against a_1 which presumably will soon lead to its elimination
from the new population. Even if the two alleles in question are less
different in kind than a_1 and a_2, it is very unlikely that their selective
values will remain unchanged (Figure 15-2).

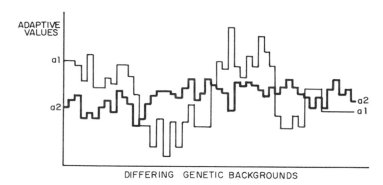

Figure 15-2. Diagrammatic presentation of the relative selective value of alleles a_1 and a_2 on many different genetic backgrounds. On some a_1 is superior, on others a_2; a_1 has a more even viability on many backgrounds; a_2 is very superior on some and very inferior on others.

One of the obvious effects of the sudden reduction of population size in the founder population will be a strong increase in the frequency of homozygotes. As a consequence, homozygotes will be much more exposed to selection and those genes will be favored which are specially viable in homozygous condition. Thus, the "soloist" is now the favorite rather than the "good mixer."

We come thus to the important conclusion that *the mere change of the genetic environment may change the selective value of a gene very considerably.* Isolating a few individuals (the "founders") from a variable population that is situated in the midst of the stream of genes that flows ceaselessly through every widespread species will produce a sudden change of the genetic environment of most loci. This change, in fact, is the most drastic genetic change (except for polyploidy and hybridization) that may occur in a natural population, since it may affect all loci at once. Indeed, it may have the character of a veritable "genetic revolution." Furthermore, this "genetic revolution," released by the isolation of the founder population, may well have the character of a chain reaction. Changes in any locus will in turn affect the selective values at many other loci, until finally the system has reached a new state of equilibrium.

THE ROLE OF THE PHYSICAL AND BIOTIC ENVIRONMENT

Focusing our attention on the decisive effects of interrupted gene flow should not make us forget the important synergistic role of selection by the new physical and biotic environment. Selection on an island differs in two ways from selection on the mainland. First, the selection pressure itself is different, since the environment (particularly the biotic one) is inevitably somewhat different. Second, this selection can express itself more directly because its effects are not continuously disturbed by the inflow of alien genes. Selection on an island will, if anything, produce even more conspicuous results than on a mainland.

The amazingly great differences among populations of adjacent islands, for example, *Tanysiptera carolinae* (Numfor) and *riedelii* (Biak), indicate that the accidents of gene assortment during the "genetic revolution" of the isolated population may be more important than the "directive" force of the similar environment of adjacent islands.

It should be emphasized that such a "genetic revolution" in the founder population is only a potentiality; it does not necessarily happen every time a population is isolated; it occurs only if the genetic constitution of the founders favors it. The amount of the genetic revolution is unpredictable since it depends on many factors. It proceeds at the most rapid rate: (1) if the parental population was particularly variable and subject to much gene-flow; (2) if the founder population contained genes of very uneven selective values in different genetic environments and particularly genes that contribute high viability in homozygous condition; (3) if the founders happened to have many genes of particularly high selective value in the new environment; and (4) if the new physical and biotic environment is capable of setting up and maintaining divergence producing selective pressures and, in particular, if it permits a shift into a vacant ecological niche somewhat different from the parental one.

The fact that neither the starling (*Sturnus vulgaris*) after its introduction into North America nor many of the introduced agricultural pests have shown indications of drastic evolutionary change confirms that the isolation of a population is not by itself a guarantee for a drastic change. Perhaps these colonies regained large population size too quickly. As has been shown by Ford and others (Ford and Ford 1930), there is an increase in variability and a relaxation of selection in a rapidly expanding population. This relaxation of selection may mitigate or even counteract at first some of the effects of the sudden isolation. Perhaps this explains why there is apparently a lag between isolation and the differentiation of the isolated populations. One might expect the isolated populations to change conspicuously within the first two or three generations, but this is usually not the case. The fact that the gene complex as a whole has to remain well integrated at all times (compare tooth elongation in fossil horses) is another retarding factor.

FRACTIONING OF A CONTINUOUS SPECIES RANGE

Not all population discontinuities in nature originate by the colonization of islands. Sometimes they arise by the contraction of a species range and the separation of a previously continuous area into separate ranges. If a continuous large species range is split, let us say, into two wide ranges, A and B, both will contain a similar rich mixture of iso-alleles, polymorph genes, pleiotropic factors, and polygenes. It will take some time before the interruption of gene flow between the two population groups will make itself felt. But eventually some of the genes in A that had continuously drifted in from B (where they are superior) will disappear and vice versa. Also, in view of the randomness of mutation, it is unlikely that the same mutations will occur in the two areas with identical frequencies. The result will be an increasing genetic divergence, accelerated by the different selection factors (of the external and genetic environment) in the two areas until the two population groups have again reached equilibrium.

If the respective environments remain rather similar and if the gene complexes that were fractioned by a secondary discontinuity were particularly well balanced, they may diverge only very slowly after separation. There are many cases known, particularly among plants, where striking geographical discontinuities have not led to much of an evolutionary change in the separated populations. Stebbins (1950) has called attention to many such cases, for instance, the American and Asiatic *Platanus,* which have not even reached reproductive isolation in spite of many millions of years of geographic isolation. It is evident from such cases that the length of the period of separation and the amount of genetic divergence are not always very closely correlated. Why some gene complexes are so stable and others evolve rapidly is still an unsolved problem. It appears very probable that differences in selection pressures are not the complete answer.

It is possible that the difference in the mode of speciation (founder population versus fractioned species range) contributes to the differences in species patterns found in nature. It is well known among systematists that in some taxonomic groups there are many very similar species (including sibling species), while in certain other groups (such as Cerambycidae) most species are so different from each other that many of them are placed in monotypic genera. It would be interesting to determine whether most speciation in the former cases proceeded via fractioning of initially large populations, while founders gave rise to many of the monotypic genera. Too many additional factors enter the picture to permit any generalization, but the possibility that the mode of speciation is one of these factors should not be entirely ignored.

GENETIC VARIABILITY

During a genetic revolution the population will pass from one well-integrated and rather conservative condition through a highly unstable period to another new period of balanced integration. The new balance will be reached after a great loss of genetic variability. There are several reasons for this loss: (1) the founders represent only a segment of the variability of the parental population; (2) during the period of rapid readjustment, alleles that had previously been of equal viability will change their relative viability, like a_1 and a_2 discussed above, and the inferior ones will become eliminated; and (3) recessives will have a much greater chance to become homozygous in the reduced population and thus become more exposed to selection. As a consequence of these gene-loss inducing factors, a population may result that is not only very different genetically from the parental population but also genetically comparatively invariable. Much that is known about island populations supports the validity of this conclusion.

The evolutionist takes, on the whole, a dim view of the future prospects of populations with depleted genetic variability. Such populations are not very plastic. If they live on an island (in the broad sense of the word), they will probably be successful as long as conditions remain stationary. However, such populations rarely have the capacity to adapt themselves to severe environmental shocks. The arrival of a new competitor or of a new enemy or a drastic change of vegetation or of the physical environment is apt to lead to extinction. It is no coincidence that even though less than 20 percent of all species of birds are island birds, more than 90 percent of all bird species that have become extinct in historical times are island species. An island bird thus has at least 50 times as great a chance to become extinct as a mainland bird. Only part of this extinction can be attributed to the small size of the range of these island species.

Permanent genetic depauperization and eventual extinction are, however, not the inevitable fate of island populations. An occasional population succeeds in making an ecological shift during the "genetic revolution" and during the period of relaxed selection accompanying the phase of rapid expansion. It may become adapted to a new ecological niche or even to a major new ecological zone. If such a population can colonize the nearest adjacent "mainland," it may find this ecological niche or zone unoccupied and be able to invade it. Once it starts spreading over wide areas it can again start accumulating additional genetic variability so that in due time it may be as full of heterozygosity and concealed variability as the ancestral species with which it may now be sympatric. This is illustrated diagrammatically in Figure 15-3.

Genetically depauperized populations have the best chance of survival in an unsaturated environment, an environment relatively free of competitors and enemies. This is the reason why speciation of this type has played such an important role on "vacant" archipelagoes, such as the Galápagos or the Hawaiian Islands, or in "empty" lakes. Similar considerations may apply to the apparent bursts of speciation encountered by the paleontologist whenever a new major "type" appeared on the scene and entered a "vacant" ecological zone.

It has been questioned whether any natural population can pass through a genetic bottleneck of reduced variability (Figure 15-3 B or C). However, there is abundant evidence that this *is* possible. Less than 20 pairs of the European starling were introduced to the United States in 1890; and only a fraction of them bred successfully. It took more than 15 years before they began to increase materially, but now (only 40

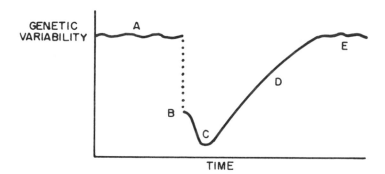

Figure 15-3. Levels of genetic variability: (*A*) in parental species, (*B*) in the founder population, (*C*) after "genetic revolution," (*D*) during recovery period, (*E*) after a new level is reached.

years after they really started to spread) they are one of the most common birds of North America, having increased to an estimated number of over 50 million individuals. The story of the house sparrow (*Passer domesticus*), the Japanese beetle, the potato beetle, or of any other kind of successful introduction is the same. But there are also many cases of spontaneous founding of new species ranges. The Australian white-eyes (*Zosterops lateralis*), a small flock of which found its way in 1856 from Australia to New Zealand, is now the most common land bird in New Zealand, having settled the outlying islands as well. Finally, it is now well established that most, if not all, of the birds of the Australian region got there by island-hopping across the Malay Archipelago. The result has been a rich fauna of successful species, genera, and even families. All of them must have gone at one stage or repeatedly through a genetically extremely depauperized condition. The adaptive radiation of birds, insects, and other animals on isolated archipelagoes, like the Galápagos Islands or Hawaii, is further proof. It is thus evident that populations can get through such a bottleneck and still become the progenitors of successful stocks.

Not enough quantitative analyses of species structure have been made to determine the average number of peripherally isolated incipient species in various groups of animals. Many species have none; others have five or six. Since most of the parental species have—speaking in terms of geological time—a long life expectancy, there is very little opportunity for replacement, unless a population succeeds in entering a novel ecological niche. Thus, in spite of the continuous budding off of peripheral populations undergoing major or minor "genetic revolutions" as described above, only few will play a role in long-term evolution, perhaps 1 in 50. The odds are very much against successfully passing through the bottleneck of reduced variability (Figure 15-3) as well as reaching a new level of high variability and an unoccupied ecological niche.

PERIPHERAL POPULATIONS AND MACROEVOLUTION

The peripherally isolated population has various attributes that are of great interest not only to the student of speciation but also to those who study major evolutionary changes. It seems to me that many puzzling phenomena, particularly those that concern paleontologists, are elucidated by a consideration of these populations. These phenomena include unequal (and particularly very rapid) evolutionary rates, breaks in evolutionary sequences and apparent saltations, and finally the origin of new "types."

Evolutionary Rates

It is now realized that the rate of evolutionary change is closely correlated with population structure. Evolutionary transformation would tend to be very slow in large panmictic populations, if such should exist in nature. A more rapid rate of evolutionary change will be induced by a different population structure. "A fine-scaled structure of partial isolation without marked environmental differences presents the most favourable condition for transformation as a single species" (Sewall Wright 1951, summarizing earlier work). It seems to me, however, that even such a system is comparatively slow and conservative because it is rich in diverse genetic factors, including multiple alleles and that, as Heuts (1951) points out, there will be a premium for constancy in such a system. Genetic factors will be selected for their ability to form viable combinations with the greatest number of other genetic factors.

It is very doubtful whether the population structure cited by Sewall Wright is favorable enough to explain the sudden, sometimes almost precipitous, changes of evolutionary rates that have been so puzzling to paleontologists. The (relative!) suddenness of these changes is unquestionable, and even Simpson, who on the whole is successful in explaining evolution within the framework of the current genetic theory, found it necessary to coin a special term, quantum evolution, for this type of rapid evolutionary change.

Two kinds of explanations for rapid evolution have most often been given previously, a genetic and an ecological one, but both are unconvincing. Previous genetic interpretations are based on the occurrence of macromutations ("systemic mutations") or on mutational avalanches. Either type of event would have exceedingly slim chances of success in a population that is part of a connected system of populations with undiminished gene flow. Equally unlikely is an ecological explanation based exclusively on a cataclysmic change of selective factors. Mountain building, shifts of climatic belts, and similar events are far too slow to account for "quantum evolution."

The genetic reorganization of peripherally isolated populations, on the other hand, does permit evolutionary changes that are many times more rapid than the changes within populations that are part of a continuous system. Here then is a mechanism that would permit the rapid emergence of macroevolutionary novelties without any conflict with the observed facts of genetics.

Phylogenetic Saltations

Many paleontologists have postulated various kinds of typostrophic "saltations" in order to explain the absence of crucial steps from the fossil record. If these changes have taken place in small peripherally isolated populations, it would explain why they are not found by paleontologists. In fact, peripheral populations are neglected even by the taxonomists of most living faunas because they are comparatively small, isolated, and often in far distant or inaccessible places. In birds, however, where such populations have been well studied, it is quite evident that they are not only incipient species but in many instances also incipient genera and higher categories. It has been pointed out earlier (Mayr 1942; Rensch 1947) that the problem of the origin of higher categories is inseparable from the problem of the origin of new species. Those who have denied this seem to be unfamiliar with the facts. It is, of course, inadmissible to apply the term speciation to subspeciation within continuously ranging populations and use this as evidence for denying that speciation could have anything to do with the origin of higher categories. As stated above, the clinal variation within continuous populations is not of the type that normally leads to the origin of major evolutionary novelties. Such are found only in isolated allopatric peripheral populations.

The drongo species *Dicrurus hottentottus* serves as an example. Though we are dealing here merely with inconsequential plumage features (Figure 15-4), it is noteworthy not only that every aberrant population is peripheral but also that several of these populations have been considered by avian taxonomists to be generically distinct. As stated above, most of these populations will eventually die out without playing a major role in the evolutionary picture. Only a very occasional one will be able to reach a vacant ecological zone. This conclusion agrees with the observed evidence, since the number of real evolutionary inventions ("new types") in the history of the earth is quite small in comparison with the total number of forms occurring at any one time.

As soon as such a population has completed its genetic reconstruction and ecological transformation, it is ready to break out of its isolation and invade new areas. Only then will it become widespread and thus likely to be found in the fossil record. But then it is already too late to record the evolutionary change through which it has gone. All the paleontologist finds is the fact that one widespread numerous species was replaced or succeeded by a rather different species that is also widespread and numerous. Contrary to the belief of many paleon-

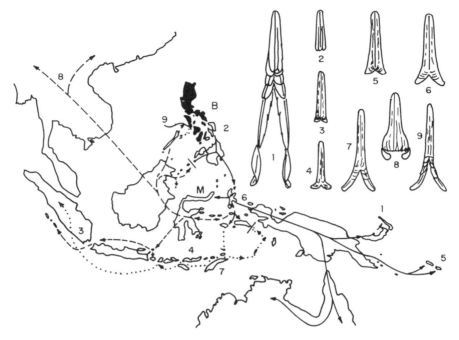

Figure 15-4. Geographic variation of the form of the tail in the polytypic drongo species *Dicrurus hottentottus.* The central populations, such as those on Sumbawa (*4*) and Halmahera (*6*), have a normal drongo tail. Nearly all peripherally isolated populations have a more or less aberrant tail, such as those on Samar (*2*), Sumatra (*3*), San Cristobal (*5*), the Asiatic mainland (*8*), Timor (*7*), Tablas, Philippines (*9*), and New Ireland (*1*). The populations of Asia (*8*), New Ireland (*1*), Tablas (*9*), and Kei, near Timor (*7*), were once considered generically distinct. (From Mayr and Vaurie 1948.)

tologists, such an apparent "jump" is consistent with the accepted genetic theory, as detailed above.

Ecological Shifts

Oceanic archipelagoes are the best place to observe the results of sudden ecological shifts of populations in living faunas. A particularly convincing case has been described by Amadon (1947) in a Hawaiian honeycreeper. Here is an allopatric population that has undergone a change not only in bill structure but also in feeding habits. It has truly switched into an entirely new ecological niche. I visualize all major evolutionary novelties as occurring in a similar manner. E. Zimmerman has described the case of a species of dragonflies isolated on the Hawaiian Islands in which the larvae have entered a totally new niche. They no longer live in water like the larvae of all other dragonflies, but in the moist humus and plant debris on the forest floor. They could well become the progenitor of an entirely new type of insect.

SUMMARY

Successful species are usually widespread and rich in genetic variability, but they tend to be rather conservative from the evolutionary point of view. Populations within such species display much ecotypic adaptation and clinal variation.

Along the periphery of the geographical range of such species there is frequent budding off of geographically isolated populations.

The genetic composition of a population that is one of a large series of contiguous populations of widespread species is continuously affected by the immigration of genes derived from adjacent or far distant populations.

In such a population there will be a selective premium on genes that do well on a great variety of genetic backgrounds and that are thus adapted to cope with the continuous influx of alien genes.

The selective value of many genes will change drastically on the altered genetic background ("genetic environment") of a newly founded peripherally isolated population. This will lead to a rapid change of gene frequencies simultaneously at many loci ("genetic revolution"), assisted by the selective effects of the change in the physical and biotic environment of the isolated area.

Furthermore, a different set of genes is apt to be superior in an area in which gene flow does not interfere with selection by the local environment.

Isolated populations are relatively invariable genetically for various reasons, and appear to become vulnerable to extinction, particularly if they live for long periods in a very uniform environment.

An occasional one of such populations may, during the period of genetic reorganization, succeed in entering a previously unoccupied ecological niche and in expanding into this niche. As such a population becomes more and more numerous, it again builds up its previously depleted genetic variability.

The period of genetic reorganization and relaxed selection pressure not only is a period permitting rapid evolutionary change, but also offers an otherwise unavailable opportunity for a drastic ecological change of a somewhat unbalanced genetic system.

Rapidly evolving peripherally isolated populations may be the place of origin of many evolutionary novelties. Their isolation and comparatively small size may explain phenomena of rapid evolution and lack of documentation in the fossil record, hitherto puzzling to the paleontologist.

REFERENCES

Amadon, D. 1947. Ecology and the evolution of some Hawaiian birds. *Evolution*, 1:63–68.

Bateman, A. J. 1950. Is gene dispersion normal? *Heredity*, 4:353–363.

Dice, L. A., and P. M. Blossom. 1937. Studies of mammalian ecology in southwestern North America, etc. *Carnegie Inst. Wash. Publ.*, 485:1–129.

Dobzhansky, T. 1950. Origin of heterosis through natural selection in populations of *Drosophila pseudoobscura*. *Genetics*, 35:288–302.

Dobzhansky, T., and H. Levene. 1951. Development of heterosis through natural selection in experimental populations of *Drosophila pseudoobscura*. *Amer. Nat.*, 85:247–264.

Dobzhansky, T., and Sewall Wright. 1947. Rate of diffusion of a mutant gene through a population of *Drosophila pseudoobscura*. *Genetics*, 32:303–324.

Ford, H. D., and E. B. Ford. 1930. Fluctuation in numbers and its influence on variation in *Melitaea aurinia*. *Trans. Roy. Ent. Soc. London*, 78:345.

Goldschmidt, R. 1940. *The material basis of evolution*. Yale University Press, New Haven.

Hayne, D. W. 1950. Reliability of laboratory-bred stocks as samples of wild populations, as shown in a study of the variation of *Peromyscus polionotus* in parts of Florida and Alabama. *Contr. Lab. Vert. Biol., Univ. Mich.*, 46:1–56.

Heuts, M. J. 1951. Les théories de l'évolution devant les données expérimentales. *Rev. Quest. Sci.*, 122:58–59.

Hooper, E. T. 1941. Mammals of the lava fields and adjoining areas in Valencia County, New Mexico. *Misc. Publ. Mus. Zool., Univ. Mich.*, no. 51:1–47.

Mayr, E. 1942. *Systematics and the origin of species*. Columbia University Press, New York.

—— 1970. *Populations, species, and evolution*. Belknap Press of Harvard University Press, Cambridge, Mass.

Muller, H. J. 1948. Evidence of the precision of genetic adaptation. *Harvey Lectures*, ser. 43(1947–1948):165–229.

Peus, F. 1950. Der Formenkreis des *Ctenophthalmus argytes* Heller (Insecta, Aphaniptera). In *Syllegomena Biologica* (Festschrift Kleinschmidt), pp. 286–318. Leipzig A. Ziemsen Verlag, Wittenberg.

Rensch, B. 1947. *Neure Probleme der Abstammungslehre. Die transspezifische Evolution*. Ferdinand Enke, Stuttgart.

Schindewolf, O. H. 1936. *Paläontologie, Entwicklungslehre und Genetik*. Borntraeger, Berlin.

Selander, R. K., W. G. Hunt, and S. Y. Yang. 1969. Protein polymorphism and genic heterozygosity in two European subspecies of the house mouse. *Evolution*, 23(3):379–390.

Spencer, W. P. 1944. Iso-alleles at the bobbed locus in *Drosophila hydei* populations. *Genetics*, 29:520–536.

Stebbins, G. L. 1950. *Variation and evolution in plants*. Columbia University Press, New York.

Wallace, B., and J. King. 1951. Genetic changes in populations under irradiation. *Amer. Nat.*, 85:209–222.

Wright, Sewall. 1931. Evolution in Mendelian populations. *Genetics*, 16:97–159.

—— 1951. The genetical structure of species. *Ann. Eugenics*, 15:323–354.

16
Geographical Character Gradients and Climatic Adaptation

The recent discussion by Scholander (1955) of the so-called ecological rules indicates a number of misunderstandings. It might be worthwhile to scrutinize the meaning and validity of these rules in some detail, since some of these misunderstandings seem to be based on widespread misconceptions, to judge from discussions with various biologists. Before the discussion a word should be said on terminology. There is no good, generally accepted collective term available for such rules as Bergmann's rule, Allen's rule, etc. The terms "climatic rules" and "ecological rules" have been used, both being rather broader than justified by the phenomena they describe. The term "ecogeographical rules," although by no means ideal, will be used in the subsequent discussions, being less inclusive than the term "ecological rules."

What do the ecogeographical rules signify? They are purely empirical generalizations describing parallelisms between morphological variation and physiogeographic features. For instance, Bergmann's rule states that "Races of warm-blooded vertebrates from cooler climates tend to be larger than races of the same species from warmer climates." As Rensch has emphasized consistently, this is a purely empirical finding that can be proven or disproven no matter to what physiological theory one might ascribe this size trend. The validity of an ecological rule, then, depends not on the validity of the physiological interpretation, but merely on the reliability of the empirical finding. To prove that an ecological rule is invalid one would have to prove that it is not valid in the majority of relevant cases. For instance, one would have to prove that races of warm-blooded vertebrates in cooler climates do not tend to be of larger average size than races in warmer climates.

This clear separation between empirical findings and interpretation has not always been observed by some of the early authors (for example, Bergmann) or in some of the textbooks. It is therefore neces-

Reprinted and revised by permission of *Evolution* from "Geographical character gradients and climatic adaptation," *Evolution*, 10, no. 1 (1956): 105–108.

sary to emphasize once more that two independent steps are involved: first, the establishment of a regularity (and these regularities may be very different in mammals, fishes, or insects) and, second, the physiological interpretation given to this regularity.

LAWS OR RULES?

A few authors, mostly those without first-hand experience with the phenomenon, have assumed that these regularities are "laws" that are invariably true. However, right from the beginning authors like Gloger (1833) and Bergmann (1849) have stressed that these rules have only statistical validity. They are true for "most" species or "many" races. Rensch in particular has emphasized this point and has devoted a series of papers (Rensch, 1936, 1938, 1939, 1940, 1948) to a determination of the percentage of cases in which the stated regularities have validity. These rules work only "other things being equal." The principal object of Rensch's investigation was not so much to prove the climatic rules, but rather to determine the degree of their validity for different kinds of animals in different regions.

DO THE ECOGEOGRAPHICAL RULES APPLY TO SPECIES AS WELL AS TO RACES?

Bergmann and other early authors applied these rules equally to species and races, but it must be pointed out that most of their "species" are considered geographic races by modern authors. It is true that these rules are sometimes valid on the species level, and Hesse, Allee, and Schmidt (1951) have listed a number of such cases. Yet nearly all contemporary supporters of the ecogeographical rules have emphasized that they have validity only for populations within species. In 1954 (see Essay 15) I stated the general viewpoint concerning the limits in the application of the ecogeographical rules as follows: "It has long been known that the validity of the so-called ecological rules, so far as it exists at all, is restricted to intraspecific variation. . . . A more northerly species is by no means always larger than its nearest more southern relative." The fact that "cold climates do not produce a fauna tending towards large-sized globular forms with small protruding parts" (Scholander 1955) is not in the least in conflict with Bergmann's or Allen's rule.

THE PHYSIOLOGICAL INTERPRETATION OF THE ECOGEOGRAPHICAL RULES

A number of recent authors have stressed that there is a difference, in degree if not in kind, between the adaptation of a population to its local conditions and the adaptation of a new kind of animal or plant to a novel ecological niche. Adaptation to local conditions has been referred to as ecotypic adaptation (Turesson) or existential adaptation (Goldschmidt). It is quite evident that the ecogeographical rules formulate empirical findings concerning the local type of adaptation. They have never been interpreted by evolutionists as playing "a detectable role in the phylogenetic engineering of warm-blooded animals for hot and cold climates" (Scholander 1955:22). Nor have evolutionists contended that the ecological rules "reflect phylogenetic pathways of heat conserving adaptation" (1955:24). They are an ecotypic, not a phylogenetic phenomenon.

The physiological interpretation will be different for each ecogeographical rule, depending on whether it concerns size, proportions, or coloration, warm-blooded or cold-blooded animals. All interpretations, however, have in common the endeavor to establish a correlation between a character gradient and a gradient in the environment. Rensch's (1939) investigation of the increase in size in *Parus montanus* in correlation with temperature gradients may be selected as an example of this method. A first attempt to correlate the size gradient within the European range of this species with the annual isotherms was unsuccessful. However, when the isotherms of the coldest month were taken, which in Europe is January, they were found to parallel closely the size cline in *P. montanus*. Similarly, ichthyologists have established for fishes a correlation between decreasing water temperatures and a tendency toward an increase in certain meristic elements. A close parallelism between character gradients and environmental gradients, if found to have a genetic basis, can be interpreted with good reason as being the result of natural selection.

If we take specifically the case of Bergmann's rule, it is entirely consistent with the known facts to assume that the parallelism between temperature gradient and size gradient is correlated with the change in the body surface–volume ratio, which is the consequence of a simple law of geometry. Thus the explanation that the increase of size to the north serves heat conservation and that the decrease of size to the south facilitates heat dissipation is entirely consistent with all the known facts. Clearly, the hypothesis that Bergmann's rule is the result of natural selection in favor of an optimal surface to volume ratio is a legitimate one. It is axiomatic in scientific methodology that a hypothesis is considered valid until it has been disproven or until a better one has been proposed.

THE ROLE OF EXCEPTIONS

The need for heat conservation is only one of many possible selection pressures affecting absolute or relative body size. It is to be expected that under certain conditions other factors may affect geographic variation in size more strongly than the need for heat conservation. For instance, if a small mammal lives in burrows during the winter, selection pressure for heat conservation will be greatly lessened, and there may be no size increase toward the north. Indeed, since an increase in size results in an increased demand for food, size may actually decrease to the north, where food is the limiting factor. This was indeed demonstrated by Stein (1951) for the European mole (*Talpa europaea*). The phenotype of an animal is the result of a compromise between many conflicting selection pressures. Every exception to the ecogeographical rules is an indication of such a conflict. These rules are now sufficiently firmly established so that the emphasis of research should be shifted to a study of the exceptions. Snow's (1954) study of trends of geographic variation in palearctic titmice illuminates the reason for some exceptions in this family. Relative bill size, for instance, decreases rapidly northward in these titmice until a minimum size is reached. It appears that the size and the amount of food taken by these birds precludes a further reduction in bill size. This minimum size, then, is the ultimate compromise of conflicting selection pressures. Exceptions to the rules in other groups of animals should be analyzed in a similar manner.

LENGTH OF BIRD WING AND BERGMANN'S RULE

In all studies on the validity of Bergmann's rule for birds the length of the wing is used as an indication of body size. Scholander (1955:21) criticizes the use of this measurement by Rensch because "there are a few short finger bones in this measurement, the rest is just wing feathers. These measurements correlate with flight, but bear no useful thermal significance." He does not appear to realize that Rensch has nowhere treated the wing as a heat-radiating body part. The wing is used as an indication of general body size, in the absence of a better measure, since "wing load" necessitates a fairly close correlation between wing length and body weight, and the latter is an indication of general size (Amadon 1943). It is unknown and badly in need of investigation how close this correlation is. A number of studies revealing exceptions are already available. It has been pointed out by several authors, including Rensch himself, that a new variable is introduced if northerly populations of an otherwise sedentary species are migratory. In most cases this leads to an attenuation and considerable lengthening of the wing of the northern population. The better a flyer the particular species is, however, the less effect the annual migration will have on wing shape and wing size, and the fact that the populations will spend both summer and winter in a warm climate will become the dominating factor. In *Dicrurus leucophaeus*, for instance, wing length in the highly migratory race of North China (*leucogenis*) (adult males average 142.8 mm) is no greater than in the more or less sedentary race of South

China (*salangensis*) (adult males average 143.0 mm) (Mayr and Vaurie 1948).

Special uses of the wing in courtship may set up a new selection pressure that entirely upsets the need for a "normal" wing load. Adult males of the regent bower bird (*Sericulus chrysocephalus*) have a wing that averages 5 mm and a tail that averages 15 mm shorter than that of immature males (Mayr and Jennings 1952). In the African weaver, *Euplectes hordaceus,* likewise the wing is much shorter in the nuptial plumage (76.3) than in the eclipse plumage (82.7) (Verheyen 1953).

These findings are additional evidence of the multiplicity of selection pressures to which an organ is exposed and of which the final phenotype is a compromise.

DOES THE EXISTENCE OF OTHER HEAT-CONSERVING MECHANISMS INVALIDATE THE ECOGEOGRAPHICAL RULES?

Scholander's criticism of the physiological interpretation of Bergmann's and Allen's rules is based on the fact that other heat-conserving mechanisms, such as an increased density of fur or plumage, fat deposits, or vascular mechanisms, are immeasurably more efficient than the slight shifts in surface to volume ratios demonstrated by the ecological rules. The facts of physiological adaptation to life in a cold climate discovered by Irving, Scholander, and other recent investigators are of the greatest interest and can be regarded as unequivocally established. Yet they do not permit the conclusions drawn by Scholander.

The philosophy of all-or-none solutions is exceedingly widespread not only in science but in all human affairs. Unfortunately, no philosophy could be worse suited for evolutionary studies. All-or-none solutions are based on typological thinking and are alien to the facts of variation. Multiple solutions for biological needs are the general rule in evolution. An animal is protected against a predator not by speed *or* an armor *or* cryptic coloration *or* poison *or* bad taste *or* by hiding *or* by nocturnal habits, but always by a combination of several of these. Simple answers are nearly always misleading. It took ornithologists a long time before they got away from such shortsighted single-aspect statements as "birds start breeding in spring because of increasing day length," or "birds migrate in fall because of a drop in temperature." Does the fact that a thicker fur or denser plumage increases protection against the cold completely eliminate any selective advantage of an improved body surface to volume ratio? Surely not!

THE TRUE MEANING OF GEOGRAPHICAL CHARACTER GRADIENTS

The regularities in the geographic variation of general size and of proportions are a simple matter of fact, as pointed out above. Vaurie, for instance, showed that all species of drongos show variation of wing length correlated with latitude except those restricted to a single island or those that live in a single restricted climatic zone (Mayr and Vaurie 1948). Adherence to the ecological rules is, however, usually not found in the case of isolated populations (Mayr 1942), a fact that is an important clue in the solution of this problem. It is becoming increasingly evident that such ecotypic adaptations as manifest themselves in the ecological rules are the means of a local population for reaching a balance between (1) the need for adaptedness to local conditions and (2) partaking at the same time of the heritage of the species as a whole, which includes all physiological mechanisms that are species specific. A species is a single large Mendelian population pervaded in all directions by gene flow. It is this cohesive force of gene flow that is primarily responsible for the validity of the climatic rules. It gives physiological unity to a species but increases the necessity for local adjustments to local conditions. That these local adjustments, reflected in the ecological rules, have only limited effectiveness is evident from the fact that even the species which obey the rules have northern and southern limits to their geographic range.

The sensitivity of this process of local adaptation is evidence of the universality of natural selection, and this is the crucial point of the entire question. Even genes with a selective advantage of only a fraction of 1 percent tend to accumulate in populations, as shown by R. A. Fisher (1930) and others. Natural selection is particularly efficient during catastrophes and other periods of great environmental stress. Let us look at a model of the effect of such a catastrophe on the survival of two alternate mechanisms adding to the survival value. Let us assume that in a population of mammals that is at rare intervals exposed to very severe winter conditions, there are 10 individuals with a superior vascular heat-conserving mechanism and also 10 with a greatly reduced tail length among the total population of 1,000 individuals. After a severe winter the 10 individuals with the superior vascular mechanism might survive, as well as 2 with the greatly reduced tail length. After an even more severe winter only one individual might survive that happens to have the combination of the superior vascular mechanism and the reduced length of tail. Students of natural selection know that such a model is quite realistic and that genes accumulate in a population, independently of each other, in accordance with the contribution they make to fitness. Problems of selection are simply statistical problems, and the prevailing phenotype of a population will be the result of a balance between opposing selection pressures. This is the meaning of the ecogeographical rules, nothing more and nothing less.

It appears to me that there is no contradiction between Scholander's interesting findings on the heat-preserving adaptations of arctic species of birds and mammals and rules, like Bergmann's and Allen's rules, that deal with adaptation of intraspecific populations to local conditions. Neither set of factors disproves the other.

REFERENCES

Amadon, D. 1943. Bird weights as an aid in taxonomy. *Wil. Bull.,* 55:164–177.

Hesse, R., W. C. Allee, and K. P. Schmidt. 1951. *Ecological animal geography.* Wiley, New York.

Mayr, E. 1942. *Systematics and the origin of species.* Columbia University Press, New York.

Mayr, E., and K. Jennings. 1952. Geographic variation and plumages in Australian bowerbirds (Ptilonorhynchidae). *Amer. Mus. Novit.,* no. 1602:1–18.

Mayr, E., and C. Vaurie. 1948. Evolution in the family Dicruridae (birds). *Evolution,* 2:238–265.

Rensch, P. 1936. Studien über klimatische Parallelität der Merkmalsausprägung bei Vögeln und Säugern. *Arch. f. Naturgesch.,* N.F., 5:317–363.

――1938. Bestehen die Regeln klimatischer Parallelität bei der Merkmalsausprägung von homöothermen Tieren zu Recht? *Arch. f. Naturgesch.,* N.F., 7:364–389.

――1939. Klimatische Auslese von Grössenvarianten. *Arch. f. Naturgesch.,* N.F., 8:89–129.

――1940. Die ganzheitliche Auswirkung der Grössenauslese am Vogelskelett. *Jour. f. Ornith.,* 88:373–388.

――1948. Organproportionen und Körpergrösse bei Vögeln und Säugetieren. *Zool. Jb. (Physiol.),* 61:337–450.

Scholander, P. F. 1955. Evolution of climatic adaptation in homeotherms. *Evolution,* 9:15–26.

Snow, D. W. 1954. Trends in geographical variation in Palaearctic members of the genus *Parus. Evolution,* 8:19–28.

Stein, G. H. W. 1951. Populationsanalytische Untersuchungen am europäischen Maulwurf. II. Über zeitliche Grössenschwankungen. *Zool. Jb. Abt. f. Syst. Ökol. u. Geog. der Tiere,* 79:567–590.

Verheyen, R. 1953. Exploration du Parc National de l'Upemba. Mission G. F. De Witte—oiseaux. *Inst. des Parcs Nat. du Congo Belge,* 19:1–687.

III
HISTORY OF BIOLOGY

Introduction

It is rather remarkable how many current controversies in biology are a continuation of long-standing arguments. The historian of science is not the only one who may profit from investigation of the historical development of these arguments. The working biologist also may benefit greatly from studying the history of such controversies because this often helps him understand the ideological background of those who first originated the controversy and, what is equally important, it often helps to clarify semantic confusions due to subtle changes in terms.

When I lectured on evolution in different countries, I became particularly aware of how history-bound concepts and arguments are. Why was the theory of evolution by natural selection developed independently by two Englishmen and why did this theory have to overcome almost insurmountable difficulties to establish itself in continental Europe? Why was Lamarck's attempt to introduce the concept of evolution into biology so unsuccessful? Why did naturalists and experimental geneticists at first differ so radically about the interpretation of evolution and why did it take 40 years to heal the breach?

Questions like these have occupied me more and more in the last 20 years and have given rise to a series of investigations, six of which are here brought together.

17
Lamarck Revisited

As long as the battle between Darwinism and Lamarckism was raging, it was quite impossible to undertake an unbiased evaluation of Lamarck. For this we are now ready, after it has been demonstrated conclusively that the various causal explanations of evolution, usually designated as Lamarckism, are not valid. Not that it really needed this final proof, but the recognition that DNA does not directly participate in the making of the phenotype and that the phenotype, in turn, does not control the composition of the DNA represents the ultimate invalidation of all theories involving the inheritance of acquired characters. This definitive refutation of Lamarck's theory of evolutionary causation clears the air. We can now study him without bias and emotion and give him the attention that this major figure in the history of biology clearly deserves.

The past study of Lamarck (1744–1829) has tended to suffer from one-sided approaches. Sometimes he was discussed under the heading, "Lamarck and Darwin": Did he anticipate Darwin? Did Darwin build on a foundation laid by Lamarck? Did Darwin fail to give him sufficient credit? It would be futile to try to reach a well-balanced evaluation of Lamarck through answering such one-sided questions. Other authors have limited their study to a comparison of Lamarck's ideas with those of such eighteenth-century philosophers as had an interest in evolution, like Diderot, Maupertuis, or Robinet. This approach utterly fails to consider the contributions made by Lamarck's professional interests. After all, he was not primarily a philosopher, but made his living as a naturalist who studied plants and animals for more than 50 years, from 1770 (when he was 26) to the 1820s (when he was in his eighties).

Some recent historians (Gillispie 1956, 1959; Wilkie 1959; Simpson 1961, 1964) have perhaps stressed too strongly the negative aspects of Lamarck's work. They have stated, not without justification, that his theory of evolution was neither new nor valid; that he did not propose a feasible mechanism of evolution; that, as an evolutionist, he was

Adapted from "Lamarck revisited," *Journal of the History of Biology,* 5, no. 1 (1972): 55–94.

hardly ever mentioned by his contemporaries; that he seemed to have had singularly little influence on the subsequent development of evolutionary thinking; and that, when later in the nineteenth century his name was applied to evolutionary mechanisms, it was usually to ones not at all stressed by Lamarck himself.

All this is largely true, and yet it does not do justice to Lamarck as a figure in the history of biology. There are two excellent evaluations of Lamarck's work in French. That of Daudin (1926) is largely limited to Lamarck's taxonomy, a subject which I will not discuss here, since I plan to treat it elsewhere. That of Guyénot (1941:415-439) is a sympathetic and reliable introduction to Lamarck's thought (and its connections with his predecessors), but it does not pretend to be a full analysis. There is also Landrieu's (1909) extensive biography of Lamarck.

A truly penetrating study of Lamarck is still a desideratum. Eventually we will want to know far more about the growth of his thought, both before and after 1800. We will want to know to what extent his system of thought consisted of widely (at his time) accepted "truths," such as the inheritance of acquired characters, and to what extent it was based on special philosophies and beliefs, such as deism, the concept of plenitude, and various aspects of the philosophies of Leibniz, Newton, or Descartes. This will require an extensive study of the thought of the eighteenth century.

Only one of Lamarck's biological publications, the *Philosophie zoologique* (1809), was ever translated into English, and even that one not until more than 100 years after its publication (Lamarck 1914). For ease of reference I am basing my analysis almost entirely on the *Philosophie zoologique*, even though I fully realize that it is a confusing and repetitive work. One sometimes has the impression that Lamarck had merely gathered together his lecture notes over many years and combined them with portions of the *Discours d'ouverture* (Lamarck 1907). How else can one explain that the identical topic is sometimes treated on 6 or 10 different pages and his conclusions stated on these pages in virtually identical sentences? On the other hand, since no Lamarck "Notebooks" are in existence, this somewhat chaotic work may permit a better insight into the workings of Lamarck's mind than the more polished final revision that was published as the introduction to the *Histoire naturelle des animaux sans vertèbres* (1815).

The emphasis in my own treatment is on three aspects of Lamarck's work. It attempts (1) to bring out as clearly as possible what Lamarck actually thought and said; (2) to show where Lamarck got himself entangled in contradictions, in part because his findings as a zoologist were in conflict with his philosophical concepts; and (3) to point out problems that are in need of further research. In other words, this essay is a prolegomenon to the study of Lamarck rather than a comprehensive work on this great naturalist. My comments are primarily those of a biologist, not those of a historian. I have made no attempt to analyze the nonevolutionary writings of Lamarck, as for instance Parts II and III of the *Philosophie zoologique*. Greene (1959) has indicated the importance of this sector of Lamarck's work. Nor have I attempted to follow in detail the changes in Lamarck's thought throughout his long life.

One of my major objectives is to help English-speaking zoologists to become better acquainted with Lamarck's work. Extensive quotations from the *Philosophie zoologique* are provided under each subject heading,[1] and I hope that they will show the reader not only how manfully (and often futilely) Lamarck struggled with facts and concepts, but also that Lamarck was a better observer and more perceptive thinker than he is usually given credit for.

PSEUDO-LAMARCKISM

Paradoxically, and as a consequence of a long tradition of historical interpretation, one must begin a discussion of Lamarck by emphasizing what the naturalist did not say. Two explanatory principles, in particular, are consistently but quite wrongly claimed to be the cornerstones of Lamarck's thinking: (1) direct effect of the environment and (2) evolution through volition.

Direct Effect of the Environment

When one checks the literature of neo-Lamarckism to find out which concept is most consistently designated as "Lamarckian," one finds that it is the belief in a direct induction of hereditary changes in organisms by the environment. Curiously, as Simpson (1961) points out, Lamarck himself emphatically rejected the existence of such an evolutionary cause. However, since he speaks again and again of "the influence of the environment," he realizes the need for an explanation:

> I must now explain what I mean by this statement: *The environment affects the shape and organization of animals,* that is to say that when the environment becomes very different, it produces in the course of time corresponding modifications in the shape and organization of animals.
>
> It is true, if this statement were to be taken literally, I should be convicted of an error; for, whatever the environment may do, it does not work any direct modification whatever in the shape and organization of animals. (p. 107)

1. Virtually all recent treatments of Lamarck merely provide an abstract of Lamarck's ideas. This is legitimate for most purposes. The particular flavor of Lamarck's reasoning, however, is lost by the mere summarizing of his ideas. I hope that the exact page references will help future students to find their way through the *Philosophie zoologique*, which is not an easy work to read. All page references (in parentheses) after quotations from the *Philosophie zoologique* refer to the Elliot translation.

He then explains how he himself envisions the effect of the environ-ment. Lamarck believed that "alterations in the environment of an-imals lead to great alterations in their needs, and these alterations in their needs necessarily lead to others in their activity" (p. 107). This in-terpretation is repeated many times throughout the *Philosophie zoolo-gique* (pp. 40, 41, 43, 45, 58, and 114).

In spite of this emphatic rejection of direct induction for higher animals that display "activity," one finds that Lamarck speaks equally freely of a modifying influence of the environment on plants and lower (for example, sessile) invertebrates. Would this not indicate that, at least in immobile organisms, he did have some belief in direct induc-tion? And to what extent did Étienne Geoffroy St. Hilaire owe his be-lief in direct induction to Lamarck? These are still unanswered questions.

The Effect of Volition

The other explanatory principle erroneously ascribed to Lamarck is the effectiveness of volition. The popular cartoon on Lamarckism de-picts an animal, preferably a giraffe, wishing to reach an objective and through this volition growing (or acquiring) the needed structure (let us say, a longer neck). The attribution of this belief to Lamarck goes far back, for even Darwin speaks of "Lamarck nonsense of . . . adaptations from the slow willing of animals" (letter of January 11, 1844, to J. D. Hooker). Actually, as pointed out by Cannon (1957), Lamarck never said anything of the sort. The error presumably arose from a mis-translation of *besoin* (need). Lyell scolded Darwin (in a letter of Octo-ber 3, 1859) for ascribing to Lamarck a theory of volition. And yet Lyell himself may have inadvertently been one of the causes of Darwin's error. In *Principles of Geology* he consistently translates Lamarck's *besoin* as the "wants" (noun) of animals. Huxley in his famous *Times* review of the *Origin of Species* (December 1859) says that according to Lamarck "the new needs will create new desires, and the attempt to gratify such desires will result in an appropriate modification." There is great danger that the hurried reader will only remember the word "desire."

LAMARCK'S EVOLUTIONISM[2]

To a modern, the word *evolution* has only a single meaning, the the-ory of common descent. In the eighteenth and early nineteenth cen-turies when essentialism (see Essay 3) was still the dominant philosophy, the word *evolution,* if used at all, was exclusively employed in the con-text of embryological phenomena, to denote the unfolding of immanent

---

2. Lamarck, so far as I know, never used the terms "evolution," "evolutionism," or "trans-formationism" later employed in the literature of English or French evolutionary biology. The term "evolutionary" is used in this essay in its modern connotation.

qualities. Ordinarily it would be applied to the embryogenesis of an individual, but increasingly often it was employed, in a somewhat metaphorical manner, to designate the "evolution" of the type. For instance, when Louis Agassiz spoke of evolution (see Essay 18), it was usually this kind of process he had in mind. It is difficult for the contemporary scientist to think in terms of eighteenth-century concepts, but we may try to describe eighteenth-century directionism as an endeavor to fuse the concept of a temporalized "scala naturae" (Lovejoy 1936) with that of the unfolding ("evolutio") of the immanent potentialities of an archetype. This postulates that the *appearance* of the type would change, but not its underlying essence. It would be, so to speak, a static evolutionism.

Some of the more extreme critics of Lamarck have implied that Lamarck's ideas on evolution were of this eighteenth-century vintage and that he did not truly propose a genuine evolutionary change, or, as we would now say, a genetic transformation. Are these critics right? Did Lamarck think only in terms of an improvement in the appearance of a fixed type without a change in its essence, or did he indeed support the idea of a genuine evolutionary change? What do Lamarck's own words tell us about this assertion?

There are numerous passages in Lamarck's writings that one might want to quote, but none of them documents the genuineness of Lamarck's evolutionary thinking better than his discussion of aquatic animals in the *Philosophie zoologique*:

> I do not doubt that . . . water is the true cradle of the entire animal kingdom.
> We still see, in fact, that the least perfect animals, and they are the most numerous, live only in water . . . that it is exclusively in water or very moist places that nature achieved and still achieves in favorable conditions those direct or spontaneous generations which bring into existence the most simple organized animalcules, whence all other animals have sprung in turn. (pp. 175-176)

> . . . After a long succession of generations these individuals, originally belonging to one species, become at length transformed into a new species distinct from the first. (pp. 38-39)

The principle of an evolutionary transformation through time could not be stated more clearly than that!

Historians concerned with the 200-year period prior to 1859 tend to go to one extreme or the other. Either they interpret any deviation from a strictly static description of the world as an anticipation of Darwin or else they regard it as a purely philosophical speculation that has nothing to do with true evolutionary thinking. The truth lies somewhere between these two extremes. As the exploration of the world accelerated in the seventeenth and eighteenth centuries and as, following Leeuwen-hoek, the study of small aquatic organisms and microorganisms became ever more popular, the Great Chain of Being (Lovejoy) acquired in-creasing substance. And with this came a belief in continuing change. After Leibniz (1693) had said that "even the species of animals have many times been transformed," it was stated more and more boldly by various eighteenth-century authors (as, for instance, by Maupertuis, Diderot, Robinet, and Bonnet) that the entire organic world is the product of evolution (see Lovejoy 1936:256-286). Most of these au-thors were philosophers with only a limited knowledge and under-standing of the world of organic beings, and their pronouncements were made rather casually and with little attempt at substantiation.

However, this cannot be said of Buffon, the foremost naturalist of his day. He might have based his entire *Histoire naturelle* on a theory of descent if he had really believed in it, but he did not. Even though Guyénot (1941:401) is convinced that Buffon must be considered "not merely a forerunner, but, indeed, a veritable founder of the theory of evolution," Lovejoy (1959) cites many good reasons why Buffon's statements must be treated with caution. They might just as well, or even more so, reflect his philosophical ideas and, in particular, his ad-herence to the principle of plenitude.

It seems to me that Lamarck has a much better claim to be designated "the founder of the theory of evolution," as indeed he has been by sev-eral French historians (such as Landrieu 1909). All others before him had discussed evolution *en passant* and incidentally to other subjects or else in poetical or metaphorical terms. He was the first author to devote an entire book primarily to the presentation of a theory of organic evo-lution. He was the first to present the entire system of animals as a prod-uct of evolution.

THE EVIDENCE FOR EVOLUTION

A specific set of observations gave rise to the evolutionary thinking of each of the founders of evolutionism. In the case of Lamarck the de-cisive observation was the perfect correlation between the structural adaptations of organisms and their environment. This was not, of course, a new observation since the very same evidence had induced the natural theologians to extol the wisdom and foresight of the Creator in creating only perfectly adapted beings. Natural theology, however, was based on the belief in an essentially static world. This dogma Lamarck was unable to accept, owing to his far better understanding of the con-tinuing geological and climatic changes on the surface of the earth and of its great age. To him, the geological, climatic, and geographic evidence indicated continuous, dynamic, and sometimes rather severe changes in the physical environment of the world. If there had been only a single organic creation and yet all the mentioned subsequent changes of the physical environment, then all organisms ought to be by now very poorly adapted. However, since we find that every organism is perfectly adapted to the particular environment in which it occurs, in its structural modifications as well as in its habits, it is obvious that organisms, in order to survive, must have the capacity for change. Anyone who wants to disprove the validity of this argument, says Lamarck (p. 127), must prove two points: (1) "that no point on the surface of the earth ever undergoes variation as to its nature, exposure, high or low situation, climate, etc."; and (2) "that no part of animals undergoes . . . any modification due to a change of environment or to the necessity which forces them into a different kind of life and activity from what has been customary to them."

I believe that the cogency of this argument of Lamarck has not been adequately stressed in the past. Organisms, in order to remain perfectly adapted at all times, have to adjust to these changes; they have to evolve (p. 106). This leads Lamarck to "the conviction that according as changes occur in the environment, situation, climate, food, habits of life, etc., corresponding changes in the animals likewise occur in size, shape, proportions of the parts, color, consistency, swiftness and skill" (p. 109).

There is a great deal of vagueness in Lamarck's discussion of this problem, perhaps largely owing to his lack of interest in species. The changes in organisms, which he describes on p. 109, must affect representatives of various species. Is it possible that these modifications of individual species will be compounded in time until they affect the very organization of the classes to which they belong? (This, indeed, as we now know, is the case). If not, what are the "units" of the classes whose improvement leads in time to the origin of higher types of organization?

The need for having to remain adapted to the constantly changing environment is for Lamarck the most powerful causal factor for evolution. It would have seemed an easy matter for Lamarck to supplement this purely inferential deduction by citing chapter and verse for such effects of the environment. Geographic variation would have been the most obvious evidence in favor of the ability of organisms to adapt. Lamarck was fully aware of the adaptation of "races" to a particular locality owing to the influence of the environment:

> Localities differ as to their character and quality, by reason of their position, construction and climate: as is readily perceived on passing through various localities distinguished by special qualities; this is one cause of variation for animals and plants living in these various places. But what is not known so well and indeed what is not generally believed, is that every locality itself changes in time as to exposure, climate, character and quality, although with such extreme slowness, according to our notions, that we ascribe to it complete stability. (p.111)

It would have seemed logical to extend this to geographically vicarious species, as Buffon had done in his comparison of certain elements of the European and North American faunas. Indeed, throughout Buffon's writings there are scattered references to the effects of geography on organisms. They stress the effects of climate rather than those of isolation, but they do stress geography. Zimmermann's volume on biogeography, Pallas's frequent comments on the subject, and particularly those of Alexander von Humboldt, culminating in his famous essay on plant geography (1805), all document the increasing preoccupation of naturalists with problems of geography at this period and show how much by 1809 this subject was "in the air." However, this aroused no response in Lamarck; perhaps the fact that he had traveled so little was responsible for his lack of interest. In this he differs strikingly from the Darwinian generation. Geography, particularly the distribution of closely related forms in space, played a decisive role in the evolutionary thinking of Darwin, Wallace, Asa Gray, Moritz Wagner, Gulick, Karl Jordan, and other leading evolutionists from 1850 on. This permitted the careful, scientific analysis of innumerable "experiments of nature" and helped, more than anything else, to remove the evolution theory from the realm of pure speculation. Lamarck's lack of interest in concrete species deprived him of the opportunity to take full advantage of the already available geographical evidence.

SPECIES AND THEIR ORIGIN

The species was the key unit in Darwin's theory of evolution. For Lamarck, whose primary emphasis was on levels of complexity in the organization of animals, the species played a subordinate role. His *Philosophie zoologique* was not an *Origin of Species*. While still a botanist, Lamarck—presumably under the influence of Linnaeus—believed in the constancy of species, and he dealt with them as if they were well delimited. After he became an invertebrate taxonomist, he still treated them in a thoroughly conventional manner in his technical taxonomic treatises, but no longer in his more philosophical discussions. Here he stated that species are not constant; they "have really only a constancy relative to the duration of the conditions in which are placed the individuals composing it" (p. 36). He therefore finds no conflict between his concept of the impermanence of species and the fact that the mummified animals found by the French expeditions to Egypt were indistinguishable from the living ones, because "the position and climate of Egypt are still very nearly what they were" 3 or 4 thousand years ago. Hence there was no cause for them to change (see below for Lamarck's thoughts about time).

His species definition is the logical consequence of these views: "It is useful to give the name of species to any collection of like individuals perpetuated by reproduction without change, so long as their environment does not alter enough to cause variation in their habits, character, and shape" (p. 44). The concept characterized by this definition does not entirely fit into any scheme. Clearly, it is closest to the nominalistic concept in its emphasis on the existence only of individuals and in the incomplete, if not arbitrary, delimitation of species against each other. However, Lamarck does not have the concept of variable populations. All the individuals that are forced by their environment to adopt similar habits will be essentially identical. Even though he does not accept the existence of an underlying essence, his species concept becomes curiously similar to that of the essentialists when he accepts the uniformity of "classes of individuals." This is particularly puzzling since, as a working taxonomist, he must have been aware of the phenomenon of individual variability. And this phenomenon was the keystone in Darwin's evolution through natural selection. Selection, as we know from the researches of Zirkle (1941) and others, was by no means an unknown concept by 1800, but it did not fit at all into Lamarck's thinking. Furthermore, since individuals will change their habits whenever their environment (their "circumstances") changes, adaptation through change of habit achieved for Lamarck all that Darwin later ascribed to the effect of natural selection.

It is remarkable how little Lamarck says about the multiplication of species,[3] considering how many years he had been working on groups of closely related species and their distribution. For him "development of new species" was simply phyletic evolution. Individuals exposed to a changed environment "after a long succession of generations . . . originally belonging to one species, become at length transformed into a new species distinct from the first" (p. 39). For the origin of new species Lamarck, therefore, does not require saltations, as does the essentialist, nor isolation combined with natural selection, as does the population thinker. Simple adaptation does the job, as far as Lamarck is concerned. Any multiplication of species would have to be sympatric speciation by ecological specialization. He describes this process as follows (p. 39): A "plant that grows normally in a damp meadow" sends out colonists into dryer and dryer terrain until "it reaches little by little the dry and almost barren ground of a mountainside. If the plant succeeds in living there and perpetuating itself for a number of generations, it will have become so altered that botanists who come across it will erect it into a separate species. The same thing happens in the case of animals that are forced by circumstances to change their climate, habits, and manner of life."

3. The term "origin of species" confounds two biologically distinct phenomena, the gradual transformation of a phyletic line into a different species and the splitting of a single species into several daughter species. It is the explanation of the latter process, also referred to as *multiplication of species* or *speciation*, which caused the early evolutionists the greatest trouble. Essentialists and population thinkers were forced to propose entirely different explanatory mechanisms (see Mayr 1970).

Lamarck, in these discussions, concentrates exclusively on two aspects of the new species, their differences ("so altered") and their occupation of a new niche (changed "habits and manner of life"). He nowhere mentions the key aspect of the modern biological species concept, the formation of a new, reproductively isolated community. Naturalists did not become conscious of the importance of this aspect of the biological species until more than 50 years later.

As an alternate mechanism Lamarck suggests, following in the footsteps of Linnaeus, that new species might originate by hybridization. Such hybridization he says will "gradually create varieties, which then become races, and in course of time constitute what we call species." Lamarck uses similar arguments to explain the origin of breeds and domestic races. This discussion is particularly interesting because Darwin uses the very same evidence as one of his major arguments in favor of the importance of natural selection. Lamarck in his discussions (pp. 110-111) does not mention the breeders' selecting activities, but ascribes the origin of the differences to the environment.

Lamarck's view that the species is largely an arbitrary aggregate of individuals exposed to "the same circumstances" and not sharply separated from other species was presumably strongly influenced by his studies of molluskan variability. There is perhaps no other group of animals with so conspicuous a variability as the land and freshwater mollusks. Indeed, they often show three superimposed kinds of variability: (1) genetic polymorphism, as in the European banded snails (*Cepaea*), where many different color types occur in a single population; (2) nongenetic modifiability, as in freshwater mussels (*Anodonta,* etc.), in which the species responds at each locality to the constellation of chemical and physical water conditions by producing a different phenotype, without necessarily changing genetically; and (3) genetically founded geographic variation, which occurs in extreme form in many land snails, particularly in the Mediterranean region and on archipelagoes. Although modern studies have shown that this variability rarely affects the sharpness of the borders between species, this was not at all evident in Lamarck's day, particularly to an author with nominalist tendencies. This leads him to describe the variation as follows: "some of these individuals have varied, and constitute races which shade gradually into some other neighboring species. Hence, naturalists come to arbitrary decisions about individuals observed in various countries and diverse conditions, sometimes calling them varieties and sometimes species" (p. 36). As the amount of material in our museums grows, "we see nearly all the gaps [between species] filled up and the lines of demarcation effaced. We find ourselves reduced to an arbitrary decision which sometimes leads us to take the smallest differences of varieties and erect them into what we call species, and sometimes leads us to describe as a variety of some species slightly different individuals which others regard as constituting a separate species" (p. 37). "These species merge more or less into one another so that there is no means of stating the small differences that distinguish them" (p. 37). On the next page he refers to some of the well-known large genera of insects and states, "these genera alone possess so many species which merge indefinably into one another" (p. 38).

In these discussions Lamarck makes no distinction between, on the one hand, individual variation or the formation of local ecotypes (neither of which normally leads to speciation) and, on the other hand, geographic variation (which—combined with isolation—may lead to speciation). Lamarck was fully aware of geographic variation. "When the observing naturalist travels over large portions of the earth's surface and sees conspicuous changes occurring in the environment, he invariably finds that the characters of species undergo a corresponding change" (p. 112). Such geographic variation is described in even more detail in his *Natural History of Invertebrates* (1815), but it is nowhere contrasted with individual variation. What is missing, in particular, is any reference to the effects of isolation. The closest Lamarck comes to this is in his description of the domestication of the dog:

> No doubt a single, original race, closely resembling the wolf, if indeed it was not actually the wolf, was at some period reduced by man to domestication. That race, of which all the individuals were then alike, was gradually scattered with man into different countries and climates; and after they had been subjected for some time to the influences of their environment and of the various habits which had been forced upon them in each country, they underwent remarkable alterations and formed various special races. (pp. 110–111)

And he continues that the various races of dogs that are now so extremely different were "formed in very distant countries." There is an intimation of the role of isolation in various of these statements, but it is nowhere clearly identified as an important factor.

Evidently not realizing that species are reproductively isolated populations and believing that the characteristics of organisms are (indirectly) the result of the environment, Lamarck could not appreciate the importance of isolation for speciation. Nor would it be important according to his theory because when a race returned to the place from which it had originated, it would have to return to its original appearance, owing to its being again exposed to the ancestral environment. Since each species is the product of its environment and since all environments merge into each other, it is only natural that all species should intergrade with each other (pp. 37, 112).

Much of what Lamarck says about intermediates between "species" is quite correct when we deal with allopatric populations. This is why the modern taxonomist so often combines groups of such allopatric "morphological species" into widespread polytypic species. What is amazing, however, is the fact that Lamarck was so totally blind to the drastic difference between the situation he described for geographically variable polytypic species and the sharp, bridgeless gaps of coexisting, sympatric species, such as the working naturalist encounters everywhere. These he never mentions nor does he anywhere refer to the isolating mechanisms that help to maintain these gaps. Since an explanation of the origin of isolating mechanisms is the very essence of an explanation of the origin of species, it is obvious how distant Lamarck was from the solution of the problem of speciation. Indeed, his entire conceptual framework was ill adapted to explain the occurrence of concrete units such as species and their origin. Though he speaks so much of "branches" among evolutionary lines, Lamarck failed to notice that evolutionary divergence, that is, the origin of such branches, must originate at the species level. It is rather evident that he was simply not thinking in terms of species, and least of all in terms of species as biological populations.

THE CAUSES OF EVOLUTIONARY CHANGE

The principal reason Lamarck failed to make more of an impression on his contemporaries and successors was that his explanatory principles were not convincing (for a detailed analysis of the reception of Lamarck's ideas by his contemporaries, see Burkhardt 1970). There were primarily two processes whose occurrence Lamarck postulated in order to account for evolutionary changes: (1) evolution toward perfection, and (2) branching evolution and adaptive radiation. However, he never supplied any proof for either. Let us consider these processes.

Evolution toward Perfection

Lamarck took it completely for granted that all classes of animals form a unique and graduated series from the simplest to the most perfect. As Daudin points out quite correctly (1926:vol. 2:111), this is "his central thesis, his master doctrine on which he insists in his lecture course and books more than on any other." According to Lamarck (p. 60), it is not difficult to place all major types of animals in a linear series based on their "affinities." Furthermore, "if one of the extremities of this series is occupied by the most perfect of living bodies, having the most complex organization, the other extremity of the order must necessarily be occupied by the most imperfect of living bodies, namely those whose organization is the simplest." Similar statements are found in the introduction (Discours Preliminaire) of his *Flore françoise*, 1778 (1779). Since the evolutionary change led from the simplest to the most complex, it is necessary in a classification to adopt the same sequence. Lamarck chides Aristotle for having chosen a descending classification: "This classification furnishes the earliest example of an arrangement, though in the opposite direction from the order of nature" (p. 62). However, he fails to notice that in Chapter V, 15, of the *Historia animalium,* when discussing the genesis of animals, Aristotle does indeed start with the simplest organisms and proceeds methodically through the animal kingdom up to the live-bearing ones.

Lamarck defends this linear sequence even though he is fully aware of Linnaeus' maplike, two-dimensional arrangement and, at least in the *Histoire naturelle* (1815-1822), of Cuvier's denial of any relationship whatsoever among the four major "embranchements" of the animal kingdom (Cuvier 1812). He particularly rejects the suggestion that the relationships of families could be presented in the form of a network (p. 58). "In each kingdom of living bodies the groups are arranged in a single graduated series, in conformity with the increasing complexity of organization and the affinities of the objects"[4] (p. 59). Instead of "complexity" he calls his principal classifying criterion sometimes "ever greater perfection," but rarely specifies by what standards he determines "perfection." When he finally does so, he adopts the traditional criterion of the *scala naturae* (p. 71): "Man . . . presents the type of the highest perfection that nature could attain to: Hence the more an animal organization approaches his, the more perfect it is." It is as simple as that!

The *scala naturae* is basically a static concept. When it is combined with the concept of evolutionism, it leads to conflicts and contradictions that Lamarck does not resolve—indeed, that he does not even seem to have been aware of. For instance, combining the image of a staircase and upward evolutionary movement leads almost by necessity to the simile of an "escalator," and this is how Gillispie (1956, 1959) quite rightly characterizes Lamarck's *scala naturae.* Yet, how does such an escalator operate when it involves only the "masses" (= major taxa), while the development of the species within each of these major taxa is regulated by "circumstances"? If one were to take the escalator simile seriously, one would have to conclude that every kind of organism moves upward to ever-greater perfection. Since the supply of the simpler organisms is continuously depleted by this process (they move up the scale and become more complex), their number has to be replenished continuously by spontaneous generation. The simplest organisms thus are the ones most recently generated, and an arrangement of organisms based on their complexity will result in a classification that will "lead us to a knowledge of the order followed by nature in bringing the various species into existence" (p. 9). When classifying, naturalists "are obliged to follow the actual order observed by nature in giving birth to her productions" (p. 29). Indeed, this is the order Lamarck adopted in Chapter VIII (pp. 128-169), entitled "of the natural order of animals,

---

4. Notice that only the major types ("groups," classes) are arranged in such an ascending sequence, not individual species (see below).

and the way in which their classification should be drawn up so as to be in conformity with the actual order of nature."

The steady upward movement of the major animal groups ("masses"), if this is indeed what Lamarck postulates, creates a conflict with certain consequences of the principle of plenitude. Either all groups of organisms move upward at the same rate, or else some "masses" move faster than others and cause the production of gaps, the existence of which Lamarck denies steadfastly (pp. 23, 33, 37, 57, 66). Yet, when speaking of man's continuing evolution, Lamarck says: "This predominant race, having acquired an absolute supremacy over all the rest, will ultimately establish a difference between itself and the most perfect animals, and, indeed, *will leave them far behind*" (p. 171; italics mine). He does not attempt to reconcile such a gap with the principle of plenitude.

When trying to account for the causes responsible for the movement of the escalator, Lamarck abandons the firm ground of experience and becomes philosophical, speculating on forces, movements, and subtle fluids. The upward movement of the scale of complexities, he says on one occasion, is simply due to the "power [pouvoir] of life." "That says no more than that the main movement is an inherent characteristic of life, which certainly explains nothing" (Simpson 1964:48).[5] Certainly not to a modern biologist. The analysis of Lamarck's theories on life and vital processes is outside the scope of the present essay; it will be dealt with in a forthcoming publication (Burkhardt in press).

In this context one should also remember that Lamarck was a deist and that this colors some of his explanatory models. He evidently believed in a "Sublime Author" or "Supreme Being," words he uses not infrequently and with apparent sincerity in the *Philosophie zoologique*. Lamarck believed in "creation" but not in a simplistic, once-and-for-all creation that would result in a static world. Ever since St. Augustine there had been two schools among the Christians with respect to creation. Lamarck evidently accepted the concept of continuing creation, later espoused by Darwin's friend Asa Gray. Consequently, he considered it rather absurd to think that the Sublime Author should have to look after each detail separately. "Could not his infinite power create an order of things which gave existence successively to all that we see as well as to all that exists but that we do not see?" "Shall I admire the greatness of the power of this first cause of everything any the less if it has pleased him that things should be so, than if his will by separate acts had occupied itself and still continued to occupy itself with the details of all the special creations, variations, developments, destructions and renewals, in short, with all the mutations which take place at large among existing things?" (p. 41). Nature can bring into existence an apparent order among organized beings "from powers conferred by the Supreme Author of all things" (p. 60; a similar statement

5. See Daudin (1926:200) on the pouvoir de la nature.

is made on p. 130). When evolution is thus interpreted as continuing creation, it can be reconciled with the belief in a supreme being.

Spontaneous Generation

One consequence of Lamarck's conception of evolution as the movement of organisms on an endless escalator is that the lower reaches of this escalator would be gradually vacated by this upward movement unless there is a steady replenishment at the lowest level. This, Lamarck says, indeed takes place through steady spontaneous generation, a phenomenon he considers so important that he devotes to it an entire chapter (Chapter VI of Part II). Spontaneous generation, according to Lamarck, does not have the power to produce higher organisms: "It is exclusively among the infusorians that nature appears to carry out direct or spontaneous generations, which are incessantly renewed whenever conditions are favorable; and we shall endeavor to show that it is through this means that she acquired power after an enormous lapse of time to produce indirectly all the other races of animals that we know." "Justification for the belief that the infusorians or most of them owe their existence exclusively to spontaneous generation is found in the fact that all these fragile animals perish during the reduction of temperature in bad seasons" (p. 103). A similar statement is found on p. 40: "Nature began and still begins by fashioning the simplest of organized bodies, and . . . it is these alone which she fashions immediately, that is to say, only the rudiments of organization indicated in the term spontaneous generation" (see also p. 130).

According to these discussions there had to be a minimum of two types of spontaneous generation, one that gives rise to infusorians and higher animals and another that originated the plant kingdom. In his famous diagram (p. 179) he admits, however, two roots even for the animal kingdom and says in the text, "in my opinion the animal scale begins by at least two separate branches . . . each of these branches derives existence only through direct or spontaneous generation" (p. 178). One of these starts with the infusorians, which in turn give rise to the polyps and these to the radiates. The other, actually more important, branch starts with "worms." More detail about this is provided in the *Histoire naturelle* (1815:455). Here Lamarck states firmly that this second line does not pass through an infusorian stage, but "with the help of particular matter (matériaux) found in the interior of already existing animals, she [nature] has given rise to spontaneous generations which are the source of the intestinal worms, among which perhaps certain ones, after having passed to the outside, have given rise to the free-living worms." The belief in the spontaneous origin of intestinal worms was thoroughly destroyed within a few decades by the brilliant researches of v. Siebold, Küchenmeister, and others on the life cycles of cestodes and trematodes.[6] The work of Pasteur and Robert Koch uprooted the

6. For a historical treatment of the discovery of the reproduction of intestinal parasites, see Foster 1965.

last remnants of a belief in the possibility of spontaneous generation, already undermined by Redi and Spallanzani prior to Lamarck's day.

The Origin of Evolutionary Novelties

Lamarck characterizes each "stage of organization" in his series of ever-greater perfection by a newly arisen structure or capacity through which it is distinguished from the next lower stage. For instance, he characterizes his Class VI (arachnids) as having "stigmata and limited tracheae for respiration; a rudimentary circulation"; his Class VII (crustaceans) as having "respiration by gills; a heart and vessels for circulation," etc. But how do the more perfect organisms acquire these new structures?

About this Lamarck is as explicit as his theory permits him to be: new environments create new needs; new needs require new efforts and habits; and these, in turn, lead to the production of new structures. Given its insistence on the primacy of behavior over structure, this would have been a remarkably modern explanation if Lamarck had not believed in the wrong mechanism. At the turn of the twentieth century the mutationists proclaimed the primacy of structure over behavior by asserting that mutations produce new structures and that thereby the use of these structures is predetermined. Some of Lamarck's contemporaries seem to have had similar ideas, for he explains: "Naturalists have remarked that the structure of animals is always in perfect adaptation to their functions, and have inferred that the shape and condition of their parts have determined their use.[7] Now this is a mistake: for it may be easily proved by observation that it is on the contrary the needs and uses of the parts which have caused the development of these same parts, which have even given birth to them when they did not exist, and which have consequently given rise to the condition that we find in each animal" (p. 113). Elsewhere he states the same principle in these words: "It is not the organs, that is to say, the nature and shape of the parts of an animal's body, that have given rise to its special habits and faculties; but it is, on the contrary, its habits, mode of life and environment that have in course of time controlled the shape of its body, the number and state of its organs and, lastly, the faculties which it possesses" (p. 114). He repeats this statement, in slightly different words, once more on pp. 127 and 174.

Once it is admitted that habits "control the shape of the body," it is only a small step to believe in the origin of entirely new structures owing to entirely new habits: "I shall now prove that the constant use of any organ, accompanied by efforts to get the most out of it, strengthens and enlarges that organ, or *creates new ones* [italics mine] to carry on functions that have become necessary" (p. 119). This is the section of the *Philosophie zoologique* that leads to the famous example of the giraffe (p. 122) that acquired a lengthened neck because it had "to

7. Could he have referred to anyone but Cuvier?

make constant efforts to reach" the leaves of trees. In the same way wading birds acquired long legs owing to their "efforts to stretch and lengthen" them (p. 119). As critics have pointed out for the last 150 years, virtually nothing is said, however, about how these modifications of the phenotype are converted into heritability.

When describing the acquisition of entirely new organs or organ systems, Lamarck often employs a language that is not too different from that which the idealistic morphologists used for their archetypes (pp. 82–83): "Nature attained to the creation of a special organ of digestion," or "she subsequently established a special organ for respiration," or "when afterwards she succeeded in producing the nervous system." From the context it is clear, however, that Lamarck had nothing in mind but the gradual, slow advance toward perfection.

His dogma that the "more perfect" animals are characterized by the acquisition of new faculties and the required executive structures demands that the newest and lowest of all animals, the infusorians, are devoid of most of the properties of higher organisms. It would be ridiculous, he says, to assume that such imperfect organisms could have any of the higher properties: "If we were to imagine that such animals possess all the organs known in other animals, but that these organs are dissolved throughout their bodies, how absurd such a supposition would be!" (p. 103). One more instance in which Lamarck made the wrong guess!

Branching Evolution and Adaptive Radiation

Lamarck, the philosopher of nature, firmly believed in a linear scale ascending toward perfection. A study of systems of organization in less perfect and more perfect animals had convinced him that all the manifold organizations could be ranked as less perfect or more perfect, and could be arranged in a single series, strictly on the basis of their complexity and without commitment on their actual derivation. It is this "unique and simple series which we are forced to adopt in order to facilitate our zoological studies" (*Histoire naturelle,* vol. 1:451). He asserts dogmatically in the *Philosophie zoologique* that "in each kingdom of living bodies the groups are arranged in a single graduated series, in conformity with the increasing complexity of organization and the affinities of the object" (p. 59).

Alas, the finely graduated scale of perfection of the philosopher does not exist in reality. As early as 1800 Lamarck (*Discours d'ouverture,* An VIII:29 [Lamarck 1907]) points out that he does not really "speak of the existence of a linear series, regular in the intervals between species and genera: such a series does not exist, rather I speak of an almost regularly graduated series of the principal groups ("masses"), such as the great families; a series which assuredly exists, among the animals as well as among the plants; but which, when the genera and particularly the species are considered, forms in many places lateral ramifications, the end points of which are truly isolated." The same idea is stated, in part with identical words, in the *Philosophie zoologique* (pp. 57-59).

The enormous diversity of living beings thus has convinced Lamarck, the naturalist, that "the genuine order of nature, that is to say, that which nature would have carried out if accidental causes had not modified her operations" (*Histoire naturelle,* vol. 1:451), does not exist in reality. In the Discours of 1800, when Lamarck had just abandoned the dogma of the fixity of species, he limits the lateral ramifications to species and genera, and the same emphasis still seems to dominate the early chapters of the *Philosophie zoologique.*

A distinct advance in Lamarck's thinking is reflected in the later chapters of the *Philosophie zoologique,* particularly in a discussion embodied in Additions to Chapters VII and VIII (pp. 173-179). All pretense that branching is limited to genera and species is now given up. Indeed, in his famous diagram (p. 179) he entirely abandons a linear classification. The infusorians, polyps, and radiarians are grouped entirely separately from the lineage that arises from the worms. Again, this line from the worms divides into two major assemblages, one terminating in the insects, arachnids, and crustaceans, the other leading to the annelids, cirripeds, and mollusks. The reptiles again lead to two branches, one giving rise to the birds, the other to the mammals. This diagram is certainly much more easily reconciled with the theory of evolution through common descent than with the *scala naturae.*

It is not only in the diagram but also in the text that Lamarck speaks quite freely of branching among the higher categories. He says of the aquatic mammals that they "were divided into three branches by reason of the diversity arising in their habits in the course of time; one of these led to the cetaceans, another to the ungulate mammals, and a third to the various known unguiculate mammals," and summarizes animal evolution as follows: "This series of animals begins with two branches where the most imperfect animals are found; the first animals therefore of each of these branches derive existence only through direct or spontaneous generation" (p. 178).

Even more branching points are admitted in the *Histoire naturelle,* particularly among the invertebrates (pp. 451-460). Here Lamarck rather frankly admits that there is no real connection between the mollusks and the fishes (although he places them adjacent to each other in his diagrams). He establishes a new group for the ascidians, which he places provisionally after the radiarians, although he confesses "that they cannot be considered their continuation nor as derived from them" (p. 451). He now admits that the "order" which nature has produced "is anything but simple; it is branched and seems to consist of several distinct series" (p. 452).

Accepting evolution thus forced Lamarck to abandon the simple series of the *scala naturae,* except as a sequence convenient for teaching and textbook writing. The rapidly increasing knowledge of the comparative anatomy of the invertebrates had simultaneously forced Cuvier to abandon the single series. Cuvier, however, replaced it by a purely creationist and static set of four "embranchements," while Lamarck's mature arrangements (particularly those of 1815) approach the later concept of the phylogenetic tree to a remarkable degree.

From 1800 on Lamarck had recognized two evolutionary processes, "progress in complexity of organization" and "anomalies due to the influence of the environment and of acquired habits" (p. 70). The latter were clearly labeled as "disturbances" in his earlier writings, due to "accidental causes." By 1815 the relative importance of the two processes seems to have become reversed in Lamarck's mind. The series of perfection is now reduced to a didactic device, based on a rational principle; the branching tree, however, reflects what is actually found in nature: "We must thus work on the composition and perfecting of two different tabulations:

> One of them represents the simple series which we must make use of in our publications and lecture courses in order to characterize, distinguish, and make known the observed animals. This series we shall base, in general, on the progression which takes place in the composition of animal organizations, considering each one in the ensemble of all of its parts and making use of the precepts which I have proposed.
>
> The other one represents the several particular series, with their simple side branches, *which nature seems to have formed when producing the different animals which actually exist. (Histoire naturelle,* vol 1:461; italics mine)

The empirical naturalist had finally gotten the better of the philosopher of the scale of perfection. More important, the disturbances and accidental causes became a dominant factor in the causation of diversity. "The reason is that nature's work has often been modified, thwarted, and even reversed by the influence exercised by very different and even conflicting conditions of life upon animals exposed to them throughout a long succession of generations" (*Philosophie zoologique,* p. 81; see also p. 107).

The causal sequence in this branching evolution is perfectly clear. The first step is invariably the entering of a new environment by an organism or a change in the environment at the traditional location. In essentially the same words it is incessantly repeated in the *Philosophie zoologique* (pp. 40, 41, 43, 45, 58, and 114) that "alterations in the environment of animals lead to great alterations in their needs, and these alterations in their needs necessarily lead to others in their activity" (p. 107).

There are special physiological mechanisms that permit conversion of the new "activities," "efforts," or "habits," as they are variously called by Lamarck, into structural changes. This is explained in the Preface (p. 2) of the *Philosophie zoologique*. Changes of habits cause changes in the movements of the fluids in the soft parts of organisms, and this results in a modification of the cellular tissues. These, as we now know, totally unsupported speculations are explained further in Part II (Physiology) of the *Philosophie zoologique*. These processes merely lead to changes in what we would now call the phenotype and are of no particular interest to the evolutionist.

Like Darwin's theory of pangenes, Lamarck's physiological interpretations are evidently influenced by his belief in the *effects of use and disuse*. Such a belief goes all the way back to antiquity (Zirkle 1946) and is still widespread in our folklore. We know what an important role use and disuse played in Darwin's thinking. As I pointed out in the preface to the facsimile edition of the first edition of the *Origin of Species* (p. xxvi), Darwin invoked use and disuse on no less than 13 pages. Lamarck does so on pp. 2, 12, 105, 107, 108, 112, 113, 115, 116, 118, 119–126, and 170 of the *Philosophie zoologique*. Much of this is based on extremely astute observations by an excellent naturalist, and that Lamarck surely was. Indeed, many of his explanations are surprisingly similar to those of Darwin. By attributing the presence of rudimentary organs to disuse (p. 115), he explains quite rationally, although incorrectly, what was such an obstacle for the argument from design. All the modern biologist has to do is to adopt natural selection as the agent through which disuse is converted into structural reduction and most of Lamarck's statements become quite reasonable.

There is little difference between Lamarck and the modern evolutionists in giving the environment an exceedingly high rating as the major causal factor in evolution, but there is a radical difference in the mechanisms by which the environment is effective. Lamarck's conceptualization provided him with no opportunity to utilize natural selection. In spite of his nominalistic emphasis on the existence only of individuals, not of species or genera, Lamarck unconsciously treated these individuals as identical, hence typologically, just as an essentialist would. All of his statements on the impact of the environment are phrased in typological language: "As changes occur in the environment . . . corresponding changes in the animals likewise occur" (p. 109).

If a given environment induces very specific needs, Lamarck postulates that different organisms entering this environment will respond with the same activities and efforts and thus acquire similar structures and adaptations. Here was an opportunity for Lamarck, the experienced zoologist, to test his theory. For instance, when he cites the sloth (*Bradypus*) as an animal that, owing to its arboreal locomotion, leaf-eating habits, and existence in the hot tropics, has acquired all sorts of adaptations, including extreme slowness of movement, he should have asked himself whether other equally tropical and arboreal mammals with similar food habits, such as the leaf-eating monkeys, had become slothlike. He would have found that they acquired entirely different adaptations and have remained quick and lively. Would he then still have maintained his conclusion that different organisms respond to the same environment in the same way?

INHERITANCE OF ACQUIRED CHARACTERS

It is not enough in evolution for an individual to acquire certain modifications, let us say, by the increased use of certain muscles. The crucial point is the transfer of his new acquisition to his offspring. This is the *sine qua non* of the evolutionary process, and it is precisely this problem that Lamarck sidestepped. Whenever (in the *Philosophie zoologique*) Lamarck invokes the principle of an inheritance of acquired characters, he remains strangely silent about the mechanism by which the transmittal takes place. The most explicit statement is found on p. 124, where he makes this claim: "Now every change that is wrought in an organ through a habit of frequently using it, is subsequently preserved by reproduction, if it is common to the individuals who unite together in fertilization for the propagation of their species. Such a change is thus handed on to all succeeding individuals in the same environment, without their having to acquire it in the same way that it was actually created." References to the inheritance of acquired characters on pp. 6, 11, 108, 109, and 175 are even less revealing. The inheritance of acquired characters is simply stated as a universally accepted principle, without a word on the mechanism of this inheritance. Lamarck considers this principle of such importance that it is stated as his Second Law (p. 113): "All the acquisitions or losses wrought by nature on individuals . . . are preserved by reproduction to the new individuals which arise, provided that the acquired modifications are common to both sexes, or at least to the individuals which produce the young."

LAMARCK'S CONCEPTUAL FRAMEWORK

It has often been remarked how different Lamarck's philosophy is from that of such contemporary essentialists as Linnaeus and Cuvier or from the later population thinking of Darwin. Indeed, it largely belongs to a seventeenth–eighteenth-century tradition which by now has been completely superseded. Many of its elements are lucidly described in Lovejoy's *Great Chain of Being*. Leibniz surely had a considerable impact on the development of this kind of thinking, but it is rather more likely that Lamarck's Leibnizism is second-hand. I wonder whether Lamarck ever read any of Leibniz's work. It will require considerable further research to determine which eighteenth-century authors, in addition to Buffon, had the greatest influence on the development of Lamarck's philosophy.

Its elements include the following:

Evolutionism (see fn. 2 above). Lamarck's rejection of a static world has been discussed above (p. 226).

Anti-essentialism. In an age which—at least on the European continent—was dominated by essentialism, Lamarck was an avowed anti-essentialist. He consistently emphasized all the things the essentialists ignored, such as continuity, the absence of gaps, change (and slow, gradual change at that!), the role of the environment, the absence of archetypes, the composition of classes (individuals, not types), the steady movement of levels of organization, and related ideas.

Uniformitarianism. His anti-essentialism was, one might say, automatically responsible for his uniformitarian views in geology (Carozzi 1964; Hooykaas 1959). Lamarck's thinking was strangely unaffected by the rising tide of catastrophism around him. Presumably there were two reasons for this. One is his faith in the potency of general laws that are responsible for events taking place in an orderly fashion and not by cataclysms. The other reason is that Lamarck, presumably following Buffon, had rejected the dogma of the recency of the earth (about 6,000 years, according to most earlier authors) and postulated an extremely high age (see below). This permitted an explanation of even the most conspicuous features of the landscape as the result of gradual and slow processes. It may well be that Lyell's uniformitarianism was more influenced by that of Lamarck than he himself realized. If this was the case, Darwin would also have been affected (via Lyell).

GRADUALNESS, CONTINUITY, AND PLENITUDE

Like many eighteenth-century authors, particularly the followers of Leibniz, Lamarck believed in the principle of *plenitude,* according to which "the range of conceivable diversity of living things is exhaustively exemplified" (Lovejoy 1936). Any type of organism that is conceivable must exist, and the diversity of types must form a complete continuum without borders or gaps. Lovejoy has shown convincingly how pervasive this concept was in the seventeenth and eighteenth centuries and what formidable difficulties it created for its adherents. After all, the world is full of evident gaps and discontinuities, which require reconciliation with the principle of plenitude. This cannot help but lead to contradictions, of which Lamarck's work provides numerous illustrations (particularly with respect to extinction, see below). On the positive side, the principle of plenitude is far more suitable as a basis for evolutionism than is essentialism, and it was clearly one of the inspirations for uniformitarianism, which through Lyell had such an impact on Darwin.

One of its important consequences is the belief that *"Natura non facit saltus,"* an insight which is responsible for Lamarck's constant reiteration of the slowness and gradualness of evolution: "With regard to living bodies, it is no longer possible to doubt that nature has done everything little by little and successively" (p. 11). Speaking of the originally aquatic animals, Lamarck says, "nature led them little by little to the habit of living in the air, first by the water's edge, etc." (p. 70).

A direct consequence of the imperceptibility of evolutionary change is that it must have been extremely slow. This, indeed, is emphasized by Lamarck again and again; for instance: "These changes only take place with an extreme slowness, which makes them always imperceptible" (p. 30); "the imperceptible changing of species" (p. 43); "it is difficult to deny that the shape or external characters of every living body whatever must vary imperceptibly, although that variation only becomes perceptible after a considerable time" (p. 45). With reference to the evolution of thinking in man, he says: "Not only is this the greatest marvel that the power of nature has attained, but it is besides a proof of the lapse of a considerable time; since nature has done nothing but by slow degrees" (p. 50). "An enormous time and wide variation in successive conditions must doubtless have been required to enable nature to bring the organization of animals to that degree of complexity and development in which we see it at its perfection" (p. 50). For nature, "time has no limits and can be drawn upon to any extent" (p. 114).

Few other authors, if any, insisted with equal persistence on the enormous time available for evolutionary changes and indeed required by them (Toulmin and Goodfield 1965). This is why Lamarck was not in the least dismayed by the discovery that the mummified animals found in the Egyptian tombs had not visibly changed in some 4,000 years. Darwin, who likewise postulated an extremely high age of the living world, is strictly in the Lamarckian tradition, as compared with the catastrophists, who tended in their chronologies to stay far closer to a literal interpretation of Genesis.

EVOLUTIONARY CONTINUITY AND THE SHARP DELIMITATION OF TAXA

According to the principle of plenitude, as well as on the basis of evolutionism, there should be no sharp discontinuities in nature. Yet every student of diversity, certainly from the seventeenth century on, observed that "bridgeless gaps" between species and higher taxa were universal. The essentialists—Linnaeus, for instance (Mayr 1957)—solved this problem by declaring all species fixed and static. The evolutionists were deeply troubled by it, and Darwin's ambivalent treatment of the species (Mayr 1959) can in part be explained by this dilemma.[8]

8. This is no longer a problem for the modern evolutionist. Theories of geographic speciation and speciation by polyploidy have demonstrated that there is no contradiction. A discontinuity between populations and incipient species can be established without an interruption of the continuity of the lines to which they belong (Mayr 1970).

The difficulty, however, is not limited to the species level. Another set of problems arises at the level of the higher taxa. Lamarck tries to ignore these difficulties as much as possible. He describes classes, orders, families, and genera as "artificial devices in natural science" (p. 20). "Nature has not really formed" such higher taxa, "but only individuals" (p. 21). Taxonomists have struggled with such conflicts for another 150 years, and it is only within the last decade or two that the true source of the trouble has been clearly recognized (Mayr 1969). When a terminological distinction is made between *taxa* (zoological groups, like birds, bats, or beetles) and *categories* (the rank of the groups, such as family, order, class), it becomes clear that taxa are often well defined and sharply delimited, while ranking in the taxonomic hierarchy is nearly always somewhat subjective, which makes the categories "artificial devices," as Lamarck had rightly said.

Even Lamarck was unable to conceal the fact that there are major gaps in nature:

> It may be said that an immense hiatus exists between crude matter and living bodies, and that this hiatus does not permit of a linear arrangement of these two kinds of bodies, nor of any attempt to unite them by link, as has been vainly attempted.
> All known living bodies are sharply divided into two special kingdoms, based on the essential differences which distinguish animals from plants, and in spite of what has been said I am convinced that these two kingdoms do not really merge into one another at any point, and consequently that there are no animal-plants, as implied by the word zoophyte . . . nor plant animals. (p. 51)

However, he claimed, when it comes to the animal kingdom as such, there are no natural gaps, no genuine lines of demarcation. All lines of demarcation "except those . . . resulting from gaps to be filled . . . always will be arbitrary and therefore changeable" (p. 57). By adopting certain principles, "it is possible to draw boundaries between each system, in such a way that there is only a small number of animals near the boundaries," even though "nature does not pass abruptly from one system of organization to another." And when discussing the 10 classes of invertebrates recognized by him (p. 66), he insists dogmatically that "races may, nay must, exist near the boundaries, halfway between two classes."

He has to make either of two possible assumptions to explain the absence of these postulated intermediates. One is that they will still be discovered in some remote part of the world (see below); the other that they are represented in the fossil state:

> It is true, especially in the animal kingdom, that several of [its

sub-] divisions appear to be really marked out by nature herself; and it is certainly difficult to believe that mammals, birds, etc., are not sharply isolated classes formed by nature. This is none the less a pure illusion, and a consequence of the limitation of our knowledge of existing or past animals. The further we extend our observations the more proofs do we acquire that the boundaries even of the apparently most isolated classes, are not unlikely to be effaced by our new discoveries. Already the *Ornithorhynchus* and the *Echidna* seem to indicate the existence of animals intermediate between birds and mammals. (p. 23)

As a consequence, all borders we establish between higher taxa are entirely artificial. "The complete series of beings making up a kingdom represent the actual order of nature . . . the different kinds of divisions which have to be set up in that series to help us distinguish objects with greater ease do not belong to nature at all. They are truly artificial" (p. 33). "Existing animals . . . form a branching series, irregularly graded and free from discontinuity, or at least once free from it. For it is alleged that there is now occasional discontinuity, owing to some species having been lost" (p. 37).

It is worth noting that all these confident statements on actual or suspected continuity between higher taxa are in the first 67 pages of the *Philosophie zoologique*. One has the impression that Lamarck is less sure of this gapless continuity in the later chapters and conspicuously less so in the *Histoire naturelle* (1815). As Carozzi (1964) has remarked for the *Hydrogéologie*,[9] Lamarck often seems to have changed his ideas while writing a book, without ever going back over the earlier chapters. There certainly are ample indications that he changed his ideas concerning the continuity of the animal series in the decade and a half after the publication of the first Discours d'Ouverture (1800).[10]

EXTINCTION

Extinction was a real problem for many of the natural philosophers of the eighteenth and early nineteenth centuries. We still lack a comprehensive treatment of the conflicting currents of thought dealing with this problem. If one believes in plenitude and in a marvelous capacity of nature for adaptation, as did Lamarck, one cannot really allow for extinction. Hence he insisted: "I am still doubtful whether the means

9. The *Hydrogeology* is the only other work of Lamarck to have been translated into English (Lamarck 1964).

10. Cuvier was in rapid ascendancy in the period from 1800 to 1815, and he and his followers presumably criticized Lamarck with an increasing feeling of superiority. It would be interesting to make a careful comparison of the *Histoire naturelle animaux sans vertèbres* of 1815 with the earlier chapters of the *Philosophie zoologique* of 1809 to see how many of his earlier ideas Lamarck had dropped in the meantime.

adopted by nature to ensure the preservation of species or races have been so inadequate that entire races are now extinct or lost" (p. 44).

There are four explanations, none of them allowing for "natural extinction," that were advanced by various naturalists during this period.

1. *Extinct animals are those that were killed by the flood or some other great catastrophe.* The most awkward aspect of this explanation was that so many of the extinct fossil species were aquatic. Lamarck, very decidedly, and for a number of reasons, rejects this explanation (p. 46). Instead, he advances several others.

2. *The supposedly extinct species might well be surviving in as yet unexplored portions of the globe.* Lamarck expresses this as follows: "There are many parts of the earth's surface to which we have never penetrated, many others that men capable of observing have merely passed through, and many others again, like the various parts of the sea-bottom, in which we have few means of discovering the animals living there. The species that we do not know might well remain hidden in these various places" (p. 44).

3. *The same species, the remains of which we find as fossils, still exist, but their outward appearance has changed so drastically that we no longer recognize them as the same.* "May it not be possible . . . that the fossils in question belonged to species still existing, but which have changed since that time and become converted into the similar species that we now actually find?" (p. 45). Louis Agassiz in his writing was a vigorous proponent of this idea. The problem requires far more research, but it seems that there may have been quite a fundamental difference between Lamarck's and Agassiz's interpretations. Agassiz was an essentialist and it gave him no trouble to conceive of a drastic alteration of the outward appearance without any change of the underlying essence. It is very obvious that Lamarck, who was not an essentialist, had a genuine evolutionary change in mind. However, since his evolution was largely a matter of the continuation of the various phyletic lines, "extinction" simply meant the preservation in fossilized condition of earlier ancestral states of still existing lines.

4. *Extinction is the work of man.* Finally, extinction might be due to man. "If there really are lost species, it can doubtless only be among the large animals which live on the dry parts of the earth; where man exercises absolute sway, and has compassed the destruction of all the individuals of some species which he has not wished to preserve or domesticate" (p. 44). These are prophetic words! Even though man has nothing to do with extinction prior to the Pleistocene, it has become increasingly apparent during the past 20 years that the extinction of many species of large mammals and flightless birds in the late Pleistocene and post-Pleistocene periods might well have been man's work (Martin et al. 1967). Man clearly was and still is an important agent of extinction. Lamarck was perhaps the first author to appreciate this possibility to its fullest extent.

THE VERDICT ON LAMARCK

At the height of the fight between neo-Darwinism and neo-Lamarck-ism, everybody was a partisan, he was either for or against Lamarck. This fight is now a matter of history since the Darwinian interpretation of the causal explanation of evolution has gained a total victory; it is now accepted by every well-informed biologist. But does this relegate Lamarck to a negligible role in the history of biology? Far from it! His contributions to the classification of the invertebrates alone secure him an honored name for all time. But merely to give him credit for his achievements as a competent descriptive zoologist and to consider him a failure as a generalizer does not do him justice at all, for a number of reasons.

First and foremost, it was Lamarck who in defiance of the *Zeitgeist* preserved and propagated a set of ideas unpopular among creationists, catastrophists, and essentialists. Curiously, Lamarck was often most right where he most differed from prevailing ideas, such as in his sup-port of evolutionism and uniformitarianism, while the errors he is best remembered for, like use and disuse, the inheritance of acquired charac-ters, and much of his physiology, were not at all original with him but represent widely held ideas merely adopted by Lamarck.

Darwin always stressed how little he owed to Lamarck (Rousseau 1969). In a letter to Lyell (October 11, 1859) he said of Lamarck's work, "I got not a fact or idea from it." This may well be true as far as the causal explanation of evolution is concerned, and also as far as evolutionary facts are concerned. Lamarck's approach was Cartesian (deductive), and he virtually never supplied detailed factual evidence in support of his sweeping generalizations. But I suspect that Darwin vastly underestimated the role Lamarck had played in preparing the intellectual climate for the subsequent Darwinian advances.

In an age still largely dominated by a belief in literal creation—a con-cept not at all unpalatable to most essentialists—Lamarck's daring and iconoclastic speculations could not help having considerable impact. Lyell's preoccupation with Lamarck is telling evidence for this. And a good case can be made for the thesis that Lyell's uniformitarianism owed much to Lamarck (see above; also Gillispie 1959:266).

Lamarck's evolutionism had numerous serious deficiencies, as point-ed out by his critics then and now. Nevertheless, it is conceivable that, if Lamarck had had the personality to found a school, his theories might have become the starting point of an improved evolutionary interpretation. What was most needed was the proposal of a feasible mechanism (or system of mechanisms). However, Lamarck had no disciples to build on the foundation that he had laid. It was also his mis-fortune to have to live in the shadow of such a giant as Cuvier (25 years his junior!). There was no common meeting ground between the evolu-tionist-uniformitarian-nominalist ideas of Lamarck and the essentialist philosophy that dominated Cuvier's thinking on all subjects (catastrophism, embranchements, etc.). A head-on collision between the representatives of these two utterly different philosophical systems was inevitable, and it was this rather than any zoological disagreements that was responsible for Cuvier's hostility toward Lamarck.

Lamarck supported rather uncritically a number of ideas that go far back in human history. These include a belief in the inheritance of acquired characters, the importance of the use or disuse of structures, and the existence in the organic world of a built-in tendency toward ever-greater perfection. Through his championship of such "folklore" ideas Lamarck had an unexpectedly large influence, particularly outside of science. Some aspects of Lysenkoism in Russia as well as the books of Teilhard de Chardin can be traced back to Lamarckian ideas. How direct this influence was and how important Lamarck's support was for the respectability of these ideas are questions that are still unanswered.

The work of Lamarck contains numerous zoological generalizations, which document what an astute and original observer he was. Originality is less evident in his philosophical ideas. It will require a thorough comparative study of the writings of the philosophers and naturalists of the seventeenth and eighteenth centuries to determine to whom Lamarck owes a particular debt for such concepts as scale of perfection (directionism), plenitude, movement of fluids, organization, and other concepts and principles that were integral components of his system. Like all major figures in the history of ideas, Lamarck was both the endpoint of a long antecedent history and the starting point of new developments.

REFERENCES

Burkhardt, R. W. 1970. Lamarck, evolution, and the politics of Science. *Jour. Hist. Biol.,* 3:275–298.

—— 1976. *The evolutionary thought of Jean-Baptiste Lamarck.* In press.

Cannon, H. G. 1957. What Lamarck really said. *Proc. Linn. Soc. London,* 168: 70–87.

Carozzi, A. V. 1964. Lamarck's theory of the earth: hydrogéologie. *Isis,* 55: 293–307.

Darwin, C. 1859. *On the origin of species by means of natural selection, or the preservation of favoured races in the struggle for life.* J. Murray, London.

Darwin, F., ed. 1888. *The life and letters of Charles Darwin.* J. Murray, London.

Daudin, H. 1926. *Cuvier et Lamarck: les classes zoologiques et l'idée de série animale* (1790–1830). Libraire Félix Alcan, Paris.

DeBeer, G. R. 1954. *Archaeopteryx lithographica: a study based upon the British Museum specimen.* British Museum, London.

Foster, W. D. 1965. *A history of parasitology.* E. and S. Livingstone, Edinburgh and London.

Gillispie, C. C. 1956. The formation of Lamarck's evolutionary theory. *Archives Int. d'Hist. Sciences,* 9:323–338.

—— 1959. Lamarck and Darwin in the history of science. In *Forerunners of Darwin, 1745–1859,* ed. B. Glass et al. The Johns Hopkins Press, Baltimore.

Gould, S. 1970. Dollo on Dollo's law, irreversibility and status of evolutionary laws. *Jour. Hist. Biol.,* 3:189–212.

Greene, J. C. 1959. *The death of Adam.* Iowa State University Press, Ames, Iowa.

Guyénot, E. 1941. *Les sciences de la vie aux XVIIe et XVIIIe siècles. L'idée d'evolution.* A. Michel, Paris.

Hooykaas, R. 1959. *Natural law and divine miracle: a historical-critical study of the principle of uniformity in geology, biology and theology.* E. J. Brill, Leiden.

Humboldt, A., and A. Bonpland. 1805. *Essai sur la géographie des plantes.* Paris.

Lamarck, J. B. 1778. *Flore françoise.* L'imprimerie royale, Paris.

—— 1809. *Philosophie zoologique, ou exposition des considerations relatives à l'histoire naturelle des animaux.* J. B. Baillière, Libraire, Paris.

—— 1815. *Histoire naturelle des animaux sans vertèbres.* Verdière, Libraire, Paris.

—— 1907. Discours d'ouverture (an VII, an X, an XI et 1806), ed. Giard. *Bull. Sci. de la France et de la Belgique,* T. 40, Paris.

—— 1914. *Zoological philosophy: an exposition with regard to the natural history of animals,* trans. Hugh Elliot. Macmillan, London.

—— 1963. *Zoological philosophy* (facsimile of 1914 edition). Hafner Publishing Co., New York.

—— 1964. *Hydrogeology,* trans. A. V. Carozzi. University of Illinois Press, Urbana.

Landrieu, M. 1909. *Lamarck, le fondateur du transformisme; sa vie, son oeuvre.* Soc. Zool., Paris.

Lovejoy, A. O. 1936. *The great chain of being.* Harvard University Press, Cambridge, Mass.

—— 1959. Buffon and the problem of species. In *Forerunners of Darwin, 1745–1859,* ed. B. Glass et al. The Johns Hopkins Press, Baltimore.

Martin, P. S., and H. E. Wright, Jr., eds. 1967. *Pleistocene extinctions: the search for a cause.* Yale University Press, New Haven.

Mayr, E. 1957. Species concepts and definitions. In *The species problem,* ed. E. Mayr. Amer. Assoc. Adv. Sci. Publ., no. 50:1–22.

—— 1959. Isolation as an evolutionary factor. *Proc. Amer. Phil. Soc.,* 103:221–230.

—— ed. 1964. *On the origin of species* (A Facsimile, the First Edition). Harvard University Press, Cambridge, Mass.

—— 1969. *Principles of systematic zoology.* McGraw-Hill, New York.

—— 1970. *Populations, species, and evolution.* Belknap Press of Harvard University Press, Cambridge, Mass.

Rousseau, G. 1969. Lamarck et Darwin. *Bull. de Muséum National d'Histoire Naturelle,* 5:1029–1041.

Simpson, G. G. 1961. Lamarck, Darwin and Butler: three approaches to evolution. *Amer. Scholar,* 30:239–249.

Toulmin, S., and J. Goodfield, 1965. *The discovery of time.* Hutchinson, London.

Wilkie, J. S. 1959. Buffon, Lamarck and Darwin: the originality of Darwin's theory of evolution. In *Darwin's biological works: some aspects reconsidered,* ed. P. R. Bell. Cambridge University Press, Cambridge.

Zirkle, C. 1941. Natural selection before the *Origin of species. Proc. Amer. Phil. Soc.,* 84:71–123.

—— 1946. The early history of the idea of the inheritance of acquired characters and of pangenesis. *Trans. Amer. Phil. Soc.,* n.s., 35:91–151.

18
Agassiz, Darwin, and Evolution

When Darwin's *On the Origin of Species* was published in 1859, there were two celebrated naturalists at Harvard University: Louis Agassiz, a zoologist, and Asa Gray, a botanist. Their reactions to the new theory of evolution were very different. Asa Gray adopted it wholeheartedly and defended it in letters, in essays, and in public debates. Louis Agassiz, on the other hand, denounced the new doctrine in the strongest terms. The reason for Agassiz's unbending opposition to Darwin's theory has always been something of a mystery. Here was a great naturalist, the world's foremost authority on fossil fishes, a man with a stupendous breadth of knowledge, founder of one of the world's great natural history museums, embryologist-anatomist-systematist-paleontologist, predestined in every way, one would think, to welcome the unifying theory of evolution with open arms. Instead, he fought it to the bitter end. Why?

Agassiz was a devout Christian and it has sometimes been claimed that his religious beliefs lay behind his opposition. Yet this cannot be the answer. Asa Gray and other contemporaries were equally devout but found nothing objectionable in the theory of evolution. It is far more likely that the different reactions of Agassiz and Gray may be explained on the basis of their utterly different educations and intellectual backgrounds. Agassiz had been educated in Switzerland and Germany during a period that was dominated by romantic ideas and by a largely metaphysical approach to nature. From childhood on Agassiz had been exposed to the *a priori* concepts of the various philosophies of the seventeenth and eighteenth centuries (as they reached into the nineteenth century), and a detailed analysis of Agassiz's writings indicates strongly that he was never able to escape this conceptual world in spite of his lifelong concern with empirical natural history. All his life Agassiz remained a child of the world of ideas of the eighteenth century. In order to understand him and his attitude toward evolution we must become familiar with the dominant concepts of that period.

In our study of the pre-Darwinian period we must guard against mis-

Revised from "Agassiz, Darwin, and evolution," *Harvard Library Bulletin,* 13(1959): 165–194.

interpreting certain verbal similarities in the stating of the "laws of nature" as evidence for a similarity of concepts. We shall presently see how utterly different from modern usage is the eighteenth-century interpretation of terms like "evolution," "progression," "higher," "type," "variety," and "species." We must never lose sight of the fact that from the Greek philosophers on to the idealistic and romantic philosophers of the eighteenth and nineteenth centuries all speculation on the nature of the world was based on the implicit assumption that the universe obeys a set of rational principles, and that one can "explain" the world by discovering these principles. As Lovejoy (1936: 327–328) has restated this point of view: "For though our intelligence is doubtless too limited" to find a specific answer to the why of "every detail of existence, it is capable of recognizing the broad principles essential to any consistent answer. By this sort of faith in the rationality of the world we live in, a great part . . . of Western philosophy and science was, for a score of centuries, animated and guided, though the implications of such a faith were seldom fully apprehended and came only slowly to general recognition."

The basic framework for such a rational interpretation of the world was provided by Plato and his successors (Aristotle and the neo-Platonists). Three ideas in particular dominated subsequent thinking, the principle of the completeness of the existing world (and of the underlying ideas), the principle of the qualitative continuity of the series of forms of natural existence, and the principle of a unilinear ascending order of excellence. It is a combination of these three principles that resulted in a conception of the universe designated by Lovejoy as the Great Chain of Being. Whether it is Aristotle or Thomas Aquinas, Leibniz or Goethe, Lamarck or Robinet, Kant or Schelling, ultimately all their concepts of the plan and the structure of the world can be traced back to the edifice originally outlined by Plato. The philosophers and scientists of the Middle Ages and of the eighteenth century may have quarreled with each other bitterly, they may have stressed continuity or discontinuity, a stationary scale of beings or a moving escalator, a progression toward goodness or initial perfection, yet their basic frame of reference had been determined by Plato.

We have to keep this clearly in mind when, in the cosmologies of writers from the Greeks to Goethe, we discover statements with an evolutionary ring, like "scale of nature" or "ladder of perfection." Actually there is not a trace in these ideologies of what a modern writer would call evolutionary thinking, a belief in "common descent." "The Chain of Being . . . was a perfect example of an absolutely rigid and static scheme of things" (Lovejoy 1936:242). Neither Aristotle nor Leibniz, nor, as is quite evident, even Goethe was an evolutionist in the biological sense. Lacking the all-important ingredient of "common descent," their concepts were in essence static.

The evolutionism that Agassiz encountered as a student and active research worker was philosophical speculation rather than scientific theory. This was true even for Lamarck, who was by far the most erudite evolutionist of the period. Evolutionary speculation of one sort or another was widespread, if not universal, during the century from 1750 to 1850. Much of that literature was such pure speculation that it must have appeared rather ridiculous to a zoologist of Agassiz's experience. Yet these ideas were so prevalent and alive that Agassiz felt duty-bound to take up the whole question of "evolution" versus an alternative interpretation of the universe and settle the problem once and for all. This he did in Volume I (1857) of his great *Natural History of the United States.* No other theme occupies so prominent a place in this work. Agassiz's familiarity with the relevant literature is evident from his frequent references to de Maillet, Buffon, Geoffroy St. Hilaire, Lamarck, Chambers, Baden Powell, and others. Their arguments in favor of evolution are taken up, one by one, and refuted. The student of the history of evolution is impressed by how many of the problems that Darwin treated in his *Origin of Species* had already been considered by Aggassiz two years earlier.

Where Agassiz argues with the philosophers of the eighteenth century, the encounter is usually very unequal, owing to Agassiz's vastly superior command of scientific facts. He was modern and forward-looking in his actual researches. And his *Natural History of the United States,* as well as his various monographs of fossil fishes and many smaller publications, is an impressive document of scholarship. His discussions in the section headed "Relations between Animals and Plants and the Surrounding World" (1857:57-63) contain some very astute observations. He was well ahead of his time in deploring the then current neglect of the study of the habits of animals and the one-sided concentration on anatomy, embryology, and taxonomy: "What does it matter to science that thousands of species more or less, should be described and entered in our systems, if we know nothing about them? . . . How interesting would be a comparative study of the mode of life of closely allied species" (1857:51, 59).

Indeed, his observations on the behavior of animals are quite remarkable and still worth reading, as, for instance, his description of the unlearned snapping behavior of the young snapping turtle: "Who could, for instance, believe for a moment longer that the habits of animals are in any degree determined by the circumstances under which they live, after having seen a little Turtle of the genus Chelydra, still enclosed in its egg-shell, which it hardly fills half-way, with a yolk bag as large as itself hanging from its lower surface and enveloped in its amnios and in its allantois, with the eyes shut, snapping as fiercely as if it could bite without killing itself?" (1857:59).

And yet, there is a radical difference between Agassiz and the modern naturalist, who constantly tests his theories against his facts. Agassiz, like the speculative philosopher of the eighteenth century, accommodates the observed facts in a preconceived framework of ideas and tends to minimize or ignore the facts that do not fit. In view of this approach of his, it becomes all important to survey the major concepts on which Agassiz bases his interpretation of nature. There are four that appear particularly important in Agassiz's thinking: (1) a rational plan of the universe, (2) typological thinking, (3) discontinuism, and (4) an ontogenetic concept of evolution. Each of these concepts is rooted in Greek philosophy, particularly that of Plato, and we will not be able to discuss the evolutionary consequences without an occasional reference to the philosophical basis.[1]

A RATIONAL PLAN OF THE UNIVERSE

I mentioned above how great a faith most philosophers from the Greeks to the nineteenth century had in the rationality of the world we live in. This basic axiom was incorporated in otherwise utterly different philosophies and theologies, ranging from rigid beliefs in a personal God who supervised and planned every moment in the life of every organism, to vague, pantheistic beliefs in a "supreme being," and finally to philosophies in which the universe, though ruled by physical forces, was yet a "Kosmos" (in the original meaning of this somewhat ambivalent Greek word). In whatever theology or philosophy the concept of a "rational world" was incorporated it permitted the development of a *Weltanschauung* of a magnificent intrinsic logic. If everything in the world follows an underlying plan, it becomes the task of the naturalist to discover the *a priori* principles of this plan. An empirical approach to an establishment of general principles would be a disorderly and futile method in a rational world. The *a priori* principle is adopted by Agassiz with deep conviction, and he rejoices in noting "to what an extraordinary degree many *a priori* conceptions, relating to nature, have in the end proved to agree with the reality, in spite of every objection at first offered by empiric observers" (1857:23).

The evidence for the rationality of the world is everywhere. To Agassiz it is irrefutable evidence for planning. Let us take, for instance, says Agassiz, the universal occurrence in nature of perfect adaptation. Whether it is a matter of the distribution of organisms in the proper environments, or the "mutual dependence of the animal and vegetable kingdoms," or the dependence of parasites on their hosts, or whatever other relations there are in nature, they could not possibly be the result

1. I would like to record here my indebtedness to Arthur O. Lovejoy's stimulating volume *The Great Chain of Being* (1936), which presents in a new light many of the concepts of the pre-Darwinian period customarily labeled "evolutionary." For the sake of simplicity I am citing this earlier literature through reference to the relevant pages of Lovejoy. For additional reading see also Guyénot 1941 and Zimmermann 1953. There is a fascinating chapter (Chapter 10, "The Effect of Natural Philosophy") in Erwin Stresemann's *Ornithology: From Aristotle to the Present* (1975) concerning the impact of the philosophical ideas of the seventeenth and eighteenth centuries on biology. This supplements in many ways Lovejoy's account of the period.

of "blind chance" or "physical causes" (1857:122-127). "All relations in nature are regulated by a superior wisdom." What simple-minded environmentalists and crude materialists have had to say on the effect of the environment does not convince Agassiz. Indeed, he has little trouble pointing out how superficial such speculations are. He challenges the environmentalist to explain why each geological period has produced different faunas and floras no matter how similar the external conditions. Conversely, he asks how exceedingly different organisms can coexist in the same environment if physical causes have a significant influence on the structure of animals. Even within a single drop of water one finds organisms as different as plants (algae), protozoans, crustaceans, and other invertebrates. The impotence of physical causes is thus proven not only by the occurrence of very different forms in the same environment but likewise by the occurrence of very similar forms in exceedingly different environments.

Is not the existence of the four types of animals, the radiates, the articulates, the mollusks, and the vertebrates, demonstrated by his great teacher Cuvier, conclusive proof for the planning of the universe? If one were to assume, says Agassiz (1857:21), that the four major types had been created in the beginning of the earth's history and then left to evolve, so to speak, on their own, it would be inconceivable how they could have maintained their mutual relationship through the ages. Systems of categories could not possibly have been maintained by individuals with their short duration of life and fleeting existence. How could "the infinitude of new animals and new plants" have evolved "cast in these four moulds, in such a manner as to exhibit, notwithstanding their complicated relations to the surrounding world, all those more deeply seated general relations, which establish among them the different degrees of affinity we may trace so readily in all the representatives of the same type?" (1857:22). The mere fact that a zoologist can distinguish major systematic categories is clear evidence to Agassiz that "all this" is "the creation of a reflective mind establishing deliberately all the categories of existence."

It is even easier for Agassiz to refute those who believe in spontaneous generation or who believe with Holbach that nature may now be "assembling in her vast laboratory the elements fitted to give rise to wholly new generations, that will have nothing in common with the species at present existing" (Lovejoy 1936:269). How could such a process possibly result in adaptation? And the same argument holds true, says Agassiz, for any new fauna and flora appearing after a previous catastrophe. "The transmutation theory furnishes no explanation of their existence. For every species belonging to the first fauna and the first flora which have existed upon earth [or after each catastrophe], special relations, special contrivances must therefore have been provided. Now, what would be appropriate for the one, would not suit the other, so that excluding one another in this way, they cannot have originated upon the same point" (1857:12–13). Since well-adapted new faunas and floras cannot be explained in terms of an origin through the direct action of the physical environment, *ergo* they must have come into being through the action of an outside creator. Agassiz's argument against the sudden creation of whole new faunas and floras out of nothing, so to speak, is indeed well taken. It is only curious that he forgets that it is not evolutionists who postulate such creations, but anti-evolutionists.

Having eliminated to his satisfaction any possible role of spontaneous events or of "physical causes," Agassiz concludes, like the philosophers of the Middle Ages, that the universe is the product of a careful, rational plan. As a consequence, "Natural History must, in good time, become the analysis of the thoughts of the Creator of the Universe, as manifested in the animal and vegetable kingdoms" (1857:135).

TYPOLOGICAL THINKING

The contemplation of the world was dominated for two thousand years by Plato's concept of the *eidos*. Different authors have stressed different aspects of Plato's Idea, some its independence from perception, others its transcendent reality, and still others its eternity and immutability. Belief in an immutable reality underlying the changeable world of appearances led to a special kind of thinking, often lately referred to as "typological thinking," that is still widespread and was virtually universal until the beginning of the nineteenth century (see Essay 3). The typologist and the populationist (one who thinks in terms of variable populations) are at opposite poles in questions of systematics and of evolution.

Agassiz was a convinced typologist. Species, genera, and the various higher classes into which the naturalist classifies the diversity of nature were to him nothing but "categories of thought." Any natural class of organisms is characterized by its "plan," its "type." The various groups of animals and plants, as delimited by the naturalist, thus have no reality to the typologist, who acknowledges only individuals and the plan ("type") upon which the differences among the individuals are founded. In Agassiz's words:

> The individuals of one species . . . exhibit characters, which . . . would require the establishment, not only of a distinct species, but also of a distinct genus, a distinct family, a distinct class, a distinct branch. Is not this in itself evidence enough that genera, families, orders, classes . . . have the same foundation in nature as species, and that the individuals living at the time have alone a material existence, they being the bearers, not only of all these different categories of structure upon which the natural system of animals is founded, but also of all the relations which animals sustain to

the surrounding world,—thus showing that species do not exist in
nature in a different way from the higher groups, as is so generally
believed? (1857:7)

The arbitrariness of the species as a category has been asserted by
virtually every philosopher from Plato down. In this even Locke is akin
to Plato: "I think it nevertheless true that the boundaries of species,
whereby men sort them, are made by men." He claims to be unable to
see "why a shock [a breed of shaggy dog] and a hound are not as distinct species as a spaniel and an elephant . . . so uncertain are the boundaries of species of animals to us" (Lovejoy 1936:229). When, at a later
period, Rousseau and Lord Monboddo asserted that man and the higher
apes were of the same species, it was evidently the species of Plato and
not that of the biologist that they were referring to (Lovejoy 1936:
235). The gradual emergence of the notion that species have a biological significance and that there are nonarbitrary criteria by which the
biological species can be defined is a fascinating chapter in the history
of science (see Essay 33).

It is evident, however, that the problem of the origin of species is, to
an adherent of Plato, either an insoluble problem or no problem at all.
This, indeed, is the basis from which Agassiz attacks Darwin. "For
many years past I have lost no opportunity of urging the idea that while
species have no material existence, they yet exist as categories of
thought, in the same way as genera, families, orders, classes, and
branches of the animal kingdom" (1860b:142). Since species are categories of thought, he continues, it is quite impossible to assume with
Darwin that they could be evolutionary stages and that members of the
same category could be of common descent. "Far from agreeing with
these views, I have, on the contrary, taken the ground that all the natural divisions in the animal kingdom are primarily distinct, founded upon
different categories of characters, and that all exist in the same way,
that is, as categories of thought, embodied in individual living forms."

It would be difficult to find a more unequivocal interpretation of
zoological classification in terms of Plato's Ideas. How can there be evolution, since no *eidos* is ever connected with any other *eidos,* either by
intermediates or by common descent? "Species are based upon relations and proportions that exclude . . . the idea of a common descent"
(1860b:143). So emphatic is Agassiz on this point that he repeats it
again and again, and yet finds it necessary to repeat it once more in
his final summary. "Species, genera, families, &c. exist as thoughts,
individuals as facts." "As the community of characters among the
beings belonging to these different categories arises from the intellectual
connection which shows them to be categories of thought, they cannot
be the result of a gradual, material differentiation of the objects themselves."

In retrospect, it is quite evident how exceedingly difficult it would be for one steeped in the tradition of Plato's philosophy to accept the idea of "common descent." This is the reason the great German zoologists of the first half of the nineteenth century failed so completely to solve the problem of evolution. They had been thoroughly indoctrinated in the concepts of idealistic philosophy, while the two "dilettantes," Darwin and Wallace in England, had spent their time watching birds, collecting insects, and reading Malthus and *Vestiges of the Natural History of Creation,* thus happily remaining unaffected by the lofty fallacies of idealistic philosophy.

The characteristics of the *eidos* are of decisive relevance for the problem of the origin of species. How can a category of thought, a type, an *eidos,* develop "a variety" and this variety eventually turn into a species, a new *eidos,* as was claimed by the evolutionists? What a series of intrinsic contradictions to a typologist! What then is a variety, and what role, if any, does it play in evolution? As we look back, it is clear that an evolutionist such as Darwin and an anti-evolutionist such as Agassiz confused two biological phenomena masquerading under the same term: the variant individual and the variant population (see Essays 10 and 11). When speaking of variety or variation Agassiz apparently always had in mind the variant individual. Such variability, he says, nowhere touches or affects the essence of the type. "Whatever minor differences may exist between the products of this succession of generations are all *individual peculiarities,* in no way connected with the essential features of the species, and therefore as transient as the individuals; while the specific characters are forever fixed" (1860b:150).

The modern reader wonders how Agassiz decided what part of the variability affects the "essential features of the species" and what part does not. It is quite evident from his own work that he had no criteria and that he based his decision in each case on, shall we say, intuition. This permitted him to maintain quite dogmatically that "varieties, properly so called, have no existence, at least in the animal kingdom" (1860a:410). He maintained to the last that "the only structural differences known between individuals of the same stock are monstrosities or peculiarities pertaining to sex, and the latter are as abiding and permanent as type itself" (1874:92). In his survey of the fishes of the Tennessee River (1854a), Agassiz encountered a number of species with high individual variability. His disbelief in the existence of "varieties" forced him to describe several "species" from schools of single species: *Lepomis megalotis* ("*sanguinolentus,*" "*inscriptus,*" "*bombifrons*"), *Aplodinotus grunniens* ("*concinnus,*" "*lineatus*"), *Fundulus notti* ("*guttatus,*" "*hieroglyphicus*"), and *Ictiobus bubalus* ("*urus,*" "*taurus*").[2] In other cases Agassiz avoided the need for the recognition of varieties by taking refuge in the criterion of intergradation. The minutes of a meeting of the American Academy of Arts and Sciences record that in a study of

2. According to information kindly supplied by Dr. Reeve M. Bailey, Museum of Zoology, University of Michigan.

echinoderms Agassiz "found a great abundance of divergent forms, which without an acquaintance with the connecting ones, and large opportunities of comparison, might be taken for distinct species, but he found that they all passed insensibly into each other" (1860c:72). This claim was not received too well by his audience, alerted by Darwin. For the minutes continue:

> Professor Parsons suggested that more extended observation might connect the received species by intermediate forms, no less than the so-called varieties.
> And Professor Gray remarked that the intermediate forms connecting, by whatsoever numerous gradations, the strongly divergent forms with that assumed as the type of the species, so far from disproving the existence of varieties, would seem to furnish the best possible proof that these were varieties. Without the intermediate forms they would, it was said, be taken for species; their discovery reduced them to varieties,—between which, but not between species (according to the ordinary view), intermediate states were to be expected.

Lest there be a misunderstanding, let me emphasize that the significance of variability was the bone of contention, not the fact as such. Agassiz, who had been working with animals for more than 30 years when these discussions took place, was far too experienced to deny variation. He stressed the frequency of variation in many of his writings: "So marked, indeed, is this individuality in many families . . . that correct descriptions of species can hardly be drawn from isolated specimens . . . I have seen hundreds of specimens of some of our Chelonians, among which there were not two identical. And truly, the limits of this variability constitute one of the most important characters of many species" (1857:58). And he reconfirms this 3 years later: "It requires extensive series of specimens accurately to describe a species, and . . . the more complete such series are, the more precise appear the limits which separate species" (1860b:144). Attempting to reconcile this extreme variability with the postulated immutability of the category species, he is induced to make this cryptic statement: "For my own part I must emphatically declare that I do not know a single fact tending to show that species do vary in any way, while it is true that the individuals of one and the same species are more or less polymorphous" (1860b:153). A contradictory statement indeed, which any Darwinian could have demolished quite easily.

And yet we must admit that on the question of the relation between variety and species Agassiz was hardly more inconsistent than Darwin himself. While Agassiz insisted on the typological fixity of species, Darwin went to the other extreme and declared species to be purely subjective inventions of the taxonomist. This exposed him to a well-deserved attack by Agassiz. "If species do not exist at all, as the supporters of the transmutation theory maintain, how can they vary? and if individuals alone exist, how can the differences which may be observed among them prove the variability of species?" (1860b:143).

These are completely valid questions. I have commented elsewhere (Essay 33) that both Linnaeus and Darwin thought that constancy and "reality" (that is, nonarbitrary delimitation) of species were merely two aspects of a single property. As a result, Darwin never solved the problem of the splitting of one species into two; indeed, he never even clearly appreciated the existence of this problem. It is a problem that can be solved only by one who thinks in terms of populations, and the species problem is one to which Darwin never consistently applied the concept of population in the way he applied it to natural selection and other evolutionary problems. The most effective refutation of Agassiz's argument would have been to say that species are composed of populations and that intermediacy among species is maintained or produced by populations, not by individuals. Not the individual, but the gene pool of the population is the real unit of evolution. This, of course, is a mode of thinking utterly alien to a typologist.

DISCONTINUISM

A consistent application of the concept of the *eidos* leads to a picture of a universe full of discontinuities. Each category of thought is separated from every other one by a clear-cut gap. No matter how much one tries to minimize these gaps by imagining intermediate types, in principle they remain discontinuities. This ideology appeared too static and disruptive to Aristotle to be acceptable, and although he did not go as far as Heraclitus, who adopted a dynamic theory of continuity, he at least proposed a static concept of continuity. This postulates that classes border each other so tightly that there is no room for a gap. Aristotle does not seem to have realized that there is an irreconcilable conflict between the principle of the *eidos,* which necessitates clear divisions and nonarbitrary classifications, and the principle of continuity. "There are not many differences in mental habit more significant than that between the habit of thinking in discrete, well-defined class-concepts and that of thinking in terms of continuity, of infinitely delicate shadings-off of everything into something else, of the overlapping of essences" (Lovejoy 1936:57). Indeed, one cannot base a system of philosophy simultaneously and equally on both concepts. The philosophers and scientists of the ensuing 2000 years can be placed almost without exception in the camp of the "continuists" or in that of the "discontinuists." And one finds that the philosophers and speculative scientists are almost invariably continuists, while the thorough and solid students of descriptive natural history confess to being discontinuists. When Aristotle describes the animal kingdom he is a discontinuist. And so were both Linnaeus and Cuvier, the two most experienced naturalists of the eighteenth and early nineteenth centuries. Agassiz, both as a disciple of Cuvier and as an avowed typologist, could not help being a discontinuist. Leibniz, the outstanding apostle of the Great Chain of Being, championed continuity: *Natura non facit saltus* (as later stated by Linnaeus). Even Kant accepted a "continuous transition from every species to every other species by a gradual increase of diversity," whereby "the diversities of species touch each other and admit of no transition from one to another *per saltum*" (Lovejoy 1936:241). These philosophical ideas rather than a study of nature were the basis of Buffon's (1749) claim that species do not exist in nature, only individuals. "It is possible to descend by almost insensible degrees from the most perfect creature to the most formless matter. . . . These imperceptible shadings are the great work of nature" (Lovejoy 1936:230). And let us remember, wrote Lamarck in 1806, that nature "knows neither classes nor orders nor genera nor species . . . and that among the living organisms, nothing exists in reality except individuals and diverse races which shade into each other in all degrees of organization" (Guyénot 1941:383). Darwin, with his often-repeated insistence on the arbitrariness of the species as recognized by the naturalists (his own work on the barnacles notwithstanding!), was clearly a standard bearer of the continuists.

Both Darwin and Agassiz, then, denied the "biological reality" of species, but for diametrically opposed reasons. Darwin denied it because he considered species an arbitrary segment in a continuous stream of individuals. Agassiz denied it because to him not the physical species as such had reality but only the category of thought that we call "species." Several more decades had to pass before the biological concept of species emerged: the great gene pool, the community of interbreeding populations in nature, reproductively isolated from other such communities (see Essay 33).

AN ONTOGENETIC CONCEPT OF EVOLUTION

Few of those who in our time use the word "evolution" realize that the term had an entirely different meaning when first introduced into biology and that until less than a hundred years ago it referred mainly to the developmental changes in ontogeny. Indeed Darwin did not use the term "evolution" in his *Origin of Species,* and Agassiz consistently spoke of the transmutation theory when referring to what we would now call evolution. And yet the original concept of evolution was a very important one in the history of ideas: it was an attempt to reconcile the observable fact of change with the concept of the fixity of the type. Does not the development of the individual from the egg to adulthood prove that one can have an "unfolding" of preexisting potentialities without any change in the essence of the type? Every individual animal, says Agassiz, goes through numerous, highly different stages from the egg to adulthood and old age, and yet its offspring revert to the typical appearance of its species. "Does this not prove that while individuals are perishable, they transmit, generation after generation, all that is specific or generic, or, in one word, *typical* in them, to the exclusion of every *individual peculiarity* which passes away with them" (1860b:151).

If one can have such an evolution of an individual without its affecting in the least the essence of its type, why not a similar evolution, that is, an unfolding of the preexisting inherent characters of a species, a genus, an order, or any higher category of animals and plants, asks the follower of Plato happily, feeling that he has thereby eliminated the apparent contradiction between the fixity of the type and the apparent changes of nature. If maggots can become flies and caterpillars butterflies, asks Leibniz (Lovejoy 1936:258), why cannot one species transform into another without any change in its essence, provided it was inherent in its essence to undergo such a change? To the adherent of this philosophy, all progressive evolution is merely an unfolding (*evolutio*) of its essence. Indeed, says Schelling (Lovejoy 1936:322), highest perfection must have existed at the very beginning if there was a supreme being at the beginning. Evolutionary change, then, is not real change, but merely a revelation, an emergence of something previously "involuted" or "folded in."

Agassiz throughout his life was a militant representative of this interpretation of evolution. Whenever he uses the term "evolution" he uses it in its eighteenth-century embryological meaning. At the very end of his life he said: "We hear so much of evolution and evolutionists that it is worth our while to ask if there is any such process as evolution in nature. Unquestionably, yes. But all that is actually known of this process we owe to the great embryologists of our century, Döllinger and his pupils K. E. von Baer, Pander, and others,—the men in short who have founded the science of Embryology" (1874:92).

It is most significant that Agassiz does not mention Darwin or Wallace in this paragraph on evolutionists written 15 years after the publication of the *Origin of Species*. And, after describing the development of the egg and of embryonic stages, Agassiz continues: "These successive stages of growth constitute evolution, as understood by embryologists, and within these limits all naturalists who know anything of Zoölogy may be said to be evolutionists. The law of evolution . . . is a law controlling development and keeping types within appointed cycles of growth, which revolve forever upon themselves, returning at appointed intervals to the same starting-point and repeating through a succession of phases the same course" (1874:92).

Reliance on the phenomena of ontogeny when making laws concerning the "development of types"—that is, the development within the higher categories of animals—has had curious consequences. It led Agassiz to espouse various "evolutionary" theories that seem quite bizarre unless one understands that phylogeny to him was the same unfolding as is ontogeny. One of these theories is that of recapitulation, or the "biogenetic law." According to it, ontogeny is nothing but a short recapitulation of phylogeny. Any organism that belongs to the same prototype will go through the same stages, but the development of the lower organisms will be arrested at an earlier stage than that of the higher organisms. This concept, usually credited to Haeckel, was not new with him. One finds traces of it in Leibniz and finds it spelled out in considerable detail by Robinet and by Bonnet. These ideas were conveyed to Cuvier through Kielmeyer and perhaps through Cuvier to Agassiz. Indeed, Agassiz reflects so many of Bonnet's ideas that it can hardly be doubted that the citizen of Neuchâtel was himself thoroughly familiar with the writings of the citizen from Geneva. That Haeckel received his ideas directly from Bonnet is suggested by the fact that he adopted from him the term "palingenesis." How completely Agassiz accepted the theory of recapitulation may be documented by a few quotations. "The phases of development of all living animals correspond to the order of succession of their extinct representatives in past geological times. As far as this goes, the oldest representatives of every class may then be considered as embryonic types of their respective orders or families among the living" (1857:115–116). "I confess I could not say in what the mental faculties of a child differ from those of a young Chimpanzee" (1857:60). If it is true that ontogeny reveals the potentialities of the type, it follows that the study of individual development should give important clues to the student of relationships:

> Embryology has . . . a wider scope than to trace the growth of individual animals . . . it ought also to embrace a comparison of these forms and the successive steps of these changes between all the types of the animal kingdom, in order to furnish definite standards of their relative standing, of their affinities, of the correspondence of their organs in all their parts. . . .
>
> I have satisfied myself long ago, that Embryology furnishes the most trustworthy standard to determine the relative rank among animals. (1857:84)

Agassiz illustrates this thesis by reference to the development of the frog, in which the earlier stages of the tadpole resemble fishes. "Next it assumes a shape reminding us more of the Tritons and Salamanders, and ends with the structure of the Frog . . . it cannot be doubted, that the earlier stages of growth of an animal exhibit a condition of relative inferiority, when contrasted with what it grows to be, after it has completed its development" (1857:84–85).

Long before Agassiz it had been suggested by Bonnet that a careful study of the ontogenetic path should make it possible to trace the history of the type, not only backward but also forward, since its future lies of course already preformed within the type. This romantic idea greatly appealed to Agassiz. He devotes an entire section of his great *Natural History of the United States* to "Prophetic Types among Animals."

> We have seen in the preceding paragraph, how the embryonic conditions of higher representatives of certain types, called into existence at a later time, are typified, as it were, in representatives of the same types, which have existed at an earlier period. . . . Certain types which are frequently prominent among the representatives of past ages, combine in their structure, peculiarities which at later periods are only observed separately in different, distinct types. Sauroid Fishes before Reptiles, Pterodactyles before Birds, Ichthyosauri before Dolphins, etc. . . . Such types, I have for some time past, been in the habit of calling *prophetic types*. . . . [They] afford . . . the most unexpected evidence, that the plan of the whole creation had been maturely considered long before it was executed. (1857:116–117)

He does not explain just why the existence of such less perfect and less successful forms as the sauroid fishes, the pterodactyls, and the ichthyosaurs should prove the existence of a mature and carefully prepared blueprint of the creation. It would seem more appropriate to describe this method of creation as a trial and error approach. This is the way it appeared to Robinet: "All the varieties intermediate between the prototype and man I regard as so many essays of Nature, aiming at the most perfect, yet unable to obtain it except through this innumerable sequence of sketches. I think we may call the collection of the preliminary studies the apprenticeship of Nature in learning to make a man" (Lovejoy 1936:280).

Again and again in his discussions Agassiz equates ontogeny with phylogeny, the development of the individual with evolutionary change. As a consequence he cannot understand why phylogenetic changes should take so much time, and this leads him to criticize Darwin: "He would have us believe that it required millions of years to effect any one of these changes; when far more extraordinary transformations are daily going on, under our eyes, in the shortest periods of time, during the growth of animals.—He would have us believe that animals acquire their instincts gradually; when even those that never see their parents, perform at birth the same acts, in the same way, as their progenitors" (1860b:146). In this confusion of ontogeny and phylogeny, Agassiz was not alone. In fact it was almost universal in the eighteenth and nineteenth centuries. The shift of the originally purely embryological term "evolution" to phylogenesis is an outstanding example. In German, the word *Entwicklungsgeschichte* was used interchangeably for both ontogeny and phylogeny throughout the nineteenth century. There is a deplorably great number of biologists left who to this very day believe that there must be a close correlation between ontogeny and phylogeny because both deal with processes occurring in the time dimension. The complete difference between the two phenomena becomes most apparent if expressed in terms of the information theory: ontogeny is the decoding of programed information, phylogeny is the creating of ever-new programs of information and the survival of the most successful ones.

The four ideas just discussed are the cornerstones of Agassiz's conceptual world. It must be admitted that it would be impossible to believe in evolution by descent from common ancestors if one were convinced of the complete rationality of the world, the existence of discontinuous types, and a mere unfolding of preformed characters both in ontogeny and in phylogeny. It would be a mistake, however, to consider Agassiz's essay entitled "The Fundamental Relations of Animals to One Another and to the World in Which They Live," which forms Chapter 1 of Part I of his *Natural History of the United States,* an anachronism or merely a set of philosophical speculations. Far from it. Helped by his phenomenal memory, his facility with languages, and his enormously broad interests, Agassiz was able to make this chapter a survey of contemporary zoology that was truly breathtaking. It presents a nearly complete survey of the literature on classification, embryology, and other branches of zoology. And in his anti-evolutionary discussions Agassiz by no means always had the worst of the argument. As the disciple of Cuvier he had no trouble proving that Lamarck and other early evolutionists were completely mistaken when trying to place all the orders and classes of animals in a single phylogenetic sequence. Triumphantly he pointed out that there are four major branches, those established by Cuvier, and that they differ so completely from each other in all their structures that no homologies can be established among them. His further demonstration, discussed above, that physical causes cannot account for the observed adaptations and that the assumption of spontaneous generation is an absurdity made him think, when he published his work in 1857, that he had struck a mortal blow against all transmutationist speculations. Little did he suspect that he had merely cut off one or two heads of a Lernaean Hydra.

THE INTERPRETATION OF THE EVOLUTIONARY EVIDENCE

Darwin and Agassiz had access largely to the same facts. It is highly interesting for the student of the scientific method to see how differently the two authors handled them. Reading Agassiz's descriptions of the "natural relations of animals," of all sorts of adaptations, and of the similarities and differences of faunas and floras, the modern student finds it almost impossible to comprehend how Agassiz could have escaped the obvious conclusion. Yet his framework of concepts established an impenetrable screen. When trying to explain a set of facts, he was severely limited in the availability of possible interpretations. The very alternatives that would seem most obvious to the modern worker were generally not included. Let us consider this in the light of some concrete examples.

Let us take, for instance, the role of the physical environment in faunal changes from one geological period to the next. Here Agassiz concludes: "The question remains simply this: When the change takes place, does it take place spontaneously, under the action of physical agents, according to their law, or is it produced by the intervention of an agency not in that way at work before or afterwards?" (1857:52).

Physical agents to him implied chaos. Observing the organic world to be orderly, to be full of the most miraculous adaptations, he regarded it as rational and planned in all of its details. The outcome of the choice between "physical causes" or "creator" was inevitable. To accept the first alternative would have meant to Agassiz the end of all zoological studies: "Enough is known to repudiate the assumption of their [organized beings'] transmutation, as it does not explain the facts, and shuts out further attempts at proper investigations" (1857:52). He was unaware that the alternative posed by him was not between creation and evolution, but between spontaneous origin of whole new faunas (a highly aberrant evolutionary theory!) by creation or by the spontaneous action of physical causes. To accept the latter alternative would indeed force one to believe in miracles and would shut out "further attempts at proper investigations." To accept evolution by gradual transformation, on the other hand, would have the opposite effect. Nothing could have been more unexpected to Agassiz than the enormous stimulation to the study of animals and plants that came in the wake of the acceptance of the evolutionary theory. The rigidity of Agassiz's conceptual framework is nowhere more in evidence than in this statement:

> All attempts to explain the origin of species may be brought under two categories: viz. 1st, some naturalists admitting that all organized beings are created, that is to say, endowed from the beginning of their existence with all their characteristics, while 2d, others assume that they arise spontaneously. This classification of the different theories of the origin of species, may appear objectionable to the supporters of the transmutation theory; but I can perceive no essential difference between their views and the old idea that animals may have arisen spontaneously. They differ only in the modes by which the spontaneous appearance is assumed to be effected. (1860b:149)

It certainly makes an objective analysis difficult if one classifies as "spontaneous" any evolutionary change, whether gradual or sudden, whether due to induction or selection, whether due to physical agents or Lamarckian forces, whether instantaneous or requiring several millions of years.

No wonder then that Agassiz missed the most obvious evidence for evolutionary change. A few examples of this may be worth quoting. He overlooks no opportunity to point out the incredible similarity in all details of structure among "related" species. But such similarity is not limited only to structure. "The more I learn upon [the mode of life of closely allied species], the more am I struck with the similarity in the very movements, the general habits, and even in the intonation of the voices of animals belonging to the same family; that is to say, between animals agreeing in the main in form, size, structure, and mode of development" (1857:59). Yet such similarity does not prove common ancestry, says Agassiz. "There is nothing parallel between the relations of animals belonging to the same genus or the same family, and the relations between the progeny of common ancestors. . . . No degree of affinity, however close, can, in the present state of our science, be urged as exhibiting any evidence of community of descent" (1860b:148). Nor does a succession of similar types in a sequence of geological strata indicate genetic relationship. To Agassiz it is merely the manifestation of a "plan." "The first representatives of each class stand in definite relations to their successors in later periods, and as their order of apparition corresponds to the various degrees of complication in their structure, and forms natural series closely linked together, this natural gradation must have been contemplated from the very beginning" (1857:24-25). This claim is repeated again and again (e.g. 1857:98-101).

The possibility of a slow, gradual molding of one type into another was so foreign to Agassiz's conceptual framework that it never seems to have entered his mind at all. For instance, he gives an excellent description of the morphological series in leg and toe reduction among the skinklike lizards. This very material is still used in zoological textbooks as an excellent illustration of evolutionary trends. But what does Agassiz say about it? "Who can look at this diagram, and not recognize in its arrangement the combinations of thought? . . . Similar series, though less conspicuous and more limited, may be traced in every class of the animal kingdom, not only among the living types, but also among the representatives of past geological ages, which adds to the interest of such series in showing, that the combinations include not only the element of space, indicating omnipresence, but also that of time, which involves prescience" (1857:45-46). Many similar statements could be cited showing with what ease Agassiz was able to explain the most obvious evolutionary evidence in terms of creation. It may not be out of place, therefore, to say a few words on Agassiz's own theory of the universe as a product of creation.

AGASSIZ ON CREATION

There is not enough space here to engage in the fascinating task of investigating the sources and details of Agassiz's theory of evolution by creation. It appears evident, however, that he was much influenced by Bonnet. Like him, he believed in a long series of cataclysms and new creations; like him, but far more definitely, he was an ovist, believing that the new life after each catastrophe arose from eggs; and, like him, he believed that the sequence of epochs and organic types indicates a progress from "lower" to "higher."

Cuvier's discovery of a seriation of fossil faunas in the Paris Basin appeared to be a brilliant confirmation of these hypotheses.[3] Yet difficulties soon arose. Cuvier had dealt with only a few faunas, requiring only a handful of creations. As stratigraphic research progressed in the first half of the nineteenth century, it became increasingly apparent that the number of successive geological horizons, each with its own fauna, was perfectly enormous. Without blinking, Agassiz stuck by his guns and insisted that each was the product of a separate creation. Nor was he shaken in this belief by the discovery that a certain percentage of forms in each fauna was identical with species known from lower strata, in fact that there were a number of "permanent types" that could be followed unchanged through vast geological epochs.

It would appear that Agassiz's theory of a never-ending series of creations was far more in conflict with the biblical account of creation than was Asa Gray's belief that God had created a universe endowed with the potency of evolving, a concept of evolution possessing biblical dignity and grandeur.

THE PUBLICATION OF ON THE ORIGIN OF SPECIES

Darwin's publication must have been a staggering blow to Agassiz—not because it was another publication upholding the detestable theory of transmutation but because it was so obviously immune to the majority of the arguments against the evolutionary theory elaborated by Agassiz just 2 years before. Cuvier's opponents—and they were mostly the same as those against whom Agassiz had been arguing—had based their theories essentially on a priori considerations of various sorts. It was easy to counter them with other a priori considerations. As naturalists they were dilettantes and no match for zoologists as erudite as Cuvier and Agassiz. That Darwin's was a totally new approach to the subject of evolution was well appreciated by Agassiz:

> Darwin has placed the subject on a different basis from that of all his predecessors, and has brought to the discussion a vast amount of well-arranged information, a convincing cogency of argument, and a captivating charm of presentation. His doctrine appealed the more powerfully to the scientific world because he maintained it at first not upon metaphysical ground but upon observation. Indeed it might be said that he treated his subject according to the

3. Contrary to common belief, Cuvier himself was not a creationist; see Potonié (1957).

best scientific methods, had he not frequently overstepped the boundaries of actual knowledge and allowed his imagination to supply the links which science does not furnish. (1874:94)

Darwin's strictly empirical approach of patiently piling fact upon fact in an almost pedestrian manner until the sheer weight of the evidence made a conclusion inevitable was received by his contemporaries with mixed feelings. The nonbiologists, in particular, felt excluded by an approach not based on "pure reason" and relieved their frustration by ridiculing Darwin's "clumsiness" as compared, for instance, with the grandiose sweep of Lamarck's *Philosophie zoologique*. It is, to put it mildly, amusing to read how Agassiz refers to Darwin's inductive method as "speculation." There is no doubt that the year 1859 ushered in a new era in the history of evolutionary biology, the era of the scientific method. It is indeed fully justifiable to refer to the entire preceding period, dominated by speculation and intuition, as the prehistory of evolutionary science. Evolutionary biology as a science started in 1859.

Agassiz was 52 years old when Darwin's work was published. He was in the midst of building his great new natural history museum, he was the most popular lecturer in America, and he had his teaching and enormous social obligations. All in all, he was unable to find the intellectual peace to undertake a critical evaluation of the foundation of his concepts and beliefs. So different was Darwin's approach that Agassiz was unable not only to understand it fully, but, where he attempted to rephrase Darwin's arguments in his own words, to do so correctly. Yet he tried manfully to refute Darwin's evolutionary proofs, particularly in two publications, one issued in 1860 and the other prepared in 1873 shortly before his death and published posthumously in 1874.

AGASSIZ'S REFUTATION OF THE EVOLUTIONARY EVIDENCE

Among Darwin's many arguments in favor of common descent by modification Agassiz singled out for criticism primarily two series, those dealing with the geological record and those dealing with the mechanisms responsible for evolutionary change. As paleontologist and embryologist Agassiz felt best qualified to deal with these two sectors.[4]

The Geological Record

The first point he takes up is the diversity of different faunas. "Before it could be granted that the great variety of types which occur at any later periods has arisen from a successive differentiation of a few still earlier types, it should be shown that in reality in former periods the types are fewer and less diversified" (1854b:318). This, however, says Agassiz, is not the case. There are 1,200 species of fossil seashells

---

4. Quotations in the following summary are sometimes taken from earlier works of Agassiz, as affording more complete statements.

known from the Eocene beds of the Paris Basin, and only 600 living
species in the Mediterranean, "affording, at once, a very striking evidence of the greater diversity and greater number of species of that geological period [the Eocene of more than 60 million years ago] when
compared even with those of a wider geographical area at the present
day" (1854b:310). Agassiz admits that the Paris fauna was so rich because it was a tropical fauna, yet he insists that its richness "is much
greater than that of any local fauna of the present period, even within
the tropics." Subsequent researches have not substantiated Agassiz's
claims. Some recent local lists of seashells record the following number
of species: Port Alfred, South Africa (nontropical), 721; west coast of
America from California to Panama, 1,600; east coast of America from
Greenland to Texas, 2,632; and Philippines, 4,152. Indeed, the Recent
fauna of marine mollusks of the Indo-Malayan area is estimated to be
6,000 to 8,000 species, far in excess of the 1,200 species from the
Eocene of Paris.[5]

   If all types of organisms have existed from the beginning, it follows
that all major types must be present in the oldest fossil-bearing rocks,
while the still earlier rocks must be void of any evidence of organic life.
This, claims Agassiz, is indeed the case. Every piece of inconvenient
evidence is eliminated by a special *ad hoc* explanation: the absence of
coelenterates ("Acalephs") in the older strata is due to the absence of
hard parts in their bodies; the absence of vertebrates in these formations is due, he says, to the incompleteness of the fossil record, for
vertebrates should be present upon "physiological grounds" (1857:24).
His major thesis, that there were as many species of animals at the very
beginning of the world as there are now, is the same that Bonnet defended so vigorously, the same that, as Lovejoy has shown, goes back
explicitly or implicitly all the way to Plato and his principle of plenitude.

   Agassiz's second argument is based on the apparent fixity of species.
Whenever animals and plants of two successive geological periods are
compared, he says, they are either completely identical or quite different. "None of those primordial forms of life, which naturalists call species, are known to have changed during any of these periods. It cannot
be denied, that the species of different successive periods are supposed
by some naturalists to derive their distinguishing features from changes
which have taken place in those of preceding ages; but this is a mere
supposition" (1857:51). As the strongest argument in favor of the
fixity of species he considers the demonstration that the animals and
plants discovered in the Egyptian tombs during Napoleon's expedition
were indistinguishable from living species, as pointed out by Cuvier and
other naturalists. It was not realized by Agassiz that the approximately
5,000 years elapsed since the entombments compare to the total length

   5. According to information kindly supplied by Dr. W. J. Clench, Museum of Comparative
Zoology, Harvard University.

of duration of a species as do about 80 days to the life expectancy of a man. It would be highly improbable that one could demonstrate an evolutionary change in such an exceedingly short fraction of the total life span of a species. Agassiz concludes the argument with the statement that he will not accept the transmutation theory "as long as no fact is adduced to show that any one well known species among the many thousands that are buried in the whole series of fossiliferous rocks, is actually the parent of any one of the species now living" (1860b:144). He nowhere states, however, what kind of evidence he would accept as proof. If one wishes to be obstinate, it is possible to claim to this very day that even the most perfect vertical series of intergrading species of fossils is nothing but proof for the "unfolding" (*evolutio*) of preformed germs in the sense of Bonnet.

Agassiz's next anti-evolutionary argument is best stated in his own words. "The supporters of the transmutation theory . . . never can make it appear that the definiteness of the characters of the class of Birds is the result of a common descent of all Birds, for the first Bird must have been brother or cousin to some other animal that was not a Bird, since there are other animals besides Birds in this world, to no one of which any bird bears as close a relation as it bears to its own class" (1860b:154). By a curious coincidence this statement was published exactly a year before the description of *Archaeopteryx*, a virtually perfect intermediate between birds and reptiles. Although numerous additional "missing links" have been discovered since that day, hardly any other of them connects two major types of animals in quite so ideal a manner as does *Archaeopteryx*. It is, of course, virtually impossible to comprehend gradual evolution if one places the diversity of the organic world in the rigid pigeonholes of "types." No doubt Agassiz would have reacted to *Archaeopteryx* as did the few remaining anti-evolutionists of a later period, who categorized it as a separate type of its own that had nothing to do with the evolution of birds from reptilian ancestors.

One of Agassiz's chief arguments is based on Cuvier's demonstration of the sharp distinction between consecutive faunas as well as the absence of any missing links between the major types. He was, therefore, particularly upset by Kowalevsky's discovery of a chorda in the ascidians that would indicate that they are a link between the mollusks (with which the ascidians had been classified by Cuvier and Agassiz) and the vertebrates (which are characterized by the chorda). It would have pleased Agassiz to learn that this discovery does not make the ascidians a missing link. It is now known that they are not to be connected with the mollusks but form the phylum chordates together with the vertebrates. The great phyla of the animal kingdom are as far apart now as they were in Agassiz's time. All the evidence indicates that they diverged from each other but that this happened in the pre-Cambrian days (more than 500 million years ago). There is no fossil record available to indicate the steps by which this divergence took place. Our ignorance concerning the origin of the major types is as great today as it was in Agassiz's time.

Finally, Agassiz raises one point concerning the fossil record by which he thinks he can inflict a mortal wound on the transmutation theory. If this theory is right, says Agassiz, the "lowest" representative of a type should be found in the lowest strata and the "highest" in the most recent strata. But this is not what one finds!

> What then are the earliest known Vertebrates? They are Selachians (sharks and their allies) and Ganoids (garpikes and the like), the highest of all living fishes, structurally speaking. . . . In all their features the Selachians, more than any other fishes, resemble the higher animals. They lay few eggs, the higher kinds giving birth only to three, four, or five at a brood, whereas the common fishes lay myriads of eggs, hundreds of thousands in some instances, and these are for the greater part cast into the water to be developed at random.   (1874:100)

In this argument Agassiz is quite oblivious to the fact that he speaks like a true Aristotelian, selecting his criteria of "high" and "low" on the basis of *a priori* considerations: "The limitation of the young is unquestionably a mark of superiority. The higher we rise in the scale of animal life the more restricted is the number of offspring. In proportion to this reduction in number, the connection of the offspring with the parent is drawn closer, organically and morally, till this relation becomes finally the foundation of all social organization, of all human civilization." The facts of internal fertilization and placenta formation among the sharks are quoted as additional evidence of their "superiority," and yet, Agassiz continues, these are the first vertebrates to be found in the fossil record, while *Amphioxus* and the lampreys (primitive chordates) are not found as fossils at all, but only in the present period, to which we ourselves belong. "This certainly does not look like a connected series beginning with the lowest and ending with the highest, for the highest fishes come first and the lowest come last." Discoveries in the fossil history of the fishes made since Agassiz have completely demolished his argument. Preceding the selachians, groups of primitive fishes have been found that appear to be directly ancestral to the lampreys and to the more advanced fishes. The bony fishes, in spite of the myriads of eggs they lay, are a comparatively recent development, derived from ganoidlike ancestors. More important, these finds show the complete invalidity of Agassiz's *a priori* criteria of what is "low" and what is "high." Many of the earliest fishes were as elaborate in their structure as any of their descendants. No wonder Agassiz finally came to the conclusion: "The whole history of geological succession shows us that the lowest in structure is by no means necessarily the earliest in time, either in the Vertebrate type or any other" (1874:101). It all depends on how we define "lowest in structure."

The Mechanisms of Evolution

Agassiz throughout his life was scornful of any theories that attempted to elaborate on the causes and mechanisms that might be responsible for evolutionary changes. I have already discussed his opinion that "physical causes" are a brute sledgehammer that cannot improve a delicate watch. Furthermore, wherever environment does have a slight effect, as in raising the milk production of a well-fed cow, it has no lasting influence, for this improvement will not be transmitted to her offspring. Agassiz was entirely right in his refutation of the inheritance of acquired characters. Indeed, most of his arguments against the environmentalists of the schools of Lamarck and Geoffroy St. Hilaire are well taken, and supported by all modern evolutionists. But where Agassiz uses strictly genetic arguments, he is a child of his times: the century of genetics had not yet arrived. Yet, on the whole, Agassiz's discussion of heredity is not much worse than Darwin's excursions in this field. Fortunately, Darwin was satisfied, particularly in the *Origin,* to take for granted the existence of genetic variability and the replenishment of genetic variability as the source material for natural selection. Indeed, as Huxley (1958) has correctly pointed out, if Darwin had been familiar with Mendel's work he might well have been misled into some sort of saltational mutationism, as were De Vries, Bateson, and other early Mendelians.

It is no longer doubted by thinking biologists that natural selection is the key mechanism of evolution. Hundreds if not thousands of objections to the universal power of natural selection have been raised during the past hundred years but they have uniformly been shown to be ill considered. It is interesting to look into Agassiz's position on this problem. Let me say at once that no typologist has ever understood natural selection, because he cannot possibly understand it. Natural selection is a population phenomenon, a shifting of statistical averages owing to differential reproduction. This is a mode of thinking so different from that of a typologist that it is bound to be incomprehensible to him (Essay 3). Agassiz was no exception. For him, "the organized beings which live now, and have lived in former geological periods, constitute an organic whole, intelligibly and methodically combined in all of its parts" (1860b:147). This, he says, cannot possibly have resulted from the play of blind physical forces. To apply the term "natural selection" to such accidental causes is a mistake, because "selection implies design; the powers to which Darwin refers the order of species, can design nothing." Here he is merely arguing against the term "selection," which Darwin had chosen in a deliberate analogy to the artificial selection of the animal and plant breeders. This, as we now see it, was an entirely legitimate terminology since, in either case, the survival into the next generation is determined by "superiority." In one case it is superiority in the eyes of the breeder, in the other case superiority of reproductive success. What Agassiz plainly missed was that no two individuals are genetically identical and, as Darwin emphasized, that not all individuals reach reproductive age and reproduce with equal success. One has a choice of only two possibilities. Either one ascribes the differences in survival and reproductive success entirely to chance or one admits that the genetic endowment of an individual may contribute to this success. If one admits the second alternative, one automatically admits natural selection. And, to express this once more in terms of "information," one can say that, owing to the genetic phenomena of mutation and recombination, every individual has a slightly different program of information controlling its development and response to the total environment. Some of these programs are more "successful" than others and therefore will contribute more than their share to the genetic reservoir of the next generation. Unsuitable programs, on the other hand, will produce less successful phenotypes and will have a smaller chance to be returned to the gene reservoir of the population. It has taken us a hundred years to reach such a sophisticated way of expressing natural selection. It is therefore understandable that Agassiz, rooted in an alien conceptual world, never really came to grips with the problem at all.

THE PASSING OF THE YEARS

Agassiz's first great outburst against the Darwinian theory came in a series of open discussions at the American Academy of Arts and Sciences, culminating in a detailed rebuttal published in the introduction to Volume III of his *Natural History of the United States* (1861) and preprinted verbatim in the *American Journal of Science and Arts* (1860b). This was the end of the scientific debate. In the next dozen years Agassiz took his case to the public. In lectures, popular articles, and books he pleaded the cause of creationism, reiterating his previous arguments in a form intelligible to the layman. However, shortly before his death, he turned once again to a more serious and systematic consideration of the question of evolution. The results of these studies he presented, in the fall of 1873, in a series of lectures, of which only the first was completed for the press, to be published posthumously in 1874. The situation with respect to Darwin had greatly changed since 1860. Darwin no longer was a maverick and rebel to whose theory Agassiz could refer as a "scientific mistake, untrue in its facts, unscientific in its method, and mischievous in its tendency" (1860b:154). The theory of evolution had by now been almost universally adopted and Darwin had become the grand old man of biology. As a consequence, Agassiz is far more gentle in his references to Darwin.

Indeed, he gives every impression of a sincere attempt to do justice

to the new theory. Yet Louis Agassiz was unable to give up old loyal-
ties. All his life he had felt himself the disciple of that great master
Cuvier, and in his old age he was not going to abandon him. As a consequence, he maintained what he had learned as a youth, even in the face
of newly discovered zoological facts. When Leuckart proposed dividing
Cuvier's radiates into coelenterates and echinoderms Agassiz protested
strongly: "The organs and the whole structural combination are the
same in the two divisions" (1874:93-94). As Leuckart had shown and
as is now known to every zoologist, this assertion is not correct: the two
phyla are as different from each other in their basic structure as any
phyla in the animal kingdom. In a similar spirit of loyalty Agassiz did
not accept even the smallest part of the transmutation theory. One may
question whether a compromise was possible. Both Darwin's theory of
common descent by modification through natural selection and
Agassiz's theory of successive special creations are so completely self-
contained and mutually exclusive that their mingling is hardly conceivable. Once one admits either a "little bit" of special creation or a "little
bit" of gradual evolution, one has no reason for not accepting all of one
or the other.

Agassiz's attitude toward the theory of evolution is an extraordinarily
interesting phenomenon in the history of the advance of scientific ideas.
It is another illustration of the familiar concept that an age has to be
ready for a new idea or a new discovery, with the corollary that contemporaries may live in different ages, some being directed more forward, others more backward. Darwin's great fortune was that he was
just enough ahead of his time to be a leader and not enough ahead to be
ignored. Agassiz's misfortune was to have absorbed in his youth a
Zeitgeist that was unsuitable for mixing with the revolutionary new
ideas. He was, one may say, a victim of the thoroughness of his education.

REFERENCES

Agassiz, L. 1854a. Notice of a collection of fishes from the southern bend of the
    Tennessee River, in the state of Alabama. *Amer. Jour. Sci. and Arts*, 2nd ser.,
    17:297-308, 353-365.
—— 1854b. The primitive diversity and number of animals in geological times.
    Amer. Jour. Sci. and Arts, 2nd ser., 17:309-324.
—— 1857. *Contributions to the natural history of the United States of America*,
    vol. 1. Little, Brown, Boston.
—— 1860a. Minutes of meeting of March 13. *Proc. Amer. Acad. Arts and Sci.*,
    4:410.
—— 1860b. Prof. Agassiz on the origin of species. *Amer. Jour. Sci. and Arts*, 2nd
    ser., 30:142-154.
—— 1860c. Minutes of meeting of October 9. *Proc. Amer. Acad. Arts and Sci.*,
    5:72.
—— 1874. Evolution and permanence of type. *Atlantic Monthly*, 33:92-101.

Guyénot, E. 1941. *L'évolution de la pensée scientifique: les sciences de la vie aux XVIIe et XVIIIe siècles. L'idée d'évolution.* Albin Michel, Paris.

Huxley, J. 1958. The emergence of Darwinism. *Jour. Linn. Soc. London, Zool.,* 45:1–14, *Bot.,* 56:1–14.

Lovejoy, A. O. 1936. *The great chain of being.* Harvard University Press, Cambridge, Mass.

Potonié, R. 1957. Zu Cuviers Kataklysmentheorie. *Paläont. Zeit.,* 31:9–14.

Stresemann, E. 1975. *Ornithology: from Aristotle to the present,* trans. H. J. and C. Epstein. Harvard University Press, Cambridge, Mass.

Zimmermann, W. 1953. *Evolution: Geschichte ihrer Probleme und Erkenntnisse.* Alber, Freiburg.

19
The Nature of the
Darwinian Revolution

The road on which science advances is not a smoothly rising ramp; there are periods of stagnation and periods of accelerated progress. Some historians of science have recently emphasized that there are occasional breakthroughs, scientific revolutions (Kuhn 1962), consisting of rather drastic revisions of previously maintained assumptions and concepts. The actual nature of these revolutions, however, has remained highly controversial (Toulmin 1966). When we look at those of the so-called scientific revolutions that are most frequently mentioned, we find that they are identified with the names Copernicus, Newton, Lavoisier, Darwin, Planck, Einstein, and Heisenberg; in other words, with one exception, all of them are revolutions in the physical sciences.

Does this focus on the physical sciences affect the interpretation of the concept "scientific revolution"? I am taking a new look at the Darwinian revolution of 1859, perhaps the most fundamental of all intellectual revolutions in the history of mankind. It not only eliminated man's anthropocentrism, but affected every metaphysical and ethical concept, if consistently applied.[1]

The earlier prevailing concept of a created, and subsequently static, world was miles away from Darwin's picture of a steadily evolving world. Kuhn (1962) maintains that scientific revolutions are characterized by the replacement of an outworn paradigm by a new one. But a paradigm

Revised from "The nature of the Darwinian revolution," *Science*, 176 (1971):981–989; copyright 1971 by the American Association for the Advancement of Science.

1. [It might be legitimate to object to speaking of *the* Darwinian revolution in the singular. Clearly, as is shown in this essay, at least two major revolutions were involved, even though Darwin himself lumped them together, referring to both of them as "my theory." One is simply the replacement of the static world of creationism by the dynamic world of evolution. Even though Lamarck had promoted this new world view from 1801 on, it was clearly Darwin's massive documentation and persuasive argumentation that made this view victorious. This is why it has so often been referred to by the historians of ideas as the Darwinian revolution. The expert evolutionist, however, means something else when he speaks of Darwinism. He means Darwin's solution of the problem of the mechanism of evolution: natural selection. Because it was a purely mechanistic explanation of adaptation and all other phenomena that previously had served as evidence for "design," natural selection was a particularly daring and altogether novel concept. Replacing design by a mechanistic process was the Darwinian revolution *sensu stricto*.]

is, so to speak, a bundle of separate concepts, and not all of these are changed at the same time. In this analysis of the Darwinian revolution, I am attempting to dissect the total change of thinking involved in the Darwinian revolution into the major changing concepts, to determine the relative chronology of these changes, and to test the resistance to these changes among Darwin's contemporaries.

The idea of evolution had been widespread for more than 100 years before 1859. Evolutionary interpretations were advanced increasingly often in the second half of the eighteenth and the first half of the nineteenth centuries, only to be ignored, ridiculed, or maligned. What were the reasons for this determined resistance?

The history of evolutionism has long been a favorite subject among historians of science (Eiseley 1958; Glan et al. 1959; Greene 1971; Ghiselin 1969; de Beer 1964; Gillispie 1951). Their main emphasis, however, has been on Darwin's forerunners, and on any and every trace of evolutionary thinking prior to 1859, or on the emergence of evolutionary concepts in Darwin's own thinking. These are legitimate approaches, but it seems to me that nothing brings out better the revolutionary nature of some of Darwin's concepts (1859) than does an analysis of the arguments of contemporary anti-evolutionists.

Cuvier, Lyell, and Louis Agassiz, the leading opponents of organic evolution, were fully aware of many facts favoring an evolutionary interpretation, and likewise of the Lamarckian and other theories of transmutation. They devoted a great deal of energy to refuting evolutionism (Cuvier 1817; Coleman 1964; Lyell 1835; Agassiz 1857) and supported instead what, to a modern student, would seem a less defensible position. What induced them to do so?

It is sometimes stated that they had no other legitimate choice, because—it is claimed—not enough evidence in favor of evolution was available before 1859. The facts refute this assertion. Lovejoy (1959), in a superb analysis of this question, asks: "At what date can the evidence in favor of the theory of organic evolution . . . be said to have been fairly complete?" Here, one can perhaps distinguish two periods. During an earlier one, lasting from about 1745 to 1830, much became known that suggested evolution or, at least, a temporalized scale of perfection (Lovejoy 1936). Names like Maupertuis (1745), de Maillet (1749), Buffon (1749), Diderot (1769), Erasmus Darwin (1794), Lamarck (1809), and E. Geoffroy St. Hilaire (1818) characterize this period. Enough evidence from the fields of biogeography, systematics, paleontology, comparative anatomy, and animal and plant breeding was already available by about 1812 (date of Cuvier's *Ossemens Fossiles*) to have made it possible to develop some of the arguments later made by Darwin in the *Origin of Species*. Soon afterward, however, much new evidence was produced by paleontology and stratigraphy, as well as by biogeography and comparative anatomy, with which only the evolutionary hypothesis was consistent; these new facts "reduced the rival hypothesis to a grotesque absurdity" (Lovejoy 1959). Yet, only a hand-ful of authors, including Meckel (1821), Chambers (1844), Unger (1852), Schaaffhausen (1853), and Wallace (1855), adopted the con-cept of evolution, while such leading authorities as Lyell, R. Owen, and Louis Agassiz vehemently opposed it.

There is not enough space here to marshal the abundant evidence in favor of evolution which existed by 1830. A comprehensive listing has been provided by Lovejoy, although the findings of systematics and biogeography must be added to his tabulation. The patterns of animal distribution were particularly decisive evidence, and it is no coincidence that Darwin devoted to them two entire chapters in the *Origin*. In spite of this massive evidence, creationism remained "the hypothesis tena-ciously held by most men of science for at least twenty years before 1859" (Lovejoy 1959). It was not a lack of supporting facts, then, that prevented the acceptance of the theory of evolution, but rather the power of the opposing ideas.

Curiously, a number of nonscientists, particularly Robert Chambers (1844) and Herbert Spencer, saw the light well before the professionals. Chambers, the author of *Vestiges of the Natural History of Creation*, de-veloped quite a consistent and logical argument for evolutionism, and was instrumental in converting A. R. Wallace, R. W. Emerson, and A. Schopenhauer to evolutionism. As was the case with Diderot and Eras-mus Darwin, these well-informed and broadly educated lay people looked at the problem in a "holistic" way, and thus perceived the truth more readily than did the professionals, who were committed to certain well-established dogmas. A view from the distance is sometimes more revealing, for the understanding of broad issues, than the myopic scrutiny of the specialist.

POWER OF RETARDING CONCEPTS

Why were the professional geologists and biologists so blind when the manifestations of evolution were staring them in the face from all direc-tions? Darwin's friend Hewett Watson put it this way in 1860 (F. Dar-win 1888:vol. 2:226): "How could Sir Lyell . . . for thirty years read, write, and think on the subject of species *and their succession*, and yet constantly look down the wrong road?" Indeed, how could he? And the same question can be asked for Louis Agassiz, Richard Owen, almost all of Lyell's geological colleagues, and all of Darwin's botanist friends from Joseph Hooker on down. They all displayed a nearly complete resistance to drawing what to us would seem to be the inevitable con-clusion from the vast amount of evidence in favor of evolution.

Historians of science are familiar with this phenomenon; it happens almost invariably when new facts cast doubt on a generally accepted theory. The prevailing concepts, although more difficult to defend, have such a powerful hold on the thinking of all investigators that they find it difficult, if not impossible, to free themselves of these ideas. To illustrate this by merely one example, I would like to quote a statement by Lyell: "It is idle . . . to dispute about the abstract possibility of the conversion of one species into another, when there are known causes, so much more active in their nature, which must always intervene and prevent the actual accomplishment of such conversions" (1835:vol.3: 162). Actually one searches in vain for a demonstration of such "known causes" and any proof that they "must" always intervene. The cogency of the argument relied entirely on the validity of silent assumptions.

In the particular case of the Darwinian revolution, what were the dominant ideas that formed roadblocks against the advance of evolutionary thinking? To name these concepts is by no means easy because they are silent assumptions, never fully articulated. When these assumptions rest on religious beliefs or on the acceptance of certain philosophies, they are particularly difficult to reconstruct. This is the major reason there is so much difference of opinion in the interpretation of this period. Was theology responsible for the lag, or was it the authority of Cuvier or Lyell, or the acceptance of catastrophism (with progressionism), or the absence of a reasonable explanatory scheme? All of these interpretations and several others have been advanced, and all presumably played some role. Others, particularly the role of essentialism, have so far been rather neglected by the historians.

NATURAL THEOLOGY AND CREATIONISM

The first half of the nineteenth century witnessed the greatest flowering of natural theology in Great Britain (Gillispie 1951:15). It was the age of Paley and the Bridgewater Treatises, and virtually all British scientists accepted the traditional Christian conception of a Creator God. The Industrial Revolution was in full swing, the poor workingman was exploited unmercifully, and the goodness and wisdom of the Creator were emphasized constantly to soothe guilty consciences. It became a moral obligation for the scientist to find additional proofs for the wisdom and constant attention of the Creator. When Chambers in his *Vestiges* (1844) dared to replace direct intervention of the Creator by the action of secondary causes (natural laws), he was roundly condemned. Although the attacks were ostensibly directed against errors of fact, virtually all reviewers were horrified that Chambers had "annulled all distinction between physical and moral," and that he had degraded man by ranking him as a descendant of the apes and by interpreting the universe as "the progression and development of a rank, unbending, and degrading materialism" (Gillispie 1951:150; Sedgwick 1845). It is not surprising that in this intellectual climate Chambers had taken the precaution of publishing anonymously. Yet the modern reader finds little that is objectionable in Chambers' basic endeavor, no matter how many errors he committed in presenting the evidence.

To a greater or lesser extent, all the scientists of that period resorted, in their explanatory schemes, to frequent interventions by the Creator (in the running of His world). Indeed, proofs of such interventions were considered the foremost evidence for His existence. Agassiz quite frankly describes the obligations of the naturalist in these words: "Our task is ... complete as soon as we have proved His existence" (1857:132). To him the *Essay on Classification* was nothing but another Bridgewater Treatise in which the relationship of animals supplied a particularly elaborate and, for Agassiz, irrefutable demonstration of His existence.

Natural theology equally pervades Lyell's *Principles of Geology*. After discussing various remarkable instincts, such as pointing and retrieving, that are found in races of the dog, Lyell states: "When such remarkable habits appear in races of this species, we may reasonably conjecture that they were given with no other view than for the use of man and the preservation of the dog which thus obtains protection" (1835:vol. 2:455). Even though cultivated plants and domestic animals may have been created long before man, "some of the qualities of particular animals and plants may have been given solely with a view to the connection which, it was foreseen, would exist between them and man" (1835:vol. 2:456). Like Agassiz, Lyell believed that everything in nature is planned, designed, and has a predetermined end. "The St. Helena plants and insects [which are now dying out] may have lasted for their allotted term" (1835:vol. 3:9). The harmony of living nature and all the marvelous adaptations of animals and plants to each other and to their environment seemed to him thus fully and satisfactorily explained.

CREATIONISM AND THE ADVANCES OF GEOLOGICAL SCIENCE

At the beginning of the eighteenth century, the concept of a created world seemed internally consistent as long as this world was considered only recently created (in 4004 B.C.), static, and unchanging. The "ladder of perfection" (part of God's plan) accounted for the "higher" and "lower" organization of animals and man, and Noah's flood for the existence of fossils. All this could be readily accommodated within the framework of a literal biblical interpretation.

The discovery of the great age of the earth (Gillispie 1951; Haber 1959) and of an ever-increasing number of distinct fossil faunas in different geological strata necessitated abandoning the idea of a single creation. Repeated creations had to be postulated, and the necessary number of such interventions had to be constantly revised upward. Agassiz was willing to accept 50 or 80 total extinctions of life and an equal number of new creations. Paradoxically, the advance of scientific knowledge necessitated an increasing recourse to the supernatural for explanation. Even such a sober and cautious person as Charles Lyell frequently explained natural phenomena as due to "creation" and, of course, a carefully thought-out creation. The fact that the brain of the human embryo successively passes through stages resembling the brains of fish, reptiles, and lower mammals discloses, "in a highly interesting manner, the unity of plan that runs through the organization of the whole series of vertebrated animals; but they lend no support whatever to the notion of a gradual transmutation of one species into another; least of all of the passage, in the course of many generations, from an animal of a more simple to one of a more complex structure" (Lyell 1835:vol. 3:20). When a species becomes extinct it is replaced "by new creations" (Lyell 1835:vol. 3:45). Nothing is impossible in creation. "Creation seems to require omnipotence, therefore we cannot estimate it" (Wilson 1970:4). "Each species may have had its origin in a single pair, or individual where an individual was sufficient, and species may have been created in succession at such times and in such places as to enable them to multiply and endure for an *appointed* period, and occupy an *appointed* space of the globe?" (Lyell 1835:vol. 3:99–100; italics mine). Everything is done according to plan. Since species are fixed and unchangeable, everything about them, such as the area of distribution, the ecological context, adaptations to cope with competitors and enemies, and even the date of extinction, was previously "appointed," that is, predetermined.

This constant appeal to the supernatural amounted to a denial of all sound scientific methods and to the adoption of explanations that could be neither proven nor refuted. Chambers saw this quite clearly. When there is a choice between two theories, either special creation or the operation of general laws instituted by the Creator, he exclaimed, "I would say that the latter [theory] is greatly preferable, as it implies a far grander view of the Divine power and dignity than the other" (1844:117). Indeed, the increasing knowledge of geological sequences, and of the facts of comparative anatomy and geographic distribution, made the picture of special creation more ludicrous every day (Lovejoy 1959:413).

ESSENTIALISM AND A STATIC WORLD

Thus, theological considerations clearly played a large role in the resistance to the adoption of evolutionary views in England (and also in France). Equally influential, or perhaps even more so, was a philosophical concept. Philosophy and natural history during the first half of the nineteenth century, particularly in continental Europe, were strongly dominated by typological thinking, often called essentialism. This presumes that the changeable world of appearances is based on underlying immutable essences, and that all members of a class represent the same essence. The enormous role of essentialism in retarding the acceptance of evolutionism was long overlooked (see Essay 3). The observed vast variability of the world has no more reality, according to this philosophy, than the shadows of an object on a cave wall, as Plato expressed it in his allegory. The only things that are permanent, real, and sharply discontinuous from each other are the fixed, unchangeable "ideas" underlying the observed variability. Discontinuity and fixity are, according to the essentialist, as much the properties of the living as of the inanimate world.

As Reiser (1958) has said, a belief in discontinuous, immutable essences is incompatible with a belief in evolution. Agassiz was an extreme representative of this philosophy (Essay 18). To a lesser extent the same can be demonstrated for all of the other opponents of evolutionism, including Lyell. When rejecting Lamarck's claim that species and genera intergrade with each other, Lyell proposes that the following laws "prevail in the economy of the animate creation. . . . Thirdly, that there are fixed limits beyond which the descendants from common parents can never deviate from a certain type; fourthly, that each species springs from one original stock, and can never be permanently confounded by intermixing with the progeny of any other stock; fifthly, that each species shall endure for a considerable period of time" (1835: vol. 2:433). All nature consists, according to Lyell, of fixed types created at a definite time. To him these types were morphological entities, and he was rather shocked by Lamarck's idea that changes in behavior could have any effect on morphology.

As an essentialist, Lyell showed no understanding of the nature of genetic variation. Strictly in the scholastic tradition, he believed implicitly that essential characters could not change; this could occur only with nonessential characters. If an animal is brought into a new environment, "a short period of time is generally sufficient to effect nearly the whole change which an alteration of external circumstances can bring about in the habits of a species . . . such capacity of accommodation to new circumstances is enjoyed in very different degrees by different species" (1835:vol. 2:464). For instance, if we look at the races of dogs, they show many superficial differences "but, if we look for some of those essential changes which would be required to lend even the semblance of a foundation for the theory of Lamarck, respecting the growth of new organs and the gradual obliteration of others, we find nothing of the kind" (1835:vol. 2:438). This forces Lyell to question even Lamarck's conjecture "that the wolf may have been the original of the dog." The fact that, in the (geologically speaking) incredibly short time since the dog was domesticated, such drastically different races as the Eskimo dog, the hairless chihuahua, the greyhound, and other extremes evolved is glossed over.

LYELL'S SPECIES CONCEPT

Holding a species concept that allowed for no essential variation, Lyell credited species with little plasticity and adaptability. This led him to an interpretation of the fossil record that is very different from that of Lamarck. Anyone studying the continuous changes in the earth's surface, states Lyell, "will immediately perceive that, amidst the vicissitudes of the earth's surface, species cannot be immortal, but must perish, one after the other, like the individuals which compose them. There is no possibility of escaping from this conclusion, without resorting to some hypothesis as violent as that of Lamarck who imagined . . . that species are each of them endowed with indefinite powers of modifying their organization, in conformity to the endless changes of circumstances to which they are exposed" (1835:vol. 3: 155–156).

The concept of a steady extermination of species and their replacement by newly created ones, as proposed by Lyell, comes close to being a kind of microcatastrophism, as far as organic nature is concerned. Lyell differed from Cuvier merely in pulverizing the catastrophes into events relating to single species, rather than to entire faunas. In the truly decisive point, the rejection of any possible continuity between species in progressive time sequences, Lyell entirely agreed with Cuvier. When he traced the history of a species backward, Lyell inexorably arrived at an original ancestral pair, at the original center of creation. There is a total absence in his arguments of any thinking in terms of populations.

The enormous power of essentialism is in part explainable by the fact that it fitted the tenets of creationism so well; the two dogmas strongly reinforced each other. Nothing in Lyell's geological experience seriously contradicted his essentialism. It was not shaken until nearly 25 years later when Lyell visited the Canary Islands (from December 1853 to March 1854) and became acquainted with the same kind of phenomena that, in the Galapagos, had made Darwin an evolutionist and that, in the East Indian Archipelago, gave concrete form to the incipient evolutionism of A. R. Wallace. Wilson (1970) has portrayed the growth of doubt that led Lyell to publicly confess his conversion to evolutionism in 1862. The adoption of population thinking by him was a slow process, and even years after his memorable discussion with Darwin (16 April 1856), Lyell spoke in his notebooks of "variation *or* selection" as the important factor in evolution in spite of the fact that Darwin's entire argument was founded on the need for *both* factors as the basis of a satisfactory theory.

LYELL AND UNIFORMITARIANISM

It is a long-standing tradition in biological historiography that Lyell's revival of Hutton's theory of uniformitarianism was a major factor in the eventual adoption of evolutionary thinking. This thesis seems to be a great oversimplification; it is worthwhile to look at the argument a little more critically (Coleman 1971). When the discovery of a series of different fossil faunas, separated by unconformities, made the story of a single flood totally inadequate, Cuvier and others drew the completely correct conclusion that these faunas, particularly the alternation of marine and terrestrial faunas, demonstrated a frequent alternation of rises of the sea above the land and the subsequent reemergence of land above the sea. The discovery of mammoths frozen into the ice of Siberia favored the additional thesis that such changes could happen very rapidly. Cuvier was exceedingly cautious in his formulation of the nature of these "revolutions" and "catastrophes," but he did admit, "The breaking to pieces and overturning of the strata, which happened in former catastrophes, show plainly enough that they were sudden and violent like the last [which killed the mammoths and embedded them in ice]" (1817:16). He implied that most of these events were local rather than universal phenomena, and he did not maintain that a new creation had been required to produce the species existing today. He said merely "that they [modern species] did not anciently occupy their present locations and that they must have come there from elsewhere" (1817:125–126).

Cuvier's successors did not maintain his caution. The school of the so-called progressionists[2] postulated that each fauna was totally exterminated by a catastrophe at the end of each geologic period, followed by the special creation of an entirely new organic world. Progressionism, therefore, was intellectually a backward step from the widespread eighteenth-century belief that the running of the universe required only occasional, but definitely not incessant, active intervention by the Creator: He maintained stability largely through the laws that He had decreed at the beginning and that allowed for certain planetary and other perturbations. This same reasoning could easily have been applied to the organic world, and this indeed is what was done by Chambers in 1844, and by many other devout Christians after 1859.

Catastrophism would not seem as great an obstacle to evolutionism as is often claimed. It admitted, indeed it emphasized, the advance each new creation showed over the preceding one. By also conceding that there had been 30, 50, or even more than 100 extinctions and new creations, it made the concept of these destructions increasingly absurd, and what was finally left, after the absurd destructions had been abandoned, was the story of the constant progression of faunas. As soon as one rejected reliance on supernatural forces, this progression automatically became evidence for evolution. The only other assumption one had to make was that many of the catastrophes and extinctions had been localized events. This was, perhaps, not too far from Cuvier's original viewpoint.

The reason catastrophism was adopted by virtually all of the truly

2. Progressionism was the curious theory according to which evolution did not take place in the organisms but rather in the mind of the Creator, who—after each catastrophic extinction—created a new fauna in the more advanced state to which His plan of creation had progressed in the meantime. This thought was promoted in Britain particularly by Hugh Miller (*Footprints*, 1847), Sedgwick (*Discourse*, 1850), and Murchison (*Siluria*, 1854), and in America by L. Agassiz (*Essay*, 1857).

productive leading geologists in the first half of the nineteenth century is that the facts seemed to support it. Breaks in fossil strata, the occurrence of vast lava flows, a replacement of terrestrial deposits by marine ones and the reverse, and many other phenomena of a similar, reasonably violent nature (including the turning upside down of whole fossil sequences) all rather decisively refuted a rigid uniformitarian interpretation. This is why Cuvier, Sedgwick, Buckland, Murchison, Conybeare, Agassiz, and de Beaumont, to mention a few prominent geologists, adopted more or less catastrophist interpretations.

Charles Lyell was the implacable foe of the "catastrophists," as his opponents were designated by Whewell (1831, 1832). In his *Principles of Geology* (1835), Lyell promoted a "steady-state" concept of the world, best characterized by Hutton's motto, "no vestige of a beginning —no prospect of an end." Whewell coined the term "uniformitarianism" for this school of thought, a term that unfortunately had many different meanings. The most important meaning was that it postulated that no forces had been active in the past history of the earth that are not also working today. Yet, even this would permit two rather different interpretations. Even if one includes supernatural agencies among forces and causes, one can still be a consistent uniformitarian, provided one postulates that the Creator continues to reshape the world actively even at the present. Rather candidly, Lyell refers to this interpretation, accepted by him, as "the perpetual intervention hypothesis" (Wilson 1970:89).

Almost diametrically opposed to this were the conclusions of those who excluded all recourse to supernatural interventions. Uniformitarianism to them meant simply the consistent application of natural laws not only to inanimate nature (as was done by Lyell) but also to the living world (as proposed by Chambers). The important component in their argument was the rejection of supernatural intervention rather than lip service to the word uniformity.

It is important to remember that Lyell applied his uniformitarianism in a consistent manner only to inanimate nature, but left the door open for special creation in the living world. Indeed, as Lovejoy (1959) states justly, when it came to the origin of new species, Lyell, the great champion of uniformitarianism, embraced "the one doctrine with which uniformitarianism was wholly incompatible—the theory of numerous and discontinuous miraculous special creations." Lyell himself did not see it that way. As he wrote to Herschel (Mrs. Lyell 1881:vol. 1:467), he considered his notion "of a succession of extinction of species, and creation of new ones, going on perpetually now . . . the grandest which I had ever conceived, so far as regards the attributes of the Presiding Mind." There is evidence, however, that Lyell did not always regard these creations as miracles, but sometimes saw them as occurring "through the intervention of intermediate causes" and thus as being "a natural, in contradistinction to a miraculous process." By July 1856, after having read Wallace's 1855 paper, and after having discussed evolution with Darwin (16 April 1856), Lyell had become completely converted to believing that the introduction of new species was "governed by laws in the same sense as the Universe is governed by laws" (Wilson 1970:123).

Only the steady-state concept of uniformitarianism was novel in Lyell's interpretation. The insistence that nature operates according to eternal laws, with the same forces acting at all times, was, from Aristotle on, the standard explanation among most of those who did not postulate a totally static world, for instance, among the French naturalists preceding Cuvier. Consequently, acceptance of uniformitarianism did not, as Lyell himself clearly demonstrated, require the acceptance of evolutionism. If one believed in a steady-state world, as did Lyell, uniformitarianism was incompatible with evolution. Only if it was combined with the concept of a steadily changing world, as it was in Lamarck's thinking, did it encourage a belief in evolution. It is obvious, then, that the statement "uniformitarianism is the pacemaker of evolutionism," is an exaggeration, if not a myth.

But what effect did Lyell have on Darwin? Everyone agrees that it was profound; there was no other person whom Darwin admired as greatly as Lyell. *Principles of Geology*, by Lyell, was Darwin's favorite reading on the *Beagle* and gave his geological interests new direction. After the return of the *Beagle* to England, Darwin received more stimulation and encouragement from Lyell than from any other of his friends. Indeed, Lyell became a father figure for him and stayed so for the rest of his life. Darwin's whole way of writing, particularly in the *Origin of Species*, was modeled after the *Principles*. There is no dispute over these facts.

But what was Lyell's impact on Darwin's evolutionary ideas? There is much to indicate that the influence was largely negative. Knowing how firmly Lyell was opposed to the possibility of a transmutation of species, as documented by his devastating critique of Lamarck, Darwin was very careful in what he revealed to Lyell. He admitted that he doubted the fixity of species, but after that the two friends apparently avoided a further discussion of the subject. Darwin was far more outspoken with Hooker, to whom he confessed as early as January 1844, "I am almost convinced . . . that species are not (it is like confessing murder) immutable" (F. Darwin 1888:23). It was not until 1856 that Darwin fully outlined his theory of evolution to Lyell (Wilson 1970:xlix). This reticence of Darwin was due not to any intolerance on Lyell's part (or else Lyell would not have, after 1856, encouraged Darwin so actively to publish his heretical views), but rather to an unconscious fear on Darwin's part that his case was not sufficiently persuasive to convert such a formidable opponent as Lyell. There has been much speculation as to why Darwin had been so tardy about publishing his evolutionary views. Several factors were involved (one being the reception of the *Vestiges*), but I am rather convinced that his awe of Lyell's opposition to the transmutation of species was a much more weighty reason than has been hitherto admitted. It is no coincidence that Darwin finally began to write his great work within 3 months after Lyell took the initiative to consult him and to encourage him. Lovejoy summarizes the effect of Lyell's opposition to evolution in these words: "It was . . . his example and influence, more than the logical force of his arguments, that so long helped to sustain the prevalent belief that transformism was not a scientifically respectable theory" (Lovejoy 1959). I entirely agree with this evaluation.

UNSUCCESSFUL REFUTATIONS OWING TO WRONG CHOICE OF ALTERNATIVES

Creationism, essentialism, and Lyell's authority were not, however, the only reasons for the delay in the acceptance of evolution; others were important weaknesses in the scientific methodology of the period. There was still a demand for conclusive proofs. "Show me the breed of dogs with an entirely new organ," Lyell seems to say, "and I will believe in evolution." That much of science consists merely in showing that one interpretation is more probable than another one, or consistent with more facts than another one, was far less realized at that period than it is now.[3]

That victory over one's opponent consists in the refutation of his arguments, however, was taken for granted. Cuvier's, Lyell's, Agassiz's, and Darwin's detailed argumentations were all attempts to "falsify," as Popper (1959) puts it, the statements of their opponents. This method, however, has a number of weaknesses. For instance, it is often quite uncertain what kind of evidence or argument truly represents a falsification. More fatal is the frequently made assumption that there are only two alternatives in a dispute. Indeed, the whole concept of "alternative" is rather ambiguous, as I shall try to illustrate with some examples from pre-Darwinian controversies.

We can find numerous illustrations in the anti-evolutionary writings of Charles Lyell and Louis Agassiz of the limitation to only two alternatives when actually there was at least a third possible choice. Agassiz, for instance, never seriously considered the possibility of true evolution, that is, of descent with modification. For him the world was either planned by the Creator or was the accidental product of blind physical causes (in which case evolution would be the concatenation of such accidents). He reiterates this singularly simple-minded choice throughout the *Essay on Classification* (1857): "physical laws" versus "plan of

---

3. Darwin's *Origin* was one of the first scientific treatises in which the hypothetico-deductive method was rather consistently employed (Ghiselin 1969). Equally important, and even more novel, was Darwin's demonstration that deterministic prediction is not a necessary component of causality (Scriven 1959). Perhaps this can be considered a corollary of population thinking, but it is further evidence for the extraordinary complexity of the Darwinian revolution.

creation" (p. 10), "spontaneous generation" versus "divine plan" (p. 36), "physical agents" versus "plan ordained from the beginning" (p. 37), "physical causes" versus "supreme intellect" (p. 64), and "physical causes" versus "reflective mind" (p. 127). By this choice he excluded the possibility of evolution not only as envisioned by Darwin, but even as postulated by Lamarck. Nowhere does Agassiz attempt to refute Lamarckian evolution. His physical causes, in turn, are an exceedingly narrow definition of natural causes, since it is fully apparent that Agassiz had a very simple-minded Cartesian conception of physical causes as motions and mechanical forces. "I am at a loss to conceive how the origin of parasites can be ascribed to physical causes" (1857:126). "How can physical causes be responsible for the form of animals when so many totally different animal types live in the same area subjected to identical physical causes?" (1857:13–14). The abundant regularities in nature demonstrate "the plan of a Divine Intelligence" because they cannot be the result of blind physical forces. (This indeed was a standard argument among adherents of natural theology.) It never occurred to Agassiz that none of his arguments excluded a third possibility, the gradual evolution of these regularities by processes that can be daily observed in nature. This is why the publication of Darwin's *Origin* was such a shock to him. The entire evidence against evolution, which Agassiz had marshaled so assiduously in his *Essay on Classification*, had become irrelevant. He had failed completely to provide arguments against a third possibility, the one advanced by Darwin.

At that period, the concept of evolution still evoked in most naturalists the image of the *scala naturae*, the ladder of perfection. No one was more opposed to this concept than Lyell, the champion of a steady-state world. Any finding that contradicted a steady progression from the simple toward the more perfect refuted the validity of evolution, he thought. Indeed, the fact that mammals appeared in the fossil record before birds, and that primates appeared in the Eocene considerably earlier than some of the orders of "lower" mammals, was, to him, as decisive a refutation of the evolutionary theory as was to Agassiz the fact that the four great types of animals appeared simultaneously in the earliest fossil-bearing strata.

The assumption that refuting the *scala naturae* would refute once and for all any evolutionary theory is another illustration of insufficient alternatives. Lyell was quite convinced that the concept of a steady-state world would be validated (including regular special creations), if it could be shown that those mechanisms which Lamarck had proposed to account for evolutionary change were improbable or impossible.

But there were also other violations of sound scientific method, for instance, the failure to see that both of two alternatives might be valid. In these cases, the pre-Darwinians arrived at erroneous conclusions because they were convinced that they had to make a choice between two processes which, in reality, occur simultaneously. For example, neither Lamarck nor Lyell understood speciation, but this failure led them to opposite conclusions. When looking at fossil faunas, Lamarck, a great believer in the adaptability of natural species, concluded that all the contained species must have evolved into very different descendants. Lyell, as an essentialist, rejected the possibility of a change in species, and therefore he believed, like Cuvier, that all of the species had become extinct, with replacements provided by special creation. Neither Lamarck nor Lyell imagined that both processes, speciation and extinction, could occur simultaneously. That the turnover of faunas could be a balance of both processes never entered their minds.

FAILURE TO SEPARATE DISTINCT PHENOMENA

A third type of violation of scientific logic was particularly harmful to the acceptance of evolutionary thinking. This was the erroneous assumption that certain characteristics are inseparably combined. For instance, both Linnaeus and Darwin assumed, as I have pointed out elsewhere (Essay 33), that if one admitted the *reality* of species in nature, one would also have to postulate their immutable *fixity*. Lyell, as a good essentialist, unhesitatingly endorsed the same thesis: "From the above considerations, it appears that species have a real existence in nature; and that each was endowed, at the time of its creation, with the attributes and organization by which it is now distinguished" (1835: vol. 3:21). He is even more specific about this in his notebooks (Wilson 1970:92). That species could have full "reality" in the nondimensional situation and yet evolve continuously was unthinkable to him. Reality and constancy of species were to him inseparable attributes.

IMPACT OF THE *ORIGIN OF SPECIES*

The situation changed drastically and permanently with the publication of the *Origin of Species* in 1859. Darwin marshaled the evidence in favor of a transmutation of species so skillfully that from that point on the eventual acceptance of evolutionism was no longer in question. But he did more than that. In natural selection he proposed a mechanism that was far less vulnerable than any other previously proposed. The result was an entirely different concept of evolution. Instead of endorsing the eighteenth-century concept of a drive toward perfection, Darwin merely postulated change. He saw quite clearly that each species is forever being buffeted around by the capriciousness of the constantly changing environment. "Never use the word(s) higher and lower" (F. Darwin 1903:114), Darwin reminded himself. By chance this process of adaptation sometimes results in changes that can be interpreted as progress, but there is no intrinsic mechanism generating inevitable advance.

Virtually all the arguments of Cuvier, Lyell, and the progressionists became irrelevant overnight. Essentialism had been the major stumbling block, and the development of a new concept of species was the way to overcome this obstacle. Lyell himself eventually (after 1856) understood that the species problem was the crux of the whole problem of evolution, and that its solution had potentially the most far-reaching consequences: "The ordinary naturalist is not sufficiently aware that, when dogmatizing on what species are, he is grappling with the whole question of the organic world and its connection with a time past and with man" (Wilson 1970:1). And, since he came to this conclusion after studying speciation in the Canary Islands, he added: "A group of islands, therefore, is the fittest place for Nature's trial of such permanent variety-making and where the problem of species-making may best be solved" (Wilson 1970:93). This is what Darwin had discovered 20 years earlier.

SPECIAL ASPECTS OF THE DARWINIAN REVOLUTION

No matter how one defines a scientific revolution, the Darwinian revolution of 1859 will have to be included. Who would want to question that, by destroying the anthropocentric concept of the universe, it caused a greater upheaval in man's thinking than any other scientific advance since the rebirth of science in the Renaissance? And yet, in other ways, it does not fit at all the picture of a revolution. Or else, how could H. J. Muller have exclaimed as late as 1959: "One hundred years without Darwinism are enough!" (1959)? And how could books such as Barzun's *Darwin, Marx, Wagner* (1941) and Himmelfarb's *Darwin and the Darwinian Revolution* (1959), both displaying an abyss of ignorance and misunderstanding, have been published relatively recently? Why has this revolution in some ways made such extraordinarily slow headway?

A scientific revolution is supposedly characterized by the replacement of an old explanatory model by an incompatible new one (Kuhn 1962). In the case of the theory of evolution, the concept of an instantaneously created world was replaced by that of a slowly evolving world, with man being part of the evolutionary stream. Why did the full acceptance of the new explanation take so long? The reason is that this short description is incomplete, and therefore misleading, as far as the Darwinian revolution is concerned.

Before analyzing this more fully, the question of the date of the Darwinian revolution must be raised. That the year 1859 was a crucial one in its history is not questioned. Yet, this still leaves a great deal of leeway to interpretation. On the one hand, one might assert that the age of evolutionism started even before Buffon, and that the publication of the *Origin* in 1859 was merely the straw that broke the camel's back.

On the other hand, one might go to the opposite extreme, and claim that not much had changed in the thinking of naturalists between the time of Ray and Tournefort and the year 1858, and that the publication of the *Origin* signified a drastic, almost violent revolution. The truth is somewhere near the middle; although there was a steady, and ever-increasing, groundswell of evolutionary ideas from the beginning of the eighteenth century on, Darwin added so many new ideas (particularly an acceptable mechanism) that the year 1859 surely deserves the special attention it has received. Two components of the Darwinian revolution must thus be distinguished: the slow accumulation of evolutionary facts and theories since early in the eighteenth century and the decisive contribution Darwin made in 1859. Together they constitute the Darwinian revolution.

The long time span is due to the fact that not simply the acceptance of one new theory was involved, as in some other scientific revolutions, but the acceptance of an entirely new conceptual world, consisting of numerous separate concepts and beliefs. And not only were scientific theories involved, but also a whole set of metascientific credos. Let me prove my point by specifying the complex nature of the revolution: I distinguish six major elements in this revolution, but it is probable that additional ones should be recognized.

The first three elements concern scientific replacements:

1. *Age of the earth.* The revolution began when it became obvious that the earth was very ancient rather than having been created only 6000 years ago (Haber 1959). This finding was the snowball that started the whole avalanche.

2. *Refutation of both catastrophism (progressionism) and of a steady-state world.* The evolutionists, from Lamarck on, had claimed that the concept of a more or less steadily evolving world was in better agreement with the facts than either the catastrophism of the progressionists or Lyell's particular version of a steady-state world. Darwin helped this contention of the evolutionists to its final victory.

3. *Refutation of the concept of an automatic upward evolution.* Every evolutionist before Darwin had taken it for granted that there was a steady progress of perfection in the living world. This belief was a straight-line continuation of the (static) concept of a scale of perfection, which was maintained even by the progressionists, for whom each new creation represented a further advance in the plan of the Creator.

Darwin's conclusion, to some extent anticipated by Lamarck, was that evolutionary change through adaptation and specialization by no means necessitated continuous betterment. This view proved very unpopular, and is even today largely ignored by nonbiologists. This neglect is well illustrated by the teachings of the school of evolutionary anthropology, or those of Bergson and Teilhard de Chardin.

The last three elements concern metascientific consequences. The main reason evolutionism, particularly in its Darwinian form, made such slow progress is that it was the replacement of one entire *Weltan-schauung* by a different one. This involved religion, philosophy, and humanism.

4. *The rejection of creationism.* Every anti-evolutionist prior to 1859 allowed for intermittent, if not constant, interference by the Creator. The natural causes postulated by the evolutionists completely separated God from His creation, for all practical purposes. The new explanatory model replaced planned teleology by the haphazard process of natural selection. This required a new concept of God and a new basis for religion.

5. *The replacement of essentialism and nominalism by population thinking.* None of Darwin's new ideas was quite so revolutionary as the replacement of essentialism by population thinking. It was this concept that made the introduction of natural selection possible. Because it is such a novel concept, its acceptance has been slow, particularly on the European continent and outside biology. Indeed, even today it has by no means universally replaced essentialism.

6. *The abolition of anthropocentrism.* Making man part of the evolutionary stream was particularly distasteful to the Victorians, and is still distasteful to many people.

NATURE OF THE DARWINIAN REVOLUTION

It is now clear why the Darwinian revolution is so different from all other scientific revolutions. It required not merely the replacement of one scientific theory by a new one, but, in fact, the rejection of at least six widely held basic beliefs (together with some methodological innovations).

Furthermore, it had a far greater relevance outside of science than any of the revolutions in the physical sciences. Einstein's theory of relativity or Heisenberg's of statistical prediction could hardly have had any effect on anybody's personal beliefs. The Copernican revolution and Newton's world view required some revision of traditional beliefs. None of these physical theories, however, raised as many new questions concerning religion and ethics as did Darwin's theory of evolution through natural selection.

In a way, the publication of the *Origin* in 1859 was the midpoint of the so-called Darwinian revolution rather than its beginning. Stirrings of evolutionary thinking preceded the *Origin* by more than 100 years, reaching an earlier peak in Lamarck's *Philosophie zoologique* in 1809. The final breakthrough in 1859 was the climax in a long process of erosion that was not fully completed until 1883, when Weismann rejected the possibility of an inheritance of acquired characters.

As in any scientific revolution, some of the older opponents, such as Agassiz, never became converted. But the Darwinian revolution differed from other scientific revolutions in the large number of workers who accepted only part of the package. Many zoologists, botanists, and paleontologists eventually accepted gradual evolution through natural causes, but not through natural selection. Indeed, on a worldwide basis, those who continued to reject natural selection as the prime cause of evolutionary change were probably well in the majority until the 1930s.

Two conclusions emerge from this analysis. First, the Darwinian and quite likely other scientific revolutions consist of the replacement of a considerable number of concepts. This requires a lengthy period of time, since the new concepts will not all be proposed simultaneously. Second, the mere summation of new concepts is not enough; it is their constellation that counts. Uniformitarianism, when combined with the belief in a static essentialistic world, leads to the steady-state concept of Lyell, but when it is combined with a concept of change, it leads to the evolutionism of Lamarck. The observation of evolutionary changes, combined with essentialist thinking, leads to various saltationist or progressionist theories, but, combined with population thinking, it leads to Darwin's theory of evolution by natural selection.

It is now evident that the Darwinian revolution does not conform to the simple model of a scientific revolution, as described, for instance, by T. S. Kuhn (1962). It is actually a complex movement that started nearly 250 years ago; its many major components were proposed at different times, and became victorious independently of each other. Even though a revolutionary climax unquestionably occurred in 1859, the gradual acceptance of evolutionism, with all of its ramifications, covered a period of nearly 250 years.

REFERENCES

Agassiz, L. 1857. *Essay on classification.* Little, Brown, Boston.

Albritton, C. C., Jr., ed. 1967. *Uniformity and simplicity.* Geol. Soc. Amer. Spec. Pap., no. 89.

Chambers, R. 1844. *Vestiges of the natural history of creation.* Wiley and Putnam, New York.

Coleman, W. 1964. *Georges Cuvier, zoologist.* Harvard University Press, Cambridge, Mass.

—— 1971. *Biology in the nineteenth century.* Wiley, New York.

Cuvier, G. 1817. *Essay on the theory of the earth,* trans. R. Jameson. Edinburgh University Press, Edinburgh.

Darwin, C. 1859. *On the origin of species by means of natural selection.* J. Murray, London.

Darwin, F., ed. 1888. *Life and letters of Charles Darwin.* J. Murray, London.

Darwin, F., and A. C. Seward, eds. 1903. *More letters of Charles Darwin.* J. Murray, London.

De Beer, G. 1964. *Charles Darwin.* Doubleday, Garden City, N.Y.

Delbrück, M. 1971. Aristotle-totle-totle. In *Of microbes and life,* ed. J. Monod and E. Borek, pp. 50–55. Columbia University Press, New York.

Eiseley, L. 1958. *Darwin's century.* Doubleday, New York.

Fruchtbaum, H. 1964. Natural theology and the rise of science. Ph.D. thesis, Harvard University.

Ghiselin, M. 1969. *The triumph of the Darwinian method.* University of California Press, Berkeley.

Gillispie, C. C. 1951. *Genesis and geology.* Harvard University Press, Cambridge, Mass.

Glass, B., O. Temkin, and W. L. Straus, Jr., eds. 1959. *Forerunners of Darwin, 1745-1859.* Johns Hopkins Press, Baltimore.

Gould, S. J. 1965. Is uniformatism necessary? *Amer. Jour. Sci.,* 263:223.

Greene, J. C. 1971. The Kuhnian paradigm and the Darwinian revolution in natural history. In *Perspectives in the history of science and technology,* ed. D. H. D. Roller. University of Oklahoma Press, Norman.

Haber, F. C. 1959. Fossils and the idea of a process of time in natural history. In *Forerunners of Darwin, 1745-1859,* ed. B. Glass, O. Temkin, and W. L. Straus, Jr., pp. 222-261. Johns Hopkins Press, Baltimore.

Hooykaas, R. 1959. *Natural law and divine miracle.* E. J. Brill, Leiden.

Hull, D. 1964. The effect of essentialism on taxonomy. *Brit. Jour. Phil. Sci.,* 15:314-326.

—— 1965. The effect of essentialism on taxonomy [continued]. *Brit. Jour. Phil. Sci.,* 16:1-18.

Kuhn, T. S. 1962. *The structure of scientific revolutions.* University of Chicago Press, Chicago.

Lovejoy, A. O. 1936. *The great chain of being.* Harvard University Press, Cambridge, Mass.

—— 1959. The argument for organic evolution before the *Origin of Species,* 1830-1858. In *Forerunners of Darwin, 1745-1859,* ed. B. Glass, O. Temkin, and W. L. Straus, Jr., pp. 356-414. Johns Hopkins Press, Baltimore.

Lyell, C. 1835. *Principles of geology,* 4th ed. J. Murray, London.

Lyell, Mrs., ed. 1881. *Life, letters and journals of Sir Charles Lyell.* J. Murray, London.

Milhauser, M. 1959. *Just before Darwin.* Wesleyan University Press, Middletown, Conn.

Miller, H. 1850. *Footprints.* Gould, Kendall and Lincoln, Boston.

Muller, H. J. 1959. One hundred years without Darwinism are enough. *School Sci. Math.,* 59:304-316.

Murchison, R. I. 1854. *Siluria.* J. Murray, London.

Popper, K. R. 1945. *The open society and its enemies.* Routledge and Kegan Paul, London.

—— 1959. *The logic of scientific discovery.* Hutchison, London.

Reiser, O. L. 1958. The concept of evolution in philosophy. In *A book that shook the world,* ed. R. Buchsbaum. University of Pittsburgh Press, Pittsburgh.

Rudwick, M. J. S. 1967. A critique of uniformitarian geology: a letter from W. D. Conybeare to Charles Lyell, 1841. *Proc. Amer. Phil. Soc.,* 111:272.

—— 1971. Uniformity and progression. In *Perspectives in the history of science and technology,* ed. D. H. D. Roller. University of Oklahoma Press, Norman.

Scriven, M. 1959. Explanation and prediction in evolutionary theory. *Science,* 130:477.

Sedgwick, A. 1845. Review of the *Vestiges of Creation. Edinburgh Review,* 82: 1-45.

—— 1850. *A Discourse on the Studies of the University of Cambridge,* 5th edition. Cambridge, England.

Simpson, G. G. 1970. Uniformitarianism. In *Essays in evolution and genetics,* pp. 43–96. Appleton-Century-Crofts, New York.

Toulmin, S. 1966. Conceptual revolution in science. *Boston Stud. Phil. Sci.,* 3:331–347.

—— 1970. Does the distinction between normal and revolutionary science hold water? In *Criticism and the growth of knowledge,* ed. I. Lakatos and A. Musgrave, pp. 39–49. Cambridge University Press, Cambridge.

Whewell, W. 1831. Review of *Principles of geology* by Charles Lyell. *Brit. Critic,* 9:180–206.

—— 1832. Lyell—*Principles of geology,* vol. 2. *Quart. Rev.,* 47:103–132.

Wilson, L. G., ed. 1970. *Sir Charles Lyell's scientific journals on the species question.* Yale University Press, New Haven.

Zimmermann, W. 1953. *Evolution, Geschichte ihrer Probleme und Erkenntnisse.* Alber, Freiburg.

20

Karl Jordan on the Theory of Systematics and Evolution

In 1955, on the occasion of the 94th birthday of Karl Jordan, the Royal Entomological Society in London published a volume of essays honoring this great entomologist. Jordan, however, was far more than a specialized insect taxonomist, and in a contribution to that volume I called attention to his magnificent contributions to biological theory.

It is little known among entomologists, and even less known among most general biologists, that Karl Jordan has made a number of highly important conceptual contributions to the fields of systematics and evolution. His name is rarely mentioned in histories of the development of these fields. Possible reasons for this neglect are manifold, but it is evident that the basic one is that Jordan was far ahead of his time and thus was destined to suffer the neglect experienced by so many pioneers. A second reason is that he worked with a material (natural populations) and with methods (nonexperimental) that were unpopular among the laboratory biologists of his time. He has always been averse to the spectacular, which, as the history of biology shows, very often monopolizes attention, no matter how soon afterward it is shown to be wrong. It is a curious paradox that the name of the investigator who finds the right answers is likely to be forgotten, because his findings are incorporated anonymously into the treasury of common knowledge and accepted theory. The subsequent analysis will reveal that this has been Jordan's fate to a considerable extent.

It has been pointed out recently that most "histories" of evolution are actually "prehistories." Their detailed treatment stops around 1860, precisely at the time when the development of modern concepts began subsequent to the publication of Darwin's *Origin of Species* in 1859. The reluctance of historians to deal with modern developments is understandable, since only a specialist can truly evaluate the significance of various contributions. With respect to systematic and evolutionary con-

Adapted from pp. 45-51, 64-65 of "Karl Jordan's contribution to current concepts in systematics and evolution," *Transactions of the Royal Entomological Society of London*, 107 (1955):45-66.

cepts there is the additional difficulty that much of the documentation is hidden away in taxonomic monographs or in rather specialized writings of naturalists. Furthermore, each author stands on the shoulders of his predecessors and with certain concepts it is hard to decide who deserves the greater credit: he who first vaguely mentioned a new idea, he who first clearly formulated it, or he who supplied final proof. Also there has been much parallelism, so that many conceptual advances have been made by several authors independently. In view of all this it would be unwise to assert too dogmatically which particular concept was pioneered by Jordan. To determine this unequivocally would require an analysis of the writings of all of Jordan's contemporaries and predecessors, an analysis that has not yet been undertaken. My evaluation of Jordan's contribution is based on the evidence now available.

The writings in which Jordan made his most important conceptual contributions were published principally during the period from 1895 to 1911. In addition to a number of shorter papers they include several major treatises that have since become classics. Among these are two general papers (on mechanical selection, 1896, and on geographical versus nongeographical variation, 1905) and a number of entomological monographs, prepared in joint authorship with Lord Rothschild, in which most of the general discussions were written and are usually signed by K. Jordan (eastern Papilios, 1895; Sphingidae, 1903; and American Papilios, 1906).

The turn of the century was intellectually an exciting period in the history of systematics and evolution. The theory of evolution had by then been almost universally accepted by biologists, yet they were still groping for the correct interpretation of the causes of evolution. As far as the "material of evolution" is concerned, the year 1900, the birthdate of the science of genetics, must be considered the starting point of our knowledge. The battles about the mode of speciation, the existence of natural selection, and the meaning of mimicry and of polymorphism were at their height. Jordan entered all these battles with enthusiasm and conviction, and one is filled with admiration bordering on awe when one compares Jordan's discussions with those of most of his contemporaries. In almost every case it is Jordan who is right, and his formulations and solutions of biological problems are very often superior to those offered by other biologists during the succeeding 50 years.

JORDAN'S BIOLOGICAL PHILOSOPHY

Jordan had the happy faculty of synthesizing new concepts of biology by fusing the best elements of various schools of thought. Through his education at the University of Göttingen, he received a training in the best tradition of German zoology. Among his teachers was Ehlers, a competent comparative anatomist of invertebrates, who brought him in contact with the battles then raging between Weismann, Haacke, Eimer,

Haeckel, and others concerning the causes of evolutionary change. Through his entomological interests Jordan came in contact with an experienced group of insect collectors and breeders. Equally, if not more, important, after his transfer to Tring, was his steady contact with Rothschild and with Hartert. Rothschild's great enthusiasm, particularly for the study of island faunas, was infectious, and Hartert at that period had become the leader of that group of ornithologists who were fighting for the consistent application of trinomials, for the downgrading of isolated allopatric populations from the rank of species to that of subspecies (wherever possible), and for a broadening of the species concept (Stresemann 1975). Hartert (1859-1934), a great naturalist and practical systematist, never wrote a major paper on principles of evolution or on the new systematics, but there is no denying his influence on Jordan, particularly during the formative years 1893-1895. Yet, a careful study of the publications of the Tring triumvirate (Rothschild, Hartert, Jordan) shows quite clearly that Jordan, perhaps as a consequence of his more complete university training and his greater interest in general biology, soon surpassed his teachers in the development of new concepts in systematics and evolution.

Jordan's entire work in the fields of entomology, systematics, and evolution is founded on a few basic beliefs. They relate to the position of systematics in biology, the application of the concept of evolution to taxonomy, the meaning of variability, and the working methods of the scientist.

Systematics and Biology

There have always been two kinds of taxonomists: those for whom species of animals and plants are like postage stamps to be collected or described merely for the sake of rarity or novelty, and those others for whom systematics is an important branch of biology. Jordan has always included himself enthusiastically in the latter camp. He expressed repeatedly (e.g., 1905:150) his regret over the short-sightedness of laboratory biologists who thought that nothing could be learned from systematics because they themselves knew nothing about this field and had contact only with the stamp collecting type of taxonomists. "Classification, as we know, has the reputation of being as dry as our cabinet specimens—if not mouldy—and of having an interest only for those who work at the special group of animals classified. There are even biologists of fame who, in their misguided wisdom, scoff at systematics and look down upon this kind of work as more or less fruitless . . . I take the opportunity . . . of stating emphatically that sound systematics are the only safe basis upon which can be built up sound theories as to the evolution of the diversified world of live beings" (1911:385.).

It cannot be denied, admitted Jordan, that systematics is unthinkable without painstaking attention to a great deal of detail. Without this there is the danger of committing glaring errors such as Darwin's statement that the birds of Madeira and the Canary Islands have not been modified in comparison with European birds. The subsequent careful studies of Hartert and other bird taxonomists have proved the error of this statement. "A precise knowledge of the lowest systematic units which only the systematist can produce is not only the foundation for a valid theory of evolution but also a necessity for the correct placing of the many units into a system" (1905:153). Throughout his career Jordan stressed the lesson he had learned as a zoology student at the university, that the taxonomist has a great deal of information of vital significance to the general biologist. For many years the attempt to convince scientists of this failed, but the increasingly closer collaboration between taxonomists and experimental biologists during the past two decades proves that the point has finally been made. Jordan has contributed his share toward the victory of the idea that every taxonomist must have a broad training in general biology and that every biologist must know and understand the important generalizations that come from the field of systematics.

Taxonomy and Evolution

Jordan was an enthusiastic evolutionist from the beginning of his career. One might go so far as to say that all his research is ultimately directed toward throwing light on the processes of evolution.

The ambiguous role of the theory of evolution in the development of the theories of systematics is often commented on. Prior to Darwin the system of animals was merely a system of similarities. But ever since 1859 there has been an argument, which has not yet been unequivocally decided, over whether the system should serve only the purely practical aims of the proper cataloguing and pigeonholing of specimens or whether, as a consequence of the theory of evolution, animals should be classified on the basis of presumed relationships. Jordan is quite emphatically of the second opinion. He cites (1905) one case after another where a modification of the classification on the basis of presumed relationships has led to a considerable improvement of the distributional picture and to a better understanding of biological phenomena. For instance, if the West African butterfly *Papilio zalmoxis* is removed from the large Oriental *Aristolochia* feeders, where this species was placed by all specialists on account of a superficial resemblance, and associated with various small African species with which it is really related, the remarkable fact is at once apparent that in spite of the presence of food plants there are no *Aristolochia* feeders in Africa. Jordan cites numerous cases from the writings of Wallace, Scharff, Eimer, and others where complex zoogeographic situations are at once clarified as soon as the respective species are classified not on the basis of superficial resemblance, but on that of relationship as revealed by the totality of characters, including internal ones. "The aim of the systematist as such is the establishment, on reasoned evidence, of the degree of relationship between the forms with which he is concerned, the evidence being furnished by the specimens and the bionomics of the species and varieties to which the specimens belong" (1911:386).

The Meaning of Variability

One of the most revolutionary changes of concept in biology has been the replacement of typological thinking by thinking in terms of populations. According to this concept, no two individuals or biological events are exactly the same and processes in biology can be understood only by a study of variation. The greatest of Jordan's papers (1905) was devoted to a discussion of the meaning of variation and the light it sheds on the interpretation of evolutionary phenomena. The study of variation is an absolute necessity in the work of the taxonomist. Jordan, partly influenced by Rothschild and Hartert, came to understand very early how important it was to study adequate samples of as many natural populations as possible: "Nowadays systematists comprehend more and more that a few specimens of each species are insufficient for a serious study, and hence try to bring together long series from every locality" (1896:447; see also 1905:182).

Methods of the Scientist

Jordan, who himself has had such a broad background, has always emphasized a broad approach and a balanced analysis. Superb master though he has become of certain groups of insects, he has always tried to remain broad in his interests. In the field of entomology he was equally at home among beetles, butterflies, moths and siphonaptera. When analyzing a given taxonomic situation he has stressed the importance of utilizing the greatest number of characters, as I shall discuss below. He believes that by analogy the same working principle should be applied to the study of evolution. He warned the evolutionist not to be deceived by an artificial system of factors just as the taxonomist must not be deceived by an artificial system of characters. "Scientific systematics utilises so far as possible all the attributes of the forms that are to be classified in order to determine their real relationship. If there is a contradiction between one organ and another, there is either an error in observation or in interpretation. The research on one organ serves as a control for the research on the other, and this permits the recognition and elimination of errors. The same method must be used in the investigation of the causes of evolution if one does not want to be satisfied with an artificial explanation of the origin of the gaps between species which corresponds to an artificial system of classification" (1905:210). The common sense and the mature balance of mind expressed in this quotation have always been characteristic of Jordan. They have been characteristic of his taxonomic decisions and of his philosophy of nomenclature. It is not my task here to discuss Jordan's contribution to the theory of nomenclature, and more specifically to the stabilization of nomenclature he so ardently desires, but mention of his attitude toward nomenclature must be made in this context. "The philosophic aspect of systematics is unfortunately much obscured . . . due in a large measure to an unduly great importance being attributed to the mere giving of names. Nomenclature is the servant of science, but has in many houses the position of master" (1911:386).

CONCEPTS OF SYSTEMATICS

It has become fashionable to apply the label "new systematics" to the set of concepts that is now prevailing in the field of taxonomy. The silent assumption is made that these concepts are new. Actually their roots go back to the middle of the nineteenth century or earlier and it would be difficult to name a single concept that was not formulated at least 30 or 40 years ago. What may come as a revelation to some of the younger students in this field is the discovery of the extent to which this "modern" philosophy is present in all of its completeness in the classic publications of Jordan. While much of it has been expressed casually or timidly by earlier authors, it is in the writings of Jordan that we find many of these concepts stated fully and in detail for the first time. His broad knowledge of the contemporary zoological literature and his contacts with students of other kinds of animals, as well as with general biologists, permitted him to develop many concepts in advance of his contemporaries. If anyone deserves to be called the father of the new systematics, it is Jordan. His influence on entomologists and taxonomists in general cannot be exaggerated. More specifically, this influence is apparent in the development of thorough taxonomic techniques, a new evaluation of subspecific categories and his contribution to the development of the modern species concept.

Taxonomic Techniques

I mentioned above that Jordan has repeatedly stressed the need for the study of large series of specimens and for the utilization of as many additional taxonomic characters, both external and internal, as can possibly be found. In his studies of butterflies he emphasized the taxonomic importance of such structural characters as scales and wing venation, and his extensive studies of the individual and geographic variation of the sexual armatures of both male and female butterflies (1896, 1905) have hardly been surpassed since. He is undoubtedly the pioneer in studying the variability of these structures in a large number of specimens from a single population and in following up their geographical variation. Moreover, he was not satisfied to use this evidence for purely taxonomic purposes, but also utilized it (see Essay 12) as an elegant proof of geographic speciation.

His concepts of the validity of taxonomic characters were far ahead of his time and many contemporary taxonomists might learn from what

Jordan said in 1905 (p. 171): "One speaks in phylogeny often of generic characters as opposed to specific ones. There is, however, no valid definition of generic morphological characters as there is none of specific ones. To call the above described variability a change in generic characters because the diagnosis of genera in our somewhat superficial taxonomy of butterflies is frequently based only on differences in wing venation, would be circular reasoning. Differences in wing venation may be either individual, specific, generic, or even a taxonomic criterion for subfamilies and higher systematic categories." He is convinced that one cannot hope to understand the variation of taxonomic characters by studying individuals. In order to understand the variability of species, it is necessary, he says, to study the offspring of a single female or "the individuals flying [he speaks of butterflies] in the same locality," in other words, local populations (1896:430).

The Subspecies

In the field of systematics the name Tring or Tring School is associated with definite concepts such as a liberal use of subspecies and an attempt to arrange natural populations into polytypic species. There is little doubt that Hartert was the original leader of this team since he had fought for these concepts in print as early as 1891 (before coming to Tring). Little of this philosophy was noticeable in Jordan's early papers on beetles; for instance, among several hundred descriptions of new beetles published in 1894, there are only a few subspecies. But as soon as Jordan started working on the species of *Papilio* with their immense geographic variation, he applied wholeheartedly the new subspecific concept as explained in his introduction (pp. 168-182) of Rothschild's revision of the Papilios of the Eastern Hemisphere (1895). Up to that time there had been great confusion as to the correct usage of the terms "subspecies," "variety," and "aberration." When the Tring triumvirate started publishing the *Novitates Zoologicae* they found it necessary to clear up this confusion in the following editorial preamble (1894, *Novit. Zool.* 1:1):

> The term "variety," especially among entomologists, has been indiscriminately used to denote an individual variation within a species as well as climatic or geographical races. We therefore, to avoid all possible errors, have determined to discard the term "variety" altogether. To denote individual variations we shall in this periodical, employ the word aberration, and for geographical forms, which cannot rank as full species, the term subspecies.
>
> *Editors*

At this period the introduction of subspecies into taxonomy was vigorously opposed by the older generation. The subspecies was still a relatively new concept and no unanimity had yet been achieved as to its definition or application. Jordan took a stand on both these points. "A subspecies is a localised group of individuals of a species, the mean of the characters of which is different from the mean of the characters of all the other localised groups, and which will, under favourable circumstances, fuse together with other groups" (1896:447). He realizes that this definition fails to make a clear distinction between populations that are sufficiently distinct to deserve to be named as subspecies and other populations that are not. "The above definition has not had regard to the degree of divergence attained by the localised form. Now, we ask, which then is the lower limit of application of the term 'subspecies'?" On the basis of cases where one sex is clearly different in two places while the other one is identical, "we shall have to use the term 'subspecies' when a localised variation is such that about half of the individuals belong to the varietal form. All lower degrees of localised variation may be termed 'localised aberration'" (1896:447–448). It is evident that at this date there was still some difficulty in distinguishing between individual and geographic variation as well as in specifying the degree of the distinctness of that half of the form which is different. In subsequent years Jordan has been particularly interested in the biological problem of the parapatric contact of two subspecies with special emphasis on a most intriguing and puzzling case among fleas he discovered (1938, 1940). As far as isolated subspecies are concerned, he adopted Hartert's viewpoint "that geographical separation of different forms cannot be an *a priori* criterion of specific distinctness, though this has often enough been alleged" (1896:431).

The Species Concept

Among all of Jordan's contemporaries, there was probably only one, E. B. Poulton, who had as advanced a concept of species and speciation as did Jordan. Indeed, his views on these subjects are of such historical importance that I have discussed them in this volume in two separate essays, Essays 12 and 32.

NATURAL SELECTION

Jordan, in contrast to virtually all the early Mendelians, was a firm believer in natural selection but was in doubt as to its precise role. Again and again he comes back to the great importance of natural selection in the evolutionary process. This is particularly evident in his repeated discussions of mimicry (1897a, 1897b, and 1911). "There is here no difficulty at all for the theory of natural selection. The theory offers a very simple explanation by assuming that the variable ancestor of the mimicking species has been gradually modified into a di- or polymorphic species by the weeding out of those intermediate forms which did not resemble protected species already existing in the country. The frequent rarity of intermediates is direct evidence for this theory"

(1911:403). He continues to say that this theory is not in conflict with other theories dealing with the origin of variation itself and those which interpret adaptation to the local environment. Jordan then presents in full detail the genetic theory of mimetic polymorphism that has since been adopted by the majority of geneticists.

How far ahead of his contemporaries Jordan was in his thinking on natural selection is evident from two notes on mimicry he published in 1897 in *Nature*. In order to account for the usual rarity of mimics as compared with the models, some students of mimicry had advanced the exceedingly anthropomorphic explanation that mimics are rare because it is to their advantage to be rare in order that the predators will not become conditioned to them. This explanation, which, as a matter of fact, is still maintained in much of the current literature, does not provide for a genetic mechanism, as was clearly realized by Jordan. Instead he advanced a theory that—anticipating the recent selection experiments of Mather, Lerner, and others—has a remarkably modern ring. If the mimic "for instance, *Papilio alcidinus*, has acquired that wonderful similarity in colour and form to its model, an Uraniid moth, in consequence of a continued selection in the one direction, it is obvious that the result of such a one-sided selection will not only be similarity to the immune model, but also physiological one-sidedness. The more rigorous the selection is, the better will the mimetic species become adapted to its model, and the more it will lose its adaptability to new biological factors. . . . Consequently, the most striking 'mimics,' in spite of, or rather in consequence of, the resemblance to immune species, are, in the long run, the less favoured in the struggle for existence, which means that they will become relatively scarce. From this consideration it is apparent to me that the selection of those specimens which are the very fittest in any *special* direction is in itself a danger to the species, and can lead to destruction" (1897a:153). This was written at a time when natural selection was being attacked by environmentalists and mutationists alike. Almost 50 years passed before the analysis of gene complexes by Mather and others led to a genetic confirmation of Jordan's views of the detrimental effects of one-sided selection.

CONCLUSIONS

The discussion here and in Essays 12 and 32 makes use of only a small portion of Jordan's life work, as presented in his early major publications. Yet I believe it is enough to show that by and large Jordan was far ahead of his times, sometimes by as much as 50 years, and that in his writings and through his contacts he exerted an influence on the development of the new systematics and on evolutionary thought that cannot be overestimated. Even though Jordan's name is not cited as widely in textbooks of evolution and zoological systematics as it deserves, there is no doubt that most of the concepts which he pioneered and for which he fought have now been accepted by virtually every worker in the field. As one of those who have benefited from Jordan's efforts to clarify evolutionary thought and to enhance the prestige of systematics, I want to record my feeling of immense gratitude.

REFERENCES

Jordan, K. 1896. On mechanical selection and other problems. *Novit. Zool.*, 3:426–525.

—— 1897a. On mimicry. *Nature*, 56:153.

—— 1897b. On mimicry. *Nature*, 56:419.

—— 1905. Der Gegensatz zwischen geographischer und nichtgeographischer Variation. *Z. wiss. Zool.*, 83:151–210.

—— 1911. The systematics of some Lepidoptera which resemble each other, and their bearing on general questions of evolution. *Int. Congr. Ent. 1* (Brussels, 1910), 2:385–404.

—— 1938. Where subspecies meet. *Novit. Zool.*, 41:103–111.

—— 1940. Where subspecies meet. *Int. Congr. Ent. VI* (Madrid, 1938), 1:145–151.

Riley, N. D. 1960. Heinrich Ernst Karl Jordan. In *Biographical Memoirs of the Fellows of the Royal Society of London*, vol. 6, pp. 107–135. Royal Society of London.

Stresemann, E. 1975. *Ornithology: from Aristotle to the present*, trans. H. J. and C. Epstein. Harvard University Press, Cambridge, Mass.

21

Where Are We?

*By the mid-1950s a peculiar myth had developed in biology, which—
with some exaggeration—can be stated as follows: Evolution is a purely
genic phenomenon; it is a change, from generation to generation, of
gene frequencies in populations. This being so, it was impossible to gain
any understanding of evolution until mathematical population genetics
was developed. Furthermore, mathematical population genetics is the
source of population thinking.*

*When I was asked to give the inaugural lecture of the 1959 Cold
Spring Harbor Symposium, I questioned this myth and presented a different picture of the actual contribution of genetics to the evolutionary
theory as I, an outsider, a nongeneticist, saw it. Perhaps not surprisingly, my interpretation did not prove particularly popular (to put it
mildly) among the leaders of the genetic establishment. It did, however,
initiate a more critical attitude toward the historiography of evolutionary genetics.*

The history of the first 100 years since the publication of the *Origin of
Species* is a fascinating one. With its many controversies, its false starts
and converging pathways toward a solution of the open problems, one
might say that we have completed a full circle and that we are closer to
Darwin now and to Darwin's original concepts than we have been at
any time during the intervening period. The science of genetics has perhaps been responsible for the greatest deviation from the original Darwinism, in the theories of the early Mendelians, but it has likewise been
responsible for a return to the original Darwinism of 1859, from which
Darwin himself had somewhat deviated in the later editions of his work.
Let us look at this history a little more closely. The nonbiologist often
overlooks that there is more to the evolutionary theory than the problem of the reality of common descent. That the living kinds of animals
and plants were descended from common ancestors was accepted by
the vast majority of biologists within two decades after 1859. By about

Adapted from "Where Are We?" *Cold Spring Harbor Symposium on Quantitative Biology,*
24(1959):409–440.

1890 essential agreement had been reached among zoologists as to the major outlines of the phylogeny of animals (very little progress in this area having been made since then). What then were the still unsolved problems of evolutionary biology? Actually there were big ones and little ones.

The most serious defect in Darwin's theory was undoubtedly its failure to account for the raw "material of evolution." Darwin was fully aware of the enormous store of genetic variability in natural populations, but had only vague notions concerning its source. Little progress was made on this problem in Darwin's lifetime, and when Darwin died, the thinking on inheritance was still one of the most backward areas in biology.

The changes that followed were dramatic and by necessity of great importance for the evolutionary theory. I would like to take this opportunity to give a short outline of the changes in genetic theory so far as it concerns evolution. Essentially one may distinguish three successive schools: Mendelism, classical population genetics, and the newer population genetics.

CHANGES IN GENETIC THEORY

Mendelism

When Weismann demonstrated the internal contradictions and improbabilities of the Lamarckian theory and emphasized the separation of soma and germ plasm, he created a new intellectual climate for genetic thinking that gave the long overlooked work of Mendel new meaning. It is not surprising that, as a consequence, after 36 years of neglect, Mendel was rediscovered simultaneously by three different authors. The school of genetics that dominated biology following the rediscovery of Mendel's laws is frequently referred to as Mendelism. Those of its representatives who were interested in evolution, particularly De Vries and Bateson, were unfortunately typological saltationists and proclaimed that the new science of genetics required a saltationist interpretation of evolution. As De Vries (1906) expressed it: "The theory of mutation assumes that new species and varieties are produced from existing forms by certain leaps." But even the more moderate Mendelians thought an organism was at the mercy of its mutations. To Morgan, as late as 1932, evolution was "due to occasional lucky mutants which happened to be useful rather than harmful. As one of Morgan's disciples, I held fairly similar views at that time." (Dobzhansky 1959:254). Consequently, the period from 1900 to about 1920 saw a sharp cleavage, an almost bridgeless gap, between the evolution-minded naturalists on the one hand and the experimental geneticists on the other hand. Reconciliation became possible when it was realized that not all genetic changes are spectacular mutations but that most mutations are small and inconspicuous (Baur, East, etc.) and that indeed there is no difference between mutations and the so-called small variations which Darwin and the naturalists had regarded as the principal material of evolution. When it was finally demonstrated by Schmidt (1917), Sumner (1924), and Goldschmidt (1920) that the genetic differences between subspecies, the incipient species of the Darwinians, are genetic in nature (contrary to the claims of some mutationists) and indicative of a genetic basis for quantitative characters, all reason for disagreement between geneticists and evolutionists was removed. The stage was set for a new genetic interpretation of evolution, for a second phase of evolutionary genetics.

Classical Population Genetics

Population genetics clearly has two separate roots. One of them, indicated by the names Fisher, Wright, and Haldane, is mathematical and its contribution to the development of population genetics has been widely emphasized. The other one, indicated by such names as Sumner, Chetverikov, Timofeeff-Ressovsky, and Dobzhansky, had its roots in population systematics and is usually ignored. I will come back to the respective contributions of the two schools.

The emphasis in early population genetics was on the frequency of genes and on the control of this frequency by mutation, selection, and random events. Each gene was essentially treated as an independent unit favored or discriminated against by various causal factors. In order to permit mathematical treatment, numerous simplifying assumptions had to be made, such as that of an absolute selective value of a given gene. The great contribution of this period was that it restored the prestige of natural selection, which had been rather low among the geneticists active in the early decades of the century, and that it prepared the ground for a treatment of quantitative characters. Yet this period was one of gross oversimplification. Evolutionary change was essentially presented as an input or output of genes, as the adding of certain beans to a beanbag and the withdrawing of others. This period of "beanbag genetics" was a necessary step in the development of our thinking, yet its shortcomings became obvious as a result of the work of the experimental population geneticists, the animal and plant breeders, and the population systematists, which ushered in a third era of evolutionary genetics.

The Newer Population Genetics

The next advance was characterized by an increasing emphasis on the interaction of genes. Not only individuals but even populations were no longer described atomistically as aggregates of independent genes in various frequencies, but as integrated, coadapted complexes. A gene is no longer considered to have one absolute selective value, but rather a wide range of potential values that may extend from lethality to high selective superiority, depending on genetic background and on the constellation of environmental factors. I have referred to this new mode of thinking as the genetic "theory of relativity" (Mayr 1955b). Dobzhansky's "balance theory" of genetic variation is one of its aspects. The thinking of this newer population genetics is in considerable contrast to that of classical population genetics and even more so to that of early Mendelism. This change of view is not always realized, even by professional geneticists or by those evolutionists who have no contact with population genetics. Numerous new problems result from it, some of which I shall presently discuss in more detail.

Let me try to place this new development into a historical perspective and determine its relationship to the preceding period. There is no doubt that the classical period of population genetics was dominated by the mathematical analyses and models of Fisher (1930), Wright (1931), and Haldane (1932). These authors, although sometimes disagreeing with each other in detail or emphasis, have worked out an impressive mathematical theory of genetical variation and evolutionary change. But what, precisely, has been the contribution of this mathematical school to the evolutionary theory, if I may be permitted to ask such a provocative question? (see also Lewontin 1971).

Some of the younger evolutionists, perhaps not too well acquainted with the earlier literature, have ascribed to this school many of the major components of the modern synthetic theory of evolution. Others, like Waddington (1957), have questioned the magnitude of its influence.

The maintenance of genetic variability in populations, expressed in the Hardy-Weinberg formula, has been known since 1908, and Gulick and other nineteenth-century authors had proposed theories of genetic drift. These earlier concepts antedate classical population genetics by many years. Where the mathematical theory made concrete contributions, as in Fisher's theory of balanced polymorphism, the mathematics is of the simplest kind. Observing all this, Waddington asks quite rightly: "What then gave the mathematical theory its undoubtedly immense importance and prestige?" It seems to me that the main importance of the mathematical theory was that it gave mathematical rigor to qualitative statements that had previously been made. It was important to realize and to demonstrate mathematically how slight a selective advantage could lead to the spread of a gene in a population. Perhaps the main service of the mathematical theory was that in a subtle way it changed the mode of thinking about genetic factors and genetic events in evolution. However, I should perhaps leave it to Fisher, Wright, and Haldane to point out themselves what they consider their major contributions. This much seems certain: that the interpretation in Mendelian terms of the inheritance of quantitative characters by Mather and the development of the more sophisticated modern views on the interaction of genetic factors, on coadaptation, and on genetic homeostasis would have been impossible without the foundation laid by the mathematical theory, as oversimplified as it may appear in retrospect.

THE CONTRIBUTIONS OF GENETICS

It seems to me that we have now reached a stage in the development of the evolutionary theory where one can look back without passion or prejudice and determine the respective contributions of experimental genetics, of systematics, of paleontology, of developmental biology, and of other branches of biology to the synthetic theory of evolution. One should do this not for purely historical reasons or to establish priorities for prestige reasons, but because the planning of future work will be helped by a clear recognition of the potential contributions that can be made by the various collaborating branches of biology. Here I shall single out the specific contributions genetics has made to the synthetic theory of evolution. In view of the erroneous statements found in much of the current literature I shall try to discuss the history of some of the concepts that together form the synthetic theory. Such an emphasis on history may be a wholesome counterweight to the exceedingly unhistorical attitude of the current age.

Mutation

The idea of an ever-continuing origin of new genetic variation goes back to folklore (Zirkle 1946). It was, as stated above, a cornerstone of Darwin's theory. The ruling interpretation of early genetics, the peculiar mutationism of De Vries, was an essentially negative contribution that retarded a real understanding of mutation for several decades. That generation of evolutionists who associated with the term "mutation" what De Vries had meant by it had to die before the term could be reintroduced into the evolutionary theory in the broader meaning of later genetics. In addition to rejecting the saltationist interpretation of mutationism, genetics has also given up its faith in "mutation pressure" as the driving force of evolutionary change, a concept widespread in the 1920s and 1930s. In both respects modern genetics has returned to the thinking of Darwin and the naturalists.

Population

The claim has been made by some population geneticists that population thinking and its application to the evolutionary theory is a contribution of genetics. This overlooks that population thinking was already strongly apparent in Darwin's own work, in particular his application of Malthus' thinking to the theory of natural selection. Population thinking was even more widespread and more concrete in the second half of the nineteenth century. Systematists, following the example set by Baird, Schlegel, and others, became increasingly interested in the analysis of local populations. This approach reached its climax in the work of Heincke (1878) and his school on populations of marine fishes. At the same period population thinking was widespread among students of birds, butterflies, beetles, and snails. Indeed, it was apparent in the writings of almost any progressive animal taxonomist in the period between 1890 and 1930. The early population geneticists (except the mathematicians) had all either started as naturalists (Chetverikov, Dobzhansky, Sumner) or been in close contact with taxonomists, as was Goldschmidt at the Munich museum when he initiated his study of *Lymantria* populations. It is of more than historical significance that population thinking came into genetics from systematics and not the reverse (Mayr 1955a).

Natural Selection

No one has claimed that the theory of natural selection is an invention of the geneticist. Yet the current generation is not fully aware of how anti-selectionist virtually all the early Mendelians were. Natural selection survived during that period in the writings of systematists like K. Jordan rather than in those of the mutationists (see Essay 20). It is, of course, entirely consistent that he who believes that mutations make new species and that mutation pressure directs the course of evolution does not have much use for natural selection.

Genetic Drift

The idea that random processes affect the genetic contents of local populations and thus control their phenotypes preceded the coining of the term "genetic drift" by many decades, if not generations, as correctly pointed out by Wright (1951). It was the principal basis of Gulick's interpretation of the variation of *Achatinella* in the Hawaiian Islands, and similar thoughts had been expressed by even earlier authors as well as later ones (e.g., Lloyd 1912, Hagedoorn 1917). What Wright did was to separate clearly the respective contributions that mutation pressure, accidents of sampling, and natural selection make to gene frequencies in populations of various sizes.

Isolating Mechanisms

The term "isolating mechanisms," coined in 1936 by Dobzhansky, is recent. That the interbreeding of species is prevented by various factors in addition to sterility was, however, known many decades earlier. One finds, for instance, discussions of these mechanisms in the writings of Darwin, Seebohm, K. Jordan, and particularly Du Rietz (1930) (see Essay 11).

Geographic Variation

The fact of geographic variation was, of course, known long before Darwin. That the differences between geographic races and the differences between individuals within a population have a similar genetic basis was established by Schmidt (1917), Sumner (1924), and Goldschmidt (1920) and was one of the main reasons for the downfall of mutationism on the one hand and of Lamarckism on the other hand. This finding permitted a selectionist interpretation of slight differences among local populations that were obviously caused by differences in the environment. A Lamarckian interpretation of geographic variation was inevitable as long as one accepted the theories of De Vries and Bateson.

I have presented this analysis not as an attempt to depreciate the contributions of genetics but rather to permit a sharper focusing on the unique contribution of genetics by eliminating false claims. As time goes by, it becomes increasingly evident that nothing compares in significance with the demonstration of the particulate nature of the genetic material. This recognition is of such overwhelming importance that it overshadows everything else. The new interpretation of natural selection, the understanding of the meaning of recombination, the new insight in the relation of gene and character, the interpretation of quantitative inheritance, and many other aspects of the genetic theory, all are ultimately inevitable consequences of the Mendelian theory.

The second reason for the great importance of genetics is that it has provided, time after time, a causal explanation for purely empirical generalizations established by systematists and other naturalists. This is true for almost every major evolutionary theory or concept. Let me illustrate it with only one or two examples.

Let us take, for instance, the theory of geographic speciation. One after the other of the reputed cases of sympatric speciation had to be reinterpreted during the past 70 years and eventually it was found in all well-analyzed cases of speciation in animals that a period of geographic isolation had preceded the acquisition of the isolating mechanisms. This was a purely empirical finding for which no immediate reason was apparent. This need for geographic isolation was finally explained by population genetics, which showed that a gene pool is such a well-integrated and coadapted system that it cannot be divided into two equally well-integrated gene pools without a period of geographic isolation of considerable duration. The purely extrinsic separation permits the independent occurrence of two separate integration processes that eventually lead to mutual reproductive isolation.

Or, let us look at the problem of the relation between embryology and evolution, which has been a source of endless disputes since the eighteenth century. Most pre-Darwinian evolutionists and many of Darwin's contemporaries regarded the development of the individual and phylogenetic change merely as two aspects of a single phenomenon. Stating this problem in terms of genetic information has greatly clarified the issue. The genetic material, handed down from generation to generation, can be considered coded norms of reaction. Individual development is the decoding of the information made available through fertilization. Evolution, on the other hand, is the generating of ever-new programs of information, of ever-new norms of reaction. With the problem stated in these terms, it is at once evident that the superficial similarity between individual and phylogenetic development is entirely spurious. One is a functional, the other a historical phenomenon.

It seems to me that the basic problem of Lamarckism, the inheritance of acquired characters, can likewise be stated far more clearly in terms of programs of information than it has been previously. If one wanted to demonstrate the occurrence of such inheritance, one would have to prove that the developmental pathway is not a one-way street and, as Weismann emphasized some 70 years ago, one would have to prove that the information from the peripheral end organs is fed back to the germ cells in the gonads and is used to reprogram the inherited information in an adaptively improved manner. I hardly need to dwell on the impossibility of such an occurrence as far as the higher organisms are concerned. Whether or not the situation is different in microorganisms is perhaps still an open question. There is at least a theoretical possibility for such an interaction, owing to the shortness of the pathway between the gene and the component of the phenotype it controls.

I have come to the end of my short historical survey of the relation between genetics and the other branches of evolutionary science. It seems evident that there is a happy symbiosis among these various fields. The naturalist has access to a vast store of observational evidence on which he bases various empirical generalizations. It is the role of the geneticist to interpret these generalizations in terms of the genetic material and to test his conclusions by experiment. I can foresee no reason for a change in this historically established pattern of cooperation. The best evidence for its success is the modern synthetic theory of evolution.

I entitled my discussion: "Where Are We?" One of the methods of fixing one's position is to look back and try to reconstruct the steps taken to reach the present position. This I have done in my historical survey. A second method is a careful scrutiny of the surroundings of our position. What do we mean by twentieth-century Darwinism? I think its essence can be characterized by two postulates: (1) that all the events that lead to the production of new genotypes, such as mutation, recombination, and fertilization, are essentially random and not in any way whatsoever finalistic; and (2) that the order in the organic world, manifested in the numerous adaptations of organisms to the physical and biotic environment, is due to the ordering effects of natural selection.

Nothing has been discovered in the decades since these principles were first clearly stated that is in any way in conflict with these basic assumptions. Yet we must realize that this is only a beginning. It has been claimed again and again in the last 30 years that evolutionary biology is exhausted as a field of research because the synthetic evolutionary theory has supplied all the answers. Let me emphasize, in a variation of Mark Twain's saying, that the news of the death of evolutionary biology is greatly exaggerated. Just how many unsolved problems there are becomes apparent as soon as one takes up individual areas and attempts to delimit the known from the unknown. Let me try to do this for the field of evolutionary genetics. Where are the pathways to the unknown?

NATURAL SELECTION

Let us begin with some problems relating to selection. In an immensely stimulating and thoughtful essay on this topic, Lerner (1959) has stated correctly: "What we have learned so far about natural selection is obviously only the beginning. What remains to be learned is immeasurably more." The power of natural selection is no longer questioned by any serious evolutionist. It is understood that it is a statistical, a populational phenomenon, the essence of which consists in differential reproduction. And as soon as this is fully understood it is evident that selection can be a creative process. Yet in spite of all these advances numerous unsolved problems remain. Let me single out only four aspects of natural selection that raise doubts in my mind.

The Selection of Genes versus the Selection of Phenotypes

Haldane (1957) has pointed out that selection places a considerable strain on populations. Too rapid a rate of simultaneous selection against too many genes might eliminate the entire population. Thus Haldane has called attention to a neglected aspect of selection. But is this really a serious problem? Is this not a situation where the new genetic theory of relativity should be applied? Haldane himself emphasizes correctly that we are dealing with relative values and that phenotypes, not genes, are exposed to selection. Those individuals which have the greatest number of minus genes have the lowest reproductive potential. Those with the greatest number of plus genes have, quite obviously, the highest reproductive potential. Natural selection thus can deal simultaneously with a great number of variable loci. The "goodness" or "badness" of a gene will be relative, depending on the fitness of the averages of all the individuals of the population. And as the population density drops, even rather poor phenotypes may survive by not being exposed to competition from better ones.

Total reproductive potential is another factor introducing a relative element. This is something we have to keep in mind when we compare *Drosophila* with man. If *Drosophila melanogaster* can withstand the impact of 300 R per generation because among the 500 fertilized eggs laid by a female, one will be a zygote with a combination of genes that has been compensated for radiation damage, it does not mean that man would be able to do likewise, even if he did step up his reproductive rate to a dozen children per family. To repeat once more, I feel that Haldane has called attention to an important problem when he asked the question: "How severe a selection pressure can a population endure at any one time?" But I am not sure that his answer is necessarily correct. And with this doubt we come to a second problem.

The Measurement of Fitness

Fitness, as we have seen, is a highly relative matter. When comparing the fitness of different genotypes, it is of crucial importance to find an objective yardstick. Fitness, as R. A. Fisher pointed out long ago, is best defined operationally as the relative contribution to the gene pool of the next generation (disregarding accidents of sampling). But how to measure it comparatively? It has become routine in population genetics to measure the fitness of whole chromosomes in homozygous condition as compared to the same chromosome in heterozygous condition paired with a balanced lethal. The chromosome then, on the basis of its viability in homozygous condition, is labeled lethal, semilethal, subvital, or viable, as the case may be. Frankly, I think this is utter nonsense, as I pointed out in 1955. Homozygous chromosomes do not occur in nature in sexual species and neither do balanced lethal tester chromosomes derived from alien populations. "To compare the relative performance of two combinations which do not occur in nature and utilize it as an index of viability for a chromosome does not appear to be particularly meaningful technique," as I put it very mildly (Mayr 1955b). Dobzhansky and Wallace have shown that there is very little relation between the fitness of a given chromosome in homozygous or in heterozygous condition. The mere fact that 20–40% of all chromosomes found in wild species of *Drosophila* are lethal when homozygous should be sufficient to indicate the inappropriateness of this terminology. According to the classical hypothesis of the genetic load of population, lethality of the homozygous chromosomes might be due to deficiencies or other lethal genes. Dobzhansky's work on synthetic lethals shows that this is not necessarily the case. It is becoming more evident from day to day, as postulated by the "balance theory" (Dobzhansky 1955, 1959), that it is the interaction of genes which determines fitness. And in outbreeding organisms, in which homozygous chromosomes do not occur, selection is for high fitness in heterozygous condition. Since every chromosome in a sexual species is unique and every diploid combination is unique, it would seem futile to determine a chromosome's fitness in any way that is meaningful. Whether we like this conclusion or not, it seems to me that there is no way of measuring the fitness of chromosomes.

And, to carry this argument one step further, what is true for chromosomes is true for individual genes. In their formulae the mathematical geneticists assign an absolute selective value S or W to a given gene, let us say gene a. Whatever usefulness such a value W may have is at once destroyed by the cautious qualification that W is to be taken as the average of all the selective values the genotype may have in different physical and biotic environments. It is obvious then that W, instead of being an absolute and uniform value, is an exceedingly relative and heterogeneous one. Even more devastating for the usefulness of the value W is the fact that the fitness of gene a is determined not only by the physical and biotic environment in which the phenotype is placed, but also by the genetic environment in which gene a is placed. Depending on the genetic background the same gene may be "good" or "bad"; indeed in an extreme case a normally valuable gene may well be a near lethal.

If there is any validity in this argument, we may have to revise many of our ideas. For instance, is the old argument really true, that virtually all mutations are deleterious, because the same loci, having mutated thousands of times in their history, must have reached the optimal genes since all inferior genes, except concealed recessives, must have long since been eliminated by natural selection? This argument makes two assumptions, both of which are demonstrably wrong, first, that the selective value of a gene is absolute, second, that the totality of environments (the physical, the biotic, and the genetic) is constant. As soon as we admit the continuous changes of the three environments, we must admit that many current mutations are far more "hopeful" than would be expected on the basis of the orthodox considerations.

But I would like to carry this argument still one step further. All evolution is a sequence of unique events. This is true not only for selective values in different genotypes and different environments, but also for selective values in segregation, crossing over, and whatever other events determine evolutionary success and evolutionary change. If we set up 10 parallel selection experiments, starting each from the same, apparently identical foundation stock, we are likely to get in the end 10 somewhat different answers. The more I study evolution the more I am impressed by the uniqueness, by the unpredictability, and by the unrepeatability of evolutionary events. Let me end this discussion with the provocative question: "Is it not perhaps a basic error of methodology to apply such a generalizing technique as mathematics to a field of unique events, such as organic evolution?"

The Population as a Unit of Selection

Haldane (1932) was the first to emphasize the importance of the population-as-a-whole as a unit of selection. There are a number of attributes of species and populations that are not of any particular selective advantage to any single individual in a population but that are of great advantage to the population as a whole. These include such factors as rate of mutation, degree of outbreeding, sex ratio, etc. It is easy to propose models by which the genetic basis of such factors can be changed. Yet such models must make certain unrealistic assumptions, such as the total isolation of each population and the total extinction of the less well adapted populations. Normal distribution patterns make these assumptions unlikely. Here again we need more thinking and perhaps more experimentation, fully taking into account the fact that most natural populations are open systems.

Reproductive Success

Natural selection owes its universal success to the fact that it applies no rigid artificial criteria. Its only criterion is success. Fitness, consequently, is quite correctly defined as the contribution to the gene pool of the next generation. And yet this eminently practical principle has one Achilles' heel: it also rewards pure reproductive superiority. A male bird of paradise who adds to the gaudiness of his plumage is rewarded in the same manner as the individual who invents a slight improvement in physiology that benefits the species as a whole, and possibly all descendant species. More and more cases are being discovered where genes have become established in a species not because they add in any way to the adaptiveness of the species (in fact they usually do exactly the opposite) but because natural selection is defenseless against them. Here is clearly a situation where the two terms "of high selective value" and "of high adaptive value" are not congruent. I strongly suspect that such selection for mere reproductive success may have frequently played a role in the extinction of species. And such extinction may be very rapid.

The problem is of more than academic interest to man and has worried eugenicists for generations. It does not require complex mathematics to figure out what would happen if the improvident low-I.Q. individual regularly had a dozen children and the prudent, superior citizen only two. I said deliberately "if" and I am not trying to create an alarmist situation. Yet one can not afford to ignore this Achilles' heel of natural selection.

Here then are four aspects of natural selection that pose unsolved problems. I could show that the same is true not only for selection but also for other major aspects of evolution, such as mutation, recombination, and errors of sampling. However, for the sake of variety, I prefer a somewhat different approach. Let me attempt, instead, to express some of the unsolved evolutionary problems in terms of the different levels of integration. Let me single out those of the individual, the population, and the species. Each of these levels has its own evolutionary problems, although nearly all of them involve mutation, recombination, selection, and errors of sampling.

THE PHENOTYPE OF THE INDIVIDUAL

What is exposed to natural selection is not the individual gene or the genotype but rather the phenotype, the product of the interactions of all genes with each other and with the environment. The central position of gene physiology in the evolutionary theory and in evolutionary happenings is not yet as much appreciated as it deserves. To say that all

quantitative characters are the product of polygenic inheritance is merely a different way of saying that they are the product of an interaction
of genes. And quantitative characters play a decisive role in contributing to fitness (*Cold Spring Harbor Symp. Quant. Biol.* 1955, Lerner
1954). The evolutionary problems connected with developmental physiology are numerous. Let me single out only one or two for more detailed discussion.

What is the relation between modifiers, polygenes, and pleiotropy?
The understanding of the interaction of genes still suffers from the
typological thinking of earlier periods. A character was thought to be
the product of a gene and if it turned out that additional genes affected
the expression of the character they were labeled modifying genes. If
it turned out that a character was the product of a considerable number
of genes, all contributing to the phenotype of this character, they were
labeled polygenes. Curiously enough, many authors have treated such
polygenes as *ad hoc* mechanisms that have no other functions in the
organism than to modify a given character. Goldschmidt (1953, 1954)
and other adherents of such a scheme argued that it would require an
impossibly large number of modifying genes. It is curious that Goldschmidt, who was so well aware of the pleiotropy of genes, was quite
unable to appreciate the probability that many polygenes are nothing
but the pleiotropic manifestations of genes that have other major functions. That this is correct is indicated by every major new mutation
which immediately encounters in its gene pool a high number of genes
which affect its dominance, its penetrance, or its expressivity. It would
be absurd to assume that every gene pool contains a large reservoir of
"reserve genes" which have no other function but to wait for the occurrence of major mutations and then mold them adaptively into the
phenotype. I would be reluctant to state these obvious facts were it not
for several recently published statements that ignore this obvious interpretation. Let me repeat once more, then, that the high degree of
pleiotropy of most genes provides abundant material for polygenic effects. Consequently, there is no reason to believe in the existence of
two sharply separated classes of genes, oligogenes and polygenes. The
same gene may behave as an oligogene for one character and as a poly-
gene for another one.

It is well known to naturalists and animal and plant breeders that unexpected or otherwise rare phenotypic traits may be revealed in certain
individuals of a population if they are exposed to exceptional conditions or to environmental shocks. A well-known, recently studied case
is the condition of cross-veinlessness in *Drosophila*, which can be revealed by temperature shocks (Waddington 1953, 1957). One aspect of
this phenomenon that is not nearly sufficiently stressed is that only a certain percentage of the individuals of a population will show this phenotypic response. Whether or not an individual responds in this manner is
not a matter of accident. Rather, the treatment reveals those individuals

which have a genotype with many of the prerequisite factors for the exceptional phenotype. Cross-veinlessness is a highly polygenic character and each of the individuals that responds to the treatment with the cross-vein phenotype will have a different combination of genes. Selecting such individuals and crossing them, continuously selecting for an increased penetrance, will lead to a gradual accumulation of such polygenes until finally the phenotype will appear as the result of normal development without need for an environmental shock. Such experiments have been made from the beginnings of genetics, as for instance Bateson's experiment on the prevention of flowering in the sugar beet (Hall 1928, from Haldane 1932). We are here simply dealing with a threshold phenomenon where numerous genes contribute to a certain phenotype but where the potentiality for that phenotype will not be pushed above the visible threshold until a sufficient number of genes have accumulated in the genotype. I entirely agree with this interpretation as presented by Milkman (1960) and Stern (1958). It seems to me that these interpretations are clearer and simpler than Waddington's interpretation. Experiments on the "revealed genotype" shed new light on the so-called Baldwin effect and indicate the possibility that a modification of the phenotype is of evolutionary significance only if it is due to concealed polygenes and will lead to their accumulation in the gene pool.

THE POPULATION

The next higher level is that of the population. That there are still many unsolved evolutionary problems at this level is true in spite of the many advances of systematics and population genetics. Let me single out just a few of these problems.

What gives the local gene pool its cohesion and coadaptation? The work of Dobzhansky and of others has revealed that there is a harmony among the genes that together make up the local gene pool. Through natural selection such genes are accumulated as produce maximal viability in their allelic and epistatic interactions. This very general statement is about all we can say about the "coadaptation," or "internal balance," of the gene pool of the local population. How it operates is still largely obscure because adequate tools have not yet been found to determine by what possible mechanisms such coadaptation, such internal balance, is maintained. Sometimes it is done by the development of supergenes, such as inversions, but there are other mechanisms as well. In particular, the process includes all the factors Dobzhansky records under the collective heading "balance hypothesis." Perhaps the system owes its harmony to the multiplicity of interactions of its components, which would automatically resist any disturbance. This would be Lerner's genetic homeostasis, although I would give, in such a homeostasis, as much weight to epistatic interactions as to allelic ones. Many otherwise puzzling evolutionary phenomena may find their explanation in epistatic inertia. Let us take, for example, the conservative nature of the frequencies of blood group genes in human races. Even though under different conditions individual blood group genes and genotypes differ in their selective values quite drastically from each other (particularly at the ABO locus), the relative frequency of the genes remains relatively stationary in a given population, apparently owing to various balance mechanisms (allelic and epistatic).

Work on the genetics of populations has revealed a far more serious population problem. It appears possible, if not probable, that the majority of the findings made on experimental populations cannot be automatically applied to natural populations. Why?

The experimental geneticist works by necessity with closed populations. Except for the rare occurrence of mutations, genetic changes in such populations are due to recombination, selection, and errors of sampling. The genetic input is negligible. It is only within recent years that we have begun to realize that these closed populations, even the large populations of the population cages, are not natural populations. Inbreeding in these closed populations is far greater than in natural populations of all species except those that have special mechanisms to secure inbreeding. In a natural population, which is a wide-open population, perhaps 40 percent or more of the members of an effective local breeding population are immigrants from the outside. The gene contents of these immigrants is, of course, largely the same as that of the local population. Yet they provide an input of new genetic factors into the population that is perhaps 100 or 1000 times greater than the input caused by mutation. No one has as yet analyzed all the consequences of this situation. I have pointed out (Essay 15) that those kinds of genes are likely to be favored in closed populations which are superior as homozygotes or in a limited number of combinations. On the other hand, those genes will be favored in open populations which can produce viable phenotypes in a vast assortment of different genetic combinations: "jack-of-all-trades genes." I would like to go one step further. Lerner (1954) and many other students of experimental populations or of domesticated animals and cultivated plants have emphasized the great importance of heterozygosity. The widespread occurrence of balanced polymorphism indicates that heterozygosity is important also in open populations. Yet there are numerous observations which indicate that the type of heterozygosity which produces heterosis is a special case and is often the product of careful selection. Epistatic interactions may contribute more to superior fitness in open populations than allelic overdominance. The relative importance of this interaction among loci is proven by the breakdown of viability in the F_2 of interpopulation crosses.

There is every reason to believe that the internal genetic structure of an open population is quite different from that of a closed one. The relative importance of most genetic processes, such as mutation, random fixation, inbreeding, and heterosis, will be entirely different in the two types of populations.

We must keep this in mind when we consider the role of mutation in natural populations. Buzzati-Traverso (1954b) has pointed out that the genetic variability of closed populations is not inexhaustible, particularly if selection pressure is high, and that a mild amount of induction of mutations by X-rays may speed up evolutionary change (see also Scossiroli 1954). The work of Wallace (1958) with largely homozygous strains of *Drosophila melanogaster* likewise indicates that in closed homozygous populations radiation-induced mutation may be beneficial because it increases heterozygosity. Let me emphasize that Wallace has not made the mistake of applying these findings to open populations. Any open population has an input of genetic novelties that may well be of the same order of magnitude as that which is induced by radiation or even greater. And yet, in the case of the open population these genetic novelties have been pretested in other populations so that all the more generally deleterious or lethal ones have already been eliminated. In open populations there is a high level of competition, in every generation, between the indigenous and the immigrant heterozygous combinations. The very imbalance of such a system makes it highly adaptive under the fluctuating environmental conditions of nature. The inbreeding closed laboratory population, living under uniform environmental conditions, is an altogether different system. We must keep this in mind when we discuss radiation damage in natural populations and in man. There are few natural populations of wild animals that have a higher rate of outbreeding than civilized man. It would be altogether unscientific to apply the findings on the effects of radiation in closed laboratory populations to a species with the wide-open population structure of modern man.

I might add that the peculiar characteristics of the open populations are not adequately coped with by mathematical geneticists, in spite of all their efforts, particularly those of Wright, to allow for the immigration factor. The mode of interaction of genes is a far more important aspect of population structure than mere gene frequency. And the immense quantitative and qualitative difference between genetic input due to mutation and that due to immigration is rarely ever mentioned. How can one appreciate the genetic dynamics of a local population in a normal outbreeding species if one does not emphasize that the genetic input through immigration may be 100 times as large as that through mutation?

There are still other aspects of the genetics of local populations which the mathematical formulations cannot represent adequately. The normal population continuum of a species consists of remarkably well-adapted local populations that nevertheless are part of a continuous system of populations that actively exchange genes with each other. Local demes (the deme is here defined in the narrowest sense as the local interbreeding population) may consist for instance of 50–200 individuals, among which 40–50 percent are immigrants. These immigrants are largely derived from neighboring demes, but an important fraction consists of long-distance immigrants. The highly skewed character of the dispersal curve is unquestionable (Bateman 1950). As I have pointed out elsewhere (Essay 15), such a system virtually demands that there are different kinds of genes. With a number of isoalleles almost certainly present at many loci, there will be a strong premium not on the genes that in certain combinations can produce the optimal genotype but rather on genes that can produce a passable genotype in the greatest number of possible combinations.

The degree of "openness" of a population will thus determine what kinds of genes are favored by selection. That the same gene may have different selective values in populations of different sizes is merely another manifestation of the genetic "theory of relativity."

THE SPECIES

The problems encountered on the level of the population are compounded on the level of the species. Nothing is less suitable as a foundation for a study of species than a typological approach. Recent work on various species of *Drosophila* has shown how much structure a species has and how little justified one is in generalizing from the genetics of one population to the genetics of another one, and from one species to another. The really difficult part to keep in mind at all times is that every population has individuality and yet that all populations are held together by gene exchange. And one suspects that the character of a species as a whole is determined by the populations near the ecological center of the species which are the most prosperous, have the highest population surplus, and are therefore the source of the greatest amount of gene flow. From this center outward to the species border, every population, to an increasing extent, has to make a compromise between coping, on the one hand, with the increasingly difficult conditions of the local environment and, on the other, with the inflow of unsuitable genes from the center of the species range. It may appear to be merely a formal difference whether one describes a species in terms of a gradient from the southernmost to the northernmost or from the most easterly to the most westerly population or, rather, as a series of more or less concentric circles around the center of the species. Yet the second alternative may actually describe the situation far more accurately.

The studies of systematists as well as those of population geneticists indicate that the peripheral populations of a species, and particularly the peripherally isolated populations, may be quite exceptional in their genetic structure. What are their special characteristics? To begin with, they live near the border of the species range, in other words, under environmental conditions that are marginal for the species. Selection will be unusually severe. Gene flow is much reduced as compared to gene flow in the more central populations of the species, and so far as it exists it is a one-way flow from the interior of the species range toward the periphery because population surplus in the peripheral populations is negligible. As I have pointed out elsewhere (Essay 15), this one-sided genetic input may well be responsible for the inability of these peripheral populations to become truly adapted to the local conditions. The simultaneous need for retaining coadaptation with the gene pool of the species as a whole and for assimilating the immigrants from the more central parts of the species range limits the opportunities for building up novel gene combinations that would permit an expansion beyond the present species range.

This limitation does not apply to the peripherally *isolated* populations, and as a result one usually finds rather deviating phenotypes in peripherally isolated populations. There are indications, although at the present time no real proof, that such populations are important in ecological shifts. Let us take, for instance, the problem of the shift in monophagous insects from one species of food plant to another. Most so-called monophagous insects are not entirely restricted to a single species of food plant but may be found occasionally on an accessory host. It is quite possible that such an accessory host may become the main host under the special environmental conditions of the peripherally isolated population. Protection from the gene flow of the parental species will permit not only the accumulation of genes leading to a better adaptation to this new host but also the acquisition of isolating mechanisms against the parental populations. If this peripherally isolated population subsequently reestablishes contact with the parental populations, it will have not only a set of isolating mechanisms against the parental species but also a different food niche and will thus be able to coexist sympatrically with the parental species.

The size of a local deme and the amount of gene exchange with its neighbors are very important. But so are numerous other factors that affect the population structure of species and the genetic system of a population. In a way, every species represents a unique system, yet certain patterns seem to emerge. Factors controlling the degree of outbreeding are particularly important in determining the population structure of a species. Obligatory inbreeders and panmictic outbreeders are the extremes of the spectrum with most species, particularly of animals, occupying a somewhat intermediate position.

To me the most impressive aspect of such diversity is that it shows the multiplicity of pathways by which a high level of adaptation can be attained. There has been much argument in the recent literature on the question whether and how one can measure the adaptedness of a species. In such arguments one should not forget that there are very different ways of achieving adaptation. Looking at numerous species of animals, I can distinguish three particularly common roads to success.

1. *The narrow specialist.* This type is particularly common among parasites and host-specific insects. The universe of such a species is narrow and confined. Yet within its universe it reigns supreme. An insect that lives as a leaf miner in the leaves of a birch tree or a fluke that lives in the liver of a particular species of vertebrates may be immensely successful and build up enormous populations. Yet its fate is closely tied to that of its host, and the balance between the parasite and the host is a precarious one that at any time can lead to the extermination of either the parasite or the host (with parasite). Nevertheless, the advantages of becoming the supreme specialist of the narrow niche are so tempting that an extraordinarily high percentage of all animals succumbs to this temptation. It is probable that more than 50 percent of all the species of animals should be classed as narrow specialists.

2. *The successful universalist.* Man, of course, is the outstanding example of this category. It contains species that can cope with many climates and with many ecological niches. There is, of course, no single species that is equally well adapted to all climates and to all habitats and it may be a matter of definition what to include in this category. Yet, by and large, we will have no difficulties in pointing out certain species in almost any group of organisms that lack conspicuous specializations and appear to be successful in several niches, species the ecologist would class as "tolerant" or "euryecous." The geneticist would suspect that such species are rich in concealed genetic variability, that they are highly heterozygous, that they may tend to chromosomal polymorphism, that they have much gene flow within the system and highly developed genetic homeostasis. Yet by all these mechanisms such species become so successful at coping with a changing environment that they tend to become conservative from the evolutionary viewpoint.

3. *The opportunist.* The third type of species is characterized, up to a point, by a combination of the two extremes previously stated. These are species with considerable geographic variability, with a successful central universalist aggregate of populations, and yet with an ability to form peripheral geographic isolates. Each of the geographic isolates is able to become specially adapted to its local environment, making the best of the local opportunities. Some of the geographic isolates are sufficiently withdrawn from the gene flow of the species to be able to find their own equilibrium without having to compensate incessantly for the disturbing effects of the immigrant genes. There is much evidence to indicate that this third type of species structure is the most important for long-range evolution. Such opportunism may be expensive but it is worth the price. Of 50 attempts to find a new ecological niche, 49 may be unsuccessful either because the new population is too small or be-

cause the new niche is not sufficiently important. Yet even if only 1 out of 50 is successful, indeed if only 1 out of 250 is successful, it would lead to a continued expansion of the total evolutionary universe. It seems to me, at least as far as animals are concerned, that much of the known evolutionary progress has been achieved by this type of species. The chance that such species are well represented in the fossil record is small and some of our difficulties in reconciling the geological record with our concepts of evolution, as based on the phenomena of genetics, may be due to the probabilities of fossil preservation. The conservative successful universalist certainly has a far greater chance to be preserved in the fossil record than the geographical isolate that switches successfully into a new niche.

CONCLUSION

In concluding my remarks I would like to make two observations. The first is that in spite of the almost universal acceptance of the synthetic theory of evolution, we are still far from a complete understanding of almost any of the more specific problems of evolution. I have tried to demonstrate this for every level from the individual to the species. There is still a vast and wide open frontier. And I would like to make a second point.

We live in an age that places great value on molecular biology. Let me emphasize the equal importance of evolutionary biology. The very survival of man on this globe may depend on a correct understanding of the evolutionary forces and their application to man. The meaning of race, of the impact of mutation, whether spontaneous or radiation-induced, of hybridization, of competition—all these evolutionary phenomena are of the utmost importance for the human species. Fortunately the large number of biologists who continue to cultivate the evolutionary vineyard is an indication of how many biologists realize this: we must acquire an understanding of the operation of the various factors of evolution not only for the sake of understanding our universe, but indeed very directly for the sake of the future of man.

REFERENCES

Bateman, A. J. 1950. Is gene dispersion normal? *Heredity*, 4:353-363.

Buzzati-Traverso, A. A. 1954a. Conclusions and perspectives. *Symp. on Genet. of Pop. Structure*, IUBS publ., ser. B, 15:126-138.

—— 1954b. On the role of mutation rate in evolution. *Proc. 9th Int. Congr. Genet.*, 1:450-462.

Chetverikov, S. S. 1926. On certain aspects of the evolutionary process from the standpoint of modern genetics. *Jour. Exper. Biol.* (Russian), A2:3-54. Eng. trans. (1961), *Proc. Amer. Phil. Soc.*, 105:167-195.

De Vries, H. 1906. *Species and varieties: their origin by mutation*, ed. D. T. MacDougal, 2nd ed. Chicago.

Dobzhansky, T. 1955. A review of some fundamental concepts and problems of population genetics. *Cold Spring Harbor Symp. Quant. Biol.*, 20:1–15.

—— 1959. Variation and evolution. *Proc. Amer. Phil. Soc.*, 103(2):252–263.

Du Rietz, G. E. 1930. The fundamental units of botanical taxonomy. *Svensk. Bot. Tidsskr.*, 24:333–428.

Eiseley, L. 1958. *Darwin's century: evolution and the men who discovered it.* Doubleday, New York.

Fisher, R. A. 1930. *The genetical theory of natural selection.* Clarendon Press, Oxford.

Goldschmidt, R. H. 1953. Pricking a bubble. *Evolution*, 7:264–269.

—— 1954. Presidential address—different philosophies of genetics. *Proc. 9th Int. Congr. Genet.*, 1:83–99.

Haldane, J. B. S. 1932. *The causes of evolution.* Longmans, Green, London.

—— 1957. The cost of natural selection. *J. Genet.*, 55:511–524.

Hagedoorn, A. C., and A. L. Hagedoorn. 1917. Rate and evolution. *Amer. Nat.*, 51:385.

Hall, A. L. 1928. Bateson's experiment on bolting in sugar beet and mangolds. *Jour. Genet.*, 20:219–231.

Heincke, F. 1878–1882. *Die Varietäten des Herings.* Berlin.

Lerner, I. M. 1954. *Genetic homeostasis.* Oliver and Boyd, Dover.

—— 1959. The concept of natural selection: a centennial view. *Proc. Amer. Phil. Soc.*, 103(2):173–182.

Lewontin, R. 1971. Genes in populations—end of the beginning. *Quart. Rev. Biol.*, 46:66–67.

Lloyd, R. E. 1912. *The growth of groups in the animal kingdom.* Longmans, Green, London.

Mayr, E. 1955a. Karl Jordan's contribution to current concepts in systematics and evolution. *Trans. Roy. Ent. Soc. London*, 107:45–66.

—— 1955b. Integration of genotypes: synthesis. *Cold Sping Harbar Symp. Quant. Biol.*, 20:327–333.

—— 1963. *Animal species and evolution.* Belknap Press of Harvard University Press, Cambridge, Mass.

Milkman, R. D. 1960. The genetic basis of natural variation. *Genetics*, 45:35–48. (Later parts in *Genetics*, 45, 46, 47, 50, 51, 53, and 54.)

Schmidt, J. 1917. Statistical investigations with *Zoarces viviparus. Jour. Genet.*, 7:105–118.

Scossiroli, R. E. 1954. Effectiveness of artificial selection under irradiation of plateaued populations of *Drosophila melanogaster. Symp. on Genet. of Pop. Structure*, IUBS publ., ser. B, 15:43–66.

Stern, C. 1958. Selection for subthreshold differences and the origin of pseudo-exogenous adaptations. *Amer. Nat.*, 93:313–316.

Stumm-Zollinger, E., and E. Goldschmidt. 1959. Geographical differentiation of inversion systems in *Drosophila subobscura. Evolution*, 13(1):89–98.

Sumner, F. B. 1924. The stability of subspecific characters under changed conditions of environment. *Amer. Nat.*, 58:481–505.

Waddington, C. H. 1953. Genetic assimilation of an acquired character. *Evolution*, 7:118–126.

—— 1957. *The strategy of the genes.* Allen and Unwin, London.

Wallace, B. 1958. The average effect of radiation induced mutations on viability in *Drosophila melanogaster. Evolution*, 12(4):532–552.

Wright, S. 1931. Evolution in Mendelian populations. *Genetics,* 16:97–159.
—— 1951. Fisher and Ford on the "Sewall Wright Effect." *Amer. Sci.,* 39(3): 452–459.
Zirkle, C. 1946. The early history of the idea of the inheritance of acquired characters and pangenesis. *Trans. Amer. Phil. Soc.,* 35(2):91–151.

22
The Recent Historiography
of Genetics

Except for the study of Darwin, there is perhaps no other area in biology the history of which has been dealt with as frequently and with as much competence as the history of genetics. In addition to reports on numerous Mendel celebrations, many books on the history of classical genetics (or certain aspects of it) have been published as well as several on the history of molecular genetics. In addition the field is particularly well supplied with source books containing collections of publications important in the history of genetics. As a result, the highly specialized literature is more readily available than is the case with the history of almost any other part of biology. What is perhaps most interesting is the fact that, with a few exceptions, all these books and articles were written by members of the genetics establishment. And since some of these geneticists belong to different schools within the field, they sometimes provide slightly different views or at least emphases concerning the same developments.

These advantages, particularly the gratifying competence of the accounts, are, however, bought at a price. The science of genetics did not emerge in a vacuum. Rather, the early history of the field was intimately connected with various other scientific endeavors, such as systematics, evolutionary biology, cytology, embryology, and mathematics. The convert to genetics cannot help but see the developments from the viewpoint of his former specialty, and this means bias even when it is subconscious.

It is a most welcome corrective for this bias that in recent years certain aspects of the history of genetics have been analyzed by historians. Two of these essays, Olby's *Origins of Mendelism* (1966) and Provine's *Origins of Theoretical Population Genetics* (1971) are excellent. Both authors have searched the original sources far more thoroughly than was done in the more broadly conceived volumes mentioned above, and both of them, as nongeneticists, have given far more emphasis to the

Revised from "The recent historiography of genetics," *Journal of the History of Biology,* 6, no. 1(1973):125–154.

legitimacy of the viewpoints of those with whom the early geneticists had their controversies. The results are most illuminating, and no one can read these volumes without learning a great deal not only about the history of genetics but indeed about the history of biology during the periods covered.

The historiography of science in some ways greatly resembles the history of science herself. Any problem that is newly solved raises additional problems, or at least places already known problems in much sharper focus. Here I shall endeavor to point out some of these problems and deal with them critically.

Historians have not given anywhere near enough attention to the importance of the scientific (and cultural) backgrounds of opposing schools. In biology, far more so than in the physical sciences, there are nearly always competing camps in every area. Genetics is no exception, particularly during the period from 1860 to 1940. Genetic variation was studied long before there was a formal science of genetics. But the naturalists viewed such variation very differently from the experimental botanists and zoologists, and the embryologists from the statisticians. Not only did they work with different material, but the very questions they asked were different. In the famous post-1900 argument about speciation "by Darwinian selection of continuous variation" or "by the saltational origin of discontinuous new mutations," the naturalists almost solidly lined up behind Darwin, while the experimental scientists favored large-scale mutations with almost equal agreement.

That such background is important is evident even in the writings of the historians themselves. In the case of Olby the influence of R. A. Fisher's genetics laboratory in Cambridge is quite conspicuous; nor has he been entirely able to escape some of C. D. Darlington's assorted biases. The claim that Darwin was a strict believer in blending inheritance is one example. In the case of Provine, the association with the University of Chicago and the effect of "two lengthy interviews with Sewall Wright" are rather noticeable in some of the interpretations. Such biases are inevitable and add spice to the discussions.

Olby's objective, as clearly stated in the title of his book, is to explore the *Origins of Mendelism*. He sees the work of Mendel as the culmination of a series of studies that began a century earlier with the classic hybridization experiments of Koelreuter. Olby not only presents facts as Stubbe, Zirkle, and Roberts had done before him, but asks forever why each investigator had done what he did and, more important, why he did not interpret his findings as we would nowadays. By this approach Olby sheds a great deal of light on previously obscure aspects of the history of genetics. There are a few conspicuous omissions in his treatment, particularly of non-English publications (e.g., Roux's brilliant conclusions of 1883 and most of the exciting cytological literature of the 1870s and 1880s), but on the whole the ground is well covered. Olby's contacts seem to have been mostly with botanists and plant geneticists, which leads him occasionally to rather one-sided conclusions. For instance, he fails to notice the drastic difference between the domestication of animals and that of plants. Hybridization has played a very subordinate role in animals, but has been involved in the origin of many of the most important crop plants. More generally, hybridization is frequent in the plant kingdom and is responsible not only for all allopolyploids, but also for much other reticulate evolution. Among animals hybridization above the subspecies level is rare and in most taxa it is negligible.

In retrospect it has become obvious that most of the great controversies in the history of genetics and of evolutionary biology were due to a failure to make precise definitions and to develop clear-cut concepts. The thesis that new facts are always responsible for the ultimate clarification of scientific problems is becoming increasingly questionable. Rather, what seems often more crucial is the superior analytical power of certain individuals who, when looking at the same facts as others, suddenly achieve a new insight. Four confusions in particular were responsible for slowing down advance in genetics in the last 200 years. The first of these was the failure to make a clear distinction between "kind," "variety," and "species," and the belief that such a distinction did not matter (see Essay 10). Even Darwin was confused on this issue. It made most of the discussions of the Mendelians (De Vries, Bateson) on the origin of species meaningless, and, as is now clear, was one of the reasons for Mendel's downfall. Mendel considered the product of a cross between two alleles (as we would now call them) in a single population just as much a hybrid as the product of a cross between two species. Naegeli would not have been able to encourage Mendel to continue with the wild goose chase of *Hieracium* crosses if Mendel had understood that this was involving him in an entirely new set of problems. To be sure, Mendel did suspect that he was dealing with a different kind of phenomenon in the cases in which he failed to get clean Mendelian segregations in the F_2–F_5 generations, and he even attempted a rather interesting explanatory model (some components of which have a strong flavor of blending inheritance). However, he never understood why some of his "hybrids" behaved in the proper Mendelian fashion while others did not. He would have had more confidence in the validity of his *Pisum* findings if he had limited himself to intraspecific crosses, but for this he would have had to have a far more sophisticated species concept.

A second confusion, which also goes back to Darwin (and in part even to Linnaeus), was the failure to distinguish between individual "varieties" and geographical "varieties" (= subspecies). In the arguments between the Mendelians and the biometricians (particularly Weldon), the opposing camps overlooked the fact that individual variants usually obey the Mendelian rules and geographical variants (being polygenic) seldom do; likewise, that individual variants (except polyploids) have nothing to do with speciation, while geographical varieties, when properly isolated, constitute incipient species.

A third confusion, which lasted into the 1950s, was the failure to realize that there are two kinds of isolation, reproductive isolation (maintained by isolating mechanisms) and geographic isolation (maintained by extrinsic barriers) (see Essay 11).

By far the most damaging, however, was the fourth confusion, the failure to understand fully that genotype and phenotype are not the same. Here, again, Mendel was ahead of his time. His terminology of "recessive" factors and some of his statements showed that he was fully aware that not all factors are expressed in the visible phenotype: "It is seen how rash it must be under such circumstances to draw from the external resemblances of hybrids conclusions as to their internal nature" (Mendel 1866:3-47). Yet those geneticists who were interested in the action of genes proceeded as if organisms were directly composed of genetic material. De Vries (1889:190) subscribed to the concept that the pangenes were the actual building material during growth and development: "The entire living protoplasm consists of pangenes; it is only they which form in it the living element." He agrees that all inherited qualities must be represented in the nuclei of the gametes, yet insists that this fact does not preclude the possibility that the cytoplasm is also composed of pangenes. With the benefit of hindsight we must concede that it was difficult if not impossible to make a strict separation of genotype and phenotype until it was discovered that the genetic material was nucleic acid rather than protein, and that it served merely as a program (a blueprint) rather than as the actual material. Even Johannsen, who coined the terms "genotype" and "phenotype," was often quite confused in his discussions, as is particularly obvious in his treatment of continuous variation. When De Vries talks about his *Oenothera* mutations, it is never clear whether he defines them in terms of the genotype or the phenotype.

This confusion affected the interpretation of virtually all genetic and evolutionary phenomena from the Darwin period to the 1940s. This is why Darwin's opponents could not understand how he could postulate a discontinuous nature of variation (i.e., pangenes) and at the same time postulate gradual and continuous evolutionary change (of the phenotype, as we would now say). This is why they accused him of believing in the blending of his pangenes when what he had primarily in mind was the blending of phenotypes. This is why Bateson and De Vries believed in a strict parallelism of the degree of genetic variation and evolutionary variation. This is why the enormously important role of recombination (and sexual reproduction), which produces most of the phenotypic variation on which selection acts, was so long ignored (Provine 1971:83). This is why so many authors of the early period included under the term "variation" the production of what we would now call new mutations as well as the numerous genotypes that genetic recombination produces in any sexual population. With De Vries using the word "mutation" (Mayr 1963:168–169) both for changed phenotypes and for changed genotypes, Weldon was not altogether wrong when he said that he could not see any difference between variation and mutation. Indeed, what Bateson had previously called discontinuous variation, De Vries now called mutation. To make matters worse, the Morgan school later expanded the application of the term mutation to include also genetic changes in the basis of continuous variation, totally in conflict with De Vries's original definition. The attacks on the evolutionary significance of mutations that were made in the 1920s and even in the 1930s reflect the utter confusion created by these shifting definitions (Allen 1968:130).

Owing to the intensive work of geneticists from 1910 on, the nature of the process of mutation was better understood, and this led—particularly in the 1920s and 1930s—to a more precise formulation of the definitions and to a clarification of concepts. A distinction was made between changes of phenotype and of genotype, between nongenetic modifications of the phenotype and genetic changes, between new variation due to mutation and that due to recombination, and between monogenic and polygenic variation. When these conceptual and terminological clarifications had been accomplished, most of the genetic controversies of the turn of the century became meaningless and the great evolutionary synthesis became possible.

THE FAILURE OF FACILE PHYSICOCHEMICAL EXPLANATIONS

Since Descartes and Newton, biologists have been told that to give a scientific explanation was to give an explanation in chemical-physical terms. It is now evident that in most cases it would have been of far greater heuristic value if, instead, the biologist had operated with black boxes, because simplistic physical explanations often had a stultifying effect. Koelreuter, for instance, explained the differences he found between the F_1 and F_2 generations strictly in chemical terms. Just as an acid and an alkali form a neutral salt, so, he believed, would the female "seed material" in a fertilized hybrid unite with the male "seed material" to form a "compound material." In the F_2 hybrids they do not necessarily combine in equal proportions, and for purely quantitative reasons this results in a variety of offspring, some of which more closely resemble one grandparent than the other. Miescher, a century later, completely missed the significance of DNA, which he had discovered, by adopting a purely mechanical interpretation of the process of fertilization.[1] The explanation of inheritance could make no advance as long

1. "Suppose the nature of the egg cell, as compared to an ordinary cell, were to be determined by the circumstance that one link were missing in the series of factors which control the active organization? Because otherwise all the essential cell constituents are found in the egg. However, during the maturation of the egg the protamin disintegrates under the formation of

as the influence of the nucleus was believed to be a purely "dynamic" one (Coleman 1965:133). In 1884 Strasburger, for instance, denied that anything material ever leaves the nucleus and postulated instead that "molecular excitations spread from the nucleus to the surrounding cytoplasm and control the processes of metabolism in the cell" (Strasburger 1884:113–114). This view was vigorously opposed by De Vries in his *Intracellular Pangenesis* (1889), where he emphasized the specificity of genetic phenomena quite correctly. For instance, he said the production of alkaloids and other metabolites is far too species-specific to be susceptible to an explanation by molecular excitation. Bateson, unfortunately, returned to a rather naive physicalism and accepted a "vibratory theory of heredity," which was one of the main reasons for his refractoriness to many post-1900 developments in genetics and in particular to the chromosome theory of inheritance (see Coleman 1970). No matter how important "movements and forces" may be at the submolecular level, as far as the interpretation of purely biological phenomena is concerned, such "dynamic" physical explanations had an appallingly stultifying effect. This, of course, is equally true in other areas of biology, as, for instance, in many of the premature physical explanations of physiological processes by Carl Ludwig, Julius Sachs, and Jacques Loeb. The historiography of biology has, up to now, concentrated on the beneficial effects of physicochemical explanations as an antidote against vitalistic notions. This mechanistic emphasis was particularly necessary in the first half of the nineteenth century when vitalism was still widespread, in part as a heritage of *Naturphilosophie*. Naive physicalism, however, became a retarding factor later in the century when it gave an oversimplified picture of the action of physicochemical processes by completely ignoring the vast complexity of the mediating biological processes (Jacob 1973).

THE IMPORTANCE OF THE RIGHT *FRAGESTELLUNG*

Historians of genetics have often commented on how surprisingly close some of Mendel's forerunners came to discovering the laws of genetics. Many had discovered 3:1 ratios, the intermediacy of the F_1, and the occurrence of dominance, but had done nothing with these facts. The usual question is: why were they so blind? Even Olby, whose treatment of Koelreuter, Gaertner, and Naudin is extraordinarily perceptive and informative, does not answer this question as directly as it deserves. The answer, it seems to me, is that the interest of these pre-Mendelians

nitrogen (N) . . . and the otherwise perfect machine is brought to a complete standstill because one screw is still missing. The spermatozoon inserts again this screw in the right position and thus restores the active organization. It does not require anything else. At the place where the chemical-physical quiescence was disturbed, the machine starts to work again, each cell produces protamin for its neighbors and thus the motion spreads according to definite laws" (letter to Boehm, 2 May 1872, Miescher 1897).

was not in the behavior of individual characters (and their genetic determination) but in the behavior of the "type" as a whole. It must be remembered that the period from 1760 to 1860 was still dominated by typological thinking and by an interest in the "nature" (= essence) of species. This is documented by the choice of the experimental material; in most cases the early hybridizers chose full species for their crosses and thus discovered relatively few cases of clear-cut Mendelian segregation in the F_2. No historian or botanist, so far as I know, has analyzed the 500 different hybridizations (involving 138 "species") Koelreuter carried out in order to determine how many of them involved true species and how many others involved intraspecific variants (alleles). To make such an analysis would seem a worthwhile task.

To a large extent the thinking of the period was influenced by certain late ideas of Linnaeus. In contrast to his assertion in younger years that species were fixed and immutable, Linnaeus proposed in later years that it might be possible to produce new species by hybridization (Olby 1966:26; Larson 1971:104–109). Much of the interest in hybridization displayed by his successors was due to their desire to prove Linnaeus wrong. Linnaeus had claimed to have produced two new species of plants by artificial cross-pollination, a goatsbeard (*Tragopogon*) and a speedwell (*Veronica*). He was so convinced that these hybrids were true-breeding species that he gave them new specific names and entered them in his catalogues as valid species. Such a conclusion was of course completely at variance with his otherwise solidly essentialist views. Here is another illustration of the phenomenon that scientists sometimes incorporate altogether incompatible observations or conclusions into their paradigms.

Koelreuter, as Olby shows convincingly, was totally opposed to this conclusion. He was not satisfied with the constancy (and intermediacy) of the new hybrid "species" in the F_1 generation. If they were new species they would have to "breed true" in subsequent generations. This, Koelreuter showed, was not the case. The F_2 hybrids of certain crosses clearly fell into three types, one of them resembling the F_1 hybrids, the others the two grandparent species. Having thus falsified Linnaeus's new theory of the origin of constant species by hybridization, Koelreuter went no further in the analysis. After all, his research problem was the nature of species and not the inheritance of characters. (The number of his other discoveries was prodigious, even if they did not include Mendel's laws.)

Olby succeeds in bringing Koelreuter alive as no one has up to this time. "The total number of experiments, judged by 18th century standards, is fantastic. Thus he carried out more than 500 different hybridizations involving 138 species, and examined the shape, color, and size of the pollen grains from over 1000 different plant species (1966:21). Koelreuter's breeding experiments indicate an excellence of planning and execution that was not approached again for more than 60 years. It is one of the tragedies of science that this brilliant investigator was almost totally ignored by his contemporaries because he lived in an era in which the discovery of new species was the all-consuming interest. Essentialism, as I have said, was the dominant philosophy of his time, and hybridization, particularly when carried over several generations and including back-crosses, simply did not fit into the framework of contemporary thinking. It was the discovery of new types that excited the imagination of his generation. The explorers of exotic floras were the heroes of the day.

Naudin and Gaertner had essentially the same approach as Koelreuter to species crossing. Even though both rediscovered F_2 segregation into three types, they merely saw in it a reversion of hybrids to the parental species. In his main work (1849) Gaertner summarizes the results of nearly 10,000 separate crossing experiments among 700 species that yielded 250 different hybrids. Darwin said of this book that "it contains more valuable matter than all other writers put together, and would do great service if better known." But what a difference between Gaertner's almost encyclopedic work and Mendel's short essay of 1866! No one has better demonstrated how important it is in scientific research to ask the right questions. By asking the same questions as his predecessors, Gaertner came up with the same answers in spite of the huge amount of material he gathered in more than 20 years of hybridizing. Neither Darwin, who studied this work carefully, nor any other contemporary saw any new laws emerge from Gaertner's facts. Perhaps one can pay him a left-handed compliment by saying that he showed so exhaustively what answers one could and what others one could not get from the traditional questions that he freed the field for an entirely new approach. We know that Mendel, who owned a copy of Gaertner's book, studied it most carefully, and it is more than likely that this was of help to him not only in selecting his experimental material but also in phrasing the new kinds of questions that resulted in his spectacular breakthrough.

Olby's careful account of the pre-Mendelians permits us to see them in a new light and far more clearly in the context of the ideas of their period. In some ways one might say that their preoccupation with the species question is of far more relevance to the history of evolutionism than to that of genetics. Minimizing this aspect, Olby occasionally arrives at evaluations that seem questionable. For instance, he expresses the curious opinion that Koelreuter's and Gaertner's findings "were later to draw Darwin's attention away from hybrids as species . . . which was unfortunate" (1966:32). On the contrary, I would say "fortunate," because hybridization is a rare and aberrant mechanism of species formation in animals, and after all Darwin was primarily a zoologist, as far as systematic competence is involved. Olby criticizes these early hybridizers for their conclusion "that results obtained from crossing of varieties has little if any relevance to the question of how new species originate." The fact of the matter is that the early hybridizers were completely right and the early Mendelians (Bateson, De Vries, and so on), who ignored them and believed the opposite, were quite wrong. Olby's further claim (p. 51) that a hostility toward the "new cytology" was responsible for Gaertner's inability to frame the right questions about inheritance seems most implausible to me. One must remember that the "new cytology" (= cell theory) was often decidedly misleading. Both Schwann and Schleiden believed that the cellular fluid was the important element (perhaps also the cell wall) and that each nucleus was newly formed by crystallization from the cellular fluid. The cell theory from the 1820s to the 1840s, the period during which Gaertner formed most of his ideas, was of no help whatsoever in the formulation of genetic theory. Furthermore, there are good reasons to believe that Gaertner would have maintained his interest in the nature of species even if he had had a far better understanding of cells. Neither Schleiden nor Virchow had anything to do with the shift of interest from species to character. For this the students of variation were responsible. The later cytology, of course, contributed decisively to the new thinking in genetics; but it must be remembered that the work on chromosomes, on mitosis and meiosis, and on fertilization from the 1870s on followed by several decades the formative periods in the lives of Gaertner and Mendel.

THE TIME SEQUENCE IN THE SOLUTION
OF SCIENTIFIC PROBLEMS

In the evaluation of Koelreuter and Gaertner it is often forgotten that certain scientific questions cannot be asked, much less answered, until some other problems have been solved. The three outstanding problems in the late eighteenth and early nineteenth centuries relating to inheritance were:

1. Does hybridization create constant new species?

2. Do both father and mother contribute to the genetic endowment of their offspring?

3. If so, are their contributions equal?

These were among the questions to which Koelreuter, Gaertner, Goss, Seton, and Knight directed most of their efforts. All uncertainty on these questions had to be removed before one could even start posing the problem of the inheritance of individual characters. As Olby narrates beautifully, it was Koelreuter who conclusively solved all three problems, refuting Linnaeus and others who had proposed different solutions.

As far as Gaertner is concerned, perhaps his most important contribution was to test an enormous number of species, showing that some were more and others less suitable for genetic research, thus making it possible for Mendel to pick the garden pea (*Pisum*) as his experimental material, a species quite exceptionally well suited for this purpose. The later history of evolutionary genetics shows what a tragedy the choice of wrong material can be. This applies equally to Naegeli and his apomict hawkweeds (*Hieracium*), to De Vries and *Oenothera* (with its balanced chromosome rings), and to Johannsen and the garden bean (*Phaseolus*), a self-fertilizing and nearly homozygous genetic material. On the basis of such aberrant material Naegeli thought he could ignore Mendel's rules, De Vries that he could establish the origin of species by single mutations, and Johannsen that he could demonstrate the impotence of natural selection.

FROM THE STUDY OF SPECIES TO THE STUDY OF VARIATION

As far as the history of genetics is concerned, perhaps the most important event was a gradual but decisive change of conceptualization in the middle of the nineteenth century. Wherever people worked with species—and this was equally true for systematists, breeders, and what we would now call evolutionists—they found that species were not the monolithic types the essentialists had believed them to be. Instead, species varied; they formed local populations, and even individuals within a population differed from each other. Population thinking was born, and interest shifted from species to variation. There was no science of genetics at that time, and the breeders were as much naturalists as were Moritz Wagner or Constantine Gloger, who had published highly original observations in the 1830s. Operationally, the study of the variation of species had to be a study of the variation of species characters. The change to population thinking and to an interest in variation came slowly, and in part subconsciously, but it was part of the new intellectual milieu of the mid-nineteenth century. To realize this helps in understanding the "Mendel phenomenon," that is, Mendel's great and seemingly premature scientific breakthrough.

MENDEL'S PIONEERING ACHIEVEMENT

Olby's treatment of Mendel is outstanding. He not only tells us a great deal about Mendel's personal life, but presents a rather plausible solution to the problem of why Mendel succeeded where all others failed. To begin with, he shows that the question "Why did this obscure monk succeed where famous scientists had failed?" is quite misleading. Although Mendel carried out his experiments in Brünn in virtual intellectual isolation, he was actually a young scientist who had spent nearly two years at the University of Vienna under the tutelage of outstanding physicists and biologists. Franz Unger, his professor of botany, was the author of an excellent textbook on the anatomy and physiology of plants, and through him Mendel had learned not only of his forerunners, but also of the modern work on cytology and embryology of plants.

Interestingly enough, as Olby points out perceptively, it was again the species question that seems to have given rise to Mendel's experiments. Unger, who had adopted the theory of evolution in 1852, had expressed the opinion that variants arise in natural populations, which in turn give rise to varieties and subspecies until finally the most distinct among them reach species level. These views contrasted strikingly with those of the early hybridizers, who believed strictly in the fixity of species. In his famous 1866 paper, Mendel states that his time-consuming experiments were necessary in order to "reach the solution of a question the importance of which cannot be overestimated in connection with the history of the evolution of organic form." With Mendel apparently wanting to test Unger's theory, he had to concentrate on variants and varieties rather than on the "essence of species," as was done by his forerunners.

This was not the only one of Mendel's intellectual departures; after all, in the 1890s Bateson and others also concentrated on variants without discovering Mendel's rules. Mendel's outstanding contribution was his method, which consisted in the careful counting of all of his observed classes and the calculation of ratios. Olby points out that Mendel's favorite high school teacher was a physicist and that physics seems to have been the major subject in his own teaching. In Vienna he took courses with the famous Doppler and with other physicists and he even served for a time as a demonstrator at the Physics Institute of the University of Vienna. It must have been here that he learned to keep careful records of his observations and to arrive at numerical generalizations. It is possible that this practice was reinforced by contacts with his fellow beekeeper, Dzierson, who is reputed to have specially emphasized the numerical approach and with whom Mendel is believed to have been in contact (both men were Silesians and Catholic priests).

Without knowing all of Mendel's reading matter, it is more difficult to determine where he acquired his hypothetico-deductive approach. Olby thinks he got it from his physics teachers, but since Darwin and Schleiden were as enthusiastic proponents of this method as the physicists, it is as likely that Mendel's inspiration came from biology (this type of theory formation had already been used in the eighteenth century, by biologists and physicists). This much is certain: Mendel, in contrast to earlier hybridizers, had formed a definite working hypothesis before he started his *Pisum* experiments. Alas, we do not have any Mendel notebooks to guide us from his vague earlier notions to the mature final theory, as published in 1866. Perhaps his first model was a purely mathematical one, a possibility mentioned to me by Coleman; and it was only after Mendel had obtained definite ratios that he asked himself what there might be within organisms that could account for the occurrence of such definite, and predictable, ratios. Perhaps it was only then that he proposed (and tested again and again) the novel hypothesis that a plant produces potentially two kinds of egg cells and two kinds of pollen grains for each of its characters. In retrospect, Mendel's achievement is even more awe-inspiring than was realized by his rediscoverers in 1900. Without any knowledge of chromosomal cytology, without the theoretical analysis of Weismann, and without the benefit of the many other seminal discoveries made between 1865 and 1900, he discovered a new way of looking at inheritance by expressing it in terms of the behavior of unit characters, and he used this interest to arrive at far-reaching generalizations. His achievement was one of the most brilliant in the entire history of science.

Olby devotes a good deal of attention to the question of why Mendel's discovery was so utterly ignored and its significance overlooked even by those who must have read his paper. Apparently Mendel was in direct touch with only a single botanist, and this contact was in every way disastrous. The Swiss botanist Naegeli, at that time a professor in Munich, has the reputation of having been a rather opinionated Herr Geheimrat. He was one of the major opponents of Darwin's views and held very reactionary ideas on the concept of species. Was it jealousy that made him deliberately suppress Mendel's views? One is almost forced to this conclusion when one reads the extensive chapter in Naegeli's 1884 book dealing with problems of inheritance, in which the experiments of numerous other hybridizers are discussed but Mendel's name is entirely omitted. Although academic arrogance may well have played a role, quite likely this is not the whole answer.

The real reason for Naegeli's rejection of Mendel seems to have escaped historians. It is, in my opinion, Naegeli's advocacy of a theory of strictly blending inheritance that forced him to reject Mendel's particulate theory, since it clearly contradicted his own. To establish this point, it is necessary to provide a short résumé of Naegeli's theory of inheritance. According to him, the idioplasm (= genetic material) consists of long strings of micelles. Each string is composed of one kind of micelles, but differs from all the other strings by the specificity of its micelles. These strings are passed without interruption from cell to cell (that is, they are not restricted to the nucleus). "All increase of idioplasm consists in a growth of these strings and takes place through an elongation of the string owing to the addition of new micelles" (p. 33). During growth "the connection of the strings of idioplasm retains its specificity and configuration and the specific quality of the idioplasm is given by this configuration" (p. 37). Naegeli then asks what happens during fertilization (pp. 199-224). Can it be a simple *"Durchdringung"* (= blending) of the soluble genetic material of the parental gametes? This Naegeli rejects as incompatible with the extraordinarily precise organization of the genome. Nor is it possible, says Naegeli, that the homologous parental strings attach themselves to each other, because this would lead after each fertilization to a doubling of the cross-section of each string. (Naegeli clearly sees the need for a reduction division.) How else can one keep the cross-section of the strings of micelles constant? "In order that the cross-section remains, on the whole, constant it is necessary to combine the idioplasm strings of the parents into new strings the length of which is the sum of the length of the parental strings. Thus a paternal and a maternal 'Anlage' becomes a filial Anlage of equal strength, that is, a group with the same number of micelles in cross-section. It is only when some parental Anlagen differ 'wesentlich' [essentially, or drastically], as occurs in the hybridization of races, varieties, and species, that the strings attach themselves laterally, resulting in an enlargement of the cross-section." Naegeli continues to say that such an enlargement of the amount of idioplasm affects only very few Anlagen. "The doubling of the length of the strings, as a result of fertilization, is nothing more than the first growth step in the new ontogeny." It is evident that the fusion of homologous parental strings is strictly blending, since the strings are qualitatively identical in their longitudinal axis.

MENDEL'S MOST SIGNIFICANT CONTRIBUTION

Olby provides a faithful presentation of Mendel's laws but he never poses the basic question as to what Mendel's most significant contribution was. Nor does he specifically refute the widespread myth that it was Mendel's concept of particulate inheritance, in opposition to the "prevailing" idea of blending inheritance. The fact of the matter is that, from Koelreuter on, virtually every author who had speculated on inheritance not only had adopted a more or less corpuscular theory but had assumed a segregation of such particles in later generations. Even in the case of species hybrids, merely a permanent attachment of the parental particles was usually postulated rather than genuine blending. Naegeli was one of the comparatively few authors to champion a true theory of blending inheritance.

Olby's account of blending inheritance is unfortunately thoroughly confused: he sometimes speaks of the blending of genetic factors, at other times of the blending of characters (components of the phenotype). His Table II is supposed to illustrate blending, but actually illustrates merely the probable percentage of genes a half-caste can expect to have received from white and from black ancestors. Contrary to Olby's claim there is abundant evidence that Darwin was not a supporter of blending inheritance. His famous letter to Huxley in 1857 is only one of the many pieces of evidence:

> I have lately been inclined to speculate, very crudely and indistinctly, that propagation by true fertilization will turn out to be a

sort of mixture, and not true fusion, of two distinct individuals, as each parent has its parents and ancestors. I can understand on no other view, the way in which crossed forms go back to so large an extent to ancestral forms.   (F. Darwin 1903:vol. 1:102-103)

Darwin's theory of pangenes, of course, was on the whole a theory of particulate inheritance, as De Vries (1889) pointed out long ago.

If the refutation of blending inheritance was not Mendel's great achievement, what then was it? This becomes apparent when we compare Mendel's theory with those of all the theorizers on inheritance in the 1880s and 1890s. They all, from Darwin on, postulated the existence of numerous identical determinants in each cell (each nucleus) for a given unit character. Weismann developed this into a major theory of embryonic development, and De Vries and Bateson (prior to 1900) used it to explain the difference between continuous and discontinuous variation. Mendel's insistence that *each character is represented in a fertilized egg cell by only two factors, one derived from the father, the other from the mother, was the new idea that revolutionized genetics.* That this must be the case is so absurdly obvious to any modern biologist that one has to go back to the confused literature prior to 1900 to appreciate the fundamental novelty of Mendel's contribution. (Wilkie, 1963:602, has also made this point.)

De VRIES

The name De Vries is usually associated with the rediscovery of Mendel's laws in 1900. All too few students of genetics are aware of the fact that in 1889 he published a brilliant book, *Intracellular Pangenesis.* Although Olby rightly credits De Vries with the discovery—independent of Mendel—of separate unit characters, he does not bring out sufficiently the pioneering character of De Vries's earlier book. While Weismann, Oscar Hertwig, and all the other theorizers on genetics during this period were concerned with developmental genetics, De Vries was the only one to be interested in transmission genetics. Furthermore, he gave a far better account of Darwin's theory of pangenesis than anyone since.

At the same time he was not uninterested in other problems of genetics, and his theory of differentiation is essentially the one to which we still subscribe:

> The material vectors of the genetic properties are conveyed from the nucleus to the cytoplasmic organelles. Most of them are inactive in the nucleus, but some can become active in the organelles of the cytoplasm. In the nucleus all characters are represented, in the protoplasm of each cell only a limited number. (1889:5-6)

De Vries has never been given full credit for developing this brilliant theory of differentiation. Equally penetrating are his arguments in favor of the study of separate characters. He decries the fact that the long-standing misconception of the species as an indivisible unit still had not yet been overcome—this, in spite of the fact that all evolutionary studies had demonstrated that the character of a species as a whole is actually composed of individual factors that are more or less independent of each other. Whether we are dealing with characters that clearly have a chemical basis, such as the color of flowers, the presence or absence of tannic acids, alkaloids, essential oils and other similar products, or more purely structural characters such as the serration at the edge of leaves, says De Vries, we always find that a given species is an extraordinarily complex composite of numerous characters:

> These factors are the units which the science of inheritance must investigate. Just as physics and chemistry go back to molecules and atoms, thus the biological sciences must penetrate to these [genetic] units in order to explain the phenomena of the living world through their combinations. (1889:9)

No one else in the 23 years since Mendel's work had understood this point.

Although at that time De Vries had already undertaken some abortive breeding experiments, it is rather apparent from numerous references in the text that he had gained much of his knowledge from discussions with plant breeders. This induced him to state that any genetic character "can be made the subject of experimentation. By appropriate selection it can be strengthened or weakened and can be brought in relation to other unchanged characters according to the breeder's intention." (Alas, by 1900 De Vries seems to have entirely forgotten this experience.) He summarizes his theory of inheritance in the statement: "Independence and mixability are the two most important characteristics of the inheritable properties (Anlagen) of all organisms. The principal task of any theory of inheritance is in my opinion to find a hypothesis that explains these two features" (1889:33).

De Vries is equally perceptive in his criticism of the work of others. Weismann would not have gone as far astray in his *Keimplasma* (1892) if he had carefully studied De Vries's refutation of his (Weismann's) theory of a complete separation of germ line and soma line and of his theory of an unequal distribution of genetic units during cell division. De Vries's inactivation theory provided a legitimate and, as we now know, very much better alternative to Weismann's theory.

In one crucial point De Vries was unable to emancipate himself from prevailing concepts. Like everyone else, he assumed that the genetic material itself provided the building material of the body: "The entire living protoplasm consists of pangenes; it is only they which form the living elements." This assumption almost inevitably leads to the postulate that a given pangene may be represented in each cell by many replicas. This would not have been too far from modern thinking if it had been restricted to the cytoplasm. But De Vries further postulated that pangenes have the capacity to multiply even within the nucleus.

After the publication of *Intracellular Pangenesis,* and with a clear hypothesis in mind, De Vries took up his breeding work seriously. Olby presents an excellent analysis of all we know about De Vries's work between 1889 and 1900. There is no doubt that he achieved many 3:1 ratios in his work, but there is no good evidence that he recognized them as such. Olby provides rather convincing evidence that De Vries came across Mendel's paper in about 1896 or 1897. It is quite possible, if not likely, that this finally convinced him of the invalidity of the concept of multiple pangenes for a single unit character and cleared the way for a straight Mendelian interpretation.

When, after 1900, De Vries realized that not he but Mendel was receiving all the credit for the genetic laws, he largely abandoned straight genetics and concentrated thenceforth on evolutionary problems. Olby pictures well De Vries's feeling of disappointment, if not bitterness, over having been scooped by Mendel (Olby 1966:129, 189). That his "origin of species by mutation" was a nearly total failure, owing to the unfortunate choice of his basic material (*Oenothera*), makes De Vries's life a double tragedy.

De Vries's strengths and weaknesses reflect his educational background. He had spent much of his earlier life in chemical and physiological laboratories doing experimental work. This was excellent preparation for the perception of—and experimenting with—unit characters. Yet De Vries lacked the experiences of a naturalist that would have prepared him for work on the origin of species. And this largely explains the failure of his mutation theory.

THE RISE OF MENDELISM

Olby's account ends with the rediscovery of Mendel's laws and with a detailed analysis of the respective contributions made by the original Mendelians. He shows graphically that method is not everything, for Bateson had proceeded along Mendelian lines well before 1900 and had obtained Mendelian ratios, yet the full explanation had eluded him.

Perhaps the most remarkable aspect of the rediscovery of Mendel's laws is the rapidity with which they were accepted. Those were the days of reliable mails and rapid publication! Correns received De Vries's paper on the morning of April 21; he finished his own on the evening of April 22; it was read at the April 27 session of the German Botanical Society and was published about May 25. Bateson in England, Cuénot in France, and numerous investigators in other countries started to test Mendel's laws almost immediately. Mendelism spread like wildfire.

Even though Provine, in his *Origins of Theoretical Population Genetics* (1971), deals only with some of the ensuing developments, designated by him as population genetics, his treatment is in many ways a continuation of Olby's story. He presents a meticulous historical account, equally invaluable for its constant endeavor to be unbiased and for its reference to innumerable obscure articles in the contemporary biological and genetics journals. As a historian he is able to be far more impartial than much of the previous secondary literature, most of it from the pens of authors who were close to the traditional genetics establishment. Provine's independence is documented by his critical evaluation of Johannsen and by his emphasis on the pioneering achievements of Castle, Yule, Harris, and others, who in spite of their major contributions are often neglected. Perhaps the outstanding section of Provine's book is his excellent treatment of the respective contributions of Fisher, Haldane, Wright, and particularly of the differences between them.

Provine, however, also has his weaknesses. Even though his account is excellently researched and almost invariably accurate in its details, he does not place as much emphasis on some of the underlying issues as would seem desirable. For instance, he fails to make it clear that the great controversy between Mendelians and biometricians owed less to their differences than to the fact that both camps made three erroneous basic assumptions:

1. There are two sharply distinguishable and quite different kinds of variation, discontinuous variation (sports, mutations) and continuous variation, a distinction going back to Darwin and earlier.

2. There is no distinction between changes in what we now call genotype and phenotype.

3. Individual and evolutionary variation are one and the same thing.

For instance, from the last-named assumption the Mendelians, led by Bateson, concluded that discontinuity of evolutionary advance would be proved if it could be shown that genetic variation was discontinuous. For the biometricians, such as Weldon, who as naturalists saw evidence for gradual evolution everywhere, it required the assumption that the evolutionarily important genetic variation also had to be "gradual" or "continuous." It seems to me that Provine makes too much of the role of personalities. To be sure, Bateson was pig-headed, intemperate, and intolerant; but his main adversary, Weldon, died in 1906 and Bateson hardly changed his ideas in the next 20 years (he died in 1926). The real reason was that Bateson was an uncompromising typologist. Like Goldschmidt and several other biologists with a background in embryology and experimental biology, he was quite incapable of understanding the nature of natural populations (in spite of his field work in Asia and his great horticultural experience).

Provine points out correctly that the argument in the Mendelian–biometrician controversy was as much over evolutionary theory as over genetic facts. The basic issue was the strictly scientific question as to whether evolution progresses gradually or by jumps. Weldon never abandoned the Darwinian tradition of gradual evolution, while Bateson had become, if he was not one already, a typological saltationist in the laboratory of the embryologist Brooks. It is important to emphasize that the drastic difference in thinking between the two schools did not result from a rediscovery of Mendelism in 1900 but antedated it by more than a decade. De Vries's saltationism, incidentally, had already been fully expressed in 1889. He selected *Oenothera* as his experimental material because it supported his already well-formed notions.

The great controversy is usually said to have ended with Bateson's complete victory after Weldon's death in 1906. This may well be true for the purely genetic argument, but it is not so for the evolutionary one. After all, the genetic argument was merely a symptom of a far deeper disagreement, the choice between saltationism and Darwin's gradual evolution through natural selection. The final reconciliation among evolutionists in the 1930s and 1940s, often designated "the great synthesis," is much more appropriately considered the real end of a controversy that had started well within Darwin's lifetime.

Provine states: "The conflict between the Mendelians and biometricians . . . drove a wedge between Mendel's theory of heredity and Darwin's theory of continuous evolution and consequently delayed the synthesis of these theories into population genetics" (1971:56). The facts hardly support this statement. It was not this conflict that delayed the synthesis but simply the typological thinking of the Mendelians and their inability to understand the population nature of species. Both Bateson and De Vries had stated their anti-Darwinian views long before the fight with the biometricians, and Bateson was still totally unable to grasp the nature of speciation in his Toronto address of 1922, 16 years after the end of his conflict with the biometricians. A clash of personalities was clearly involved, but conceptual confusion on both sides was more fundamental and far more important.

THE BENEFIT OF CONTROVERSIES

One can argue whether such controversies as that between the Mendelians and biometricians are or are not of benefit to science. It seems to me that this differs from case to case. I rather suspect that Bateson's stubborn resistance to the chromosome theory resulted in much effort by members of the Morgan school that could have been devoted to new frontiers of genetics. As for the controversy between Mendelians and biometricians, it likewise led to a hardening of the viewpoints on both sides and to a neglect of the correct components in the views of the other side. With Weldon claiming that even discontinuous variation did not obey Mendel's laws, it is not surprising that the study of continuous variation was pushed into the background, not to be taken up seriously until Mather and his school started to work in the 1940s. In some instances, however, such a controversy is beneficial because it forces both sides to analyze their argument and to marshal supporting facts. I doubt that I would have written my *Systematics and the Origin of Species* (1942) if I had not been provoked by Goldschmidt's claims in his *Material Basis of Evolution* (1940) that geographic speciation is an irrelevant evolutionary process.

Owing to Weldon's strategy and his ultimate defeat, typological thinking prevailed in genetics for many more decades. Even in the literature of the 1930s and 1940s one can still find numerous references to "the wild type" and "the mutation," as if the wild type were a "type" and the mutation a "new type." This is why the thinking of Chetverikov and his school was so refreshing.

INSTANTANEOUS OR GRADUAL EVOLUTION?

Provine brings out, far better than the previous literature, that the arguments of the biometricians were not nearly as silly as they are sometimes made out to be, or at least that the arguments of the Mendelians, as far as evolution is concerned, were equally silly.

The development of the conflict between Bateson and Weldon was tragicomical. When Bateson concluded that Mendelian inheritance supported discontinuous evolution, Weldon (who fully understood that evolution was Darwinian—that is, continuous) fell into Bateson's trap and thought he would have to disprove the discontinuous nature of inheritance. If Weldon had associated himself with some of the leading naturalists of his time, like Sir Edward Poulton and Karl Jordan (and less with Karl Pearson), he would have had little difficulty in demolishing Bateson's claims that evolution always takes place by "discontinuous mutations" and that natural selection is irrelevant (see below).

While Bateson forever stressed the radically different nature of continuous and discontinuous variation,[2] going so far as stating that the "swamping effect of intercrossing" obliterated gradual variation (the Darwinian material for natural selection), Weldon was much closer to the truth when he rejected such an absolute difference. He asserted, for instance, that when regression "is better understood than it is at present such naturalists as Professor De Vries and Mr. Bateson will abandon their attempts to distinguish between 'variation' and 'mutations.'" This is, of course, what the Morgan school eventually did, by extending the applicability of the term "mutation" even to the smallest variants, including those involved in Galton's regression. Weldon, however, because he had adopted the same three basic misconceptions as the Mendelians (see above), was unable to see the consequences of his own suggestion. Pearson and Johannsen likewise did everything to em-

2. Professor Provine has informed me that Bateson did not always insist on a complete difference between continuous and discontinuous variation, particularly in his later years (see, for instance, Provine 1971:115, n. 48).

phasize the difference between continuous and discontinuous variation. There was one outstanding exception at that period, the biometrician Yule. As Provine points out, Yule recognized not only "that Mendelism and biometry were compatible but also even more crucial, that Mendelism and Darwin's idea of continuous evolution were compatible." Yule, alas, was too far ahead of his time to be appreciated and it required many more years of tortuous argument before others arrived at the same conclusion.

THE UPHILL STRUGGLE OF NATURAL SELECTION

By the turn of the century natural selection had fallen upon hard times. Except for Weismann, the naturalists (from Alfred Russel Wallace to Poulton, Karl Jordan, and David Starr Jordan), and the biometricians (Pearson and Weldon), nearly everybody rejected natural selection as the principal mechanism of evolution. This was particularly true for the laboratory scientists; indeed I am not aware of a single experimental biologist of the period who championed natural selection.[3] There were manifold reasons for this rejection, reasons that have never been properly tabulated. These include:

1. Typological thinking, which denied the significance of minor deviations from the "type." Johannsen, for instance, tended to consider such variation nongenetic.

2. The belief that continuous variation had no selective value and was therefore of no evolutionary consequence.

3. The belief that the small differences of which continuous variation is composed would be "swamped by intercrossing" (Bateson).

4. The complete neglect of the role of recombination in the production of variability. This led Galton to consider selection ineffective because the deviations would soon be obliterated by regression. The same disregard of recombination induced Jennings and Johannsen to consider selection merely a process isolating pure lines.

5. The curious misconception that selection was a "chance process"— that is, that it did not obey any natural laws, and was therefore not reputable scientifically.

6. The failure of selection experiments.

Provine is fully aware of the crucial importance of the selection problem and provides a rich and perceptive analysis. He shows how poorly

3. This statement may seem contradicted by the fact that Bateson and other Mendelians talked a great deal about the importance of natural selection. However, when we look closely at the discussions of these authors, we discover that under "natural selection," they understood the *elimination* of deleterious deviations from the type. It is the same concept of selection that we find in the writings of Blyth and other pre-Darwinian authors. Evolutionary advance, for Bateson, was caused by the pressure of favorable mutations, and not by creative selection in the Darwinian sense.

the early selection experiments were designed, most of them, owing to various misconceptions, not at all testing the effectiveness of selection. For instance, what Johannsen proved was that selection cannot create new genetic variation, but by working with virtually homozygous, self-fertilizing stocks, he proved nothing about the power of selection since he never tested its action when operating on an ever-new supply of mutations and recombinants. As a matter of fact, as Provine shows, Johannsen's results by no means exclude the possibility that selection had indeed occurred in some of his own tests.

This weakness of Johannsen's argument was noted a long time ago by Pearson, Weldon, and Yule. The pure line work of Johannsen (and of those who followed him) shows better than anything else how little the Mendelians understood natural selection.

The various attempts to refute the power of selection were truly a comedy of errors. Bateson tried to do so (on his Russian trips) by establishing a correlation between phenotype and environment, Johannsen by selecting pure lines, still others by using long-lived, slow-breeding species. None of those who denied the efficacy of selection so insistently bothered to visit the animal breeders, who at this very same period made constant advances in the productivity of their flocks and herds, owing to deliberate selection. The only author who made use of the material of the breeders, Raymond Pearl, found "no evidence whatever that there is a cumulative effect of the selection of fluctuating variations." It is still uncertain how he could have arrived at such an obviously erroneous conclusion (Lerner 1950:20–22).

Provine supplies an excellent analysis of the reasons why anti-selectionism finally collapsed. The first was the discovery by Morgan and his school (also by Jennings, Baur, East, and others) of mutations with small phenotypic effects. The second was the Mendelian interpretation of continuous variation by Nilsson-Ehle and East, and the third was Castle's successful selection experiments. The latter were particularly important because they proved three facts that had been vigorously denied by the anti-selectionists: the genetic nature of continuous variation, the response of continuous variation to selection, and the ability of selection to push variation beyond the limits of the "pure lines" that had previously existed in these populations. All this had been achieved by 1920, but there was still a wide gap between the views of the naturalists and those of the laboratory geneticists. The latter still refused to acknowledge the immense genetic variability of natural populations; they still talked about the wild type; they still placed mutation pressure above selection; and they still virtually ignored epistatic effects, as well as the action of natural selection in wild populations. Indeed, their approach was still largely typological. Many of the Mendelians, among them Pearl and Punnett, continued to be strongly opposed to the Darwinian interpretation. All this did not change until population thinking was brought into genetics by biologists who, having been trained as systematists, had been exposed to the immense variation of natural populations and had adopted population thinking.

THE ORIGINS OF THEORETICAL POPULATION GENETICS

Even though this is the title of Provine's volume, almost three quarters of the work is devoted to the earlier history of evolutionary genetics, a period when geneticists simply did not think in terms of populations. The convention of designating the work connected with the names of Fisher, Haldane, and Wright as population genetics is, in my opinion, completely legitimate, because they were the first to ask questions concerning the frequency of genes in populations and the factors that affect this frequency. However, it was the great confusion of the first two decades of the century that revealed the need for the development of population genetics and that dominated the choice of the questions to which the mathematical population geneticists devoted their early attention. Provine gives an excellent treatment of the respective contributions of the three great geneticists and particularly of the differences in their questions. Basically, all three set out to prove that there is no contradiction between Mendelian inheritance and the observed facts of variation or the effectiveness of natural selection. Fisher's *Genetical Theory of Natural Selection* was a magnificent achievement when it was published (1930), and Provine is fully justified in claiming that it was the most substantial contribution to the synthesis of Mendelism and Darwinism. Yet, as I shall point out presently, it was not quite so exclusively responsible for this synthesis as one might gather from reading Provine.

Fisher's and, to a lesser extent, Haldane's backgrounds were mostly in mathematics, while Sewall Wright had started out as a laboratory geneticist (under Castle) and was for years associated with the animal-breeding program of the U.S. Department of Agriculture. Provine remarks very perceptively that this made Sewall Wright emphasize certain factors (e.g., population size and interaction of genes) that, at least in the beginning, were of no particular interest to Fisher and Haldane. He was, therefore, better prepared to cope with the increasing complexities of the phenomena of population genetics than Fisher, who for the sake of manageable mathematics made all sorts of simplifying assumptions, such as: population size large, epistatic effects and linkage negligible, accidents of sampling unimportant, effects of individual genes usually slight. As necessary as all of these assumptions were during the infancy of population genetics, they contained the germ of much of the trouble that plagued the field during the ensuing 40 years.

Provine essentially ends his story with the period 1930–1932, which saw the major contributions of the three great mathematical geneticists. The later history of population genetics has not yet been written. If Provine had surveyed the recent events in this field more closely, he might have modified some of his evaluations. It is also evident that some of his sources are one-sided. He had "two lengthy interviews" with Sewall Wright, but he might have seen things in a somewhat different light if he had also interviewed Dobzhansky, Michael Lerner, and Bruce Wallace.

To me it has always seemed that there are two major schools in population genetics. One is represented by Fisher and Haldane and by those workers of the Morgan tradition (like Muller and Crow) who are interested in population genetics. The background of this school is in mathematics or in the genetics laboratory. The other school, the beginnings of which can be equated with the publication of Chetverikov's pioneering paper of 1926, had its roots in natural history and animal breeding (Adams 1968:23-29). Provine decided not to include the contributions of the Russian school because "by the time Chetverikov's paper in theoretical population genetics was known in England and the United States, the theoretical construct erected by Fisher, Haldane, and Wright had progressed beyond it" (Provine 1971:10). Unfortunately, Provine thereby indicates how little he understands the contribution of the Russian school, later continued by Dobzhansky and some of his associates and students (including Wallace and Lerner). Their special emphasis is on a high variability of natural populations (originally postulated by Chetverikov and repeatedly confirmed by the Russian school), on gene interaction and coadaptiveness, and on recombination rather than on mutation. Many other basic concepts of modern population genetics—indeed, true population thinking—owe much more to the thinking of the Russian school, as well as to the taxonomists and to the breeders, than to the mathematicians and laboratory geneticists. Sewall Wright held a position somewhat intermediate between the two schools, and one can list him as a supporter of either side, depending on which paragraphs of his papers one wants to quote. The history of the conflicting views of the two schools and the rapid developments after 1966 has not yet been written (Mayr 1970; Lewontin 1971).

One major development of the 1915-1930 period is entirely ignored by Provine, evidently because it was not connected with any of the major genetics laboratories. I am referring to the work on the genetics of geographic races. Such work was conducted by zoologists and by botanists, in Europe as well as in the United States. On the zoological side it includes the work of Johannes Schmidt on the fish *Zoarces*, of Richard Goldschmidt on the moth *Lymantria dispar*, and of Francis Sumner and Lee Dice on mice of the genus *Peromyscus*. There are numerous equivalent studies of plants, beginning with the work of R. Wettstein, but culminating in the work of E. Baur on snapdragons (*Antirhinum*). These studies dealing with natural populations did far more to convince the "Darwinians" of the Mendelian nature of selectively important "natural" variation than either the "artificial" (as they called them) *Drosophila* mutations or mathematical calculations. All of these authors demonstrated the genetic nature of much of geographic variation and, since it was often closely correlated with climatic factors, the importance of natural selection. This work was a crucial factor in making the "new synthesis" of the 1930s and 1940s possible. It should not be ignored in a history of population genetics because it had an important impact on theory.

SUMMARY

It is evident how much Olby and Provine have contributed to a better understanding of the emergence of genetics. It is equally evident, I believe, how many obscure issues still remain to be elucidated. Indeed, their volumes have raised as many new questions as they have answered old ones. In particular, the role of constructive as well as retarding contemporary concepts in the development of new generalizations still requires far more analysis. The somewhat independent trends of various national schools and the influence of neighboring fields (e.g., statistics, animal husbandry, systematics) are other areas deserving further study. All these influences contributed, one way or another, to the growth and maturation of genetics.

REFERENCES

Adams, M. B. 1968. The founding of population genetics: contributions of the Chetverikov school, 1924–1934. *Jour. Hist. Biol.,* 1:23–29.

Allen, G. 1968. Thomas Hunt Morgan and the problem of natural selection. *Jour. Hist. Biol.,* 1:113–139.

Coleman, W. 1965. Cell, nucleus, and inheritance: an historical study. *Proc. Amer. Phil. Soc.,* 109:133.

—— 1970. Bateson and chromosomes: conservative thought in science. *Centaurus,* 15:228–314.

Darwin, F., ed. 1903. *More letters of Charles Darwin.* J. Murray, London.

De Vries, H. 1889. *Intracelluläre Pangenesis.* Fischer Verlag, Jena. Trans. by C. S. Gager as *Intracellular pangenesis,* Open Court Publishing Co., Chicago, 1910.

Gaertner, C. F. 1849. *Versuche und Beobachtungen über die Bastarderzeugung im Pflanzenreich.* K. F. Herring, Stuttgart.

Jacob, F. 1973. *The logic of life.* Pantheon, New York.

Larson, J. L. 1971. *Reason and experience.* University of California Press, Berkeley.

Lerner, M. 1950. *Population genetics and animal improvement.* Cambridge University Press, Cambridge.

Lewontin, R. 1971. Genes in populations—end of the beginning. *Quart. Rev. Biol.,* 46:66–67.

Mayr, E. 1963. *Animal species and evolution.* Belknap Press of Harvard University Press, Cambridge, Mass.

—— 1970. *Populations, species, and evolution.* Belknap Press of Harvard University Press, Cambridge, Mass.

Mendel, G. 1866. Versuche über Pflanzen-Hybriden. *Verh. Naturf. Ver. Brünn,*

4:3–47. Trans. as *Experiments in plant hybridization,* Harvard University Press, Cambridge, Mass., 1965.

Miescher, F. 1897. *Die histochemischen und physiologischen Arbeiten,* 2 vols. Vogel, Leipzig.

Naegeli, K. v. 1884. *Mechanisch-physiologische Theorie der Abstammungslehre.* Oldenbourg, Munich.

Olby, R. C. 1966. *Origins of Mendelism.* Constable Press, London.

Provine, W. B. 1971. *The origins of theoretical population genetics.* University of Chicago Press, Chicago.

Strasburger, E. 1884. *Neue Untersuchungen über den Befruchtungs-vorgang bei den Phanerogamen als Grundlage für eine Theorie der Zeugung.* Fischer, Jena.

Wilkie, J. S. 1963. Commentary. In *Scientific change,* ed. A. C. Crombie, pp. 597–603. Heinemann, London.

IV
PHILOSOPHY OF BIOLOGY

Introduction

At American universities, philosophy is not a required subject of study for a scientist. It was, however, required at many German universities when I was a student, and for my Ph.D. at the University of Berlin, I had to take an examination in philosophy. At that time I carefully studied the positivists, while in my post-doctoral years I concentrated on Bergson and Driesch. This, as well as my reading of philosophical works in later years, left me altogether dissatisfied. None of this philosophy even approached the requirements for a philosophy of biology. The vitalists, rather openly, operated with metaphysical forces which I rejected—even more so after coming to the United States than before—because I was firmly convinced that all living processes and phenomena have to be consistent with the established laws of chemistry and physics and that there are no unexplained vital forces.

I was equally dissatisfied with the naively and coarsely mechanistic interpretations of the reductionists. I have some six or seven volumes on the shelves of my library, each purporting to be a philosophy of science. Alas, they were all written either by physicists or ex-physicists and are nothing but philosophies of physics, or else (worse) by logicians and deal with operational problems instead of with subject matter. Until quite recently no volume existed that came even close to being a philosophy of biology. One should perhaps not blame physicists and logicians for not understanding the very special properties of enormously complex and unique biological systems or the teleonomic capacities of historically acquired programs of information. There is nothing like this in the inanimate world, and smashing these complex systems into fragments, as some reductionists naively attempt, inevitably destroys the specific properties of these systems.

It is really up to the biologist to develop a philosophy of biology (Mayr 1969). Such a philosophy rejects vitalism, by firmly endorsing the conclusion that all biological processes obey the laws of the physical sciences, but it also rejects reductionism by showing that complex biological systems have numerous properties that one simply does not find in inanimate nature. In recent years more and more biologists have *become aware of this challenge and have begun to work on philosophical problems. Simpson's essay "Biology and the Nature of Science" (1963) was a pioneering endeavor in this cause. Happily, the younger generation of philosophers has also in recent years discovered the special problems raised by the philosophy of biology, as exemplified by the valuable contributions made by David Hull, Michael Ruse, Ronald Munson, William C. Wimsatt, Stephen Toulmin, and others. A number of symposia have dealt specifically with problems of the philosophy of biology (Ayala and Dobzhansky 1974; Mendelsohn 1969; Waddington 1968–1969).*

My own contributions started over 15 years ago when I discussed various philosophical questions raised by evolutionary problems (Essays 4 and 9). In 1961 I attempted to clarify the difference between proximate and ultimate causes in biological phenomena (a difference that does not exist in the inorganic world) and I showed how different the problems and the methodology are in the two largely separate fields of biology (Essay 23). Recently, I attempted a far more detailed analysis of the problem of teleology in biology and the universe in general (Essay 26). Two shorter essays (24 and 25) deal with problems of definition and methodology. Essays 29 and 30, presented in the section "Theory of Systematics," also raise problems of interest to the philosopher. My hope is that these writings have contributed to the recognition that a philosophy of biology is an integral component of a philosophy of science.

REFERENCES

Ayala, F. J., and T. Dobzhansky. 1974. *Studies in the philosophy of biology.* University of California Press, Berkeley.

Mayr, E. 1969. Footnotes on the philosophy of biology. *Phil. Sci.*, 36:197–202.

Mendelsohn, E., ed. 1969. Explanation in biology. *Jour. Hist. Biol.*, 2:1–268.

Simpson, G. G. 1963. Biology and the nature of science. *Science*, 139:81–88.

Waddington, C. H. 1968–1969. *Towards a theoretical biology*, vols. 1 and 2. Edinburgh University Press, Edinburgh.

23
Cause and Effect
in Biology

Being a practising biologist I feel that I cannot attempt the kind of analysis of cause and effect in biological phenomena that a logician would undertake. I would instead like to concentrate on the special difficulties presented by the classical concept of causality in biology. From the first attempts to achieve a unitary concept of cause, the student of causality has been bedeviled by these difficulties. Descartes' grossly mechanistic interpretation of life, and the logical extreme to which his ideas were carried by Holbach and de la Mettrie, inevitably provoked a reaction leading to vitalistic theories, which have been in vogue, off and on, to the present day. I have only to mention names like Driesch (entelechy), Bergson (élan vital), and Lecomte du Noüy among the more prominent authors of the recent past. Though these authors may differ in particulars, they all agree in claiming that living beings and life processes cannot be causally explained in terms of physical and chemical phenomena. It is our task to ask whether this assertion is justified, and, if we answer this question with "no," to determine the source of the misunderstanding.

Causality, no matter how it is defined in terms of logic, is believed to contain three elements: (1) an explanation of past events (*a posteriori* causality), (2) prediction of future events; and (3) interpretation of teleological—that is "goal-directed"—phenomena. The three aspects of causality (explanation, prediction, and teleology) must be the cardinal points in any discussion of causality and were quite rightly singled out as such by Nagel (1965). Biology can make a significant contribution to all three of them. But before I can discuss this contribution in detail I must say a few words about biology as a science.

FUNCTIONAL BIOLOGY AND EVOLUTIONARY BIOLOGY

The word *biology* suggests a uniform and unified science. Yet recent developments have made it increasingly clear that biology is a most

Reprinted from "Cause and effect in biology," *Science,* 134 (1961):1501–1506; copyright 1961 by the American Association for the Advancement of Science.

complex area—indeed, that the word *biology* is a label for two largely separate fields that differ greatly in method, *Fragestellung*, and basic concepts. As soon as one goes beyond the level of purely descriptive structural biology, one finds two very different areas, which may be designated functional biology and evolutionary biology. To be sure, the two fields have many points of contact and overlap. Any biologist working in one of these fields must have a knowledge and appreciation of the other field if he wants to avoid the label of a narrow-minded specialist. Yet in his own research he will be occupied with problems of either one or the other field. We cannot discuss cause and effect in biology without first having characterized these two fields.

The functional biologist is vitally concerned with the operation and interaction of structural elements, from molecules up to organs and whole individuals. His ever-repeated question is "How?" How does something operate, how does it function? The functional anatomist who studies an articulation shares this method and approach with the molecular biologist who studies the function of a DNA molecule in the transfer of genetic information. The functional biologist attempts to isolate the particular component he studies, and in any given study he usually deals with a single individual, a single organ, a single cell, or a single part of a cell. He attempts to eliminate or control all variables, and he repeats his experiments under constant or varying conditions until he believes he has clarified the function of the element he studies. The chief technique of the functional biologist is the experiment, and his approach is essentially the same as that of the physicist and the chemist. Indeed, by isolating the studied phenomenon sufficiently from the complexities of the organism, he may achieve the ideal of a purely physical or chemical experiment. In spite of certain limitations of this method, one must agree with the functional biologist that such a simplified approach is an absolute necessity for achieving his particular objectives. The spectacular success of biochemical and biophysical research justifies this direct, although distinctly simplistic, approach.

The evolutionary biologist differs in his method and in the problems in which he is interested. His basic question is "Why?" When we say "why" we must always be aware of the ambiguity of this term. It may mean "how come?" but it may also mean the finalistic "what for?" It is obvious that the evolutionist has in mind the historical "how come?" when he asks "why?" Every organism, whether an individual or a species, is the product of a long history, a history that dates back more than 2000 million years. As Max Delbrück (1949:173) has said, "a mature physicist, acquainting himself for the first time with the problems of biology, is puzzled by the circumstance that there are no 'absolute phenomena' in biology. Everything is time-bound and space-bound. The animal or plant or micro-organism he is working with is but a link in an evolutionary chain of changing forms, none of which has any permanent validity." There is hardly any structure or function in an organism that can be fully understood unless it is studied against this historical background. To find the causes for the existing characteristics, and particularly adaptations, of organisms is the main preoccupation of the evolutionary biologist. He is impressed by the enormous diversity of the organic world. He wants to know the reasons for this diversity as well as the pathway by which it has been achieved. He studies the forces that bring about changes in faunas and floras (as in part documented by paleontology), and he studies the steps by which have evolved the miraculous adaptations so characteristic of every aspect of the organic world.

We can use the language of information theory to attempt still another characterization of these two fields of biology. The functional biologist deals with all aspects of the decoding of the information contained in the DNA program of the fertilized zygote. The evolutionary biologist, on the other hand, is interested in the history of these programs of information and in the laws that control the changes of these programs from generation to generation. In other words, he is interested in the causes of these changes.

Many of the old arguments of biological philosophy can be stated far more precisely in terms of these genetic programs. For instance, as Schmalhausen, in Russia, and I have pointed out independently, the inheritance of acquired characteristics becomes quite unthinkable when applied to the model of the transfer of genetic information from a peripheral phenotype to the DNA of the germ cells.

But let us not have an erroneous concept of these programs. It is characteristic of them that the programing is only in part rigid. Such phenomena as learning, memory, nongenetic structural modification, and regeneration show how "open" these programs are. Yet even here there is great specificity, for instance, with respect to what can be "learned," at what stage in the life cycle "learning" takes place, and how long a memory engram is retained. The program, then, may be in part quite unspecific, and yet the range of possible variation is itself included in the specifications of the program. Therefore, the programs are in some respects highly specific; in other respects they merely specify "reaction norms" or general capacities and potentialities.

Let me illustrate this duality of programs by the difference between two kinds of birds with respect to species recognition. The young cowbird is raised by foster parents—let us say in the nest of a song sparrow or warbler. As soon as it becomes independent of its foster parents it seeks the company of other young cowbirds, even though it has never seen a cowbird before. In contrast, after hatching from the egg, a young goose will accept as its parent the first moving (and preferably also calling) object it can follow and become imprinted to. What is programed is in one case a definite *Gestalt*, in the other merely the capacity to become imprinted to a *Gestalt*. Similar differences in the specificity of the inherited program are universal throughout the organic world (see Essay 2).

CAUSALITY

Let me now get back to the main topic and ask: Is *cause* the same thing in functional and evolutionary biology?

Max Delbrück (1949:173) has reminded us that as recently as 1870 Helmholtz postulated "that the behavior of living cells should be accountable in terms of motions of molecules acting under certain fixed force laws." Now, says Delbrück correctly, we cannot even account for the behavior of a single hydrogen atom. As he also says, "any living cell carries with it the experiences of a billion years of experimentation by its ancestors."

Let me illustrate the difficulties of the concept of causality in biology by an example. Let us ask: What is the cause of bird migration? Or more specifically: Why did the warbler on my summer place in New Hampshire start its southward migration on the night of August 25? I can list four equally legitimate causes for this migration.

1. *An ecological cause.* The warbler, being an insect eater, must migrate, because it would starve to death if it should try to winter in New Hampshire.

2. *A genetic cause.* The warbler has acquired a genetic constitution in the course of the evolutionary history of its species that induces it to respond appropriately to the proper stimuli from the environment. By contrast, the screech owl, nesting right next to it, lacks this constitution and does not respond to these stimuli. As a result, it is sedentary.

3. *An intrinsic physiological cause.* The warbler flew south because its migration is tied in with photoperiodicity. It responds to the decrease in day length and is ready to migrate as soon as the number of hours of daylight have dropped below a certain level.

4. *An extrinsic physiological cause.* Finally, the warbler migrated on August 25 because a cold air mass, with northerly winds, passed over our area on that day. The sudden drop in temperature and the associated weather conditions affected the bird, already in a general physiological readiness for migration, so that it actually took off on that particular day.

Now, if we look over the four causations of the migration of this bird once more, we can readily see that there is an immediate set of causes of the migration, consisting of the physiological condition of the bird interacting with photoperiodicity and drop in temperature. We might call these the *proximate* causes of migration. The other two causes, the lack of food during winter and the genetic disposition of the bird, are the *ultimate* causes. These are causes that have a history and that have been incorporated into the warbler's system through many thousands of generations of natural selection. It is evident that the functional biologist would be concerned with analysis of the proximate causes, while the evolutionary biologist would be concerned with analysis of the ultimate causes. This is the case with almost any biological phenomenon we might want to study. There is always a proximate set of causes and an ultimate set of causes: both have to be explained and interpreted for a complete understanding of the given phenomenon.

Still another way to express these differences would be to say that proximate causes govern the responses of the individual (and its organs) to immediate factors of the environment, while ultimate causes are responsible for the evolution of the particular DNA program of information with which every individual of every species is endowed. The logician will, presumably, be little concerned with these distinctions. Yet the biologist knows that many heated arguments about the "cause" of a certain biological phenomenon could have been avoided if the two opponents had realized that one of them was concerned with proximate and the other with ultimate causes. I might illustrate this by a quotation from Loeb (1916): "The earlier writers explained the growth of the legs in the tadpole of the frog or toad as a case of adaptation to life on land. We know through Gudernatsch that the growth of the legs can be produced at any time, even in the youngest tadpole, which is unable to live on land, by feeding the animal with the thyroid gland."

Let us now get back to the definition of cause in formal philosophy and see how it fits with the usual explanatory cause of functional and evolutionary biology. We might, for instance, define cause as "a non-sufficient condition without which an event would not have happened," or as "a member of a set of jointly sufficient reasons without which the event would not happen." Definitions such as these describe causal relations quite adequately in certain branches of biology, particularly in those that deal with chemical and physical unit phenomena. In a strictly formal sense they are also applicable to more complex phenomena, and yet they seem to have little operational value in those branches of biology that deal with complex systems. I doubt that there is a scientist who would question the ultimate causality of all biological phenomena —that is, that a causal explanation can be given for past biological events. Yet such an explanation will often have to be so unspecific and so purely formal that its explanatory value can certainly be challenged. In dealing with a complex system an explanation can hardly be considered very illuminating that states: "Phenomenon A is caused by a complex set of interacting factors, one of which is b." Yet often this is about all one can say. We will have to come back to this difficulty in connection with the problem of prediction. However, let us first consider the problem of teleology.

TELEOLOGY

No discussion of causality is complete that does not come to grips with the problem of teleology. This problem had its beginning with Aristotle's classification of causes, one of the categories being the "final" causes. This category is based on the observation of the orderly and purposive development of the individual from the egg to the "final" stage of the adult, and of the development of the whole world from its beginnings (chaos?) to its present order. Final cause has been defined as "the cause responsible for the orderly reaching of a preconceived ultimate goal." All goal-seeking behavior has been classified as "teleological," but so have many other phenomena that are not necessarily goal-seeking in nature.

Aristotelian scholars have rightly emphasized that Aristotle—by training and interest—was first and foremost a biologist, and that it was his preoccupation with biological phenomena which dominated his ideas on causes and induced him to postulate final causes in addition to the material, formal, and efficient causes. Thinkers from Aristotle to the present have been challenged by the apparent contradiction between a mechanistic interpretation of natural processes and the seemingly purposive sequence of events in organic growth, in reproduction, and in animal behavior. Such a rational thinker as Bernard (1885) has stated the paradox in these words:

> There is, so to speak, a preestablished design of each being and of each organ of such a kind that each phenomenon by itself depends upon the general forces of nature, but when taken in connection with the others it seems directed by some invisible guide on the road it follows and led to the place it occupies.
>
> We admit that the life phenomena are attached to physico-chemical manifestations, but it is true that the essential is not explained thereby; for no fortuitous coming together of physico-chemical phenomena constructs each organism after a plan and a fixed design (which are foreseen in advance) and arouses the admirable subordination and harmonious agreement of the acts of life. ... Determinism can never be [anything] but physiochemical determinism. The vital force and life belong to the metaphysical world.

What is the x, this seemingly purposive agent, this "vital force," in organic phenomena? It is only in our lifetime that explanations have been advanced which deal adequately with this paradox.

The many dualistic, finalistic, and vitalistic philosophies of the past merely replaced the unknown x by a different unknown, y or z, for calling an unknown factor *entelechia* or *élan vital* is not an explanation.

I shall not waste time showing how wrong most of these past attempts were.

Where, then, is it legitimate to speak of purpose and purposiveness in nature, and where is it not? To this question we can now give a firm and unambiguous answer. An individual who—to use the language of the computer—has been "programed" can act purposefully. Historical processes, however, can*not* act purposefully. A bird that starts its migration, an insect that selects its host plant, an animal that avoids a predator, a male that displays to a female—they all act purposefully because they have been programed to do so. When I speak of the programed "individual" I do so in a broad sense. A programed computer itself is an "individual" in this sense, but so is, during reproduction, a pair of birds whose instinctive and learned actions and interactions obey, so to speak, a single program. The completely individualistic and yet also species-specific DNA program of every zygote, which controls the development of the central and peripheral nervous systems, of the sense organs, of the hormones, of physiology and morphology, is the *program* for the behavior computer of this individual.

Natural selection does its best to favor the production of programs guaranteeing behavior that increases fitness. A behavior program that guarantees instantaneous correct reaction to a potential food source, to a potential enemy, or to a potential mate will certainly give greater fitness in the Darwinian sense than a program that lacks these properties. Again, a behavior program that allows for appropriate learning and the improvement of behavior reactions by various types of feedbacks gives greater likelihood of survival than a program that lacks these properties.

The purposive action of an individual, insofar as it is based on the properties of its genetic program, therefore is no more or less purposive than the actions of a computer that has been programed to respond appropriately to various inputs. It is, if I may say so, a purely mechanistic purposiveness.

We biologists have long felt that it is ambiguous to designate such programed, goal-directed behavior "teleological," because the word *teleological* has also been used in a very different sense for the final stage in evolutionary adaptive processes. When Aristotle spoke of final causes, he was particularly concerned with the marvelous adaptations found throughout the plant and animal kingdoms. He was concerned with what later authors have called design or plan in nature. He ascribed to final causes not only mimicry and symbiosis but all the other adaptations of animals and plants to each other and to their physical environment. The Aristotelians and their successors asked themselves what goal-directed process could have produced such a well-ordered design in nature.

It is now evident that the terms *teleology* and *teleological* have been applied to two entirely different sets of phenomena. On the one hand, there are the production and perfecting throughout the history of the animal and plant kingdoms of ever-new genetic programs. On the other hand, there is the testing of these programs throughout the lifetime of each individual. There is a fundamental difference between, on the one hand, end-directed behavioral activities or developmental processes of an individual or system, which are controlled by a program and, on the other, evolutionary adaptation controlled by natural selection, leading to a steady improvement of genetic programs.

In order to avoid confusion between the two entirely different types of end direction, Pittendrigh (1958:394) has introduced the term *teleonomic* as a descriptive term for all end-directed systems "not committed to Aristotelian teleology." Not only does this negative definition place the entire burden on the word *system,* but it makes no clear distinction between the two teleologies of Aristotle. It would seem useful to restrict the term *teleonomic* rigidly to systems operating on the basis of a program of information. Teleonomy in biology designates "the apparent purposefulness of organisms and their characteristics," as Julian Huxley expressed it (1960:9).

Such a clear-cut separation of teleonomy, which has an analyzable physicochemical basis, from teleology, which deals more broadly with the overall harmony of the organic world, is most useful because these two entirely different phenomena have so often been confused with each other.

The development or behavior of an individual is purposive; natural selection is definitely not. When MacLeod (1957:477) stated, "What is most challenging about Darwin, however, is his reintroduction of purpose into the natural world," he chose the wrong word. The word *purpose* is singularly inapplicable to evolutionary change, which is, after all, what Darwin was considering. If an organism is well adapted, if it shows superior fitness, this is not due to any purpose of its ancestors or of an outside agency, such as "Nature" or "God," that created a superior design or plan. Darwin "has swept out such finalistic teleology by the front door," as Simpson (1960:966) has rightly said.

We can summarize this discussion by stating that there is no conflict between causality and teleonomy, but that scientific biology has not found any evidence that would support teleology in the sense of various vitalistic or finalistic theories (Simpson 1950:262; Koch 1957:245; Simpson 1960:966). (For a more detailed discussion of teleonomy, see Essay 26.)

THE PROBLEM OF PREDICTION

The third great problem of causality in biology is that of prediction. In the classical theory of causality the touchstone of the goodness of a causal explanation was its predictive value. This view is still maintained in Bunge's modern classic (1959:307): "A theory can predict to the extent to which it can describe and explain." It is evident that Bunge is a physicist: no biologist would have made such a statement. The theory of natural selection can describe and explain phenomena with considerable precision, but it cannot make reliable predictions, except through such trivial and meaningless circular statements as, for instance, "the fitter individuals will on the average leave more offspring." Scriven (1959:477) has emphasized quite correctly that one of the most important contributions to philosophy made by the evolutionary theory is that it has demonstrated the independence of explanation and prediction.

Although prediction is not an inseparable concomitant of causality, every scientist is nevertheless happy if his causal explanations simultaneously have high predictive value. We can distinguish many categories of prediction in biological explanation. Indeed, it is even doubtful how to define "prediction" in biology. A competent zoogeographer can predict with high accuracy what animals will be found on a previously unexplored mountain range or island. A paleontologist likewise can predict with high probability what kind of fossils can be expected in a newly accessible geological horizon. Is such correct guessing of the results of past events genuine prediction? A similar doubt pertains to taxonomic predictions, as discussed in the next paragraph. The term *prediction* is, however, surely legitimately used for future events. Let me give you four examples to illustrate the range of predictability:

1. *Prediction in classification.* If I have identified a fruit fly as an individual of *Drosophila melanogaster* on the basis of bristle pattern and the proportions of face and eye, I can "predict" numerous structural and behavioral characteristics that I will find if I study other aspects of this individual. If I find a new species with the diagnostic key characters of the genus *Drosophila*, I can at once "predict" a whole set of biological properties.

2. *Prediction of most physicochemical phenomena on the molecular level.* Predictions of very high accuracy can be made with respect to most biochemical unit processes in organisms, such as metabolic pathways, and with respect to biophysical phenomena in simple systems, such as the action of light, heat, and electricity in physiology.

In examples 1 and 2 the predictive value of causal statements is usually very high. Yet there are numerous other generalizations or causal statements in biology that have low predictive values. The following examples are of this kind.

3. *Prediction of the outcome of complex ecological interactions.* The statement "An abandoned pasture in southern New England will be replaced by a stand of gray birch (*Betula populifolia*) and white pine (*Pinus strobus*)" is often correct. Even more often, however, the re-

placement may be an almost solid stand of *P. strobus*, or *P. strobus* may be missing altogether and in its stead will be cherry (*Prunus*), red cedar (*Juniperus virginianus*), maples, sumac, and several other species.

4. *Prediction of evolutionary events.* Probably nothing in biology is less predictable than the future course of evolution. Looking at the Permian reptiles, who would have predicted that most of the more flourishing groups would become extinct (many rather rapidly) and that one of the most undistinguished branches would give rise to the mammals? Which student of the Cambrian fauna would have predicted the revolutionary changes in the marine life of the subsequent geological eras? Unpredictability also characterizes small-scale evolution. Breeders and students of natural selection have discovered again and again that independent parallel lines exposed to the same selection pressure will respond at different rates and with different correlated effects, none of them predictable.

As is true in many other branches of science, the validity of predictions for biological phenomena (except for a few chemical or physical unit processes) is nearly always statistical. We can predict with high accuracy that slightly more than 500 of the next 1000 newborns will be boys. We cannot predict the sex of a particular unborn child.

Without claiming to exhaust all the possible reasons for indeterminacy, I can list four classes. Although they overlap somewhat, each deserves to be treated separately.

1. *Randomness of an event with respect to the significance of the event.* Spontaneous mutation, caused by an "error" in DNA replication, illustrates this cause for indeterminacy very well. The occurrence of a given mutation is in no way related to the evolutionary needs of the particular organism or of the population to which it belongs. The precise results of a given selection pressure are unpredictable because mutation, recombination, and developmental homeostasis are making indeterminate contributions to the response to this pressure. All the steps in the determination of the genetic contents of a zygote contain a large component of this type of randomness. What I have described for mutation is also true for crossing over, chromosomal segregation, gametic selection, mate selection, and early survival of the zygotes. Neither underlying molecular phenomena nor the mechanical motions responsible for this randomness are related to their biological effects.

2. *Uniqueness of all entities at the higher levels of biological integration.* In the uniqueness of biological entities and phenomena lies one of the major differences between biology and the physical sciences. Physicists and chemists often have genuine difficulty in understanding the biologist's stress on the unique, although such an understanding has been greatly facilitated by the developments in modern physics. If a physicist says "ice floats on water," his statement is true for any piece of ice and any body of water. The members of a class usually lack the

individuality that is so characteristic of the organic world, where all individuals are unique; all stages in the life cycle are unique; all populations are unique; all species and higher categories are unique; all inter-individual contacts are unique; all natural associations of species are unique; and all evolutionary events are unique. Where these statements are applicable to man, their validity is self-evident. However, they are equally valid for all sexually reproducing animals and plants. Uniqueness, of course, does not entirely preclude prediction. We can make many valid statements about the attributes and behavior of man, and the same is true for other organisms. But most of these statements (except for those pertaining to taxonomy) have purely statistical validity. Uniqueness is particularly characteristic for evolutionary biology. It is quite impossible to have for unique phenomena general laws like those that exist in classical mechanics.

3. *Extreme complexity*. The physicist Elsässer stated in a recent symposium: "[an] outstanding feature of all organisms is their well-nigh unlimited structural and dynamical complexity." This is true. Every organic system is so rich in feedbacks, homeostatic devices, and potential multiple pathways that a complete description is quite impossible. Furthermore, the analysis of such a system would require its destruction and would thus be futile.

4. *Emergence of new qualities at higher levels of integration*. It would lead too far to discuss in this context the thorny problem of "emergence." All I can do here is to state its principle dogmatically: "When two entities are combined at a higher level of integration, not all the properties of the new entity are necessarily a logical or predictable consequence of the properties of the components." This difficulty is by no means confined to biology, but it is certainly one of the major sources of indeterminacy in biology. Let us remember that indeterminacy does not mean lack of cause, but merely unpredictability.

All four causes of indeterminacy, individually and combined, reduce the precision of prediction.

One may raise the question at this point whether predictability in classical mechanics and unpredictability in biology are due to a difference of degree or of kind. There is much to suggest that the difference is, in considerable part, merely a matter of degree. Classical mechanics is, so to speak, at one end of a continuous spectrum, and biology is at the other. Let us take the classical example of the gas laws. Essentially they are only statistically true, but the population of molecules in a gas obeying the gas laws is so enormous that the actions of individual molecules become integrated into a predictable—one might say "absolute"—result. Samples of 5 or 20 molecules would show definite individuality. The difference in the size of the studied "populations" certainly contributes to the difference between the physical sciences and biology.

CONCLUSIONS

Let me now return to the initial question and try to summarize some of my conclusions on the nature of the cause-and-effect relations in biology.

1. Causality in biology is a far cry from causality in classical mechanics.

2. Explanations of all but the simplest biological phenomena usually consist of sets of causes. This is particularly true for those biological phenomena that can be understood only if their evolutionary history is also considered. Each set is like a pair of brackets that contains much that is unanalyzed and much that can presumably never be analyzed completely.

3. In view of the high number of multiple pathways possible for most biological processes (except for the purely physicochemical ones) and in view of the randomness of many biological processes, particularly on the molecular level (as well as for other reasons), causality in biological systems is not predictive, or at best is only statistically predictive.

4. The existence of complex programs of information in the DNA of the germ plasm permits teleonomic purposiveness. On the other hand, evolutionary research has found no evidence whatsoever for a "goal seeking" of evolutionary lines, as postulated in that kind of teleology which sees "plan and design" in nature. The harmony of the living universe, so far as it exists, is an *a posteriori* product of natural selection.

5. Finally, causality in biology is not in real conflict with the causality of classical mechanics. As modern physics has also demonstrated, the causality of classical mechanics is only a very simple, special case of causality. Predictability, for instance, is not a necessary component of causality. The complexities of biological causality do not justify embracing nonscientific ideologies, such as vitalism or finalism, but should encourage all those who have been trying to give a broader basis to the concept of causality.

REFERENCES

Bernard, C. 1885. *Leçons sur les phénomènes de la vie*, vol. 1. Librairie J.-B. Bailliere et Fils, Paris.

Bunge, M. 1959. *Causality*. Harvard University Press, Cambridge, Mass.

Delbrück, M. 1949. A physicist looks at biology. *Trans. Conn. Acad. Arts Sci.*, 38:173-190.

Huxley, J. 1960. The openbill's open bill: a teleonomic enquiry. *Zool. Jb. Abt. Anat. u. Ontog. Tiere*, 88:9-30.

Koch, L. F. 1957. Vitalistic-mechanistic controversy. *Sci. Monthly*, 85:245-255.

Loeb, J. 1916. *The organism as a whole*. Putnam Press, New York.

MacLeod, R. B. Teleology and theory of human behavior. *Science*, 125:477-480.

Nagel, E. 1965. Types of causal explanation in science. In *Cause and effect*, ed. D. Lerner. Free Press, New York.

Pittendrigh, C. S. 1958. Adaptation, natural selection, and behavior. In *Behavior and evolution,* ed. A. Roe and G. G. Simpson, pp. 390–416. Yale University Press, New Haven.

Scriven, M. 1959. Explanation and prediction in evolutionary theory. *Science,* 130:477–482.

Simpson, G. G. 1950. Evolutionary determinism and the fossil record. *Sci. Monthly,* 71:262–267.

—— 1960. The world into which Darwin led us. *Science,* 125:966–974.

24
Explanatory Models
in Biology

An interest in scientific explanation goes back to the Greek philosophers. It is, however, only in the last 200 years that this concern has become a major branch of philosophy. Darwin was vigorously attacked by several philosophers for having violated the canons of good science, as well documented by Hull in Darwin and His Critics *(1973). In this century most of the dialogue on this subject has been among philosophers, but increasingly often biologists have entered the discussion. An outstanding example is Ghiselin's* Triumph of the Darwinian Method *(1969).*

My own interest in the area is expressed in Essays 24 and 25, each of which originated as a set of comments I was asked to present as chief discussant of a major paper. The following essay deals with the impact of physical models on concept formation in physiology early in the nineteenth century. It is a set of comments on Everett Mendelsohn's analysis of the attitudes of Johannes Müller and his students, particularly Theodor Schwann, toward Naturphilosophie *and toward reductionist explanations (* Boston Studies Phil. Sci., 2[1965]:127–150*).*

Dr. Mendelsohn has analyzed with sensitivity and great insight the emergence of a new system of explanation in nineteenth-century biology. He presented a well-balanced selection of quotations from the writings of Johannes Müller, Theodor Schwann, and their associates, and traced the origin of the new ideas. Since he knows the literature of that period so much better than I, it will be best if I limit my comments to the interpretation of the currents and counter-currents of that period. As in any other historical event, there are, of course, several interpretations possible concerning the reasons for the shift in Schwann's thinking. For the sake of argument, I shall attempt an interpretation that differs somewhat, at least in emphasis, from Dr. Mendelsohn's analysis. I would not want to claim that it is a superior explanation, but merely that it is a legitimate alternative. In view of our uncertainty

Reprinted from "Comments" (on "Explanation in nineteenth century biology," by Everett Mendelsohn), *Boston Studies in the Philosophy of Science,* 2(1965):151–156.

about the intellectual currents in the first half of the nineteenth century, it would seem worthwhile in future researches to make a comparative exploration of both alternatives.

It is extraordinary how many new movements, indeed perhaps all of them, start as a rebellion against that which preceded them. Since there is no doubt that a new interpretation of physiological phenomena emerged in the minds of Müller's students, Schwann, Du Bois-Reymond, and Helmholtz (with Schwann recognized as their leader), we must ask: "What did Müller's students rebel against?" Clearly it was something in Müller's world of ideas, but what? Was it really his attachment to *Naturphilosophie,* as it might appear on the surface? Personally, I rather doubt it. Müller himself had rejected this influence at an early age, and his inaugural lecture on October 19, 1824, was distinctly anti-Schelling and hence anti-*Naturphilosophie.* To be sure, Müller was still pro-Goethe and hence pro–idealistic morphology, but this is something rather different from *Naturphilosophie.* Before taking up what the new generation rebelled against, let me add an aside on *Naturphilosophie.*

When we look at the scene contemporary with Müller, we do not seem to find a great following for *Naturphilosophie* among scientists. Indeed, it is my impression that, except for Oken and Carus, there was not a single respectable anatomist, physiologist, zoologist, or botanist whom one could truly label a disciple of Schelling or follower of Oken, not even Meckel. I admit that I have not made a special study of their writings, but from what little I know of their work, all the leaders of contemporary science, not only Rudolphi, but also Retzius, Rathke, Tiedemann, Ehrenberg, and Purkinje were rather sober, descriptive, and uncommitted scientists.

Romanticism had a greater impact on poetry, literature, painting, and music in Germany than in any other country in the world. *Naturphilosophie* was the expression of Romanticism in philosophy, but its application to science was singularly unsuccessful and, I add, unimportant. Many of the ideas often ascribed to *Naturphilosophie* originated in the eighteenth century (Bonnet) or earlier, even though they were later incorporated into *Naturphilosophie.* Still others, such as vitalism, have really nothing to do with *Naturphilosophie.* I shall come back to this later.

What, then, in Müller's thinking was Schwann opposed to? I think there are two points, both of them appropriately stressed by Dr. Mendelsohn. The first is any and all mixing of philosophy and science. This is clear from Schwann's correspondence in 1835 with his brother Peter and from Virchow's later statement. Idealistic morphology and other "interpretations" borrowed from philosophy were for Schwann as objectionable as *Naturphilosophie.*

More specifically, the rebels objected to the recognition of any kind of vital force, later often referred to as "vitalism." Vitalistic ideas existed long before the days of *Naturphilosophie* and have persisted long after the demise of *Naturphilosophie*. Indeed, a belief in vital forces has been part of the philosophy of many a biologist who had not a trace of connection with *Naturphilosophie*. It is extraordinarily difficult to discuss vitalism in an objective, detached manner. The closer we come to the recent period, the less successful have been the attempts to revive a vitalistic interpretation of phenomena of life. Vitalism as a possible theory of biology has now been dead for some 40 or 50 years, as has been the entire argument of mechanism versus vitalism. The situation was quite different in the days of Johannes Müller and Claude Bernard. In those days, a belief in vital forces was, curiously enough, in many cases a sophisticated reaction to crude theories of mechanism. It was not a special philosophy and it had no separate manifesto. Usually it consisted merely in the statement of a working biologist that he was willing to study the phenomena of life in a detached, objective manner, and if he found an inexplicable residue, an unresolved *x*, so to speak, exclusively connected with living organisms, he would call it a vital force, or whatever terminology he used. This was actually a rather sober attitude, and it characterized Claude Bernard and Müller in his mature days, as well as many other reputable biologists in the middle of the last century. This approach was quite different from the later ones of Bergson and Driesch, who used vitalism as a broad platform on which to erect the large edifice of a whole philosophy of organisms. The vitalism of the working physiologists in the first half of the nineteenth century was a philosophy of resignation and skepticism. It was based on an admission that there was a residue that could not be resolved.

I believe it was this resignation in Müller, the consequence of his disappointment with *Naturphilosophie*, against which his young students rebelled. It may sound paradoxical, but the writings of Schwann and his associates give one the impression that they were actually romantics at heart. To be sure, they were not romantics of mysticism and obscurantism, like Schelling and Oken, but reductionist romantics, romantics who dreamed of reduction of all biological phenomena to physical and chemical phenomena and laws, who were not willing to compromise, who were not willing to set aside an unknown or an unknowable. However, it was not for very long that the reductionists held the field. Dr. Mendelsohn has quoted quite perceptively Virchow's statement of 1855 in which he castigates "the scientific prudery of insisting that all organic processes be reduced to the resultant of inherent molecular forces." Du Bois-Reymond's later "Ignoramus, ignorabimus" is another illustration of the ultimate resignation that caught up also with this rebellious generation.

Dr. Mendelsohn asked the question why there was such a difference between, on the one hand, the reserved approach of the French physiologists, of the mature Müller, and of Virchow, and on the other hand, the radical, uncompromising approach of the young German physiologists. No doubt there are a number of reasons for this difference, but the most important one probably is the dominant philosophy of the period preceding these scientists. If there was a radical movement first, the followers would be moderates. Where moderates had been dominant, radicals followed them. In the case of the French physiologists, moderation was the reaction to the extreme mechanism (reductionism) of Lamettrie and Baron Holbach. In the case of Müller, it was the reaction to *Naturphilosophie;* in the case of the young German physiologists, it was the reaction to the compromising moderation of the middle-aged Müller, now thoroughly cured of his *Naturphilosophie* fantasies. Perhaps the most fascinating aspect of the developments at that period is the rapidity of reactions and counter-reactions. Within a period of less than 10 years, we have Müller's reaction to *Naturphilosophie* and Schwann's reaction to Müller's resignation and skepticism. Not much later follows Virchow's reaction to the youthful reductionists.

One aspect of the reductionist approach fascinates me particularly. In our day we see clearly that there is no vital force and that it is possible to interpret even teleonomic phenomena in terms of physics and chemistry as the product of genetic information programed in the DNA of the germ cells. As a consequence, we are *a priori* biased against any author admitting vital forces. Yet, if we look back at explanations in the first half of the nineteenth century, we find that the reductionists in their crude materialism ("muscle action like that of any elastic substance") are actually much farther from the truth than their vitalist opponents. All one has to do is to replace the x (= vital force) in the formula of the vitalist with certain newer findings and one has essentially an interpretation that can still stand. There is an obvious parallel here with the situation in evolutionary biology during the first two decades of the twentieth century.

Their differences in philosophy affected the scientific methodologies of the physiologists. The reductionists did not care whether they worked on a muscle inside an organism or on an isolated "dead" muscle. The vitalists, in order to remove the factor x from their variables, took great pains to do their experiments only on parts of living organisms. By eliminating the complicating factor of death, they were much better able to study physiological phenomena per se. How wise Claude Bernard and Müller were in their insistence on this method is indicated by the fact that in principle this is still the prevailing method in physiology, except at the strictly molecular level.

You may feel that I have gone too far in turning the tables and labeling the reductionists as romantics and the vitalists as the sober, skeptical scientists. Perhaps I have, but this different emphasis is necessary in order to bring home an important point, and in order to help to establish a genuine balance in our interpretation. It sometimes takes a real effort to put oneself back into thinking of a long-past period, and yet one has to do this in order to make just value judgments. To judge an 1830s scientist in the hindsight of 1975 may add much to an analysis, but cannot give complete balance.

Let me add one final point. When I stress the deficiencies of Schwann's reductionist approach, I am not trying to devalue either its uncompromising honesty or its heuristic power. I think it can be shown again and again in the history of biology that the oversimplified, often crude, often atomistic approach almost invariably produces more results and produces results more rapidly than the more sophisticated, more holistic approach, which is perhaps most useful after the reductionist approach has exhausted the pathways that are accessible to it. The immense success of the reductionist approach of Schwann, Du Bois-Reymond, and Brücke is a matter of history.

REFERENCES

Ghiselin, M. 1969. *The triumph of the Darwinian method.* University of California Press, Berkeley.

Hull, D. 1973. *Darwin and his critics.* Harvard University Press, Cambridge, Mass.

Mendelsohn, E. 1965. Physical models and physiological concepts: explanation in nineteenth century biology. *Boston Studies Phil. Sci.,* 2:127–150.

25
Theory Formation in Developmental Biology

*In 1968 June Goodfield gave a lecture on theoretical entitites and func-tional explanation in biology (*Boston Studies Phil. Sci., *5:421–449). In particular she considered the gene, the organizer, messenger RNA, and the repressor. I commented as follows on her interpretation.*

The philosophy of science has been dominated by physics to such an extent during past decades that any discussion of theory formation in biology must start out with the similarities between physics and biol-ogy. June Goodfield has, therefore, quite rightly concentrated on aspects of biology that have a good deal in common with the physical sciences, such as theoretical entities and functional processes. It is good strategy to start out with functional biology, the part of biology where indeed physical entities (e.g., molecules) and physical processes, such as energy and information transfer, often play a decisive role. It is surely no coincidence that a biological entity, the gene, conforms so closely to the physical entities as described by the philosophers of physics, be-cause the gene is simultaneously a macromolecule and thus as much an object of the physical as of the biological sciences. This still leaves the question open to what extent the simplistic notions of physical theory formation are useful in the more complex areas of biology.

Classical genetics dealt with the performance of a given gene. From here investigators have moved in two directions, both concerned with the interaction of genes. I will not pursue the problem of evolutionary interactions in the gene pool of a species, a problem of intense interest to the evolutionist and ecologist. Instead, I shall make a few comments on the interaction of genes during development, the major object of June Goodfield's analysis. This is one of the two fields of biology into which molecular biologists have moved in large numbers during the past few years. It is the problem of differentiation, so lucidly discussed and analyzed by June Goodfield, which excites the functional biologist as

Reprinted from "Comments on 'Theories and hypotheses in biology,'" *Boston Studies in the Philosophy of Science,* 5(1968):450–456.

perhaps no other problem. The challenge consists in the paradox between the identity of the genetic program in the chromosomes of all cells, on the one hand, and the increasing divergence of cells during development, on the other hand. What happens when the fertilized egg cleaves into two daughter cells, and sequentially into more and more cells until finally there are nerve cells, gland cells, epidermal cells that make hair, and cells that give rise to gametes? What is responsible for this tremendous differentiation, all based on the same genotype?

The problem of differentiation has been with us for several hundred years, and it is quite fascinating to observe the changes in the conceptual framework within which it has been investigated. No matter how unbiased and uncommitted an investigator believes himself to be, we know that he makes all sorts of silent assumptions. This, perhaps, was Spemann's greatest problem. Ever since C. F. Wolff, developmental biology has been dedicated to the routing out of the last traces of the preformism that had previously dominated embryology for so many centuries. Particulate inheritance suddenly seemed to restore preformism, and this was one of the sources of intense antagonism against Mendelism (shared up to 1910 by T. H. Morgan) during the early decades of the century. Spemann had been raised in an intellectual climate of extreme anti-preformism. Every modern biologist knows that every attribute of an organism is affected by genes, and yet, when an embryologist discusses a given developmental process, he acts to an amazing degree as if genes did not exist and as if all development was the result of an endogenous system of developmental gradients: once the fertilized egg starts developing, then the further course of development is prescribed by the total field of development. Consciously or unconsciously, the embryologist felt that there was a conflict between his epigenetic conceptual framework and the Mendelian theory according to which development is controlled by individual genes. Theory formation in embryology was preoccupied for several decades with attempting to resolve the apparent conflict between these two interpretations.

June Goodfield has focused attention on the resulting change in conceptualization among embryologists by posing the problem of differentiation in the form of two questions, the classical one of the epigenetic embryologist and the modern one. The classical question is: How does an apparently homogeneous, small ball of yolk-laden cytoplasm, with a nucleus, turn into a large, complicated, highly organized adult with fully functioning organs? The modern one is: How does the *encoded* structure, compressed into the final developmental stages of a specialized, maternal organ—the egg—become transformed into the *realized* structure of an adult organism? The two questions represent two extremely different viewpoints. Those asking the classical question assumed that all the complexities of an organism arise *de novo* in an extreme epigenetic sense. The new question takes it for granted that much of the complexity is preformed in the blueprint of the genetic program and is only translated into the development.

I wonder whether the conceptual revolution represented by these two questions is not more fundamental than admitted by June Goodfield. She says: "Even after decades of brilliant studies, we are conceptually not much further on [than the embryology of 25 years ago]. There is, as yet, no general theory of development." As a matter of fact, I myself feel that there is such a theory even though it may not turn out to be correct. It all hinges on the interpretation of the gene as the determinant of characters. June Goodfield emphasized how cautiously Johannsen expressed himself about the characteristics of the gene. However, all geneticists for the next 40 years or so (at least until 1944) presumed that whatever else the gene was, it was a protein. No one saw what difficulties this assumption raised for the genotype-phenotype problem. Nor has it ever been pointed out, I believe, how completely this problem changed when the determinants became a genetic program that merely serves as a blueprint *but does not, as such, participate in the development.* The genetic program is first translated into polypeptides, which are the agents of development. The objection might be raised that a distinction between the nucleic-acid program and the protein executor is irrelevant because there is a straight one-to-one relationship owing to the messenger RNA. This objection is invalid because there is no one-to-one relationship between gene and character. Since the genes themselves stay outside the developmental process, and since the same enzyme may participate in the developmental process at different places at different times, always governed by the genetic program, it becomes evident why the sharp separation of the genetic program and the translated proteins is so tremendously important for the causal interpretation of the developmental process.

Is it not possible, indeed is it not probable, that the machinery of translating the genetic program provides the "general theory of development" which embryologists are looking for so assiduously? Is not the seeming conflict between the views of the classical, epigenetic embryologists and the molecular biologists eliminated by the finding of diffusable repressors and inducers? The Monod-Jacob model is undoubtedly an oversimplification, but does it not seem to solve previously existing difficulties, at least in principle?

Development, it seems to me, can be compared with the activities of a symphony orchestra. The musical score tells the musicians what to produce and when. The conductor reinforces and synchronizes the "turning on" and "turning off" of the activities of individual musicians. The similarity between this and the "turning on" and "off" of the activities of individual gene loci by inducers and repressors is remarkably close. The activity of an orchestra, including that of its conductor, is just as much controlled by the score it is playing as the development of an organism is controlled by its genetic program. There are good reasons for believing that even most feedbacks and other homeostatic mechanisms are encoded in the genetic program. It is very probable, for instance, that even the time is programed when each locus in a given cell becomes active and the conditions under which it again becomes deactivated. This, of course, allows for all sorts of feedbacks and diffusable repressors.

Mind you, I am not claiming that we now understand development. All I am saying is that it is quite possible that we already have, in principle, a "general theory of development." But development takes place in organisms, and this introduces a level of complexity that is altogether unknown in the inorganic world. A higher organism has enough DNA for about 5 million genes and the complexity introduced by the actual and potential interaction of these 5 million genes during development is an awe-inspiring consideration. Biological systems are of an order of complexity that places them in a class by themselves. This leads me back to some more basic considerations.

Biologists still suffer from the dead hand of the past. We are only two or three decades removed from the period when all theory formation in biology was dominated by the controversy between vitalism and mechanism. Only two alternatives seemed to be available at that time. You could either be a woolly vitalist (finalist or teleologist) or a naive physicochemical reductionist. Even though most respectable biologists rejected vitalism in all of its forms, they felt that a primitive reductionism was singularly unsatisfactory for biological interpretation, because it was inapplicable to all truly biological phenomena. Now that vitalism has been dead for some 30 years and every biologist admits that "processes in living systems" are not "any less material or less physical in nature" than processes in nonliving systems (Simpson 1963), it has become possible and respectable to study biological phenomena for which there are no equivalents in inorganic nature (if for no other reason than their complexity). In order not to be misunderstood, I would like to quote Simpson once more: "Insistence that the study of organisms requires principles additional to those of the physical sciences does not imply a dualistic or vitalistic view of nature."

A discussion of theory formation in biology would be incomplete that does not consider certain biological phenomena which, although not in conflict with any of the laws of physics, are nevertheless not elucidated by them. This point can be made particularly well by recalling that functional biology is only part of the whole of biology (see Essay 23). It is the combination of functional biology and evolutionary biology that gives us the whole of biology. The way questions are asked and theory is formed is rather different in these two parts of biology. In functional biology, as we have seen, one deals with the decoding of genetic programs. All of it ultimately involves chemical reactions guided by the genetic program and regulated by complex feedback systems. It is quite conceivable that functional biology could be reduced to physics and chemistry, at least in principle, even though the complexity of the interactions between the genetic program and the resulting proteins produces a system of such diversity that a complete analysis is probably unattainable. The most common question we ask in evolutionary biology is how something came about. Every phenomenon, as correctly stated by Max Delbrück (1948), is time bound and space bound. Everything we now see is the result of a long evolution. Every genetic program has a history of billions of years and has incorporated a great deal of teleonomic purposiveness. Except for a few analogies between the phenomena of evolutionary biology and some of the phenomena of cosmological evolution, there is hardly any equivalence between evolutionary biology and anything found in the physical sciences.

In her closing words, June Goodfield pointed out two reasons why theoretical entities, so popular in physical explanation, are so rare in biology. One is the enormous variability in the biological universe, and the other is that the entities with which the biologist works, from macromolecules and cellular components on up, are either visible or otherwise clearly demonstrable and not merely a matter of inference. Both of these reasons explain, in part, why the needs for theory formation are so different in physics and in biology. They also help to explain why the needs of biology are not met by theories specifically developed to satisfy the needs of physics. To go back to my earlier argument, it is no longer necessary to stress that biological phenomena are consistent with the laws of chemistry and physics. The time has now come in biological theory formation to stress those "material and physical phenomena" (Simpson) that are exclusive to biology and not encountered when one deals with inanimate objects.

Let me start with the question of uniqueness. I recently attended a lecture on elementary particles by a physicist who stressed *identity* as a crucial characteristic of physical entities. A pi(π)-meson, regardless of its source, the earth or any other galaxy in the universe, always has identical properties, and this is true for any of the basic units in the physical sciences. In biology such identity is found only at the chemical level. It is quickly replaced by uniqueness when we get into more complex systems. No two individuals are identical in sexually reproducing species nor are any two populations. No two species or natural communities are identical. Uniqueness is the inevitable consequence of the incredible complexity of biological systems. It is this complexity that is part of the reason for the slow progress and incomplete analysis in behavioral biology, ecology, and population biology. Generalizations made when one deals with unique phenomena are probability statements with empirically determined statistics. Fortunately, since the rise of quantum mechanics, an area of agreement between physics and biology has developed in the interpretation of laws as statistical laws. Yet, when we look at individual biological phenomena like gene pools, biological species, biological classification, the process of speciation, the concept of fitness, most behavioral phenomena like learning, imprinting, or innateness, or such concepts of ecology as competition, niche, food pyramid, or balanced diversity, we realize how little they have to do with the conventional entities of physical theory.

This raises the more fundamental question to what extent theory formation in the physical sciences is equivalent to theory formation dealing with highly complex biological systems, especially biological phenomena controlled by historically evolved genetic programs. I am not at all certain what the answer to this question is.

Can we escape from this dilemma? Biology is obviously so much broader in its dimensions than the sciences of inanimate objects that it can only be stultifying to squeeze it into the straitjacket of physical terminology and theory. Biologists must have the courage to formulate theories and define terms that are meaningful and heuristic as far as biology is concerned (regardless of their applicability to physics). I have attempted elsewhere (1969) to do this for certain biological terms and concepts. Actually, biological theory formation is ultimately quite simple, even though the detailed interpretation and analysis of any biological system is appallingly complex. The reductionist approach has paid off handsomely in the physical sciences. It has been equally productive in molecular biology. Yet carrying the reductionist approach too far can lead only to trivialities (Simpson 1963). When one dissects complex biological systems down to the level of atoms and elementary particles, one loses everything that is legitimately called biological. At the same time one must avoid, as June Goodfield has shown so beautifully, the acceptance of unanalyzed terms, like the term "organizer," which only inhibit analysis. The art of theory formation in biology is to stay away from either of these two extremes.

REFERENCES

Delbrück, M. 1948. A physicist looks at biology. *Trans. Conn. Acad. Arts Sci.*, 38:173–190.

Goodfield, J. 1968. Theories and hypotheses in biology. *Boston Studies Phil. Sci.*, 5:421–449.

Mayr, E. 1969. Footnotes on the philosophy of biology. *Phil. Sci.*, 36(2):197–202.

Simpson, G. G. 1963. Biology and the nature of science. *Science*, 139:81–88.

26
Teleological and Teleonomic:
A New Analysis

Teleological language is frequently used in biology in order to make statements about the functions of organs, about physiological processes, and about the behavior and actions of species and individuals. Such language is characterized by the use of the words *function, purpose,* and *goal,* as well as by statements that something exists or is done "in order to." Typical statements of this sort are "It is one of the functions of the kidneys to eliminate the end products of protein metabolism," or "Birds migrate to warm climates in order to escape the low temperatures and food shortages of winter." In spite of the long-standing misgivings of physical scientists, philosophers, and logicians, many biologists have continued to insist not only that such teleological statements are objective and free of metaphysical content, but also that they express something important which is lost when teleological language is eliminated. Recent reviews of the problem in the philosophical literature (Nagel 1961; Beckner 1969; Hull 1973; to cite only a few of a large selection of such publications) concede the legitimacy of some teleological statements but still display considerable divergence of opinion as to the actual meaning of "teleological" and the relations between teleology and causality.

This confusion is nothing new and goes back at least as far as Aristotle, who invoked final causes not only for individual life processes (such as development from the egg to the adult) but also for the universe as a whole. To him, as a biologist, the form-giving of the specific life process was the primary paradigm of a finalistic process, but for his epigones the order of the universe and the trend toward its perfection became completely dominant. The existence of a form-giving, finalistic principle in the universe was rightly rejected by Bacon and Descartes, but this, they thought, necessitated the eradication of any and all teleological language, even for biological processes, such as growth and behavior, or in the discussion of adaptive structures.

Adapted from "Teleological and teleonomic: a new analysis," *Boston Studies in the Philosophy of Science,* 14 (1974):91–117.

The history of the biological sciences from the seventeenth to the nineteenth centuries is characterized by a constant battle between extreme mechanists, who explained everything purely in terms of movements and forces, and their opponents, who often went to the opposite extreme of vitalism. After vitalism had been completely routed by the beginning of the twentieth century, biologists could afford to be less self-conscious in their language and, as Pittendrigh (1958) has expressed it, were again willing to say "a turtle came ashore to lay her eggs," instead of saying "she came ashore and laid her eggs." There is now consensus among biologists that the teleological phrasing of such a statement does not imply any conflict with physicochemical causality.

Yet the very fact that teleological statements have again become respectable has helped to bring out uncertainties. The vast literature on teleology is eloquent evidence for the unusual difficulties connected with this subject. This impression is reenforced when one finds how often various authors dealing with this subject have reached opposite conclusions (e.g., Braithwaite 1954; Beckner 1969; Canfield 1966; Hull 1973; Nagel 1961). They differ from each other in multiple ways, but most importantly in answering the question: What kind of teleological statements are legitimate and what others are not? Or, what is the relation between Darwin and teleology? David Hull (1973) has recently stated, "evolutionary theory did away with teleology, and that is that," yet, a few years earlier McLeod (1957) had pronounced, "what is most challenging about Darwin, is his reintroduction of purpose into the natural world." Obviously the two authors must mean very different things.

Purely logical analysis helped remarkably little to clear up the confusion. What finally produced a breakthrough in our thinking about teleology was the introduction of new concepts from the fields of cybernetics and new terminologies from the language of information theory. The result was the development of a new teleological language, which claims to be able to take advantage of the heuristic merits of teleological phraseology without being vulnerable to the traditional objections.

TRADITIONAL OBJECTIONS TO THE USE
OF TELEOLOGICAL LANGUAGE

Criticism of the use of teleological language is traditionally based on one or several of the following objections. In order to be acceptable, teleological language must be immune to these objections.

Teleological Statements and Explanations Imply the Endorsement
of Unverifiable Theological or Metaphysical Doctrines in Science

This criticism was indeed valid in former times, as for instance when natural theology operated extensively with a strictly metaphysical teleology. Physiological processes, adaptations to the environment, and all forms of seemingly purposive behavior tended to be interpreted as being due to nonmaterial vital forces. This interpretation was widely accepted among Greek philosophers, including Aristotle, who discerned an active soul everywhere in nature. Bergson's (1910) élan vital and Driesch's (1909) *Entelechie* are relatively recent examples of such metaphysical teleology. Contemporary philosophers reject such teleology almost unanimously. Likewise, the employment of teleological language among modern biologists does not imply adoption of such metaphysical concepts (see below).

The Belief That Acceptance of Explanations for Biological Phenomena That Are Not Equally Applicable to Inanimate Nature Constitutes Rejection of a Physicochemical Explanation

Ever since the age of Galileo and Newton it has been the endeavor of the "natural scientists" to explain everything in nature in terms of the laws of physics. To accept special explanations for teleological phenomena in living organisms implied for these critics a capitulation to mysticism and a belief in the supernatural. They ignored the fact that nothing exists in inanimate nature (except for manmade machines) which corresponds to DNA programs or to goal-directed activities. As a matter of fact, the acceptance of a teleonomic explanation (see below) is in no way in conflict with the laws of physics and chemistry. It is not in opposition to a causal interpretation, and it does not imply an acceptance of supernatural forces in any way whatsoever.

The Assumption That Future Goals Were the Cause of Current Events Seemed in Complete Conflict with Any Concept of Causality

Braithwaite (1954) stated the conflict as follows: "In a [normal] causal explanation the explicandum is explained in terms of a cause which either precedes it or is simultaneous with it; in a teleological explanation the explicandum is explained as being causally related either to a particular goal in the future or to a biological end which is as much future as present or past." This is why some logicians up to the present distinguish between causal explanations and teleological explanations.

Teleological Language Seemed to Represent Objectionable Anthropomorphism

The use of words like *purposive* or *goal-directed* seemed to imply the transfer of human qualities, such as intent, purpose, planning, deliberation, or consciousness, to organic structures and to subhuman forms of life. Intentional, purposeful human behavior is, almost by definition, teleological. Yet I shall exclude it from further discussion because the words *intentional* and *consciously premeditated,* which are usually used in connection with such behavior, may get us involved in complex controversies over psychological theory, even though much of human behavior does not differ in kind from animal behavior. The latter, although usually described in terms of stimulus and response, is also highly "intentional," as when a predator stalks its prey or when the prey flees from the pursuing predator. Yet seemingly "purposive," that is, goal-directed, behavior in animals can be discussed and analyzed in operationally definable terms, without recourse to anthropomorphic terms like "intentional" or "consciously."

As a result of these and other objections teleological explanations were widely believed to be a form of obscurantism, an evasion of the need for a causal explanation. Indeed some authors went so far as to make statements such as "Teleological notions are among the main obstacles to theory formation in biology" (Lagerspetz 1959:65). Yet biologists insisted on continuing to use teleological language.

The teleological dilemma, then, consists in the fact that numerous and seemingly weighty objections to the use of teleological language have been raised by various critics, and yet biologists have insisted that they would lose a great deal, methodologically and heuristically, if they were prevented from using such language. It is my endeavor to resolve this dilemma by a new analysis, and particularly by a new classification of the various phenomena that have traditionally been designated "teleological."

THE HETEROGENEITY OF TELEOLOGICAL PHENOMENA

One of the greatest shortcomings of most recent discussions of the teleology problem has been the heterogeneity of the phenomena designated "teleological" by different authors. To me it would seem quite futile to arrive at rigorous definitions until the medley of phenomena designated "teleological" is separated into more or less homogeneous classes. To accomplish this will be my first task. Furthermore, mingling a discussion of teleology with consideration of such extraneous problems as vitalism, holism, and reductionism only confuses the issue. Teleological statements and phenomena can be analyzed without reference to major philosophical systems.

By and large all the phenomena that have been designated in the literature as teleological can be grouped into three classes: (1) unidirectional evolutionary sequences (progressionism, orthogenesis), (2) seemingly or genuinely goal-directed processes, and the so-called (3) teleological systems. The ensuing discussion will serve to bring out the great differences between these three classes of phenomena.

UNIDIRECTIONAL EVOLUTIONARY SEQUENCES
(PROGRESSIONISM, ORTHOGENESIS)

Beginning with Aristotle and other Greek philosophers, and becoming increasingly widespread in the eighteenth century, was a belief in an upward or forward progression in the arrangement of natural objects. This was expressed most concretely in the concept of the *scala naturae,* the scale of perfection (Lovejoy 1936). Originally conceived as something static (or even descending, owing to a process of degradation), the Ladder of Perfection was temporalized in the eighteenth century and merged almost unnoticeably into evolutionary theories such as that of Lamarck. Progressionist theories were proposed in two somewhat different forms. The steady advance toward perfection was directed either by a supernatural force (a wise creator) or, rather vaguely, by a built-in drive toward perfection. During the flowering of Natural Theology the "interventionist" concept dominated but after 1859 it was replaced by the so-called orthogenetic theories, which were widely held by biologists and philosophers (see Lagerspetz 1959:11–12 for a short survey). Simpson (1949) refuted the possibility of orthogenesis with particularly decisive arguments. Actually, as Weismann had said long ago (1909), the principle of natural selection solves the origin of progressive adaptation without any recourse to goal-determining forces.

It is somewhat surprising how many philosophers, physical scientists, and occasionally even biologists still flirt with the concept of a teleological determination of evolution. Teilhard de Chardin's (1955) entire dogma is built on such a teleology and so are, as Monod (1971) has stressed quite rightly, almost all of the most important ideologies of the past and present. Even some serious evolutionists play, in my opinion rather dangerously, with teleological language. For instance, Ayala (1970:11) says:

> the overall process of evolution cannot be said to be teleological in the sense of directed towards the production of specified DNA codes of information, i.e., organisms. But it is my contention that it can be said to be teleological in the sense of being directed toward the production of DNA codes of information which improve the reproductive fitness of a population in the environments where it lives. The process of evolution can also be said to be teleological in that it has the potentiality of producing end-directed DNA codes of information, and has in fact resulted in teleologically oriented structures, patterns of behavior, and regulated mechanisms.

To me this seems a serious misinterpretation. If "teleological" means anything it means "goal-directed." Yet natural selection is strictly an *a posteriori* process that rewards current success but never sets up future goals. No one realized this better than Darwin, who reminded himself "never to use the words higher or lower." Natural selection rewards past events, that is, the production of successful recombinations of genes, but it does not plan for the future. This is precisely what gives evolution by natural selection its flexibility. With the environment

changing incessantly, natural selection—in contradistinction to orthogenesis—never commits itself to a future goal. Natural selection is never
goal oriented. It is misleading and quite inadmissible to designate such
broadly generalized concepts as survival or reproductive success as
definite and specified goals.

The same objection can be raised against certain arguments presented
by Waddington (1968:55–56). Like so many other developmental
biologists, he is forever looking for analogies between ontogeny and
evolution. "I have for some years been urging that quasi-finalistic types
of explanations are called for in the theory of evolution as well as in
that of development." Natural selection "in itself suffices to determine,
to a certain degree, the nature of the end towards which evolution will
proceed, it must result in an increase in the efficiency of the biosystem
as a whole in finding ways of reproducing itself." He refers here to
completely generalized processes, rather than to specific goals. It is
rather easy to demonstrate how ludicrous the conclusions are which
one reaches by overextending the concept of goal direction. For
instance, one might say that it is the purpose of every individual to die
because this is the end of every individual, or that it is the goal of every
evolutionary line to become extinct because this is what has happened
to 99.9 percent of all evolutionary lines that have ever existed. Indeed,
one would be forced to consider as teleological even the second law of
thermodynamics.

One of Darwin's greatest contributions was to have made it clear that
goal-directed processes involving only a single individual are of an entirely different nature from evolutionary changes. The latter are controlled by the interplay of the production of variants (new genotypes)
and their sorting out by natural selection, a process that is quite decidedly not directed toward a specified distant end. A discussion of
legitimately teleological phenomena would be futile unless evolutionary
processes are eliminated from consideration.

SEEMINGLY OR GENUINELY GOAL-DIRECTED PROCESSES

Nature (organic and inanimate) abounds in processes and activities
that lead to an end. Some authors seem to believe that all such terminating processes are of one kind and "finalistic" in the same manner
and to the same degree. Taylor (1950), for instance, if I understand him
correctly, claims that all forms of active behavior are of the same kind
and that there is no fundamental difference between one kind of movement or purposive action and any other. Waddington (1968) gives a
definition of his term "quasi-finalistic" as requiring "that the end state
of the process is determined by its properties at the beginning." Further
study indicates, however, that the class of "end-directed processes" is
composed of two entirely different kinds of phenomena. These two
types of phenomena may be characterized as teleomatic processes in
inanimate nature or teleonomic processes in living nature.

Teleomatic Processes in Inanimate Nature

Many movements of inanimate objects as well as physicochemical processes are the simple consequences of natural laws. For instance, gravity provides the end state for a rock that I drop into a well. It will reach its end state when it has come to rest on the bottom. A red-hot piece of iron reaches its "end state" when its temperature and that of its environment are equal. All objects of the physical world are endowed with the capacity to change their state and these changes follow natural laws. They are "end-directed" only in a passive, automatic way, regulated by external forces or conditions. Since the end state of such inanimate objects is automatically achieved, such changes might be designated *teleomatic*. All teleomatic processes come to an end when the potential is used up (as in the cooling of a heated piece of iron) or when the process is stopped by encountering an external impediment (as when a falling stone hits the ground). Teleomatic processes simply follow natural laws, that is, lead to a result consequential to concomitant physical forces, and the reaching of their end state is not controlled by a built-in program. The law of gravity and the second law of thermodynamics are among the natural laws that most frequently govern teleomatic processes.

Teleonomic Processes in Living Nature

Seemingly goal-directed behavior in organisms is of an entirely different nature from teleomatic processes. Goal-directed "behavior" (in the widest sense of this word) is extremely widespread in the organic world; for instance, most activity connected with migration, food getting, courtship, ontogeny, and all phases of reproduction is characterized by such goal orientation. The occurrence of goal-directed processes is perhaps the most characteristic feature of the world of living organisms.

Definition of the Term "Teleonomic"

For the last 15 years or so the term *teleonomic* has been used increasingly often for goal-directed processes in organisms. In 1961 I proposed the following definition for this term: "It would seem useful to restrict the term 'teleonomic' rigidly to systems operating on the basis of a program, a code of information" (Essay 23). Although I used the term "system" in this definition, I have since become convinced that it permits a better operational definition to consider certain activities, processes (like growth), and active behaviors as the most characteristic illustrations of teleonomic phenomena. I therefore modify my definition as follows: *A teleonomic process or behavior is one that owes its goal directedness to the operation of a program.* The term "teleonomic" implies goal direction. This, in turn, implies a dynamic process rather than a static condition, as represented by a system. The combination of "teleonomic" with the term system is, thus, rather incongruent (see below).

All teleonomic behavior is characterized by two components. It is guided by a "program" and it depends on the existence of some end point, goal, or terminus that is foreseen in the program that regulates the behavior. This end point might be a structure, a physiological function, the attainment of a new geographical position, or a "consummatory" (Craig 1918) act in behavior. Each particular program is the result of natural selection, constantly adjusted by the selective value of the achieved end point.

My definition of "teleonomic" has been labeled by Hull (1973) a "historical definition." Such a designation is rather misleading. Although the genetic program (as well as its individually acquired components) originated in the past, this history is completely irrelevant for the functional analysis of a given teleonomic process. For this it is entirely sufficient to know that a "program" exists which is causally responsible for the teleonomic nature of a goal-directed process. Whether this program originated through a lucky macromutation (as Richard Goldschmidt conceived possible) or through a slow process of gradual selection, or even through individual learning or conditioning, as in open programs, is quite immaterial; the mere existence of a program, whatever its origin, is enough to classify a process as "teleonomic." However, a process that does not have a programed end does not qualify to be designated as teleonomic (see below for a discussion of the concept "program").

All teleonomic processes are facilitated by specifically selected executive structures. The fleeing of a deer from a predatory carnivore is facilitated by the existence of superlative sense organs and the proper development of muscles and other components of the locomotory apparatus. The proper performing of teleonomic processes at the molecular level is made possible by highly specific properties of complex macromolecules. It would stultify the definition of "teleonomic" if the appropriateness of these facilitating executive structures were made part of it. However, it is in the nature of a teleonomic program that it does not induce a simple unfolding of some completely preformed *Gestalt,* but that it always controls a more or less complex process which must allow for internal and external disturbances. Teleonomic processes during ontogenetic development, for instance, are constantly in danger of being derailed, even if only temporarily. There exist innumerable feedback devices to prevent this or to correct it. Waddington (1957) has quite rightly called attention to the frequency and importance of such homeostatic devices, which virtually guarantee the appropriate canalization of development.

We owe a great debt of gratitude to Rosenblueth et al. (1943) for their endeavor to find a new solution for the explanation of teleological phenomena in organisms. They correctly identified two aspects of such phenomena: (1) they are seemingly purposeful, being directed toward a goal; and (2) they consist of active behavior. The background of these authors was in the newly developing field of cybernetics, and it is only natural that they should have stressed the fact that goal-directed behavior is characterized by mechanisms which correct errors committed during the goal seeking. They considered the negative feedback loops of such behavior its most characteristic aspect and stated, "teleological behavior thus becomes synonymous with behavior controlled by negative feedback." This statement emphasizes important aspects of teleological behavior, yet it misses the crucial point: *The truly characteristic aspect of goal-seeking behavior is not that mechanisms exist which improve the precision with which a goal is reached, but rather that mechanisms exist which initiate, that is, "cause" this goal-seeking behavior.* It is not the thermostat that determines the temperature of a house, but the person who sets the thermostat. It is not the torpedo that determines toward what ship it will be shot and at what time, but the naval officer who releases the torpedo. Negative feedbacks improve the precision of goal seeking, but they do not determine it. Feedback devices are only executive mechanisms that operate during the translation of a program. Therefore, it places the emphasis on the wrong point to define teleonomic processes in terms of the presence of feedback devices. They are mediators of the program, but as far as the basic principle of goal achievement is concerned, they are of minor consequence.

Recent Usages of the Term "Teleonomic"

The term "teleonomic" was introduced into the literature by Pittendrigh (1958:394) in the following paragraph:

> Today the concept of adaptation is beginning to enjoy an improved respectability for several reasons: it is seen as less than perfect; natural selection is better understood; and the engineer-physicist in building end-seeking automata has sanctified the use of teleological jargon. It seems unfortunate that the term "teleology" should be resurrected and, as I think, abused in this way. The biologists' long-standing confusion would be more fully removed if all end-directed systems were described by some other term, like "teleonomic," in order to emphasize that the recognition and description of end-directedness does not carry a commitment to Aristotelian teleology as an efficient [*sic*] causal principle.

It is evident that Pittendrigh had the same phenomena in mind as I do,[1] even though his definition is rather vague and his placing the term

[1] This is quite evident from the following explanatory comment I have received from Professor Pittendrigh by letter (dated February 26, 1970): "You ask about the word 'teleonomy.' You are correct that I did introduce the term into biology and, moreover, I invented it. In the course of thinking about that paper which I wrote for the Simpson and Roe book (in which the

"teleonomic" in opposition to Aristotle's "teleology" is unfortunate. As we shall see below, most of Aristotle's references to end-directed processes refer precisely to the same things that Pittendrigh and I would call teleonomic (see also Delbrück 1971).

Other recent usages of the term that differ from my own definition are the following. B. Davis (1962), believing that the term denotes "the development of valuable structures and mechanisms" as a result of natural selection, considers the term virtually synonymous with adaptiveness. The same is largely true for Simpson (1958:520–521), who sees in "teleonomic" the description for a system or structure that is the product of evolution and of selective advantage:

> The words "finalistic" and "teleological" have, however, had an unfortunate history in philosophy which makes them totally unsuitable for use in modern biology. They have too often been used to mean that evolution as a whole has a predetermined goal, or that the utility of organization in general is with respect to man or to some supernatural scheme of things. Thus these terms may implicitly negate rather than express the biological conclusion that

term is introduced) I was haunted by that famous old quip of Haldane's to the effect that 'Teleology is like a mistress to a biologist: he cannot live without her but he's unwilling to be seen with her in public.' The more I thought about that, it occurred to me that the whole thing was nonsense—that what it was the biologist couldn't live with was not the illegitimacy of the relationship, but the relationship itself. Teleology in its Aristotelian form has, of course, the end as immediate, 'efficient', cause. And that is precisely what the biologist (with the whole history of science since 1500 behind him) cannot accept: it is unacceptable in a world that is always mechanistic (and of course in this I include probabilistic as well as strictly deterministic). What it was the biologist could not escape was the plain fact—or rather the fundamental fact—which he must (as scientist) explain: that the objects of biological analysis are organizations (he calls them organisms) and, as such, are end-directed. Organization is more than mere order; order lacks end-directedness; organization *is* end-directed. [I recall a wonderful conversation with John von Neumann in which we explored the difference between "mere order" and "organization" and *his* insistence (I already believed it) that the concept of organization (as contextually defined in its everyday use) always involved "purpose" or end-directedness.]

"I wanted a word that would allow me (all of us biologists) to describe, stress or simply to allude to—without offense—this end-directedness of a perfectly respectable mechanistic system. Teleology would not do, carrying with it that implication that the end is causally effective in the current operation of the machine. Teleonomic, it is hoped, escapes that plain falsity which is anyhow unnecessary. Haldane was, in this sense wrong (surely a rare event): we can live without teleology.

"The crux of the problem lies of course in unconfounding the mechanism of evolutionary change and the physiological mechanism of the organism abstracted from the evolutionary time scale. The most general of all biological 'ends,' or 'purposes,' is of course perpetuation by reproduction. *That* end [and all its subsidiary "ends" of feeding, defense and survival generally] is in some sense effective in causing natural selection; in causing evolutionary change; but not in causing itself. In brief, we have failed in the past to unconfound causation in the historical origins of a system and causation in the contemporary working of the system. . . .

"You ask in your letter whether or not one of the 'information' people didn't introduce it. They did not, unless you wish to call me an information bloke. It is, however, true that my own thinking about the whole thing was very significantly affected by a paper which was published by Wiener and Bigelow with the intriguing title 'Purposeful machines.' This pointed out that in the then newly-emerging computer period it was possible to design and build machines that had ends or purposes without implying that the purposes were the cause of the immediate operation of the machine."

organization in organisms is with respect to utility to each separate species at the time when it occurs, and not with respect to any other species or any future time. In emphasis of this point of view, Pittendrigh [above] suggests that the new coinage "teleonomy" be substituted for the debased currency of teleology.

Monod (1971) likewise deals with teleonomy as if the word simply meant adaptation. It is not surprising, therefore, that Monod considers teleonomy "to be a profoundly ambiguous concept." Furthermore, says Monod, all functional adaptations are "so many aspects or fragments of a unique primary project which is the preservation and multiplication of the species." He finally completes the confusion by choosing "to define the essential teleonomic project as consisting in the transmission from generation to generation of the invariance content characteristic of the species. All these structures, all the performances, all the activities contributing to the success of the essential project will hence be called teleonomic." What Monod calls "teleonomic" I would designate as of "selective value." Under these circumstances it is not surprising that Ayala (1970) claims that the term "teleonomy" was introduced into the philosophical literature in order "to explain adaptation in nature as the result of natural selection." If this were indeed true, and it is true of Simpson's and Davis's cited definitions, the term would be quite unnecessary. Actually, there is nothing in my 1961 account that would support this interpretation, and I know of no other term that would define a goal-directed activity or behavior that is controlled by a program. Even though Pittendrigh's discussion of "teleonomic" rather confused the issue and has led to the subsequent misinterpretations, he evidently had in mind the same processes and phenomena that I denoted as *teleonomic*. It would seem well worthwhile to retain the term in the more rigorous definition, which I have now given.

The Meaning of the Term "Program"

The key word in my definition of "teleonomic" is the term "program." Someone might claim that the difficulties of an acceptable definition for teleological language in biology had simply been transferred to the term "program." This is not a legitimate objection, because it fails to recognize that, regardless of its particular definition, a program (1) is something material and (2) exists prior to the initiation of the teleonomic process. Hence, it is consistent with a causal explanation.

Nevertheless, it might be admitted that the concept "program" is so new that the diversity of meanings of this term has not yet been fully explored. The term is taken from the language of information theory. A computer may act purposefully when given appropriate programed instructions. Tentatively, "program" might be defined as *coded or pre-*

arranged information that controls a process (or behavior) leading it toward a given end. As Raven (1960) has remarked correctly, the program contains not only the blueprint but also the instructions of how to use the information of the blueprint. In the case of a computer program or of the DNA of a cell nucleus, the program is completely separated from the executive machinery. In the case of most manmade automata the program is part of the total machinery.

My definition of "program" is deliberately chosen in such a way as to avoid drawing a line between seemingly "purposive" behavior in organisms and in manmade machines. The simplest program is perhaps the weight inserted into loaded dice or attached to a "fixed" number wheel so that they are likely to come to rest at a given number. A clock is constructed and programed in such a way as to strike at the full hour. Any machine programed to carry out goal-directed activities is capable of doing this "mechanically."

The programs that control teleonomic processes in organisms are either entirely laid down in the DNA of the genotype ("closed programs") or constituted in such a way that they can incorporate additional information ("open programs") acquired through learning, conditioning, or through other experiences (Essay 2). Most behavior, particularly in higher organisms, is controlled by open programs. Once the open program is filled in, it is equivalent to an originally closed program in its control of teleonomic behavior (see Essay 47).

Open programs are particularly suitable for demonstrating the fact that the mode of acquisition of a program is an entirely different matter from the teleonomic nature of the behavior controlled by the program. Nothing could be more purposive, more teleonomic, than much of the escape behavior in many prey species (in birds and mammals). Yet in many cases the knowledge of which animals are dangerous predators is learned by the young who have an open program for this type of information. In other words, this particular information was not acquired through selection and yet it is clearly in part responsible for teleonomic behavior. Many of the teleonomic components of the reproductive behavior (including mate selection) of species that are imprinted for mate recognition is likewise only partially the result of selection. The history of the acquisition of a program, therefore, cannot be made part of the definition of "teleonomic."

The origin of a program is quite irrelevant for the definition. It can be the product of evolution, as are all genetic programs, or it can be the acquired information of an open program, or it can be a manmade device. Anything that does *not* lead to what is at least in principle a predictable goal does not qualify as a program. Even though the future evolution of a species has severe limits set on it by its current gene pool, its course is largely controlled by the changing constellation of selection pressures and is therefore not predictable. It is not programed inside the contemporary gene pool.

The entire concept of a program of information is so new that it has received little attention from philosophers and logicians. My tentative analysis may, therefore, require considerable revision when subjected to further scrutiny.

How does the program operate? The philosopher may be willing to accept the assertion of the biologist that a program directs a given teleonomic behavior, but he would also like to know how the program performs this function. Alas, all the biologist can tell him is that the study of the operation of programs is the most difficult area of biology. For instance, the translation of the genetic program into growth processes and into the differentiation of cells, tissues, and organs is at the present time the most challenging problem of developmental biology. The number of qualitatively different cells in a higher organism almost surely exceeds 1 billion. Even though all (or most) cells have the same gene complement, they differ from each other owing to differences in the repression and derepression of individual gene loci and owing to differences in their cellular environment. It hardly needs stressing how complex the genetic program must be to be able to give the appropriate signals to each cell lineage in order to provide it with the mixture of molecules that it needs in order to carry out its assigned tasks.

Similar problems arise in the analysis of goal-directed behavior. The number of ways in which a program may control a goal-directed behavior activity is legion. It differs from species to species. Sometimes the program is largely acquired by experience; in other cases it may be almost completely genetically fixed. Sometimes the behavior consists of a series of steps, each of which serves as reinforcement for the ensuing steps; in other cases the behavior, once initiated, seems to run its full course without need for any further input. Feedback loops are sometimes important, but their presence cannot be demonstrated in other kinds of behavior. Again, as in developmental biology, much of the contemporary research in behavioral biology is devoted to the nature and the operation of the programs that control behavior and more specifically teleonomic behavior sequences (Hinde and Stevenson 1970). Almost any statement one might make is apt to be challenged by one or another school of psychologists and geneticists. It is, however, safe to state that the translation of programs into teleonomic behavior is greatly affected both by sensory inputs and by internal physiological (largely hormonal) states.

TELEOLOGICAL SYSTEMS

In the philosophical literature, the word "teleological" is particularly often combined with the term "system." Is it justified to speak of "teleological systems"? Analysis shows that this combination leads to definitional difficulties.

The Greek word *telos* means end or goal. Teleological means end directed. To apply "teleological" to a goal-directed behavior or process would seem quite legitimate. I am perhaps a purist, but it bothers me to apply the word *teleological*, that is *end-directed*, to a stationary system. Any phenomenon to which we can refer as teleomatic or teleonomic (discussed above) represents a movement, a behavior, or a process that is goal directed because it has a determinable end. This is the core concept of "teleological," the presence of a *telos* (an end) toward which an object or process moves. Rosenblueth et al. (1943) have correctly stressed the same point.

However, extending the term "teleological" to cover static systems leads to contradictions and illogicalities. A torpedo that has been shot off and moves toward its target is a machine showing teleonomic behavior. But what justifies calling a torpedo a teleological system when, with hundreds of others, it is stored in an ordnance depot? Why should the eye of a sleeping person be called a teleological system? It is not goal directed at anything. Part of the confusion is due to the fact that the term "teleological system" has been applied to two only partially overlapping phenomena. One comprises systems that are potentially able to perform teleonomic actions, like a torpedo. The other comprises systems that are well adapted, like the eye. To refer to a phenomenon in this second class as "teleological," in order to express its adaptive perfection, reflects just enough of the old idea of evolution leading to a steady progression in adaptation and perfection to make me uneasy. What is the telos toward which the teleological system moves?

The source of the conflict seems to be that "goal directed," in a more or less straightforward literal sense, is not necessarily the same as purposive. Completely stationary systems can be functional or purposive, but they cannot be goal directed in any literal sense. A poison on the shelf has the potential of killing somebody, but this inherent property does not make it a goal-directed object. Perhaps this difficulty can be resolved by making a terminological distinction between functional properties of systems and strict goal directedness, that is, teleonomy of behavioral or other processes. However, since one will be using so-called teleological language in both cases, one might subsume both categories under teleology.

R. Munson (1971) has recently dealt with such adaptive systems. In particular, he studied all those explanations that deal with aspects of adaptation but are often called "teleological." He designates sentences "adaptational sentences" when they contain the terms "adaptation," "adaptive," or "adapted." In agreement with the majority opinion of biologists he concludes that "adaptational sentences do not need to involve reference to any purpose, final cause, or other non-empirical notion in order to be meaningful." Adaptational sentences simply express the conclusion that a given trait, whether structural, physiological, or behavioral, is the product of the process of natural selection and thus favors the perpetuation of the genotype responsible for this trait.

Furthermore, adaptation is a heuristic concept because it demands an answer to the question in what way the trait adds up to the probability of survival and does so more successfully than another conceivable trait. To me, it is misleading to call adaptational statements teleological. "Adapted" is an *a posteriori* statement, and it is only the success (statistically speaking) of the owner of an adaptive trait that proves whether the trait is truly adaptive (= contributes to survival) or is not. Munson summarizes the utility of adaptational language in the sentence: "To show that a trait is adaptive is to present a phenomenon requiring explanation, and to provide the explanation is to display the success of the trait as the outcome of selection" (1971:214). The biologist fully agrees with this conclusion. Adaptive means simply: being the result of natural selection.

Many adaptive systems, as for instance all components of the locomotory and of the central nervous systems, are capable of taking part in teleonomic processes or teleonomic behavior. However, it only obscures the issue to designate a system "teleological" or "teleonomic" because it provides the executive structures of a teleonomic process. Is an inactive, not-programed computer a teleological system? What "goal" or "end" is it displaying during this period of inactivity? To repeat, one runs into serious logical difficulties when one applies the term "teleological" to static systems (regardless of their potential) instead of to processes. Nothing is lost and much is to be gained by not using the term "teleological" too freely and for too many rather diverse phenomena.

It may be necessary to coin a new term for systems that have the potential of displaying teleonomic behavior. The problem is particularly acute for biological organs that are capable of carrying out useful functions, such as pumping by the heart or filtration by the kidney. To some extent this problem exists for any organic structure, all the way down to the macromolecules, which are capable of carrying out autonomously certain highly specific functions owing to their uniquely specific structure. It is this ability that induced Monod (1971) to call them teleonomic systems. Similar considerations have induced some authors, erroneously in my opinion, to designate a hammer as a teleological system, because it is designed to hit a nail (a rock, not having been so designed, but serving the same function, not qualifying!).

The philosophical complexity of the logical definition of "teleological" in living systems is obvious. Let me consider a few of the proximate and ultimate causes to bring out some of the difficulties more clearly. The functioning of these systems is the subject matter of regulatory biology, which analyzes proximate causes. Biological systems are complicated steady-state systems, replete with feedback devices. There is a high premium on homeostasis, on the maintenance of the *milieu interieur*. Since most of the processes performed by these systems are programed, it is legitimate to call them teleonomic processes. They are "end directed" even though very often the "end" is the maintenance of the status quo. There is nothing metaphysical in any of this because, so far as these processes are accessible to analysis, they represent chains of causally interrelated stimuli and reactions, of inputs and of outputs.

The ultimate causes for the efficiency and seeming purposefulness of these living systems were explained by Darwin in 1859. The adaptiveness of these systems is the result of millions of generations of natural selection. This is the mechanistic explanation of adaptiveness, as was clearly stated by Sigwart (1881).

Proximate and ultimate causes must be carefully separated in the discussion of teleological systems (Essay 23). A system is capable of performing teleonomic processes because it was programed to function in this manner. The origin of the program that is responsible for the adaptiveness of the system is an entirely independent matter. It obscures definitions to combine current functioning and history of origin in a single explanation.

THE HEURISTIC NATURE OF TELEONOMIC LANGUAGE

Teleological language has been employed in the past in many different senses, some of them legitimate and some of them not. When the distinctions outlined above are made, the teleological *Fragestellung* is a most powerful tool in biological analysis. Its heuristic value was appreciated by Aristotle and Galen, but neither of them fully understood why this approach is so important. Questions that begin with "What" and "How" are sufficient for explanation in the physical sciences. In the biological sciences no explanation is complete until a third kind of question has been asked: "Why?" It is Darwin's evolutionary theory that necessitates this question: No feature (or behavioral program) of an organism ordinarily evolves unless it is favored by natural selection. It must play a role in the survival or in the reproductive success of its bearer. Given this premise, it is necessary for the completion of causal analysis to ask for any feature, why it exists, that is, what its function and role in the life of the particular organism is.

The philosopher Sigwart (1881) recognized this clearly:

A teleological analysis implies the demand to follow up causations in all directions by which the purpose [of a structure or behavior] is effected. It represents a heuristic principle because when one assumes that each organism is well adapted it requires that we ask about the operation of each individual part and that we determine the meaning of its form, its structure, and its chemical characteristics. At the same time it leads to an explanation of correlated subsidiary consequences which are not necessarily part of the same purpose but which are inevitable by-products of the same goal-directed process.

The method, of course, was used successfully long before Darwin. It was Harvey's question concerning the reason for the existence of valves in the veins that made a major, if not the most important, contribution to his model of the circulation of blood. The observation that during mitosis the chromatic material is arranged in a single linear thread led Roux (1883) to question why such an elaborate process had evolved rather than a simple division of the nucleus into two halves. He concluded that the elaborate process made sense only if the chromatin consisted of an enormous number of qualitatively different small particles and if their equal division could be guaranteed only by lining them up linearly. The genetic analyses of chromosomal inheritance during the next 60 years were, in a sense, only footnotes to Roux's brilliant hypothesis. These cases demonstrate most convincingly the enormous heuristic value of the teleonomic approach. It is no exaggeration to claim that most of the greatest advances in biology were made possible by asking "Why?" questions. They ask for the selective significance of every aspect of the phenotype. The former idea that many if not most characters of organisms are "neutral," that is, that they evolved simply as accidents of evolution, has been refuted again and again by more detailed analysis. It is asking why such structures and behaviors evolved that initiates such analysis. Students of behavior have used this approach in recent years with great success. It has, for example, led to questions concerning the information content of individual vocal and visual displays (Smith 1970; Hinde 1972).

As soon as one accepts the simple conclusion that the totality of the genotype is the result of past selection, and that the phenotype is a product of the genotype (except for the open portions of the program that are filled in during the lifetime of the individual), it becomes one's task to ask about any and every component of the phenotype what its particular functions and selective advantages are.

It is now quite evident why all past efforts to translate teleonomic statements into purely causal ones were such a failure: a crucial portion of the message of a teleological sentence is invariably lost by the translation. Let us take, for instance, the sentence: "The wood thrush migrates in the fall into warmer countries *in order to* escape the inclemency of the weather and the food shortages of the northern climates." If we replace the words "in order to" by "and thereby," we leave the important question unanswered as to *why* the wood thrush migrates. The teleonomic form of the statement implies that the goal-directed migratory activity is governed by a program. By omitting this important message the translated sentence is greatly impoverished as far as information content is concerned, without gaining in causal strength. The majority of modern philosophers are fully aware of this and agree that "cleaned-up" sentences are not equivalent to the teleological sentences from which they were derived (Ayala 1970; Beckner 1969).

One can go one step further. Teleonomic statements have often been maligned as stultifying and obscurantist. This is simply not true.

Actually the nonteleological translation is invariably a meaningless platitude, while it is the teleonomic statement that leads to biologically interesting inquiries.

ARISTOTLE AND TELEOLOGY

No other ancient philosopher has been as badly misunderstood and mishandled by posterity as Aristotle. His interests were primarily those of a biologist and his philosophy is bound to be misunderstood if this fact is ignored. Neither Aristotle nor most of the other ancient philosophers made a sharp distinction between the living world and the inanimate. They saw something like life or soul even in the inorganic world. If one can discern purposiveness and goal direction in the world of organisms, why not regard the order of the Kosmos-as-a-whole also as due to final causes, that is, as due to a built-in teleology? As Ayala (1970) said quite rightly, Aristotle's "error was not that he used teleological explanations in biology, but that he extended the concept of teleology to the non-living world." Unfortunately, it was this latter teleology that was first encountered during the scientific revolution of the sixteenth and seventeenth centuries (and at that in the badly distorted interpretations of the scholastics). This is one of the reasons for the violent rejection of Aristotle by Bacon, Descartes, and their followers.

Although the philosophers of the last 40 years acknowledge quite generally the inspiration that Aristotle derived from the study of living nature, they still express his philosophy in words taken from the vocabulary of Greek dictionaries that are hundreds of years old. The time would seem to have come for the translators and interpreters of Aristotle to use a language appropriate to his thinking, that is, the language of biology, and not that of the sixteenth-century humanists. Delbrück (1971) is entirely right when he insists that it is quite legitimate to employ modern terms like "genetic program" for *eidos* where this helps to elucidate Aristotle's thoughts. One of the reasons why Aristotle has been so consistently misunderstood is that he uses the term *eidos* for his form-giving principle, and everybody took it for granted that he had something in mind similar to Plato's concept of *eidos*. Yet the context of Aristotle's discussions makes it abundantly clear that *his eidos* is something totally different from Plato's *eidos* (I myself did not understand this until recently). Aristotle saw with extraordinary clarity that it makes no more sense to describe living organisms in terms of mere matter than to describe a house as a pile of bricks and mortar. Just as the blueprint used by the builder determines the form of a house, so does the *eidos* (in its Aristotelian definition) give the form to the developing organism, and this *eidos* reflects the terminal *telos* of the full-grown individual. There are numerous discussions in many of Aristotle's works reflecting the same ideas. They can be found in the *Analytika* and in the *Physics* (Book II), and particularly in the *Parts of Animals* and in the *Generation of Animals*. Much of Aristotle's discussion becomes remarkably modern if one inserts modern terms to replace obsolete sixteenth- and seventeenth-century vocabulary. There is, of course, one major difference between Aristotle's interpretation and the modern one. Aristotle could not actually *see* the form-giving principle (which, after all, was not fully understood until 1953) and assumed therefore that it had to be something immaterial. When he said, "Now it may be that the Form *(eidos)* of any living creature is soul, or some part of soul, or something that involves soul" (P. A. 641a 18), it must be remembered that Aristotle's psyche (soul) was something quite different from the conception of soul later developed in Christianity. Indeed, the properties of "soul" were to Aristotle something subject to investigation. Since the modern scientist does not actually "see" the genetic program of DNA either, it is for him just as invisible for all practical purposes as it was for Aristotle. Its existence is inferred, as it was by Aristotle.

As Delbrück (1971) points out correctly, Aristotle's principle of the *eidos* being an "unmoved mover" is one of the greatest conceptual innovations. The physicists were particularly opposed to the existence of such a principle by

> having been blinded for 300 years by the Newtonian view of the world. So much so, that anybody who held that the mover had to be in contact with the moved and talked about an "unmoved mover," collided head on with Newton's dictum: *action equals reaction.* Any statement in conflict with this axiom of Newtonian dynamics could only appear to be muddled nonsense, a leftover from a benighted prescientific past. And yet, "unmoved mover" perfectly describes DNA: it acts, creates form and development, and is not changed in the process. (Delbrück 1971:55)

As I stated above, the existence of teleonomic programs—unmoved movers—is one of the most profound differences between the living and the inanimate world, and it is Aristotle who first postulated such a causation.

KANT AND TELEOLOGY

The denial of conspicuous purposiveness in living organisms and the ascription of their adaptive properties to the blind accidental interplay of forces and matter, so popular in the eighteenth century, was not palatable to discerning philosophers. No one felt this more keenly than Immanuel Kant, who devoted the second half of his *Critique of Judgment* to the problem of teleology. It is rather surprising how completely most students of Kant ignore this work, as if they were embarrassed that the great Kant had devoted so much attention to such a "soft" subject. Yet, as in so many other cases, Kant was more perceptive than his critics. He clearly saw two points: first, that no explanation of nature is complete that cannot account for the seeming purposiveness of much of the development and behavior of living organisms and, second, that the purely mechanical explanations available at his time were quite insufficient to explain teleological phenomena. Unfortunately, he subscribed to the prevailing dogma of his period that the only legitimate explanations are purely mechanical ("Newtonian") ones, which left him without any explanation for all teleological phenomena. He therefore concluded that the true explanation was out of our reach and that the most practical approach to the study of organisms was to deal with them "as if they were designed." Even though he was unable to free himself from the design–designed analogy, he stressed the heuristic value of such an approach: it permits us to make products and processes of nature far more intelligible than trying to express them purely in terms of mechanical laws.

Kant's interest was clearly more in the explanation of "design" (adaptation) than in teleonomic behavior, yet he thought that an explanation of design was beyond the reach of human intellect. Just 69 years before the publication of the *Origin of Species* Kant (1790) wrote as follows:

> It is quite certain that we can never get a sufficient knowledge of organized beings and their inner possibility, much less explain them, according to mere mechanical principles of nature. So certain is it, that we may confidently assert that it is absurd for men to make any such attempt, or to hope that maybe another Newton will some day arrive to make intelligible to us even the production of a blade of grass according to natural laws which no design has ordered. Such insight we must absolutely deny to mankind. (Quoted from McFarland 1970)

Darwin removed the roadblock of design, and modern genetics introduced the concept of the genetic program. Between these two major advances the problem of teleology has now acquired an entirely new face. A comparison of Kant's discussion with our new concepts provides a most informative insight into the role of scientific advances in the formulation of philosophical problems. Equally informative is a comparison of three treatments of Kant's teleology, roughly separated by 50-year intervals: Stadler 1874; Ungerer 1922; and McFarland 1970.

CONCLUSIONS

1. The use of so-called teleological language by biologists is legitimate; it implies neither a rejection of physicochemical explanation nor a noncausal explanation.

2. The terms "teleology" and "teleological" have been applied to highly diverse phenomena. I have made an attempt to group these phenomena into more or less homogeneous classes.

3. It is illegitimate to describe evolutionary processes or trends as goal directed (teleological). Selection rewards past phenomena (mutation, recombination, etc.), but does not plan for the future, at least not in any specific way.

4. Processes (behavior) whose goal directedness is controlled by a program may be referred to as *teleonomic*.

5. Processes that reach an end state caused by natural laws (e.g., gravity, second law of thermodynamics) but not by a program may be designated *teleomatic*.

6. Programs are in part or entirely the product of natural selection.

7. The question of the legitimacy of applying the term "teleological" to stationary functional or adaptive systems requires further analysis.

8. Teleonomic (i.e., programed) behavior occurs only in organisms (and manmade machines) and constitutes a clear-cut difference between the levels of complexity in living and in inanimate nature.

9. Teleonomic explanations are strictly causal and mechanistic. They give no comfort to adherents of vitalistic concepts.

10. The heuristic value of the teleological *Fragestellung* makes it a powerful tool in biological analysis, from the study of the structural configuration of macromolecules up to the study of cooperative behavior in social systems.

REFERENCES

Ayala, F. J. 1970. Teleological explanations in evolutionary biology. *Phil. Sci.,* 37:1–15.

Baer, K. E. von. 1876. Über den Zweck in den Vorgängen der Natur. In *Studien, etc.,* pp. 49–105, 170–234. St. Petersburg.

Beckner, M. 1969. Function and teleology. *Jour. Hist. Biol.,* 2:151–164.

Bergson, H. 1907. *Evolution créative.* Alcan, Paris.

Braithwaite, R. D. 1954. *Scientific explanation.* Cambridge University Press, Cambridge.

Canfield, J. V., ed. 1966. *Purpose in nature.* Prentice-Hall, Englewood Cliffs, N.J.

Craig, W. 1918. Appetites and aversions as constituents of instincts. *Biol. Bull.,* 34:91–107.

Davis, B. D. 1961. The teleonomic significance of biosynthetic control mechanisms. *Cold Spring Harbor Symp. Quant. Biol.,* 26:1–10.

Delbrück, M. 1971. Aristotle-totle-totle. In *Of microbes and life,* ed. J. Monod and E. Borek. Columbia University Press, New York.

Driesch, H. 1909. *Philosophie des Organischen.* Quelle und Meyer, Leipzig.

Hinde, R. A., ed. 1972. *Non-verbal communication.* Cambridge University Press, Cambridge.

Hinde, R. A., and J. G. Stevenson. 1970. Goals and response controls. In *Development and evolution of behavior,* ed. L. R. Aronson et al. W. H. Freeman, San Francisco.

Hull, D. 1973. *Philosophy of biological science.* Prentice-Hall, Englewood Cliffs, N.J. (I consulted the ms. of this book.)

Kant, I. 1790. *Kritik der Urteilskraft,* part 2.

Lagerspetz, K. 1959. Teleological explanations and terms in biology. *Ann. Zool. Soc. Vanamo,* 19:1–73.

Lehman, H. 1965. Functional explanation in biology. *Phil. Sci.,* 32:1–20.

Lovejoy, A. O. 1936. *The great chain of being.* Harvard University Press, Cambridge, Mass.

MacLeod, R. B. 1957. Teleology and theory of human behavior. *Science,* 125:477.

Mainx, F. 1955. Foundations of biology. *Foundations of the Unity of Science,* 1(9):1–86.

McFarland, J. D. 1970. *Kant's concept of teleology.* University of Edinburgh Press, Edinburgh.

Monod, J. 1971. *Chance and necessity.* Knopf, New York.

Munson, R. 1971. Biological adaptation. *Phil. Sci.,* 38:200–215.

Nagel, E. 1961. The structure of teleological explanations. In *The Structure of Science.* Harcourt, Brace and World, New York.

Pittendrigh, C. S. 1958. Adaptation, natural selection, and behavior. In *Behavior and Evolution,* ed. A. Roe and G. G. Simpson. Yale University Press, New Haven.

Raven, C. P. 1960. The formalization of finality. *Folia Biotheoretica,* 5:1–27.

Roe, A., and G. G. Simpson, eds. 1958. *Behavior and evolution.* Yale University Press, New Haven.

Rosenblueth, H., N. Wiener, and J. Bigelow. 1943. Behavior, purpose, and teleology. *Phil. Sci.,* 10:18–24.

Roux, W. 1883. *Über die Bedeutung der Kerntheilungsfiguren. Eine hypothetische Erörterung.* W. Engelmann, Leipzig.

Sigwart, C. 1881. "Der Kampf gegen den Zweck." In *Kleine Schriften,* vol. 2: 24–67. Mohr, Freiburg.

Simpson, G. G. 1949. *The meaning of evolution: a study of the history of life and of its significance for man.* Yale University Press, New Haven.

Smith, W. J. 1969. Messages of vertebrate communication. *Science,* 165:145–150.

Sommerhoff, G. 1950. *Analytical biology.* Oxford University Press, London.

Stadler, H. 1874. *Kant's Teleologie und ihre erkenntnistheoretische Bedeutung.* F. Dümmler.

Taylor, R. 1950. Comments on a mechanistic conception of purposefulness. *Phil. Sci.,* 17:310–317.

Teilhard de Chardin, P. 1955. *Le phénomène humain.* Editions de Seuil, Paris.

Theiler, W. 1925. *Zur Geschichte der Teleologischen Naturbetrachtung bis Aristoteles.* O. Füssli, Zurich and Leipzig.

Ungerer, E. 1922. *Die Teleologie Kants und Ihre Bedeutung für die Logik der Biologie.* Bornträger, Berlin.

Waddington, C. H. 1957. *The strategy of the genes.* Allen and Unwin, London.

—— 1968. *Towards a theoretical biology.* University of Edinburgh Press, Edinburgh.

Weismann, A. 1909. The selection theory. In *Darwin and modern science,* ed. A. C. Seward. Cambridge University Press, Cambridge.

V
THEORY OF SYSTEMATICS

Introduction

The diversity of life is one of the most spectacular aspects of this earth. Its existence and even more so its hierarchical arrangement has supplied what is perhaps the most convincing piece of evidence for evolution. Owing to the uniqueness of all forms of life, generalizations about diversity are possible only by classifying organisms into groups of more or less related forms, and this endeavor raises numerous problems, from the definition of species to the theory of classification. It is the task of the systematist to cope with these problems and this occupation has given rise to the disciplines of taxonomy and systematics. Taxonomy is the theory and practice of classifying organisms. The science of organic diversity in a broader, more biological sense is systematics.

Man classified organisms long before the origin of writing, but it is only in recent decades that the theory of classification has matured and that the full scientific importance of biological diversity has been recognized. Systematics has become an important branch of biology. My own interest in this field goes back to my student days and particularly to the influence of E. Stresemann and B. Rensch. It is, however, only in the last 25 years that I have published extensively in this area. My interests are reflected by the four essays that follow.

27

The Challenge of Diversity

If one wanted to single out the theme that more than any other brings systematists and evolutionists together, one would have to say it is the study of organic diversity, because it poses the same challenging scientific problems to both disciplines. What is more satisfying for a scientist than to be challenged by scientific problems and to work in a field that continually poses new ones?

The idea that all phenomena of nature somehow reflect an underlying order goes back to the ancient Greeks, and as a matter of fact even further back to the old religions and to the mythology of primitive man. The study of this order—and the endeavor to explain it—is one of the tasks of science. But when we look at the explanations for this order advanced by various scientists, we become shockingly aware of the heterogeneity of science. Until a few years ago, when an evolutionist or systematist opened a book on the philosophy of science and read about the basic concepts, methods, and objectives of science, he was bound to be distressed to discover how little all this had to do with his own particular endeavor. The reason for this incongruity is that these books were written either by logicians or by physicists. These authors did not realize what Simpson has pointed out so beautifully, that the physical sciences are a very specialized branch of science. Its ideal is to explain everything under a few general laws and to subordinate all diversity under a limited number of broad-based generalizations. Such an approach is primarily interested in processes and functions and attempts to elucidate these with the help of experiments.

Perhaps the outstanding aspect of the physical sciences is the identity of the entities with which it deals. A sodium atom is a sodium atom no matter where you encounter it and what its chemical history might have been. It always has exactly the same properties. The same is true for the elementary particles, the protons, electrons, mesons, etc., or for the aggregates of atoms, the molecules. It is the sameness of these entities that permits the determination of extremely precise constants

Reprinted from "The challenge of diversity," *Taxon*, 23, no. 1(1974):3–9.

for all the properties of these constituents as well as their inclusion in general laws.

How different is the material of the systematist and evolutionist! Its outstanding characteristic is uniqueness. No two individuals in sexually reproducing populations are the same (not even identical twins), no two populations of the same species, no two species, no two higher taxa, nor any associations of species, such as two forests, two ponds, the biota of two mountains, and so on, *ad infinitum*. Instead of the sameness encountered in the physical sciences, we are dealing with uniqueness throughout the organic world.

Such uniqueness does not mean that generalizations are impossible, but it does mean that generalizations and predictions are achieved in a rather different way. Sometimes they are formulated by determining mean values, and by proceeding afterward as if these mean values correspond to the true constants of the identical entities of physical science. A more valid, and usually far more interesting, approach is to study diversity spectra of classes composed of unique entities.

It is precisely the study of such diversity spectra that is now one of the most active areas of research in evolutionary, environmental, and populational biology. Examples are comparisons of the diversity of tropical and temperate zone biota, of continental and insular biota, of the diversity of fossil biota in different geological periods, or of diversity differences in different ecological niches and adaptive zones.

LEVELS OF DIVERSITY

Diversity is, of course, not altogether absent from the world of the physical sciences. There is the diversity of the elementary particles and there is the diversity of atoms and molecules. Certain aspects of any study of diversity, such as classification and the endeavor to explain evolution, also are encountered in the physical sciences. There is diversity in the kinds of stars and in many of the phenomena studied by geophysics, but such studies of diversity play a rather subordinate role in the total realm of the physical sciences.

The reason for the dominant role of diversity in the living world is that it occurs at so many levels of integration. Let me remind you of some of these levels. For instance, there are at least 10,000 different kinds of macromolecules in a higher organism (some estimates going indeed much higher than that); counting all the different states of repression and derepression there are millions, indeed perhaps billions, of different cells in a higher organism; there are thousands of different organs, glands, tissues, muscles, neurocenters, etc.; no two individuals in any sexually reproducing species are the same, not only because they are genetically different, but also because they differ by age, sex, and by having accumulated different types of information in their open memory programs.

What is particularly fascinating is that one can ask very similar questions at each of these levels of diversity, such as the extent of the diversity, the estimated mean value and variance of the diversity spectrum, the origin of the diversity, its functional role, and its selective significance. As is characteristic for so much in the biological sciences, the answers to many of these questions are of a qualitative, rather than of a quantitative, nature.

Whatever level of diversity one is dealing with, the first step in its study is obviously that of inventory taking. It is the discovery and description of different kinds of species in taxonomy, of different kinds of bones, muscles, glands, and nerves in anatomy, of different kinds of normal and abnormal cells in cytology and pathology, of different kinds of associations and biota in ecology and biogeography, to mention just a few areas of diversity, that form the basis of all subsequent studies. It is obvious that, as science matures, this descriptive basement level becomes less and less popular, and yet it is the foundation on which all else is built. This has been discovered even by the molecular biologist, who is constantly discovering and naming new macromolecules, for instance new enzymes. In systematics this first level in the study of diversity is often referred to as "alpha-taxonomy."

It is at once apparent that an inventory is useless as an information storage system if the information is not classified. Ever since the time of Aristotle and Theophrastus naturalists have been concerned with classification. The history of this endeavor from the medieval herbals through Cesalpino, Tournefort, Linnaeus, Adanson, and Cuvier is utterly involved and most fascinating. The approach of these pioneers was purely static, particularly the classifying procedure of those who proceeded by the method of logical division. Centuries of groping for a more meaningful approach came dramatically to an end in 1859 with the publication of the *Origin of Species.* Not only was the cause of the endless organic diversity explained by Darwin but also the reason for its orderliness, because each taxon, lower as well as higher, is composed of the descendants of a common ancestor.

As I have just told the story it may sound as if systematics had been groping in the dark, finally to be rescued by evolutionary biology. Actually in many ways one might say that the opposite was true. The evolutionary theory was quite literally a by-product, so to speak, of the classifying endeavors of the systematists.

Darwin in his autobiography recounts quite graphically how satisfying the explanatory power of the evolutionary theory was when confronted by puzzling facts of diversity. It answered for him the intriguing question of why the fossil mammals of the Argentinian pampas were so closely related to the living mammals of South America rather than to the fossil faunas of the Old World. It explained why the faunas of the temperate zone of South America were related to those of the tropical Amazonian forests rather than to the faunas of the climatically similar parts of Africa or eastern Asia and Australia.

After 1859 there was never again any doubt about the correctness of the evolutionary explanation of organic diversity. But to apply it to classification remained a difficult task: how are we to know which groups of species are the joint descendants of a common ancestor? It is sometimes forgotten that Darwin actually worked out the guidelines for the making of classifications remarkably well in the brilliant thirteenth chapter of the *Origin*. No systematist or evolutionist can afford not to have read this masterly analysis. Here Darwin clearly points out that a classification must be more than a genealogy, it must also reflect such phylogenetic events as the invasion of new adaptive zones or periods of accelerated divergence, and so on. Darwin even provides detailed instructions as to how one can avoid being misled by convergence or by superficial specializations.

It is by reading this chapter that one understands the true meaning of the old saying that classifying is an art. But what is art? To be sure, a superior classification provides a genuine esthetic pleasure, but the word "art" in the old saying is used in a somewhat different sense. As in the word "artisan" it refers to a craft, to a professional competence that can be acquired only through years of practice.

Such a view is, of course, somewhat unfashionable in this modern day of technology. Two schools of taxonomy have emerged in the last generation as a reaction to what some taxonomists called the subjective and arbitrary approach of the evolutionary method. These are the schools of cladistics and phenetics. Although otherwise diametrically opposed to each other, they both think they have found the ideal of a purely objective method of achieving a "natural" classification (see Essays 29 and 30). But let me return to the main theme.

THE ORIGIN OF DIVERSITY

Perhaps the most challenging of all problems of diversity is that of the origin of *new* diversity. This is a problem which has a different solution for each level of diversity. For instance, the origin of new genotypes, that is, of genetically different individuals, has been extremely well studied by geneticists and has been well explained in terms of mutation, recombination, cross-over inhibition, and so on.

Population geneticists and population systematists have devoted half a century to the solution of the problem of the origin of new populations. This includes the formation of new ecotypes, climatic adaptation through geographic variation, and all the related problems such as gene flow, the role of founder populations, and so forth.

The origin of diversity at the species level was singled out by Darwin for the title of his classic, *On the Origin of Species*. Here also the basic mechanisms are now well understood, but the relative contribution of various factors still remains controversial. In the period of the 1930s to 1950s the main emphasis was on the study of reproductive isolation (see Essays 13–15). Even today there is some mild disagreement on the relative importance in speciation of chromosomal repatterning versus purely genic mechanisms such as Carson has described for homosequential speciation in Hawaiian *Drosophila*.

The second component of speciation, the acquisition of ecological compatibility, has only recently received full attention after generations of neglect. The crucial question here is "How can a new species carve out a new niche for itself?" The importance of this problem, at least for animals, is documented by the frequency of parapatric distributions on the horizontal plane and by strict altitudinal replacements on the vertical scale. The determining factors of species diversity in plant associations are even more difficult to determine, although some good beginnings have now been made here also.

Both systematists and evolutionists are interested in the origin of new diversity not only at the level of the species but also at that of the higher categories. Speciation is not merely a purely quantitative process leading to an ever-increasing number of new species but also a qualitative one. It often leads to invasion of a new niche, occasionally to invasion of a new adaptive zone; in short it may lead to the origin of new types of animals and plants, to the origin of new taxa ranked in the higher categories.

In the post-Darwinian period most of those interested in these problems of macroevolution were anti-Darwinians, like Cope, H. F. Osborn, Schindewolf, and Goldschmidt, to mention some outstanding representatives. The origin of new diversity at the macroevolutionary level was greatly neglected in the first third of this century and particularly by population genetics and by the new systematics. It was G. G. Simpson who showed in 1944 in his *Tempo and Mode in Evolution* that there is no conflict between the findings of genetics and those of macroevolutionary phenomena, and new attention was directed to this field by various other paleontologists, by zoologists like Rensch and myself, and by evolutionary anatomists like Dwight Davis and Walter Bock. In recent years the importance of this aspect of the study of diversity has been particularly emphasized by G. Ledyard Stebbins.

There is now little argument about two aspects of the origin of higher taxa: first, that this process is nothing but an extension of evolution at the species level and, second, that it is always, so to speak, a response to the availability of a previously vacant adaptive zone.

These conclusions still leave many questions unanswered. For instance, are unspecialized or highly specialized types more often successful in making a major adaptive shift? How many successive shifts can a given evolutionary line undertake? Does a successful shift narrow the future potential? Any evolutionist can at once suggest scores of other similar questions. The study of diversity at the macroevolutionary level is clearly still a somewhat neglected field of biology. Before leaving it let me refer to one particularly challenging problem in this area.

One of the great puzzles of diversity is the rapid origin of multicellular plants and of all major types of animals in the late Pre-Cambrian and earliest Cambrian at an enormously accelerated evolutionary rate. This, for a long time, seemed to be a totally refractory problem, but real progress has been made in recent years. First of all, the paleobotanical researches of Barghoorn, Cloud, and W. Schopf have filled numerous gaps in our knowledge of plant evolution, from the first record of prokaryotes (about 3200 million years ago) to the first record of multicellular algae (probably not more than about 800 million years ago). Simultaneously, the researches of the geochemists and geophysicists have reasonably well documented—and nailed down chronologically—the changes in the earth's atmosphere from a reducing to an oxidizing one. Finally, ecological speculations have entered the fray and have suggested how the origin of predatory protozoans could have stimulated the diversity of the food web and led to an enormous acceleration in adaptive radiation.

It would be an exaggeration to say that the solution of the problem of metazoan diversity is in hand, but I think we can claim at least that this is no longer a hopeless problem and that the ultimate solution can already be dimly seen.

DIVERSITY AND THE PROBLEM OF EVOLUTIONARY PROGRESS

For Lamarck and most nineteenth-century evolutionists evolution represented a steady upward progress toward ever-greater perfection. Darwin warned against too easy an acceptance of such a view and it has come increasingly under attack in recent decades, as a reaction against a naive belief in a steady upward evolution. But perhaps this reaction has now gone too far.

If we study the record of life on earth from its very beginning, we do note a progression. To be sure, the evolution of any progression of diversity on earth has not been a movement on a smooth upward ramp. No matter how much speciation might have occurred among bacteria and blue-green algae, the spectrum of organic diversity would have remained rather meager if evolution had not advanced beyond the early prokaryote level. At some time in evolution a major step of cellular reorganization occurred that brought about the distinctive organization of the eukaryotes. Subsequently, a handful of other major advances in biological organization occurred during organic evolution. This includes the widespread adoption of autotrophy, sexuality, multicellularity, and certain more specialized developments like vascularity and angiospermy in plants, and terrestriality and homoiothermy in animals. Each of these evolutionary inventions provides an entirely new platform for the development of evolutionary diversity and each of them represents what Simpson called a "quantum step" in evolution. Interestingly, most of these evolutionary novelties did not lead to an extermination of the previous level of organization.

DIVERSITY AND ECOLOGY

In recent decades diversity has moved more and more into the center of attention of the ecologists. One cannot open an ecological journal or even a journal devoted to evolution such as the *American Naturalist* or *Evolution* without encountering several articles on species diversity, its meaning, and its evolution. Ever since Evelyn Hutchinson's brilliant paper of 1959, "Homage to Santa Rosalia, or Why Are There So Many Kinds of Animals?" there has been a flood of investigations on the relation between latitude, altitude, area, habitat, vegetation, and species diversity. The partitioning of the factors that contribute to diversity is still controversial. How important is the heterogeneity of the habitat or of the substrate as compared with such other factors as the amount and the diversity of food resources, the seasonal stability of the environment, or the presence and diversity of predators, to mention merely a few factors? How different is the relative importance of these factors in terrestrial and aquatic animals, how different for sessile or mobile communities, for microscopic and for large organisms, for animals and for plants? Clearly this is one aspect of diversity in the study of which the law of diminishing returns has not even begun to make itself felt.

CONCLUSIONS

I am sure you will believe me that I could go on and on, pointing out still other fascinating aspects of the study of diversity. Instead I want to call attention to a rather different meaning of diversity. To be sure, the study of diversity is a most important area of *research* in many branches of biology. But it has also a far broader significance. This unique importance consists in the fact that it develops a kind of thinking which is not generated by other branches of science. The study of diversity has perhaps made its most important contribution to the development of new human conceptualizations, to a new approach in philosophy. More than anything else, it is the study of diversity which has undermined essentialism, the most insidious of all philosophies. By emphasizing that each individual is uniquely different from every other one, the students of diversity have focused attention on the importance of the individual, and have developed population thinking, a type of thinking that is of the utmost importance for the interaction of human subgroups, human societies, and human races. By showing that each species is uniquely different from every other species and thus irreplaceable, the student of evolution has taught us a reverence for every single product of evolution, one of the most important components of conservation thinking.

By stressing the importance of the individual, by developing and applying population thinking, by giving us a reverence for the diversity of nature, systematic and evolutionary biology have supplied a dimension to human conceptualization that had been largely ignored, if not denied, by the physical sciences. And yet it is a component that is crucial for the well-being of human society and for any planning of the future of mankind. This is properly emphasized in Stephen Toulmin's great new book *Human Understanding*, where a new philosophy of science is developed on the basis of the principles that emerge from the study of organic diversity.

Considering the far-reaching importance of biological diversity, and acknowledging its stimulating influence on all of biology, I might, with equal justification, have called this paper "In praise of diversity."

28
The Role of Systematics
in Biology

What do we mean by "systematics"? To answer this question meaning-fully requires an excursion into the history as well as the philosophy of biology. The ancient Greeks saw a natural order in the world that, they thought, could be demonstrated and classified by certain logical pro-cedures. They tried to discover the true nature (essence) of things and approached classification through the methods of logic. Indeed, Aris-totle, the first great classifier, was also the father of logic. The philoso-phy of essentialism (see Essay 3) dominated the thinking of taxonomists up to and including Linnaeus. Taxonomic nomenclature and the so-called typological thinking of taxonomists right up to our day have been permanently affected by the Greek legacy.

HISTORY OF TAXONOMY

During the early history of biology, the Aristotelian influence was no great handicap. Botany and zoology, to state it in a highly oversimpli-fied manner, arose during the sixteenth century as applied sciences, at-tached to medicine. Botany began as a broadened study of medicinal herbs, and early botanical gardens were herb gardens. With but one or two exceptions, all the great botanists and herbalists from the sixteenth to the eighteenth century (Linnaeus included) were professors of medi-cine or practicing physicians. Zoology arose in connection with human anatomy and physiology. When botany and zoology became indepen-dent sciences, the first concern of the two fields was to bring order into the diversity of nature. Taxonomy was therefore their dominant con-cern and, indeed, in the eighteenth and early nineteenth centuries botany and zoology were virtually coextensive with taxonomy. More-over, by sheer necessity, taxonomy at that period was essentially the technique of identification.

Reprinted from "Introduction: the role of systematics in biology," pp. 4–15 in *Systematic biology*, publication 1692 of the National Academy of Sciences–National Research Council (Washington, D.C., 1969).

The middle third of the nineteenth century was a period of decisive change to which many separate streams of development contributed. Increasing professionalism was one. Increasing specialization was another, to mention just two. Taxonomy itself accelerated the change by introducing several new concepts into biology. The greatest unifying theory in biology, the theory of evolution, was largely a contribution made by students of diversity, as we might call the taxonomists. It is no coincidence that Darwin wrote his *Origin of Species* after encountering taxonomic problems during the voyage of the *Beagle* and after 8 years of concentrated work on barnacle taxonomy. Comparison of different kinds of organisms is the core of the taxonomic method and leads at the same time to the question of how these differences originated. The findings of explorer taxonomists, paleotaxonomists, and comparative anatomists inexorably led to the establishment and eventual acceptance of the theory of evolution.

One might have expected that the acceptance of evolution would result in a great flowering of taxonomy and enhancement of its prestige during the last third of the nineteenth century. This was not the case—in part for almost purely administrative reasons. The most exciting consequences of the findings of systematics were studied in university departments, while the very necessary but less exciting descriptive taxonomy, based on collections, was assigned to the museums. Furthermore, most taxonomists were satisfied to use evolutionary concepts for rather practical purposes, such as a source of evidence on which to base inferences on classification. As a consequence, evolutionary biology did not contribute as much to strengthening the bridge between taxonomy and other branches of biology as might have been expected. The great contributions to biology made by taxonomists during this period, such as population thinking, the theory of geographic speciation, and the biological species concept, were incorporated into biology anonymously and in such a way that taxonomy did not receive due credit.

Biology is no exception to the well-known phenomenon that there is a continuous change of fashions and frontiers in science. Since the 1870s there has been one breakthrough after another, beginning with the improvements of the microscope and the exciting discoveries of cytology. Perhaps the dominant trend during this period was an increasing interest in biological mechanisms and in the physicochemical explanation of biological functions. This led to the flourishing of various branches of physiology, of endocrinology, of genetics, of embryology, of immunology, of neurophysiology, of biochemistry, and of biophysics. Taxonomy, the oldest, the most classical branch of biology, inevitably suffered in competition with all these brilliant developments. Whenever there was an interesting new growing point in taxonomy, it quickly became independent and left behind a rather descriptive, static, and sometimes almost clerical residue. In the early part of this century taxonomy was regarded by most biologists as an identification service.

Some of the best universities in this country refused to accept Ph.D. theses in taxonomy. The Guggenheim Foundation was the only granting agency that considered taxonomy worthy of support. Under these circumstances it is not surprising that only the most dedicated naturalists chose taxonomy as their life's work, and we must pay tribute here to some inspired teachers who attracted gifted youngsters into the field.

Even today, systematists feel that they are not getting their full share of recognition, of financial support, and of superior graduate students, yet one must recognize that circumstances have changed for the better in the last 20 or 30 years. This change has had many causes, but for some aspects it is not easy to say what is cause and what is effect. Taxonomists played a decisive role in the development of the synthetic theory of evolution, and this fact is being increasingly recognized by the leaders of biology. Julian Huxley and others have emphasized that taxonomy is a vital branch of biology. Simultaneously, we have witnessed a steady improvement in the scientific training of taxonomists. In order to obtain a position it is no longer sufficient that the young taxonomist know how to describe new species; he is now expected to have acquired an adequate training in, and understanding of, genetics, statistics, animal behavior, biochemistry, and other branches of experimental–functional biology. The bridge between museums and universities is being broadened and strengthened in many places, and the strong barriers between a narrowly defined taxonomy and the adjacent branches of biology are being obliterated. This new generation of taxonomists is no longer satisfied to work on preserved specimens. This new breed of naturalist-taxonomists insists on studying taxa as living organisms and pursues its investigations in the field and in the experimental laboratory, wherever effort will be most productive.

The ultimate result of these developments has been general recognition that the universe of the taxonomist is far greater than was previously envisioned. Taxonomists now take an ever-increasing interest in evolutionary, ecological, and behavioral research and, indeed, have assumed leading roles in these fields. Until very recently the terms "taxonomy" and "systematics" were generally considered to be synonymous. In view of current developments, it seems advantageous to restrict the term "taxonomy" to the theory and practice of classifying, more narrowly defined, and to use the term "systematics" for the study of organic diversity, more broadly defined. This new viewpoint is represented by Simpson's definition (1961): "Systematics is the scientific study of the kinds and diversity of organisms and of any and all relationships among them." In short, *systematics is the study of the diversity of organisms.*

THE POSITION OF SYSTEMATICS IN BIOLOGY

When we look at biology as a whole we see that systematics occupies a unique position. In 1961 (Essay 23) I pointed out that there are basically two biologies. One deals with functional phenomena and investigates the causality of biological functions and processes; the other, evolutionary biology, deals with the historical causality of the existing organic world. Functional biology takes much of its technique and *Fragestellung* from physics and chemistry, and is happiest when it can reduce observed biological phenomena to physicochemical processes. Evolutionary biology, dealing with highly complex systems operated by historically evolved genetic programs, must pursue a very different strategy of research in order to provide explanations. Its most productive method is the comparative method, for which the taxonomists have laid the foundation. Indeed, I can hardly think of an evolutionary problem that has not developed out of some finding of taxonomy.

One can express these basic concerns also in a somewhat different manner. At one extreme, biology is preoccupied with the ultimate building stones and ultimate unit processes that are the common denominators in living organisms. This has largely been the concern of molecular biology, from the structure of macromolecules to such functional unit processes as the Krebs cycle. As legitimate as the reductionist methodology is when applied to functional problems, it quickly carries us down to a level where we leave behind most of what is most typically biological, and we are left with a subject matter that is essentially physicochemical. This is surely true for the chemistry and physics of the ultimate building stones and unit processes of living organisms. If this were the only level of integration in biology, it would be quite legitimate to combine biology with chemistry or physics.

At the other extreme is preoccupation with the level of biology that deals with whole organisms, with uniqueness, and with systems. It is a matter of historical record that taxonomists are among those biologists who have been most consistently concerned with whole organisms and who have most consistently stressed the organismic, the systems approach to biology.

No one will question the immense importance of molecular phenomena, but they are not the only aspect of biology. As Michael Ghiselin has stated so perceptively, just as architecture is more than the study of building materials, so biology is more than the study of macromolecules. In systematics, in evolutionary biology, and in much of organismic biology, one normally deals with hierarchical levels of biological integration that are many orders of magnitude above the molecular. Each level has its specific problems and its appropriate methods and techniques. That there is such a difference in levels of integration is taken completely for granted in the physical sciences. No one would expect the aeronautical engineer to base the design of airplane wings on the study of elementary particles. But a unique role for each level is even more evident at the different levels of biological integration. Lest I be misunderstood, I see no conflict between molecular biology and organismic biology (including systematics), but it must be emphasized that each level of integration poses its own specific problems, requires its own methods and techniques, and develops its own theoretical framework and generalizations.

The role of systematics should now be quite clear: It is one of the cornerstones of all biology. It is the branch of biology that produces most of our information on the levels of integration designated as natural populations and higher taxa. It supplies urgently needed facts and, more important, it cultivates a way of thinking—a way of approaching biological problems—that is alien to the reductionist but is of great importance for the balance and well-being of biology as a whole.

THE CONTRIBUTIONS OF SYSTEMATICS TO BIOLOGY

Many biologists do not appreciate the magnitude of the contributions made by systematics. And yet these achievements are extraordinary, even if we adopt the narrowest definition of taxonomy. They include the description of about 1 million species of animals and a half million species of higher and lower plants, as well as their arrangement in a system. This classification, though we continue to modify it in detail, is, on the whole, amazingly logical, internally consistent, and stable. It is an immensely useful system of information storage and retrieval. All the comparative work of morphologists, physiologists, and phylogenetically inclined molecular biologists would be meaningless if it were not for the classification.

Taxonomists supply a desperately needed identification service for taxa of ecological significance and for the determination of fossil species needed for work in stratigraphy and geological chronology. In all areas of applied biology good taxonomy is indispensable: in public health, in the study of vector-borne diseases and of parasites; in the study of the relatives of cultivated plants and of domestic animals; and in the study of insect pests and of their biological control. Much work in conservation, wildlife management, and the study of renewable natural resources depends for its effectiveness on the soundness of taxonomic research. The faunas, floras, handbooks, and manuals prepared by taxonomists are indispensable in many branches of biology and are widely used by the general public.

As important as these descriptive and service functions of taxonomy are, they are only part of the contribution of systematics, to many of us the least important part. I have already pointed out that evolutionary biology was founded on the work of taxonomists. They also supplied the solutions of many individual evolutionary problems, including the role of isolation, the mechanism of speciation, the nature of isolating mechanisms, rates of evolution, trends of evolution, and the problem of the emergence of evolutionary novelties. Taxonomists (including paleontologists), more than any other kind of biologists, have made significant contributions to all these subjects.

There is hardly a taxonomic operation during which the systematist does not have to face basic biological questions. In order to assign specimens to species he must study variability, particularly poly-morphism, and quite often he has to undertake a rather complete population analysis, including the study of life cycles. In the study of polytypic species he concerns himself with geographic variation and its meaning, he studies the adaptation of local populations, and he tests the validity of climatic rules. When studying the population structure of species, he examines isolates and belts of hybridization. Indeed, tax-onomists have developed in the last two generations a veritable "science of the species," in the same sense in which cytology is the science of the cell and histology is the science of tissues. At every step the sys-tematist must think about the adaptation of populations, their past histories, and the magnitude of dispersal (gene exchange between populations).

Many new concepts that arose out of the work of taxonomists have since diffused broadly into genetics, ecology, physiology, and other areas of biology. By far the most important of these, as I have often stressed in the past, is population thinking. Biology, like all other sci-ences, was permeated by typological thinking until late in the nine-teenth century, and still is to a certain extent. When the learning psy-chologist speaks of "The Rat" or "The Monkey," or when the racist speaks of "The Negro," these are instances of typological thinking. The early Mendelians were pure typologists. A mutation changed "The Wildtype," and the result was a new type of organism—according to De Vries, a new species. I have pointed out elsewhere (Mayr 1963) that taxonomists began as early as the 1840s and 1850s to collect large series of individuals—population samples as we would now say—and to describe the variation of these samples. From this purely pragmatic operation eventually emerged a wholly new way of thinking that re-placed typological essentialism. From taxonomy, population thinking spread into adjacent fields and was in part instrumental for the devel-opment of population genetics and population cytology. Population thinking now has spread into the behavior field—into physiology and ecology. This one conceptual contribution has been of such great benefit to vast areas of biology that it alone justifies support for systematics.

As the interests of the systematists broaden, it is becoming more and more true that systematics has become, as stated by Julian Huxley (1940), "one of the focal points of biology." Although he may not be able to solve these problems himself, it is the systematist who frequent-ly poses the problems that are of concern to the population geneticist, the physiologist, the embryologist, and the ecologist. For instance, systematics poses the problems in the area of ecology that deals with the phenomena of species diversity, the differences in the richness of faunas and floras in different climatic zones and habitats, and so on. A succession of prominent taxonomists have led the study of species competition, niche utilization, and structure of ecosystems.

Environmental physiology owes much to systematics. Zoological systematists like C. L. Gloger, J. A. Allen, and Bernhard Rensch have made major contributions to the discovery of adaptive geographic variation and the establishment of climatic rules. Up to the 1920s it was almost universally believed that geographic differences among populations of a species were nongenetic modifications of the phenotype and were of no evolutionary interest. As it was then stated, "the type of the species is not affected." It was zoologists with taxonomic competence who demonstrated the genetic basis for adaptive differences between geographic races. The stress on unique characteristics of individuals, the recognition of differences between populations, and the emphasis on the phenotype as a compromise between multiple selection pressures all represent thinking that came directly from evolutionary systematics, and that has exercised and is continuing to exercise a profound influence on environmental physiology.

Taxonomic principles as applied to the interpretation of man's evolution by Simpson, LeGros Clark, Mayr, and Simons have decisively added to our understanding of man's evolution and of hominid classification. The chaos of 29 generic names and more than 100 specific names caused by the earlier typological approach was replaced by a biologically oriented classification in which three genera, *Paranthropus, Australopithecus,* and *Homo* (the latter with two or three species) are recognized.

Whole branches of biology could not exist without systematics. Biological oceanography is one example, biogeography another. The latter field has been so traditionally the domain of the taxonomists that it is unnecessary to stress the contribution of systematics. Cytogenetics and bioacoustics are other areas of biology that derive much of their inspiration from systematics. Systematists have contributed enormously to ethology through their studies of comparative behavior, particularly of insects.

These contributions must be stressed for two reasons. One is that those persons who have come into biology from the outside (for example, from physics or chemistry) simply do not know this aspect of its history. The other is that there has been a tendency, even among those who know the situation, to give credit to the neighboring fields—population genetics, ecology, or ethology—even when the advances were made by practicing taxonomists and were made possible only by the experience they had gained as taxonomists. It is totally misleading to limit the labels "taxonomy" or "systematics" to purely clerical, descriptive operations and to give a different label to the broader findings and concepts that are the direct result of more elementary operations. Regrettably, even some taxonomists have supported the myth that all the more biologically interesting activities and findings of the taxonomist are not a part of taxonomy. In this connection it will be of some importance, in order to clarify the situation, to add a few words on the structure of systematics.

THE STRUCTURE OF SYSTEMATICS

Earlier I described how systematics, as we now understand it, emerged from essentialism and nominalism (by rejecting these concepts) and came to be based on the fact of evolution. Systematists began to study organic diversity as the product of evolution and to recognize that every classification is a scientific theory with the properties of any scientific theory: that is, it is explanatory, because it explains the existence of natural groups as the products of common descent, and it is predictive, because it can make highly accurate predictions about the pattern of variation of unstudied features of organisms and the placing of newly discovered species. Finally, systematics established many new contacts with other areas of biology by adopting the thesis that the characteristics of the living organism are as important (or more so) for classification as those of preserved specimens.

How did these profound changes in the science of systematics affect its working procedure? In some ways not at all, because the needs for sound classification have not changed. There is still the same need to order the diversity of nature into its elementary units—the biological species. Sorting variable individuals and populations into species (and naming and describing them) is sometimes referred to as alpha taxonomy. There is still need for some alpha taxonomy, even in as mature a branch of systematics as bird taxonomy. New species, new subspecies, and all sorts of new taxa of birds are still being found. We still discover occasionally that a species given in the literature is nothing but a variant of another species. In ornithology we still are in need of compilations, checklists, and descriptive works of various sorts that fall under the designation of alpha or beta taxonomy. Yet even in these relatively elementary procedures of taxonomy there is a drastic difference between doing them in a typological (essentialist) fashion or in a biological-evolutionary fashion.

The typologist acts as if he were dealing with the "essential natures" of created types. He stresses morphotypes and discontinuities; variation is treated as a necessary evil to be ignored as much as possible. The biological systematist knows that he is dealing with samples of variable natural populations, and he is interested in the biological meaning of this variation. He knows that he is dealing with living organisms and wants to study all their attributes, whether they concern morphology, behavior, ecology, or biochemistry.

An understanding of the biological meaning of variation and of the evolutionary origin of groups of related species is even more important for the second stage of taxonomic activity—the sorting of species into groups of relatives (taxa) and their arrangement in a hierarchy of higher categories. This activity is what the term "classification" denotes; it is also referred to as beta taxonomy. No matter how interested a taxonomist is in the evolutionary and ecological aspects of the taxa he studies, he must also devote a major share of his time to alpha and beta taxonomy, not only because so much work still remains to be done, but also because the more interesting biological problems are found only through research in alpha and beta taxonomy.

THE FUTURE OF SYSTEMATICS

I would be rather pessimistic about the future of taxonomy if it were only an identification service for other branches of biology, as some with little imagination think. But anyone who realizes that systematics opens one of the most important doors toward understanding life in all its diversity cannot help but feel optimistic. Environmental biology, behavioral biology, and even molecular biology are all paying increasing attention to organic diversity. The most exciting aspect of biology is that, by contrast with physics and chemistry, it is not possible to reduce all phenomena to a few general laws. Nothing is as typically biological as the never-ending variety of solutions found by organisms to cope with similar challenges of the environment. Nothing is more intriguing than the study of differences between related organisms and the challenge to explain these differences as the result of natural selection. Even in cases where the ultimate solution may come from genetics or biochemistry, it is usually the systematist who poses the challenging questions. The opportunities for exciting research are virtually unlimited. This is becoming clearer and more widely appreciated every year.

These opportunities are not without obligations. Let us remember at all times that every taxonomist is a spokesman for systematics. He must carry out his activities in such a way as to reflect favorably on his field. Let us remember that taxonomy is not a kind of stamp collecting but a branch of biology. Let systematists desist from all practices that are injurious to the prestige of systematics, as for instance indulging in nomenclatural practices that lower the value of scientific nomenclature as an information storage and retrieval system. Finally, let us remember that in virtually every taxonomic finding certain generalizations that are of value and broad interest to biology as a whole are implicit. Systematists should make these findings known. They are bound to have impact well beyond the limits of systematics.

It is my sincere belief that systematics is one of the most important and indispensable, one of the most active and exciting, and one of the most rewarding branches of biological science. I know of no other subject that teaches us more about the world we live in.

REFERENCES

Huxley, J. A. 1940. *The new systematics.* Clarendon Press, Oxford.

Mayr, E. 1963. *Animal species and evolution.* Belknap Press of Harvard University Press, Cambridge, Mass.

—— 1969. *Principles of systematic zoology.* McGraw-Hill, New York.

Simpson, G. G. 1961. *Principles of animal taxonomy.* Columbia University Press, New York.

Theory of Biological Classification

Taxonomy experienced a vigorous rejuvenation during the 1930s and 1940s. In this period the analysis of the population structure of species was emphasized, and the resulting "population systematics" made important contributions to the shaping of the synthetic theory of evolution (Mayr 1942). The new systematics, as this movement has also been called (Huxley 1940), rather neglected the consideration of the meaning of higher taxa and categories and of the whole theory of classification. Even though individual authors had long shown an interest in the theory of classification (Simpson 1945; Rensch 1945), it was not until the publications of Hennig (1950), Remane (1952), Cain (1958, 1959), Sokal and Sneath (1963), Simpson (1961), and some other authors in the 1950s and 1960s that a vigorous revival of interest in macrotaxonomy occurred.

Before describing the nature of these new developments and the controversies that emerged from them, it is necessary to stress one basic consideration. Theory formation in taxonomy, as elsewhere in biology, has been greatly handicapped by endeavors to squeeze biological concepts into the straitjacket of concepts and theories developed by logicians and philosophers of physics. Many of the generalizations derived from the facts of the physical sciences are irrelevant when applied to biology. More important, many phenomena and findings of the biological sciences have no equivalent in the physical sciences and are therefore omitted from philosophies of science that are based on physics (see Essay 23). It is not only the enormous complexity of biological systems which requires concepts that have no analog in the physical sciences but, in particular, the fact that organisms contain a historically evolved genetic program in which the results of 3 billion years of natural selection are incorporated. The uniqueness of almost all biological situations above the level of molecules (for example, individuals, populations, species, and the like) is another phenomenon requiring an emphasis in biology rather different from that customary and appropriate in the physical sciences.

Revised from "Theory of biological classification," *Nature,* 220, no. 5167 (1968):545–548.

This is not the place to analyze this difference in detail, but it has to be mentioned, because many of the terms used in the formation of theories in biology (for example, relationship, species, classification, population, and the like) have also been used (but often with an entirely different meaning) in the formation of theories in nonbiological sciences. As long as biologists attempted to use the definitions of these terms that had been customary in physical science, and as long as they stayed strictly within the framework of theory that was appropriate for physical science, it was impossible to accommodate the special demands of the biological situation. There was no hope for the development of a sound theory of classification until taxonomists abandoned the erroneous notion that they had to find a theory of taxonomy that was equally applicable to artifacts and to evolved organisms. Emancipation from the definitions and concepts of the physical sciences began in 1859 and has greatly accelerated in recent years. The development of the biological species concept is one of the manifestations of this emancipation, and the evolutionary theory of biological classification, which is utterly different in underlying philosophy from any theory of classification prevalent before 1859, is another.

What, then, is biological classification? Unhappily, no agreement on the answer to this question exists yet among biologists. Indeed, quite aside from the philosophical disagreements, the word *classification* is actually used traditionally in two different senses. It denotes, first, the act of classifying, that is the ordering of organisms into groups, and, second, the finished product of the classifying activity, like the classification of mammals, of butterflies, or of any other group of organisms. Fortunately, this terminological ambiguity is not troublesome for my discussions because the main points I plan to discuss are equally pertinent to the activity of classifying and the product of such activity. On the whole, when I speak of classification, I refer to the act of classifying.

The difference between the classification of organisms and of inanimate objects is not total. There are some basic principles which apply equally well to the classification of inanimate objects and of organisms. One of these is that objects placed in a given class in a classification must have certain attributes in common. The other is the principle of hierarchy, that is, that classes of objects, in turn, can be combined into larger classes. In a library, the categories American History, French History, Modern History, Ancient History, and the like, can be combined into the higher category History, just as the taxa Diptera, Coleoptera, and Hymenoptera can be combined with others into the higher taxon Insecta.

But this is about as far as the congruence goes between biological classifications and classifications of artifacts. There are certain principles relating to the classification of artifacts that are simply inapplicable to organisms and vice versa. For instance, two or three entirely independent classifying criteria may be used for the classification of books in a single library. For instance, in the library of my own institution, the following four classifying principles are used: (1) year of publication, (2) subject matter, (3) size, (4) author's name. Each of these criteria is purely arbitrary and utilitarian. All utilitarian classifications try to reduce the heterogeneity of classes of objects and to break them down into manageable parcels. Purely arbitrary-utilitarian classifications exist also in biology, but only for very special purposes. A limnologist, for example, may divide his faunas into such categories as freshwater, brackish water, and marine.

These are special cases. In general, when a zoologist speaks of animal classification or a botanist of plant classification, he has something very different in mind, something for which the basis has been provided by nature herself, something not arbitrarily made by the taxonomist. There are very few equivalents in nature to this nonarbitrary biological classification, the periodic table of the elements and the system of elementary particles being among the few conspicuous exceptions. Darwin was the first to stress the difference between arbitrary and biological classifications. The decisive point is that penguins, bats, or beetles are not groups arbitrarily made by the operations of our mind, like the categories in a library, but are groups produced by evolution because the members of the groups are descendants of a common ancestor. The crucial point, as so convincingly demonstrated by Simpson (1961), is that we do not base the taxon on the similarity of the included species, but the included species are similar to each other because they are descendants from a common ancestor. The causal relationship corresponds to that of the similarity of monozygous twins. Two individuals are not twins because they are very similar, but they are so similar because a single zygote gave rise to both of them. To understand the causal connection between similarity and taxon delimitation is of the utmost importance for a sound theory of classification. I shall come back to this again and again in my subsequent discussion, particularly in the analysis of the reasoning of the nominalists and their contemporary representatives, the pheneticists.

There is one other basic difference between purely utilitarian classifications of inanimate objects and a biological classification. Every biological classification is a scientific theory (Mayr 1969). Classifications have the same properties as all theories in science. A given classification is explanatory: it asserts that a group of organisms grouped together consists of descendants of a common ancestor. A good classification, like a good scientific theory, has a high predictive power with respect to the assignment of newly discovered species and the pattern of variation of previously unused characters. That classification is the best which is least affected by such new discoveries. Like all theories, classifications are provisional and may have to be modified in the light of new discoveries.

We are now ready to attempt a definition of the term "classification," a definition that can serve as the basis of further analysis. Most taxonomists define it approximately this way: "Classification is the ordering of species into groups and their ranking on the basis of similarity and relationship." Disagreement on the interpretation of the terms "similarity" and "relationship" has been the reason for continuing controversies and has given rise to the various principal theories of biological classification (Mayr 1969).

Taxonomists have been, traditionally, quite inarticulate when asked to explain their theory of classification. They would say that they attempt to establish "natural groups" or groups showing "natural affinity." For the rest they left it to logicians and other philosophers to work out the principles of taxonomy. This situation, however, has changed decisively within the past 20 years. Owing to the publications of Hennig, Remane, Cain, Simpson, and Sokal and Sneath, there is now a lively interest in this area and most taxonomists give at least some thought to the theory of classifying. As a result, the pages of our journals are filled with controversy. Looking over this entire field, it seems to me that the many conflicting endeavors of various taxonomists can be assigned to five basic theories of taxonomy. Three of these go back to the pre-Darwinian period, while two are based on the concept of evolution. The five theories which I distinguish are essentialism, nominalism, empiricism, cladistics, and evolutionary classification.

ESSENTIALISM

Essentialism considers it the task of pure knowledge to discover the hidden nature (or form, or essence) of things. When applied to organic diversity, it believes that all members of a taxon share the same essential nature; they conform to the same type. This is why essentialist ideology is also referred to as typology. Classification of organic diversity for the essentialists consists in assigning the variability of nature to a fixed number of basic types at various levels. Variation is considered a trivial and irrelevant phenomenon. Essentialism has a long history in philosophy, and it dominated biological classification from Aristotle to Linnaeus (Cain 1958,1959).

The fatal flaw of essentialism is that there is no way of determining what the essential properties of an organism are and why these and no other properties of an organism are essential (Simpson 1961; Hull 1965). It is simply not true that groups of things share an underlying essence and that there is a fixed number of discrete, sharply separated essences. This philosophy might have seemed defensible as long as the creationist dogma prevailed, but to the modern biologist it is a strangely inappropriate concept. Adherence to essentialism led to great arbitrariness in taxonomy because there are no objective or nonarbitrary methods of determining what the essential natures of organisms are. This is particularly well illustrated by the complete difference in the choice of the classifying criteria adopted by the various pre-Linnaean plant taxonomists adhering to essentialist principles. Nothing further needs to be said about the essentialist theory of classification because it now has virtually no adherents.

NOMINALISM

According to nominalism only individuals exist. All groupings, all classes, are artifacts of the human mind since universals do not exist. Such things as birds or snakes are not real but are only names invented by man and arbitrarily attached by him to groups of individuals considered to be similar. I have already pointed out the fallacy of applying to organisms a philosophy which has much merit when one deals with inanimate objects but which is inappropriate for evolved organisms. The basic fallacy of the nominalists is their misinterpretation of the causal relationship between similarity and relationship. Nominalism arose in opposition to essentialism. It played a considerable part during the eighteenth century among the French opponents of Linnaeus, but faded out toward the end of that century.

It would seem appropriate, at this point, to discuss a modern school of taxonomy whose operations are based on nominalist principles. This is the school of the pheneticists, one of the branches of numerical taxonomy (Sokal and Sneath 1963). They deliberately set out to "make" taxa on the basis of calculated overall similarity. This procedure usually leads to a classification which is not very different from one based on the evolutionary approach. The reason for this is that, usually, two organisms will be the more similar the more closely they are related by descent. The phenetic approach, so far as it adopts the nominalist philosophy, has, however, a number of serious weaknesses, which have been pointed out by Simpson (1961) and many other recent authors (Mayr 1969). The greatest is, of course, the demonstrably false claim that groups in nature are the product of the human mind (or of the computer!) rather than of evolution. That such groups are the product of evolution is self-evident for all well-defined natural taxa and particularly for species, with their genetically programed isolating mechanisms that safeguard their reproductive isolation. They are not an arbitrary, subjective, man-made phenomenon.

The phenetic approach has been most useful when applied to groups with immature classifications, particularly single-character classifications, and to those with numerous nonredundant characters. In such taxa it has produced some groupings that are clearly superior to traditional ones. It faces its greatest difficulties in the actual ranking procedure, because even slight changes in the method of calculation may result in rather different phenograms. To summarize an extremely large literature, I might say that a purely phenetic approach, disregarding all considerations of phyletic weighting, has been largely a failure when applied to higher organisms. One can expect, however, that computer methods such as have been pioneered by the pheneticists will become increasingly important in taxonomy when combined with the philosophy of evolutionary taxonomy and with the proper weighting of characters.

EMPIRICISM

The hundred years between the tenth edition of the *Systema Naturae* (1758) and the publication of Darwin's *Origin of Species* (1859) was a period of transition. Deductive principles, whether based on essentialism or nominalism, were increasingly rejected, and taxonomists to an increasing extent based their taxa on the totality of characters. This approach was started by Adanson, but a strongly empirical philosophy characterized virtually all the leading taxonomists of that period. The term "natural" acquired a new meaning during this period, signifying a classification unbiased by *a priori* considerations and based on a consideration of totality of characteristics. By their uncommitted labors, the empiricists prepared the way for coming developments. If there was anything wrong with the approach of the empiricists, it was that they could not supply a reason for the existence of natural groups. This was, of course, precisely the deficiency that was repaired by the theory of evolution.

All pre-Darwinian theories of biological classification lost their meaning in 1859. If I have discussed them at all, it is because so much in our taxonomic ritual, such as binominal nomenclature, goes back to these earlier theories. Also, as in the case of phenetics, the older theories are sometimes revived under a new guise. One might think that the arrival of the theory of evolution would have terminated all further arguments, but this is not the case. The principal current argument is between two different theories of classification, both claiming to be the true and only theory of evolutionary or phylogenetic classification. Let me take them up in the reverse chronological order.

CLADISTICS

Hennig, who is the most articulate spokesman for cladistics, insists that all organisms be ranked and classified exclusively according to "recency of common descent" (1950). Categorical status, according to this theory of classification, depends on the position of the branching points on the phylogenetic tree. The merits and shortcomings of this approach are discussed in Essay 30.

EVOLUTIONARY CLASSIFICATION

The theory of evolutionary classification, first proposed by Darwin, delimits taxa on the basis of two considerations—common ancestry and subsequent divergence. Its method is to infer relationship on the basis of an *a posteriori* weighting of similarity. It would be going too far to discuss such methods of weighting (Mayr 1969; 217–228), but they are essentially those which the great masters of taxonomy have practiced for more than a hundred years. To me it seems that the evolutionary approach combines the best features of the phenetic and of the cladistic approaches. By not being committed to any one-sided dogmas, such as that all characters have equal weight or that there is only one process in evolution (the splitting of branches), it is able to evaluate all available evidence and arrive at balanced conclusions. This is tremendously important for another reason.

Evolutionary taxonomists are sometimes falsely accused of believing that the prime and ultimate aim of taxonomy is to discover and embody the phylogenetic tree in one single system of biological classification. As I have said earlier, the idea of one ideal natural system is a phantom. Even if we had a perfect understanding of phylogeny, it would be possible to convert it into many different classifications. According to the evolutionary taxonomists, taxa have to meet only one qualification: they must consist of clusters of species inferred to be more closely related to each other, that is, to be genetically more similar to each other, than to species of other clusters. This permits innumerable adjustments in the delimitation of taxa and their ranking, in order to facilitate information retrieval.

Two authors, one of whom is a splitter, the other a lumper, will produce two entirely different classifications both based on the same evolutionary principles. This shows that the grouping together of descendants of common ancestors still leaves an enormous number of degrees of taxonomic freedom and that among these there surely are some that best serve the practical needs of a classification. I have reread recently what Bather (1927) said in criticism of evolutionary classifications. I was rather astonished to discover, contrary to my recollections, that he did not criticize evolutionary classification at all, and that what he did criticize were two bad practices, one an excessive amount of splitting and the other one, quite interestingly, the cladistic approach, which, however, he erroneously referred to as the phylogenetic approach.

It cannot be emphasized too strongly that the object of an evolutionary classification is precisely that of any good classification, that is, to store information in the most reliable manner and to permit retrieval of this information most efficiently. The evolutionary taxonomist is convinced that a classification based on the evolutionary principles stated here is indeed more practical than any classification based on one-sided considerations.

If we study the current literature, we find that it reflects a tremendous shift of interest from that which prevailed in the 1930s. Population taxonomy has been supplemented by what one might call macro-taxonomy, a taxonomy that is chiefly concerned with the theory and practice of classifying higher taxa. Taxonomists, up to now, have proceeded almost entirely on empirical principles. The emergence of a vigorous interest in the theory of classification is one of the most gratifying developments of the past 20 years (Takhtajan 1968). Now that a far better understanding exists of what a sound theory of classification ought to be, one can hope that future progress will be not only more rapid but also surer. Empiricism alone is not enough; a healthy advance of taxonomy depends on a sound theoretical foundation.

REFERENCES

Bather, F. A. 1927. Biological classification past and future. *Quart. Jour. Geol. Soc.,* 83:58–64.

Cain, A. J. 1958. Logic and memory in Linnaeus's system of taxonomy. *Proc. Linn. Soc. London,* 169:144–163.

Cain, A. J. 1959. The post-Linnaean development of taxonomy. *Proc. Linn. Soc. London,* 170:234–243.

Hennig, W. 1950. *Grundzüge einer Theorie der Phylogenetischen Systematik.* Deutscher Zentralverlag, Berlin.

Hull, D. 1965. The effect of essentialism on taxonomy. *Brit. Jour. Phil. Sci.,* 15: 314–326, 16:1–18.

Huxley, J., ed. 1940. *The new systematics.* Clarendon Press, Oxford.

Mayr, E. 1942. *Systematics and the origin of species.* Columbia University Press, New York.

—— 1969. *Principles of systematic zoology.* McGraw-Hill, New York.

Remane, A. 1952. *Die Grundlagen des natürlichen Systems, der vergleichenden Anatomie und der Phylogenetik.* Geest und Portig, Leipzig.

Rensch, B. 1947. *Neuere Probleme der Abstammungslehre,* 1st ed. Enke, Stuttgart.

Simpson, G. G. 1945. The principles of classification and a classification of mammals. *Bull. Amer. Mus. Nat. Hist.,* 85:1–339.

—— 1961. *Principles of animal taxonomy.* Columbia University Press, New York.

Sokal, R. R., and P. H. Sneath. 1963. *Principles of numerical taxonomy.* Freeman Press, San Francisco.

Takhtajan, A. 1968. Classification and phylogeny, with special reference to the flowering plants. *Proc. Linn. Soc. London,* 179:221–227.

30
Cladistic Analysis or
Cladistic Classification?

*Ein besonderes Anliegen ist es mir, zu betonen, dass
die kritische Auseinandersetzung mit einem Autor
dessen Verdienste nicht herabsetzen möchte. Das
Gegenteil ist richtig. An unwesentlichen Arbeiten
lohnt es sich nicht, Kritik zu üben.*

Hennig 1969

The choice of method in a scientific discipline depends to a large extent
on the objectives of that discipline. If one wants to determine which of
several methods of classifying animals and plants is most productive,
one must first clarify one's concept of systematics. Quite rightly, therefore, Hennig begins his *Grundzüge einer Theorie der phylogenetischen
Systematik* (1950) with a discussion of the concept of systematics (pp.
1–12). Systematics, he says, is the ordering of the diversity of nature
through constructing a classification which can serve as a general reference system. "Creating such a general reference system, and investigating the relations that extend from it to all other possible and necessary systems in biology, is the task of systematics" (Hennig 1966:7,
1950:10).

The task of the creator of classifications, thus, is to find the best
possible "general reference system." However, one can and should be
more specific: a classification, in contradistinction to an identification
scheme, functions as a biological theory (with all the explanatory, predictive, and heuristic properties of a theory) (Mayr 1969:79–80); it
must provide a sound foundation for all comparative studies in biology,
and it must be able to serve as an efficient information storage and
retrieval system (Mayr 1969:229–244).

Generalizations in large parts of biology are derived from comparisons. Comparisons of groups, however, are meaningful in evolutionary studies only when such groups are correctly formed, that is,
consist of "related" elements. The construction of efficient classifications is thus, as stated by Hennig, a prerequisite for sound work in large
parts of biology. Warburton (1967) has attempted to specify the criteria on which some classifications can be judged to be superior to
others in fulfilling the demand to serve as "general reference systems,"
as sound biological theories, and as efficient information storage and
retrieval systems.

Reprinted from "Cladistic analysis or cladistic classification?" *Zool. Syst. Evol.-forsch.*
12 (1974):94–128; reprinted by permission of Verlag Paul Parey, Hamburg.

The 1930s and 1940s were dominated by the so-called new systematics. Taxonomists concentrated their attention on the level of species and populations (*microtaxonomy*), which was also the area of principal concern of the newly emerging field of population genetics. The problems relating to the classification of higher taxa (*macrotaxonomy*) were largely neglected. There was, however, a significant minority of workers, particularly among the paleontologists and comparative anatomists, who felt that the seemingly simple Darwinian credo that classifications should reflect "relationship" or "common descent" raised many unanswered questions. This is evident from several contributions to the volumes edited by Huxley (1940), Heberer (1943), and Jepsen, Mayr, and Simpson (1949), and more specifically from the writings of Simpson (1945) and Rensch (1947). The intellectual ferment of this period led to the formulation of three competing theories of classification during the 1950s and 1960s, each of them claiming to be more objective and a better general reference system than the other two. These three theories, referred to by Günther (1971:76) and described in more detail by Mayr (1969:68–77) will now be characterized. (Unfortunately, a considerable number of taxonomists have hardly any theory at all and deal with species and higher taxa purely descriptively, regarding classifications simply as identification systems.)

THE THREE CURRENT THEORIES OF CLASSIFICATION

Phenetic Systematics (Phenetics)

Organisms are classified, according to phenetics, on the basis of "overall similarity." Similarity is calculated from the presence or absence of numerous unweighted characters or character states (Sokal and Sneath 1963). This method does not establish groups by inspection, but orders the lowest taxonomic units (usually species) into groups with the help of standardized procedures.

The methods and principles of phenetics have been critically analyzed elsewhere (Mayr 1965, 1969; Johnson 1970; Hull 1970).

Cladistic Systematics (Cladistics)[1]

Organisms are classified and ranked, according to the theory of cladistics, exclusively on the basis of "recency of common descent." Membership of species in taxa is recognized by the joint possession of derived ("apomorphous") characters. Grouping and ranking are given simultaneously by the branching points. (See below for a statement of the reasons why the designation "phylogenetic systematics" for this theory is misleading).

---

1. The nouns *cladism* and *cladistics* have been used interchangeably. Since the ending "-ics" corresponds to that of phenetics, systematics, and genetics I now prefer to use cladistics.

Evolutionary Systematics

Organisms are classified and ranked, according to evolutionary systematics, on the basis of two sets of factors: (1) phylogenetic branching ("recency of common descent," retrospectively defined) and (2) amount and nature of evolutionary change between branching points. The latter factor, in turn, depends on the evolutionary history of a respective branch, for instance, whether or not it has entered a new adaptive zone and to what extent it has experienced a major radiation. The evolutionary taxonomist attempts to maximize simultaneously in his classification the information content of both types of variables (1 and 2 above).

The synthetic or evolutionary method of classification thus combines components of cladistics and of phenetics. It agrees with cladistics in the postulate that as complete as possible a reconstruction of phylogeny must precede the construction of a classification since groups that are not composed of descendants of a common ancestor are artificial and of low predictive value. More generally, it agrees also with cladistics in the careful weighting of characters. It rejects, however, the "divisional" process of classification ("downward" classification), which is most evident in the cladists' definition of "monophyletic." Evolutionary classification rejects most of the conceptual axioms of phenetics, but agrees with it in the actual procedure of grouping by a largely phenetic approach. However, in contrast to the unweighted approach of the pheneticists, it bases its conclusions on the careful weighting of characters.

The method in which cladistic and phenetic components are combined was originated by Darwin (see below).

IS CLADISTICS THE BEST THEORY OF CLASSIFICATION?

The cladists are sincerely convinced that their theory produces the best classifications. Hennig, for instance, states "that the claim of phylogenetic systematics for primacy among all possible forms of biological systematics has never been refuted even in the slightest" (1971:9). Günther (1971:38) likewise states: "W. Hennig has elaborated and substantiated his theory of phylogenetic systematics to such an extent that it can be considered as irrefutable." Günther (p. 76) furthermore claims that, among the three prevailing conceptualizations of biological systematics, it is only the consistently phylogenetic (genealogical) concept which permits drawing phylogenetically unequivocal conclusions. Similar statements can be found in the writings of Brundin, Crowson, Nelson, Schlee, and other cladists. Griffiths, for instance, states that Hennig's method "provides the only theoretically sound basis for achieving an objective equivalence between the taxa assigned to particular categories in a phylogenetic system" (1972:9).

Given this conviction of the superiority of their method, cladists are genuinely puzzled as to "why there are nevertheless so many systematists who have not (or only with reservations) committed themselves to phylogenetic systematics" (Hennig 1971:9). Hennig answers his own question by implying that it is simply insufficient familiarity with the objectives and methods of cladistics which has prevented a more general adoption. Günther believes that it is the unavailability or neglect of three sets of facts which has prevented the more general application of cladistics: (1) the lack of sufficient available distinguishing characters, (2) the uncertainty as to which characters are ancestral and which derived, and (3) the difficulty of a clear recognition of convergences (1971:77). In other words, both of these authors feel that empirical rather than conceptual reasons are responsible for the delay in adopting cladistics.

Is this conclusion really justified? Does a purely genealogical arrangement answer the demands of a "best classification"? Indeed, how do we determine which of several alternate classifications is the best? There has long been agreement among the theoreticians of classification that in most cases those classifications are "best" which allow the greatest number of conclusions and predictions. Mill (1874:466–467) expressed this, a hundred years ago, in the statement: "The ends of scientific classification are best answered when objects are formed into groups respecting which a greater number of general propositions can be made, and those propositions (being) more important, than could be made respecting any other groups into which the same things could be distributed." The opponents of cladistics claim that cladistic classifications do not satisfy Mill's criterion of a "best classification." The number of evolutionary statements and predictions that can be made for many holophyletic[2] groups (like birds and crocodilians) is often quite minimal, consisting ultimately only of a list of the synapomorphies. Indeed, the cladistic theory of classification would seem to suffer from several fundamental conceptual flaws. The argument cannot be settled without a searching analysis of the theory (including all the underlying assumptions) on which cladistics is based.

There are many indications that most cladists have never given serious consideration to alternative theories of classification, particularly to the theory of evolutionary taxonomy. For how else could Hennig (1971:7) have classified the evolutionary taxonomists in the same group as those taxonomists who work without any theory at all (see also Brundin 1972:111)? Other cladists, in their arguments, proceed as if phenetics (classification simply based on similarity) is the only alternative to cladistics. Even Griffiths (1972:18), who clearly distinguishes between the three methods of classifying, argues in his actual defense of cladistics only against "morphological-phenetic classifica-

---

2. A holophyletic group contains all the descendants of a stem species.

tions." Objections to the cladistic theory are being brushed aside as being due to inconsistencies or as having a purely psychological basis (Günther 1971:38).

There can be no hope for a meeting of the minds until the cladists face up to the criticisms of their opponents and attempt to refute them, point by point. Griffiths (1972) is the only cladist who has even attempted such a refutation.

In contrast to the flood of defenses of cladistics published in recent years (by Bigelow, Brundin, Cracraft, Crowson, Griffiths, Günther, Hennig, Kiriakoff, Nelson, Rosen, Schlee, and others) there has been only a limited amount of critical analysis of their theory. Simpson (1961) and myself (1969) have dealt with it *en passant* in major textbooks. There have been several short critical book reviews, and certain specific points (like the definition of "monophyly") were criticized by Ashlock, Colless, Farris, Gutmann, Johnson, Michener, Peters, and others. But Darlington's papers were the only serious attempt at a broad criticism of cladistics. And even his criticism deals more with the application of cladistics to biogeography than with the underlying basic assumptions. Indeed, Darlington himself states that his criticism "is not a general consideration of cladism" (1970:1). It is the objective of the present critical analysis to fill a serious gap in the taxonomic literature.

THE COMPONENTS OF CLADISTICS

It is most important for the understanding of cladistics to realize that it actually consists of two quite different sets of operations: (1) the reconstruction of the branching pattern of phylogeny through *cladistic analysis* and (2) the construction of a *cladistic classification* based on this branching pattern. The first of these two operations is important and largely unobjectionable. It is the second one which has encountered widespread criticism and will be carefully analyzed in the ensuing pages, with particular attention to the claim of the cladist that a classification should be a mirror image of the branching pattern of the phylogeny.

Reconstruction of the Branching Pattern of Phylogeny (Cladistic Analysis)

Cladistic analysis starts from the basic assumption that a sound classification cannot be constructed without a thorough understanding of the phylogeny of the given group. The evolutionary taxonomist agrees, on the whole, with this assumption. All phylogeny, except in cases of reticulate evolution, is strictly genealogical. Hennig is quite right when he states: "Phylogenetic research as biological science is possible only if it adopts the discovery of the genealogical relation of species as its first objective" (1969:33).

But how is one to proceed if one wants to reconstruct the phylogeny of a group? As Hennig has stated (1969:19), this method rests, in the

last analysis, on the simple realization "that all differences and agree-
ments between various species originated in the course of phylogeny.
During the splitting of a species its characters are transmitted to the
daughter species either changed or unchanged." Hennig is fully aware
that all that can be inferred by this method is the sequence of splits;
their absolute chronology must be determined in some other way.
Hennig has emphasized, and quite rightly so, that a phylogeny does not
need to be based on fossils but can be inferred from a careful comparative analysis of morphological characters. This thesis is well substantiated by the classification of the Recent mammals. Our ideas of their
relationships, based on a study of their comparative anatomy, have in
no case been refuted by subsequent discoveries in the fossil record.
However, the fossil record is entirely indispensable for the determination of absolute chronologies.

The most important step in cladistic analysis is the attempt to separate characters into ancestral (plesiomorphous) and derived (apomorphous) characters. Only the latter are considered legitimate evidence for
relationship, and taxa are therefore based on the joint possession of
derived characters (synapomorphies). (For a consideration of the value
of symplesiomorphies in the process of ranking see below.) Neither
Hennig nor any of his followers has claimed that this important principle was new when proclaimed by Hennig: "The observation that taxa
should only be characterized by apomorphous (derivative) conditions in
their ground plan is, of course, by no means new and to many people
seems self-evident" (Griffiths 1972:21). To mention only one example,
Tillyard's classification of the Perlaria (1921:35-43) was based on this
principle. In fact, one can say that most of the better taxonomists of
former eras applied this principle, as is quite evident from a study of
their classifications.

Nevertheless, Hennig deserves great credit for having fully developed
the principles of cladistic analysis. The clear recognition of the importance of synapomorphies for the reconstruction of branching sequences
is Hennig's major contribution. The cladograms that can be constructed
with the help of this method are as important for the evolutionary
taxonomist as they are for the cladist. I have previously (Mayr 1969:
212-217) called attention to the extreme value of this method for the
delimitation of taxa. The relative time sequence of the various branching points that the cladogram provides is of great value in many studies,
particularly in zoogeographic ones, as Hennig himself has demonstrated
for the diptera of New Zealand (Hennig 1960).

The formidable operational difficulties encountered by a cladistic
analysis are discussed below.

Cladistic Analysis and Cladistic Classification

There is little argument between cladists and evolutionary taxonomists about the cladogram that results from the cladistic analysis. The

argument arises over the relationship of such a cladogram to the classification that is to be based on it. Cladists assume that a one-to-one relationship exists between cladogram (phyletic diagram) and classification. The cladogram, once it is constructed, "automatically" provides also the classification. A cladogram and a classification are for the cladist merely two sides of the same coin. The evolutionary taxonomist, on the contrary, believes that a mere branching pattern cannot convey nearly as much interesting information as an evolutionary classification that takes additional processes of evolution into consideration (see below).

Traditionally, the first step in the classification of animals taken by the practicing taxonomist has always been the delimitation by inspection of seemingly "natural" groups. At first these are frankly based on the apparent "similarity" of the included species, that is, on phenetic criteria. When Hennig first proposed the cladistic method (1950) virtually all major higher taxa of animals were already known. He therefore automatically adopted the traditional method of taxonomy, of ranking and regrouping animal taxa that other authors had previously delimited. The validity of these provisional groups is subsequently tested in traditional taxonomy against a whole series of additional criteria, such as the homology of characters (similar and dissimilar ones), the presence of synapomorphies, the chronological relation to similar groups, an absence of conflict with the fossil record (when available), an absence of convergence (= spurious similarities), a meaningful geographic distribution, and so on. The more experienced a taxonomist is, the more quickly and thoroughly he can undertake these tests. (The classifying procedure of the pheneticist is drastically different.) Once established, a classification is thus constantly improved by the process designated by Hennig (1950) as "reciprocal illumination," which, as Hull (1967) has shown, does not involve circular reasoning. Indeed, the method is nothing more than another application of the hypotheticodeductive approach (Popper 1959, 1963) so commonly used in all branches of science and particularly (since Darwin) in biology. As soon as a new (or improved) classification is proposed, it will generate new information that, in turn, will lead to a reanalysis and possibly an improvement of the classification. This traditional approach to classification was followed without serious criticism during much of the nineteenth and twentieth centuries.

Unfortunately, this approach by trial and error is sometimes rather inefficient and has led to frequent changes in classifications. Again and again, therefore, taxonomists called for a more reliable approach. The better taxonomists agreed on two minimal conditions, the crucial importance of the right choice of characters (a point Darwin already had emphasized) and the necessity of basing taxa on numerous characters. But this still left much uncertainty.

It was Hennig's novel proposal simply to translate the cladogram into a hierarchical classification and thus do away with all the previous uncertainty. The proposal to construct a classification directly from the cladogram, however, does not convey the entire theory of cladistic classification. For this reason it would be highly desirable to present a detailed exposition of the entire theory in all of its ramifications, but this is rather difficult. Not only is Hennig's original work (1950) written in a rather turgid style, but some of his earlier theses seem to be no longer maintained in more recent publications. Furthermore, some of his followers, like Brundin, Schlee, and Griffiths, seem to have added postulates which, although perhaps implicit in the original theory, were not explicitly made by Hennig himself. Nevertheless, the major theses and postulates of cladistics have been stated sufficiently often to permit their enumeration. I shall try to list the more important ones, preferably by direct quotations from the works of cladists. I suspect that my listing is not complete and that some of the postulates are not adopted by every cladist. However, I hope that this list can serve as a convenient basis for the ensuing analysis.

These are the more important postulates of cladistics:

1. All taxa should be "monophyletic," with this term redefined in a novel way, in conflict with its traditional definition.

2. The term "phylogenetic" should be restricted to the branching (cladistic) component of phylogeny.

3. Relationship should be measured in terms of "recency of common descent," that is, narrowly genealogically.

4. "There is only one dimension in phylogeny and that is the time dimension" (Brundin 1966). Consequently, the splitting of phyletic lines (as reconstructed from the joint acquisition of derived characters) is admitted as the only legitimate evidence in the construction of classifications. To consider also similarities or the relative amount of ancestral (plesiomorphous) characters would lead to a "syncretistic system" that "robs the combination of any scientific value" (Hennig 1966:77). Hennig quotes with approval Bigelow's (1956) statement: "Classification must be based on one or the other (on overall resemblance *or* recency of common ancestry . . .), not on both, if philosophical confusion is to be avoided" (Hennig 1966).

5. The categorical rank of a taxon is automatically given by the absolute geological age of the stem species, or (in a less rigorous formulation) by the "relative age" of the stem species. (See also Crowson 1970:251 and disclaimer by Griffiths 1972:10, 16.)

6. Species can be delimited in time by two successive events of speciation.

7. The splitting of lines is always a dichotomy, resulting in the production of two sister groups.

8. "Homology . . . is usually defined in terms of common origin in time" (Griffiths 1972:17). (This is simply not true. Except for the ancestor-descendant relationship, the concept of homology is com-

pletely independent of the time dimension. No other cladist has made such a claim. Hennig himself adopts Remane's [1952] concept of homology.)

9. "Basically" all classifications should be horizontal classifications, valid only for a given time period (Hennig 1950:259) and therefore the same taxon might be given different categorical rank in different geological periods.

OBJECTIONS TO THE THEORY OF CLADISTIC CLASSIFICATION

The basic postulate of the cladistic theory, a complete congruence of cladogram and classification, can be satisfied only by making numerous assumptions and redefinitions and by ignoring numerous facts of evolution and of phylogeny (broadly defined). This results in major theoretical and practical shortcomings that will now be analyzed, point by point, in three major and a number of subordinate sections.

Arbitrary Decisions

In order for their method of classification to work, cladists have to make a number of arbitrary decisions, involving redefinition of well-known terms, reinterpretation of adaptive evolution, the proposal of a new species definition, and an unrealistic mode of origin of higher taxa. When these arbitrary decisions are rejected, very little support for cladistic classification remains.

Redefinition of Well-Known Terms

A large part of the controversy between cladistics and its opponents can be ascribed to the fact that the cladists have given an entirely new meaning to a number of widely used evolutionary terms which in a rather different sense had been in essentially consistent usage for about 100 years. The transfer of well-known and universally understood terms to entirely new concepts cannot fail to produce confusion. This is particularly true for three terms: *phylogeny, relationship,* and *monophyletic.*

Phylogeny
Since the days of Darwin and Haeckel the term "phylogeny" has been applied to all aspects of descent. Any article or book in the last 100 years that has used the term "phylogeny" (and its adjective phylogenetic) has used it comprehensively for all phenomena revealed by *Stammbaumforschung.* The Collegiate Dictionary of Zoology, for instance, defines it as "1. Evolutionary relationships and lines of descent in any taxon. 2. The origin and evolution of higher taxonomic categories" (Pennak 1971:395). But Hennig now attempts to restrict this term to a single aspect of phylogeny, that of branching. He states, "We will call 'phylogenetic relationship' the . . . [genealogical] relations

between different sections (in the diagram), each bounded by two cleavage processes in the sequence of individuals that are connected by tokogenetic relations" (1966:20). Or, "We have defined the phylogenetic relationships . . . as those segments of the stream of genealogical relationships that lie between two processes of speciation" (p. 29). Schlee (1971) defines phylogeny as "the origin of taxa, that is that part of evolution which is designated as its cladistic component."

Hennig's definition and use of the term "phylogenetic" is clearly in conflict with the previously universal use of the term. His diagrams are cladograms and not at all phylogenetic trees, which by the lengths and angles of their branches convey far more information than a cladogram. It would only aggravate the confusion if Hennig's specialized theory of classification would continue to be designated as "phylogenetic" classification.

Gisin (1964) has proposed designating Hennig's method by the term "genealogical" classification, since genealogical (kinship) relationships are an important aspect of the Hennig method. Unfortunately, the designation "genealogical" does not differentiate cladistics from certain other types of classification, because any classification in which each taxon consists of species that are derived from a common ancestor is a genealogical classification. Darwin was, therefore, quite right in stating that evolutionary classifications are genealogies. In any phylogeny in which hybridization does not occur (and this is the case in nearly all animal phylogenies), there can be only one genealogy, because each speciational event is unique. This unequivocality of the branching component of phylogeny is rightly emphasized by the cladists. However, even an unequivocal genealogy can usually be converted into a number of different classifications. And this is where evolutionists and cladists come to a parting of the way. Those who allow for a different weighting of different adaptational processes and events (for example, by giving greater weight to the occupation of a major adaptive zone) may arrive at a very different classification from someone who uses branching as his only basis (as do the cladists), even though the genealogies on which both base their classifications are identical. Since the term "genealogical" does not discriminate between evolutionary and cladistic classifications, I prefer the term "cladistic" for Hennig's method because it applies to it unequivocally and not to any other system of classification. Furthermore, it conforms to the terminology proposed by Rensch (1947) and Cain and Harrison (1960). It is the only term which accurately conveys the emphasis of the Hennig method on branching and on branching alone.

When employed in phylogeny the term "genealogical" is obviously used in a somewhat generalized sense, with "taxa" corresponding to the "individuals" of a conventional genealogy. The only alternative would be to classify "generations" of individuals. This indeed, would be logically impeccable, but useless for purposes of biological classification.

Relationship

The term "relationship" has been used in the systematic literature in many different senses. Recourse to a dictionary is of no help, since it lists extremely divergent definitions. The term "relationship" (or "affinity") was widely used in the eighteenth-century taxonomic literature, long before the evolutionary theory was adopted. For most authors, at that period, it simply meant similarity. Yet, even today, the term "relationship" is being defined in many different ways.

Pheneticists, in the beginning, were operating on the basis of the assumption that the phenotype accurately reflects the genotype and that an unweighted determination of "overall similarity" allows a correct determination of relationship. They are no longer as dogmatic about this as they were in 1963, but relationship still means similarity to them. The shortcomings of this interpretation have been pointed out by Mayr (1969) and many others.

The cladists go to the other extreme and restrict the term "relationship" to kinship in a strictly genealogical sense. According to Hennig, "the measure of phylogenetic relationship is the relative recency of common ancestry" (1966:74). Relationship between two species is measured by the number of branching steps which separate them from the common ancestor (Hennig 1950:129). When Cracraft (1972:381) claims that the cladistic method "is the best one available for determining relationships in a relatively unambiguous fashion," he falls victim to a circular argument because he uses the highly specialized definition of relationship of the cladists. To repeat, relationship for the cladist is genealogical kinship. But cladistic kinship alone, for an evolutionist, is a completely one-sided way of documenting relationship, because it ignores the fate of phyletic lines subsequent to splitting.

Let me explain. Since a person receives half of his chromosomes from his father, and his child again receives half of his chromosomes from him, it is correct to say that a person is genetically as closely related to his father as to his child. The percentage of shared genes (genetic relationship), however, becomes quite unpredictable, owing to the vagaries of crossing over and of the random distribution of homologous chromosomes during meiosis, when it comes to collateral relatives (siblings, cousins) and to more distant descendants (grandparents and grandchildren, etc.). Two first cousins (even two brothers, for that matter) could have 100 times more genes in common with each other than they share with a third first cousin (or brother) (among the loci variable in that population). The more generations are involved, the greater becomes the discrepancy between genealogical kinship and similarity of genotype, even though all these relatives still derive their genes from the gene pool of a single species. In phylogeny, where thousands and millions of generations are involved—that is, thousands and millions of occasions for a change in gene frequencies owing to stochastic processes, recombination, selection, and genetic revolutions—it becomes quite meaningless to express relationship only in terms of genealogical kinship.

In addition to the cytogenetic processes, there are also numerous aspects of selection that can result in a highly unequal degree of genetic change in different lines of descent. One of several phyletic sister lines may enter a new adaptive zone and there become exposed to severe novel selection pressures. As a result it will diverge dramatically from its cladistically nearest relatives and may become genetically so different that it would be biologically misleading to continue calling the sister groups near relatives. Yet, being the joint descendants of a stem species, they must be designated sister groups. And being sister groups they must be coordinate in rank, that is, according to cladistic theory, they must have the same absolute categorical rank in the hierarchy (Hennig 1966:139). This decision ignores the fact that one is still very much like the stem species while the other has evolved in the meantime into a drastically different type of organism.

This situation is best illustrated by a diagram (Figure 30-1). There will be a maximal genetic difference of 25 percent between the genomes of B and C, but of 60 to 70 percent between C and D. The cladist will say that C is more nearly related to D than to B, the evolutionist and the pheneticist that C is much closer to B than to D.

This independence of adaptive shifts from phyletic splitting is the reason why the evolutionary taxonomist adopts a very different definition of relationship. To him relationship means the inferred amount of shared genotype; it means gene content rather than purely

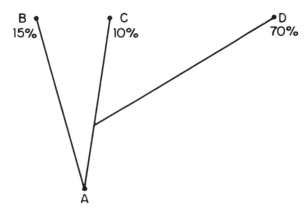

Figure 30-1. Inferred percentual difference from ultimate ancestor (*A*). Taxon C is more closely related to *B* than to *D*, even though it shares a more recent common ancestor with *D*.

formalistic kinship. For what is of primary interest in a taxon, its evolutionary role, its system of adaptations, and all the correlations in its structure and characters, is ultimately encoded in its genotype.

Since distant relatives cannot be analyzed genetically and since a purely additive analysis of a complex system of adaptations would be quite meaningless anyhow, it is necessary to infer degree of genetic relationship on the basis of indirect evidence. In this approach use is made of every available clue, but primarily of the combined evidence from phyletic branching and from a carefully weighted phyletic analysis. The evolutionary taxonomist believes that an approach which superimposes a carefully weighted phenetic analysis on a preceding cladistic analysis is better able to establish degree of relationship than either a purely cladistic or an unweighted phenetic approach. And a classification based on such a multiple-based determination of relationship will be more reliable and more predictive than one based on one-sided criteria. The criteria that have to be employed to make such weighting meaningful have been discussed elsewhere (Mayr 1969: 217–228). These criteria permit measuring something that is more than mere "overall similarity."

Convergence, parallelism, and mosaic evolution, all these evolutionary phenomena underline the importance in evolution of the underlying invisible genotype. Figure 30–2 will illustrate the difficulties created by the concealed potential of the genotype. If species C, owing to its concealed genotype, acquires by parallel evolution very much the same characters as D and E, even though it branched off from the line leading to A, it would seem legitimate to classify it with D + E. The

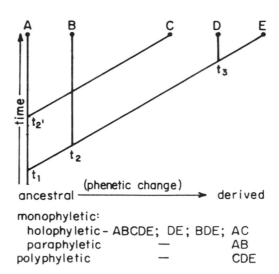

Figure 30-2. Taxon C, through parallel evolution, has become more similar to D and E than to A, with which it shares a nearer common ancestor. The converse is true for B. (After Ashlock 1969.)

cladist would presumably call a taxon composed of C + D + E polyphyletic, as if the similarity were due to convergence. Phenotypically, indeed,
C + D + E form a polyphyletic group. Many evolutionary phenomena
indicate the existence of a concealed genotype that cannot be read out
completely and directly from the visible phenotype. Most of our
difficulties with apparent polyphyly are due to the manifestations of
such concealed genotypes.

Monophyletic

The traditional procedure of taxonomy is to recognize a higher taxon
"intuitively," that is, on the basis of shared characteristics, evolutionary
role, and so on. As Hennig stated it (1966:146): "Taxonomy can begin
its grouping task with the assumption that the degree of similarity
between species corresponds with the degree of their phylogenetic kinship," an assumption to be thoroughly tested subsequently. There is
thus agreement between Hennig and the traditional taxonomist that the
characterization and delimitation of groups has primacy in the classifi-
catory operation. This primacy is confirmed by Griffiths (1972:7), who
proposes "that the phylogenetic system should be expressed by revision
of the traditional Linnaean system rather than by proposal of a separate
classification."

Not until a taxon is established provisionally does the taxonomist ask
whether it is truly a "natural" taxon, composed of species that are each
other's closest relatives. One of the ways to satisfy this condition is to
ask whether all members of the taxon have descended from a common
ancestor, that is, whether the taxon is monophyletic. The term "monophyletic" is a qualifying adjective for a noun, the noun *group* (or
taxon). Among possible groups (and taxa) there are some we can
identify as being monophyletic. It is a distinctly "retrospective" term
(Mayr 1969:75). For this reason the term "monophyletic," ever since
Haeckel, has been applied to groups that satisfied two conditions: (1)
the component species, owing to their characteristics, are believed to be
each other's nearest relatives, and (2) they are all inferred to have
descended from the same common ancestor. The second qualification is
needed to exclude unnatural groupings due to convergence. When this
traditional definition of monophyletic is applied to taxa, difficulties are
rarely encountered: birds are monophyletic, crocodilians are monophyletic, and reptiles are monophyletic. The concept as such is entirely
unambiguous, even though its application encounters occasional
difficulties, as in the therapsid–mammalian transition (Simpson 1961:
124–125; but see Crompton and Jenkins 1973).

Hennig has created enormous confusion by adding to the traditional
definition of a monophyletic taxon the following qualification: ". . .
and which includes all species descended from this stem species." This
definition is the inevitable consequence of the elimination of all consideration of adaptive evolution from Hennig's concept of phylogeny.

Since birds and crocodilians (excluding all other living reptiles) are derived from a common ancestor, his method forces the cladist to recognize a taxon for birds and crocodilians together, even though this is a useless assemblage. Hennig has transferred the qualification "monophyletic" from the taxon to the mode of descent. From a retrospective principle he has made monophyly a prospective criterion. This ignores, indeed it quite deliberately conceals, the most interesting aspects of evolution and phylogeny, those of adaptive radiation and the invasion of new adaptive zones (further discussed below under "Grades").

His definition of monophyletic forces Hennig to add another term to the phyletic terminology: *paraphyletic*.[3] A taxonomic group is paraphyletic if it has given rise to specialized sidelines that are not considered part of the group. For instance, the Reptilia are—for Hennig—a paraphyletic group, because certain Reptilia were the stem mothers both of other reptilia and of the birds (and the same is true for the stem species that gave rise to the mammalian branch). To designate groups as paraphyletic strikes me as a purely formalistic approach. It is of no relevance whatsoever for the relationship between crocodilians and other reptiles to know that the branch leading to the crocodilians (the archosaurian lineage) produced an offshoot that eventually became the class Aves. The animal taxonomist does not classify a logician's schemata or diagrams, but concrete groups of organisms. It is of no relevance for our judgment on the biological classification of the crocodilians whether or not a side branch gave rise to a drastically modified daughter group.

To take the traditional term "monophyletic" and transfer it to a new concept for which it had never been used before is contrary to sound language practices and to principles of scientific terminology. It strikes me as ludicrous when Hennig's adherents criticize their opponents for an "illogical usage of the word monophyletic." If one wants to have a term for "the aggregate of all groups descended from a common ancestor," one must coin a new term. Ashlock (1971:65, 1972) has recognized this quite clearly and has proposed for it the term "holophyletic." This corresponds to "monophyletic" as used by Hennig in contrast to the traditional usage.

No cladist seems to have noticed some of the consequences of the redefinition of the term "monophyletic." It forces him to abandon upward classification as traditionally practiced by the empirical taxonomists ever since Darwin and even earlier, and to replace it by "downward" classification. Although starting from entirely different premises, the cladist has methodologically returned to the "divisional" method of classification that was dominant from Cesalpino to Linnaeus. His

---

3. Hennig designates a group as *paraphyletic* if the similarity of the composing taxa is based on symplesiomorphy. For instance, the Reptilia are a paraphyletic group, in contrast to the archosaurians (crocodilians and birds) and the therapsids (mammallike reptiles and mammals), which are holophyletic.

criterion of division is of course very different from that of the adherents of Aristotle's logical division, but the principle of classifying of both schools (cladists and logicians) is very much the same.

It may sound like a platitude to say that when classifying one ought to deal with entities which one has in front of himself. The pheneticist and evolutionist classify species and genera in this manner. Not so the cladist, who deals with the unknown quantities produced by phylogenetic splits. It is implicit in his principles that he is forced to make the prediction that sister lines derived from a stem species will have sufficiently similar evolutionary fates so that the resulting sister groups can be ranked at the same categorical level (= are coordinate). The case of the birds and crocodilians is a particularly convincing illustration of the thousands of occasions where this prediction does not come true. It is the abandonment of the principle of upward classification, dominant since Darwin, and its replacement by Aristotle's downward classification which is the fatal flaw in the philosophy of cladistic classification.

Those who would adopt the three terms *phylogeny, relationship,* and *monophyletic* in their new, aberrant Hennigian definitions are forced to adopt drastic changes in the whole theory and practice of phylogeny and classification. It is quite true that one can operate in an entirely logical manner within the framework set by these new definitions. However, as Ghiselin has pointed out so often, one can operate in an entirely logical manner on the basis of totally wrong premises. Many, if not most, of the claims of the cladists go back to the consequences of their new definitions of these three terms.

The Neglect of the Dual Nature of Evolutionary Change

Darwin saw clearly that speciation involves two independent processes. One is the acquisition of reproductive isolation, a prerequisite of prevention of the hybridization between the two incipient species. The other one is the acquisition of niche differences resulting in "divergence of character" in order to overcome the effects of competition (Darwin 1859:111).

What is true at the level of the species is equally true in macroevolution. We can distinguish two processes of evolution, that of the splitting of phyletic lines and that of the invasion of new adaptive niches and major adaptive zones by phyletic lines. Any theory of classification which pays no attention to the tremendous range of difference between shifts of phyletic lines into minor niches and into entirely new adaptive zones is bound to produce classifications that are unbalanced and meaningless. But such a neglect of different kinds of phyletic evolution is precisely what the cladistic method demands.

The cladist proceeds in the construction of his classifications as if the splitting of lineages were the only phylogenetic process and as if all such splits were equivalent. All splits have equal weight for the cladist, just as all characters have equal weight for the pheneticist. The cladist's exclusive preoccupation with splitting has been confirmed by Hennig

on several occasions, but will be documented here only by the following quotation: "Decisive is the fact that processes of species cleavage
are the characteristic feature of evolution; they are the only positively
demonstrable historical processes that take place in supra-individual
organism groups in nature." (Hennig 1966:235). This leads Hennig to
the claim that his method is the only one that gives historically correct
answers.

The cladist states openly that branching is the only aspect of phylogeny of interest to him. That some of the resulting lineages may enter
entirely new adaptive zones and then become extraordinarily different
from other, more conservative lineages is regarded by him as irrelevant;
if removed from the common ancestor by the same number of speciational steps, they all will have to be given the same taxonomic rank.
Cladistics treats any apomorphous character like any other. Those taxa
are combined which have the greatest number of joint apomorphies (as
being derived from the same stem species). So far as I can judge from
the cladistic literature, no weighting of apomorphies is undertaken, and
derived characters that signify entrance into an entirely new adaptive
zone (as in the case of the birds) are given no more weight than the
joint apomorphies of birds and crocodiles (which distinguish these
living archosaurians from other living reptiles). The acquisition of minor
specializations is given the same weight as major adaptive innovations.
That some events in adaptive evolution are far more important than
others is completely ignored. This is, perhaps, the most significant difference between cladistic and evolutionary classification.

It is evident that the cladist reveals great ambivalence in the treatment of divergence. He pays lip service to the fact that there are differences in the rate of evolution in various communities of descent, but
does not draw any of the obvious conclusions from this observation.
Neither Hennig nor Brundin nor any of their younger followers (Schlee,
Nelson, Griffiths) pays even the slightest attention to these differences
in rates mentioned by Hennig when constructing classifications. They
likewise entirely ignore various other important evolutionary phenomena, such as the existence of "grades" and of highly specialized side
lines, all the phenomena of mosaic evolution, and all causal factors in
evolution. How rapidly a new branch diverges, how it changes in relation to the "sister group," how many additional characters it acquires,
which new adaptive zone it has invaded, all these are questions that the
cladist hardly ever mentions. By considering only genealogical distance,
the cladist acts as if he assumed that all lines diverge in an equivalent
manner and that genealogical distance corresponds to genetic distance.
By claiming that branching is the only historical process of consequence, he denies that other aspects of evolutionary change, such as
rate of evolution, adaptive radiation, the occupation of new adaptive
zones, mosaic evolution, and many other macroevolutionary phenomena, are eligible for the term "historical process."

Both components of phylogeny are potentially of equal importance for the evolutionary taxonomist, and both must be judiciously considered in the construction of classifications. Splitting as well as phyletic change occur simultaneously in evolution, but in most groups either one or the other process predominates at a certain geological period. Whenever there is massive splitting, such as, for instance, during the speciation of the 50,000 or 100,000 species of weevils (Curculionidae) or of the several thousand species of *Drosophila*, phyletic divergence is relatively insignificant. Among the invertebrates, and more specifically among the arthropods, there are numerous higher taxa in which abundant speciation has occurred without any impact on the basic morphology and without any shift into a new adaptive zone. All the species of these assemblages are repeated variations on a theme. This is in strong contrast to such memorable episodes in the history of the world as the origin of the vascular plants, the angiosperms, the chordates, the vertebrates, the terrestrial tetrapods, the reptiles, and the birds.

Rensch (1947), Huxley (1942, 1958), and Simpson (1959b, 1961) have particularly emphasized the importance of these levels of adaptation, designated by Huxley as *grades*. All members of a grade are characterized by a well-integrated adaptive complex. The successful evolution of a phyletic lineage toward and into a new adaptive zone is characterized by the stepwise acquisition (mosaic evolution) of a series of novelties to adapt it (and its descendants) to its new position in the ecosystem. Subsequently, the basic new type of this phyletic line may undergo little evolutionary change but experience instead abundant adaptive radiation as a result of bountiful speciation and various modifications in the basic adaptive theme of the grade. In the history of the vertebrates we know many such cases of the formation of successful new grades, such as the sharks, the bony fishes, the amphibians, reptiles, birds, and mammals. Each of these is characterized by a certain type of adaptation to the environment (Bock 1965), regardless of the amount of cladistic break-up within the grade. It results in a great deal of loss of information to ignore the adaptive component of evolution expressed by the concept of grade and to limit one's attention only to the splitting of lines. But this is precisely what the cladists are doing.

Actually, the existence of minor and major grades is one of the most interesting phylogenetic phenomena, even though it is a phenomenon which we are still unable to understand adequately. Why is there so often such a uniformity of type within a higher taxon? There is a rich diversity of species of parrots, but all of them from the smallest pygmy parrot or lorikeet to the largest cockatoo or macaw are characteristically parrots. And so it is in many, if not most, higher taxa. The reptiles represent a well-defined grade between the amphibian level and that of the two derivatives of the reptiles, the birds and the mammals.

Simpson, in his various publications (1953, 1959b, 1961) has repeatedly discussed the contrast between clades and grades. Crocodiles have a more recent common ancestor with birds than with lizards. They belong, thus, to the same clade as birds but they do not belong to the avian grade, but rather to that of the reptiles. To which of these two aspects of evolution shall we give primacy? There are literally thousands of similar dilemmas in the evolution of animals and plants. For instance, the African apes (*Pan*) have a more recent common ancestor with man (*Homo*) than with the orang (*Pongo*). However *Pan* belongs to the same grade as *Pongo*, very different from that typified by man.

The better the fossil record becomes known, the more often one encounters such dilemmas. To the evolutionary taxonomist the existence of grades seems often more significant and more meaningful biologically than the mere splitting of phyletic lines. How little some of the cladists appreciate the biological significance of grades is illustrated in a comment by Brundin (1972:111), who designates groups such as the reptiles as "timeless abstractions."

One senses two reasons for the deliberate neglect of evolutionary divergence by cladists. One is that this factor cannot be measured precisely and unequivocally. Indeed, rates and degrees of evolutionary divergence can usually be inferred only by extrapolation or other indirect approaches. Yet, by appropriate weighting (Mayr 1969:220–228) one can draw meaningful probabilistic inferences, which, although not entirely precise, are far more valuable than the advice to ignore evolutionary divergence altogether. A second reason is that cladists seem to think that they have to make a choice, in the delimitation of taxa, between basing them *either* on branching points *or* on degrees of evolutionary divergence. They fail to appreciate the added amount of information one gains by utilizing *both* sources of evidence.

Cladists, when criticized for the neglect of evolutionary divergence, try to defend themselves by referring to Hennig's *deviation rule:* "When a species splits, one of the two daughter species tends to deviate more strongly than the other from the common stem species" (1966:207). The establishment of this rule, say the cladists, proves that they do not ignore phyletic evolution. Several aspects of this rule are remarkable. First of all, it is in flat contradiction to Hennig's assertion that splitting is the only historical process in phylogeny. Unequal deviation is a historical process that—as such—is independent of splitting. Second, although the deviation rule was pronounced by Hennig in 1950 (p. 111) and confirmed in 1966 and 1969 (p. 43), and although it is in principle adopted by most cladists (for example, Brundin 1972:108), it seems to play no role whatsoever in the construction of any of their classifications. The dendrogram illustrating the deviation rule (1950:Fig. 25) is one of Hennig's very few diagrams in which the angles of all clades are not the same. One almost gets the impression that it is the whole purpose of the deviation rule to permit cladists to defend themselves against the accusation of having ignored evolutionary divergence altogether. For the consequences of the deviation rule are completely neglected. They would become obvious if the process of unequal deviation were to occur after each step of speciation. Then it would become evident how important phyletic divergence is. For instance, if one of the newly arisen "sister groups" is very much like the parental group or even identical with it (as implied by Hennig 1966:59) while the other deviates strongly, the terminology "sister groups" is no longer applicable since one now deals with a continuing parental group from which a daughter group has split off (Hennig 1966:Figs. 14, 15). But beyond this formal objection, there is the much more serious one that a greatly accelerated rate of divergence in one branch (one sister group), while subsequent branches of the other sister group diverge only slightly, would lead in time to such an unbalance of the system as to destroy completely the usefulness of the dichotomous cladogram. Such asymmetric branching happens in evolution very frequently and is easily accommodated in the classifications of the evolutionary taxonomists but fits only very awkwardly (if at all) into the cladistic classifications.

Some of Hennig's followers have recognized these contradictions. Schlee (1971:5, 30, 37) sees rather clearly that the deviation rule is dispensable: "It forms neither an argument for the justification of Hennig's method nor a prerequisite for the work with Hennig's principle." Schlee adds a mysterious interpretation by Hennig himself (p. 6): "that the 'deviation rule' must be understood in a special genealogical sense, but not in a morphological–biological sense." Actually, the opposite is true: in a strict genealogical sense, there can be no deviation. If there is an unequal deviation, it must be in "a morphological–biological sense." Such morphological deviation is the normal situation in phylogeny. Whenever there is a splitting of phyletic lines, almost invariably one line will diverge more rapidly and more widely than the other; in fact, one of them may not change at all. The complete neglect of the frequent occurrence of this process is one of the fatal errors in the translation of the cladistic analysis into a classification.

Since the consideration of grades in classification is sometimes referred to in the cladistic literature as a typological approach, I would like to call attention to a somewhat different usage of the term "typological" (typologisch) in the American and the German literature. "Typological" in American usage is a straight synonym for essentialistic, referring to the abstraction of an underlying *eidos* (essence) and the neglect of the existence and importance of variation. In the German phylogenetic literature *Typus* or *Bauplan* is also an abstraction, representing either the inferred ancestral "type" or the "ideal" *Bauplan* of a major taxon. To recognize the reptiles as a legitimate taxon means recognizing a generalized reptilian *Bauplan,* and this procedure is then referred to as a typological approach. The emphasis in the German usage is on the typological philosophy of idealistic morphology. "Typological" in the German usage often also implies "phenetic" (broadly defined). The confusion between the two concepts of "typological" is well illustrated in a discussion by Schindewolf (1967).

A Purely Formalistic Species Definition

Hennig's species definition in his original treatise (1950) is concerned only with the delimitation of species in the time dimension. A species is simply the distance between two branching points of the phylogenetic tree (p. 111). This concept is retained in his 1966 book (pp. 56–65). For instance: "The limits of the species in a longitudinal section through time would consequently be determined by two processes of speciation: the one through which it arose as an independent reproductive community, and the other through which the descendants of this original population ceased to exist as a homogeneous reproductive community" (p. 58). Although the last words of this definition have a slight flavor of the biological species definition, the diagrams in which it is illustrated (particularly his Fig. 15, p. 60; see Figure 30–3) show how purely formalistic Hennig's species concept is. For instance, species B and D_1 differ in no way from the stem species A, but must be called different species, because C and E had in the meantime branched off this stem. In contrast species D_2 is different from species D_1 ("morphologically" says Hennig, but his argument would be the same if they were "reproductively" isolated), but must be called the same species, because no branch had budded off from this stem in the meantime.

I am calling attention to Hennig's species concept, not because it is the decisive capstone in the Hennig theory (it is *not*, Hennig's own claims notwithstanding), but merely to give another example of the arbitrariness and purely formalistic nature of the major components of Hennig's cladistic theory. Hennig's species concept is so obviously

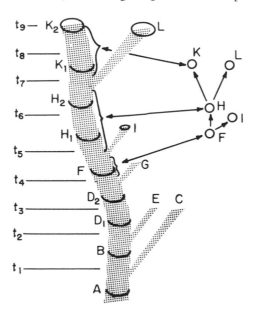

Figure 30–3. The relation, according to Hennig, between branching and speciation. (From Hennig 1966:fig. 15.)

unbiological and unrealistic that it has been rejected by numerous recent critics. Peters, for instance, refutes Hennig's formulations very effectively (1970:28–30), showing that among all the possible properties of species their duration in the time dimension is the least meaningful. Cain (1967:412) also exposes the biological meaninglessness of Hennig's formalism.

Hennig's concept of the process of speciation is strongly affected by his species concept. The various possibilities implied in Hennig's discussions can be formulated as three alternatives:

A. Parental species eliminated by
　1. the splitting of the mother species into two daughter species (dichotomy), or
　2. the simultaneous splitting of the mother species into more than two daughter species, or
B. Parental species continuing
　3. after the origin of one of several side branches from the essentially unchanged phyletic mainline.

In his 1950 book Hennig allows only for alternative 1, and among cladists this is still the favored alternative because it fits most easily into the cladistic scheme. Cladists are not unaware of alternative 3 but find a formal solution that does away with this inconvenient phenomenon. If a species A throws off a species C, species A will have to be called species B from the branching point on, in order to satisfy the cladistic postulates, even when B is biologically quite indistinguishable from A. In kinship studies, writes Hennig (in Schlee 1971:28): "the question of the biological identity of different species from different chronological horizons becomes totally irrelevant." This illustrates well to what extremes Hennig carries dogmatic formalism. Because, no matter what he says, if A and B are biologically identical, then they simply are not different species but the same species from which species C has branched off at some time. And this is of critical importance for the discrimination between sister and daughter groups. Hennig's solution is thoroughly misleading. The production of a side branch, a new phyletic lineage, does not change the parental species.

Dichotomy or not. Modern speciation studies permit us to determine at which relative frequency alternatives 1 and 2 occur. All the indications are that a simple dichotomy into two daughter species is not the rule. Polytypic species almost invariably have more than two subspecies. Far better evidence is provided by superspecies that consist of groups of allospecies. In North America at least 48 (= 40 percent) of 126 superspecies of birds have more than 2 allospecies (Mayr and Short 1970). Among 94 Northern Melanesian superspecies 61 (= 65 percent) are nondichotomous, containing 3 to 13 allospecies. The dichotomous "standard" is also refuted by all species-rich genera and by the frequency of clusters of sibling species. What is usually found in speciation studies is that the maternal species undergoes relatively little evolutionary change while numerous daughter species bud off at the periphery. Hennig himself is not unaware of this situation, as is shown by his presentation of geographic variation in the snake *Dendrophis pictus,* in which the large central population is surrounded by 6 peripherally isolated populations (Hennig 1966: Fig. 16, p. 61). It is increasingly realized by biologists (Mayr 1942, 1963) that peripheral budding is the most frequent process of speciation, even though most of these daughter species are extremely short-lived. Under these circumstances I fail to comprehend the logic of Nelson's (1971:374) assertion: "The use of dichotomous speciation as a methodological principle is required before an hypothesis of multiple speciation is even tentatively acceptable."

The difference between splitting and budding might seem a purely semantic one, if it were not for the fact that the cladists base such far-reaching conclusions on the postulate of consistent dichotomy. Darlington (1970:2–4) has presented an incisive critique. Even some cladists are beginning to abandon the principle of obligatory dichotomy (for instance, Schlee 1971:27), but they have not yet faced up to the consequences this poses for their theory of sister groups. If a non-dichotomous split has produced three or four independent phyletic lines, it would mean that each of the resulting groups has two or three different sister groups rather than a single one. Many of the discussions of sister groups found in the cladistic literature would become meaningless under these circumstances.

In his *Stammesgeschichte der Insekten* (1969) Hennig recognizes a number of higher taxa that consist of more than two sister groups. However, he emphasizes that this is a purely provisional arrangement, acceptable only "as long as the exact relationships are still uncertain."

The role of *extinction* is greatly underestimated in the cladistic constructions. Pairs of related taxa (Hennig's sister groups) are indeed frequently encountered. Their existence, however, is usually due to the extinction of numerous intermediate phyletic lines rather than to a peculiar process of speciation, that is, the splitting of a mother species into two daughter species. There is little evidence that this occurs frequently, and none whatsoever that this is the universal process of phylogeny.

The Mode of Origin of Higher Taxa

His phylogenetic theory forces the cladist to propose an unrealistic mode of origin for higher taxa. Since he recognizes branching as the only phylogenetic process, he has to give his branching points two properties: they are the origin of new species and also of new higher taxa. This arbitrary assumption in no way corresponds to the facts, as correctly pointed out by Darlington (1970:2). Speciation—that is, the

acquisition of reproductive isolating mechanisms between populations—and the acquisition of phylogenetically significant new
apomorphous characters, are two largely independent processes. The
study of groups of sibling species and of most species-rich genera shows
that the acquisition of reproductive isolation often (if not usually) has
no effect on the morphological criteria that a taxonomist or evolutionary biologist would associate with the origin of new higher taxa. It
is the exception rather than the rule that one of the daughter species
acquires during speciation a character which is of potential significance
for the characterization of a new higher taxon. The appearance of new
apomorphous characters is correlated with the invasion of new niches
and adaptive zones rather than with speciation (Simpson 1956b; Bock
1965). The occurrence of phylogenetic dichotomy thus becomes more
plausible when it is divorced from speciation, because the probability
that several daughter lines will shift simultaneously into the same new
adaptive zone is small. It is, however, not nil, since a change of climate
and vegetation or the arrival of a new predator may indeed cause a
simultaneous identical shift in several related lines.

Probably more frequent and potentially more troublesome is the
occurrence of the simultaneous acquisition of the same apomorphous
characters in different lines in a cluster of sister groups, let us say, the
descendants of a highly polytypic superspecies. Extinction often provides a practical solution to this dilemma. The evolution of a new
higher taxon from the line initiated by the new species is a later event,
representing the second phylogenetic process, evolutionary divergence,
and has nothing to do with splitting as such. It is a misleading formulation to say that higher taxa split. The evidence for this assertion is this:
A higher taxon is a collective assemblage that comprises numerous
species and lesser evolutionary lines. Very few of them ever diverge to
the extent that they form a separate higher taxon. But every once in
a while one of these lesser branches diverges to such an extent that it
must be removed eventually as a separate higher taxon.

Two aspects are important: The "Adam" of the new phyletic line
almost invariably belongs to the ancestral taxon. The first member of
the phyletic line that eventually led to the birds (long before *Archaeopteryx*) was presumably an otherwise rather conventional dinosaur, but
with feathers or featherlike scales. Is a single derived character enough
to throw such an "Adam" into the new taxon that will eventually
emerge from his lineage? Paleontologists have long been concerned
about this problem, which has been discussed in great detail by
Simpson (1961 and elsewhere).

Ashlock (1971:67) makes the sensible suggestion that the delimitation of the ancestral taxon against the derived taxon should be arrived
at as follows: "The unique innovations found in the living members
should be traced through the available fossils and the break placed
somewhere between the first appearance of one of these characters and

the first appearance of all of them ... The specific assignment of the boundary should be phenetically determined, weighting characters if appropriate, so as to establish attainment of the grade of the derived group." He admits that lack of preservation of the soft parts and the absence of a fossil record may make this task difficult, if not impossible, but in principle this is certainly the appropriate procedure.

The other important conclusion is that the origin of a side branch is of no evolutionary consequence for the main branch (except for possible competition). For instance, the class Reptilia, a well characterized grade of tetrapods, has existed since the Carboniferous and survives in four living orders. Some time in the Triassic one of the numerous reptilian side branches (the cynodonts among the therapsids) evolved into the mammals and a little later another one gave rise to the birds (Figure 30-4). A rigid application of their dogma forces the cladists to break up the reptilian grade into many separate "classes" and to designate particular reptilian lineages as the "sister groups" of birds and mammals. The fact that no one would place the crocodilians outside the reptiles if birds did not exist reveals how artificial and arbitrary this procedure is. The essential unity of the reptiles is best illustrated by the continuing argument among paleontologists as to which particular orders of reptiles are most closely related to which others.

One of the major sources of the cladistic difficulty is the assumption that the origin of any new higher taxon requires the disappearance of

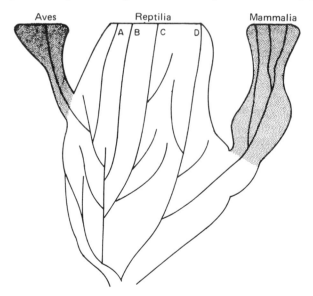

Figure 30-4. The independence of the emergence of the avian and mammalian grades from the branching pattern of the reptilian grade. *A* (= crocodilians) belongs cladistically with the Aves, but is still a characteristic member of the reptilian grade. The origin of birds and mammals does not affect the categorical status of the reptilian branches from which they arose.

the parental taxon. Even though there is massive evidence that many, if not most, higher taxa are lateral derivatives (side branches) of other taxa that had existed long previously and have continued to exist long after the split, his particular formalism forces the cladist to make such unrealistic claims as: "The stem species belongs neither to one nor to the other [daughter] group [to which it has given rise]: it cannot be assigned to either of the two" (Hennig 1969:33).

How unsuitable this cladistic approach is can be demonstrated best by the study of a richly diversified group which is still in existence and actively speciating. The modern family of Drosophilidae represents such a case (Figure 30–5). In this higher taxon one has such rich information from morphology, chromosomes, behavior, and other characteristics that one can reconstruct the probable phylogeny with a considerable degree of reliability. The resulting dendrogram shows that several specialized genera originated from various locations within the *Drosophila* dendrogram. The origin of a specialized side line, to which one accredits separate generic status, in no way affects the taxonomic status of the main line of *Drosophila*, which continues as the genus *Droso*

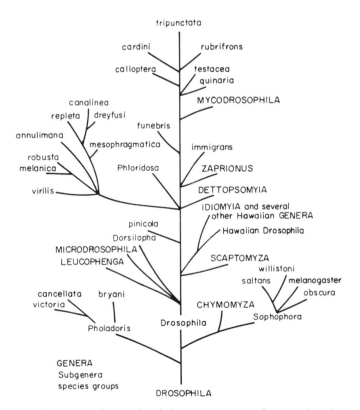

Figure 30–5. The origin of specialized derivative genera from within the *Drosophila* phyletic tree. (From Throckmorton 1965.)

phila. This is not a unique case. The origin of multiple side lines, each specializing in a somewhat different way while the major stem group continues essentially unchanged, is a very common occurrence in phylogeny.

A Misleading Conceptualization of Ranking

The process of classification consists essentially of two steps: (1) the *grouping* of lower taxa (usually species) into higher taxa and (2) the assignment of these taxa to the proper categories in the taxonomic hierarchy (*ranking*). The considerations that govern these two processes are quite different from each other (Mayr 1969:chap. 10). The cladists, following the erroneous assumption that phylogeny is a unitary process (consisting only of branching), assert that classifying likewise is a single-step procedure and that the grouping of taxa simultaneously also supplies their rank. Traditionally the rank of a taxon is determined by such criteria as degree of difference, uniqueness of the occupied adaptive zone, or amount of adaptive radiation within (Mayr 1969: 233). According to cladistics, rank is given automatically by time of origin, and the same rank must be given to sister groups.

This erroneous conclusion is reinforced by Hennig's frequent confusion of the terms "taxon" and "category." In 1950 such confusion was excusable because up to that time no terminological distinction had been made. By 1966, when the difference had been made abundantly clear by numerous writers, the confusion was no longer defensible. Yet, even in 1966 (pp. 77–83) Hennig speaks of the "reality and individuality" of categories, when he means taxa, and this confusion thoroughly obfuscates his discussion of the ranking procedures. The whole process of classification is based on the clear discrimination of *taxon* and *category*: zoological groups (taxa) have "reality" in nature because taxa names, like birds, butterflies, or bats, are unambiguously names for clearly distinguishable groups. It is equally evident that the ranking of these taxa in the Linnaean hierarchy of taxonomic categories is rather arbitrary and often highly controversial. What one author considers a tribe, a second may call a subfamily, and a third a family or even superfamily (see below).

Hennig, unfortunately, believes (or at least originally believed) that a knowledge of branching points would permit him to determine the categorical rank of a taxon. He utilizes two criteria: the geological age of the branching point (leading to the nearest sister group) and the number of subsequent branching points. He has stated this without reservation at numerous places in his writings. For instance, the taxa in the hierarchy "are subordinated to one another according to the temporal distance between their origins and the present; the sequence of subordination corresponds to the 'recency of common ancestry' " (1950:83). "In the phylogenetic system . . . the absolute rank order cannot be independent of the age of the group since . . . the coordination and sub-

ordination of groups is by definition set by their relative age of origin"
(1966:160).

Both of Hennig's criteria have been frequently criticized. Peters
(1970), for instance, recognizes that among two evolutionary lines that
evolve at the same rate one might give off a large number of side
branches, while the other might maintain a monolithic singularity. It
would make no sense to grant a higher rank to the frequently branching
line than to the nonbranching line. After all, branching and evolutionary divergence are two independent processes. Hennig and some of
his followers now speak occasionally of "relative chronological age"
instead of "absolute geological age," but in 1966 Hennig still maintained that the location of the branching point on the geological time
scale determines categorical rank (Table 30-1). Accordingly, taxa that
originated from a split in the pre-Cambrian are ranked as phyla; between Cambrian and Devonian as classes; between Mississippian and
Permian as orders; between Triassic and Lower Cretaceous as families;
between Upper Cretaceous and Oligocene as tribes; and in the Miocene
as genera. Hennig continues: "then the mammals would have to be
called an order ... the Marsupialia and Placentalia would have to
be downgraded to families, and the 'orders' of the Placentalia would be
tribes" (p. 187). Similar statements can be found in the writings of
other cladists. For instance: "Species are to be ranked 1. according to

Table 30-1. Hennig's assignment of categorical rank on the basis of the
absolute geological age of the stem species. (After Hennig 1966.)

Geological time periods		Categorical rank
VI.	Miocene	Genus
V.	Oligocene	Tribe
	Upper Cretaceous	
IV.	Lower Cretaceous	Family
	Triassic	
III.	Permian	Order
	Mississippian	
II.	Devonian	Class
	Cambrian	
I.	Precambrian	Phylum

their relative time of origin or 2. such that sister groups are given equal rank" (Nelson 1972:366). "Relative age" is a term frequently referred to in the cladistic literature but no one has yet proposed an operational method of determining relative age.

If one would really rank taxa on the basis of their (geological) time of origin, one would be forced to adopt an entirely unbalanced system. Since the genus *Lingula* originated earlier than either the class Aves or the class Mammalia, one would have to place *Lingula* (and other contemporary, still surviving genera) into higher categories than birds and mammals. Even a cladist would presumably rather be inconsistent than go to such extremes. The discussions of Griffiths (1972:10) and Crowson (1970:250–254) reveal their difficulties.

Curiously, Hennig has cited the classification of parasites as support for the cladistic method (1950:261–269). He confirms that "host group and group of parasites must be given the same systematic rank . . . when a group of parasites, which is restricted to a certain host group, originated simultaneously with it, perhaps in the manner that the stem species of the group of parasites lived as parasites in the host group" (p. 265). And yet, as Osche (1961) has pointed out, the chronology of parasites and their hosts actually refutes the cladistic method. It is highly probable that many vertebrate parasites originated at the same geological period as their host taxa. This would necessitate, if one were to follow Hennig strictly, that a genus of cestodes be raised to the rank of a family or order because it parasitizes a family or order of vertebrates or reciprocally that the host taxon be reduced to the rank of a genus. It would also mean that the superfamily Ascaroidea of the nematodes be given the same categorical rank as the class Cestoda because the stem species of both taxa are believed to have invaded the vertebrates at the same time.

The more carefully phylogenies are studied the more difficulties for the Hennig principle of ranking are discovered. Various auxiliary devices proposed by Hennig do not improve the situation. By assuming branching to be a regular process, with an approximately even rate, Hennig believes that the number of species contained in a taxon provides an approximate measure for its age: "Monophyletic groups with large numbers of species cannot be very young" (1966:182). Recent evolutionary research has clearly demonstrated that this conclusion does not hold. The correlation between age of a taxon and number of contained species is extremely loose. The Hawaiian drosophilids, a clearly monophyletic group, are not only very recent (probably less than 4 million years old) but also extremely rich in species (at least 600 to 800). Most of the species flocks of African cichlid fishes are likewise very young, a product of the last couple of million years. Most families of rodents, rich in species, are geologically much younger than most of those families of mammals that contain only 2 or 3 genera. The monotremes with 4 to 5 species are as old as or older than the eutherians with more than 3,000 species (Peters 1970:31). Well-known types of marine invertebrates go back 300 to 400 million years without ever having produced rich species flocks.

Hennig and his supporters claim that the great merit of the cladistic method is that it provides nonarbitrary definitions of the higher categories. Griffiths (1972:16), for instance, praises Hennig for having proposed a logically unobjectionable definition "in terms of the age of origin of the stem species." I have shown above how unrealistic this proposal is and how utterly it fails to provide a sound procedure of ranking. Invertebrate paleontologists, who can demonstrate extremely different evolutionary fates for different phyletic lines (derived from the same stem species), have frequently pointed out to what unbalanced classifications the cladistic method of ranking would lead. Simpson (1961:142–144) likewise has described what the application of this principle would do to vertebrate classification. Crowson (1970:260) admits, "If phylogenetic classification proceeds usually by dichotomous divisions, and very unequal ones at that, it will necessitate the usage of many more categories [and taxa names] than were needed for older, 'formal,' systems." Simpson is fully justified in calling such classifications "completely impractical." Hennig himself (1968:10) now realizes that this method leads to an absurd scale of ranking and has abandoned in his book on the phylogeny of insects any attempt to provide categorical ranks for the higher taxa. But what method of classification is this which cannot rank higher taxa?

Operational Neglect of Evident Facts

Many aspects of phylogeny create difficulties for cladistics. Hennig and his followers are not unaware of these facts and occasionally discuss them freely. However, they ignore these difficulties in the construction of their classifications, or at least they offer no operational instructions on how to cope with them. I have referred to this already in the discussion of the "deviation rule," in which the unequal divergence of phyletic lines is acknowledged, but without drawing the obvious consequences for classification. Other facts that are neglected are the following.

The Difficulty of Determining the Direction of Evolutionary Sequences

Taxa are constructed by Hennig on the basis of synapomorphies (derived characters). The direction of the evolutionary change must therefore be determined for each character sequence. Sometimes it is easy to make such a decision; in many other cases it is quite difficult. Hennig is well aware of this problem and has formulated four rules (1950:172ff.) that help to determine which of several alternate characters are more ancestral and which derived. Most of these rules, reformulated by Peters and Gutmann (1971:242), are based on practices that go back to Darwin or even earlier periods. Peters and Gutmann

(p. 256) emphasize correctly that any formalism in phylogenetic research (including cladistics) tends to lead to superficial and often unreliable conclusions. It must be replaced by a far more biological attitude toward the morphological evidence. Phylogenetic reconstruction must be based on an analysis of function and adaptive significance, such as is practiced in evolutionary morphology (Bock 1965, 1969). I refer to the essay of Peters and Gutmann (1971) for a very perceptive statement of this weakness of cladistics. Their considerations make it obvious that the magnitude of an adaptive innovation is of the utmost importance in classification, a fact consistently ignored by the cladists.

The Discrimination between Parallelism and Convergence

The common possession of the same character in two different taxa may have one of four possible causes (Mayr 1969:202):

1. *Plesiomorphous similarity:* the sharing of characters with an ancestor (see the cladistic literature for more precise specifications).

2. *Synapomorphous similarity:* the unique sharing of characters derived from a stem species, for example, all birds, but no other organisms, have feathers.

3. *Similarity due to parallelism:* characteristics produced by a shared genotype inherited from a common ancestor.

4. *Similarity due to convergence:* these can be either convergent acquisitions (wings in pterodactyls and bats) or convergent losses (leglessness in snakes and worm lizards).

To discriminate between these four possibilities is not nearly as easy as it may sound, because all that is accessible to the student of phylogeny are phenotypes, while phylogeny actually consists of a change in genotypes.

The traditional recipe for the discrimination between parallelism and convergence is the analysis of homology. Cases where a homology between characters can be established, that is, a derivation from the equivalent character in the common ancestor, are the result of parallel evolution. Similarities in nonhomologous characters are the result of convergence (Bock 1963, 1967). To follow this recipe in practice is, however, very difficult, not only because the establishment of homology is often fraught with difficulties (Hull 1967), but also because in distant relatives the line between parallelism and convergence is often not sharp (for a discussion of parallelism see also Simpson 1961: 103–106). Osche (1965) has given a perceptive analysis of the difficulties caused by the potentialities of the hidden genotype.

Cladists pay little attention to the various possibilities: "In deciding whether corresponding characters of several species are to be regarded as synapomorphies, convergences, homologies, or parallelisms we must determine whether the same character was already present in a stem species that is common only to the bearers of the identical characters" (Hennig 1966:120). Contrary to Hennig's claim, this does not permit a discrimination between the four possibilities but only between syn-apomorphy on the one hand and the other three on the other hand. Here and elsewhere (for instance, 1966:121) Hennig is quite outspoken in his lack of interest in making a distinction between parallelism and convergence (see also Peters 1972:168). And yet such a distinction is of crucial importance in deciding degree of relationship between taxa.

Taxa are classified in cladistics on the basis of the presence or absence of derived characters. Indeed, Hennig has emphasized re-peatedly that nothing counts in classification but such characters. Un-fortunately, only part of the genotype is expressed in the visible pheno-type and yet the hidden part of the genotype is often as important for the future evolution of a phyletic line as that which is revealed in the visible phenotype. Hennig (1950:176) himself has shown that a poten-tial for stalked eyes is widespread among the acalyptrate dipterans but is realized only in scattered species and genera. A secondary jaw articu-lation originated at least 14 times independently in the class Aves (Bock 1969). In the genus *Drosophila* the repeated manifestation of concealed potentialities is extremely frequent (Throckmorton 1962, 1965). There is hardly a higher taxon known in which such derived characters do not occur scattered through the system. In any particular case the presence or absence of the character is not in the least determined by the presence or absence of the character in the common ancestor. To be sure, the genotype of the common ancestor has the potential for the development of this character, but its realization is unpredictable.

When such an independent realization of incipient tendencies occurs with several characters, there is no way to determine in what sequence these derived characters were acquired in different lines and in what sequence the various related lines branched from each other. This can be shown in a diagram (Figure 30–6). The only way the cladist can cope with this problem is to say that it will disappear if one takes enough characters. Unfortunately, this is not necessarily true. Usually there are not enough such characters available and, when only few are available, the decision will have to be made on the basis of careful weighting, often hardly distinguishable from being arbitrary. To imply that there really is not much difference between convergence and parallelism (Hennig 1966:121) reveals a peculiar reluctance to come to grips with one of the major weaknesses of the cladistic system. These are not purely theoretical problems. The recent effort by Inger (1967) and of Kluge and Farris (1969) to arrive at a satisfactory phylogeny of the anurans depends to a considerable extent on decisions concerning the relative primitiveness of various characters.

Griffiths (1972:24–26) has a lengthy discussion on the question of whether or not convergence poses a serious problem for cladistics. He is surely on safe ground in his belief that highly complex characters or character combinations are not apt to be acquired independently in unrelated phyletic lines. (For a critique of the complexity criterion see

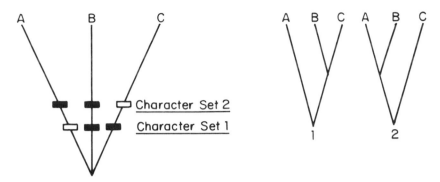

Figure 30-6. Contradictory information provided by different sets of characters owing to mosaic evolution. According to character set *1*, the lineages *B* and *C* form a sister group to *A*; according to character set *2*, the lineages *A* and *B* form a sister group to *C*.

Peters 1972:168–170.) However, characters (even rather complex ones) can be lost independently in different lines and rather simple characters can be acquired convergently. Unfortunately, Griffiths misinterprets Darwin in the following sentence: "Doubts have been raised about the validity of Darwin's distinction between 'adaptive' and 'non-adaptive' characters for purposes of evolutionary evaluation, and I therefore do not employ such a criterion" (p. 25). Actually Darwin makes no such distinction. He speaks only of *ad hoc* specializations, which indeed have low phyletic weight and can be acquired convergently. This is what Cracraft (1972) has overlooked in his recent discussion of ratite evolution. If several groups of running birds lose their power of flight independently, acquire large size, and specialize entirely in running, one would expect them to acquire the ratite complex of characters even if the stem species of this assemblage did not have these characters. This consideration is quite independent of the question whether or not the families of birds which lost their power of flight were a closely knit group of related genera or only distantly related to each other. Potentially the same objection applies to the acquisition of diving adaptations by *Hesperornis,* grebes and loons. The arguments of the cladistic school do not weaken the validity of Darwin's warning against relying too much on *ad hoc* specializations (see also Mayr 1969:220, 223).

Hennig believes that one can distinguish convergence (and parallelism) from synapomorphy by "taking into account as many characters as possible" (1966:121). This is true in principle, but often impossible to implement. Schlee (1971:23) lists some additional criteria which, indeed, are quite helpful. However, when well-established phylogenies are carefully analyzed, it becomes obvious that an unequivocal decision can only rarely be reached. Owing to a very scanty fossil record of insects, as compared, for instance, with that of mammals, the insect taxonomists are not nearly so aware as they ought to be of the frequency of parallelism. For instance, "within the most advanced group (of therapsid reptiles), the cynodonts, several mammalian features (e.g., dentary-squamosal jaw articulation, loss of alternate tooth replacement, complex occlusion, and double-rooted check teeth) are known to have evolved independently in several phyletic lines" (Crompton and Jenkins 1973:138). Most workers also accept that the incorporation of the quadrate and articular into the middle ear must have occurred independently in therian and nontherian mammals. The frequency of the independent acquisition of identical adaptations in birds, for instance, a secondary jaw articulation (Bock 1959) or various specializations in different lines of woodpeckers (Bock and Miller 1959), and the frequency of independent phyletic advances in *Drosophila* (Throckmorton 1965) highlight the difficulties (see Bock 1967 and Peters 1972:168).

More disturbing is the fact that cladists, when actually constructing classifications, never seem to come to grips with the difficulties caused by parallelism and convergence. They simply ignore them.

The Information Content of Ancestral Characters

Cladists rightfully criticize the adoption of "mere similarity" as the only criterion of taxon delimitation, but so have evolutionary taxonomists for many years. Ghiselin (1969b), Simpson (1961), and I (1969) have consistently pointed out the pitfalls of basing a classification entirely on similarity, particularly (as proposed by the pheneticists) on unweighted similarity. However, this does not justify going to the other extreme by rejecting all consideration of similarity in the construction of a classification. To do so would result in the loss of a great deal of taxonomically important information. After all, similarity, when properly evaluated, is an important index of the amount of shared genotype and is the basis for the determination of homology. Cladists completely ignore this and as a result overlook the fact that the retention of a large number of ancestral characters is just as important an indicator of "relationship" (traditionally defined) as the joint acquisition of a few "derived" characters.

Two extremes have been proposed with respect to the relative importance of conservative (ancestral) and advanced (derived) characters. The cladists consider only the latter, while it is sometimes said that it is the study of conservative characters which is most apt to reveal relationship. The evolutionary taxonomist is convinced (and has acted on this basis for the last 100 years) that one must evaluate information from both types of characters in order to be able to construct a sound classification. Nothing could be further from the truth than the claim that "primitive similarities contain no phylogenetic information" (Cracraft 1972:383).

Let me illustrate this with two examples: There are three major families of living gallinaceous birds. Among these the Megapodiidae have the greatest number of primitive characters, while the Phasianidae have the greatest number of derived characters. The South American Cracidae are intermediate. They share a few derived characters with the advanced Phasianidae but a far greater number of primitive characters with the Megapodiidae. Traditionally, because they seemingly share to such a large extent the same genotype, the Cracidae have been said to be more closely related to the megapodes than to the Phasianidae. As a cladist, however, Cracraft (1972:283) insists that the Cracidae are "more closely related" to the Phasianidae because they share with them a few derived characters. I consider this to be a misleading statement. In the aggregate of their characters Cracidae are obviously much closer to the Megapodiidae than to the Phasianidae. To know that the Phasianidae branch off the same branch that leads to the Cracidae is important, but only part of the evidence that leads to a classification.

Let me cite another example which Hennig has recently used in order to illustrate the superiority of the cladistic approach (1971:12–14) (see Figure 30–7). In the Canadian amber of the Upper Cretaceous two fossil diptera were found which show relationship to two modern families: the Phoridae, a cosmopolitan family with some 2500 species, and the Sciadoceridae with two Recent species, both in the southern continents. In which of the two Recent families should one place these fossils? The original describers of the fossils (McAlpine and Martin 1966) acknowledged distant relationship with the Phoridae but placed the fossils in the Sciadoceridae because they are far more similar to species of that family than to any recent Phoridae.

The two Canadian fossils (*Sciadophora* and *Prioriphora*) differ from

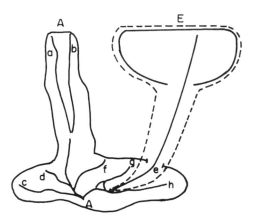

Figure 30-7. Two different interpretations of the taxonomic assignment of fossil flies. According to the traditional interpretation (solid outline) the fossils (*e*) belong to the Sciadoceridae (*A*); according to the cladistic interpretation (broken outline) they belong to the Phoridae (*E*). Letters *a-h* = sublines of the Sciadoceridae with varying evolutionary potentials, most of which (*c, d, f, g, h*) have become extinct.

the living Sciadoceridae consistently only in a single apomorph character, the absence of a discal cell. In two other apomorph characters (absence of anal cell, dorsal arista) the recent Sciadoceridae are variable. One of the two Canadian fossils (*Prioriphora*) acquired some additional phoridlike apomorph characters (loss of proscutellum, apically enlarged palpus, insertion of R_{4+5} far from the wing tip, m_1 not joining m_2), but in all these characters the other Canadian fossil (*Sciadophora*) retains the plesiomorph (Sciadoceridae-like) condition. Although the Canadian fossils (as a group) show one or two derived characters, they agree in the majority of their characters far better with the living Sciadoceridae than with the Phoridae. In particular, they lack the additional fusion and reduction in the length of wing vein r_3 and the loss of the second basal cell. Also in typical phorids the coxae and femora are stouter and the whole habitus is stockier.

In this case, and in several others I have studied, the entire cladistic reclassification rests on one or two characters. This is implicitly nothing but a return to a "single-character classification." In spite of the cladists' exhortations to base classifications on the holomorph (= totality of characters), in practice virtually all dichotomies in classifications are based on exceedingly few characters, often a single character pair. This has been rightly criticized by Darlington (1970:17).

Mosaic Evolution

As far as ancestral versus derived character states are involved, cladistics assumes that once "evolution" has "decided" to give one phyletic line a primitive character state and the "sister group" the derived character state, this difference will be perpetuated forever. In many instances, this is indeed the case. In many others, and this is completely ignored by the cladists, the potential for the derived character exists in all daughter lines of the original ancestor, as I have just pointed out, but is realized in the various daughter lines irregularly and at different rates, leading to parallel evolution. The claim "if derived character states can be identified, then monophyletic lineages can be constructed" (Cracraft 1972:381), is clearly not justified.

Cladists have not been able to overcome the difficulties caused by mosaic evolution. Yet it is not true (as has been claimed) that they believe that dichotomies create primitive and advanced *groups*. They realize, like all good taxonomists, that most groups possess a mixture of primitive and derived characters. Even the most primitive group of living mammals, the monotremes, has a number of derived characters.

However, a difficulty is created by the fact that newly acquired characters are sometimes lost again in subsequent evolution. Such a double apomorphy (= secondary primitiveness) would be masquerading as a plesiomorphy. Although some cladists are aware of this possibility, I do not recall that it was ever taken into consideration in the construction of cladistic classifications.

DARWIN AND CLASSIFICATION

Evolutionary taxonomists have long been convinced that they strictly follow Darwinian principles of classification by giving equal consideration to branching *and* to phyletic change. As Simpson (1961:52) has said: "Evolutionary taxonomy stems explicitly and almost exclusively from Darwin" (see Simpson 1959a for a more detailed discussion of Darwin's theory of classification). In recent years cladists have tried to claim Darwin for their side and Nelson has gone so far as to state: "If, indeed, there is a cladistic school Darwin is its founder and chief exponent" (1971:375). He has referred to cladistics as "the Darwin-Hennig classification" (1972:370). Is there any justification for this claim?

Darwin's *Origin* (1859) was the first major publication to propose evolution through common descent, at that time an entirely novel concept (quite different from Lamarck's concept of evolution). Genealogical language quite naturally played an important role in this volume. Like the modern evolutionary taxonomist, Darwin started out on a strictly genealogical basis: "The arrangement . . . must be strictly genealogical in order to be natural." Since any group of animals can have only one genealogy, as pointed out above, Darwin's postulate is an axiom for any evolutionary taxonomist. However, this was only the first step for Darwin, because he continues: ". . . but that the amount of difference in the several branches or groups, though allied in the same degree in blood to their common progenitor, may differ greatly, being due to the different degrees of modification which they have undergone; and this is expressed by the forms being ranked under different genera, families, sections, or orders" (1859:420; see also Ghiselin 1969a:84). It is significant that Darwin at this point refers back to his famous diagram in the fourth chapter (opposite page 116). Here he shows that 3 congeneric Silurian species (A, F, and I) evolved into 15 modern genera. These represent 3 "sister groups" derived from the 3 Silurian genera. The descendants of A and of I now constitute distinct families or even orders (p. 125). "But the existing genus F 14 may be supposed to have been but slightly modified; and it will then rank with the parent genus F; just as some few still living organic beings belong to Silurian genera" (p. 421). No more explicit statement could be wished for to refute the claim that Darwin was an exponent of cladistic classification.

Ghiselin and Jaffe (1973) have shown how frequently Darwin in his classification of the Cirripedia deviated from a cladistic classification: he places *Alcippe* in the Thoracica even though it is on the Abdominalia stem; *Pachylasma,* on the branch leading to the Balaninae, is included with the Chthamalinae; *Pollicipes* gives rise to the Lepadidae (with which it is included) and biphyletically to the stalkless cirripedes (Verrucidae and Balanidae). He could have adopted none of these classifications if he had followed the cladistic definition of monophyly. In each case Darwin established what the cladists would call a paraphyletic group.

In contrast to Nelson's is Brundin's claim (1972:107) that the Hennig method had revealed the "weaknesses of current neo-Darwinistic theory." No other cladist makes such claims. There are several indications in Brundin's writings that he fails to comprehend the Darwinian theory. Does he perhaps believe in some sort of orthogenesis, as implied by the statement: "The evolutionary process is far more orderly than admitted by the neo-Darwinists of today" (p. 119)?

CLADISTIC OBJECTIONS TO THE METHODS OF EVOLUTIONARY (COMBINED) CLASSIFICATION

Griffiths (1972:16) has stated that the "combined" (= evolutionary) grouping should be rejected because it "raises serious logical difficulties." Hennig, furthermore, has claimed repeatedly that a "synthetic" (1971:11) or "syncretistic" (1966:77) systematics is unable to elaborate a consistent system ("robs the combination of any scientific value") and leads to serious error. Griffiths (1972:15–17), however, is the only author who has tried seriously to enumerate and classify what he considers to be the shortcomings of the evolutionary method. In order to minimize future misunderstandings, let me attempt to refute some of his objections.

Griffiths states that I reject the endeavor to achieve "an unequivocal correspondence between phylogeny and classification above the species level" (1972:16). On the contrary, I object to Hennig's narrow definition of phylogeny, which forces the cladist to neglect at least half of the information phylogeny provides. It is my contention that the traditional method of classification reflects phylogeny to a far greater extent than Hennig's.

Griffiths further states that it is impossible to measure rates of evolutionary change accurately. Correct! Perhaps it is inappropriate to say a classification should reflect "rates of evolution." However, a classifier should not and cannot ignore the *results* of highly uneven rates of evolution. Even if it should be impossible to compare numerically the (slow) rate of evolution between the stem species and the modern crocodilians with the (rapid) rate between the same stem species and the modern birds, every beginner can see how much more drastically the birds differ from the common ancestor than the crocodilians. To ignore this altogether, because it cannot (yet) be measured accurately, would seem a poor escape from a difficulty.

A third objection is that the evolutionists fail to provide an objective definition of the categories above the species level. Their definition compares very unfavorably, claims Griffiths (1972:7–8), with "Hennig's proposal to define categories above the species level in terms of the origin of the stem species of the member taxa." Griffiths forgets that Hennig has traded biological meaning for a hoped-for logical consistency. As Darwin showed, the modern descendants of two Silurian sister species may be, biologically speaking, respectively an order and a genus. To give them the same categorical rank to satisfy Hennig's formalistic definition may be logically impeccable, but is simply wrong biologically. Hennig himself, in the meantime, has abandoned the claim to be able to give a nonarbitrary definition of the higher categories and refrains from placing higher taxa of insects into categories (see above). Griffiths' objections to the evolutionary criteria of ranking are based on the special cladistic definitions of monophyly and phylogeny. They become invalid and irrelevant when these definitions are rejected.

It is sometimes stated that the evolutionary taxonomists fail to provide definitions for categories. This is not correct. All I have emphasized is that the species category is the only category for which a nonarbitrary definition is possible. Both Simpson (1961) and I (1969) have provided formal definitions for the higher categories, even though for reasons we have stated such definitions have limits to their usefulness.

These are Griffiths' specified criticisms. More generally the cladist criticizes the evolutionary taxonomist for failing to provide simple criteria for making decisions in classification. It would seem to me that the nature of the material precludes a simplistic approach. The number of variables that must be considered in the construction of a classification is so large that simple methods will not work. This is the reason why the evolutionary taxonomist carefully weighs the evidence and uses his judgment in arriving at conclusions. He asks what role a higher taxon plays in the economy of nature. He considers the nature of the adaptive breakthrough which gave rise to the taxon (Peters 1972). In short, he insists on approaching his material as a biologist and evolutionist, rather than looking for automatic answers. In the short run this may create difficulties and uncertainties, but who would want to question that a classification which utilizes all potentially available information is more informative, more predictive, and indeed more truly reflective of past evolution than a classification which arbitrarily restricts itself entirely to the information provided by the branching pattern?

CONCLUSIONS

A sound classification of a group of organisms cannot be devised without a well-considered reconstruction of its phylogeny. One component of such a reconstruction is an establishment of the branching pattern of the various phyletic lines and the design of a cladogram. Hennig has demonstrated that this can be done in a relatively unequivocal manner by classifying characters into apomorphous (derived) and plesiomorphous (ancestral) characters.

The high information content of derived (apomorph) characters was appreciated by many taxonomists long before Hennig, but never sufficiently stressed, and, in fact, entirely ignored by some authors. The emphasis on the proper weighting of synapomorphies, under the impact of Hennig's cladistic theory, has been a healthy development in systematics.

It would require an impartial analysis to determine how many recent improvements in the classification of fishes, insects, and other groups were the result of a rigorous application of cladistic analysis. Cracraft (1972) has recently asserted that application of the cladistic method would revolutionize avian taxonomy. However, after 12 pages of discussion, he failed to produce even a single case in which he could demonstrate that the currently accepted classification was wrong. All he was able to show was that a classification based only on branching points is sometimes different from a classification in which phylogenetic divergence is given primacy. The case of the fossil Sciadoceridae (discussed above) is another illustration of a change in classification, but not necessarily an improvement. The reasons why the cladistic method cannot show more achievements should be obvious from the previous discussion.

Cladistic grouping encounters many difficulties even if we exclude unacceptable decisions in ranking. For instance, very often there are not enough apomorphous characters available, or else there is doubt as to which of two alternative character states is ancestral and which is derived, and, finally, owing to mosaic or parallel evolution, there may be conflict between the information provided by different apomorphous characters.

As valuable as the cladistic analysis is, it does not automatically provide a classification. By making use of only one of the two available sets of phylogenetic data, cladistics produces classifications which are less able to serve as general reference systems than evolutionary classifications, because they have a poorer information content. Evolutionary taxonomists, by weighting the information from both sources, arrive at classifications which may be criticized as being more "subjective," but which reflect evolutionary history more accurately and are therefore more meaningful biologically. The cladist believes that the simple adding up of synapomorphies will provide the correct classification, so to speak, automatically. The evolutionary taxonomist, in contrast, feels that only a careful weighing of all the evidence will reveal meaningful degrees of relationship, in the sense of inferred genetic relationship. He feels, furthermore, that a classification must pay attention to major adaptive events in evolutionary history, like becoming terrestrial or airborne, since these are of greater importance for the ranking of taxa than the mere splitting of phyletic lines.

Johnson (1970) and Michener (1970:20–22) are other systematists who have recently stated the case in favor of evolutionary (synthetic) systematics.

Griffiths (1972:17) has proposed that if one wants to express the effects of different degrees of evolutionary divergence in different phyletic lines, one should for this purpose construct an entirely separate classification, in other words, that one should have several (or at least two) sets of classifications. This proposal strikes me as altogether impractical. Peters and Gutmann (1971:256) likewise reject the purely formalistic approach of the Hennig school and demand that it be replaced by a biological attitude toward the morphological evidence. This includes giving due consideration to the size of adaptive breakthroughs.

An eclectic classification which considers with equal care the branching points in phylogeny and all aspects of phylogenetic divergence would seem the best way to generate biologically meaningful classifications, permitting the greatest number of broad generalizations. Darwin's advice to use both these sources of information has been adopted by the most successful classifiers of the last 100 years.

In conclusion, it is evident that, no matter how useful cladistic analysis is, it cannot be automatically translated into a classification.

REFERENCES

Ashlock, P. 1971. Monophyly and associated terms. *Syst. Zool.*, 20:63–69.

——1972. Monophyly again. *Syst. Zool.*, 21:430–438.

——1974. The uses of cladistics. *Ann. Rev. Ecol. Syst.*, 5:81–99.

Bigelow, R. 1956. Monophyletic classification and evolution. *Syst. Zool.*, 5:145–146.

Bock, W. 1959. Preadaptation and multiple evolutionary pathways. *Evolution*, 13:194–211.

——1963. Evolution and phylogeny in morphologically uniform groups. *Amer. Nat.*, 97:265–285.

——1965. The role of adaptive mechanisms in the origin of higher levels of organization. *Syst. Zool.*, 14:272–287.

——1967. The use of adaptive characters in avian classification. *Proc. 14th Int. Ornith. Congr.* (1966), pp. 61–74.

——1968. Review of Hennig, *Phylogenetic systematics. Evolution*, 22:646–648.

——1969. Comparative morphology in systematics. In *Systematic Biology*, pp. 411–448. National Academy of Sciences, Washington, D.C.

Bock, W., and W. Miller. 1959. The scansorial foot of the woodpeckers, with comments on the evolution of perching and climbing feet in birds. *Amer. Mus. Novit.*, 1931.

Brundin, L. 1966. Transantarctic relationships and their significance, as evidenced by chironomid midges. With a monograph of the subfamilies Podonominae and Aphroteniinae and the austral Heptagyiae. *Kungl. svenska Vetensk. Akad. Handl.*, 4:11.

——1972. Evolution, causal biology, and classification. *Zool. Scripta*, 1:107–120.

Cain, A. J. 1967. One phylogenetic system, a review of *Phylogenetic systematics* by W. Hennig. *Nature,* 16:412–413.

Cain, A. J., and G. A. Harrison. 1960. Phyletic weighting. *Proc. Zool. Soc. London,* 135:1–31.

Cracraft, J. 1972. The relationships of the higher taxa of birds: problems in phylogenetic reasoning. *Condor,* 74:379–392.

Crompton, A. W., and F. A. Jenkins. 1973. Mammals from reptiles: a review of mammalian origins. In *Annual review of earth and planetary sciences,* pp. 131–155.

Crowson, R. A. 1970. *Classification and biology.* Heinemann, London.

Darlington, P. J. 1970. A practical criticism of Hennig-Brundin *Phylogenetic systematics* and antarctic biogeography. *Syst. Zool.,* 19:1–18.

Darwin, C. 1859. *On the origin of species by means of natural selection, or the preservation of favoured races in the struggle for life.* J. Murray, London.

Farris, J. 1966. Estimation of conservatism of characters by consistency within biological populations. *Evolution,* 20:587–591.

——1967. The meaning of relationship and taxonomic procedure. *Syst. Zool.,* 16:44–51.

Ghiselin, M. T. 1969a. *The triumph of the Darwinian method.* University of California Press, Berkeley.

——1969b. The principles and concepts of systematic biology. In *Systematic Biology,* pp. 45–55. National Academy of Sciences, Washington, D.C.

Ghiselin, M. T., and L. Jaffe. 1973. Phylogenetic classification in Darwin's monograph on the subclass Cirripedia. *Syst. Zool.,* 22:132–140.

Gisin, J. 1964. Synthetische Theorie der Systematik. *Z. f. zool. Syst. Evolut.-forsch.,* 2:1–17.

Greenwood, P. H., and D. E. Rosen, 1971. Notes on the structure and relationship of the Alepocephaloid fishes. *Amer. Mus. Novit.,* 2473:1–41.

Griffiths, G. D. C. 1972. The phylogenetic classification of Diptera Cyclorrhapha with special reference to the structure of the male postabdomen. W. Junk, N.V., The Hague.

Günther, K. 1956. Systematik und Stammesgeschichte der Tiere. *Fortschr. Zool.,* 10:37–55.

——1971. Abschliessende Zusammenfassung der Vorträge und Diskussionen. In *Erlanger Forschungen,* ed. R. Siewing. Erlangen.

Heberer, G., ed. 1943. *Die Evolution der Organismen: Ergebnisse und Probleme der Abstammungslehre.* G. Fischer, Jena.

Hennig, W. 1950. *Grundzüge einer Theorie der Phylogenetischen Systematik.* Deutscher Zentralverlag, Berlin.

——1960. Die Dipteran-Fauna von Neuseeland als systematisches und tiergeographisches Problem. *Beitr. z. Ent.,* 10:15–329.

——1966. *Phylogenetic systematics.* University of Illinois Press, Urbana.

——1969. *Die Stammesgeschichte der Insekten.* Senckenberg-Buch 49, Frankfurt.

——1971. Zur Situation der biologischen Systematik. In *Erlanger Forschungen,* ed. R. Siewing. Erlangen.

Hull, D. 1967. Certainty and circularity in evolutionary taxonomy. *Evolution,* 21:174–189.

——1970. Contemporary systematic philosophies. *Ann. Rev. Ecol. Syst.,* 1:19–54.

Huxley, J., ed. 1940. *The new systematics.* Clarendon Press, Oxford.

——1942. *Evolution: the modern synthesis.* Allen and Unwin, London.

——1958. Evolutionary processes and taxonomy, with special reference to grades. *Uppsala Univ. Arsskr.,* 6:21–39.

Illies, J. 1961. Phylogenie und Verbreitungsgeschichte der Ordnung *Plecoptera*. *Verh. Deutsch. Zool. Gesellsch. Bonn Zool. Anz. Suppl.*, 25:384–394.

Inger, R. 1967. The development of a phylogeny of frogs. *Evolution*, 21:369–384.

Jepsen, G., E. Mayr, and G. G. Simpson. 1949. *Genetics, paleontology, and evolution*. Princeton University Press, Princeton.

Johnson, L. A. S. 1970. Rainbow's end: the quest for an optimal taxonomy. *Syst. Zool.*, 19:203–239.

Kiriakoff, S. G. 1959. Phylogenetic systematics versus typology. *Syst. Zool.*, 8:117–118.

Kluge, A., and J. Farris. 1969. Quantitative phyletics and the evolution of Anurans. *Syst. Zool.*, 18:1–32.

Mayr, E. 1942. *Systematics and the origin of species*. Columbia University Press, New York.

———1963. *Animal species and evolution*. Belknap Press of Harvard University Press, Cambridge, Mass.

———1965. Numerical phenetics and taxonomic theory. *Syst. Zool.*, 14:73–97.

———1969. *Principles of systematic zoology*. McGraw-Hill, New York.

Mayr, E., and L. Short, 1970. *Species taxa of North American birds*. Nuttall Ornith. Club, Cambridge, Mass.

McAlpine, J., and J. Martin. 1966. Systematics of Sciadoceridae and relatives with descriptions of new genera and species from Canadian amber and erection of family Ironomyiidae (Diptera: Phoroidea). *Canad. Ent.*, 98:527–544.

Michener, C. D. 1970. Diverse approaches to systematics. *Evol. Biol.*, 4:1–38.

Mill, J. S. 1874. *A system of logic, ratiocinative and inductive, being a connected view of the principles of evidence and the methods of scientific investigation*, 8th ed. Longmans, Green, London.

Nelson, G. 1971. Cladism as a philosophy of classification. *Syst. Zool.*, 20:373–376.

———1972. Comments on Hennig's *Phylogenetic systematics* and its influence on ichthyology. *Syst. Zool.*, 21:364–374.

Osche, G. 1961. Aufgaben und Probleme der Systematik am Beispiel der Nematoden. *Verh. Deutsch. Zool. Gesellsch. Bonn*, 1960:329–384.

———1965. Über latente Potenzen und ihre Rolle im Evolutionsgeschehen. *Zool. Anz.*, 174:411–440.

———1971. Discussion comments. In *Erlanger Forschungen*, ed. R. Siewing. Erlangen.

Pennack, R. 1964. *The collegiate dictionary of zoology*. Ronald Press, New York.

Peters, D. S. 1970. *Über den Zusammenhang von biologischem Artbegriff und phylogenetischer Systematik*. Aufsätze u. Red. Senckenberg naturforsch. Ges. Waldemar Kramer, Frankfurt.

———1972. Das Problem konvergent entstandener Strukturen in der anagenetischen und genealogischen Systematik. *Z. f. zool. Syst. Evolut.-forsch.*, 10:161–173.

Peters, D. S., and W. F. Gutmann, 1971. Über die Lesrichtung von Merkmals- und Konstruktions-Reihen. *Z. f. zool. Syst. Evolut.-forsch.*, 9:237–263.

Popper, K. R. 1959. *The logic of scientific discovery*. Hutchinson, London.

———1963. Conjectures and refutations. Routledge and Kegan Paul, London.

Remane, A. 1952. *Die Grundlagen des natürlichen Systems, der vergleichenden Anatomie und der Phylogenetik*. Geest and Portig, Leipzig.

Rensch, B. 1947. *Neuere Probleme der Abstammungslehre. Die transspezifische Evolution*. F. Enke, Stuttgart.

Schindewolf, O. H. 1967. Über den "Typus" in morphologischer und phylogenetischer Biologie. *Abh. Akad. Wiss. Lit. Math. Naturw. Klasse* (Mainz), n.s., 4:57–131.

Schlee, D. 1971. *Die Rekonstruktion der Phylogenese mit Hennigs Prinzip.* Aufsätze u. Red. Senckenberg naturforsch. Ges. Waldemar Kramer, Frankfurt.

Siewing, R., ed. 1971. Methoden der Phylogenetik. In *Erlanger Forschungen,* Reihe B, 4. Erlangen.

Simpson, G. G. 1945. The principles of classification and a classification of mammals. *Bull. Amer. Mus. Nat. Hist.,* 85:1–350.

———1953. *The major features of evolution.* Columbia University Press, New York.

———1959a. Anatomy and morphology: classification and evolution, 1859 and 1959. *Proc. Amer. Phil. Soc.,* 103:286–306.

———1959b. The nature and origin of supraspecific taxa. *Cold Spring Harbor Symp. Quant. Biol.,* 24:255–271.

———1961. *Principles of animal taxonomy.* Columbia University Press, New York.

Sokal, R., and P. Sneath, 1963. *Principles of numerical taxonomy.* Freeman Press, San Francisco.

Throckmorton, L. 1962. The problem of phylogeny in the genus *Drosophila. Studies in Genet.,* (University of Texas), 2:207–343.

———1965. Similarity versus relationship in *Drosophila. Syst. Zool.,* 14:221–236.

Tillyard, R. 1921. A new classification of the order Perlaria. *Canad. Ent.,* 53:35–43.

Warburton, F. E. 1967. The purposes of classification. *Syst. Zool.,* 16:241–245.

VI
THE SPECIES

Introduction

Perhaps no other scientific problem has occupied my attention for as many years as the species problem. The species is a crucially important unit in taxonomy, in evolution, and in ecology. Naturalists have wrestled with the species problem for more than 300 years. It would be gratifying if one could say that this attention has led to complete clarity. Unhappily, this is not the case. A glance at the recent literature reveals how much uncertainty and disagreement still exists.

The picture, however, is not all black. Compared with the understanding of species at the time of Darwin, our present understanding is a decisive advance. Not only do we see much more clearly what the difficulties are, but we can also say that some previously existing confusions have been definitely removed.

As far back as the late nineteenth century, there have been authors who had a remarkably sound perception of what a species is. Seebohm, K. Jordan, and Poulton were outstanding among these. Their species concept was based on their systematic studies and represented the first formulations of what we now call the biological species concept. However, they articulated only insufficiently the underlying reasons for their own species concept and for the rejection of previously prevailing concepts. For this reason their advanced thinking did not have as much of an impact as it deserved.

The last 30 to 40 years have witnessed a concerted effort to clarify the conceptual basis of what the biologist calls species. One might perhaps single out four areas in which decisive advances were made within this period:

1. The recognition that the word *species* may refer to a taxon or to a category. *This means that two sets of difficulties exist, those of the delimitation of particular species taxa and those of the definition of the word* species, *that is, of the species category. Furthermore, it means that in order to be able to delimit species taxa intelligently, one must first have a yardstick that permits decisions in doubtful cases, that yardstick being the species concept. All recent attacks on the biological species concept contained an appalling confusion of these two appli-*

cations of the word species. *Difficulties in the delimitation of species taxa do not necessarily signify confusion about the species category.*

2. The recognition that species taxa always consist of populations. *The designation of certain species as polytypic refers to species taxa and has nothing to do with the species concept. Populations in sexual species are the temporary manifestations of gene pools, and these gene pools or Mendelian populations are the basic components of species rather than individuals. Species clearly are not classes of individuals such as exist for inanimate objects. Indeed, each gene pool can be regarded as an individual.*

3. The species concept is a relational concept. *Like the word* brother *(which has meaning only in relation to other sibs), the species concept does not refer to an intrinsic property but to the relation of a species population to others, the relation being that of reproductive isolation. This relationship can be defined nonarbitrarily. All the higher categories are intrinsically defined, that is, the taxa contained in the higher categories are characterized by certain properties. Since there is no nonarbitrary method for determining which properties justify higher categorical rank, higher categories, in contrast to species, cannot be defined nonarbitrarily. This is why there is so much uncertainty about higher categories and so much disagreement among specialists about the delimitation of higher taxa and their ranking.*

4. The biological significance of species is twofold: *(a) Each species is a reproductive community, as specified under point 3 above; (b) each species plays a highly specific role in the household of nature; it occupies a species-specific niche and plays a definite role in the ecosystem. This duality in the biological nature of species is important in the consideration of the process of speciation.*

This summary is my current evaluation of the problem of the nature of species. The ensuing essays, covering a period of more than 20 years, reflect my continuous struggle with this problem and the gradual approach to the position I now hold. I suspect that these presentations not only are of historical interest but also may appeal to various readers who do not go along with me all the way to my present position.

Toward a Modern Species Definition

What is a species?

This question has bothered taxonomists and biologists alike ever since biologists, especially Linnaeus, firmly established the species concept, but the confusion has grown steadily with the constant refinement of taxonomic technique and with the advent of evolutionary thought. Unfortunately, we cannot dispose of the question simply by dodging the definition as irrelevant, because, for example, a good many controversies are primarily due to the different species concepts held by the opponents (see Goldschmidt 1937).

The difficulties of and possible objections to a species definition can best be illustrated by analyzing and criticizing a number of the definitions given by well-known taxonomists or geneticists. To Linnaeus the species was a unit that could be defined on a morphological basis. Consequently, he described in numerous cases males and females, young and adult, as different species, because they showed well-definable morphological differences. Since that time, but particularly during the last 30 to 40 years, the species concept in most taxonomic groups has undergone an almost revolutionary change. Painstaking taxonomic work, particularly by ornithologists, lepidopterologists and malacologists, has shown that gliding intergradation connects most geographically representative species, so that they had to be reduced to subspecies. In Darwin's day the discovery of such transitions was hailed as proof of the change of one species into another. Nowadays we know that these intergradations are the normal condition and that there are relatively few "good" species that are not actually composed of groups of "subspecies." A few taxonomists, especially entomologists, are left who still insist on calling species the lowest systematic categories. The best-known advocate of this is Kinsey (1937), whose point of view is as follows:

Confusion will be avoided if we call the basic taxonomic unit

Adapted from pp. 250–256 of "Speciation phenomena in birds," *American Naturalist*, 74(1940):249–278.

the species. It is the unit beneath which there are in nature no subdivisions, which maintain themselves for any length of time or
over any large area. The unit is variously known among taxonomists as the species, subspecies, variety, *Rasse* or geographic race.
It is the unit directly involved in the question of the origin of species, and the entity most often indicated by non-taxonomists
when they refer to species. Systematists often introduce confusion
into evolutionary discussions by applying the term to some category above the basic unit.

Goldschmidt (1937) has already voiced some objections to these
claims, but I might add a few others. It is not true that "there are in nature no subdivisions" below the species (of Kinsey) "which maintain
themselves for any length of time." Actually there are all degrees of
distinctness between "the effective breeding population" within a
continuous array of populations and the subspecies that is completely
isolated by geographical barriers. Recent genetic work by Dobzhansky
and others, as well as Kinsey's own taxonomic work, shows this quite
clearly. Furthermore, the lowest category is not "the entity most often
indicated by non-taxonomists when they refer to species." When the
layman or nontaxonomic biologist speaks of "the song sparrow," he is
not concerned with the numerous subspecies of this species, as for example the Atlantic, the Eastern or the Mississippian race. He means the
total sum of all these races, or else the particular local race wherever he
meets it. Nor do we call the human races species, although they are the
basic taxonomic units of *Homo sapiens.* Timofeeff-Ressovsky (1940)
overcomes most of the difficulties of other definitions, but does not
adequately deal with the problem of reproductive isolation. He writes:
"A species is a group of individuals that are morphologically and
physiologically similar (although comprising a number of groups of the
lowest taxonomic category) which has reached an almost complete
biological isolation from similar neighbouring groups of individuals inhabiting the same or adjacent territories. Under biological isolation we
understand the impossibility or nonoccurrence of normal hybridisation
under natural conditions." Most geographically isolated forms would
be species under this definition. Dobzhansky (1937) wants to overcome
this difficulty (which he realizes very clearly) when he defines species
as "that stage of the evolutionary process at which the once actually or
potentially interbreeding array of forms becomes segregated in two or
more separate arrays which are physiologically incapable of interbreeding." This definition is an excellent description of the process of speciation, but it defines what a group of species is (that is, two or more
separate arrays), not what *one* species is. My principal objection to
Dobzhansky's definition is that he leaves difficult situations of practical
importance undecided. For example, how does his definition deal with
those cases where a species (in the widest sense, not in Dobzhansky's)
consists of a long geographical chain of different populations or subspe-

cies, of which each one intergrades with its neighbors and is not separated by any physiological isolating mechanisms, but where the final members of the chain are completely sterile when brought together. There are several such cases known from beetles (Epilachna, Carabus) and a number of less drastic ones from other groups. Second, how can it be practically tested that two forms are "physiologically incapable of interbreeding"? There are probably more than a million species of animals in existence and in less than 1/10 of 1 percent have crosses with one another been attempted. Should we wait until the other 99 9/10 percent have been crossed before these species can be fully established?

Before I try to give my own species definition, I might analyze very briefly what criteria are generally used to define species.

Morphological characters. Descriptive characters, such as structure, proportions, color patterns, and so on, are the conventional means used to define a species. It is therefore only natural if even modern species definitions state that a species is composed of "groups of individuals with similar morphological characters." However, geographical races differ in certain families (Phasianidae, Paradisaeidae, Trochilidae) by "stronger" morphological characters than good species in others (Apodidae, Tyrannidae, Zosteropidae). Perfectly good species are often remarkably similar—I only need to remind you of *Drosophila pseudoobscura* and *D. miranda*—while in other cases the extreme links of a long chain of subspecies are more different from each other than are most related species. Rensch (1929, 1938, 1939) has pointed out that nearly every morphological character which has been used to separate species may also vary geographically within the species. Furthermore, morphological characters often vary independently from fertility and are therefore of no value in the all-important borderline cases. Morphological characters are, therefore, of no decisive value in a species definition, because *there is no difference in morphological characters between subspecies and species.*

Genetic distinctness. In the earlier years of the science of genetics, great stress was laid on the genetic distinctness of two species. Now we know not only that all subspecies are genetically different, but also that populations within subspecies are genetically different (Dobzhansky for *D. pseudoobscura,* Sumner and Dice for *Peromyscus,* Goldschmidt for *Lymantria dispar,* etc.). In fact, every individual is a different biotype. We readily admit this in regard to man and domestic animals and plants, but it is equally true for all other animals. Genetic distinctness, being a *sine qua non* condition, is therefore of little value in a species definition.

Lack of hybridization. This criterion is only of limited value. If no hybridization is possible between two neighboring populations, then there is little doubt that they are good species. Occurrence of some hybridization does not, however, necessarily mean that the respective populations are not good species. Many good species of animals are capable of producing hybrids in captivity, but never interbreed in nature.

A successful species definition should not lay too much stress on any one of the three above-listed criteria. Of all the definitions that have come to my attention within recent years, the one proposed by Sewall Wright (1940) seems to show the fewest flaws. According to him, species are "groups within which all subdivisions interbreed sufficiently to form intergrading populations wherever they come in contact, but between which there is so little interbreeding that such populations are not found." This definition again omits reference to those "subdivisions" which have the misfortune of not being able to come in contact with other subdivisions because they are spatially or ecologically isolated.

I might therefore propose the following emended definition: *A species consists of a group of populations which replace each other geographically or ecologically and of which the neighboring ones intergrade or hybridize wherever they are in contact or which are potentially capable of doing so (with one or more of the populations) in those cases where contact is prevented by geographical or ecological barriers."* Such a definition is applicable in practically all difficult cases of bird taxonomy except where the terminal links of a chain of races overlap. It remains to be seen how useful it is when applied in other groups, although Remane (1927), who is not an ornithologist, arrived at a very similar formulation. As a taxonomist, I am, of course, interested in a practical definition, and a definition like Dobzhansky's is of little use in taxonomic work. Even the formulation given above is at best only an approach. In many cases of interrupted distribution, it is necessary to leave it to the judgment and the systematic tact of the individual taxonomist to decide whether or not two particular forms are "potentially capable" of interbreeding, in other words, whether they are species or subspecies.

REFERENCES

Dobzhansky, T. 1937. *Genetics and the origin of species.* Columbia University Press, New York.

Goldschmidt, R. 1937. Cynips and Lymantria. *Amer. Nat., 71*:508–514.

Kinsey, A. C. 1937. Supra-specific variation in nature and in classification. *Amer. Nat., 71*:206–222.

Remane, A. 1927. Art und Rasse. *Verh. Ges. Phys. Anthrop.,* pp. 2–23.

Rensch, B. 1929. *Das Prinzip geographischer Rassenkreise und das Problem der Artbildung.* Borntraeger Verlag, Berlin.

—— 1938. Some problems of geographical variation and species formation. *Proc. Linn. Soc. London, 150*:275–285.

—— 1939. Typen der Artbildung. *Biol. Rev., 14*:180–222.

Wright, Sewall. 1940. The statistical consequences of Mendelian heredity in relation to speciation. In *The new systematics,* ed. J. Huxley, pp. 161–183. Clarendon Press, Oxford.

Timofeeff-Ressovsky, N. W. 1940. Mutations and geographical variation. In *The new systematics,* ed. J. Huxley, pp. 73–136. Clarendon Press, Oxford.

Karl Jordan and the
Biological Species Concept

Again and again in his writings Karl Jordan comes to grips with the species concept. He realizes clearly that this concept is not only fundamental for the practical work of systematics, but equally crucial for studies of speciation and evolution, as is so painfully evident from the writings of those who have ignored the species problem in their discussions. In order to understand and evaluate Jordan's own viewpoint and his contributions to the subject, it will be necessary to outline the current status of this problem. It is rather evident from the fact that the controversy is still continuing (Gregg 1950; Burma 1954; Burma and Mayr 1949) that a full understanding of the species problem has not yet been reached. After a thorough study of most of the literature on this subject, I have come to the startling conclusion that the disagreement is due to the fact that there are in existence not merely one but actually three entirely different species concepts. All past arguments and discussions have suffered from the fact that an author has either championed one of the three concepts against the others or has wavered between two of these without realizing it.

The first of these concepts is the typological one, which goes back to the *eidos* of Plato. Such a species is a "different thing." Implicit in this concept is that variation as such is unimportant since it represents only the "shadows" of the *eidos*. Translated into biology, the typological concept becomes the morphological species concept. Certain individuals are a different species "because they are different." When a mineralogist speaks of "species of minerals" or a physicist of "nuclear species" he has this typological species concept in mind. This concept is still widespread in certain branches of invertebrate zoology and in paleontology. The objections to this concept are manifold and have been stated by Jordan with great clarity and vigor. First of all, it is a strictly subjective concept. "It would almost appear, in fact, as if a 'species' is that which a respective author chooses to consider a 'species'" (1896:426). Inevitably, this concept cannot be applied without all sorts

Adapted from pp. 51–57 of "Karl Jordan's contributions to current concepts in systematics and evolution," *Transactions of the Royal Entomological Society of London,* 107(1955): 45–66.

of inconsistencies. "If it is the presence of morphological difference which leads us to split up in the one case, and the absence of such difference in the other case to unite, why then are not *Distomum, Redia,* and *Cercaria; Rhabdomena* and *Rhabditis; Vanessa levana* and its offspring *prorsa,* the same species? Morphological difference alone is not a criterion of specific distinctness" (1896:434). After citing numerous other cases that show the worthlessness of a typological-morphological species concept, Jordan comes "to the conclusion that morphological differences of any kind and degree are not decisive criteria as to specific distinctness; the systematist actually sinks his species in spite of distinguishing characters as soon as it is proved that the morphologically different forms appear among the offspring of the same female. The most general case of bodily difference which is not regarded as being specific is the difference between males and females; notwithstanding the great dissimilarity which the sexes so often exhibit, not only in the reproductive organs, but also in other morphological characters, the systematist puts male and female together in one species, and hence makes at once the concession that his term 'species' is not a purely morphological one, but that the higher criterion of the term is of a physiological kind" (1896:436). One has only to read the papers of De Vries and other contemporary plant breeders and early Mendelians to appreciate how far ahead of them Jordan was in his thinking. Having laid to rest the morphological concept so decisively, Jordan was ready to investigate what the physiological criteria are that characterize species. This led him to a consideration of the second species concept.

This concept is altogether different from the typological species. It does not deal with things, describing their degree of difference, but specifies a relationship. It is a concept like the word *brother,* which has a meaning only with reference to some other object. This "biological" or "nondimensional" species concept describes the relationship of two natural populations that coexist at the same locality and specifies this relationship as "noninterbreeding." The concept has been designated as "objective" because it can be defined as objectively as the word "brother" and many similar designations of relationship. It has nothing to do with the objectivity of the concept itself that its *application* to taxonomic situations may occasionally run into difficulties, just as it is occasionally difficult to establish the father of a baby even though there is nothing subjective about the biological concept "father" itself.

The revolt against the typological species concept and the attempt to replace it by a more biological concept had started long before Jordan. Indeed, it is this concept of the nondimensional species, expressing the relationship of noninterbreeding among sympatric populations of a single locality, which was the foundation of the original biological species concept of John Ray (1686), of that of Linnaeus, and of that of all local naturalists since. Although they had arrived at this concept empirically, they had not thought it through sufficiently to describe it and define it unequivocally. That a species is a reproductively isolated population had also been previously expressed. Eimer, for instance, had spoken of species as "groups of individuals modified in such a manner that interbreeding between them and other groups no longer takes place or is not possible with success indefinitely" (1889:16). Poulton (1903:94) had said that "the idea of a species as an interbreeding community, as syngamic, is I believe the more or less acknowledged foundation of the importance given to transition" in geographic variation. Yet in all my search through the literature I have not found a single statement by any other author which indicates an understanding equal to that of Jordan. He was the first to point out the completely objective nature of the species concept as a measure of relationship in a nondimensional situation (1905:157): "Individuals connected by blood relationship form a single faunistic unit in an area, to which unit we must add all the other individuals of the area which resemble them. This is the cornerstone of the building of the systematist and the starting point for an exact analysis of the correctness of the theory of evolution." And speaking of the single locality of the neighborhood of Goettingen, his alma mater, he continues: "The three common *Pieris* of the gardens of Goettingen, the carab beetles of the Hainberg, the physopodes in the flowers of the Botanical Garden, the mice on the fields of the Weend, the bembids on the sand of the shore of the Leine, they all prove that the living inhabitants of a region are not a chaotic mass of intergrading groups of individuals, but that they are composed of a finite number of distinct units which are sharply delimited against each other and each of which forms a closed unit . . . The units, of which the fauna of an area is composed, are separated from each other by gaps which at this point are not bridged by anything. This is a fact which can be tested by any observer. Indeed all faunistic activity begins with the searching out of these units. A list of the species that occur in a region is an enumeration of such independent units which with Linnaeus we call species" (1905:157). This masterly statement, published in 1905, is as true today as it was then. If it had not been ignored by so many taxonomists and nontaxonomists since that time, we could have saved ourselves much useless argument.

The third species concept is that of the multidimensional species. While the nondimensional species defines a relationship of two populations, the multidimensional species is a grouping of populations. It is thus essentially the same as any higher category, a grouping of units of the next lower category, as the genus, for instance, is a grouping of species. This third species concept, then, is a collective concept and, like all collective concepts, it lacks in principle the precision and objectivity which is generally inherent in relational concepts. If we group a number of geographically representative populations under the heading of a single polytypic species, we inevitably have to make a number of subjective judgments. As long as these populations are interbreeding and intergrading with each other there is usually little conceptual difficulty. In such cases it is generally granted that the combined populations conform to the criterion of the nondimensional species concept and form a single reproductive community. As soon as populations are added to this polytypic species that are separated in space or time, the judgment becomes truly subjective. The classifying taxonomist is faced here with the same situation as he is on the generic level when he must make a decision on the status of an aberrant species that is considered congeneric by some authors and generically different by others.

It was an inevitable consequence of the great period of geographical exploration that the nondimensional species of the local naturalist had to be expanded into a multidimensional species, a geographically variable species. Early authors were confused about what to do with geographical varieties nomenclaturally. Some, like Pallas, described the differing populations of new localities as "varieties," and so did Linnaeus himself in some cases. Other authors, limiting themselves rigidly to the classical binominal system of Linnaeus, called each of these different populations a separate species. Eventually a system of triple names won out, with the geographical varieties listed as subspecies. That this new method produced a basic conflict of two very different species concepts was overlooked, either consciously or unconsciously. The first author, I believe, who clearly realized that the nondimensional species of Linnaeus and the multidimensional species that was so rapidly gaining favor in the second half of the nineteenth century are not one and the same thing was O. Kleinschmidt. He restricted the term species to the nondimensional relation and coined for the collective species the term "Formenkreis," later emended by Rensch into "Rassenkreis." This action, though inconvenient for practical purposes, was entirely logical. It also presented the historical developments correctly. The recent attempt to restrict the term "species" to the multidimensional concept and to apply the group symbol of symbolic logic to the concept species, as was done by Gregg (1950), is based on a misunderstanding of the historical developments.

It is only natural that Jordan, as a member of the Tring team, endorsed wholeheartedly the application of the concept of polytypic species to natural populations. Hartert had earlier come to the conclusion that isolated populations should not be excluded from polytypic species if they seemed to belong to them on the basis of corroborating evidence. Jordan repeatedly explained why such a decision is inevitable. Since it is known from experiments and observation that strikingly different individuals may be members of a single interbreeding population, "it is *a priori* evident that also geographically separated forms, in spite of their being morphologically distinct and in spite of their not being connected with one another by intergradations, can very well be subspecies of one species, i.e., that they can under favorable circumstances fuse into one form. The actual proof of specific distinctness the

systematist as such cannot bring . . . we work with the mental reservation that the specific distinctness of our species novae deduced from
morphological differences will be corroborated by biology" (1896:
450–451). The viewpoint of the Tring school was at first not well
received among taxonomists, and Jordan found it necessary to return to
this argument in 1905. He uses a discussion of the geographic variation
in the genitalic structures and other taxonomic characters of the butterfly *Papilio dardanus* to prove, in a brilliant, closely reasoned argumentation, that complete continuity of populations is not necessary in order
to permit combining geographically isolated subspecies into polytypic
species (1905:196–197). The tremendous simplification of classification these arguments have made possible is now a matter of history.
Yet there are some contemporary biologists who still fail to make a
distinction between the *reproductive* isolation of sympatric populations, which proves that they are different species, and the *geographic*
isolation of allopatric populations, which does not necessarily prove it.
I recommend to them reading Jordan's arguments, which have as much
weight today as they had when he first stated them.

In retrospect we can understand why taxonomists have had so much
difficulty during the past 100 years in applying the species concept to
the purely practical task of classifying natural populations. They did
not realize that there is no such thing as "the species concept" but
rather three different concepts, each of which may permit a different
conclusion in a particular situation. Jordan himself was no exception in
this respect. In his various attempts at a species definition he wavered
between the three stated concepts and finally attempted to include in
his species definition criteria taken from all three. It is still of interest
to study his discussion of the species definition as given in 1896
(p. 438). His definition then reads "a species is a group of individuals
which is differentiated from all other contemporary groups by one or
more characters, and of which the descendants which are fully qualified
for propagation form again under all conditions of life one or more
groups of individuals differentiated from the descendants of all other
groups by one or more characters." It is evident that as a taxonomist he
feels that in his practical work he cannot proceed with a species definition that omits reference to morphological criteria. "The question of
specific distinctness or non-distinctness is therefore two-fold: first, one
of morphological, and second, one of physiological difference" (1896:
442). He continues by pointing out that in most cases the systematist is
not able to test by experiments the presence of physiological distinction and that therefore he can never prove with certainty from
specimens alone whether the distinguishing morphological characters
are of specific value or not. However, experience shows that physiological isolation and degree of morphological difference are so closely
correlated that one is permitted, in the case of isolated populations, to
determine their taxonomic status by comparing their morphological

distinctness with that between related species and subspecies. Therefore, "If in a given case we have to decide whether A and B, which live
together, are two different species, or two forms of one species, the
morphological characters of A compared with those of B and the geographical representatives of B will have to guide us in our judgement"
(1896:453). "The same kind of evidence we may employ when we have
to come to a decision as to the specific distinctness of geographically
separated forms which are not connected by intergradation. We must
accept as a general law that forms which are connected by all intergradations, or forms which overlap in characters, are specifically identical." This statement, made in 1896 (p. 454), is still somewhat vague. By
1905 Jordan had considerably clarified his ideas. He states (1905:157)
that the working taxonomist is usually faced with two potential difficulties. The first is to decide whether a group of variable specimens
from one locality belongs to one species or to several, and the other is
whether somewhat different samples from two different localities
belong to one or to two species. How can we tell, he asks, whether a
certain number of specimens collected at one locality belong to one
species or to several? Unfortunately, he says, it is quite impossible to
give a general answer. Species criteria must be determined anew for
each group of organisms. In some cases only the breeding in the laboratory can supply the final proof, yet, much as the specific criteria change
from group to group, there is no evidence that the species concept as
such varies from one taxonomic group to another. The study of living
populations and the breeding in captivity indicate that the morphological species criteria are correlated with a physiological condition.
"We find that the morphological gap between individuals of two different species is accompanied by a physiological difference which is missing in the case of the morphological gap between individuals of one and
the same species. This physiological difference has two consequences,
namely:

"First, that individuals of a species, no matter how different from
each other morphologically, will produce only individuals of their own
species and,

"Second, that different species can co-exist in the same area without
fusing into a single species" (1905:159).

He then continues to summarize his viewpoint in one of the most
important passages in the history of biology. "The criterion of the
concept species is thus a triple one and each of its three aspects can be
investigated: A species has certain morphological characters, does not
produce individuals belonging to a different species, and does not fuse
with other species. As in 1896 I place great emphasis on this latter
point. The non-fusion is the explanation for the immense number of
existing species. Nothing keeps the species of an area separated but
their own organisation. Individuals of one species live side by side with
those of other species so genetically independent as if there had never

been any genetic connection between them, as if each species has been created separately. The discovery of this fact led, during Linnaeus' time, to the dogma of the constancy of species. The experience which the observer of the individuals of his district and of his time made was erroneously extended to the individuals of all times and all districts." Here again Jordan emphasizes the validity of the species concept for the nondimensional system. He continues: "If non-fusion is the principal criterion of specific difference it follows that one can prove specific distinctness only for those related forms which co-exist in the same area. The work of the systematist must begin with these and when dealing with all presumptive species from different regions he must ask himself whether the differences between them justify the conclusion that these forms could co-exist in the same area" (1905:159–160).

There is one among the conventional species criteria in insects to which Jordan paid special attention, difference in the structure of the genitalic armatures. Dufour had asserted that the male genital armatures were different in every species and had proposed the hypothesis that this difference serves as a mechanism to prevent hybridization between different species and thus preserves the purity of each species. During the 50 years preceding Jordan's paper (1896) on mechanical selection, the idea had become quite universal among entomologists that each species could be diagnosed on the basis of its genitalia and that there was no individual variability with respect to this character. Only a few authors, such as Kolbe (1887), Perez (1894), and Edwards (1894), had disagreed with this hypothesis. Jordan, however, had an *a priori* doubt about this concept "because if the concept of evolution is correct then species differences in the genital organs must have evolved, and the beginning of such differences must be visible already within the species" (1905:164). It is the main object of his important 1896 paper to demonstrate individual and geographical variation in these structures. An analysis of 27 species of *Papilio* proved conclusively the existence of geographic variation. Jordan illustrates this on 4 plates with 189 figures. Each of the examined species of *Papilio* had a diagnostic difference in the structure of the genitalia from every other species of the genus. In 1896 Jordan was therefore convinced that Dufour was at least in part right. However, further studies convinced him that species-specific difference in genital morphology is not a necessity. For instance, among 698 studied species of hawk moths (Sphingidae) there were 48 that did not differ in their genitalic structures; at least Jordan was not able to establish differences on the basis of the available material (1905). Jordan thus was the first to prove conclusively that reproductive isolation between species of insects was not necessarily supported by mechanical isolating mechanisms and that genitalic structures are subject to the same laws of variability as all other taxonomic characters. The aspect of this investigation most pleasing to a modern worker is the method of approach. Jordan did not single out one convenient example to demonstrate his case; no, he made a complete statistical analysis of all the available species of a whole family. Jordan's work on the variation of genitalic structures in insects is outstanding in its formulation of the problem, its procedure, and its conclusions.

Although Jordan left a few problems open, such as the relation between a nondimensional species concept and multidimensional species taxa, it is evident that he had already developed in broad outlines all the modern views. Substantially no progress was made in the 30 years after Jordan's 1905 paper. In fact, in the entire genetic literature I have not found a single discussion of the nondimensional species that is as penetrating as that of Jordan.

REFERENCES

Burma, B. H. 1954. Reality, existence, and classification: a discussion of the species problem. *Madroño* (S. Francisco), 12:193–209.

Burma, B. H., and E. Mayr. 1949. The species concept: a discussion. *Evolution,* 3:369–373.

Edwards, W. H. 1894. Notes on "A revision of the genus Oeneis" (Chionobas). *Canad. Ent.,* 26:55–64.

Eimer, G. H. T., 1889. *Artbildung und Verwandtschaft bei Schmetterlingen.* Jena.

Gregg, J. R. 1950. Taxonomy, language and reality. *Amer. Nat.,* 84:419–435.

Jordan, K. 1896. On mechanical selection and other problems. *Novit. zool.,* 3:426–525.

—— 1905. Der Gegensatz zwischen geographischer und nichtgeographischer Variation. *Z. wiss. Zool.,* 83:151–210.

Kolbe, H. J. 1887. Carabologische Auseinandersetzung mit Herrn Dr. G. Kraatz. *Ent. Nachr.,* 13:132–144.

Perez, J. 1894. De l'organe copulateur mâle des Hymenoptères et de sa valeur taxonomique. *Ann. Soc. Ent. Fr.,* 63:74–81.

Poulton, E. B. 1903. What is a species? *Proc. Ent. Soc. London,* 1903:xciv.

Ray, J. 1692. *Historia Plantarum.*

33
Species Concepts
and Definitions

The importance of one fact of nature is being recognized to an ever-increasing extent: that the living world is comprised of more or less distinct entities which we call species. Why are species so important? Not just because they exist in huge numbers, and because each species, when properly studied, turns out to be different from every other, morphologically and in many other respects. Species are important because they represent an important level of integration in living nature. This recognition is fundamental to pure biology, no less than to all subdivisions of applied biology. An inventory of the species of animals and plants of the world is the base line of further research in biology. Whether he realizes it or not, every biologist—even he who works on the molecular level—works with species or parts of species and his findings may be influenced decisively by the choice of a particular species. The communication of his results will depend on the correct identification of the species involved and, thus, on its taxonomy.

Considering this importance of the species, one would expect to find that naturalists and other biologists have long been concerned with questions such as: What is a species? How is it to be defined? What is its significance in the household of nature? This expectation has indeed been fulfilled. No other problem has preoccupied the leading students of the diversity of nature, from John Ray (1686) on, to a greater extent than the species problem. Recognizing its pivotal role in evolution, Darwin perceptively entitled his great work *On the Origin of Species*. One would further expect that more than 250 years of attention to this problem would have cleared away all obscurities so that the species concept of the modern biologist would be as well defined and clear as a crystal. Alas, *this* expectation has not been fulfilled. A glance at the recent evolutionary literature reveals how much uncertainty and disagreement there still is. There is even a modern textbook of evolution

Adapted from "Species concepts and definitions," pp. 1–22 in *The species problem,* ed. E. Mayr, publication no. 50 of the American Association for the Advancement of Science (Washington, D.C., 1957); copyright 1957 by the American Association for the Advancement of Science.

in which the authors attempt to interpret evolution without mentioning species. Other authors, such as Monod (1974), have designated the species the unit of evolution. It is quite obvious that we are still a long way from having such a clear understanding of species that unanimity among systematists and evolutionists is automatic.

One way to reach an understanding is to go over the past history of the species problem and attempt to isolate and define the disagreements and misunderstandings. Who was the first to realize that there is a species problem and what was his proposed solution? What were the subsequent developments? Space does not permit a thorough coverage of the field, but even a glance at the highlights is revealing. If we open a history of biology, the two names mentioned most prominently under the heading of "Species" will be Linnaeus and Darwin. Linnaeus will be cited as the champion of two characteristics of the species, their constancy and their sharp delimitation (their "objectivity"). One of the minor tragedies in the history of biology has been the assumption during the 150 years after Linnaeus that constancy and clear definition of species are strictly correlated and that one must *either* believe in evolution (the "inconstancy" of species) and then have to deny the existence of species except as purely subjective, arbitrary figments of the imagination, *or,* as most early naturalists have done, believe in the sharp delimitation of species but think that this necessitates denying evolution. We shall leave the conflict at this point and merely anticipate the finding made more than a hundred years after Linnaeus that there is no conflict between the fact of evolution and the fact of the clear delimitation of species in a local fauna or flora.

The insistence of Linnaeus on the reality, objectivity, and constancy of species is of great importance in the history of biology for three reasons. First, it meant the end of the belief in spontaneous generation as far as higher organisms are concerned, a belief which at that time was still widespread. Lord Bacon and nearly all leading writers of the pre-Linnaean period, except Ray, believed in the transmutation of species ("heterogony"), and the Linnaean conception "of the reality and fixity of species perhaps marks a necessary stage in the progress of scientific inquiry" (See Poulton 1903:lxxxiv–lxxxvii for further references on the subject). "Until about 1750 almost no one believed that species were stable. Linnaeus had to show that species were not erratic and ephemeral units before organic evolution as we know it could have any meaning" (Conway Zirkle in litt.). The idea that the seed of one plant could occasionally produce an individual of another species was so widespread that it died only slowly. We all know that it has raised its ugly head once more in recent years (Lysenko). In spite of Redi's and Spallanzani's experiments, spontaneous generation was still used in 1851 by the philosopher Schopenhauer as an explanation for the origin of higher categories. Linnaeus thus did for the higher organisms what Pasteur did a hundred years later for the lower.

A second reason why Linnaeus' emphasis was important is that it took the species out of the speculations of the nominalist philosophers, who stated that "only individuals exist. The species of a naturalist is nothing but an illusion" (Robinet 1768). I shall return later to the point why species are more than merely an aggregate of individuals.

A third reason why the insistence on the sharp delimitation of species in the writings of Linnaeus is of historical importance is that it strengthened the viewpoint of the local naturalist and established the basis for an observational and experimental study of species in local faunas and floras, of which Darwin took full advantage.

Linnaeus was too experienced a botanist to be blind to the evidence of evolutionary change. Greene (1912) gathered numerous citations from Linnaeus' writings that clearly document his belief in the common descent of certain species, and Ramsbottom (1938) and Sirks (1952) have traced how Linnaeus expressed himself more and more freely on the subject as his prestige grew (see also Larson 1971 and Stafleu 1971). Paradoxically, Linnaeus did more, perhaps, to lay a solid foundation for subsequent evolutionary studies by emphasizing the constancy and objectivity of species than if he, like Darwin, had emphasized the opposite.

Darwin looked at the species from a viewpoint almost directly opposite to that of Linnaeus. As a traveler naturalist and particularly because of his studies of domesticated plants and animals, he was impressed by the fluidity of the border between species and the subjectivity of their delimitation. The views of both Linnaeus and Darwin underwent a change during the life of each. With Linnaeus the statements on the constancy of species became less and less dogmatic through the years. For Darwin, the more firmly the idea of evolution became fixed in his mind, the stronger grew his conviction that this should make it impossible to delimit species. He finally regarded species as something purely arbitrary and subjective. "I look at the term species as one arbitrarily given for the sake of convenience to a set of individuals closely resembling each other, and that it does not essentially differ from the term variety which is given to less distinct and more fluctuating forms. . . . The amount of difference is one very important criterion in settling whether two forms should be ranked as species or variety." And finally he came to the conclusion that "in determining whether a form should be ranked as a species or a variety, the opinion of naturalists having sound judgment and wide experience seems the only guide to follow" (1859). Having thus eliminated the species as a concrete unit of nature, Darwin had also neatly eliminated the problem of the multiplication of species. This explains why he made so little effort in his classical work to solve the problem of speciation.

The 75 years following the publication of the *Origin of Species* (1859) saw biologists rather clearly divided into two camps, which we might call, in a somewhat oversimplified manner, the followers of

Darwin and those of Linnaeus. The followers of Darwin, who included the plant breeders, geneticists, and other experimental biologists, minimized the "reality" or objectivity of species and considered individuals to be the essential units of evolution. Characteristic for this frame of mind is a symposium held in the early Mendelian days that endorsed unanimously the supremacy of the individual and the nonexistence of species. Statements made at this symposium (Bessey 1908) include the following: "Nature produces individuals and nothing more. . . . Species have no actual existence in nature. They are mental concepts and nothing more. . . . Species have been invented in order that we may refer to great numbers of individuals collectively." Taxonomists, one of the speakers claimed, did not merely name the species found in nature but actually "made" them. "In making a species the guiding principle must be that it shall be recognizable from its diagnosis." Left over from this period is a statement by a well-known geneticist: "Distinct species must be separable on the basis of ordinary preserved material."

It is a curious paradox in the history of biology that the rediscovery of the Mendelian laws resulted in an even more unrealistic species concept among the experimentalists than had existed previously. They either let species saltate merrily from one to another, as did Bateson and De Vries, defining species merely as morphologically different individuals, or they denied the existence of species altogether except as intergrading populations. Whether these early Mendelians regarded species as continuous or discontinuous units, they all agreed in their arbitrariness and artificiality. There is an astonishing absence of any effort in this school to study species in nature, to study natural populations.

A study of natural populations had become the prevailing preoccupation in an entirely independent conceptual steam, that of the naturalists, which ultimately traces back to Linnaeus. The viewpoint of the naturalist was particularly well expressed by Jordan (1905): "The units of which the fauna of a region is composed are separated from each other by gaps which, at a given place, are not bridged by anything. This is a fact which can be checked by any observer. Indeed, the activity of a local naturalist begins with the searching out of these units which with Linnaeus we call species." Although this was the prevailing viewpoint among taxonomists, it was completely ignored by the general biologists, by whom, as a result of Darwin's theory, "species were mostly regarded merely as arbitrary divisions of the continuous and ever changing series of individuals found in nature . . . of course, active taxonomists did not overlook the existence of sharply and distinctly delimited species in nature—but as the existence of those distinct units disagreed with the prevailing theories, it was mentioned as little as possible" (Du Rietz 1930). The two streams of thought are still recognizable today, even though most geneticists, under the leadership of Dobzhansky, Huxley, Ford, and others, have swung into Jordan's camp. The principal opponents of the concept of objectively delimitable species are today found among philosophers and paleontologists. Publications maintaining this viewpoint are those of Gregg (1950), Burma (1949, 1954), Yapp (1951), and Arkell (1956). These are only a few titles from a vast literature, some of which is cited in the references.

The point that is perhaps the most impressive when one studies these voluminous publications is the amount of disagreement that has existed and still exists. The number of possible antitheses that have been established in this field may be characterized by such alternate views, to mention only a few, as follows:

Subjective *versus* objective
Scientific *versus* purely practical
Degree of difference *versus* degree of distinctness
Consisting of individuals *versus* consisting of populations
Only one kind of species *versus* many kinds of species
To be defined morphologically *versus* to be defined biologically

As interesting as this chapter in the history of human thought is, the detailed presentation of the gropings and errors of former generations would add little to the task before us. Let me concentrate therefore on the gradual emergence of the ideas that we, today, consider central and essential. Three aspects are stressed in most modern discussions of species: (1) they are based on distinctness rather than on difference and are therefore to be defined biologically rather than morphologically; (2) they consist of populations, rather than of unconnected individuals, a point particularly important for the solution of the problem of speciation; (3) they are more succinctly defined by isolation from non-conspecific populations than by the relation of conspecific individuals to each other. The crucial species criterion is thus not the fertility of individuals, but rather the reproductive isolation of populations. Let me try to trace the emergence of these and related concepts.

It is not surprising that species were considered merely "categories of thought" by many writers in periods so strongly dominated by philosophy as were the eighteenth and nineteenth centuries. Thoughts like Robinet's "only individuals exist" and the species "is nothing but an illusion" were echoed by Agassiz, Mivart, and particularly by those paleontologists who considered their task merely the classification of "objects" (= fossil specimens). In opposition to this, an increasingly strong school developed that regarded species as "definable," "objective," "real." Linnaeus was, of course, the original standard bearer of this school, to which also belonged Cuvier, de Candolle, and many taxonomists in the first half of the nineteenth century. Some of them supported their case with purely morphological arguments; others used a more biological argument, as I will discuss below.

What is unexpected for this pre-Darwinian period, however, is the frequency with which "common descent" is included in species definitions. When such an emphatically anti-evolutionary author as von Baer (1828) defines the species as "the sum of the individuals that are united by common descent," it becomes evident that he does not refer to evolution. What is really meant is more apparent from Ray's species definition (1686) or a statement by the Swedish botanist Oeder (1764) that characterizes species "dass sie aus ihres gleichen entsprungen seien und wieder ihres gleichen erzeugen." Expressions like "community of origin" or "individus descendants des parents communs" (Cuvier) are frequent in the literature. These are actually attempts at reconciling a typological species concept (with its stress of constancy) with the observed morphological variation. Constancy was a property of species taken very seriously not only by Linnaeus and his followers but curiously enough also by Lamarck and by Darwin himself: "The power of remaining constant for a good long period I look at as the essence of a species" (letter to Hooker, Oct. 22, 1864). Such constancy in time was the strongest argument in favor of a morphological species concept, but it could be proved only by the comparison of individuals of different generations. Different morphological "types" that are no more different than mother and daughter or father and son can safely be considered conspecific. They are "of the same blood." It is obvious that this early stress of descent was essentially the consequence of a morphological species concept. Yet this consideration of descent eventually led to a genetic species definition.

Virtually all early species definitions regarded species only as aggregates of individuals, unconnected except by descent, as is evident not only from the writings of Robinet, Buffon, and Lamarck, but also of much more recent authors (e.g., Britton 1908; Bessey 1908). The realization that these individuals are held together by a supraindividualistic bond, that they form populations, came only slowly. Illiger (1800) spoke of species as a community of individuals that produce fertile offspring (Mayr 1968). Brauer (1885) spoke of the "natural tie of blood relationship" through which the "individuals of a species are held together," and which "is not a creation of the human mind ... if species were not objective, it would be incomprehensible that even the most similar species mix only exceptionally and the more distant species never." Plate (1914) was apparently the first to state explicitly the nature of this bond: "The members of a species are tied together by the fact that *they recognize each other as belonging together* and reproduce only with each other. The systematic category of the species is therefore entirely independent of the existence of Man." Finally, in the language of current population genetics, this community becomes the "coadapted gene pool," again stressing the integration of the members of the population rather than the aggregation of individuals (a viewpoint that is of course valid only for sexually reproducing organisms).

The growth of thinking in terms of populations went hand in hand with a growing realization that species were less a matter of difference than of distinctness. "Species" in its earlier typological version meant merely "kind of." This, as far as inanimate objects are concerned, is measured in terms of difference. But one cannot apply this same standard to "kinds of" organisms, because there are various biological "kinds." Males and females may be two very different "kinds" of animals. Jack may be a very different "kind" of person from Bill, yet neither "kind" is a species. Realization of the special aspects of biological variation has led to a restriction of the term "species" to a very particular "kind," namely the kind that would interbreed with each other (Buffon 1749). Other early authors who state this clearly are Voigt (1817), "Man nennt Spezies . . . was sich fruchtbar mit einander gattet, fortpflanzt"; Oken (1830), "Was sich scharet und paaret, soll zu einer Art gerechnet werden"; and Gloger (1833), "What under natural conditions regularly pairs, always belongs to one species" (he stated that by stressing "regularly" he wanted to eliminate the complications due to occasional hybridization). Gloger later (1856) gave a different, but similar, definition: "A species is what belongs together either by descent or for the sake of reproduction." It is interesting how completely all these definitions omit any reference to morphological criteria. They are obviously inapplicable to asexually reproducing organisms.

This is an exceedingly short outline of some of the trends in the development of a modern species concept. More extensive treatments can be found in the publications of Geoffroy St. Hilaire (1859), Besnard (1864), de Quatrefages (1892), Bachmann (1905), Plate (1914), Uhlmann (1923), Du Rietz (1930), Kuhn (1948), and other authors cited in the references. Several conclusions are evident. One is that biological, or so-called modern, species criteria were used by authors who published more than a hundred years ago, long before Darwin. Another is that there has been steady progress toward clarification, but that there is still much uncertainty and dissent. One has a feeling that there is a hidden reason for so much disagreement. Perhaps it is due to the fact that there is more than one kind of species and that we need a different definition for each of these species. Many attempts have been made during the last hundred years to distinguish several kinds of species. Camp and Gillis (1943), for instance, recognized no less than 12 different kinds of species. Yet, a given species in nature might fit into several of their categories, and in view of this overlap no one has adopted either this elaborate classification or any of the simpler schemes proposed before or afterward.

SPECIES CONCEPTS

An entirely different approach to the species problem stresses the kaleidoscopic nature of any species and attempts to determine how many different aspects a species has. Depending on the choice of criteria, it leads to a variety of "species concepts" or "species definitions." At one time I listed five species concepts, which I called the practical, morphological, genetic, sterility, and biological (Mayr 1942). Meglitsch (1954) distinguishes three concepts, the phenotypic, genetic, and phylogenetic, a somewhat more natural arrangement. Two facts emerge from these and other classifications. One is that there is more than one species concept and that it is futile to search for *the* species concept. The second is that there are at least two levels of concepts. Such terms as "practical," "sterility," and "genetic" signify concrete aspects of species that lead to what one might call "applied" species concepts. They specify criteria that can be applied readily to determine the status of discontinuities found in nature. Yet they are secondary, derived concepts, based on underlying philosophical concepts, which might also be called primary or theoretical concepts. I believe that the analysis of the species problem would be considerably advanced if we could penetrate through such empirical terms as phenotypic, morphological, genetic, phylogenetic, and biological to the underlying philosophical concepts. A deep, and perhaps widening, gulf has existed in recent decades between philosophy and empirical biology. It seems that the species problem is a topic where productive collaboration between the two fields is possible.

An analysis of published species concepts and species definitions indicates that all of them are based on just three theoretical concepts. An understanding of these three philosophical concepts is a prerequisite for all attempts at a practical species definition. And all species criteria or species definitions used by the taxonomist in his practical work trace back ultimately to these basic concepts.

The Typological Species Concept

The typological concept is the simplest and most widely held species concept. It merely means "kind of." There are languages, for instance German, where the term for "kind" (*Art*) is also used for "species." A species in this concept is "a different thing." This concept is very useful in many branches of science and it is still used by the mineralogist who speaks of "species of minerals" (Niggli 1949) or the physicist who speaks of "nuclear species." This simple concept of everyday life was incorporated in a more sophisticated manner in the philosophy of Plato. Here, however, the word *eidos* (*species,* in its Latin translation) acquired a double meaning that survives in the two modern words *species* and *idea,* both of which are derived from it. According to Plato's thinking, objects are merely manifestations, "shadows," of the *eidos.* By transfer, the individuals of a species, being merely shadows of the same type, do not stand in any special relation to each other, as far as a typologist is concerned. Naturalists of the "idealistic" school endeavor to penetrate through all the modifications and variations of a species in order to find the "typical" or "essential" attributes. Typological thinking finds it easy to reconcile the observed variability of the individuals of a species with the dogma of the constancy of species because the variability does not affect the essence of the *eidos,* which is absolute and constant. Since the *eidos* is an abstraction derived from individual sense impressions, and a product of the human mind, according to this school, its members feel justified in regarding a species as "a figment of the imagination," an idea. Variation, under this concept, is merely an imperfect manifestation of the idea implicit in each species. If the degree of variation is too great to be ascribed to the imperfections of our sense organs, more than one *eidos* must be involved. Thus, species status is determined by degrees of morphological difference. The two aspects of the typological species concept, subjectivity and definition by degree of difference, therefore depend on each other and are logical correlates.

The application of the typological species concept to practical taxonomy results in the morphologically defined species; "degree of morphological difference" is the criterion of species status. Species are defined on the basis of their observable morphological differences. This concept has been carried to the extreme where mathematical formulas were proposed (Ginsburg 1938) that would permit an unequivocal answer to the question whether or not a population is a different species.

Most systematists have found this typological-morphological concept inadequate and have rejected it. Its defenders, however, claim that all taxonomists, when classifying the diversity of nature into species, follow the typological method and distinguish "archetypes." At first sight there seems an element of truth in this assertion. When assigning specimens either to one species or to another, the taxonomist bases his decision on a mental image of these species that is the result of past experience with the stated species. The utilization of morphological criteria is valuable and productive in the taxonomic practice. To assume, however, that this validates the typological species concept overlooks a number of important considerations. To begin with, the mental construct of the "type" is subject to continuous revision under the impact of new information. If it is found that two archetypes represent nothing more than two "kinds" within a biological species, they are merged into a single one. It was pointed out above that males and females are often exceedingly different "kinds" of animals. Even more different are in many animals the larval stages, or in plants sporophyte and gametophyte, or in polymorph populations the various genotypes. A strictly morphological-typological concept is inadequate to cope with such intraspecific variation. It is equally incapable of coping with another difficulty, namely, an absence of visible morphological differences between natural populations that are nevertheless distinct and reproductively isolated, and therefore to be considered species. The frequent occurrence of such "cryptic species" or "sibling species" in nature has been substantiated by various genetic, physiological, and ecological methods. They form another decisive argument against defining species on a primarily morphological basis. Any attempt in these two situations to define species "by degree of difference" is doomed to failure. Degree of difference can be specified only by a purely arbitrary decision.

More profound than these two essentially practical considerations is the fact that the typological species concept treats species merely as random aggregates of individuals that have the "essential properties" of the "type" of the species and "agree with the diagnosis." This static concept ignores the fact that species are not merely classes of objects but are composed of natural populations that are integrated by an internal organization and that this organization (based on genetic, ethological, and ecological properties) gives the populations a structure that goes far beyond that of mere aggregates of individuals. Even a house is more than a mere aggregate of bricks or a forest more than an aggregate of trees. In a species an even greater supraindividualistic cohesion and organization is produced by a number of factors. Species are a reproductive community. The individuals of a species of higher animals recognize each other as potential mates and seek each other for the purpose of reproduction. A multitude of devices insures intraspecific reproduction in all organisms. The species is an ecological unit that, regardless of the individuals of which it is composed, interacts as a unit with other species in the same environment. The species, finally, is a genetic unit consisting of a large, intercommunicating gene pool, whereas each individual is only a temporary vessel holding a small portion of this gene pool for a short period of time. These three properties make the species transcend a purely typological interpretation or the concept of a "class of objects."

The very fact that a species is a gene pool, with numerous devices facilitating genic intercommunication within and genic separation from without, is responsible for the morphological distinctness of species as a by-product of their biological uniqueness. The empirical observation that a certain amount of morphological difference between two populations is normally correlated with a given amount of genetic difference is undoubtedly correct. Yet it must be kept in mind at all times that the biological distinctness is primary and the morphological difference secondary. As long as this is clearly understood, it is legitimate and indeed very helpful to utilize morphological criteria. This caution has been exercised, consciously or unconsciously, by nearly all proponents of the morphological species concept. As pointed out by Simpson (1951) and Meglitsch (1954), they invariably abandon the morphological concept when it comes in conflict with biological data. This was true for Linnaeus himself and for his followers to the present day.

The typological species concept has a certain amount of operational usefulness when applied to inanimate objects. Ignoring the population structure of species, however, and incapable of coping with the facts of biological variation, it has proved singularly inadequate as a conceptual basis in taxonomy. Much of the criticism directed against the taxonomic method was provoked by the application of the typological concept by taxonomists themselves or by other biologists who mistakenly considered it the basis of taxonomy.

The Nondimensional Species Concept

The essence of the nondimensional species concept is the relationship of two coexisting natural populations in a nondimensional system, that is, at a single locality at the same time (sympatric and synchronous). This is the species concept of the local naturalist. It was introduced into the biological literature by the English naturalist John Ray and confirmed by the Swedish naturalist Linnaeus. It is based not on difference but on distinction, and this distinction in turn is characterized by a definite mutual relationship, namely, that of reproductive isolation. The word *species* is here best defined in combination with the word *different*. The relationship of two "different species" can be objectively defined as reproductive isolation. We have, thus, an objective yardstick for this species concept, something that is absent in all others. Philosophers have objected to the use of the terms "objective" or "real" for species, and it may be more neutral to use the terms arbitrary or nonarbitrary (Simpson 1951). Presence or absence of interbreeding of two populations in a nondimensional system is a completely nonarbitrary criterion.

This species concept seems so self-evident to every naturalist that it is only rarely put in words. That the species is more than an aggregate of individuals held together by a biological bond has long been realized, as was pointed out in the historical survey above. The interbreeding within the species is more conspicuous, and it was thus more often emphasized than is the reproductive isolation against other species. Eimer, as early as 1889 (p. 16), defined species as "groups of individuals which are so modified that successful interbreeding [with other groups] is no longer possible." But the nondimensional species concept was first stated in its full extent and implication in 1905 by Jordan (see Essay 32).

In spite of its theoretical superiority, the nondimensional species has a number of serious drawbacks, particularly its limitation to sexually reproducing species and to such without the dimensions of space and time. (These difficulties are discussed in Mayr 1969:30–35.)

The Multidimensional Species Concept

In contradistinction to the other two concepts, of which one is based on a degree of difference, the second one on the completeness of a discontinuity, the multidimensional, or polytypic, species concept is a collective one. It considers species groups of populations, namely, such groups as interbreed with each other, actually or potentially. Thus this species concept is a concept of the same sort as the higher categories, genus, family, or order. Like all collective categories it faces the difficulty, if not impossibility, of clear demarcation against other similar groupings. What this species gains in actuality by the extension of the nondimensional situations in space and time, it loses in objectivity. As unfortunate as this is, it is inevitable, because the natural populations encountered by the biologist are distributed in space and time and cannot be divorced from these dimensions. Thus, this species concept likewise has its good and its bad points.

SPECIES DEFINITIONS

All our reasoning in discussions of "the species" can be traced back to the three primary concepts discussed above. As concepts, of course, they cannot be observed directly, and we refer to certain observed phenomena in nature as "species" because they conform in their attributes to one of these concepts or to a mixture of them. From these primary concepts, we come thus to secondary concepts, based on particular aspects of species. I have already mentioned the so-called morphological species concept, which, in most cases, is merely an applied typological concept, using morphological criteria. The case of the so-called genetic species concept shows that all three of the basic concepts can be expressed, on this level, in genetic terms. Some geneticists, for instance, subscribed to the typological concept and defined species by the degree of genetic difference, as did Lotsy or De Vries; others stressed the genetic basis of the isolating mechanisms between species, thereby endorsing the nondimensional species concept; still others finally emphasized the gene flow among interbreeding populations in a multidimensional system, thus adopting the multidimensional collective species concept. All three groups of geneticists thought they were dealing with a uniquely "genetic species concept," yet they were merely observing secondary manifestations of the primary concepts.

It is evident from the analysis of the morphological and genetic species concepts that such derived concepts are attempts to deal directly with the discontinuities in nature. In the past, almost every taxonomist worked with his own personal yardstick based on a highly individual mixture of elements from the three basic concepts. As a consequence, one taxonomist might designate as a species every polymorph variant; a second one, every morphologically different population; and a third one, every geographically isolated population. Such lack of standards, which is still largely characteristic for the taxonomic literature, has been utterly confusing to taxonomists and other biologists alike. It has therefore been the endeavor of many specialists within recent decades to find a standard yardstick on which there could be general agreement. A historical study of species definitions indicates clearly a trend toward acceptance of a synthetic species definition, often referred to as a "biological species" definition. It is essentially based on the nondimensional ("reproductive gap") and the multidimensional ("gene flow") species concepts. Nearly all species definitions proposed within the last 50 years incorporate some elements of these two concepts. This is evident from the species definitions of Jordan (see Essay 32), Stresemann (1919), and Rensch (1929). Du Rietz (1930) called the species "a syngameon . . . separated from all others by . . . sexual isolation." Dobzhansky (1935) was apparently the first geneticist to define species in the terms customary among naturalists and taxonomists, namely, interbreeding and reproductive isolation; other recent definitions are variants of the same theme. Mayr (1940) defined species as "groups of actually or potentially interbreeding natural populations which are reproductively isolated from other such groups." Simpson (1943) gave the definition "a genetic species is a group of organisms so constituted and so situated in nature that a hereditary character of any one of these organisms may be transmitted to a descendant of any other," and Dobzhansky (1950) defined the species as "the largest and most inclusive . . . reproductive community of sexual and cross-fertilizing individuals which share in a common gene pool."

Finally, it might be useful to mention some qualifications that are often, though needlessly, included in species definitions. Anything that is equally true for categories above and below species rank should be omitted, since there is no sense burdening a species definition with features that do not help discrimination between species and intraspecific populations.

1. Species characters are adaptive. This component of Wallace's (1889) species definition was correctly rejected by Jordan (1896). Adaptiveness is not diagnostic for species characters and not even necessarily true. Not every detail of the phenotype needs to be adaptive as long as the phenotype as a whole is adaptive and as long as the genotype itself is the result of selection.

2. Species are evolved and evolving. Again this is true for the entire organic world from the individual to the highest taxa and adds nothing to the species definition.

3. Species differ genetically. This is only the morphological species concept expressed in genetic terms. It does not permit discriminating species from intraspecific populations or from individuals.

4. Species differ ecologically. This qualification is unnecessary and misleading for the same reasons as the genetic one. Ecological differences exist for all ecotypes within species and in general for all geographical isolates. Conspecific populations are sometimes more different ecologically than are good species.

A yardstick such as the biological species concept is not automatic. To apply it properly requires skill and experience. This is particularly true in the recognition of situations where it cannot be applied directly, for one reason or another, and where the worker has to fall back on the criterion of "degree of difference."

REFERENCES

Arkell, W. J. 1956. The species concept in paleontology. *Syst. Assoc. Publ.*, 2:97–99.

Bachmann, H. 1905. Der Speziesbegriff. *Verh. schweiz. naturforsch. Ges.*, 87: 161–208.

Baer, K. E. von. 1828. *Entwicklungs-Geschichte der Thiere.* Königsberg.

Besnard, A. F. 1864. Altes und Neues zur Lehre über die organische Art (Spezies). *Abhandl. zool. mineral. Ver. Regensburg*, 9:1–72.

Bessey, C. E. 1908. The taxonomic aspect of the species question. *Amer. Nat.*, 42:218–224.

Brauer, F. 1885. Systematisch-zoologische Studien. *Sitzber. Akad. Wiss. Wien*, 91 (Abt. 1): 237–413.

Britton, N. L. 1908. The taxonomic aspect of the species question. *Amer. Nat.*, 42:225–242.

Buffon, G. L. LeClerq. 1749. *Histoire naturelle, générale et particulière, avec la description du Cabinet du Roi.* Paris.

Burma, B. H. 1949a. The species concept: a semantic review. *Evolution*, 3:369–370.

———1949b. The species concept: postscriptum. *Evolution*, 3:372–373.

Camp, W. H., and C. L. Gillis. 1943. The structure and origin of species. *Brittonia*, 4:323–385.

Darwin, C. 1859. *On the origin of species by means of natural selection, or the preservation of favoured races in the struggle for life.* J. Murray, London.

Dobzhansky, T. 1935. A critique of the species concept in biology. *Phil. Sci.*, 2:344–355.

———1950. Mendelian populations and their evolution. *Amer. Nat.*, 84:401–418.

Doederlein, L. 1902. Über die Beziehungen nahe verwandter "Thierformen" zu einander. *Z. Morphol. Anthropol.*, 26:23–51.

Dougherty, E. C. 1955. Comparative evolution and the origin of sexuality. *Syst. Zool.*, 4:145–169.

Du Rietz, G. E. 1930. The fundamental units of botanical taxonomy. *Svensk. Bot. Tidsskrift*, 24:333–428.

Eimer, G. H. T. 1889. *Artbildung und Verwandtschaft bei Schmetterlingen.* Jena.

Geoffroy Saint Hilaire, I. 1859. *Histoire naturelle générale des règnes organiques.* Paris.

Ginsburg, I. 1938. Arithmetical definition of the species, subspecies and race concept, with a proposal for a modified nomenclature. *Zoologica*, 23:253–286.

Gloger, C. L. 1833. *Das Abändern der Vögel durch Einfluss des Klimas.* Breslau.

———1856. Über den Begriff von "Art" ("Species") und was in dieselbe hinein gehört. *Jour. f. Ornith.*, 4:260–270.

Greene, E. L. 1912. Linnaeus as an evolutionist. In *Carolus Linnaeus.* C. Sower, Philadelphia.

Gregg, J. R. 1950. Taxonomy, language and reality. *Amer. Nat.*, 84:419–435.

Huxley, J. 1942. *Evolution: the modern synthesis.* Allen and Unwin, London.

Illiger, J. C. W. 1800. *Versuch einer systematischen vollständigen Terminologie für das Thierreich und Pflanzenreich.* Helmstedt.

Jordan, K. 1896. On mechanical selection and other problems. *Novit. Zool.,* 3:426–525.

———1905. Der Gegensatz zwischen geographischer und nichtgeographischer Variation. *Z. wiss. Zool.,* 83:151–210.

Kuhn, E. 1948. Der Artbegriff in der Paläontologie. *Eclogae Geolog. Helv.,* 41:389–421.

Lorkovicz, Z. 1953. Spezifische, semispezifische und rassische Differenzierung bei *Erebia tyndarus* Esp. *Rad. Acad. Yougoslave,* 294:315–358.

Mayr, E. 1940. Speciation phenomena in birds. *Amer. Nat.,* 74:249–278.

———1942. *Systematics and the origin of species.* Columbia University Press, New York.

———1949. The species concept: semantics versus semantics. *Evolution,* 3:371–372.

———1951. Concepts of classification and nomenclature in higher organisms and microorganisms. *Ann. N.Y. Acad. Sci.,* 56:391–397.

———1968. Illiger and the biological species concept. *Jour. Hist. Biol.,* 1(2): 163–178.

———1969. *Principles of systematic zoology.* McGraw-Hill, New York.

Mayr, E., and C. Rosen. 1956. Geographic variation and hybridization in populations of Bahama snails (*Cerion*). *Amer. Mus. Novit.,* 1806:1–48.

Mayr, E., E. G. Linsley, and R. L. Usinger. 1953. *Methods and principles of systematic zoology.* McGraw-Hill, New York.

Meglitsch, P. A. 1954. On the nature of the species. *Syst. Zool.,* 3:49–65.

Monod, J. 1974. Préface. In *Populations, espèces et évolution,* by E. Mayr. Hermann, Paris.

Niggli, P. 1949. *Probleme der Naturwissenschaften (Der Begriff der Art in der Mineralogie).* Basel.

Plate, L. 1914. Prinzipien der Systematik mit besonderer Berücksichtigung des Systems der Tiere. In *Die Kultur der Gegenwart,* vol. 3, pp. 92–164. B. G. Teubner, Leipzig.

Poulton, E. B. 1903. What is a species? *Proc. Ent. Soc.* London, 127–166.

Prosser, C. L. 1957. The species problem from the viewpoint of a physiologist. In *The Species Problem,* ed. E. Mayr, pp. 339–369. American Association for the Advancement of Science, Washington, D.C.

Quatrefages, A. de. 1892. *Darwin et les précurseurs français.* Paris.

Ramsbottom, J. 1938. Linnaeus and the species concept. *Proc. Linn. Soc. London,* 150th session:192–219.

Ray, J. 1686. *Historia Plantarum.*

Rensch, B. 1929. *Das Prinzip geographischer Rassenkreise und das Problem der Artbildung.* Bornträger, Berlin.

Robinet, C. 1768. *De la nature.* Amsterdam.

Schopenhauer, A. 1851. *Parerga und Paralipomena: kleine philosophische Schriften.* Berlin.

Simpson, G. G. 1943. Criteria for genera, species, and subspecies in zoology and paleozoology. *Ann. N.Y. Acad. Sci.,* 44:145–178.

———1951. The species concept. *Evolution,* 5:285–298.

Sirks, M. J. 1952. Variability in the concept of species. *Acta Biotheoretica,* 10:11–22.

Spring, A. F. 1838. Über die naturhistorischen Begriffe von Gattung, Art und Abart und über die Ursachen der Abartungen in den organischen Reichen. Leipzig.

Spurway, H. 1955. The sub-human capacities for species recognition and their correlation with reproductive isolation. Acta 11th Congr. Int. Ornith. (Basel, 1954), pp. 340–349.

Stafleu, F. A., 1971. Linnaeus and the Linnaeans. A. Oosthoek's Vitgerersmaatschappij, Utrecht.

Stresemann, E. 1919. Über die europäischen Baumläufer. Verh. Ornith. Gesellsch. Bayern, 14:39–74.

Sylvester-Bradley, P. C. 1956. The species concept in paleontology. Syst. Assoc. Publ., no. 2: introduction.

Thomas, G. 1956. The species concept in paleontology. Syst. Assoc. Publ., no. 2:17–31.

Uhlmann, E. 1923. Entwicklungsgedanke und Artbegriff in ihrer geschichtlichen Entstehung und sachlichen Beziehung. Z. Naturw. (Jena), 59:1–114.

Voigt, F. S. 1817. Grundzüge einer Naturgeschichte als Geschichte der Entstehung und weitern Ausbildung der Naturkörper. Frankfurt.

Wallace, A. R. 1889. Darwinism: an exposition of the theory of natural selection, with some of its applications. J. Murray, London.

Yapp, W. B. 1951. Definitions in biology. Nature, 167:160.

34
Sibling or Cryptic Species
among Animals

During my discussion with W. Thorpe in 1947 (Essay 13) about the prevalence of sympatric speciation in animals, it became quite apparent that much of the difference of opinion was based on differences in the species concepts. At that time most applied zoologists, particularly entomologists and parasitologists, still had a morphological species concept and described as biological races morphologically indistinguishable populations that lived on different hosts or otherwise occupied different niches. They considered these populations incipient species. I differed from them by considering these "races" species, in spite of the absence of morphological difference, since they were reproductively isolated. These taxa perfectly fitted my definition of sibling species, a term I had introduced (1942) for previously existing equivalent terms in the French and German literature.

At that period there was great uncertainty about how frequent such cryptic species were, and I therefore undertook an analysis of a considerable portion of the zoological literature to determine how often such sibling species had originally been described as races and, more broadly, how common such sibling species were in different higher taxa of the animal kingdom. The following essay contains part of the results of that investigation. The effect of my analysis was that after 1949 the designation of "biological race" for sibling species became unfashionable. Changes in the ranking of the American orthopterans (crickets, grasshoppers, etc.) are a good illustration for the change in conceptualization that took place in 1948.

Pairs or groups of morphologically nearly or completely identical species have been termed sibling species (Mayr 1942:151).[1] Their existence is usually revealed only if they coexist at the same locality, and in my

Adapted from pp. 227–231 of "The bearing of the new systematics on genetical problems: the nature of species," *Advances in Genetics,* 2(1948):205–237.

1. The literature references for this essay are not given because they are of interest only to a few specialists. They can be found in the article from which this essay was excerpted.

original definition I therefore limited the definition to sympatric species. This is, however, not necessary since it is feasible with modern techniques to test the conspecificity of allopatric populations. Morphologically similar pairs or groups of species have been described in the recent literature with increasing frequency. They are found in nearly all taxa, although they seem to be decidedly more frequent in some than in others.

The discovery of sibling species is possible only in groups that are either very well known taxonomically, such as birds, or to which particularly refined methods have been applied. Such methods may be either biometric (as particularly among fish), or cytogenetic (as, for example, in *Sciara* and *Drosophila*), or combined taxonomic-ecological (as, for example, in *Anopheles*). Elaborate studies of this sort are generally carried out only in groups that are medically (e.g., *Anopheles*) or otherwise of special significance. It is for this reason as yet impossible to state how widespread is the occurrence of sibling species, or to give the approximate percentage of sibling species in various orders. Furthermore, many situations have been published in the literature which cannot yet be classified because they are still incompletely analyzed.

Physiological races have been described particularly often among insects. However, as A. Emerson (1945) points out, most of these so-called physiological races answer all the criteria of good species. This is true for the species of the *Anopheles maculipennis* complex and for the tree crickets of the *Nemobius fasciatus* group. Recent work in several orders of insects fully confirms these conclusions.

In the Hymenoptera, differences between closely related species of ants are often particularly slight. This is one of the reasons for the complication of ant nomenclature. In the past, myrmecologists did not dare to recognize barely distinguishable sympatric forms as full species and were thus forced to resort to a complicated quadrinomial nomenclature of races, subspecies, and varieties. E. V. Gregg (1945) carefully compared two of the American species of the *Lasius niger* complex. She comes to the conclusion that the two so-called varieties *neoniger* and *americanus* of *Lasius niger* are good species. In addition to a number of ecological differences between the two forms, she found that no morphological overlap existed. All of 20 workers of *neoniger* had hair on the scape of the antennae and on the tibiae where not one hair was found in *americanus*. The differences in the number of hairs on coxae, trochanters, and femora were statistically significant. At least in the Chicago area, where this investigation was made, there was no sign of intergradation between the two species. The nature of the isolating mechanisms is not known in this case. Season and time of day of swarming seem identical. Evidence for specific distinctness was also found for two European "varieties" (*niger* and *alienus*) of the *Lasius niger* complex (Diver 1940).

Among the wasps of the genus *Polistes*, Bequaert (1940) recognized in the species *fuscatus* 18 "varieties" that differ in color, but not in structure. Near Kirkwood, Missouri, Rau (1942, 1946) found three of these "varieties," *pallipes, variatus,* and *rubiginosus,* but they behave like three good species. Each of the three is clearly characterized by distinctive color patterns and intermediates are absent. The color characters are correlated with 7 distinct behavior or ecological differences. The naturalist Rau finally exclaimed: "I, for my part, am only too happy to meet and greet *pallipes* and *variatus* as two distinct species, but after all, I am not a taxonomist." The two Californian "varieties" of *Polistes fuscatus,* recognized by Bequaert, differ not only in color, habits, and ecology, but also in having different strepsipteran parasites; the parasites are frequent in one, rare in the other (Bohart 1942). To apply the term "variety" to such obvious species is possible only for adherents of an extremely rigid morphological species concept.

The order of Diptera is another group of insects that is rich in sibling species. McCarthy (1945) found that specimens identified by a *Sciara* specialist as belonging to *Sciara fenestralis* actually belonged to two exceedingly similar species. The two species failed to produce hybrids and the banding patterns of the salivary chromosomes did not match. Females of species A produce regularly both male and female offspring with the males in excess (bisexual reproducer), while a given female of species B produces exclusively offspring of one sex, male or female (unisexual producer). Morphological differences were negligible. A similar case involving *Sciara ocellaris-reynoldsi* had previously been described.

The number of known sibling species in the genus *Anopheles* is rapidly increasing, although there is still a tendency among the medical entomologists to label such noninterbreeding natural populations as varieties. This is evident in Aitken's revision (1945) of the western North American species and other recent works. However, more and more authors follow Bates's logical and consistent step (Mayr 1942) of raising such "varieties" to the rank of full species. This was done for *Anopheles melas,* a sister species of *A. gambiae* (Ribbands 1944).

It is fortunate for the study of sibling species that they are very common in the genus *Drosophila.* There is every intergradation between pairs of species that in spite of their similarity can be distinguished by the specialist, like *Drosophila melanogaster* and *D. simulans,* and pairs, like *Drosophila willistoni* and *D. equinoxialis* (Dobzhansky 1946), that are morphologically entirely indistinguishable, except for a slight average size difference. Flies of *D. willistoni* reach sexual maturity more quickly, since at the age of 24 hours after emergence from the pupae half of the females contain sperm, while in *D. equinoxialis* this proportion of inseminated females is reached at 48 hours. Cross matings are completely sterile and not a single hybrid was produced. The widespread occurrence of morphologically very similar species of *Drosophila* has been clearly established, chiefly through the work of Patterson and his coworkers. Such sibling species are *Drosophila macrospina* and *D. subfunebris* (Mainland 1942); the *Drosophila virilis* group, consisting of *D. virilis, americana, texana, nova-mexicana, montana,* and *lacicola* (Patterson, Stone, and Griffen 1942; Patterson and Griffen 1944); the *mulleri* group, consisting of *mulleri, aldrichi, mojavensis, arizonensis,* and *buzzattii* (Crow 1942); the *cardini* group (Streisinger 1946); and *Drosophila pseudoobscura* and *D. persimilis* (Dobzhansky and Epling 1944). Additional sibling species complexes from Europe (*obscura* complex) and South America have been mentioned in preliminary publications. The significance of these findings will be discussed at the end of this section.

Only a few sibling species have been described among beetles. In the rice weevil, *Calandra oryzae,* two strains have been described by Zumpt, Richards, and Birch (1946) that differ in size and various physiological characters, but not morphologically. A few F_1 hybrids are produced in single choice crosses. F_1 females lay eggs and are inseminated by males of the large strain, but all eggs laid by such females are unfertilized. The indication is that sibling species are involved. Two borers of the genus *Cyllene* are also very similar, but have different hosts and breeding seasons (King 1943).

Sibling species are common among moths, although careful studies of the genital armatures have shown that the differences between the species are often more pronounced than is apparent from the external morphology. Most of the common food-pest moths have turned out to be groups of sibling species (Corbet 1943). Of the four species of the *Tinaea granella* complex, only two have been found to infest grain, and of these *T. granella* prefers wheat, *T. infimella* rye. As in mosquitoes, the larvae of these moths are sometimes more distinct than the adults.

Many of the species of insects that attack forest trees also seem to be really groups of sibling species. Brown and Mackay (1943) showed that the budworm (*Cacoelia* [*Archips*] *fumiferana,* Tortricidae) that lives on balsam fir and spruce is different from the one that lives on jack pine, although there is no clear-cut morphological distinction. Either species may occasionally infest the "wrong" host without any change of characters.

The work of Cantrall (1943) has shed much light on the so-called biological races of tree crickets of the *Nemobius fasciatus* group. The three species *N. fasciatus, N. socius,* and *N. tinnulus* occur together in the George Reserve in Michigan. In the laboratory any species will mate with any other, but only the *fasciatus* x *tinnulus* cross is fertile. Both species have very distinctive songs and F_1 hybrids have an intermediate song, which is not heard in nature. In addition to the song differences there are very definite differences in habitat preference. *N. fasciatus* lives in dry grasslands, *N. socius* in marshes, and *N. tinnulus* in sunny oak-hickory forests. Some entomologists still refer to these species as "physiological races."

It is impossible to make statements concerning the presence and frequency of sibling species in groups that are not well known taxonomically. This is true not only for most families of insects, but also for mollusks (e.g. genus *Pisidium*) and most internal parasites. Among marine organisms also only a few sibling species complexes have been studied. A study of the *Gammarus pulex* complex (crustaceans) (Sexton 1942) is a notable exception. Others occur in limpets, for example, in the genera *Patella* (Eslick 1940) and *Acmaea* (Test 1946).

What corresponds to sibling species in the higher animals exists also among the protozoa. An earlier summary of the work was published by Kimball (1943), but there is a more detailed study by Sonneborn and Dippell (1946) on the situation in *Paramecium aurelia*. The seven known so-called "varieties" of *P. aurelia* belong to two groups, A and B. No variety of group A reacts sexually with a variety of group B, nor any variety in group B with any other variety. The four varieties of group A do not have complete sexual isolation, as believed earlier, but there are potent bars to the exchange of genes. Sexual isolation in nature may well be complete. Although these "varieties" in *Paramecium* have all the biological characteristics of good species, it would serve no good end to give them distinct species names until the analysis has been carried a good deal further.

Morphologically indistinguishable so-called biological races occur commonly among pathogenic protozoans, particularly in *Plasmodium* and *Trypanosoma* (Hoare 1943). However, it seems to be unknown whether they are reproductively isolated and the question cannot yet be answered whether or not they correspond to the sibling species of higher animals.

This review shows that it has been clearly established during recent years that morphologically similar or indistinguishable species are not an exception but seem to occur in all well-studied groups of animals, although with greater frequency in some orders and families. Furthermore, in principle there is no difference between sibling species and strikingly distinct congeneric species. In fact, there is a complete gradation between the cases in which congeneric species are conspicuously different, only slightly different, and completely identical in external morphology. However, even these latter sibling species are reproductively isolated from each other and there is no evidence of a transfer of genes from one species to the next. In short, they satisfy all the requirements of a biological species definition.

It has been asked how the evolution of sibling species can be explained (Huxley 1943). Two explanations seem feasible. One is that they are species *in statu nascendi* and that they are of such recent origin that they have had no time yet to acquire significant morphological differences. A corollary of this hypothesis would be to assume that reproductive isolation had been acquired through one or only a few steps and that sibling species are on the whole genetically nearly identical. All the known facts, particularly those derived from a study of *Drosophila pseudoobscura-persimilis,* the *Anopheles maculipennis* complex, and the *Paramecium aurelia* varieties, indicate that this assumption is incorrect. Nearly every physiological and ecological character of these sibling species that has been carefully studied has been found to be genetically distinct in the various sibling species. Even though I do not agree with Epling's hypothesis (Epling 1944) of possible early Tertiary origin of the species *Drosophila pseudoobscura* and *D. persimilis,* the genetic differences between them and their failure to hybridize in nature where their ranges overlap indicate clearly that they are *not* species *in statu nascendi.*

The alternative explanation would be that the morphological characters of sibling species are genetically so well integrated that they are not visibly affected by the mutational steps that have produced ecological and reproductive isolating mechanisms. If such were the case it might be interpreted as indicating that in certain genera (e.g. *Drosophila, Anopheles*) a certain standardized morphology has a definite selective value that prevents visible changes. What is particularly puzzling is the fact that those sibling species that have been tested genetically, like *Drosophila pseudoobscura* and *D. persimilis,* seem to have normal mutation rates and the expected share of visually conspicuous mutations. The problem of the similarity of sibling species seems akin to that of phylogenetically conservative characters and of the characters of the higher categories (Mayr 1942). The taxonomist cannot do much about the solution of these problems except to call them to the attention of the geneticist.

[The comments in the last three paragraphs were directed against the opinion, widespread in the 1940s and 1950s, that sibling species are species *in statu nascendi* and in quite a different class from ordinary species. Hence my emphasis on their species character, that is, their reproductive isolation, and my further emphasis on the fact that they are part of a complete spectrum from extremely different to morphologically indistinguishable species. I also stressed that the integration of the genotype in certain genera like *Anopheles* and *Drosophila* was so well balanced that the genetic changes correlated with the acquisition of reproductive isolation did not result in a visible change of morphological structure, whereas morphology in other genera was far more labile.

The electrophoretic analysis of protein polymorphism in recent years has permitted for the first time a quantitative determination of genetic differences among species. Comparisons of species in different higher taxa are rather meaningless, but when comparisons are made between species of the same genus, for example *Drosophila,* it is found that morphologically similar species, for instance sibling species, are less different from each other genetically than morphologically dissimilar species. Yet the two groups grade into each other.]

35
The Biological Meaning
of Species

There is perhaps no other subject in biology for which one can document as long-standing a controversy as the species concept. If one gathered together all that has ever been written about the species, it would easily fill several shelves in a library. What then is the reason for so much confusion?

There are actually many reasons, but it would not be worthwhile to pursue them all. Many of the difficulties have been removed in recent years. Let me start with a most elementary linguistic consideration. It is necessary to make a distinction between categories and taxa. The concept *tree* is a category, but actual trees such as willows, oaks, and pines are taxa that we place in the category tree. The categories employed by the taxonomist are species, genus, family, order, and so forth, but the words robin, blackbird, chiff-chaff, and blue tit signify taxa to be placed in the species category. We see here at once that there are two levels of difficulties, the delimitation of taxa and their ranking in the proper category.

Let me illustrate this with a human example. There was a widespread theory in the early nineteenth century that the human races had descended from the different sons of Noah and were actually different species. On the taxon level, this posed the problem whether an intermediate population, let us say the North Africans, should be placed in the taxon of the Caucasians or that of the Negroes. This, then, was one species problem. The second problem was whether the proper category for each of these human types was that of the species or the subspecies. This second decision depends entirely on the concept of species adopted, whereas the placing of the North Africans with either the white or the Negro race has nothing to do with the species concept as such. Much of the argument about the species concept has been due to the confusion between these two classes of problems, those having to do with the assignment of populations to taxa and those having to do with the ranking of these taxa in categories. This will all become clearer as we go more deeply into the arguments.

Reprinted from "The biological meaning of species," *Biological Journal of the Linnean Society,* 1(1969):311–320.

Let us start with a historical survey of different species concepts. Considering the reams of paper devoted to the subject, it comes as somewhat of a surprise to learn that all the countless species definitions can be assigned to no more than three basic concepts of the category of the species.

THE TYPOLOGICAL OR ESSENTIALIST SPECIES CONCEPT

According to the typological concept, the observed diversity of the universe reflects the existence of a limited number of underlying "universals," or types. Individuals do not stand in any special relation to each other, being merely expressions of the same type. Variation is the result of imperfect manifestations of the idea implicit in each species. This species concept, going back to the philosophies of Plato (his *eidos*) and Aristotle, was the species concept of Linnaeus and his followers. This school of philosophy is now usually referred to as *essentialism,* following Karl Popper, and its species concept as the essentialist species concept. According to it, species can be recognized by their essential natures or essential characters, and these are expressed in their morphology. In its practical application, this species concept is usually called the morphological species concept.

In retrospect, it becomes obvious that not even Linnaeus and his followers had a strictly morphological species concept. For instance, Linnaeus described the male and the female mallard duck as two different species. When it was realized that the two so-called species were nothing but male and female, they were without hesitation combined into a single species even though there had been no change in the degree of morphological difference.

Even though morphological evidence is still used as a basis for inferences on the delimitations of biological species, a morphological species concept is no longer maintained by the modern biologist. In addition to the various conceptual reasons for its rejection are two practical ones. First, individuals are frequently found in nature that are clearly conspecific with other individuals in spite of striking morphological differences owing to sexual dimorphism, age differences, polymorphism and other forms of individual variation. An essentialist species concept is helpless in the face of caterpillar and butterfly, or sporophyte and gametophyte among plants, or whatever other drastic forms of intraspecific variation are found in nature. It is equally helpless in the face of so-called sibling species, that is, perfectly good genetic species which lack conspicuous morphological differences. Its theoretical as well as its practical weaknesses are the reasons why the essentialist species concept is now universally abandoned.

THE NOMINALISTIC SPECIES CONCEPT

The nominalists (Occam and his followers) deny the existence of "real" universals. For them only individuals exist, while species are manmade constructs. The nominalistic species concept was popular in France in the eighteenth century and has some adherents to the present day, particularly among botanists. Bessey (1908) expressed this viewpoint particularly well: "Nature produces individuals and nothing more. . . . Species have no actual existence in nature. They are mental concepts and nothing more. . . . Species have been invented in order that we may refer to great numbers of individuals collectively."

When I read statements such as this, I always remember an experience I had 40 years ago when I lived all alone with a primitive tribe of Papuans in the mountains of New Guinea. These superb woodsmen had 136 names for the 137 species of birds I distinguished (confusing only two nondescript species of warblers). That primitive Stone Age man recognizes the same entities of nature as western university-trained scientists refutes rather decisively the claim that species are nothing but a product of human imagination. The same, of course, is true for the sharp definition of animal species in our neighborhood. When you study the birds in your woods and gardens, do you ever find intermediates between blue tits and great tits, or between thrushes and blackbirds, or between jackdaws and rooks? Of course you do not. Every species of bird, mammal, or other higher animal is extraordinarily well defined at a given locality, and hybridization or intermediacy is a rare exception. Species are the product of evolution and not of the human mind. However, the nominalist species concept may well be legitimate when one deals with inanimate objects and particularly with human artifacts. It ignores, however, the fact that there is a fundamental difference between classes of *objects* that are the product of the human mind, like kinds of furniture, and classes of *organisms* that are the product of evolution rather than of human imagination. As Simpson has emphasized correctly, the basic fallacy of the nominalists is their misinterpretation of the causal relation between similarity and relationship. Members of a species taxon are similar to each other because they share a common heritage. It is not true that they belong to this taxon because they are similar, as is claimed by the nominalists. The situation is the same as with identical twins. Two brothers are identical twins not because they are so extraordinarily similar, but they are so similar because they are both derived from a single zygote, that is, because they are identical twins. Incidentally, it is this same misinterpretation of the connection between similarity and relationship that is the fatal weakness of numerical phenetics. Anyone who believes in evolution must reject the nominalistic species concept.

THE BIOLOGICAL SPECIES CONCEPT

It began to be realized in the late eighteenth century that neither of these two medieval species concepts, the essentialistic and the nominalistic, was applicable to biological species. An entirely new species concept began to emerge after about 1750, but it took another 150 years before it had been thought through in all of its consequences. This third concept differs quite drastically from the concept of inanimate species. It rejects the idea of defining the species typologically as a "class of objects." Indeed, it breaks with all philosophical traditions by defining species purely biologically, as follows: *Species are groups of interbreeding natural populations that are reproductively isolated from other such groups.*

A species, owing to the properties mentioned in this definition, has three separate functions. First, it forms *a reproductive community.* The individuals of a species of animals (the situation is somewhat different in plants) recognize each other as potential mates and seek each other for the purpose of reproduction. The species-specific genetic program of every individual ensures intraspecific reproduction. Second, the species is also *an ecological unit* that, regardless of the individuals composing it, interacts as a unit with other species with which it shares the environment. The species, finally, is *a genetic unit* consisting of a large intercommunicating gene pool, whereas the individual is merely a temporary vessel holding a small portion of the contents of the gene pool for a short period of time. In all three characteristics the biological species is nonarbitrarily defined, and differs quite drastically from so-called species of inanimate objects. It is called "biological" not because it deals with biological taxa, but because the definition is biological. It utilizes criteria that are meaningless as far as the inanimate world is concerned.

The species has two properties that distinguish it completely from all other taxonomic categories, let us say the genus. First of all, it permits a nonarbitrary definition—one might even go so far as to call it a self-operational definition—by stressing that it is defined by the noninterbreeding with other populations. Second, while all other categories are intrinsically defined, by having certain visible attributes, species are relationally defined. The word *species* corresponds very closely to other relational terms such as, for instance, the word *brother.* A given person is not a brother on the basis of certain intrinsic properties of his, but only in relation to someone else. A population is a species only with respect to other populations. To be a different species is not a matter of degree of difference but of relational distinctness.

The relational definition of the species is both the strength and the weakness of the biological species concept. It permits nonarbitrary decisions with respect to all other coexisting populations, that is, synchronic and sympatric species populations. This is where the concept is needed most frequently by the biologist and where its application faces the fewest difficulties. This is the situation sometimes referred to as the nondimensional species. The more distant two populations are in space and time, the more difficult it becomes to test their species status in relation to each other, but the more irrelevant biologically this also becomes.

Before entering into a discussion of the biological significance of species, let me say a few words on the dimensions of this universe. Few nontaxonomists have any conception of the magnitude of biological diversity. More than a million species of animals have already been described and nearly half a million species of plants. However, our knowledge is highly uneven. Only about 3 new species of birds are described annually, a very small addition to the 8600 species previously recorded. But let us look at some other groups. I still remember the days when many papers were published in the genetic literature giving the name of the organism simply as *Drosophila*. This was implicitly considered to be synonymous with *D. melanogaster*. Now more than 1000 species of *Drosophila* are recognized, and almost as many new species were discovered in the last 17 years as in the 170 years preceding 1950. I want to give you another statistic. One group of mites, the chiggers (Trombiculidae), are now known to be of great medical importance as vectors of scrub-typhus and other rickettsial diseases. Only three species were known in 1900, 33 in 1912, 517 in 1952, and about 2250 in 1966. It is estimated that several hundred thousand species of mites in the many different families of this order still await description. What the total of species of animals is no one knows. It may be 3 million, it may be 5 million, and it may even be 10 million. Most taxonomists nowadays partition their time, devoting part of it to the more classical operations of taxonomy, the describing and classifying of species, and the other part to a study of the biological aspects of species. For nothing could be more discouraging than devoting one's life entirely to the endless collecting, describing, and naming of new species. To do only that would be nothing but stamp collecting. Describing does not make a scientist; a scientist wants to understand and explain. He wants to determine the causes of the multitude of phenomena and relations at the species level.

What are the kinds of question for which we look for a causal answer? Let me single out six major problems.

Discontinuity

"Why is variation in nature organized in the form of species rather than being continuous?" To be very frank, we have a descriptive answer to this question, perhaps I should say an empirical answer, but the complete causal analysis has only begun. To make clear what we are after, let us imagine a universe without species. Every individual in such a world may, during reproduction, exchange genetic material with any other individual. What would happen under this set of rules of the game? Every once in a while mutation and recombination would produce an individual that would be particularly successful in utilizing the resources of the environment. Alas, during the next reproductive period, this unique combination of genetic factors would be broken up and its genotype lost forever.

There are two ways of preventing this and nature has adopted both. One is to abandon sexual reproduction and maintain the superior genotype through asexual reproduction as long as the environmental situation lasts for which this genotype is specially adapted. The other solution, of course, is the "invention" of the species, if I may express myself that way, that is, the acquisition of a genetic program which will permit reproduction and genetic recombination only with such other individuals as are genetically similar, that is, conspecific.

The division of the total genetic variability of nature into discrete packages, the so-called species, that are separated from each other by reproductive barriers, prevents the production of too great a number of disharmonious, incompatible gene combinations. This is the basic biological meaning of species, and this is the reason why there are discontinuities between sympatric species. We do know that genotypes are extremely complex epigenetic systems. There are severe limits to the amount of genetic variability that can be accommodated in a single gene pool without producing too many incompatible gene combinations. We still do not understand why, on the whole, hybrids are not only far more frequent but also apparently less handicapped in plants than in animals.

The mechanisms that guarantee the discreteness of species are called the *isolating mechanisms*. There is a great diversity of such mechanisms, the sterility barrier being only one, and as far as animals are concerned, one of the less essential ones. Behavioral barriers are the most important class of isolating mechanisms in animals. It is necessary to emphasize that it is coded in the genetic program of every species to what signals an individual should respond during the reproductive period. The study of isolating mechanisms has become one of the most important and fascinating areas of biology, and every textbook of evolutionary biology, cytology, genetics, or behavior now deals with them quite extensively.

Multiplication of Species

The second great problem of species is that stated in the title of Darwin's great book *The Origin of Species.* How do species multiply? The answer to this question can now be stated in much more meaningful terms than was possible a generation ago. Species originate when populations acquire isolating mechanisms. A few special cases excepted, species multiply either by polyploidy (a process largely restricted to the plant kingdom) or by geographic speciation, that is by the genetic reconstruction of spatially isolated gene pools. The subject having been dealt with exhaustively in several recent books, I will say nothing further about it.

However, I would like to mention three sets of unsolved problems of speciation.

1. How frequent are exceptional situations, such as the sympatric

evolution of host races into full species or the essentially sympatric origin of species through disruptive selection?

2. What role does chromosomal reorganization play during speciation? And how often does the acquisition of isolating mechanisms occur purely through genic mutation without any additional chromosomal reorganization?

3. To what extent does the acquisition of genic isolating mechanisms entail a reorganization of the entire epigenetic system?

Some of these questions may seem peculiar to someone who has not followed the recent genetic literature. However, unless I am very much mistaken I am discerning at the present time the emergence of a new area in genetics that constitutes a third set of problems related to the biological meaning of species.

The Genetics of Species

In the 1920s when I was a student and when the battle between the mutationists and the biometricians had not yet completely died down, there was a widespread idea that Mendelian factors controlled only the variation of intraspecific characters and that species differences were controlled by genetic factors in the cytoplasm. This idea has, of course, been dead for 40 years and even the discovery of DNA in mitochondria and other cellular organelles is not likely to lead to its revival. However, a number of phenomena have been discovered in recent years that indicate that our concept of an organism as a bag full of genes is an oversimplification. One of these phenomena is the remarkable phenotypic uniformity of most species over vast distances, a uniformity difficult to explain as the result of gene flow. I postulated in my 1963 book that such populations are held together by sharing in a single system of epistatic interactions or, as Waddington would call it, a single system of canalizations, but evidence for the existence of such a system is indirect and entirely based on inference. The study of the distribution of enzyme systems by Hubby, Lewontin, and others is now beginning to open the door to an entirely new realm of research. It seems that the same enzyme loci are variable in many populations of *Drosophila pseudoobscura* and often even have similar allele frequencies. The only exception was a peripherally isolated population. If these findings are confirmed for other species, it would bring us back to the idea that indeed a species may have a species-wide epistatic genic system on which geographic variation and other types of polymorphism are superimposed. It further suggests that this basic epistatic system undergoes a genetic revolution in connection with speciation. It is, of course, far too early to base sweeping conclusions on preliminary results, and the only reason I am mentioning them at all is because they fit so extremely well with some previous postulates.

Let me now cite another door that has been opened onto a *terra incognita*. Until quite recently, when one asked a *Drosophila* specialist how many genes *Drosophila* has, he might have said about 10,000 or, if he was in a very generous mood, 50,000; a mouse geneticist would have given similar answers. Yet, if one measures the amount of DNA in a single mammalian cell nucleus, one finds that it contains enough DNA for about 5 million cistrons, that is, for 5 million genes. It is a great puzzle what the other 4,950,000 genes are doing. We still do not know, but recent studies by Britten and his group in the Carnegie Institution, and by Walker (in Edinburgh) and his group show that there is great heterogeneity in the nuclear DNA and, in particular, that certain genomes may contain large quantities of identical DNA sequences. If such special DNA's should be species-specific, as some of the evidence indicates, it would raise an entirely new set of problems. The reason I am referring to this research is to make it clear how little we still understand what I have previously referred to as the *genetics* of *species,* the very particular genetic structure of species. It is quite possible, if not probable, that the acquisition of isolating mechanisms is merely a coincidental by-product of a far more fundamental genetic event, a genetic restructuring of populations. Only the simultaneous study of several loci or several characters will give us the kind of answers we are looking for. Any day may bring further exciting new discoveries in this area.

Let me now go back to some more classical problems.

The Role of Species in Evolution

The biologist, when he contemplates large-scale evolution, speaks of trends, adaptations, specializations, and invasions of new adaptive zones and niches. The explanation of these phenomena has, however, suffered owing to the fact that the most important part of the story, the role of the agents of these evolutionary phenomena, was omitted. Actually, in each case, it is a species or a group of species that is responsible for the evolutionary events. *Species are the real units of evolution,* they are the entities which specialize, which become adapted, or which shift their adaptation. And speciation, the production of new gene complexes capable of ecological shifts, is the method by which evolution advances. The species truly is the keystone of evolution.

The role of species is to some extent comparable to the role of mutations. Most mutations are irrelevant or deleterious, but whenever there *is* any genetic improvement, it is due to the incorporation of a new mutation into an improved genotype. It is the same with species. Recent taxonomic studies have shown how frequent incipient species are. Speciation obviously is a prolific process, but the majority of new species have a short life expectancy—they become extinct sooner or later. But one out of 100, 1000, or 10,000 makes an evolutionary invention and is able to occupy a novel adaptive zone. Birds, bats, vertebrates, insects, they all ultimately go back to one particular, unusually successful species. Every species is a new evolutionary experiment. Most of them are failures, but an occasional one is a spectacular success. Even when we look at a group of closely related species, we find almost invariably one or the other with an unusual specialization or adaptation. In most cases this merely leads into an evolutionary dead-end street, but occasionally it opens the door to an entirely new world. To repeat, the species plays an enormously important role in evolution.

Species and Ecosystems

One of the unsolved problems relating to species is why some of them, in fact the vast majority, are so narrow in their ecological specialization, while a few species seem to have an extraordinary ecological tolerance. For instance, we can say descriptively that a certain plant host-soecific moth is so specialized that its larvae can live, as leaf miners, only in the leaves of one particular species of plants, while another species of moths has such broad tolerance that it can feed on the leaves of all the species in, let us say, 8 or 10 families of plants. Or to give another example, one species of ants always has small colonies, and only a few of these colonies per unit of area, while another species of the same or a related genus may be extremely successful and become a tramp species that is carried all over the world, establishing colonies wherever it goes. Extremely little is known so far about the reasons for such differences among species. Carson has examined this problem for the genus *Drosophila* and he found that the so-called garbage species, those that can live successfully in most countries, in most climates, and on many sources of food, do, indeed, on the average differ in their karyotype from the less successful, less common, more localized species. But this is only a beginning, and the actual truth of the matter is that we have very little understanding so far of the genetic basis for the tremendous differences in ecological tolerance between different species. It is rather obvious that the classical method of trying to describe species differences in terms of gene frequencies and the fitness of individual genes will not get us very far in such an analysis. The genetics of the ecological role of species is still at its very beginning.

Species and Species Diversity

Up to now I have focused attention on a single species at a time, but there is another aspect to species, and that is the total diversity of species in a given region. To be sure, the total diversity of species at a given place at a given time is the product of the characteristics of all the individual species of which the total is composed. Nevertheless, as in the case of many complex systems, the analysis of the system as a whole gives us new insights into the properties of the component parts, just as the study of the water molecule reveals certain properties of the elements hydrogen and oxygen that the study of these elements in their pure state would not or not easily reveal.

The study of species diversity is one of the most active frontiers of

ecology, and the number of unanswered questions is legion. For instance, what is the ecological interaction of species? I think here, in
particular, not of such rather simple-minded matters as food chains, but
of the far more complex problems indicated by such words as *niche,
competition,* and *exclusion.* The niche concept is an old one and even
Darwin referred to the "place an animal or a plant occupies in nature"
and used other similar expressions. Originally, niche was quite rightly
defined as the requirements of a species. In other words, it was designed
from the animal or plant outward, as something that the species requires in order to survive and prosper. Unfortunately, there has been an
increasing tendency to look at nature as a huge old-fashioned roll-
topped desk with an enormous number of pigeonholes, each one the
niche of a species. This interpretation leads to many difficulties. There is
a far better way of looking at niches, namely, by defining them in terms
of the genetic potential of a species to utilize certain components of the
environment. The niche, then, is no longer a static property of the environment, but a reflection of the contained species. As soon as we do
this, we can understand how the niche utilization can be broadened
when a species invades a new area, or in another case how it can be
narrowed when the area is invaded by a new species that is more efficient in utilizing certain resources of the environment. There is nothing
new in this way of looking at the niche problem, but a great deal of
rather sterile controversy could have been avoided by regarding niches
as the outward projections of the genetic potential of species. This also
helps us understand differences in species diversity between different
latitudes. Where violent seasonal fluctuations make high demands on
the genetic potential, comparatively few species can cope with the situation, and this is one of the reasons why there are so many fewer species in the temperate zone than in the tropics.

It has always been stated, and qutie rightly so, that successful
speciation depends not only on the acquisition of isolating mechanisms,
but also on an ability to utilize certain resources of the environment
more successfully than any competitor. The species thus is one of the
important units of ecology, this importance being due to the fact that
any given gene pool has only a limited ecological competence.

CONCLUSION

I am afraid this has been a rather rapid survey of an immensely wide
field. The naturalist, the student of local faunas and floras, has understood the importance of the species as a biological unit for hundreds of
years. However, the ill-conceived essentialistic and nominalistic philosophies and their translation into arbitrarily defined morphological species taxa has long prevented the full appreciation of the great biological
importance of species. This lack of appreciation, I feel, is now a matter
of the past. Students of animal behavior and particularly students of

species-specific isolating mechanisms have helped in demonstrating the nonarbitrariness of species. More recently, this is being further supported by the kind of genetic studies I mentioned earlier and also by the studies of biochemists and immunologists. I think there can be little doubt that the species represents a level of integration—in the hierarchy of levels from the subcellular to the community—that is of the utmost importance in all branches of biology, particularly in physiology, behavior, ecology, and evolution. It is fully as legitimate to study species as it is to study molecules; indeed, for the healthy integration of all knowledge in biology, it is vitally important that this particular level of biological integration not be neglected. For it is the study of species, more than anything else, that provides a joint interest for some otherwise very different branches of biology, and thus contributes to the unity of biology as a whole.

REFERENCES

Bessey, C. E. 1908. The taxonomic aspect of the species question. *Amer. Nat.*, 42:218–224.

Mayr, E. 1963. *Animal species and evolution.* Belknap Press of Harvard University Press, Cambridge, Mass.

VII
MAN

Introduction

Evolution is of such great interest to many because most findings of evolutionary biology are of immediate applicability to man. Even in this volume, only a few of the essays do not have an immediate bearing on man and the meaning of the world in which he lives. This has not always been understood. Anthropology, the science of man, was traditionally rather self-contained and thus isolated. Biological anthropology, in particular, became very one-sided, almost entirely morphological, and purely descriptive at that. It was forgotten, as obvious as this would seem to be, that man can be fully understood only if all branches of biology make their contribution. How fruitful such a collaboration can be was demonstrated at the Fifteenth Cold Spring Harbor Symposium of Quantitative Biology in 1950, when anatomists, anthropologists, paleontologists, systematists, population biologists, and geneticists all applied their special competence to the problem of man's evolutionary history. Not only did the participants educate each other, but they achieved a new synthesis that served as the basis, in the ensuing years, for increasing research in human evolution.

36

Taxonomic Categories
in Fossil Hominids

The classification of fossil hominids had become completely chaotic by the 1940s. As Campbell (1965) showed, no less than 29 generic and more than 100 specific names had been given to various hominid fossils recovered in different parts of the world. The diversity of putative human ancestors and their relatives had become so bewildering that it had become virtually impossible to attempt a reconstruction of human evolution. The approach of the anatomists who had been responsible for most of these names had been purely typological. It seemed very tempting, in this situation, to try to apply the concepts of the new systematics to hominid classification and see whether this would produce clarity. This I attempted in a preliminary paper in 1944, followed in 1950 by the more formal proposal reprinted here.

My strategy was to apply the principle of Occam's razor. I asked: What is the simplest arrangement of taxa that can be reconciled with the plethora of names? The result was startling: I came up with a classification of only one genus (possibly two) and only three species. As will be shown in a postscript, this turned out to be an oversimplification, but it was surely far closer to the truth than the arrangements I tried to displace. It permitted the posing of concrete research problems and introduced an entirely new approach to the study of fossil man: thinking in terms of populations instead of in types.

It is one of the most fruitful procedures of modern science to bring specialists of various fields together to discuss the problems that concern the zone of overlap of their fields. Since I have no first-hand knowledge of paleoanthropology, my own contribution to the question of the taxonomic categories of fossil man will be that of a systematist. Significant progress has been made within recent years among biologically thinking taxonomists in the understanding of the categories of subspecies, species, and genus, and it is my hope that this knowledge may help in a better understanding of fossil man.

Revised from "Taxonomic categories in fossil hominids," *Cold Spring Harbor Symposia on Quantitative Biology*, 15 (1951):109–118.

The whole problem of the origin of man depends, to a considerable extent, on the proper definition and evaluation of taxonomic categories. But there is less agreement on the meaning of the categories species and genus in man and the primates than perhaps in any other group of animals. Some anthropologists, in fact, imply that they use specific and generic names merely as labels for specimens without giving them any biological meaning. Weidenreich, for example, stated that in anthropology "it always was and still is the custom to give generic and specific names to each new type without much concern for the kind of relationship to other types formerly known." Broom (1950) likewise states, "I think it will be much more convenient to split the different varieties [of South African fossil ape-man] into different genera and species than to lump them." The result of such standards is a simply bewildering diversity of names. In addition to various so-called species of *Homo,* I have found the following names for various hominid remains in the literature: *Australopithecus, Plesianthropus, Paranthropus, Eoanthropus, Giganthopithecus, Meganthropus, Pithecanthropus, Sinanthropus, Africanthropus, Javanthropus, Paleoanthropus, Europanthropus,* and several others. No two authors agree either in nomenclature or in interpretation. It seems to me that an effort should be made to give the categories species and genus a new meaning in the field of anthropology, namely, the same one that in recent years has become the standard in other branches of zoology.

A reevaluation of the nomenclature of hominid taxonomy is facilitated by the fact that in recent years a magnificent body of new data has been accumulated by anthropologists, based partly on comparative anatomical studies and partly on significant new discoveries of fossil hominids in southeast Asia and in eastern and southern Africa.

The nomenclatorial difficulties of the anthropologists are chiefly due to two facts. The first is a very intense occupation with only a very small fraction of the animal kingdom that has resulted in the development of standards that differ greatly from those applied in other fields of zoology. The second is the attempt to express every difference of morphology, even the slightest one, by a different name and to do this with the limited number of taxonomic categories that are available. This difference in standards becomes very apparent if we, for example, compare the classification of the hominids with that of the *Drosophila* flies. There are now about 600 species of *Drosophila* known, all included in a single genus. If individuals of these species were enlarged to the size of man or of a gorilla, it would be apparent even to a lay person that they are probably more different from each other than are the various primates and certainly more so than the species of the suborder Anthropoidea. What in the case of *Drosophila* is a genus has almost the rank of an order or, at least, suborder in the primates. The discrepancy is equally great at lower categories, as we shall presently see. It is not mere formalism to try to harmonize the categories of anthropology with those of the rest of zoology. Rather, the evaluation of human evolution depends to a considerable extent on the proper determination of the categories of fossil man.

There are two recent developments in general systematics that will be particularly helpful in our efforts. The first one is that the biological meaning of the categories species and genus is now better understood than formerly. The second is that, in the attempt to close the gap between the complexity of nature and the simplicity of categories, the number of existing categories has been augmented by intermediate and group categories, such as "local population" or "local races" and "subspecies groups." The adoption of these intermediate categories facilitates classification without encumbering nomenclature.

THE TAXONOMIC CATEGORIES

The work in the new systematics has led to a far-reaching agreement among zoologists on the meaning of the categories subspecies, species, genus, and family. In the following analysis I shall attempt to see how far the current usage of these categories can be extended to fossil hominids and what such a reclassification means in terms of human evolution.

The Genus

The genus is a taxonomic category for a group of related species. It is usually based on a taxonomic group that can be objectively defined. However, the delimitation of these groups against each other, as well as their ranking, is frequently subjective and arbitrary. A conventional definition of the genus would read about as follows: "A genus consists of one species, or a group of species of common ancestry, that differ in a pronounced manner from other groups of species and are separated from them by a decided morphological gap."

Recent studies indicate that the genus is not merely a morphological concept but that it has a very distinct biological meaning. Species that are united in a given genus occupy an ecological situation which is different from that occupied by the species of another genus, or, to use the terminology of Sewall Wright, they occupy a different adaptive plateau. It is part of the task of the taxonomist to determine the adaptive zones occupied by the various genera. The adaptive plateau of the genus is based on a more fundamental difference in ecology than that between the ecological niches of species.

Unfortunately, there is no such thing as a recognized or absolute generic character. Early taxonomists knew this; indeed, Linnaeus stated, "It is the genus that gives the characters, and not the characters that make the genus." The genus is a group category and it defeats the object of binomial nomenclature to place each species into a separate genus, as has been the tendency among students of primates.

The acceptance of the new concept of biologically defined polytypic species (see below) necessitates the upward revision of all other categories (Mayr 1942). Often what was formerly designated a group of allopatric species is now called a single polytypic species with numerous subspecies. To leave each of these polytypic species in a separate genus deprives the genus of its significance as a truly collective category. I shall illustrate this need for the combining of genera by an example. Gorilla and chimpanzee are two excellent species that, as Adolph Schultz has shown, differ from each other by a wealth of characters. At one time several species of gorillas and of chimpanzees were recognized, but the allopatric forms within the two species are now considered subspecies. Being left with one species of gorilla and one species of chimpanzee, we are confronted by the question whether or not they are sufficiently different to justify placing them in different genera. A specialist on anthropoids impressed by the many differences between these species may want to do so. Other zoologists will conclude that the differences between the two species are not indicative of a generic level of difference when measured by the standards customary in most branches of zoology. To place these two anthropoids into two separate genera defeats the function of generic nomenclature and conceals the close relationship between gorilla and chimpanzee as compared with the much more different orang and the gibbons. Recognizing a separate genus for the gorilla would necessitate raising the orang and the gibbon to subfamily or family rank, as has indeed been done or suggested. This only worsens the inequality of the higher categories among the primates.

The same is true for the fossil hominids. After due consideration of the many differences between Modern man, Java man, and the South African ape-man, I did not find any morphological characters that would necessitate separating them into several genera. Not even *Australopithecus* has unequivocal claims for separation. This form appears to possess what might be considered the principal generic character of *Homo,* namely, upright posture with its shift to a terrestrial mode of living and the freeing of the anterior extremity for new functions that, in turn, have stimulated brain evolution. Within this type there has been phyletic speciation resulting in *Homo sapiens.*

The claim that the many described genera of hominids and Australopithecines have no validity, if the same yardstick is applied that is customary in systematic zoology, is based on two major points. Both of these are admittedly somewhat vulnerable. One is the overall picture of morphological resemblance with a deliberate minimizing of the brain as a decisive taxonomic character. To this point I shall return presently. The other point is the assumption that all these forms, including *Australopithecus,* are essentially members of a single line of descent. Additional finds might easily disprove this. However, taking the currently available evidence all together, it seems to me far more logical and consistent at the present time to unite the hominids into a single genus than to continue the current multiplicity of names.

This reevaluation of the generic status of the fossil hominids forces us to consider also the categories above the genus. Does *Homo* belong to a separate family Hominidae? The morphological differences between *Pan*, the genus to which the chimpanzee and gorilla belong, and *Homo* are so slight that there seems to be no justification for placing them in separate families. There is even less justification for placing South African man in a separate subfamily, the Australopithecinae. The most primitive known hominids, those of South Africa, combine certain typical hominid characters, such as upright posture, with others that are usually considered simian, such as small size of brain and protruding face. It is noteworthy, however, as pointed out by several investigators, that these hominids, even at this primitive stage, lack certain other simian features that were formerly considered primitive: powerful canines, large incisors, a sectorial form of the first lower premolar, an exaggerated development of the supra-orbitals, a simian shelf, and powerful brachiating arms. It now appears probable that many of these characters are functional specializations that were acquired by the anthropoid apes after the hominid line had branched off.

The fact that the hominids lack these specializations has been used by some authors as evidence to postulate a very early human origin and a very isolated position of the hominid branch. This is by no means the only possible interpretation. Rather, it seems to me that most of these typical characters of the living anthropoids may well be a single character complex evolved in response to a highly arboreal mode of living. It now appears probable that the African anthropoids, the orang, and the gibbons may have acquired most of these characters independently and are therefore, in a sense, a polyphyletic group. The available evidence seems to indicate that man may be more closely related to the gorilla–chimpanzee group than this group is either to the orang or to the gibbons. The degree of similarity in certain morphological traits cannot necessarily be used to measure degree of phylogenetic relationship. The arboreal, brachiating large anthropoids are exposed to a similar type of selection and will therefore evolve in a parallel, if not convergent, manner. When the *Homo* line acquired upright posture it entered a completely different adaptive zone and became exposed to a severely increased selection pressure. This must have resulted in a sharp acceleration of evolutionary change, leading to the well-known differences between man and the living anthropoids. This factor must be taken into consideration when the phylogeny of man and the anthropoids is reconstructed. It would therefore appear to be misleading from the purely morphological-phylogenetic point of view to separate man from the anthropoid apes as a special family. It would be equally misleading to go to the other extreme and to use the evidence of the somewhat independent evolution of man and the various anthropoids as a means of denying their close relationship.

Denying the genus *Homo* family rank is based on purely morphological considerations. It does not take into account man's unique position in nature. Man has undoubtedly found an adaptive plateau that is strikingly different from that of any other animal. There are some who feel that there is only one way in which to emphasize this uniqueness of man, namely, to place *Homo* in a separate family. The conventional standards of taxonomy are insufficient to decide what is correct in this case.

From the purely biological point of view, man is certainly at least as different as a very good genus. We have thus the evolution of a new higher category in the geologically short period of 1 to 2 million years. This is another significant illustration of the rapidity by which one major taxonomic entity can be transformed into another one, without any jumps.

The Subspecies

Before we can attempt to answer the question how many species of fossil man have existed, we must say a few words on infraspecific categories. The species of the modern systematist is polytypic and multidimensional. It has the geographical dimensions of longitude and latitude and also the time dimension. It is polytypic because it is composed of lower units, such as subspecies and local populations. Customarily in anthropology, distinct local populations have been referred to as races, and a similar custom exists in some branches of zoology, for example, in ichthyology.

The amount of geographical variation and the degree of difference between the geographical subdivisions of a species are different from case to case. Some species appear quite uniform throughout their entire range; other species have a few or many more or less well defined subspecies. For instance, the two African forest anthropoids, chimpanzee and gorilla, show only a moderate amount of geographical variation, although both have well-defined subspecies, and some investigators have proposed splitting the chimpanzee into two species. Geographical variation is much more pronounced in the gibbons and even more so in some of the South American monkeys, where geographical races are often different enough to be considered full species by the majority of authors.

Modern man is comparatively homogeneous because there is much interbreeding between different tribes and races. Still, we find living close to each other such strikingly different races as bushmen and Bantus in South Africa, or the Congo pygmies and Watusi in central Africa, or the Wedas and Singhalese in Ceylon. There is much indirect evidence that primitive man was much more broken up into small scattered tribes with little contact with each other, intensely subject to local selective factors.

In addition to this much greater geographical variation of primitive man, there is evidence also of greater individual variation (including sexual dimorphism). The variability of Mt. Carmel man has been commented upon in the literature. It seems possible, if not probable, that the various South African finds, *Australopithecus, Plesianthropus,* and *Paranthropus,* might well be age or sex stages of a few related tribes, notwithstanding Broom's (1950) assertions to the contrary.

Differences between young and adult and between male and female appear to be greater in the gorilla and orangutan than they are in modern man. Variability may increase or decrease in the course of evolution. Abundant proof for this statement can be found in the paleontological literature. I interpret the available literature to indicate that primitive man showed more geographical as well as individual variation than modern man.

Why primitive man should have been more variable than modern man is not entirely clear. A study of the family structure of anthropoids might shed some light on this problem. Perhaps there was a greater functional difference between male and female than in modern man. Perhaps the ancestral hominids had a system of polygamy that favored the selection of secondary sex characters in the male. We don't know. Whatever the reasons, we should not use the variability within populations of modern man as a yardstick by which to judge the probable variability of extinct populations.

This point is important because it bears on the question whether or not more than one species of hominid has ever existed on the earth at any one time. Indeed, all the evidence now available can be interpreted as indicating that, in spite of much geographical variation, never more than one species of man existed on the earth at any one time. I shall come back to this point later.

The Species

As described in several publications, the concept of the species has undergone a considerable change during recent years. The morphological and typological species of the early taxonomists has been replaced by a biological species. The species is now defined as "a group of actually or potentially interbreeding natural populations that is reproductively isolated from other such groups." When this concept is applied to man, it is at once obvious that all living populations of man are part of a single species. Not only are they connected everywhere by intermediate populations, but even where strikingly distinct human populations have come in contact, such as Europeans and Hottentots, or Europeans and Australian aborigines, there has been no sign of biological isolating mechanisms, only social ones.

The problem of species delimitation is much more difficult with respect to fossil man. How shall we determine which populations were "actually or potentially interbreeding"? It is evident that we must use all sorts of indirect clues. The first concrete problem is what types of fossil man should be included in the species *Homo sapiens*. Cro-Magnon man is so nearly identical with *Homo sapiens* that its inclusion in that species is not doubted by any serious student.

The problem of Neanderthal man is much more difficult. Should he be included in the same species as modern man or not? When the first finds of Neanderthal man were made there seemed to be no problem. These fossils were characterized by distinct morphological features and were clearly replaced by modern man in Europe on a distinct chronological level. There is no morphological or cultural intermediacy. Additional finds, however, have caused various difficulties. The Mt. Carmel finds in Palestine belong to a population that combines some features of Neanderthal man with some of modern man. It is immaterial whether we interpret this as a hybrid population, as an intermediate population, or as a population ancestral to both. The fact remains that Mt. Carmel man makes the delimitation of modern man from Neanderthal man exceedingly difficult, if not impossible, as pointed out by Dobzhansky (1944). Weidenreich supported the theory that modern man was a direct descendant of Neanderthal man. Boule and others have raised serious objections to this theory. But how can we reconcile the apparently incompatible views that modern man and Neanderthal man are conspecific and that modern man is *not* a descendant of typical European Neanderthal man? A possible clue is furnished by the hominids that were widespread in Europe in the mid-Pleistocene. The skulls of Steinheim, Swanscombe, and Fontéchevade combine features of modern man and of Neanderthal man, together with primitive and specialized features of their own. They lived apparently in interglacials and were more closely linked with a warm climate than Neanderthal man.

If I understand the evidence correctly, it is possible to interpret these early European fossils as remains of populations of *Homo* that were ancestral both to *sapiens* and to "classical" Neanderthal man and from which these two forms evolved by geographical variation. Tentatively, the working hypothesis can be made that Neanderthal man in its classical form was a geographical race that occurred in central Europe and was represented in Africa by Rhodesian man and in Java by Njandong man, while a more *sapiens*-like population occurred at the same period as some of these Neanderthaloids either in north Africa or western Asia or in some other area that has not yet yielded remains of fossil man. When *sapiens* began to expand and spread, he eliminated the other contemporary races, just as the white man drove out the Australian aborigines and the North American Indians. The process of elimination of the Neanderthal characters in mixed populations was presumably helped by selection preference in favor of the characters of modern man.

It is very probable that additional finds will make the delimitation of *sapiens* against Neanderthal even more difficult. It seems best to follow Dobzhansky's suggestion and to consider the two forms, as well as the ancestral group that seems to combine their characters, a single species.

Homo erectus

Java and Peking man are distinct enough from modern man that they have to be considered a separate species, which must be called *Homo erectus*. This is true regardless of the fact that on Java, at least, Njandong and Wadjak man may have formed a practically unbroken chain of hominids leading from Java man to modern man. Peking man (*Homo erectus pekinensis*) is, on the whole, so similar to Java man that it should be considered merely subspecifically distinct, as I proposed previously (Mayr 1944).

Homo africanus

South African ape-man again is one level further back and is sufficiently far removed from Java man to be considered a full species. Actually, no less than three genera and five species of South African ape-man were described, which, in Broom's terminology, have the following names: *Australopithecus africanus* 1925 (Taung), *A. prometheus* 1947 (Makapan), *Plesianthropus transvaalensis* 1936 (Sterkfontain), *Paranthropus robustus* 1938 (Kromdraai), and *Paranthropus crassidens* 1949 (Swartkrans). Since one of these names was based on a child, another on an adult female, and a third on an adult male, an enumeration of diagnostic differences is virtually impossible. The extant skulls are somewhat altered in shape owing to crushing, and the fact that the cephalic index in the Taung child is 62.4 while it is 83.5 in the Sterkfontein male is therefore not as significant as Broom thinks. Nor is the fact that the finds are associated with different faunas. Contemporary modern man can be found associated with okapis or elephants or tigers or kangaroos or South American edentates or polar bears. The various finds of South African man are presumably not contemporary, but there is nothing in the evidence that has so far been presented (e.g., Broom 1950) that would prove that more than one species is involved.

Until a real taxonomic distinction has been established, it will be safer and more scientific to refer to the different South African fossils by vernacular names. There is no danger of confusion if we speak of the Sterkfontain or Makapan finds, while it implies an obviously erroneous conclusion, namely that of generic distinctness, if we refer to them as *Plesianthropus* and *Australopithecus*. New discoveries are still being made in these cave deposits and many of those that have already been made have not yet been fully worked out. There is good reason to believe that it will be firmly established in the not-too-distant future how many different tribes, temporal subspecies, or even species of South African ape-man once existed. To consider them all one species is the simplest solution that is consistent with the available evidence.

A more important question is whether the South African hominids are ancestral to modern man or rather represent a specialized or aberrant side line. The exact dating of these fossils has not yet been achieved but they are believed to be very early Pleistocene or latest Pliocene; in fact, they presumably ranged over a considerable period of time. There is thus no definite chronological reason why the South African ape-man could not be considered a possible ancestor of modern man. The principal objection that has been raised is that South African man shows a combination of characters that "should not" occur in an early hominid. This argument is based on typological considerations. Adherents of this concept believe that missing links should be about halfway between the forms they connect and that they should be halfway in every respect. This undoubtedly is not the case with *Australopithecus*. It is apparently amazingly like modern man in its upright posture, structure of the pelvis, and other features, while it is very simian in its massive mandibles, large molars, prognathism, and small brain. *Australopithecus* lacks those specializations that stamp gorilla, orang, and gibbon as typical anthropoids.

The peculiar combination of characters that is found in *Australopithecus* is due to the fact that during the evolution of man different characters evolved at different rates. If we set the point where the human line branched off from the other anthropoids as zero and the *Homo sapiens* stage as 100, we might arbitrarily give the following points to the various organs of *Australopithecus:* pelvis, 90; premolars, 75; occipital condyles, 80; incisors, 55; the setting of the brain case, 70; shape of the tooth row, 70; the profile of the jaw, 30; the molar teeth, 40; the brain, 35; and so on. It is obvious that one type does not change into another type evenly and harmoniously; rather, some features run way ahead of the others.

The Simplified Nomenclature of Fossil Man

Reducing the bewildering assortment of genera and species of hominids to one genus with three species not only results in simplicity but also makes certain conclusions obvious that were previously not apparent. Before discussing these conclusions, however, I might point out some of the disadvantages of such a simplified classification.

There have been two trends in human evolution, as, indeed, there are in the evolution of all organisms. First of all, there is a continuous evolutionary change in time, the so-called phyletic evolution, starting in the hominids with the most simian forms and ending with modern man. Simultaneously a centrifugal force has been operating, namely, geographical and other local variation, which tries to break up the uniform human species. This geographic variation leads to the formation of races and subspecies, and, if this trend should go to completion, to the formation of new separate species. There are all sorts of intermediate stages in both these trends, and it is obvious that all the many possible differences and gradations between the various kinds of hominids cannot be expressed completely in the simple nomenclature of species, genus, and subspecies.

For instance, man as he exists today has pronounced racial groups, such as the Whites, Negroes, and Mongoloids, which might well deserve subspecific recognition. But there are minor racial differences within each of these subspecies. Furthermore, preceding modern man there have been types of *Homo sapiens* that are now extinct, like Cro-Magnon man and his contemporaries. These, no doubt, are a level of subspecies different from those of living man. Neanderthal man is a third level, and the pre-Neanderthal man, who combines certain features of *sapiens* and Neanderthal, is a fourth level. It is unsatisfactory for biological, as well as for practical, reasons to treat each of these levels as a separate species. However, combining them into a single species conceals the pronounced differences between these levels and reduces the taxonomic difference between Neanderthal and modern man to the level of difference between White Man and Negro. How can this be avoided?

First of all, we must realize that no system of classification and nomenclature can ever hope to express adequately the complicated relationships of natural populations. However, by giving species and genus the well-defined meanings that we have assigned to these categories, we make at least an attempt to standardize taxonomic categories and make them comparable. A possible solution of our particular difficulties may come from a refinement of the levels of infraspecific categories. In addition to the subspecies we may use such infrasubspecific categories as "race" and "local population," as well as the suprasubspecific category of the "subspecies group." Hence, we should be guided by the following practical rules:

1. Not to assign a formal name to any local population or race that does not deserve subspecific rank.

2. To give trinomials to all forms that do not deserve higher than subspecies rank.

3. To group together as subspecies groups all those subspecies within a species that form either geographical or chronological groups. Such subspecies groups in *Homo sapiens,* for instance, might be: (a) modern man, (b) Neanderthal group, and (c) pre-Neanderthal group.

4. Not to give formal generic and specific names to new fossil finds that are not sufficiently known. Vernaculars, such as "Steinheim man" or "Njandong man," are just as useful and much less misleading. The formal application of generic and specific names simulates a precision that often does not exist. To give the impression of an unjustified precision is as much a methodological error as to make calculations to the fifth decimal when the accuracy of the original data extends only to the first decimal.

Anthropologists should never lose sight of the fact that taxonomic categories are based on populations, not on individuals. Different names should never be given to individuals that are presumably members of a single variable population.

CONCLUSIONS

The arranging of all finds of fossil hominids into a single genus with three species helps to focus attention on the following conclusions.

The Question of the "Missing Link"

Ever since there has been an appreciation of man's anthropoid origin there has been a search for the "missing link." Some anthropologists may disclaim this and say that they realize the gradual evolution of mankind, but the fact remains that accurate criteria of humanhood are elaborated even in the most recent literature, such as Sir Arthur Keith's criterion of the brain volume of 750 cc.

The analysis of this problem will be facilitated by the realization that it is an oversimplification to use in this case the uninomial alternative "ape" versus "man." Taxonomists know by experience the inadequacy of uninomialism. Once we classify man binomially as *Homo sapiens,* it immediately becomes apparent that we must look for two missing links, namely, that which connects *sapiens* with his ancestor and that which connects *Homo* with his ancestor. Or, to express this differently, the two points of interest are the first on the phyletic line of man where he reached the *sapiens* level and second the place where the *Homo* line branched off from the other primates.

Let us look more closely at these two problems of the origin of man. The branching off of *Homo* from the other anthropoids was a case of orthodox speciation distinguished only by the fact that the new species simultaneously reached a new adaptive plateau. It is now evident, as has been stated by many authors, that a change in the mode of locomotion and a corresponding alteration of the entire organization of the body, in other words, the assuming of the upright posture, were the essential steps that led to the evolution of *Homo*. This evolutionary trend apparently affected first the pelvis and posterior extremities, and soon afterward the anterior extremities. The corresponding reorganization of the skull apparently lagged behind. It is therefore singularly difficult to localize in both time and space this important evolutionary step of the attainment of the upright posture with the help of jaw and tooth fragments, such as constitute most of the primate and anthropoid remains in eastern Africa during the Pliocene and Miocene.

To determine the exact point in the phyletic evolution of *Homo* where the *sapiens* level was reached is quite impossible. It was a very gradual process leading from *erectus* to *sapiens* and no particular form can be singled out as the missing link. However, there is a lower level in the phyletic evolution of *Homo* that is of special evolutionary interest,

namely, the level at which the hominids first displayed those intellectual qualities that are considered distinctly human rather than
simian.

Attempts have been made to measure the attainment of this *Homo*
level in terms of brain size. This method is fraught with difficulty. First
of all, brain size is to some extent correlated with body size. If, for
instance, a large gorilla should have a brain of 650 cc., this is not at all
necessarily equivalent to the brain of a fossil hominid of 650 cc., if that
hominid were much smaller than a gorilla. If the brain of the gorilla
averages one-fourth larger than that of the chimpanzee, it does not
mean that he is on the average 25 percent more intelligent. The correlation between brain size and intelligence is very loose. There is good
evidence that the brain size of late Pleistocene man may have averaged
larger than that of modern man. If true, this does not mean necessarily
that there has been a deterioration of man's intelligence since the Pleistocene, for intelligence is not determined by brain size alone. It is, of
course, still unknown what neurological structures affect intelligence,
but the folding of the cortex and all sorts of specializations within the
cortex appear to be as important as size. It is therefore dangerous, in
fact outright misleading, to use size as an absolute criterion and to say
that the *Homo* stage was reached when brain size reached a level of 700
or 750 cc.

It has been suggested that we measure the attainment of the human
level by some cultural achievement, such as the use of fire, rather than
by an anatomical standard like brain size. This is unquestionably a
superior approach, but it entails the practical difficulty that the first
moment of fire making was not fossilized and can never be dated
accurately. However, the first making of fire may have occurred not
much after the first use of tools by hominids, and some lucky finds
may some day shed light on the period when that occurred. South
African man was presumably a user of tools, and the first use of tools
may be coincident with the evolution of South African man.

Speciation in Man

In the strict sense of the word, speciation means the origin of
discontinuities through the origin of reproductive isolating mechanisms.
How often has man speciated? The answer is that he has speciated only
once if our assumption is correct that never more than one species of
man existed on the earth at any one time. This single event of speciation was the branching off of *Homo* from the anthropoid stock. That
some fairly distinct hominid remains have been found in approximately
contemporary deposits does not prove their specific distinctness. The
subdivision of the human species into independent tribes favors diversification. If fossils of Congo pygmies and of Watusi were to be found in
the same deposit by a paleontologist a million years hence, he might
well think that they belonged to two different species. As stated

previously, the known diversity of fossil man can be interpreted as being the result of geographic variation within a single species of *Homo*. Apparently, this variation nowhere led to the simultaneous occurrence of several species of *Homo*. What is the cause for this puzzling trait of the hominid stock to stop speciating in spite of its eminent evolutionary success? It seems to me that the reason is man's great ecological diversity. Man has, so to speak, specialized in despecialization. Man occupies more different ecological niches than any known animal. If the single species man occupies successfully all the niches that are open for a *Homo*-like creature, it is obvious that he cannot speciate. This conforms strictly to Gause's rule. Also, man is apparently slow in establishing isolating mechanisms. This is indicated by the numerous instances of incomplete speciation in the history of the hominids. In no case was this speciation completed, the reason being that the segregating populations were either absorbed by intermarriage or exterminated. Man is apparently particularly intolerant of competitors. The wiping out or absorption of primitive populations by culturally more advanced or otherwise more aggressive invaders, which we have witnessed so many times during the eighteenth and nineteenth centuries in Australia, North America, and other places, has presumably happened many times before in the history of the earth. The elimination of Neanderthal man by the invading Cro-Magnon man is merely one example.

There is one striking difference between man and most of the animals. In animals whenever there is competition between two subspecies the one that is better adapted for a specific locality seems to win out. Man, who has reached such a high degree of independence from the environment, is less dependent on local adaptation, and a subspecies of man can quickly spread into many geographically distant areas if it acquires generalized adaptive improvements such as are described by the social anthropologist. Such improvements do not need and probably often do not have a genetic basis. The authors who have claimed that man is unique in his evolutionary pattern are undoubtedly right. Even though the phyletic evolution of man will continue to go on, the structure of the human species at the present time is such that there appears to be very little chance for speciation, that is, for the division of the single human species into several separate species.

POSTSCRIPT, 1975

The 25 years since the publication of this analysis have witnessed unprecedented discoveries of new material on fossil man. Outstanding in this field are the finds made by Louis and Mary Leakey in Olduvai (Tanzania) and by Richard Leakey east of Lake Rudolf. Other discoveries were made in Kenya, Ethiopia, Uganda, North Africa, Iran,

China, Australia, and many other places. New methods of dating geological strata were discovered and applied. The research resulting from all these discoveries indicates that the picture I painted was over-simplified. Unfortunately, the new material has perhaps raised more questions than it has answered. It is quite clear now that there were at least two australopithecine lineages, a gracile one, represented by *"Australopithecus" africanus,* and a robust one, represented by *"Paranthropus" robustus,* the latter a collateral line not involved in the human ancestry. What is still uncertain is whether and to what extent *A. africanus* lived side by side with *"Homo" habilis,* a large-brained hominid discovered at various sites in Tanzania, Kenya, and Ethiopia. How much geographic variation was there in *A. africanus,* and was *H. habilis* derived from one of the races of polytypic *A. africanus*? Was *H. habilis* the direct ancestor of *H. erectus,* and did the evolution of the latter result in the extinction of *P. robustus* and of the remaining remnants of *A. africanus*? All such questions still remain to be answered.

The time scale has been thoroughly revised. *Australopithecus* is now known as far back as 5-½ million years ago, and other rather large-brained, *Homo*-like fossils as far back as at least 2-½ million years. Tool making has also been pushed far back, and the discovery of tool use and tool making (although not making of stone tools!) among chimpanzees makes it probable that tool making may have been an ability of man's ancestors for more than 10 million years. The whole story of brain evolution has to be rewritten in the light of these new discoveries. Correcting brain size for body size rather drastically changes the numerical values of the rate of brain evolution (Pilbeam and Gould 1974). Where *Meganthropus* of Java (scraps of jaw) fits into all this is still a complete mystery.

When it comes to more recent forms, not much has been discovered that drastically alters our concepts. Everything confirms the fact that Cro-Magnon man *(Homo sapiens sapiens)* invaded Europe about 35,000 years ago (completely displacing Neanderthal man), but where he came from is as great a mystery as ever. I have discussed some of these problems elsewhere (Mayr 1963).

Much of the uncertainty concerns details. This much remains clearly established: the replacement of thinking in terms of anatomical types by thinking in terms of geographically variable populations has led to a complete revolution in our thinking on human evolution.

REFERENCES

Broom, R. 1950. The genera and species of the South African fossil ape-man. *Amer. Jour. Phys. Anthrop.,* 8:1–13.

Campbell, B. 1965. *The nomenclature of the Hominidae.* Occasional paper no. 22, Royal Anthropological Institute, London.

Dobzhansky, T. 1944. On species and races of living and fossil man. *Amer. Jour. Phys. Anthrop.*, 2:251–265.

Mayr, E. 1942. *Systematics and the origin of species.* Columbia University Press, New York.

——1944. On the concepts and terminology of vertical subspecies and species. *Nat. Res. Council Bull.*, 2:11–16.

——1963. The taxonomic evaluation of fossil hominids. In *Classification and human evolution*, ed. S. L. Washburn, pp. 332–346. Aldine Press, Chicago.

Pilbeam, D., and S. J. Gould. 1974. Size and scaling in human evolution. *Science*, 186:892–900.

Straus, W. L., Jr. 1949. The riddle of man's ancestry. *Quart. Rev. Biol.*, 24:200–223.

VIII
BIOGEOGRAPHY

Introduction

My Ph.D. thesis dealt with a zoogeographical problem, the rapid invasion of large parts of Europe by the serin finch (Mayr 1926). Since that time, zoogeography in all its aspects has remained one of the focal points of my interest.

The following nine essays deal with the biogeographical questions that were of the greatest concern to me: Is it better to describe distributions in terms of regions or faunas? What factors determine species number on an island? How many colonizations account for the total number of species on an archipelago? What particular balance between colonization and extinction accounts for the size of an island fauna? What factors are responsible for the differences in colonizing ability of different species?

Many of these questions are now being investigated again. My great advantage was that I had access to the material of the Whitney South Sea Expedition, which for more than a dozen years had visited nearly every island in the tropical Pacific, taking a complete inventory of all the island avifaunas. This information permitted a quantitative analysis such as had never before been possible. Indeed, much of this material has been found useful also in MacArthur and Wilson's Island Biogeography (1967).

In the first half of the century, postulating land bridges as the pathway of colonization was at its height, and the few authors, like Gulick, Simpson, and myself, who opposed the facile invocation of land bridges were voices crying in the wilderness. Even such a perceptive and farsighted author as Bernhard Rensch assumed automatically that distribution patterns in the Indo-Australian archipelago had to be based on former land connections (Rensch 1936). My early biogeographic papers were therefore devoted to showing not only that land bridges in the western Pacific had no geological basis, but also that the actual composition of the island faunas reflected over-water colonization rather than what one would have predicted from a colonization over a land bridge.

Continental drift was invoked by a few authors. In the 1930s and 1940s it was not taken seriously by most zoogeographers, not so much *because the geophysicists opposed it unanimously, but rather because drift was invoked to explain very recent distribution patterns, patterns that must have been due to faunal movements in the later Tertiary and Quaternary. Postulating recent drift led to glaring contradictions and inconsistencies. The revised chronology provided by plate tectonics requires no revision of the previously adopted explanation of patterns of colonization by bird faunas in the western Pacific.*

The new generation of biogeographers is beginning to ask new questions, questions that did not occur to us in the 1930s, 1940s, and 1950s. Nevertheless, these questions could not have been asked until those questions had been answered satisfactorily that are posed in the following nine essays.

Additional zoogeographical studies of mine not included in this collection are:

1926. *Die Ausbreitung des Girlitz (*Serinus canaria serinus *L.). Jour. f. Ornith., 4:571–671.*

1930. *Theoretisches zur Geschichte des Vogelzuges. Der Vogelzug, 1(4):149–172. (With W. Meise.)*

1931. *Birds collected during the Whitney South Sea Expedition. XIV. With notes on the geography of Rennell Island and the ecology of its bird life (by H. Hamlin). Amer. Mus. Novit., no. 488:1–29.*

1933. *Die Vogelwelt Polynesiens. Mitt. Zool. Mus. Berlin, 19:306–323.*

1941. *Wanderung oder Ausbreitung? Zoogeographica, 4(1):18–20.*

1941. *Borders and subdivision of the Polynesian region as based on our knowledge of the distribution of birds. Proc. 6th Pacific Sci. Congr., 4:191–195.*

1944. *Timor and the colonization of Australia by birds. Emu, 44:113–130.*

1945. *Symposium on age of the distribution pattern of the gene arrangements in* Drosophila pseudoobscura. *Introduction and some evidence in favor of a recent date. Lloydia, 8(2):69–82.*

1951. *Bearing of some biological data on geology. Bull. Geol. Soc. America, 62:537–546.*

1952. *Introduction (p. 85) and Conclusion (pp. 255–258) from "The problem of land connections across the South Atlantic with special reference to the Mesozoic." Bull. Amer. Mus. Nat. Hist., vol. 99.*

1954. *On the origin of bird migration in the Pacific. Proc. 7th Pacific Sci. Congr., 4:387–394.*

1963. *The fauna of North America, its origin and unique composition. Proc. 16th Int. Congr. Zool. (Washington, D.C.), 4:3–11.*

1965. *Avifauna: turnover on islands. Science, 150:1587–1588.*

1967. *The challenge of island faunas. Australian Nat. Hist., 15(12): 369–374.*

1967. *The origin of the bird fauna of the south Venezuelan highlands. Bull. Amer. Mus. Nat. Hist., 136:269–327. (With W. Phelps.)*

1972. Continental drift and the history of the Australian bird fauna. Emu, *72(1):26-28.*

1972. Geography and ecology as faunal determinants. Proc. 15th Int. Ornith. Congr., *pp. 549-561.*

REFERENCES

MacArthur, R., and E. Wilson. 1967. *The theory of island biogeography.* Princeton University Press, Princeton.

Rensch, B. 1936. *Die Geschichte des Sundabogens.* Borntraeger, Berlin.

37
What is a Fauna?

Interest in the study and analysis of faunas has perhaps reached its nadir in the present generation. The word *fauna* is not even included in a recent glossary of important zoogeographical words. Yet hardly another concept in zoogeography is as controversial and as frequently misunderstood as that of fauna. It was an exciting concept to former generations and a scattering of perceptive faunal analyses in recent decades shows that this topic has lost nothing of its challenge and excitement, at least for some evolutionary biologists.

When the young Darwin boarded the *Beagle* in 1831, he took it for granted that the floras and faunas of all regions were the "products" of these regions and that faunas owed their characteristics to the local physical environment. But what he found and never tired of emphasizing was totally at variance with his preconceptions. For instance: "In the southern hemisphere, if we compare large tracts of land in Australia, South Africa, and western South America, between latitudes 25° and 35°, we shall find [these] parts extremely similar in all their conditions, yet it would not be possible to point out three faunas and floras more utterly dissimilar. Or again, we may compare the productions of South America south of Lat. 35° with those north of 25°, which consequently inhabit a considerably different climate, and they will be found incomparably more closely related to each other than they are to the productions of Australia or Africa under nearly the same climate" (1859:347).

These observations, and others to be discussed presently, destroyed for Darwin the idea, widespread since Buffon and earlier, that flora and fauna are "the product of a country" or, as we might say today, that factors of the physical environment determine the composition of faunas and floras. It became evident to Darwin that faunas could not have been created *in situ* but must be a product of history. It is clear from Darwin's letters, from his autobiography, and from many discussions in his *Origin of Species*, particularly the very first paragraph, that

Revised from "What is a fauna?" *Zool. Jb. Syst.*, 92(1965):473–486.

these faunistic observations more than anything else were the source of his evolutionary ideas. As stated by Darwin in his autobiography: "During the voyage of the Beagle I had been deeply impressed by discovering in the Pampean formation great fossil animals covered with armour like that on the existing armadillos: secondly, by the manner in which closely allied animals replace one another in proceeding southwards over the continent: thirdly, by the South American character of most of the productions of the Galapagos Archipelago."

All three sets of observations demonstrate the primacy of historical geography over the physical environment: (1) the temperate fauna of South America has to a large extent the same types as the tropical portions of that continent rather than as the temperate zones of other continents; (2) the fossil faunas contain to a considerable extent the same types as the recent faunas rather than the same types as the contemporary fossil faunas of other continents; and, finally, (3) the faunas of islands are very similar to those of nearby continents, even though, to quote Darwin with respect to the Galapagos, "there is nothing in the conditions of life, in the geological nature of the islands, in their height or climate, or in the proportions in which the several classes are associated together, which resembles closely the conditions of the South American coast: in fact there is a considerable dissimilarity in all these respects" (Darwin 1859:398). As Wallace said in 1876, "It was long thought, and is still a popular notion, that the manner in which the various kinds of animals are dispersed over the globe is almost wholly due to diversities of climate and vegetation."

I shall not rehearse the history of zoogeography in the last 100 years. I shall merely remark that the faunal and historical approach favored by Darwin tended to recede into the background as the geographical approach of Sclater and Wallace came to the fore and as an increasing number of authors expended their energies in trying to determine the borders between geographic regions and in subdividing these regions into subregions and biotic provinces. For those whose interest is strictly geographical, it is easy to answer the question: "What is a fauna?" They will give the simple, static, and purely descriptive definition: "The animals found in a given region." This surely is a correct answer as far as it goes, and yet confined to the stated definition, the word *fauna* would denote a strangely sterile and purely descriptive concept.

Conspicuously missing from this definition is any reference to what one might call faunal "structure." As a result of various historical forces, a fauna is composed of unequal elements, and no fauna can be fully understood until it is segregated into its elements and until one has succeeded in explaining the separate history of each of these elements.

The exaggerated emphasis on regions and on a purely static definition of faunas was the reason for a rebellion by E. R. Dunn (1922), the leader of a new movement in biogeography. Like many reformers,

Dunn perhaps went too far when he said, ". . . the zoogeographical realms are nothing save and except the great land masses with lines drawn to correspond to the physiographic barriers. There is a great philosophical difference between such terms as Holarctic Fauna and Holarctic Region. In the first case we speak of zoological matters in terms of zoology, in the second of geographical matters in terms of mythology." Yet the new look recommended by Dunn unquestionably led to a deeper analysis and encouraged a revival of the historical approach Darwin had found so useful. Dunn (1931), Simpson (1943), and I (Essay 38) found that attempts to reconstruct faunas in terms of faunal elements supplemented most fruitfully the paleontological approach, particularly for groups with an inadequate fossil record.

METHODS OF INFERENCE IN ZOOGEOGRAPHY

The validity of such reconstructions depends entirely on the soundness of the methods of inference employed by the zoogeographer. It is therefore necessary to say a few words on method. Inferences in zoogeography are based on the proper evaluation of three sets of information: (1) the relative age of various taxa, (2) the determination of the dispersal capacity of taxa, and (3) the distribution of related taxa. Let me illustrate these with some examples.

Relative Faunal Age of Various Taxa

When we study the birds of Australia, we find that they range all the way from exceedingly old endemics, of which no close relatives are known, to recent immigrants from Asia or the tropical archipelagoes that have not even differentiated subspecifically. In a tentative analysis, I arranged the Australian bird fauna some years ago in five groups (see Table 37-2, p. 557) on the basis of differentiation and inferred period of immigration in Australia (Mayr 1944:188–189). A similar determination of relative age is possible for most islands, such as Madagascar, New Zealand (Fleming 1962), New Caledonia, and Fiji, where, in almost all cases, endemic families and genera can be inferred to have been older immigrants than endemic species or subspecies. In the case of continents that are in contact with other continents, as are North America and Eurasia across the Bering Straits, a fairly reliable ranking of the relative age of cross-colonists can usually be undertaken. It cannot be questioned, for instance, that the Wheatear (*Oenanthe*) and the leaf warbler (*Phylloscopus*) have colonized Alaska very recently from Eastern Asia.

The total distribution pattern of a given taxon, as well as its categorical rank, thus often permits reliable conclusions on the relative age of the taxon in a given region. Even though it will not be possible, in the absence of a fossil record, to determine the absolute date of colonization of a given group, nevertheless it will often be possible to infer whether a taxon arrived earlier or later than certain other taxa.

The Determination of Dispersal Capacity

Of prime importance in reconstructing the history of faunas is determination of the dispersal capacity of various types of animals. As a broad generalization, one can assert that the less capable a taxon is of crossing water barriers, the more valuable it becomes as a zoogeographic indicator. Primary freshwater fishes rank perhaps highest on this basis, and many groups of mammals also respect water barriers extremely well. On the other hand, some groups of small mammals seem to raft readily, for instance the marsupial genus *Phalanger*. It is widespread on oceanic islands in the Papuan region, as far out as the Talaud Islands, the Admiralty Islands, and the Solomon Islands.

Owing to their power of flight, birds are generally assumed to be highly successful in crossing water barriers, and as a broad generalization, this is correct. But it has not been sufficiently stressed that different species, genera, and families of birds differ strikingly from each other in dispersal ability. The mesomyodean Passeres (Suboscines) of South America, with the exception of a single family (Tyrannidae), are almost totally unable to cross water gaps (Essay 39). This is in strong contrast to the nine-primaried Oscines of North America, which undertook numerous colonizations of South America across the water gap that existed through most of the Tertiary. Sixty families of birds occur in New Guinea, but only 16 of these families supply most of the Polynesian bird fauna (Essay 45).

One concludes that different faunal elements must be given different weight in faunal analyses. In an attempt to determine former land connections, the weight of a taxon is inversely correlated to its dispersal ability.

The Distribution of Related Taxa

Much can be learned about the former history of taxa from consideration of the distribution of related taxa. This source of information is particularly important for all groups in which a fossil record is absent. For instance, the nine-primaried Oscines are well distributed over North and South America. Indeed, more species of this group are found in the Neotropical than in the Nearctic region. Yet, a study of the relationship of these families with each other and with other families in Eurasia and South America indicates rather decisively that the original center of radiation of this group of families must have been North America (Essay 39). A similar analysis can be made for genera and subfamilies of parrots, pigeons, corvids, thrushes, and so on.

All inferences based on these three sets of information are vulnerable owing to extinction, which can sometimes lead to gross distortion of faunal composition. In view of this potential weakness every effort must be made to confirm one's inferences with the evidence of the fossil record. Matthew, Simpson, and other students of fossil mammals have proved how much detail is possible in a reconstruction of faunal history when there is an abundant fossil record. The fossil record of such avian groups as the Cracidae, Old World Vultures, Gruidae, and Cariamae proves how wrong inferences may be that are entirely based on current distributions (Essay 39). It would be discouraging if such cases were frequent. Fortunately, inferences from living faunas and information derived from fossils are usually in good agreement.

TYPES OF FAUNAL ORIGIN

Combining our knowledge of past and present components of faunas, on the one hand, and of the size and former connections of continents, on the other, we may achieve an understanding of faunas that will permit their classification. Many classifications of faunas are possible: the one adopted here places particular stress on the historical aspects of faunal origin (Table 37–1). On the basis of six classifying criteria, an overlapping of the several subdivisions is unavoidable. As a result, most existing faunas are composites of several faunal types.

1. Autochthonous Adaptive Radiation

Perhaps the simplest type of faunal origin, and one that would result in the most homogeneous fauna, is the complete isolation of a segment of a previously existing fauna. Let us imagine, for instance, that North and South America once had a single homogeneous fauna. When a water barrier separated the two continents, the South American segment began to evolve independently from the North American (and vice versa), and the more successful elements in the newly isolated southern continent had an opportunity for adaptive radiation. Eventually they evolved into a distinct fauna, differing in many ways from that of North America.

Any new island or archipelago that receives its occupants essentially at one time has a similar opportunity for developing an autochthonous fauna. Indications of such a development are noticeable in every isolated land area or ecologically distinct geographical subdivision of a continent. Superimposed on it, however, will be a continuous input of new arrivals owing to insufficient geographic isolation of the area.

Types of faunas that owe their origin to a continuous input of new elements will be discussed under headings 2, 3, and 5.

Table 37–1. Types of faunal origin.

1. Autochthonous adaptive radiation
2. Continued single-origin colonization
3. Continued multiple-origin colonization
4. Fusion of two faunas
5. Successive adaptation
6. Composite origin through various combinations of 1–5

2. Continued Single-Origin Colonization

This type of faunal origin, according to my interpretation, is well illustrated by the Australian bird fauna. Biogeographers have long argued about the history of the Australian biota. I will not enter the argument whether and to what extent continental drift or antarctic connections have played a role. The Australian bird fauna, with the possible exception of the Ratites (Mayr 1944), gives no indication of such connections, but can be interpreted as the result of a steady and continuous colonization from Asia. On the basis of relative age, five major layers of colonists can be distinguished arbitrarily, but it must be remembered that there are actually no clearly defined discontinuities (see Table 37-2). Not much can be said about the origin of the oldest of these five layers, except that these old elements are definitely not more closely related to South American than to Asiatic elements. In all other cases, relationship is unequivocally with southeast Asia. The conclusion is clear. Australia has always been accessible to avian colonists from Asia and, so far as we can determine, from no other place. The earlier the colonization, the more distinct the resulting element and the more likely it has secondarily undergone adaptive radiation in Australia. There must have been an old bird fauna in Australia when it was still part of the Antarctic continent and in contact with South America. It is most puzzling that no remnants of this fauna can be found in Australia (with the possible exception of the emu-cassowary group).

A similar pattern is displayed by the Madagascar mammals (Simpson 1940). Omitting bats and the extinct hippopotamus, this fauna is the product of four colonizations, presumably all from Africa, resulting in a rich fauna of lemurs, insectivores, civet cats, and rodents. The birds and mammals of South America fall likewise into this category (Essay 39).

3. Continuous Multiple-Origin Colonization

The birds of Madagascar are derived from two sources. More than 90 percent came from Africa, but at least 5 percent from India (Rand 1936). The terrestrial fauna of New Zealand is in part derived from Australia, yet an appreciable component reached the island more directly from the Papuan region across the Melanesian island arcs. The bird

Table 37-2. Composition of the Australian bird fauna.
(From Mayr 1944.)

1. Strongly endemic families and subfamilies, about 15
2. Less isolated indigenous families and subfamilies, about 8
3. Endemic genera of nonendemic families, about 30
4. Endemic species of nonendemic genera, about 60
5. Recently immigrated species, not differentiated at all, or only subspecifically, about 40

fauna of Pantepui (Venezuelan Highlands) consists of 42 percent segregates from the tropics and 58 percent immigrants from the Andes and other mountain chains (Mayr and Phelps 1967).

The method of colonization plays an important role, as is well demonstrated by the fauna of the West Indies. Most aerial invaders, for instance the birds, are derived from Central America (Bond 1963). However, those colonists, such as mammals and reptiles, that depend on rafts, are likely to come from big rivers. For the West Indies this means the Magdalena, the Orinoco, and other rivers of the South American mainland. Islands that lie between two continents, for instance the Hawaiian Islands (Essay 44) and the Tristan da Cunha group (Rand 1955), almost invariably have faunal elements from both source areas.

4. Fusion of Two Faunas

Whenever the barriers between two faunal regions break down, after a lengthy preceding isolation, a new fusion fauna develops in the area of secondary contact. As far as birds are concerned, the two most spectacular fusion faunas in existence are found in "Wallacea," between the Indian and Australo-Papuan regions, and in Central America. I have analyzed both faunas on previous occasions (Mayr 1944; Essays 39 and 42). The most characteristic aspect of fusion faunas is that they are dominated by certain families that are exceptionally good colonizers. The ecological aspects of the secondary interdigitation of two faunas has in no case been well investigated. Fusion faunas are particularly interesting as suitable material for the testing of zoogeographic methods. Two problems of special importance in an analysis of fusion faunas are competition and habitat specialization. Let us first take up competition.

It is widely believed among biogeographers, going back to Darwin and earlier, that members of some faunas are competitively superior to members of other faunas. Matthew (1915), in particular, frequently emphasized that faunal elements of the great Holarctic land mass were competitively superior to elements of the southern continents. Indeed the faunas of the southern continents seemed to consist of series of layers of northern invaders. The interpretation is often made that new invaders are competitively superior to the older indigenous elements and frequently cause their extinction. Parts of this orthodox interpretation are surely legitimate, but matters are not so simple as they first appeared. Patterson and Pascual (1972) have shown that much of the extinction of Tertiary South American mammals took place prior to the invasion of northern elements, which followed the joining of the continent.

As far as birds are concerned, the situation is rather complex. Northern elements have been far more successful than South American elements in crossing the Tertiary water gap between North and South America. Many of the presumably originally northern elements, the tanagers for instance, have had an extremely successful adaptive radiation in South America, so that they can now be considered truly South American elements. While only two families of originally southern origin, the Trochilidae and the Tyrannidae, have successfully invaded North America, virtually all of the families of presumed North American origin are now well represented in South America (Essay 39). Among birds, however, unlike mammals, many of the original South American elements are still the dominant types, as indicated by such groups as the Suboscines, Tinamidae, Ramphastidae, and Galbulidae. The fossil record does not reveal wholesale extinction of South American avian groups equivalent to the sweeping extinction of mammalian types.

Perhaps the most interesting aspect of the fusion of North and South American elements is the difference between habitat types—the uniformity of life in the northern and southern rain forests compared with the heterogeneous arid region faunas. The present bird fauna in the tropical rain forest of Central America, from Chiapas and Guatemala southward, is remarkably similar in composition to that of the Amazonian forest (see Table 39–4. p. 598). What caused the obliteration of the differences that must have existed between the northern and southern rain forest faunas during the Tertiary separation of the two continents? Two somewhat interrelated causes might be suggested. First, the area of Central America suitable for tropical rain forest became increasingly restricted during the period of Tertiary mountain building, heavy volcanism, and climatic deterioration that culminated in the Pleistocene. The fauna occupying this life zone was correspondingly decimated. When the climate ameliorated again, during the warm spells of the Pleistocene and the post-Pleistocene warm and humid period, the vast faunal reservoir of tropical South America released a flood of colonists into the now expanding rain forest of Central America.

A second reason for the apparent success of South American rain forest elements in colonizing Central America is the relative size of the populations involved. The North American tropical elements were probably restricted to refuges and may have consisted of numerous more or less isolated and frequently quite small populations. This strongly attenuated fauna was apparently inferior in competition with the aggressive South American invaders derived from the vast reservoir of the South American rain forests. Whatever North American elements could compete successfully invaded South America in turn, thus completing the homogenization of the fauna in the tropical rain forest of the Americas. This homogeneity was favored, as Griscom (1950) stressed, by the geographical continuity of the rain forest from Central to South America.

In contrast, the disruption of the arid habitats has contributed to the much greater differences between the arid faunas of North and South

America. The distribution pattern of the arid tropical, subtropical, and warm temperate faunal elements is quite different from that of the humid tropical elements. The area in North America available for these faunas expanded greatly during the Miocene and Pliocene. The opening niches and habitats were invaded by autochthonous North American elements, which became increasingly well-adapted to these life zones. When finally the land bridge to South America became established, it had remarkably little effect on the faunal composition of these arid subtropical or warm temperate habitats.

The evidence derived from birds concerning the Central American fusion fauna agrees quite well with that derived from fishes (Myers 1963) and from reptiles and amphibians (Schmidt 1946). In all cases, the Central American fauna consists primarily of a tropical North American element to which tropical South American elements have been added in varying degrees across the Panamanian filter bridge.

5. Successive Adaptations (Specialized Habitat Faunas)

The four types of faunas so far discussed characterize geographic regions, or at least originated in them. Another type of fauna characterizes what one might call "ecological regions." The faunas of the deep sea, of the mountains, or of the deserts owe their composition to the ability of their members to meet the specific demands of specialized habitats.

No deserts, mountains, or even deep seas existed from the very beginning. They originated at some period during the earth's history, and expanded and contracted in subsequent periods. All existing deserts, for instance, are relatively recent in terms of the earth's history, and their faunas and floras did not exist prior to the Tertiary. Most of the high mountain areas of the earth are young, as is the Arctic. How, then, did the faunas of these areas originate? According to one view, the preexisting fauna was segregated within the developing new zone and gradually acquired its adaptation as a whole and *in situ*. This interpretation is, perhaps, the prevailing concept right now. My own studies indicate that it is not correct. The other interpretation is that a newly developing adaptive zone functions as a vacuum that attracts enterprising colonists. Because such colonists continue to invade the partially empty zone throughout geological time, any such specialized "ecological area"—the deep sea, the deserts, high mountains, or the Arctic—should show a complete gradient between old endemics and very recent immigrants. And this, indeed, is what my analysis reveals.

As an example of a mountain area, we might consider Pantepui, the Venezuelan Highlands. Of its bird fauna, we can determine degree of endemicity as well as probable point of origin. The 96 species that make up the characteristically montane element of Pantepui fall into a graded series of decreasing endemicity (Table 37-3). As far as origin is

Table 37–3. Levels of endemicity of Pantepui birds.

Species in endemic genera	2
Endemic species in nonendemic genera	23
Endemic species in widespread superspecies	5
Total endemic species	30
Nonendemic species with only endemic subspecies	48
Nonendemic species with endemic and nonendemic subspecies	7
Nonendemic species without endemic subspecies on Pantepui	11
Total nonendemic species	66
Total species of birds	96

concerned, 42 percent of the species are derived from the avifauna of the surrounding tropical lowlands (by ecological shift) and 58 percent by immigration from more distant mountains (Mayr and Phelps 1967). The latter category again is heterogeneous, several different mountain ranges having served as source areas: the coastal ranges of northern Venezuela, the Andes, and even the Brazilian highlands. There is not a shadow of a doubt that the mountain avifauna of Pantepui is the product of a steadily continuing colonization from several source areas.

Australian zoogeographers recognize three major faunal elements in Australia (Serventy and Whittell 1962), the Eyrean of the arid interior, the Bassian of the cool temperate districts of southeast and southwest Australia, and the Torresian of the Tropical Zone. It is implied in the literature that each of these elements has had a different faunal history. To test this suggestion, I tried to determine the age of the typically Eyrean, Bassian, and Torresian elements. In order to guarantee complete objectivity, I asked Dr. D. L. Serventy to prepare for me a list of all the Australian bird species that he considered typically Eyrean, Bassian, and Torresian. The list, which he most kindly supplied, served as the basis of my calculations. The age of these elements is based, in turn, on a tabulation I had prepared earlier (1944:188–189) (Table 37–2), adjusted only to allow for a few recent advances in the classification of Australian birds. The result of these calculations is given in Table 37–4. Although the Eyrean fauna has the highest percentage of old and the lowest percentage of new elements, the differences among the Australian faunas are merely of degree. Indeed, the basic similarity of the three faunas with respect to degrees of endemicity is astonishing. The faunas in the specialized life zones of Australia fully support the thesis that they are the product of a continuous adaptation of new elements to specialized climatic conditions.

Studies of the arid zone faunas in the Americas, in Africa, and in Asia show exactly the same phenomenon, and this is evident also for

Table 37-4. Faunal elements among Australian birds.

Species	Eyrean	Bassian	Torresian
Species in endemic[a] families and subfamilies	60 (63.8%)	44 (49.4%)	92 (46%)
Species in endemic genera of nonendemic families	25 (26.6%)	25 (28.1%)	49 (24.5%)
Endemic species in nonendemic genera	9 (9.6%)	13 (14.6%)	37 (18.5%)
Species found also outside Australo-Papuan area	0 (0.0%)	7 (7.9%)	22 (11%)
Total	94 (100%)	89 (100%)	200 (100%)

[a]"Endemic" refers to the Australo-Papuan region.

the bird fauna of the Arctic (Johansen 1960) and other climatically specialized regions. In all cases the faunas are heterogeneous in degree of endemicity, indicating a continued steady process of colonization.

The deep sea is sometimes described as a refuge of ancient types that have elsewhere succumbed to the competition of more modern types. The actual analysis of deep sea faunas has not confirmed this image. To be sure, there are some ancient types in deep waters, such as *Neopilina,* but equally ancient types are found in shallow waters and even in the tidal zone. Barnard, in a discussion of deep sea amphipods, concludes that the "deep sea is largely populated with species belonging to genera successful in shallower waters and that abyssal generic endemism is quite low." Among the 132 genera that have abyssal species, only 28 (21 percent) are endemic abyssal genera. There are no endemic abyssal families of amphipods; indeed, even the three or four families best represented in the abyssal fauna are only slightly richer in the deep sea than they are in certain other areas, particularly the shallower cold waters of the Arctic and Antarctic. In view of the 28 genera that are abyssal endemics (some of them not clearly related to any shallow-water genus), in addition to the much larger recent abyssal element, it can be asserted with conviction that there have been successive invasions of faunal elements into the abyssal zone (Barnard 1962:19–21). Clarke (1962:291–306) has shown that the deep sea mollusk fauna likewise contains ancient as well as recent types. No families or higher groups of mollusks appear to have evolved in the deep sea, but a number of orders are better developed below 2000 meters than in shallow water. Clarke concludes that most of the present deep sea molluskan fauna "was first derived from various shallow water faunas in the relatively recent geological past, i.e., probably during the late Mesozoic era or the Cenozoic era."

6. Composite Origin

In addition to these five definite types of faunas (Table 37-1), we must recognize that the fauna of any extended area is actually a composite of all five types. Yet only recognition that such types exist will permit their analysis.

CONCLUSIONS

Returning to our original question. "What is a fauna?" we find unsatisfactory the purely static definition: "A fauna is the totality of species living in an area." An expanded definition is far more rewarding: "A fauna consists of the kinds of animals found in an area as a result of the history of the area and its present ecological conditions." One might even include in the definition a reference to the isolation that is a prerequisite for the development of a truly distinctive fauna. Or one might stress the fact that a fauna consists of faunal elements which differ in origin, age, and adaptation. Perhaps we need more faunal analyses before a satisfactory definition can be attempted.

It is evident that the undisturbed development of a well-balanced, adaptively well-diversified, and reasonably homogeneous fauna is possible only under conditions of at least partial isolation from other areas in which other faunas evolve. Obviously, then, the longer a land mass is isolated from all others, the more pronounced will become the distinguishing characters of its fauna. Australia is the outstanding illustration of this, South America somewhat less so. Elements of a fauna may eventually spread to other areas, but will, at least at first, be aliens in the new area of colonization.

Elements of a fauna radiate, in due time, into all the available habitats and niches, no matter how forbidding. This radiation involves adaptation to functional niches (arboreal, terrestrial, climbing, running, burrowing, etc.), as well as to highly diverse habitats, such as forest, savannah, deserts, mountains, and aquatic habitats.

Wherever two previously isolated land masses come into contact with each other, they will exchange components of their respective faunas. As a result, the fauna of an area will nearly always consist of an old, autochthonous element plus such other elements as have later been added by immigration from neighboring or more distant areas. It is one task of the biographer to sort out in a fauna the old, autochthonous element (if any) from the various kinds and "layers" of later immigration.

The listing of species is thus only the first step in the study of a fauna. The analysis of faunal components and the study of the climatic, vegetational, and geographical conditions under which the various components colonized the area are equally important parts of zoogeography. Indeed, such analyses are among the most interesting tasks of evolutionary biology.

REFERENCES

Barnard, J. L. 1962. South Atlantic abyssal amphipods collected by R. V. Vema. In *Abyssal Crustacea*, pp. 1–78. Columbia University Press, New York.

Bond, J. M. 1963. Derivation of the Antillean avifauna. *Proc. Acad. Nat. Sci.*, 115: 79–98.

Clarke, A. H., Jr. 1962. On the composition, zoogeography, origin and age of the deep-sea mollusk fauna. *Deep-Sea Research*, 9:291–306.

Darwin, C. 1859. *On the origin of species by means of natural selection, or the preservation of favoured races in the struggle for life.* J. Murray, London.

—— 1958. *The autobiography of Charles Darwin*, ed. N. Barlow. Collins, London.

Dunn, E. R. 1922. A suggestion to zoogeographers. *Science*, 56:336–338.

—— 1931. The herpetological fauna of the Americas. *Copeia*, 1931:106–119.

Fleming, C. A. 1962. History of the New Zealand land bird fauna. *Notornis*, 9: 270–274.

Griscom, L. 1950. Distribution and origin of the birds of Mexico. *Bull. Mus. Comp. Zool.*, 102:341–382.

Johansen, H. 1960. Die Entstehung der arktischen Vogelfauns. *Proc. 11th Int. Ornith. Congr.* (1958), 1:358–362.

Matthew, W. D. 1915. Climate and evolution. *Ann. N.Y. Acad. Sci.*, 24:171–318.

Mayr, E. 1944. The birds of Timor and Sumba. *Bull. Amer. Mus. Nat. Hist.*, 83: 123–194.

Mayr, E., and W. H. Phelps, Jr. 1967. The origin of the bird fauna of the South Venezuelan Highlands. *Bull. Amer. Mus. Nat. Hist.*, 136:269–328.

Myers, G. S. 1963. The fresh-water fish fauna of North America. *Proc. 16th Int. Congr. Zool.*, 4:15–20.

Patterson, B., and R. Pascual. 1972. The fossil mammal fauna of South America. In *Evolution, mammals, and southern continents*, ed. A. Keast, pp. 247–309. State University of New York Press, Albany.

Rand, A. L. 1936. The distribution and habits of Madagascar birds. *Bull. Amer. Mus. Nat. Hist.*, 72:143–499.

Schmidt, K. P. 1946. On the zoogeography of the Holarctic Region. *Copeia*, 1946: 144–152.

Serventy, D. L., and H. M. Whittell, 1962. *Birds of western Australia*, 3rd ed. Paterson Brokensha Pty., Perth.

Simpson, G. G. 1940. Mammals and land bridges. *Jour. Wash. Acad. Sci.*, 30: 137–163.

——1943. Turtles and the origin of the fauna of Latin America. *Amer. Jour. Sci.*, 241:413–429.

Wallace, A. R. 1876. *The geographical distribution of animals.* Harper and Bros., New York.

38
History of the North American Bird Fauna

The acceptance of the theory of plate tectonics in the 1960s caused such a revolution in our concepts of the earth's history that one might expect it to have made all previously published papers in historical biogeography obsolete. This is, however, not always the case. For instance, most of the pattern of avian distribution is sufficiently recent to have become established after the continents had reached positions near the present ones. The conclusions of the following essay on the North American bird fauna, written in 1946, are, therefore, still largely valid. In presenting excerpts from that essay, I have omitted much descriptive detail and have attempted to point out where the modern interpretation is different from the original (1946) one. It is primarily the interpretation of the holarctic faunal element that has changed because of the impact of plate tectonics.

The prevailing geological dogma in 1946, when I prepared this history of the North American bird fauna, was that of the permanence of continents and oceans. Now, almost 30 years later, the newer geophysical evidence has clearly established the former crustal unity of North American with Europe and of South America with Africa. As far as South America is concerned, the separation is sufficiently ancient to have left remarkably little imprint on the composition of its bird fauna: hardly any other two bird faunas are more fundamentally different than those of Africa and South America. The story is different with North America. In 1946 one took it for granted—since the Atlantic was accepted as an ancient permanent ocean—that all faunal exchange between the New World and the Old World had to have taken place across the Bering Straits bridge. Now it is obvious that much of the older holarctic element dispersed over a North Atlantic land connection.

Omissions and insertions were made during the editing of this essay to allow for the new findings of geological science. Fortunately, by far

Adapted from "History of the North American bird fauna," *Wilson Bulletin*, 58, no. 1 (1946):3–41.

the greater part of the analysis is as valid now as it was in 1946, despite the advances of plate tectonics.

The bird student cannot help becoming envious on observing with what accuracy and amazing detail the student of mammals reconstructs the history of that class. Rich finds of fossils have enabled the paleo-mammalogist to determine the probable region of origin not only of families but also of genera, sometimes even of species, and to trace past modifications in their ranges. The student of birds is far less fortunate. Bird bones, being small, brittle, and often pneumatic, are comparatively scarce in fossil collections. The majority of Tertiary species of birds described from North America belong to zoogeographically unimportant families of water birds. Even fewer fossil birds are known from South America. The absence of certain families or orders from the fossil record of either North or South America proves nothing as far as birds are concerned. Furthermore, the history of birds is more difficult to reconstruct than that of mammals for two other reasons. First, birds seem to be a more ancient group than the mammals, many or most of the Recent families having already been in existence at the beginning of the Tertiary. And second, since birds cross water gaps more easily than mammals, the isolation of a land mass does not necessarily result in the isolation of its bird fauna. On these premises it would seem almost impossible to trace the history of the components of a local bird fauna, but this is by no means the case. Indirect methods of faunal analysis lead to fairly reliable results, since most families of birds are rich in genera and species. A quantitative analysis is, of course, impossible in small families, and their place of origin (as, for example, that of the limpkins) can be determined only with the help of fossils. In a paper read in 1926 before the International Ornithological Congress at Copenhagen, Lönnberg (1927) demonstrated the productivity of the indirect method by applying it in an investigation of the origin of the present North American bird fauna. Although most of Lönnberg's conclusions are still valid today, so much additional knowledge has accumulated during the past 20 years that a fresh analysis seems timely.

FAUNAL AND REGIONAL ZOOGEOGRAPHY

Trends and fashions have occurred in the science of zoogeography as in any other science. The zoogeography of the nineteenth century—the classical zoogeography of Schmarda (1853), Sclater (1858), and Wallace (1876)—was descriptive, essentially regional, and nondynamic. It was based on the premise that different parts of the world are inhabited by different kinds of animals; and each of these major areas was called a zoogeographical region. This method seemed successful while knowledge of the distribution of animals was still incomplete. As far as the boundaries between these regions were concerned, it was recognized that they "depend upon climatic conditions, which are in a measure determined or modified by features of topography" (Allen 1893:120). However, as the various parts of the world became better known, it became evident that the various regions proposed were of unequal value. This led to the proposal of new regions or to the fusion of previously separated regions into larger units. It is impossible to give here the history of the never-ending attempts to find a "perfect" zoogeographical classification. For example, it was soon found that the fauna of North America was somewhat intermediate between that of Asia and that of South America, which resulted in conflicting proposals concerning the zoogeographic position, or rank, of North America.

According to one school, North America was only part of a larger region combining North America, Europe, and north Asia. Gill (1875: 254) called this region the Arctogaean, while Heilprin (at the suggestion of Newton) called it the Holarctic (Heilprin 1883:270). This region (with the Palearctic and Nearctic as subregions) is perhaps even today the most frequently adopted zoogeographical classification of the northern hemisphere. Reichenow (1888:673ff.) took emphatic exception to this classification. He showed that, as far as birds were concerned, North America was much closer to the "Neotropical" than to the Old World, and that North and South America should be combined in a "Western Zone" or "New World Region." This point is well substantiated by his statistics. J. A. Allen (1893:115) showed that the Old World element in the warm temperate parts of North America amounted to only 23 to 37 percent of the genera, but he did not draw any conclusions from these figures. Subsequent writers almost completely ignored Reichenow's conclusions. Heilprin (1883) went to the opposite extreme. He refused to recognize the Nearctic even as a subregion. He drew a zoogeographic boundary right across North America, putting the northern half into the "Holarctic Region," the southern half in the "Neotropical Region." Wallace himself thought (1876:66) that it was a question "whether the Nearctic Region should be kept separate, or whether it should form part of the Palaearctic or of the Neotropical regions." The literature, particularly of the 1880s and 1890s, was filled with discussions of this question.

Eventually it was realized that the whole method of approach—the *Fragestellung*—of this essentially static zoogeography was wrong. Instead of thinking of fixed regions, it is necessary to think of fluid faunas. As early as 1894, Carpenter said: "No zoological region can be mapped with the hard and fast line of a political frontier, and the zoologist must always think more of faunas than of geographical boundaries" (1894:57). The faunal approach made slow but steady progress in Europe and in America. In Europe it has led to such excellent studies as those of Stegmann (1938a) on the birds of the Palearctic and of Stresemann (1939) on the birds of the Celebes. In America it was E. R. Dunn who was the pioneer of this concept. In a spirited attack on the older, static, regional zoogeography, he stated (1922: 336):

> There has been a constant search for some sort of scheme whereby ranges of animals might be reduced to a common denominator.
> . . .
> By far the most generally used of these philosophical methods is that of Realms, Regions and Zones. These are all based on the idea that large numbers of species have the same range, and that by picking out some of the conspicuous forms and mapping their ranges one has *ipso facto* a set of regions, to which other ranges may be referred, and with which other ranges should agree.
> This is, in some degree, true, but in nearly every case in which the ranges of any two species agree, the agreement is due to the geographic factors and not to the zoologic factors.
> It is obvious that the zoogeographical realms are nothing save and except the great land masses with lines drawn to correspond to the physiographic barriers. There is a great philosophical difference between such terms as Holarctic Fauna and Holarctic Region. In the first case we speak of zoological matters in terms of zoology, in the second of geographical matters in terms of mythology.
> The Palearctic fauna is an aggregate of species and may invade (in fact *has* invaded) Australia without forfeiting its name.

Following up these thoughts, Dunn (1931:107) analyzed the reptile fauna of North America and found that it could be classified into the following three groups: "(1) A northern, circumpolar, modern element. This would be truly *Holarctic.* (2) A more southern, older element, which I shall call *Old Northern.* . . . (3) A still more southern, still older element, the original fauna of South America, with its analogues in the Australian or Ethiopian regions. This I shall call *South American,* as I wish to avoid the term Neotropical. . . ."

I have attempted in the following sections to classify the North American bird fauna in a similar manner. This classification, tentative as it is under the circumstances, is very useful as a test of the various arrangements proposed by regional zoogeographers. It provides at least provisional answers to such questions as: "Is it justifiable to recognize a neotropical fauna and a nearctic fauna?" "Is the nearctic fauna, if it exists, part of a New World or of a holarctic fauna?" "Does North America have a fauna of its own, or is it merely an area of intergradation between the Eurasian and the South American faunas?" "Are the faunas of given geographical areas sufficiently homogeneous to justify the recognition of zoogeographic regions, or does the delimitation of zoogeographic regions convey an erroneous impression?"

We are in a much better position today to answer these questions than was Lönnberg in 1926. First, there has been a general advance in the whole field of zoogeography—a complete change in the concept of the functions of the science—signaled by the important publications of Simpson, Stegmann, and Stresemann. Classical zoogeography asked: What are the zoogeographic regions of the earth, and what animals are found in each region? The modern zoogeographer asks when and how a given fauna reached its present range and where it originally came from; that is, *he is interested in faunas rather than in regions.* In the light of this new concept of the science, such familiar terms as "holarctic," "nearctic," and "neotropical" acquire completely new meaning. Second, there have been many very specific recent additions to our knowledge, contributed partly by the paleontologist and partly by the taxonomist, that permit a more accurate analysis than Lönnberg could give.

Recent Contributions of the Paleontologist

The number of important discoveries of fossil birds has been greatly augmented in recent years, the Californian school and Alexander Wetmore having made the most valuable contributions. [For the latest summary see Brodkorb (1971).] Finds of particular zoogeographic significance concern the following groups (Wetmore 1940): (1) The Aramidae. The limpkin (*Aramus*) is the only living representative of this family; and, as Lönnberg said (1927:24), "if one has to judge only from the present distribution, [it] would certainly be regarded as South American"; but the fact that there are two extinct Tertiary genera (*Badistornis* and *Aramornis*) in North America favors a North American origin for the family. (2) The Old World vultures (Aegypiinae), which are now restricted to the Old World. Nobody would suspect the former occurrence in the New World of this subfamily of the Accipitridae if fossil remains of three extinct genera had not been found in the Miocene (*Palaeoborus*), Pliocene (*Palaeoborus, Neophrontops*), and Pleistocene (*Neogyps, Neophrontops*) of North America. No conclusion can be drawn, however, as to the origin of the family. (3) The New World vultures (Cathartidae), which Lönnberg (1927:22) listed as a South American family. The fact that Wetmore (1940 and 1944) has found several striking genera in the early Tertiary of North America indicates either a North American or pre-Tertiary origin for the family. (4) The Cracidae (curassows and guans), whose present center of distribution is in South America, where the vast majority of the species occur and where most of the genera are endemic. Even though seven Recent species occur in Central America and two genera are endemic there (*Penelopina* and *Oreophasis*), this family would surely be considered a comparatively recent arrival in North America were it not for the occurrence of two species in the Tertiary of North America (*Ortalis tantala* in the lower Miocene; *O. phengites* in the lower Pliocene) and for the occurrence in the Wyoming Eocene of the related (fossil) family Gallinuloididae.

Recent Contributions of the Taxonomist

Unsound classifications have caused much confusion in zoogeography, as ably pointed out by Simpson (1940b) in a discussion of the so-called evidence for an antarctic land bridge. Of particular zoogeographic significance are the following recent changes in the classification of birds.

"New World Insect Eaters"

From a study of a number of South American genera it would seem that the tanagers (Thraupidae)—including the South American swallow-tanagers (Tersinidae), honeycreepers (Coerebidae), wood warblers (Parulidae—formerly "Compsothlypidae"), vireos (Vireonidae)—including the shrike-vireos (Vireolaniidae) and the pepper-shrikes (Cyclarhidae), blackbirds and troupials (Icteridae), and some of the finches (the subfamily Emberizinae) are closely related, constituting a single superfamily [now designated the nine-primaried Oscines], perhaps the New World equivalent of the Old World family Muscicapidae of recent authors (J. T. Zimmer, verbal information).

Troglodytidae

Sharpe's Hand-list (vol. 4, 1903) and other older taxonomic works included among the wrens a considerable number of south Asiatic genera (*Pnoëpyga, Elachura, Spelaeornis, Sphenocichla,* and sometimes *Tesia*). Lönnberg (1927:9–10) consequently had considerable difficulty in proving an American origin for this family. Recent taxonomic work has clearly established the fact that none of the listed Asiatic genera (superficially wrenlike babbling thrushes and Old World warblers) belongs to the Troglodytidae and that *Troglodytes troglodytes* is the only wren that occurs in the Old World. The strictly American character of the wren family is now beyond dispute.

"Chamaeidae"

The wren-tit (*Chamaea*) is not the sole representative of a separate family, but a member of the Paradoxornithinae (parrot bills and suthoras), and possibly not even generically separable from *Moupinia* of southwest China.

Fringillidae

The so-called finches are an assemblage (probably highly artificial) of seed-eating birds with cone-shaped bills. Three major groups can be distinguished within the fringillids that are established in North America: (1) Carduelinae—the cardueline finches; (2) Emberizinae—

certain buntings and American sparrows; and (3) Cardinalinae—the cardinals, or South American finches. (See Sushkin 1924 or 1925.) There is little doubt that the Carduelinae are Old World in origin; the Emberizinae North American, although some species are found in the Old World; the Cardinalinae South American, although some genera have become thoroughly established in North America. (It should be noted that no final decision can be reached on the last two groups until it has been determined whether certain South American genera belong to the Emberizinae or to the Cardinalinae. A discussion of the characters of the fringillid subdivisions, as well as an incomplete listing of the genera, can be found in Sushkin.)

THE GEOLOGICAL HISTORY OF NORTH AMERICA

The North America of today is connected with South America by an isthmus and is separated from Asia only by a narrow oceanic strait. These connections with the two adjoining faunal areas are of the greatest importance, and a study of their history, both geologically and climatically, is a prerequisite to a full understanding of the faunal history of North America. At the present time there is only a tenuous direct connection with Europe through the arctic-subarctic islands of the North Atlantic (Greenland, Iceland). The wheatear (*Oenanthe oenanthe*) is one of the few birds that has come to us via this bridge. As I shall presently discuss, this connection was far more important in the early Tertiary.

The theory of the permanence of continents and oceans, still accepted when this essay was first published, has now been replaced by the theory of plate tectonics, as mentioned earlier. Although the new theory does not drastically affect the reconstruction of the major components of the continents (the "plates" of current geology), it does reject the former belief in the permanence of the ocean basins. It revives Alfred Wegener's theory of continental drift, although with a rather different, that is, very much earlier, chronology. Of the three major potential land connections of North America, with Asia, with South America, and with Europe, only the last one has required an entirely new interpretation as the result of the acceptance of plate tectonics. To the best of our current knowledge, the history of the three continental connections was as follows.

1. Asia and North America were in continental connection with each other across what is now the Bering Straits at various times during the later Mesozoic and the Tertiary. The students of mammals (Simpson 1947) have shown that periods of very active faunal exchange have alternated with periods of little or no exchange. Presumably an oceanic strait like the present Bering Straits opened and closed repeatedly, just as the Bering Straits did during the Pleistocene. The botanical as well as the mammalian evidence indicates that climate and vegetation on this

bridge was temperate or at best cool subtropical, but was rather unsuited for the passage of tropical elements.

2. Our ideas on the firm connection between North America and South America through Central America have not been affected by the adoption of plate tectonics. It is still believed that this connection is, geologically speaking, rather recent. The isthmus of Panama established connection with Colombia only about 5-6 million years ago through the closing of the Atrato portal. Prior to that event the two continents were separated throughout the Tertiary and Mesozoic by oceanic gaps of various extents. North America reached originally only as far south as Guatemala and Honduras. There are still some unsolved puzzles, however. For instance, there are indications in the pattern of mammalian distributions that North and South America were nearer to each other at the end of the Mesozoic or in the early Tertiary than in the middle Tertiary. Yet I believe there is no solid geological evidence for an actual connection in the Eocene. Nor is much known about the number and size of the islands that served as stepping stones, particularly for such aerial colonists as birds. The faunal exchange across the gap is described in detail in Essay 39.

3. The North Atlantic is not an old ocean, as used to be believed. Indeed, North America and northern Europe were still broadly connected in the early Tertiary. Exactly when the two continents parted is still controversial. The middle or late Eocene (about 45 million years ago) is a reasonable guess. At any rate, there were periods in the early Tertiary when the mammalian faunas of North America were virtually identical with those of Europe, but quite distinct from those of Asia. Europe and Asia were two distinct land masses at that time, separated by an arm of the ocean near the present Ural Mountains (Turgai Straits). It will be pointed out presently what faunal elements seem to belong to this old Euramerican fauna.

Most important for an understanding of the origin of the North American fauna is the fact, emphasized by Lönnberg (1927), Dunn (1931), and Simpson (1943b), that the whole southern half of North America was subtropical or tropical during most of the Tertiary, while it was separated from South America by oceanic gaps. Even in the later Tertiary, a tropical climate prevailed in the southernmost section of North America. This means that (with the exception of those animals that cross water gaps easily) there was not merely one tropical American fauna, the "Neotropical," but two quite distinct ones: one south of the ocean gaps, the other north of them. F. M. Chapman (1923) showed that the motmots (Momotidae), usually referred to as a "typically Neotropical" family, had actually originated in Middle America, "where the ancestral forms of the existing genera were possibly developed during the Oligocene when this region consisted of scattered islands which would afford the isolation favorable to differentiation" (p. 58). Lönnberg (1927:12) states correctly that the same would probably be found to be true in other families if they were examined as "thoroughly and masterfully" as the Momotidae were by Chapman. In the meantime, Dunn (1931), Simpson (1943b:428), and Hubbs (1944:271) have emphasized the importance of this Middle American (that is, tropical North American) element among reptiles and fishes.

The mid-Tertiary fauna of North America was probably not only highly peculiar but also far more homogeneous than it is now. To visualize its composition, one must look at the South America of today. The temperate zone of South America, which admittedly is rather small because of the continent's southern attenuation, does not have a fauna that is basically different from that of the tropical areas. It has its share of endemic species and even genera, but its fauna (although poorer) is composed more or less of the same families as that of the warmer portion. A similar faunal homogeneity was perhaps true for North America during Tertiary times, the faunas of the tropical, of the subtropical, and of the warm-temperate zones being very much alike in composition. The present-day contrast between the fauna of tropical-subtropical Central America and that of temperate North America has two causes: (1) the climatic deterioration in the late Tertiary and Pleistocene which eliminated most tropical elements then existing in North America, and (2) the invasion (from South to North America) of a new tropical element after the closing of the Central American water gaps and particularly at the end of the Pleistocene. This faunal mixing during the late Pliocene and the Pleistocene led to a complete reshuffling of faunal elements (see Essay 39). Obviously, it would be a zoogeographical error to classify families that were originally North American with the truly autochthonous[1] South American families. Yet, nearly all the older zoogeographical treatises classify as "Neotropical" what is really a mixture of North and South American faunal elements. An effort has been made in the following classification to avoid this error. (In this paper zoogeographical North America is considered to extend southward to the edge of the tropical rain forest.)

CLASSIFICATION OF THE FAUNAL ELEMENTS OF THE AMERICAS[2]

Three Tertiary land masses are the primary contributors to the present fauna of the Americas: South America, North America, and Eurasia. It would therefore appear that the simplest classification of faunal elements would be into the same categories: South American,

1. In this paper I have used the terms "endemic" and "autochthonous" as follows: Endemic = restricted to a given region; not found elsewhere. Autochthonous = having originated in a given region; now sometimes found beyond the borders of that region.
2. The account of the analysis of the faunal elements here presented is considerably shortened. For further detail see the original and Essay 39.

North American, and Eurasian (or "Old World"). These three classes undoubtedly must be recognized, but they are not sufficient to cover all families and genera of birds. First, an additional category must be recognized for groups that cannot be analyzed for one reason or another (to be stated below). Second, there are certain groups ("holarctic," or "pan-boreal," elements) that have moved back and forth across the Bering Straits so freely that they cannot be assigned with certainty to either continent. Others ("pan-American") crossed the Central American water gaps sufficiently freely to obscure their center of origin. Finally, there is an old tropical element ("pan-tropical") that is of such similar composition in the Old World and New World tropics that it is impossible at the present time to determine the original home. It is into these categories (Figure 38–1) that I have tried to classify all the families of birds known to occur in the Americas, whenever possible carrying the analysis even further: to subfamilies, genera, and occasionally to species. This is particularly necessary in the case of families that originated outside of North America, for parts of which North America

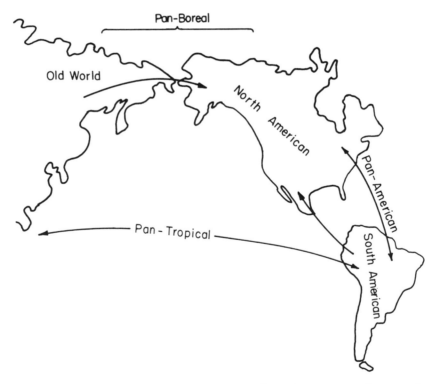

Figure 38–1. Diagram of the faunal elements of North America. The unanalyzed element, whose geographical origin cannot be determined, is of course omitted from the map. [Although the pan-tropical element is here indicated as trans-Pacific, it is now clear that its connection was across the Atlantic.]

became a secondary center of evolution (e.g., quails, jays, thrushes), and of those other families that reached North America repeatedly at different geologic periods (e.g., the swallows).

Unfortunately, the bird geographer has, as stated above, relatively few fossils to guide him in his analysis. He is therefore forced to utilize indirect evidence, which is often difficult to evaluate (For an analysis of these criteria see Essay 37). These indirect methods are fully reliable only in richly developed families. The value of the evidence is uncertain in regard to families consisting of only one or merely a few species.

The Unanalyzed Element

The separation of land masses, which is responsible for the divergent development of terrestrial faunas, has little bearing on the evolution of sea bird faunas. Roughly, the oceanic birds can be classified into (1) a southern group: penguins (Spheniscidae) and sheath-bills (Chionidae); (2) a tropical group: tropic-birds (Phaëthontidae), boobies and gannets (Sulidae), frigate-birds (Fregatidae); (3) a northern group: skuas and jaegers (Stercorariidae); and (4) a worldwide group: albatrosses, shear-waters, fulmars, and petrels (Tubinares), gulls and terns (Laridae). A further analysis and determination of the point of origin of these sea birds is outside the scope of this paper.

Equally obscure is the place of origin of the partly oceanic, partly freshwater families of the pelicans (Pelecanidae) and the cormorants (Phalacrocoracidae). Among the true freshwater groups, a number of families are so evenly distributed in the Old and New World as to make determination of their centers of origin impossible. These include the grebes (Podicipedidae), herons and bitterns (Ardeidae), storks and jabirus (Ciconiidae), ibises and spoonbills (Threskiornithidae), fla-mingoes (Phoenicopteridae), ducks, geese, and swans (Anatidae), and rails, coots, and gallinules (Rallidae). With most of these, it is not simply the family as a whole that is widespread, but also the sub-families, many of the genera, and frequently even individual species. This point is well illustrated by the duck family, of which an up-to-date classification is available (Delacour and Mayr 1945). Of the nine recog-nized tribes (or "subfamilies"), only the monotypic torrent duck tribe (Merganettini) is restricted to a single continent. Of the 40 genera, no less than 18 are found on two or more continents. Many species are circumtropical or at least very widespread. For example, the white-faced whistling duck (*Dendrocygna viduata*): South America, Africa, Madagascar; the fulvous whistling duck (*Dendrocygna bicolor*): America, Africa, India; the superspecies *Tadorna ferruginea* (which in-cludes the four species formerly separated as *"Casarca"*): Europe, Asia, South Africa, Australia, New Zealand; the black duck-mallard group of river ducks (*Anas platyrhynchos-fulvigula*): spread over most of the world except South America; and other species groups in the genera *Aythya, Cairina, Mergus,* and *Oxyura*: South America, Africa, Australia. Widespread genera and species are typical also of most other families of freshwater and shore birds.

Among the strictly terrestrial birds, there are 8 families that are so widespread or so evenly distributed as to make analysis difficult at the present time. These families are the hawks and eagles (Accipitridae), the osprey (Pandionidae), falcons and caracaras (Falconidae), nightjars (Caprimulgidae), swifts (Apodidae), woodpeckers (Picidae), and swallows (Hirundinidae). The evidence indicates that all of these families originated at such an early date (Eocene or Cretaceous) that subsequent shifts in distribution have obliterated most of the clues. Most of them are also excellent transoceanic colonizers.

The Pan-Tropical Element

While representatives of the hawks, owls, and swifts are found in several climatic zones, there are certain other families that are also widespread but only within the tropical belt. For five families of freshwater birds (in some cases, partly marine), the area of origin is difficult to fix because each of them is found both in the Old World and New World tropics, though represented only by a single species, or merely a few species. These families are the snake-birds (Anhingidae), sun-grebes (Heliornithidae), jacanas (Jacanidae), painted snipes (Rostratulidae), and the skimmers (Rynchopidae). All of them now have widely disrupted ranges, as can easily be seen from the map of the sun-grebes (Figure 38–2). It is also remarkable that the Recent Old World and New World representatives are often the members of a single species or superspecies (*Anhinga, Rostratula benghalensis, Rynchops*). This would

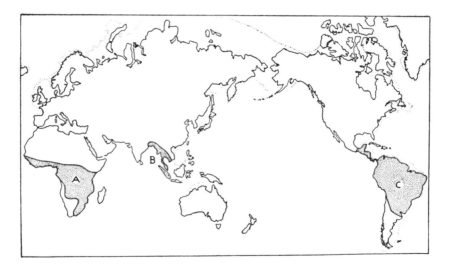

Figure 38–2. Present distribution of the sun-grebes (Heliornithidae), a typical family of the pan-tropical group. *A = Podica, B = Heliopais, C = Heliornis.*

indicate either extremely slow evolution or an enormous capacity for transoceanic dispersal.

Among the land birds, three families are pan-tropical. The barbets (Capitonidae) and the trogons (Trogonidae) have a notably similar distributional pattern. The ranges of both families are restricted to the humid tropics, and are bounded in the east by Wallace's Line. Fossil trogons have been found in the Eocene of France, and this fact, together with the scarcity of trogons in South America, has led most authors to assume an Old World origin for the family. However, trogons are much more diversified in Central America than in the Old World tropics; in fact, all the African and Indian species could be included in a single genus. Tropical North America is the most likely place of origin. The barbets, with a similar distributional picture, are so much more richly developed in the Old World tropics than in the New that an Old World origin is probable (see Ripley 1945:543–544).

The distribution of the parrots (Psittacidae) is considerably more extensive than that of the barbets and trogons. The parrots, with about 315 species, are one of the richest of all bird families, but about an equal number are found in the Old and the New World. However, most of the more aberrant types, such as the lories (Loriinae), cockatoos (Cacatuinae), and pigmy parrots (Micropsittinae), are found in the Old World, more specifically in the Australian region. It is therefore probable that the Psittacidae originated in the Old World, but the great number of endemic genera and species in America indicates a very early arrival in the New World. The pathway of dispersal of this pan-tropical element will be discussed below.

The Pan-Boreal Element

The loons (Gaviidae) among the freshwater birds, the phalaropes (Phalaropodidae) among the shore birds, and the auk family (Alcidae) among the sea birds are typical of a large class of circumboreal birds. All three families are distributed in the arctic or in the north temperate zone and are about equally well represented in the Old and the New World. The auk family and the loons are known from the Tertiary of both North America and Europe. The temperate zones of Eurasia and America were in such direct contact for a good part of the Tertiary (by means of the Bering bridge) that it will be very hard to determine which of the two land masses was the giver and which the taker of the members of this temperate zone group. Among genera and species, this circumboreal element is much stronger than among families. Well over 80 percent of the species of the circumboreal tundra zone belong to it, and it is impossible to determine their ultimate source. Stegmann (1938a) believes that Asia, more particularly Siberia, has probably made the greatest contribution to the group because it is the largest land mass in the temperate zone.

The Asian Element

It is generally admitted that the connection between Asia and North America across the Bering Strait is very ancient (pre-Tertiary). As far as birds are concerned, a more or less active faunal exchange probably took place right through the Tertiary, even during periods when the two land masses were separated by water. This long-standing accessibility of North America to Asian immigrants is reflected in the taxonomic composition of this element in America. According to the date of their immigration, these birds have (1) not changed at all, e.g., the Alaska yellow wagtail (*Motacilla flava alascensis*), the red-spotted bluethroat (*Luscinia* ["*Cyanosylvia*"] *suecica robusta*), and the wheatear (*Oenanthe oe. oenanthe*); or (2) have become subspecifically distinct, e.g., Kennicott's willow warbler (*Phylloscopus* ["*Acanthopneuste*"] *borealis kennicotti*), the northern shrike (*Lanius excubitor borealis*), the brown creeper (*Certhia familiaris americana*); or (3), if they arrived very early, they have evolved into separate species, genera, or even subfamilies—that is, America has become for them a secondary center of evolution.

The third case is true of the Old World pheasant family (Phasianidae), which has produced the American quails (subfamily Odontophorinae). And it is probably true of the cuckoos (Cuculidae). In this family, Peters (Check-list, vol. 4, 1940) recognizes six subfamilies. Three of these, the Cuculinae, the Couinae (Madagascar), and the Centropodinae, are restricted to the Old World; the Crotophaginae are American; the Neomorphinae have five genera in the New World, one in the Old; and the Phaenicophaeinae have nine in the Old World, three in the New. The evidence points toward an Old World origin of the family, and to tropical North America as a secondary center of evolution for the three subfamilies. [In all the families in which the center of distribution is tropical an early Tertiary trans-Atlantic colonization is more probable than a late Tertiary Bering Straits route.]

The pigeons (Columbidae) are worldwide in distribution—which indicates their great age. However, the rich development of the family in the Australian region, where the most aberrant members of the family occur (e.g., *Caloenas, Goüra, Otidiphaps,* and *Didunculus*), and the fact that most American species belong to just a few phyletic lines prove an Old World origin. It seems probable that some species reached South America as early as the middle Tertiary and established a second evolutionary center (see also Goodwin 1967).

Both the crow family (Corvidae) and the thrushes ("Turdidae") are examples of Old World groups that have established minor secondary evolutionary centers in North America, particularly in the tropical part. For the Corvidae, Amadon (1944:16–20) has presented detailed evidence. The blue jay group (*Cyanocitta*) developed in America, but since there is not a single endemic genus in South America, it is obvious that the jays reached there only after the closing of the Central American water gaps in the late Tertiary. The genera *Corvus, Nucifraga,* and *Perisoreus* represent separate later invasions of the Corvidae into North America. In view of the early arrival of the jay group, it seems conceivable that some of the palearctic genera (*Perisoreus, Nucifraga, Garrulus*) evolved in America and crossed back to Asia by the Bering Straits, but it would be impossible to prove this.

The thrush subfamily Turdinae (see Mayr 1941:106) presents a very similar distributional pattern and probably had a similar history.

The kingfishers (Alcedinidae) are a rich Old World family of which only one branch (Cerylinae) has reached the New World. This colonization cannot have been very recent, since a few species (the neotropical group *Chloroceryle*) are sufficiently distinct from their nearest Old World relatives to be considered by most authors a separate genus.

The cardueline subfamily of the Fringillidae is an Old World group, but one of the lines seems to have arrived in America rather early, since it has produced a number of endemic South American species (*"Spinus"*).

The Paridae (titmice) are a mainly Eurasian family that has repeatedly invaded North America, where it has even developed two endemic genera, verdins (*Auriparus*) and bush-tits (*Psaltriparus*). But the latter genus is closely related to the Asiatic genera *Aegithaliscus* and *Psaltria,* while the other American titmice are still more closely related to Asiatic species; some are even conspecific. They must have crossed the Bering Straits during or after the late Pleistocene.

Six additional Old World families (or subfamilies) have colonized the Americas so recently, and the New World representatives are still so similar to the Old World forms (congeneric or even conspecific), that North America cannot be considered a secondary evolutionary center for them. These are: barn owls (Tytonidae), larks (Alaudidae), nuthatches (Sittidae), creepers (Certhiidae), Old World warblers and kinglets (Sylviinae), and shrikes (Laniidae). The Old World origin of most of these groups has been discussed by Lönnberg (1927) and earlier authors. Only two of them (the larks and barn owls) have reached South America, and they have arrived so recently that the South American representatives are no more than subspecifically distinct.

The North American Element

The fauna that developed in North America during the Tertiary, while this continent was separated from South America and connected with Asia only by the Bering Strait bridge, is of great zoogeographical importance. It was much neglected in the past, when some of its components were labeled "Holarctic," others "Neotropical." The greater part of the Tertiary North American continent had a subtropical or tropical climate, and it is therefore not surprising that tropical families and genera are well represented in this North American element.

The fact that there is an important autochthonous element in the bird fauna of North America was clearly recognized by Lönnberg (1927). He considered that the thrashers and mockingbirds (Mimidae), the vireos (Vireonidae), the wood warblers (Parulidae), the waxwings (Bombycillidae) with their relatives the silky flycatchers (Ptilogonatidae), and the wrens (Troglodytidae) and motmots (Momotidae) are also North American in origin. The monotypic family palm-chats (Dulidae) also belongs to this group. Subsequent research has fully confirmed his conclusions (the evidence is given in more detail in Essay 39). The most important North American element are the so-called nine-primaried Oscines, the Vireonidae, Thraupidae, Parulidae, Cardinalinae, Emberizinae, Icteridae, and related genera. Many of them, like the Thraupidae and Icteridae, spread to South America very early, and this is where they had their major radiation (see below under "Pan-American Element"). A North American origin is indicated also for the New World vultures (Cathartidae), the limpkins (Aramidae), turkeys (Meleagridae), dippers (Cinclidae), and the families mentioned by Lönnberg, particularly the wrens (Troglodytidae) and mockingbirds (Mimidae). (For further details on the North American element see the original essay and Essay 39.)

North America became a secondary center of evolution for several originally Old World groups, such as the American quails (Odontophorinae), the *Cyanocitta* group of jays of the family Corvidae, the *Myadestes–Catharus–Hylocichla* group of thrushes, and some others. In particular, the Odontophorinae, a whole subfamily restricted to North America, and known there as far back as the Miocene, well deserve to be included among the typically North American fauna. Part of the pan-American element (certain Icteridae), discussed below, has also now become sufficiently well established in North America to be considered part of the North American element.

The Pan-American Element

The water gaps that existed between North and South America up to the late Pliocene produced an almost complete separation of the mammalian faunas of the two continents (Simpson 1940a:157–163). The intervening chain of islands permitted colonization by only a few groups especially adapted to "island hopping." On the whole, the geographical picture of this line of islands was apparently very similar to that of the Malay Archipelago, where colonization by mammals was almost completely prevented, even though the islands were more numerous and the water gaps comparatively small. For birds, these interisland straits of the Malay Archipelago were much less of a barrier (Mayr 1944:171–194). The same is true for the inter-American island belt. It explains many of the difficulties of the bird geographer. There are quite a number of American families that are so rich, in both North and South America, in endemic genera and species that it is difficult to determine their primary country of origin without fossil evidence. It is rather obvious that these are the families able to utilize islands as stepping stones from one continent to the other. During the greater part of the Tertiary, the whole southern part of North America was apparently more humid, and certainly warmer, than it is today. It would have been more difficult for many of the species that developed in this climatic zone to enter the more temperate parts of North America than to cross into tropical South America. In the reverse direction, the same was true for species of tropical South America. This is one of the reasons that the contrast between the North and the South American Tertiary faunas is much less pronounced in birds than in mammals, and much less than one would expect on the basis of the length of separation of the two continents.

[In the original version of this 1946 essay all the families that were well represented both in North and South America were grouped together under the term "pan-American element." Eighteen years later I analyzed this element again, partitioning it into an originally southern element (Trochilidae, Tyrannidae) and several kinds of presumably originally northern elements (nine-primaried Oscines). For further detail, see Essay 39.]

All of these families have endemic genera in both North and South America. These are too distinct to have developed in the rather short time since the establishment of the Panamanian land connection. They must have descended from birds with the faculty of trans-oceanic colonization.

The South American Element

[The 1946 essay also included a listing and discussion of the South American element (pp. 25–26). This is here omitted, since the subject is covered in more detail in Essay 39.]

It is most remarkable that none of the families of clearly South American origin succeeded in advancing beyond North America by crossing the Bering Straits. Old World families, on the other hand, have sent many branches into South America. This contrast in the potential of the two faunal elements suggests that a temperate zone family can become adapted to the tropics more easily than a tropical family to a temperate climate.

An African Element?

[During the first half of the Mesozoic the South Atlantic had not yet opened up and Africa and South America were part of a large southern continent. It would therefore be conceivable that North America received some African elements by way of South America. There is, however, little evidence for this. Even though the connection was probably broken some time in the Cretaceous, faunal interchange between Africa and South America was presumably possible for good fliers to the end of the Cretaceous and perhaps into the early Tertiary.

Yet, hardly any two other bird faunas are more different than those of Africa and South America. There has been no faunal sharing between South America with its suboscines, hummingbirds, toucans, tinamous, and Africa with its hornbills, honeyguides, bee-eaters, Musophagidae, Eurylaemidae, Nectariniidae, Alaudidae, and Ploceidae.

It is possible that some of the Pan-tropical element crossed not from Europe to North America but from Africa to South America, but there is no definite evidence for this. Up to now, except for a few water birds, it is impossible to identify in the North American avifauna any element that might have gotten there from Africa via South America.]

COMPARISON OF BIRDS WITH OTHER ANIMALS
AND WITH PLANTS

On a walk through the woods in temperate North America, one encounters flowers and trees that differ but little from species found in temperate Asia. The admixture of tropical South American elements is negligible. The same is true for mammals. The porcupine and the armadillo are apparently the only South American elements in the present North American mammal fauna, compared with a 13 to 20 percent South American element in the bird fauna, except at the northern fringe (Table 38-1). I do not know of any exact published figures, but I gather from the writings of mammalogists that more than half of the temperate North American mammals are of Old World origin. In birds (again excepting the northern fringe), only a third or less originated in the Old World.

There are mainly two reasons why the Old World element is so much weaker among North American birds than among most other animal groups—or perhaps I should better say: why the South American and warm North American elements in temperate America are so much stronger in birds than in other animal groups. One of these reasons is the ability of birds to cross water gaps. Thus, while the indigenous mammals were imprisoned in South America during the Tertiary separa-

Table 38-1. Analysis by geographical origin of the breeding passerine species of several districts of North America.

Area	South American	North American	Old World
Yakutat Bay, southeast Alaska (Hudsonian Zone)	3%	39%	58%
Oregon	14	47	39
Nipissing area, southern Ontario, 46°N (Canadian Zone)	13	57	30
New Jersey	14	63	23
Florida	20	59	21
Sonora, Mexico	27	52	21

tion of the two continents, several groups of South American birds crossed the water gap into the northern continent (see Essay 39). Some of these genera and generic groups must have arrived in North America at a very early date. Preempting many ecological niches, the 40 or 50 species of such originally South American groups as hummingbirds and tyrant flycatchers have helped stem the influx of Old World species.

A second and more important factor is bird migration. It enables many tropical or semitropical birds to include in their breeding range areas of the temperate zone that have a hot summer season by moving to these areas in the summer and then flying back to their tropical homes when the cool season begins. An analysis of the mid-winter avifauna of temperate eastern North America shows that it is composed almost entirely of Old World elements. The difference in migratory behavior between the autochthonous and the Old World elements is illustrated in the following statistics. Among the 28 species of permanent residents (excluding water birds and unanalyzable species) listed by Cruickshank (1942:25–26) for the New York region, no less than 23 (82.1 percent) are of Old World origin. By contrast, among 67 analyzable species of summer residents (which migrate south in the fall), only 8 (11.9 percent) are of Old World origin. If the 95 species of the two categories are combined, it is found that of the 12 species of the South American element only 1 (8.3 percent) is a permanent resident, and of the 52 species of the North American element only 4 (8.3 percent) are permanent residents, while of the 31 species of the Old World element no less than 23 (76.7 percent) are permanent residents. The Old World element, which, as Stegmann (1938a) has shown, developed for the most part in the always cold land mass of northern Siberia, is so thoroughly adapted to the cold that it can survive in this latitude without migration, whereas the autochthonous American element, most of which developed in a warm zone, survives the winter by avoiding it.

The combination of these two factors has resulted in the peculiar composition of the contemporary North American bird fauna. It is, therefore, obvious that no general zoogeographic scheme can be based on the distribution of birds, and that zoogeographical classifications based on the distribution of mammals or reptiles are inapplicable to birds. This difference between birds and other animal groups is the reason for much of the "New World" versus "Holarctic" controversy. Those who wanted to unite North and South America into a single "New World" based their conclusion mainly on a study of birds. Those who wanted to include North America with Eurasia in a "Holarctic" region based their conclusions on studies of mammals or reptiles.

THE HISTORY OF THE PAN-TROPICAL ELEMENT

In a previous section I discussed a number of families that are more or less restricted to the tropics, but are found in the Old as well as in the New World. A similar distribution has been documented for various families and subfamilies of turtles (Simpson 1943b) and other reptiles (Dunn 1931), as well as for mammals (e.g., tapirs) and other groups. Various explanations have been advanced to account for this type of distribution. In a few exceptional cases, for example, the white-faced and fulvous whistling ducks (*Dendrocygna viduata* and *D. bicolor*) and the southern pochard (*Netta erythrophthalma*), as well as some species of herons, it is reasonably certain that trans-oceanic colonization is the answer. This explanation is, however, exceedingly improbable for most of the other groups that have closely related representatives in the tropics of both the Old and the New World, for example, some of the snake-birds (Anhingidae), the sun-grebes (Heliornithidae), jacanas (Jacanidae), barbets (Capitonidae), trogons (Trogonidae), and parrots (Psittacidae) among the birds that I have classified with the pan-tropical element; as well as some of the storks (ciconiidae), ibises (Threskiornithidae), flamingos (Phoenicopteridae), nightjars (Caprimulgidae), woodpeckers (Picidae), and hawks (Accipitridae and Falconidae). A different explanation must be found for their movement from one continent to another.

Three different kinds of explanations have been advanced to account for this close relationship of certain faunal elements in the New World and Old World tropics. The earliest explanation was to postulate the former existence of "land bridges" across various parts of the Atlantic and Pacific. Since no geological evidence for a former existence of now submerged land bridges could be found—and also for zoogeographic reasons—the land-bridge theories were eventually completely rejected (Matthew 1915).

Subsequently, the theory of a permanence of continents and oceans was most widely accepted, defended particularly convincingly by the students of fossil mammals. Matthew (1915), Simpson (1943a:9, 1947), and others postulated that throughout the Tertiary there was an active faunal exchange between Eurasia and North America across the Bering Straits bridge. When the Bering Straits were open, the exchange was reduced; when they were closed (owing to a lowering of the sea level), the exchange was active. It was postulated by these zoogeographers that not only arctic and temperate zone elements participated in these movements, but also subtropical and tropical ones. Stegmann (1938b) objects to this solution. He quotes considerable evidence from the field of paleobotany and paleoclimatology that indicates (p. 485) "that the climate in the region of Bering Strait was at times warmer than it is now, but never reached tropical temperatures. Indeed it is quite certain that in northwestern America and in nearly all of Siberia the climate was never tropical or even subtropical during the entire Cenozoic and Cretaceous. . . . The Bering region was thus far outside the tropics during the entire period that needs to be taken into consideration for the evolution of Recent birds, so that it is without the slightest significance as a 'land bridge' for tropical groups." The records of American plant paleontologists support this contention. Chaney (1940) shows that as far back as the Eocene only a temperate climate existed in the countries east and west of the Bering Strait bridge (see Figure 38-3). One has to go as far south as the state of Washington on the American side, and to China on the Asiatic side, to find fossil plants that indicate even a subtropical climate.

Formerly the opinion was widespread among paleogeographers that a uniformly tropical climate prevailed all over the world during past geological periods. Reputed finds of Tertiary palms in Greenland seemed to strengthen this theory. However, these botanical reports have since been found to be erroneous; furthermore, geophysicists have made it abundantly clear that climatic zones must always have existed on the earth. This is a corollary of the earth's curvature. Less radiated heat from the sun will reach a given area in the higher latitudes than will reach the equatorial districts, where at noon the sun is nearly overhead during the greater part of the year. Furthermore—and this is a factor strangely neglected in books on past climates—the axis of the earth is inclined at an angle of 23½° to the perpendicular to the plane of the ecliptic. This inclination causes our seasons. The northern hemisphere is turned away from the sun during the winter and turned toward the sun during the summer. Geophysicists believe that this angle of the ecliptic has not changed significantly during the recent geological past. This means that north of the Arctic Circle an Arctic winter night has existed at all times. The Arctic Circle goes exactly through the Bering Straits, and there can be little doubt that an arctic "winter" (in terms of daily sunlight) must have existed at least as far south as the

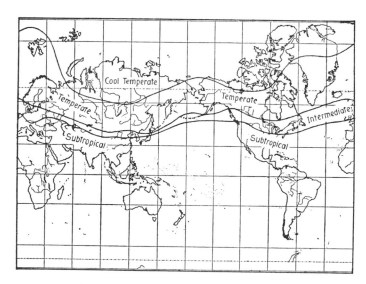

Figure 38-3. Eocene climatic zones as indicated by fossil plants. (Based on Chaney 1940.)

Aleutians, in other words, beyond the southern edge of the Bering shelf. Surely this would not be a favorable condition for tropical faunas and floras to pass freely back and forth between Asia and America. Nor is the argument weakened by the movements of the poles, believed to have taken place during the last 100 million years.

Yet, the close relationship between the Old and New World members of the pan-tropical element, whose ranges are now widely discontinuous, proves that such a faunal exchange must have taken place, and this places the zoogeographer in a real quandary. The customary solution for the problem is to ignore it. Stegmann (1938b:492) and other authors of the Russian school (e.g., Wulff 1943:173–196) were the first to suggest solving it by postulating a modified Wegenerian land connection across the North Atlantic lasting at least until the middle of the Tertiary. Simpson (1943a:20–22), however, objected to this proposal on the basis of the small number of early Tertiary mammalian forms that were common to Europe and North America. Similar objections came from the field of botany.

[This is how matters stood in 1946. There seemed to be no way out of the dilemma. The Bering Straits rather clearly were not the colonization route for the pan-tropical elements, yet every other alternative was at that time even more unsatisfactory (as shown in my discussions on pp. 36–37 of the original essay, here omitted). The development of the theory of plate tectonics raised new hopes for a solution of the problem. It firmly established that the final separation of North America from Europe did not take place until the early Tertiary (45–50 million years ago). Since Greenland and other islands could serve as steppingstones, there was probably active faunal trans-Atlantic interchange until Oligocene times and for good fliers even later.

This new insight forces us to take a second look at the history of the holarctic element. It is no longer permissible to assume that all the exchange took place across the Bering Straits. On the contrary, a trans-Atlantic exchange was more probable during the early Tertiary than a Bering Straits exchange. Only a new analysis can establish which pathway was used by various genera in the following families: Caprimulgidae, Apodidae, Picidae, Hirundinidae, Gruidae, Columbidae, Cuculidae, Strigidae, Corvidae, Alcedinidae, Turdinae, and Carduelinae.

The question remains whether the trans-Atlantic connection explains the origin of the pan-tropical element. The answer, unfortunately, is that it does not. The Eocene connection was in the far north Atlantic, leading across Greenland. Greenland, in turn, was separated from the American continent by the Labrador Straits, which had apparently opened up in the early Cretaceous. Even though the climate on this bridge was warm temperate, perhaps even subtropical, it was not at all suitable for the passage of strictly tropical elements.

How then did the tropical elements move between the Old World and the New World? There is the possibility of a much earlier passage (?

early Cretaceous) when the bulge of North Africa was still within reach of North America. To me this solution seems unsatisfactory since it creates more difficulties (rates of evolution, dispersal abilities of trogons and barbets) than it removes. In short, at the present time there is still not a satisfactory explanation for the dispersal route of the pan-tropical elements.]

REFERENCES

Allen, J. A. 1893. The geographical origin and distribution of North American birds, considered in relation to faunal areas of North America, *Auk*, 10:97–150.

Amadon, D. 1944. *The genera of Corvidae and their relationships.* Amer. Mus. Novit., no. 1251.

Brodkorb, P. 1971. Origin and evolution of birds. In *Avian biology*, vol. 1:19–55. Academic Press, New York.

Carpenter, G. 1894. Nearctic or Sonoran? *Nat. Sci.*, 5:53–57.

Chaney, R. 1940. Tertiary forests and continental history. *Bull. Geol. Soc. Amer.*, 51:469–488.

Chapman, F. 1923. The distribution of the motmots of the genus Momotus. *Bull. Amer. Mus. Nat. Hist.*, 48:27–59.

Cruickshank, A. 1942. Birds around New York City. *Amer. Mus. Nat. Hist. Handbook Ser.*, No. 13.

Delacour, J., and E. Mayr, 1945. The family Anatidae. *Wilson Bull.*, 57:3–55.

Dunn, E. 1922. A suggestion to zoogeographers. *Science*, 56:336–338.

———1931. The herpetological fauna of the Americas. *Copeia*, 1931:106–119.

Gill, T. 1875. On the geographical distribution of fishes. *Ann. and Mag. Nat. Hist.*, ser. 4, 15:251–255.

Goodwin, D. 1967. *Pigeons and doves of the world.* Brit. Mus. (Nat. Hist.), London.

Heilprin, A. 1883. On the value of the "Nearctic" as one of the primary zoological regions. *Proc. Acad. Nat. Sci.* (Philadelphia), 1883:266–275.

Hubbs, C. 1944. Review of "Studies in the genetics of Drosophila. III." *Amer. Nat.*, 78:270–271.

Lönnberg, E. 1927. Some speculations on the origin of the North American ornithic fauna. *Kungl. Svenska Vetenskapsakad. Handl.*, ser. 3, 4(6):1–24.

Matthew, W. D. 1915. Climate and evolution. *Ann. N.Y. Acad. Sci.*, 24:171–318.

Mayr, E. 1941. *List of New Guinea birds.* Amer. Mus. Nat. Hist., New York.

———1944. The birds of Timor and Sumba. *Bull. Amer. Mus. Nat. Hist.*, 83:123–194.

Peters, J. 1940. *Check-list of birds of the world*, vol. 4. Harvard University Press, Cambridge, Mass.

Reichenow, A. 1888. Die Begrenzung zoogeographischer Regionen vom ornithologischen Standpunkt. *Zool Jahrb., Abtheilung für Systematik, Geographie und Biologie der Thiere*, 3:671–704.

Ripley, S. 1945. The barbets. *Auk*, 62:542–563.

Schmarda, L. 1853. *Die geographische Verbreitung der Thiere*, vol. 2. Vienna.

Sclater, P. 1858. On the general geographical distribution of the members of the class Aves. *Jour. Proc. Linn. Soc. London*, Feb. 1858, pp. 130–145.

Sharpe, R. 1903. *A hand-list of the genera and species of birds*, vol. 4. Brit. Mus. (Nat. Hist.), London.

Simpson, G. G. 1940a. Mammals and land bridges. *Jour. Wash. Acad. Sci.*, 30: 137–163.

——1940b. Antarctica as a faunal migration route. *Proc. 6th Pacific Sci. Congr.* (1939), 2:755–768.

——1943a. Mammals and the nature of continents. *Amer. Jour. Sci.*, 241:1–31.

——1943b. Turtles and the origin of the fauna of Latin America. *Amer. Jour. Sci.*, 241:413–429.

——1947. Holarctic mammalian faunas and continental relationships during the Cenozoic. *Bull. Geol. Soc. Amer.*, 58:613–688.

Stegmann, B. 1932. Die Herkunft der paläarktischen Taiga-Vögel. *Arch f. Naturg,* n.s., 1:355–398.

——1938a. Principes généraux des subdivisions ornithogeographiques de la région paléarctique. In *Faune de l'URSS*, n.s. 19, *Oiseaux*, vol. 1, no. 2.

——1938b. Das Problem der atlantischen Landverbindung in ornithogeographischer Beleuchtung. *Proc. 8th Int. Ornith. Congr.* (1934), 476–500.

Stresemann, E. 1939. Die Vögel von Celebes. Zoogeographie. *Jour. f. Ornith.*, 87: 312–425.

Sushkin, P. 1924. Résumé of the taxonomical results of morphological studies of the Fringillidae and allied groups. *Bull. Brit. Ornith. Club*, 45:36–39.

——1925. A preliminary arrangement of North American genera of Fringillidae and allied groups. *Auk*, 42:259–261.

Szidat, L. 1940. Die Parasitenfauna des Weissen Storches und ihre Beziehungen zu Fragen der Oekologie, Phylogenie und der Urheimat der Störche, *Z. Parasitenkunde*, 11:563–592.

Wallace, A. 1876. *The geographical distribution of animals*, vol. 1. Harper & Bros., New York.

Wetmore, A. 1940. *A check-list of the fossil birds of North America.* Smiths. Misc. Coll., 99, no. 4.

——1944. A new terrestrial vulture from the upper Eocene deposits of Wyoming. *Ann. Carnegie Mus.*, 30:57–69.

Wulff, E. 1943. *An introduction to historical plant geography.* Chronica Botanica Co., Waltham, Mass.

Inferences Concerning the Tertiary American Bird Faunas

North and South America were separated during much of the Mesozoic as well as during almost the entire Tertiary. The size and location of the ocean gap between the two continents varied through geological time. It was apparently largest in the Cretaceous and earliest Tertiary but fluctuated in extent between the early Cretaceous and the mid-Pliocene. The mid-Eocene period was more favorable to north-south colonization than the Oligocene and Miocene (Haffer 1970). At its widest, the gap ranged from Guatemala or Honduras southward and included much of northwestern Colombia. (Childs and Beebe 1963; Darlington 1957; Harrington 1962; Woodring 1954).[1] Various islands served as stepping-stones for transoceanic colonists in the area now occupied by Nicaragua, Costa Rica, and Panama; nevertheless, the isolation between the northern and southern continents was sufficiently drastic to permit the development of different faunas north and south of the gap.

The degree of difference between these two Tertiary faunas depends largely on the dispersal ability of the respective animal groups. On the whole, it is unexpectedly great, when one considers that there was little climatic difference between South America and the southern half of North America, at least during the early Tertiary. At the end of the Eocene, for instance, the tropics of South America were matched by a

Adapted from "Inferences concerning the tertiary American bird faunas," *Proceedings of the National Academy of Sciences*, 4 (1941):197–216.

1. It must be emphasized, opposing suggestions (Darlington 1957) notwithstanding, that there is no evidence in favor of the hypothesis that Central America was—during the Tertiary—a continent separate from North America. But even if it had been, its tropical fauna would not have been restricted, because the area north of Tehuantepec covered by tropical vegetation in the early Tertiary was far larger than Tertiary Central America south of Tehuantepec. Central America, in spite of its current climatic and vegetational distinctness, was not a separate region in the Tertiary, but merely a southern peninsula of North America. The isthmus between North and South America arose apparently as a series of islands that in due time fused with the Central American peninsula and with each other, from the north southward. The last portal to close was apparently that between Panama and Colombia. Panama already had a North American mammalian fauna in the Miocene. The term North America is, therefore, here used for any part of Tertiary North America, including Central America as far south as Honduras.

large tropical zone in North America extending 10-12 degrees farther north than now. Fossil remains of subtropical forests have been found in Tennessee, Missouri, and coastal Oregon. Vegetation classified as subtropical extended at that time at least as far north as the Middle Atlantic States in the east and British Columbia in the west (Chaney 1947; Scott 1954; Dorf 1960).[2] Consequently, the huge area from about Honduras to the Canadian border was available in the early Tertiary for the development of a tropical-subtropical North American fauna.

Our knowledge of the faunas that lived during the Tertiary in the two tropical Americas is scanty. It is only for mammals that we have satisfactory evidence. Simpson (1950) has enumerated the older and more recent elements in the South American mammal fauna (Table 39-1). During the same period of time, North America had a rich evolution of Artiodactyla, Perissodactyla, Carnivora, rodents, insectivores, and other groups of mammals. When the closing of the gap between North and South America was completed at the end of the Pliocene, the two faunas came into direct contact with each other, which resulted in an acceleration of extinction (particularly of South American groups), as well as the spreading of South American elements into North America (armadillo, opossum, porcupine—to name only the surviving forms) and of North American elements into South America (camels, tapirs, carnivores, artiodactyls, etc.). This has been excellently described by Simpson and other authors.

The reason the evolution of the two mammalian faunas, north and south of the gap, can be described in so much detail is the availability of an abundant fossil record. Only scanty fossil remains or none at all are available for most other kinds of organisms, for instance, birds (Wetmore 1956).

Birds create a second difficulty for a zoogeographic analysis through their ability of flight. They can and do cross water gaps far more easily than mammals. Fortunately, different groups of birds differ drastically from each other in this capacity. Most large birds, like herons, ducks, and hawks, pay little attention to water barriers, and a single species may have, even today, a virtually worldwide distribution. Other families, particularly small, forest-inhabiting land birds, may respect zoogeographic barriers extremely well. Such considerations underlie the criteria, stated in Essay 38, by which families can be selected that are suitable for zoogeographic analysis.

Since the need for a knowledge of the Tertiary composition of the

2. The picture is rather complex when studied in detail. Even in early Tertiary some areas were drier and others more humid. Tropical floras extended much further north along the Pacific coast (at least to 49° latitude) than in the Mississippi embayment (37°). The relationship of much of the tropical North American plant element was as much or more with tropical Old World than with South America. Since fruits and seeds are more diagnostic than leaves, but less frequently found, there is still much uncertainty about the floristics of tropical North America in early Tertiary.

Table 39-1. Mammalian colonizations of South America
by North American immigrants.

Across water gaps
1. Late Cretaceous–Paleocene
     Marsupials
     Condylarths and derived ungulates (pyrotheres, astrapotheres, xenungulates,
        litopterns, etc.); possibly several colonizations
     Notoungulates
     Edentates
2. Late Eocene–Oligocene
     Caviomorph rodents
     Primates
3. Late Miocene or early Pliocene
     Carnivores (one group of procyonids)

Over land
4. Late Pliocene and Quaternary
     North American rodents (squirrels, cricetids)
     Carnivores (dogs, bears, weasels, cats)
     Ungulates (peccaries, deer, tapirs, camels, horses)
     Proboscidians
     Lagomorphs (rabbits) (one genus with two species)
     Insectivores (shrews) (one species, marginally)
     (Bats omitted from this tabulation)

avifaunas is considerable, some method must be found to reconstruct distributions in past geological periods, in spite of the two stated handicaps. Such a method, known since the very beginning of the science of biogeography, consists in an evaluation of the present pattern of distribution (in relation to dispersal facilities) and in a study of the distribution of near relatives. Direct proof is impossible by this method, but it allows for inferences with varying degrees of probability.

The object of this paper is the testing of various sets of data for possible inferences on the approximate composition of the bird fauna on the two American land masses during the Tertiary. Excluded from consideration are all families containing fewer than six living species, to guard against the danger of interpreting a relict distribution as indication of a center of origin.

Applying these criteria, we are left with 8 families of non-Passeres and 15 families of Passeres that are available for analysis. Since these families were selected for their inability or unwillingness to cross water barriers, we can justifiably infer that families in which the genera have a prevailingly South American distribution are of South American origin, and families in which the genera have a prevailingly North American distribution are North American in origin. Species of such genera, however, may have actively spread into the other continent after the closing of the Panamanian gap. On the basis of these assumptions one would infer that among the selected non-Passeres, 2 families are northern and 6 families are southern (See Table 39–2). The hummingbirds (Trochilidae), although clearly centering in South America, are known to be facile crossers of water gaps. As a result, they have experienced a secondary radiation in North America, and 15 genera are now restricted to North America, north of Panama.

All but 2 of the 15 passerine families that are suitable for analysis belong to two large assemblages of closely related families, the American mesomyodean Passeres and the American "nine-primaried Oscines." The American mesomyodean Passeres or Suboscines (suborder Tyranni) include 10 families or subfamilies (number of Recent species in parentheses): the Rhinocryptidae (27), Conopophagidae (10), Formicariidae (221), Furnariidae (212), Dendrocolaptidae (47), Tyrannidae (365), Oxyruncidae (1), Pipridae (60), Cotingidae (90), and Phytotomidae (3). All of these are very poor in the crossing of water gaps (see Table 39–2A and Figure 39–1A) except the Tyrannidae (tyrant flycatchers), which have had secondary radiations in North America like the hummingbirds. The South American origin of the Suboscines can hardly be questioned.

The other group consists of the so-called nine-primaried Oscines (not all of them actually have nine primaries). The families and subfamilies at present included in this group are the Vireonidae, Thraupidae, Coerebidae, Parulidae, Cardinalinae, Emberizinae, and Icteridae. The group as a whole seems more nearly related to several groups of Old World birds (Muscicapidae *sensu lato* and particularly Fringillidae *sensu stricto*) than to any strictly South American group. This indicates an original North American home for this group. The secondary radiation of the Emberizinae in the Old World (Eurasia and Africa) strengthens the probability of an originally North American center of the nine-primaried Oscines. In spite of this inferred northern origin of the group, the constituent families cannot necessarily be designated as northern. The Thraupidae (tanagers) clearly have had their center of radiation in South America (Table 39–2), and even the Icteridae and Emberizidae have had at least as much of a radiation in South as in North America. There can be little doubt that most of the nine-primaried Oscines crossed the water gap rather early in the Tertiary, and quite likely repeatedly in both directions.

Secondary radiation is the rule among the older immigrants into South America, as it is among the mammals (Table 39–1). In this respect South America is quite like Australia, which also became a secondary center of radiation for immigrant elements, in this case from Asia (see Mayr 1944).

For the sake of this analysis it might be useful to classify the South American element among the Passeres into four categories: (1) primary or autochthonous South American (p.s.), consisting of the Suboscines (except the Tyrannidae); (2) expanding South American (e.s.), Tyran-

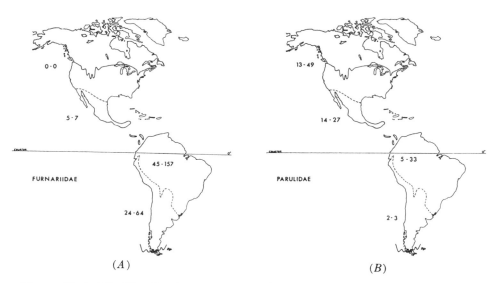

Figure 39-1. (*A*): The distribution pattern of a primarily South American family, the ovenbirds (Furnariidae). (*B*): The distribution pattern of a primarily North American family, the wood warblers (Parulidae). On both maps, the broken lines separate a tropical-subtropical from a temperate climatic region on each continent (see text). On both maps, the first figure gives the number of genera now breeding in the area, the second figure the number of species. The solid lines indicate the estimated border between the subtropical and temperate zones at the end of the Eocene.

nidae; (3) secondarily South American (s.s.), Thraupidae; and (4) pan-American (p.a.), Emberizidae and Icteridae.

The remaining two passerine families (Troglodytidae, Mimidae) are believed to have their nearest relatives in the Muscicapidae (*sensu lato*) assemblage of the Old World. A North American origin is also favored by their distribution pattern, even though some wrens (Troglodytidae) seem to have entered South America prior to the closing of the Pana-manian gap ("expanding North American").

The conclusions of this analysis are well supported by the evidence derived from the current distribution of genera and summarized in Table 39-2. A widely distributed genus that is not clearly either north-ern or southern is recorded as 0.5 under both northern and southern elements, so that the totals agree with the known number of genera.

A related but somewhat independent approach is to count the num-ber of genera and species now living in each of four regions (not in-cluding the area from northwestern Colombia to Nicaragua, which was once largely submerged): (1) North America north of the Mexican border,[3] (2) subtropical–tropical North America from Mexico to Hon-

3. The Mexican border, an admittedly artificial and imprecise line, was chosen deliberately as the border between temperate and tropical-subtropical North America to avoid the bias that is inevitable when one attempts to draw a more precise line between subtropical and temperate Mexico.

Table 39–2. Present distribution of American bird families belonging to different faunal elements (number of genera).

Families	Essentially northern, occurring		Essentially southern, occurring	
	Not south of Panama	Also south of Panama	North beyond Nicaragua	Not north of Nicaragua
A. Primary South American families				
Non-Passeres				
Tinamidae	0	0	2	7
Nyctibiidae	0	0	1	0
Bucconidae	0	0	1	9
Galbulidae	0	0	0	5
Ramphastidae	0	0	3	2
Passeres				
Dendrocolaptidae	0	0	7	6
Furnariidae	0	0	5	54
Formicariidae	0	0	10	43
Rhinocryptidae	0	0	0	10
Conopophagidae	0	0	0	2
Cotingidae	0	0	8	24
Pipridae	0	0	4	17
Total (12 families)	0	0	41	179

B. Old World (O.W.) and primarily North American families

Trogonidae (O.W.)	1	1	1	0
Momotidae	3	1	2	0
Troglodytidae	6	3.5	2.5	3
Mimidae	5	0.5	0.5	1
Parulidae	15	4	2	0
Vireonidae	2	1	3	0
Total (6 families)	32	11	11	4

C. Families with readily colonizing genera

Expanding South American

Trochilidae	15	3.5	7.5	84
Tyrannidae	2	5	20	79

Secondarily South American

Thraupidae	1	1.5	9.5	44

Pan-American

Icteridae	5	5	6	17
Emberizidae	20	5	6	21
Total (5 families)	43	20	49	255

duras, (3) tropical-subtropical South America, and (4) temperate and arid South America (including the higher Andes). The distribution pattern of a typical South American family is shown in Figure 39–1A, that of a North American family in Figure 39–1B. The number of genera and species now living in each of the four regions is recorded in Table 39–3 for 19 families and subfamilies. As one might have predicted, the results tabulated in Tables 39–2A and 39–3A, Tables 39–2B and 39–3B, and Tables 39–2C and 39–3C, respectively, are in fairly close agreement with each other. The inference to be drawn from these tabulations is evident; there appears to be a high degree of probability that the families listed in Tables 39–2A and 39–3A are autochthonous South American elements, while those of Tables 39–2B and 39–3B are basically North American elements.

Still another line of evidence is provided by the radiation of faunal elements into various vegetation zones. Although the richest (in species) bird fauna is found in the tropical rain forest, which had a much wider distribution in the early Tertiary than now, species and species groups of this fauna continued throughout the Tertiary to radiate into subtropical and arid tropical habitats. When the contact between North and South America was established late in the Pliocene, the bridge apparently was highly diversified ecologically. Active volcanism and the Pleistocene fluctuations of sea level provided for a wide distribution of successional plant formations. The fact that members of the camel family and other dry-habitat forms (like horses) were able to cross the isthmus of Panama shows that the isthmus at that time was by no means solidly and continuously covered by rain forest. Tropical forest, however, was far more extensive south of the Panama bridge than north of it. The arid tropical habitats of Central America (and to some extent the subtropical habitats) were better protected against mass invasion of South American elements than the tropical forest and therefore reflect the composition of the Tertiary North American fauna more accurately than the humid tropical. The number of recent South American elements is much smaller in the arid habitats of Central America than in the tropical rain forest (Table 39–4). Conversely, the arid habitats of subtropical and temperate South America have a remarkably high percentage of South American elements.

There are some notable differences between recent and old colonists. The wood warblers (Parulidae) have their center of diversification in North America, where 21 genera and 70 species occur. Only two genera, *Myioborus* and *Basileuterus,* have succeeded in evolving endemic species in South America. Indeed, both genera actively speciated after their ancestors had reached the new continent,[4] *Myioborus* now

4. There is enough uncertainty among the nine-primaried Oscines about the allocation of genera to families to permit the suggestion that the two genera are not true wood warblers, but evolved independently in South America from tanagers (E. Eisenmann, verbal communication). A similar uncertainty concerning allocation is true for many genera among the various families of the nine-primaried Oscines.

Table 39–3. Present distribution of American families in four climatic-vegetational zones.[a]

Families	Temperate N. Amer.	Trop.-subtr. N. Amer.	Trop.-subtr. S. Amer.	Temperate S. Amer.
A. South American families				
Dendrocolaptidae	0-0	7-14	11-43	4-4
Furnariidae (Fig. 39–1A)	0-0	5-7	45-157	24-64
Formicariidae	0-0	9-10	53-222	2-2
Conopophagidae and Rhinocryptidae	0-0	0-0	8-27	6-9
Cotingidae	0-0	9-11	29-84	1-3
Pipridae	0-0	4-4	21-51	0-0
Total	0-0	34-46	167-584	37-82
B. North American families				
Trogonidae (O.W.)	0-0	3-8	2-12	1-1
Momotidae	0-0	5-6	3-3	0-0
Troglodytidae	6-9	10-28	9-38	2-2
Mimidae	4-9	4-11	2-5	1-4
Parulidae (Fig. 39–1B)	13-49	14-27	5-33	2-3
Vireonidae	1-10	6-16	4-22	2-2
Total	24-77	42-96	25-113	8-12
C. Readily colonizing families				
Trochilidae	4-8	25-44	89-231	15-18
Tyrannidae	8-21	30-46	93-259	34-64
Icteridae	9-16	12-28	26-50	11-18
Thraupidae	1-4	12-25	53-175	5-6
Emberizinae and Cardinalinae	25-53	31-60	43-147	26-55
Total	47-102	110-203	304-862	91-161

[a]The four zones are shown in Figure 39–1. The first figure in each column indicates number of genera, the second, number of species.

Table 39–4. Faunal composition of various habitats in Central and South America (in percent).

Habitat	Faunal elements			
	Old World, N. Amer., and expanding northern (Table 39–2B)	Pan-Amer. (Table 39–2C)	Expanding and secondarily S. Amer. (Table 39–2C)	Primarily southern (Table 39–2A)
Humid habitats (tropical rain forest)				
Central America	12.7	14.3	34.9	38.1
Amazonian Forest	9.0	10.5	33.0	47.5
Arid habitats				
Middle America	49.1	15.1	27.3	8.5
Northern South America	29.5	13.6	45.5	11.4
Southern South America	23.1	17.9	7.7	51.3

Table 39–5. Attenuation of a South American family in Central America exemplified by the Formicariidae. (From Griscom 1950.)

	Genera	Species	Species per genus
Total number of taxa in the family	55	233	4.2
Ranging north to western Panama	18	27	1.5
Ranging north to Nicaragua	17	21	1.2
Ranging north to Guatemala	11	11	1.0
Ranging north to Mexico	8	8	1.0

having 11 South American species and *Basileuterus* about 20. Just how long ago the colonization occurred that led to this active speciation is uncertain, but it surely preceded the closing of the Panama gap. Similarly, early invasions of the Trochilidae and Tyrannidae into North America led to an almost explosive speciation of certain genera in the new continent. Early colonizations (in either direction) are found in all pan-American groups, but also in some pan-tropical (see Essay 38) and North American groups.

More recent colonizations, particularly in the late Pleistocene or post-Pleistocene, have led to a very different distribution pattern. Griscom (1950), in a discussion of the birds of Mexico, correctly emphasized the recency of most of the South American elements in Central America. Primarily South American genera have 148 species that penetrate as far north as Mexico and "in only 38 cases has specific change taken place north of the Colombian Andes." He illustrates graphically the steady and steep decrease of South American elements for a typically South American family, the ant birds (Formicariidae) (Table 39–5).

It is thus apparent that there is a considerable difference between the older elements, which often have experienced a secondary radiation in the newly colonized continent, and the newer elements,[5] which mostly have not even changed into different species since arrival from the continent of origin. This finding poses some questions concerning the so-called neotropical element in Central America.

SUMMARY

Present distribution patterns permit certain inferences on the history of American bird families of which there is only an insufficient fossil

5. The complexity of the analysis is indicated by the probability that some tropical North American elements that had presumably colonized South America just prior to or during the establishment of the isthmus were subsequently exterminated in North America as a result of the Pleistocene climatic deterioration, and recolonized Central America in the later Pleistocene and post-Pleistocene as part of the wave of tropical elements that moved northward in the wake of the climatic amelioration.

record. The now-existing bird fauna of the American tropics is a mixture of old North American, old South American, and secondarily South American elements. There is a component of ("pan-American") families which crossed the Tertiary water gap in both directions so freely that they are difficult to assign faunally. South American bird families appear to have been more successful in competing with invading North American elements than were their counterparts among the mammals. In part this may be due to the less effective isolation of the birds during most of the Tertiary and the resulting continuous selection for an ability to compete with northern hemisphere invaders. In part it is due to the specialization of South American elements for life in the humid forest, an ecological zone that was available to a far greater extent in South America than north of Panama.

REFERENCES

Chaney, R. W. 1947. Tertiary centers and migration routes, *Ecol. Monogr.*, 17: 140–148.

Childs, O. E., and B. W. Beebe, eds. 1963. *Mem. Amer. Assoc. Petr. Geol.*, vol. 2.

Darlington, P. J. Jr. 1957. *Zoogeography*. Wiley, New York.

Dorf, E. 1960. Climatic changes of the past and present. *Amer. Sci.* 48:341–364.

Griscom, L. 1950. Distribution and origin of the birds of Mexico. *Bull. Mus. Comp. Zool. Harv.*, 103:341–382.

Haffer, J. 1970. Geologic-climatic history and zoogeographic significance of the Uraba Region in northwestern Colombia. *Caldasia*, 10:603–636.

Harrington, H. J. 1962. Paleogeographic development of South America. *Bull. Amer. Assoc. Petr. Geol.*, 46:1773–1814.

Mayr, E. 1944. The birds of Timor and Sumba. *Bull. Amer. Mus. Nat. Hist.*, 83: 123–194.

Patterson, B., and R. Pascual. 1972. The fossil mammal fauna of South America. In *Evolution, mammals, and southern continents*, ed. A. Keast, pp. 247–309. State University of New York Press, Albany.

Simpson, G. G. 1950. History of the fauna of Latin America. *Amer. Sci.*, 38: 361–389.

Scott, R. A. 1954. Fossil fruits and seeds from the Eocene clarno formation of Oregon, *Palaeontographica*, 96:66–97.

Wetmore, A. 1956. *A check-list of the fossil and prehistoric birds of North America and the West Indies*. Smiths. Misc. Coll., no. 131.

Woodring, W. 1954. Caribbean land and sea through the ages. *Bull. Geol. Soc. Amer.*, 65:719–732.

The Origin and History of the Polynesian Bird Fauna

Elsewhere (Mayr 1941) I have attempted to show that the islands of the Polynesian subregion have a fairly homogeneous avifauna, and that the various islands and archipelagoes can be divided into four districts, each of which has an even more uniform bird life. It is, of course, not satisfactory to stop at this purely descriptive treatment of the bird fauna; the zoogeographer wants to go beyond that, he wants to interpret his findings. He wants to know when and in what manner the birds reached the various islands and whence they came. To answer these questions we must have an accurate picture of the present-day distribution, and fortunately we possess that. No other group of animals is nearly so well known as the birds, and the distributional facts are extensive enough to permit a wealth of conclusions. Following Rensch's advice (1936), I have attempted to separate carefully the purely zoological from the purely geological evidence. Too many of the current zoogeographic hypotheses have been proposed by zoologists who combined their incomplete knowledge of distribution with equally insufficient geological data. It is much sounder to develop a hypothesis exclusively on the basis of data derived from one field and to use the data of the other field only for the ultimate checking.

Of course, even if we should succeed in elucidating fully the distribution of birds, we would have no complete picture of the zoogeography of Polynesia. There are two reasons for this. The first is that the means of dispersal are different in every taxonomic group; the second, that the period of expansion may be different in each group. The reptiles and some of the invertebrates, such as primitive mollusks, had their day in the Mesozoic age, whereas birds and mammals did not reach their peak of development until after the mid-Tertiary. It is therefore possible that some groups may show a distributional pattern different from that of the birds. Nothing definite can be said, however, until the distribution of these other groups is better known.

Adapted from "The origin and the history of the bird fauna of Polynesia," *Proceedings of the Sixth Pacific Science Congress*, 4(1941):197–216.

THE VALUE OF BIRDS IN ZOOGEOGRAPHICAL STUDIES

It is sometimes questioned whether it is admissible to make use of the distribution of birds for zoogeographical studies. Birds are excellent flyers and thus capable of rapid and active spreading. As a matter of fact, it cannot be denied that birds are capable of crossing considerable stretches of the open sea to settle in new territories. There is abundant evidence for this, such as the resettlement of Krakatau Island (Dammerman 1948), the recent arrival of Australian birds on New Zealand (Fleming 1962), and the colonization of unquestionably oceanic islands (Mayr 1931; Mayr and deSchauensee, 1939).

What birds lose as indicators of paleogeographical phenomena by their capacity for active spreading, most other groups of animals lose by their tendency to be passively transported. Wallace pointed out many years ago (1876, 1880) that the distribution of animals can be fully understood only if we know completely their means of dispersal. This is frequently forgotten by those zoogeographers who naively build land bridges wherever they find discontinuous distributions of island forms. Although we still know very little about the methods of dispersal, we are beginning to realize that most animals, except the larger mammals, some amphibia, and true freshwater fishes, have a much greater faculty of being transported passively than was thought until recent years.

The possibility of transport by floats or in logs is not to be underestimated. Many tropical currents have a speed of at least 2 knots, that is, about 50 miles a day, or 1000 miles in 3 weeks. It is probably not a great task for a wood-boring insect to survive 3 weeks in a drifting log. Air currents are, however, of incomparably greater importance than sea currents. Even slight winds are of great influence on the distribution of floating and flying animals, as recent investigation has shown. It is astonishing how rich the "aerial plankton" is, even up to altitudes of 1000 meters and more (Burton 1927–1935:vol. 9; Hardy and Milne 1938; Glick 1939). Normal winds would, of course, not account for the spreading of mollusks, flightless insects, and other small invertebrates. However, most of the islands with which we are concerned are situated within the zone of tropical hurricanes, the lifting force of which is quite extraordinary. Whole houses weighing several tons are sometimes blown away. Small objects are sometimes sucked thousands of feet up, and it is quite credible that as they get into the upper air currents they may be carried with the hurricane for hundreds or even thousands of miles. The fact that there are small mollusks and flightless insects on such typical oceanic islands as Easter Island, Juan Fernandez, and Saint Helena is almost unassailable proof that such a method of dispersal is a reality (Gulick 1932; Murphy 1936). This has convinced most of the doubting Thomases that the Galápagos Islands, the Hawaiian Islands, and most of the Polynesian islands are also oceanic, as had already been assumed by most of the geologists. The results of the recent surveys in the Hawaiian Islands, the Marquesas, and on Samoa indicate that there are indeed very few animals that cannot be transported across considerable stretches of the sea by wind, waves, other animals, or man. The absence of the larger mammals, of true freshwater fishes, and of certain amphibia is therefore rather good evidence that an island is oceanic, provided there is no other very strong evidence to the contrary.

I hope I have brought out in this discussion on the means of dispersal that, in spite of their capacity for active spreading, birds are quite as suitable criteria as most other animals or even more so.

Ornithologists have pointed out that most species of birds, particularly on tropical islands, are extraordinarily sedentary. A comparison of the bird faunas of neighboring islands brings out this point very clearly. Of the 265 species of land birds which are known from that part of New Guinea which is opposite New Britain, only about 80 species have a representative on New Britain. In other words, the 45-mile-wide stretch of water which separates the two islands has prevented the crossing over of 70 percent of the New Guinea species.

Even more conspicuous is a situation in the western Papuan Islands. The straits between the islands of Salawati and Batanta, which are less than two miles wide, have prevented the crossing of 17 full species of Salawati birds to Batanta, and of 5 species from Batanta to Salawati. Furthermore, a considerable number of species is represented on the two islands by different subspecies. Literally hundreds of similar instances could be listed from the distribution of birds in the Indo-Australian archipelagoes, all of them indicating the sedentary habits of their avian inhabitants.

These examples illustrate the point that the majority of species of birds use their ability to fly less for active spreading than to return to their home island in heavy storms during which insects would be blown away for considerable distances.

THE GEOLOGICAL HISTORY OF POLYNESIA

There is hardly another region in the entire world about which so little is known geologically as Polynesia. The zoogeographer who wished to reconstruct the paleography of this region was therefore forced to rely more or less on his imagination, and it is no wonder that no two authors agreed on the number and extent of the land bridges believed necessary to explain the faunal phenomena. The confusion was aggravated through misunderstandings of certain highly important terms, such as "oceanic" and "continental" islands. It is necessary to clarify this terminology before we go any further in the discussion.

Oceanic and Continental Islands

Darwin was the first to direct attention to the importance of oceanic islands. He showed that with very few exceptions all the remoter islands of the great oceans were lacking indigenous mammals or amphibia, and he maintained that none of these islands had ever formed part of a continent. Wallace (1880:234) extended these ideas, and finally arrived at a definition of oceanic islands; they are "islands of volcanic and coralline formation, usually far from continents and always separated from them by very deep sea, entirely without indigenous land mammalia or amphibia, but with abundance of birds and insects, and usually with some reptiles."

This definition has been accepted rather generally and is now in wide use. Actually it is not a very good definition, because it includes a number of qualifications which are not essential for the "oceanic" character of such islands. After all, the term "oceanic" island, as opposed to "continental" island, was created by a zoogeographer, and has primarily a zoogeographical, not a geological, significance. It is time to draw attention to the fact that there is a difference between the geologist's and the zoogeographer's definition of an "oceanic" island. To the geologist it is "an island, usually of volcanic or coralline origin, which is not situated on a continental shelf"; to the zoogeographer it is "an island which received its fauna across the sea," or, to put it negatively, "which did not receive its fauna by way of a land bridge."

The two definitions overlap, but only partly. Curiously enough, nobody seems to have pointed this out before. Every "oceanic" island of the geologist is obviously also an "oceanic" island for the zoogeographer, but the contrary relation is not necessarily true. Countless islands are known which are situated on continental shelves but have never been in complete "dry" connection with the neighboring mainland, at least not during the more recent periods of the earth's history, which alone are of interest to the zoogeographer. Other islands have at one time been in connection with the mainland but have subsequently become submerged under the ocean so that their entire fauna was drowned. After the next emergence they were truly "oceanic" islands to the zoogeographer although still retaining their status as "continental" for the geologist. The confusion about these two totally different kinds of "oceanic" islands has had disastrous results in discussions of the zoogeography of the western Pacific. I hardly need mention that in the subsequent discussion here the term "oceanic island" will be used in its restricted zoogeographical meaning: "an oceanic island is an island which has received its fauna across the sea and not by way of land bridges." The only complication which might occur in connection with this formula is that some islands might be "continental" for their older elements and "oceanic" for their more recent elements. I consider an island continental if it can be shown conclusively that it has received even part of its fauna by way of a land bridge. It is obvious that some of the most "continental" islands, like England, have received part of their more recent fauna across the sea.

The Geology of the Western Pacific

[The conservative interpretation of geology at the time this essay was written was to postulate a considerable stability of the ocean floor. I say "conservative" because this was at the end of a period of the most reckless land-bridge building by F. Sarasin, H. Crampton, and Skottsberg, a movement that had succeeded in recruiting many adherents. It must be pointed out that this section was written more than 25 years before the establishment of plate tectonics and ocean floor spreading. Considering this fact, it is rather remarkable to what extent my conclusions of 1939 are still valid today.]

Among the conflicting statements of the geological literature one concept seems to be generally accepted. The Pacific (between Australia and the Americas) can be divided into an eastern and a western part. The eastern part consists of the "great Pacific basin," one of the old oceanic basins with little change in geological history. The western part consists of the area situated west of a line running from Fiji to New Zealand. The principal feature of this area is a series of 4 or 5 arcs, more or less concentric with Australia, which begin in the north in the New Guinea area, the Solomon Islands, and Fiji, and which can be traced, at least in part, as far as New Caledonia and the New Zealand region. The exact age of these arcs is not known, but we do know that they have seen considerable ups and downs during the Tertiary. There is no evidence in the geological literature that the various arcs were at one time connected with one another, nor is there any evidence that at any period the entire arcs had completely emerged above the ocean. On the contrary, many of the existing facts speak in favor of the idea that parts of these arcs were always submerged and others exposed as islands. This brings us to a discussion of the land bridges.

Land Bridges

There was a period early in this century when most zoogeographers were busy manufacturing land bridges whenever they found it convenient to explain certain difficulties of faunal distribution. It is, of course, fascinating to be able to tell the uninitiated: "Here are two islands with similarities in their faunas; consequently they must once have been connected! (Isn't science wonderful?)." These efforts culminated, so far as the Polynesian islands are concerned, in the work of F. Sarasin (1925), who constructed a whole network of land bridges and raised and lowered the sea level by 2 and 3 thousand fathoms in quick succession. But even much more conservative authors have always maintained the continental nature of the faunas of New Caledonia, the New

Hebrides, Fiji, and other neighboring groups. I myself grew up in this belief and was rather surprised when my faunistic studies did not support the contention. They showed, rather, that even the just-mentioned islands are "oceanic." The ornithological evidence will be given in a subsequent section, and I shall content myself at present with a few more theoretical considerations that speak against the existence of the postulated land bridges.

1. According to geological theory, land connections must have been along the arcs, if they existed at all. Zoogeography, however, shows that there is very little relationship between islands situated on the same arcs, for example, Fiji and Tonga with the Kermadec Islands and New Zealand, or the Louisiade Archipelago and New Caledonia.

2. All arcs are concentric with Australia, they run more or less from the northwest to the southeast, and they are separated by deep-sea troughs. The postulated land bridges of the land-bridge builders, however, run from Fiji to the Solomon Islands, or from New Caledonia to Australia, in other words from east to west, radially to the arcs. Such connections are very unlikely from the geologist's point of view.

3. The depths between the islands that have to be connected by land bridges are very considerable, which means that the entire area would have to be raised and the resulting land masses would have to be of very great size. This actually is the usual conclusion. However, the islands which are the leftovers of these hypothetical land masses differ in so many of their elements that separate (and very selective) land bridges would have to be postulated for each island. This is clearly an impossibility.

The means of dispersal of most plants and animals are much more extensive than was formerly realized, and even rather irregular distributions can be explained without the help of land bridges. Dispersal across the sea is, of course, most obvious for birds, and ornithologists were among the first to accept the ideas of the permanency of continents and oceans. Most entomologists are also beginning to realize that they can solve most of their distributional difficulties without land bridges. Many conchologists, however, still postulate continental connections between all or nearly all the islands where land shells exist. It seems to me that the wide acceptance of land bridges by conchologists is chiefly due to three reasons: (1) our almost complete ignorance of the means of dispersal of snails, (2) our lack of knowledge of the speed of speciation in snails, and (3) faulty classification, particularly generic classification. A. Gulick (1932) has already directed attention to the presence of snails on most oceanic islands. They were unquestionably carried there by some unknown means of transportation. Occasionally we must accept this even for larger snails. If one (or several) species of the large snail *Placostylus* are found in northern New Zealand, I would not, as Hedley did, create a continent embracing all the areas where the genus *Placostylus* is found (New Zealand, New Caledonia, New Hebrides,

Solomon Islands, and eastern New Guinea), because the acceptance of such a land mass is contrary to all the other evidence. To me it seems incomparably simpler to assume a still unknown method of transportation than a land bridge that is unsupported by any other fact.

THE COMPOSITION OF THE POLYNESIAN BIRD FAUNA

[Pp. 201 to 212 of the original essay contained a detailed analysis of the species of birds found in the four major divisions of the Polynesian subregion: (1) Micronesia (Palau, Marianas, Carolines), (2) central Polynesia (Fiji, Tonga, Samoa, and adjacent islands), (3) eastern Polynesia (Austral Islands, Tuamotus, Society Islands, Marquesas, etc.), and (4) southern Melanesia (New Caledonia, Loyalty Islands, New Hebrides, and Santa Cruz group). The major conclusion of my analysis was that the pattern of distribution was in each case consistent with cross-oceanic colonization, but clearly incompatible with a colonization on land bridges. The evidence for this conclusion is too specialized to be presented in full. To illustrate the nature of my argumentation I shall reproduce here part of my analysis of the composition of the central Polynesian bird fauna.]

The islands I group in central Polynesia form the core of Oceania, and their fauna differs by several characters from those of the other subdivisions of Polynesia. First of all, it is a comparatively rich fauna, considering the distance from the nearest continent, and, second, there is very little "pollution" by foreign elements, such as are found in Micronesia (Palearctic and Philippine elements) and southern Melanesia (Australian and Papuan elements).

Within this district we have to distinguish between two types of islands, quite different in their geological history and their ecology. One type comprises the old, big, and mountainous islands in the Fijian and Samoan groups, and the other, all remaining islands, mostly coralline. To the first belong, in the Fiji group, Kandavu, Viti Levu, Ovalau, Vanua Levu, and Taviuni, and possibly also Ngau and the Koro Islands; and in the Samoa group, the two islands Upolu and Savaii. These are the old islands of central Polynesia and the only ones that have a notable endemic fauna. All the other islands have a much impoverished fauna and lack distinctive endemic elements. The islands of Fiji and Samoa have what is generally considered the most typically Polynesian fauna. Elements that have developed in that region have subsequently spread into Micronesia, eastern Polynesia, and southern Melanesia. It is therefore very important to make a more detailed analysis of the birds of this region. The results of this analysis are shown in the list below.

This list does not include widespread species, such as *Anas superciliosa, Rallus philippensis, Porzana tabuensis, Porphyrio porphyrio, Ducula pacifica, Tyto alba, Petroica multicolor, Turdus poliocephalus,* and *Aplonis tabuensis,* all of which occur on both Fiji and Samoa, as well as on a great many other islands and archipelagos.

The following species and subspecies of birds occur on the large islands of Fiji (excluding Kandavu) and Samoa.

Fiji (Viti Levu, Ovalau, Vanua Levu, Taviuni)	Samoa (Upolu, Savaii)
Butorides striatus subsp.	—
Accipiter rufitorques	—
Falco peregrinus ernesti	—
Nesoclopeus pœcilopterus	—
—	*Pareudiastes pacificus*
Ptilinopus porphyraceus porphyraceus	*Ptil. porph. fasciatus*
Ptilinopus perousii mariae	*Ptil. perousii perousii*
Chrysœna victor, viridis, etc.	—
Ducula latrans	—
Gallicolumba stairii subsp.	*Gallicolumba stairii* subsp.
Columba vitiensis vitiensis	*Columba vit. castaneiceps*
—	*Didunculus strigirostris*
Phigys solitarius	—
—	*Vini australis*
Charmosyna aureicincta	—
Prosopeia tabuensis	—
Prosopeia personata	—
Tyto longimembris oustaleti	—
Cacomantis pyrrhoph. simus	—
Halcyon chloris	*Halcyon (Todirhamphus) recurvirostris*
Lalage maculosa woodi	*Lalage maculosa maculosa*
—	*Lalage sharpei*
Myiagra vanikorensis	*Myiagra albiventris*
Myiagra (Lophomyiagra) azureicapilla	—
Rhipidura spilodera	*Rhipidura nebulosa*
Lamprolia victoriæ	—
Vitia ruficapilla	—
Ortygocichla (Trichocichla) rufa	—
Clytorhynchus nigrogularis	—
Clytorhynchus vitiensis	—(*powelli* on Manua Is.)
Mayrornis lessoni	—
Pachycephala pectoralis (*graeffei*-group)	*Pachycephala flavifrons*
Artamus leuc. mentalis	—
—	*Aplonis atrifusca*
Amoromyza viridis	*Amoromyza samoensis*
Foulehaio car. procerior	*Foulehaio car. carunculata*
—	*Myzomela cardinalis nigriventris*
Myzomela jugularis	—
Zosterops explorator	—
—	*Zosterops samoensis*
Zosterops lateralis flaviceps	—
Erythrura cyanov. pealei	*Erythrura cyanov. cyanovirens*
Erythrura (Rhamphostr.) kleinschmidti	—

It is possible to test various zoogeographical working hypotheses with the help of the material presented in this list. The following are the three principal hypotheses.

1. "Fiji, Tonga, and Samoa are remainders of a large land mass; the faunas of the three archipelagos are therefore essentially the same or at least quite similar."

The facts do not conform to this hypothesis. There are 47 species listed in the tabulation. Of these, only 7 species (15 percent) are found on both island groups, but always (*Gallicolumba stairii* being a doubtful exception) in different forms. Five representative pairs of species (*Halcyon, Myiagra, Rhipidura, Pachycephala,* and *Amoromyza*) are found on both islands (10 species = 21 percent). The remaining 30 species (= 64 percent) are restricted either to Fiji or to Samoa. The element common to both groups (except for the above-mentioned widespread species) comprises only 36 percent of the tabulated species.

Of the 7 species found on Samoa, but not on the main islands of Fiji, 5 are endemic on Samoa (*Pareudiastes* pacificus, *Didunculus* strigirostris, Lalage sharpei, Aplonis atrifusca,* and *Zosterops samoensis*). Of the 23 species found in Fiji, but not on the main islands of Samoa, not less than 16 are endemic on Fiji: *Accipiter rufitorques, Nesoclopeus poecilopterus, Chrysœnas* victor* (etc.), *Ducula latrans, Phigys* solitarius, Charmosyna aureicincta, Prosopeia* tabuensis, Prosopeia* personata, Myiagra† azureicapilla, Lamprolia* victoriœ, Vitia ruficapilla, Ortygocichla† rufa, Mayrornis lessoni, Myzomela jugularis, Zosterops explorator,* and *Erythrura† kleinschmidti.* (Asterisk denotes endemic genera; dagger, endemic subgenera.) The other 7 species have a more or less wide distribution outside the Fiji group.

These figures show clearly that the bird faunas of Fiji and Samoa are very different, at least so far as species and genera are concerned. The Samoa group, consisting of relatively small islands, has only a rather poor endemic bird fauna, the only outstanding member of it being the toothed pigeon, *Didunculus*. A comparison of the floras of Fiji and Samoa leads to exactly the same conclusions, as does an analysis of the insects. I fully agree with Buxton (1927–1935:vol. 9:26), who says: "I fail to find any evidence supporting the view which has been enunciated that Fiji, Tonga, and Samoa are the remains of an old land mass."

2. "Samoa is an oceanic island; Fiji, on the other hand, is continental, having received a considerable part of its fauna by way of a land bridge."

I have already pointed out that this hypothesis is not supported by any geological evidence. But what does the faunal analysis show? If Fiji is continental, it must have some element which reached it by land bridge and which could not reach the oceanic Samoa. To find this element we must deduct from the list of Fijian birds, first, all the widespread Polynesian species, second, all the species that occur jointly on Fiji and Samoa, and third, all those that have arrived rather recently from Australia by way of the New Hebrides. If we analyze the remaining species, we find that they all belong to families (such as rails, pigeons, parrots, etc.) of genera that are notorious for their colonizing faculties. The only apparent exceptions are three undergrowth dwellers, *Lamprolia, Vitia,* and *Ortygocichla (Trichocichla).* However, *Vitia* has also settled on the easternmost of the Solomon Islands, and the genus *Ortygocichla* has a notoriously spotty range, the only other species being found on New Britain. This leaves the aberrant genus *Lamprolia,* of unknown relationship, possibly a remnant of a more widely distributed group. The same basic agreement in regard to family relationships is found in other taxonomic groups. Berland (1934) writes, for example: "If one examines the list of species of spiders known from Fiji, one notices a very marked parallelism with that of Samoa: the same families predominate, and the same ones are absent. There are only two remarkable exceptions." The same absence jointly from Fiji and Samoa of widespread Papuan genera and families is, incidentally, also a marked phenomenon in birds.

All this evidence indicates clearly that both Fiji and Samoa are oceanic islands that were populated by the same elements, which possessed unusual colonizing abilities. The Fiji group, consisting of much larger islands and being much closer to the Papuan region, received, of course, a considerably larger number of immigrants.

Those who believe in the continental character of Fiji usually advance as their last and most convincing argument that there are frogs on Fiji, which could not possibly have arrived there except on a land bridge. However, this argument is not valid. Dr. G. K. Noble (personal communication, July 19, 1939), the well-known specialist on Amphibia, kindly gave me the following information on the two Fijian genera of frogs:

> I place both genera in the subfamily Cornuferinae of the family Ranidae. Both genera seem to have arisen directly from *Rana,* which is widely distributed in the Northern Hemisphere and invades the Southern Hemisphere at various points. In other words, *Cornufer* and *Platymantis* are of very recent origin. The fact that both genera are in Fiji means merely that they have been successful in crossing over the water. A close parallel is found in Madagascar. Most of the Madagascar fauna is distinctive, but there are three or four ranids there that are closely allied or identical to the African ranids which have invaded the area. I assume this to mean that they arrived by flotsam-jetsam methods or perhaps were transported by native boats.
>
> All Fijian frogs are closely allied to Solomon Island frogs. The distribution of the genera is as follows: *Cornufer* is found in Burma, the Philippines, Borneo, and the Solomon and Fiji Islands. *Platymantis* is found in the Philippines, Halmahera, Kei Islands, New Guinea, New Britain, the Solomons, and Fiji Island.

I agree with you thoroughly that, on the basis of these frogs, we cannot assume that Fiji is continental.

3. "Both Fiji and Samoa are zoogeographically oceanic islands that derived their faunas across the sea."

This is, in my opinion, the most plausible interpretation of the distributional phenomena in central Polynesia. The reason why it is not more generally accepted is threefold. First, the means of dispersal of various animal groups are not fully appreciated; second, the geology of Fiji has been misunderstood; and third, the differences in the faunas of Fiji and Samoa have been interpreted as being differences of kind instead of degree.

The question of origin of the Fijian bird life is not particularly involved. The Malayan Region has not participated directly. The Papuan Region is by far the largest contributor, either via the Solomon Islands, or via Santa Cruz Island or the New Hebrides. All the endemic elements belong to this fauna. There is a weak but significant Australian element consisting of 4 open country or grassland species: *Tyto longimembris, Cacomantis pyrrhophanus, Artamus leucorhynchos,* and *Zosterops lateralis,* and of two species more partial to forest, *Myiagra vanikorensis* and *Petroica multicolor.* It is obvious that these forms are recent immigrants, because all of them, with the exception of *Myiagra,* are only slightly different from their next Australian relatives. Most of these species (except *Cacomantis* and *Petroica*) are absent from the Solomon Islands, but all of them (except *Tyto*) are found in the New Hebrides. It is obvious that these recent immigrants from Australia had access to Fiji only after the New Hebrides had emerged from the sea to serve as a stepping-stone for transoceanic settlers. The only species of this group to reach Samoa is *Petroica multicolor,* one of the forest dwellers. Lack of proper habitat on Samoa, more than the distance, prevented colonization of the grassland and open country species.

SOME GENERAL CONSIDERATIONS

Derivation of the Polynesian Avifauna

Aside from a small Palearctic element in Micronesia, and apart from a not very strong Australian element in southern Melanesia and central Polynesia, all the Polynesian forms are derived from the west. This is essentially the same conclusion reached by workers in other taxonomic groups. Buxton (1927–1935:vol. 9:90) says, for example: "It is clear, and generally admitted, that the fauna of Samoa is essentially Indo-Malayan, and that the fauna of Oceania, or at least the greater part of it, has spread out from the west, passing through Melanesia and Fiji into Polynesia, and becoming poorer with the passage of each area of sea." Chopard, a specialist of Orthoptera, also calls the Polynesian fauna "une faune malaise appauvrie à l'extrême."

Although there is complete agreement concerning the western origin of the Polynesian fauna, it would be quite incorrect, as far as birds are concerned, to call the Polynesian fauna "a Malayan fauna." The two faunas hardly have an element in common. The Polynesian avifauna derived from the New Guinea region, a very important evolutionary center for birds.

One more point must be discussed in this connection, namely, the question of a "Polynesian fauna." In an analysis of the flora of New Guinea, botanists sometimes distinguish a Malayan, an Australian, and a "Polynesian" element (e.g., Lam 1934). To call this last-mentioned element "Polynesian" may be historically correct, because a great majority of the genera and species in question were discovered in the Polynesian Islands before they were found in the Papuan region. However, if the conditions in the distribution of the plants correspond to those in the birds, it is biogeographically incorrect to call this element "Polynesian." Actually it is of "Papuan" origin.

To prove this for birds we might look at the six genera of birds that have evolved in Polynesia more than 20 forms: *Zosterops* (33), *Aplonis* (31), *Halcyon* (28), *Acrocephalus* (25), *Pachycephala* (25), *and Ptilinopus* (23).

Zosterops. This is a widespread genus, extending from Africa to central Polynesia. Endemic species of the related genera of this family are found in Malaysia (3), Lesser Sunda Islands (4), Philippines (2), Celebes (1), Moluccas (3), Palau (1), Ponapé (1), Santa Cruz (1), and Rennell Island (1). All the forms of *Zosterops* found in southern Melanesia and central Polynesia are, apparently, relatively recent arrivals. The genus has not reached eastern Polynesia. Obviously, Polynesia cannot be the center of origin.

Aplonis (+ relatives). This group is found from Malaysia as far east as the Cook Islands. Most of the Polynesian forms belong to one superspecies ("*tabuensis*"), but endemic species have developed on Samoa (*atrifusca*), Santo, New Hebrides (*santovestris*), on Kusaie (*corvina*), and on Ponapé (*pelzelni*). The New Guinea area has four species (*cantoroides, mysolensis, metallica,* and *mystacea*), and all the related genera are found in the area of New Guinea and the Indo-Malayan Archipelago. Polynesia cannot be considered the center of origin of this group.

Halcyon (28). All the Polynesian forms (with the exception of the rather recent *farquhari*) are geographical representatives of the widespread *Halcyon chloris*. *Halcyon* is the only genus of kingfishers in Polynesia, while in the Papuan area a great number of related genera occur, in addition to several species of *Halcyon*. Clearly again, the center of origin is not Polynesia.

Acrocephalus (= *Conopoderas*). This is a recent arrival from the Palearctic region.

Pachycephala. All the Polynesian *Pachycephala* belong to the *Pachy-*

cephala pectoralis group, while on New Guinea there are not less than 11 different specific groups, in addition to 4 related genera.

Ptilinopus. All the Polynesian fruit doves belong to two ancestral lines (formerly subgenera), both of which are also represented on New Guinea, in addition to four additional subgenera.

These six examples should be enough to illustrate my point. I have not been able to find a single widely distributed genus for which a primary Polynesian center of origin could be assumed. This statement does not include the endemic genera, most of which are, however, monotypic.

Colonizing Powers

Considering the haphazard manner by which these oceanic islands receive their populations, it is rather astonishing how similar the faunas of the various islands are. Berland, on the basis of the distribution of the spiders, has come to the conclusion that the fauna of all Polynesia is so uniform as to suggest that these islands are but the fragments of a single land mass. This view is similar to Pilsbry's (1939), founded on Mollusca. Actually, this paradox of the similarity of the faunas of oceanic islands is solved in a quite different manner. Of all the possible families, genera, and species of the Papuan region that are theoretically in a position to colonize, only a small fraction will actually avail themselves of the opportunity. The faculty to reproduce after the completion of the long colonization voyage, and other ecological factors, are of greater importance than actual distances (see Essay 45). The Hawaiian honeyeaters *Chaetoptila* and *Moho* have their next relative in *Amoromyza* of central Polynesia (Fiji-Samoa) and probably conquered a distance of 2500 miles at the time they colonized Hawaii. The same is true for the Hawaiian flycatcher *Chasiempis,* which is related to the Polynesian genera *Pomarea* and *Mayrornis.* There is irrefutable evidence for any number of colonizing flights of 1500 miles and more. However, this capability of making successful flights across the sea is utterly lacking in other families. Not a single bird of paradise has crossed the narrow straits (45 miles) between New Guinea and New Britain. Any number of genera, for example *Poecilodryas* and *Sericornis,* have been unable to make the short jump from the Vogelkop or from Japen Island to Biak Island. The open sea is like a filter which lets pass only certain specially adapted elements, and that, in my opinion, is the reason for the relative uniformity of the animal life of Polynesia.

There is good evidence that many species of birds pass through periods of active expansion but lose this faculty again at later periods of their evolutionary history. I have discussed this problem in connection with the spreading of the serin finch (*Serinus serinus*) (Mayr 1926). There is a good deal of evidence that such expansive periods have also occurred in South Sea Island species. The Polynesian ranges of *Acrocephalus, Turdus,* and *Petroica* offer good illustrations. Others are the wide distribution, even on small and very recent coral islets, of such genera as *Aplonis, Myzomela, Lalage, Clytorhynchus, Foulehaio*, etc., not to mention such notoriously vagrant fruit eaters as *Ducula pacifica* and the *Ptilinopus* group. It is in these genera that we find most of the double invasions of one island (see Mayr 1933).

Factors Controlling Dispersal

The composition of the bird fauna of the Polynesian islands suggests that dispersal depends primarily upon the following factors.

1. *The age of the island.* The oldest islands have, if all other conditions are the same, the richest faunas. Evidence supporting this view has been quoted repeatedly (Fiji Islands, New Hebrides), but it will not be possible to work out a more detailed correlation until the geological evidence on the age of the islands is more complete.

2. *The size of the island (and the number of available habitats).* New Caledonia, the largest of the islands included by me in Polynesia, has the largest number of species. Viti Levu, the second largest island, has the second largest number. In the New Hebrides the island of Espiritu Santo has more species than all the other islands of this group, which is probably due to a combination of age and area. There is a possibility that size is even more important than age. Daly (1924) holds that Tutuila is the oldest of the Samoan Islands, Upolu intermediate, and Savaii the most recent. On the basis of this age gradation one would expect an equivalent correlation of the number of species. Actually just the reverse order is true. Savaii has the highest number, Upolu has a few less, and Tutuila has a much poorer fauna. This latter gradient, however, runs exactly parallel to the size of the islands. Savaii has 730 square miles, Upolu 430, and Tutuila only 54. There are probably three reasons for this importance of size. First, a large island is more easily reached by a trans-oceanic colonizer than a small one; second, a large island usually has more habitats and ecological niches than a small one; and third, the larger the island, the larger the effective breeding population and the smaller the vulnerability to factors causing extermination (both genetic and ecological factors).

3. *The distance from the nearest land mass.* The gradual diminution in the number of species on islands as one moves farther and farther away from continents has been stressed by many authors. One aspect, however, is frequently neglected. One can classify the inhabitants of oceanic islands into colonizers and strays, although there is, of course, not a sharp line between the two. The colonizers comprise those species which often go deliberately in search of new territories (balloon spiders, aphids, many fruit-eating birds, etc.), and which spread rapidly from island to island once they reach a new archipelago. In contradistinction to these aggressive species are the strays, which apparently are passively transported and show little inclination to spread once they reach their new destination. They are often restricted to that particular island of an archipelago which is closest to the nearest land mass.

A few examples may illustrate this point. I have already mentioned above that several Papuan species reach the Loyalty Islands from the New Hebrides, but do not touch New Caledonia. In the Fiji Islands we have a similar situation with respect to Kandavu. A number of species blown to that exposed island from somewhere in the west or southwest survived there but did not succeed in spreading thence to the rest of the Fiji group (*Rhipidura personata, Meliphacator,* etc.). The Palearctic elements in the Marianne Islands, and the Philippine (viz. Papuan) elements in the Palau group, which failed to spread beyond, belong also to this group. In the Solomon Islands we have particularly interesting conditions on San Cristobal Island. There are probably half a dozen species that were blown to that island by the southeast trade wind from somewhere in southern Melanesia or central Polynesia, but that have not succeeded in going any farther (*Rhipidura flabellifera, Lalage naevia, Rallus philippensis,* probably also *Vitia parens,* and *Edithornis sylvestris,* etc.). These strays are particularly characteristic of marginal areas and do not form a significant portion of eastern Polynesia.

4. *Climatic conditions.* Although the Polynesian Islands are geographically accessible to immigrants from the north (Palearctic), the west (Malayan, Papuan, and Australian regions), and from the south (New Zealand), most of these sources have not materialized. The reason there are no New Zealand elements and very few from either the Palearctic or Australia is primarily a climatic one. Nearly all Polynesia is within the tropics or just barely reaching into the subtropical zone. Most of the regions just mentioned are, however, in the temperate zone or on the border of subtropical and temperate zones. That the presence or absence of grassland has influenced the distribution of some forms has already been mentioned.

Extinction. Not only have the small and usually rather isolated islands of Polynesia not been new centers of evolution, like the Galápagos or Hawaiian Islands, but there is good evidence that many of them are "traps." Species that reach these islands are doomed to extinction. This highly interesting problem has never received the attention it deserves, and this is not the place to go into all the detail. The fact that island birds are very vulnerable is generally known among ornithologists. Of all the birds that have become extinct during historical times more than 90 percent have been island species. It has been rather generally assumed that this extermination was due to human interference, such as destruction of habitat or introduction of new predators. This is unquestionably true for some, but it is not the whole answer, and for some species the correlation is not at all established.

We must not lose sight of the fact that the extinction process must have been going on for rather long geological periods. Let us examine just the island of New Caledonia. It has been dry land ever since the Oligocene. In its early days the island may not have had as rich a fauna and flora as it has now, but surely it should have been able to support 20 to 30 species, compared to the 64 species of today. Actually, there is only 1 bird left on New Caledonia, the Kagu (*Rhynochetos*), the immigration of which we might date as far back as the Oligocene. The other 4 endemic genera of New Caledonia are not sufficiently distinct from their next relatives to give them an age that goes back much farther than Pliocene or late Miocene. In other words, only 1 of the 20 or 30 species of New Caledonia's first bird fauna survives, and only 5 of the 30 to 40 species of the second fauna.

The recent considerations of Sewall Wright have given us a possible key to this curious phenomenon. Apparently, in these isolated populations there is more gene loss than gene mutation. The species are therefore adjusted to an exceedingly narrow limit of environmental conditions. They are unable to respond to any major change of conditions and must die if such a change occurs, or are crowded out if new competitors arrive.

[My thesis that the size of an island fauna is the result of a balance between colonization and extinction was long ignored, but is now accepted, after it was transformed into graphs and mathematical formulae. It is the basic thesis of MacArthur and Wilson's *Island Biogeography* (1967).]

REFERENCES

Berland, L. 1934. Les araignées du Pacifique. *Mem. Soc. Biogeogr.*, 4:154–180.

Buxton, P. A. 1927–1935. *Insects of Samoa*. Brit. Mus. (Nat. Hist.), London.

Chasen, F. N. 1937. On a collection of birds from the Krakatau group of islands, Sunda Strait. *Treubia*, 16:245–259.

Dammerman, K. W. 1948. The fauna of Krakatau. *Verhandel Koninkl. Ned. Akad. Wetenschap. Afdel. Naturk.*, sec. 2, 44:1–594.

Fleming, C. A. 1962. History of the New Zealand land bird fauna. *Notornis*, 9:270–274.

Glick, P. A. 1939. *The distribution of insects, spiders and mites in the air*. U.S. Dept. Agr. Tech. Bull., no. 673.

Gulick, A. 1932. Biological peculiarities of oceanic islands. *Quart. Rev. Biol.*, 7:415.

Hardy, A. C., and P. S. Milne. 1938. Studies in the distribution of insects by aerial currents. *Jour. Anim. Ecol.*, 7:199–229.

Lam, H., 1934. Materials towards the study of the flora of the island of New Guinea. *Blumea*, 1:135, 141–144.

MacArthur, R. H., and E. O. Wilson. 1967. *The theory of island biogeography*. Princeton University Press, Princeton, New Jersey.

Mayr, E. 1926. Die Ausbreitung des Girlitz (*Serinus canaria serinus* L.). *Jour. f. Ornith.*, 4:571–671.

—— 1931. The relationship and origin of the birds of Rennell Island. *Amer. Mus. Novit.*, 488:7–11.

—— 1933. Die Vogelwelt Polynesiens. *Mitt. Zool. Mus. Berlin*, 19:322–323.

—— 1941. Borders and subdivisions of the Polynesian region. *Proc. 6th Pacific Sci. Congr.*, 4:191–195.

Mayr, E., and R. M. de Schauensee. 1939. The birds of the island of Biak. *Proc. Acad. Nat. Sci. Phila.*, 91:1–37.

Murphy, R. C. 1936. *Oceanic birds of South America.* Amer. Mus. Nat. Hist., New York.

Pilsbry, H. 1939. *Land mollusca of North America (north of Mexico).* George W. Carpenter Fund, Philadelphia.

Rensch, B. 1936. Geschichte des Sundabogens. Borntraeger, Berlin.

Sarasin, F. 1925. Über die Tiergeschichte der Länder des Südwestlichen Pazifischen Ozeans auf Grund von Forschungen in Neu-Caledonien und auf den Loyalty-Inseln. *Nova Caledonia,* 4:160.

Stresemann, E. 1931. Aves. In *Handbuch der Zool.,* ed. W. Kükenthal, vol. 7, pt. 2, pp. 640–642. Walter de Gruyter, Berlin and Leipzig.

Wallace, A. R. 1876. *Geographical distribution of animals.* Harper & Bros., New York.

—— 1880. *Island life.* Macmillan, London.

41
Land Bridges and Dispersal Faculties

The zoogeographers of island regions belong to two rather well-defined schools. The one I like to call (with a bit of malice) the "naive" school takes it for granted that two islands or other disconnected land masses must have been united previously by a land bridge if parts of their faunas of flightless animals are identical or closely related. This concept provided a great stimulus to faunistic research, since it implied that a careful analysis of faunal differences and relationships would permit the reconstruction of paleogeographic maps showing the distribution of land and sea in former ages. Conversely, it was assumed that a knowledge of former land connections, as provided by geology, would explain the phenomena of present distribution.

In recent years an ever-increasing body of facts has accumulated proving that this simple hypothesis is not correct. In this, geologists and biologists have cooperated. Land bridges are now readily admitted only for such islands as Britain, the Greater Sunda Islands, the Aru Islands, and Formosa, which prove by their well-balanced faunas that they once must have been part of a continent. The zoogeographical evidence is well supported by geological data in all the cases mentioned, since all these islands are situated on continental shelves. However, the existence of former land bridges must be questioned in the case of all islands with an impoverished or unbalanced fauna. These new findings and interpretations effected a considerable change of opinion among zoogeographers; in fact they resulted in the development of a new school. The proponents of this school of zoogeography point out that the dispersal faculties of island animals have been vastly underrated in the past (Gulick 1932; Darlington 1938; Mayr 1941; Zimmerman 1942). This is true not only of flying animals, like bats, birds, and most insects, but of nearly all groups. There is ample evidence that even Anura and true freshwater fishes can occasionally overcome ocean gaps. It has been found that even strictly oceanic islands in the Pacific may have rather

Excerpted from pp. 181–186 of "The birds of Timor and Sumba," *Bulletin of the American Museum of Natural History*, 83(1944):127–194; copyright 1944 by *The Bulletin of The American Museum of Natural History*.

rich faunas of snails, flightless insects, lizards, and other animals that were formerly believed to be able to spread only with the help of land bridges. The geological evidence against land bridges is so unequivocal and convincing in these instances that no doubt remains that these species must have been deposited on these islands by hurricanes or marine currents or by any other means except land bridges. If gaps of 2000 and 3000 kilometers can be spanned, how much more potent must be the factor of passive dispersal where gaps of only 20 or 100 kilometers are involved! The importance of rafts for this dispersal has been stressed by Matthew (1915), M. A. Smith (1943), and others. Dispersal through the air seems to be even more important, particularly for those organisms and their eggs which are unable to withstand long exposure to salt water.

Birds, bats, and Lepidoptera, being flying animals, have always been credited with a considerable ability to cross water barriers. However, it has been pointed out by Stresemann (1927–1934, 1939), by Rensch (1936), by Mayr (1941; see Essay 40), and others that birds, for example, are as useful as zoogeographic indicators as are strictly terrestrial groups. The most successful colonizers among the beetle fauna of eastern Polynesia are small flightless species that breed in dead twigs. Some of the small species of land snails seem to be particularly adapted to interisland transport. The factors that promote such colonization, like rafts and hurricanes, have been discussed by so many recent authors (Gulick 1932; Darlington 1938; Zimmerman 1942) that nothing more needs to be said.

Why certain species and genera are so successful in overcoming water barriers and others are not is a point about which we require considerably more information. The relative dispersal faculty of various bird species and genera has been discussed by me repeatedly (Mayr and de Schauensee 1939; Mayr 1941; see Essay 40). Nearly all the species that are found in the Malay Archipelago in the island belt between Sunda and Sahul shelves are good colonizers. Parrots, pigeons, honeyeaters, starlings, and white-eyes have been particularly successful colonizers in Polynesia, and the reason seems to be that they all travel occasionally in flocks. If such a flock is blown out to sea, it has an infinitely better chance of establishing itself on a previously unoccupied island than a single individual. Woodpeckers, on the other hand, are among the most solitary of the birds. This may account for the fact that they have made so little progress in colonizing the Malay Archipelago. They got into the Philippines, two species reached Celebes, and one the Lesser Sunda Islands (east to Alor). This distribution compares with 18 species on Borneo and more than 20 on Sumatra.

Mammals are, on the whole, poor colonizers of islands. Simpson (1940) points out how few successful colonizations of Madagascar and the West Indian Islands have been made. The very impoverished mammal fauna of Celebes and of the Lesser Sunda Islands shows that the same is true of the Malay Archipelago. The factors that control the distribution of a particular genus are often rather complex. The marsupials of the genus *Phalanger* have unusually well-developed dispersal faculties. They reached the Solomon Islands, the Admiralty Islands, the Timor group, the Moluccas, and Celebes. Yet they have been unable to jump the small gap between Celebes and Borneo. It seems likely to me that this gap was actually bridged repeatedly, but that the numerous predators on Borneo have prevented the establishment of *Phalanger*.

The fact that even mammals can cross water gaps is demonstrated particularly well by the mammals of the Kei Islands. A land bridge between the Kei Islands and New Guinea has frequently been postulated because there are six species of marsupials on the Kei Islands. Still this is a very poor marsupial fauna, compared with that of New Guinea with 50 to 100 species. However, the Kei Islands fauna is rich compared to that of Halmahera (only *Phalanger* and *Petaurus*) and the southern Moluccas (only *Phalanger* and the endemic *Rhynchomeles prattorum*). Noticeable also is the slight degree of endemism among the six species of Kei Islands marsupials. The answer is presumably ecological. The Kei Islands are only 100 kilometers distant from the shore of Pleistocene Sahul-land and near the place where the large rivers that drain the Snow Mountains must have entered the sea. These rivers, which probably compared favorably in size with the present-day Fly, Sepik, and Mamberano rivers, must have been a great source of tree rafts, and this may be an explanation for the rich mammalian fauna of the Kei Islands. Halmahera and Seran, on the other hand, although they are closer to New Guinea, are situated opposite a mountainous coast, along which only small streams enter the sea. There was much less opportunity for colonization by rafts.

Some additional evidence regarding a land bridge between the Kei Islands and New Guinea can be summarized as follows: All the geological data are in conflict with the theory of a former land bridge (at least during geologically recent times). The composition of the freshwater fish fauna is also opposed to such an explanation. There is not a single primary freshwater fish to be found on Kei, and the secondary freshwater fish show a greater affinity with the Indian than with the Australian fauna. It is for all these reasons that it is to be concluded that no late Tertiary or Pleistocene land bridge existed between Aru and Kei Islands and that the six species of marsupials reached the Kei Islands on rafts. The slight or absent racial differentiation indicates that this colonization took place quite recently, possibly while the edge of Sahul shelf was near the Kei Islands during the height of the Pleistocene glaciations.

Different dispersal factors must be taken into consideration for each island and for each group of animals. It is quite possible, for example, that it will be shown eventually by comparative studies that the Kei Islands have a particularly high Papuan element in all groups that depend on rafts for dispersal, while Halmahera has a particularly high Papuan element in the groups that are carried through the air. More faunistic and taxonomic work needs, however, to be done before such studies can be undertaken.

Amphibians are another group of animals that are frequently regarded as incontrovertible proof of former land bridges. This idea also has become very doubtful in recent years. Most amphibians, whether as eggs, larvae, or adults, are very susceptible to injury by salt water. It is improbable that they would survive lengthy oceanic voyages on rafts. Therefore, if frogs are found on islands with otherwise oceanic faunas, like the Fijis or the Caribbean Islands, one must assume that they were transported there by hurricanes, clinging to palm fronds or other wind-borne vegetation. In the island region of the Malay Archipelago the species of the arboreal genus *Rhacophorus* (in particular *leucomystax*) have a much wider distribution than the terrestial, that is, aquatic, species. Of about 35 species of Java frogs and toads, only 9 or 10 reached Lombok, and only two Timor. For most frogs even narrow ocean straits are dispersal barriers of considerable magnitude; still, the Australian species *Hyla rubella* managed to reach both Timor and Timorlaut. These fragmentary notes are offered merely because dispersal faculties in amphibians have never been critically analyzed.

It cannot be my task in the present paper to provide a comparative treatment of dispersal faculties in various groups of animals. The time for this has not yet come. But possibly these remarks will stimulate specialists of various groups to correlate the ecological characteristics of each species with its distributional pattern.

The mammalian fauna of the Kei Islands represents only a single specific instance in which the hypothesis of trans-oceanic chance dispersal is more successful in explaining the facts than is the assumption of a former land bridge. It seems necessary, in view of the widespread belief in land bridges, to cite additional evidence in favor of trans-oceanic colonization. Distribution patterns in the Malay Archipelago supply such evidence. There are, for example, four sets of phenomena that have puzzled the land-bridge builders and that fit exactly the hypothesis of chance dispersal.

1. *Faunal relationships within the Malay Archipelago are independent of the submarine contours (below the 200-meter line) but are closely correlated with the distances of the islands from each other.* Timor is zoogeographically very close to Wetar, although both islands are separated by an oceanic deep. Equally, the faunal relationship with Sumba is not very great, although both islands are on the same submarine ridge. Good zoogeographic connections also exist between Aru–Kei, between Seran–New Guinea, and between Seran–northern Moluccas, although in all these cases there are no submarine banks that might be interpreted as drowned land bridges. This is supplemented by the surprisingly small zoogeographical relationship of some islands which are

situated on the same submarine ridge, such as Sumba–Timor, Alor–Wetar, Timor–Timorlaut–Seran. It can be stated, and there are only a few exceptions to this rule, that the fauna of each island of the Indo-Malayan island belt (except on the continental shelves) is the direct product of its accessibility to over-the-water immigrants from other neighboring islands, provided age, size, and ecology are comparable. Islands that are separated by straits of less than 200 meters in depth may have been in continental connection during recent glaciations and as a result may show greater faunal affinity.

2. *Slight faunal affinities exist even between geologically unrelated regions.* De Beaufort (1926:160–161) is rather puzzled about the faunal connections between the northern Celebes and Halmahera, between the Philippines and Halmahera, and between the Philippines and New Guinea. The geologist finds no possibility for a land connection between these islands during the late Tertiary or the Pleistocene, which is the period when the faunal exchange must have taken place. It is now obvious that the animals that occur jointly on the island groups mentioned have the ability to cross water gaps of considerable width.

3. *The small percentage of endemic species.* If each island or group of islands received its fauna by one or a few land bridges and was then completely isolated, one would expect that most of the species would evolve into endemics during this isolation. It is found, however, that with the exception of a few old islands the number of endemics is rather low. Rensch (1936:261–262) is perplexed by the fact that so few of the animals of Flores, Wetar, Timor, or the other eastern Sunda Islands are restricted in their distribution to the Sunda arc. This is true for only 17.5 percent of the Flores species, for 27 of 297 (=9.1 percent) of the analyzed Timor species, and for 7 of 161 (=4.3 percent) of the Wetar species. If there had been a continuous land bridge along the inner Banda arc, the percentage should be much higher. What these figures really indicate is that any species that can jump the water gaps of Lombok Strait and of the strait between Alor and Wetar is equally likely to get across to Celebes, to the Moluccas, to Timorlaut, or to Australia. As soon as this happens it will, of course, no longer be restricted to the Sunda arc. The low percentage of species that are widespread on the Sunda arc, but restricted to it, proves the superior dispersal faculty of the species of which the fauna of the Lesser Sunda Islands is composed.

4. *The spread of discontinuously distributed mountain animals can be understood only if a great ability to overcome topographical barriers is assumed.* Stresemann (1939:379–384) points out in his study of the birds of Celebes that many of the species of the mountain forest have an unexpected ability not only to cross stretches of lowland of considerable extent, but even to embark on lengthy trans-oceanic flights. Such species as *Muscicapula melanoleuca, Zosterops montana, Dendrobiastes hyperythrus, Eumyias panayensis, Turdus poliocephalus, Phyllergates*

cucullatus, Phylloscopus trivirgatus, Bradypterus montis, Urosphena subulata, Seicercus montis, and *Dicaeum sanguinolentum* must have made large jumps to get from one mountain top to the next one. Wallace and Rensch have attempted to explain the discontinuous distribution of mountain animals by assuming that the habitats of these species had been continuous during the climatic depression of the Pleistocene. Paleobotanical and paleoclimatic research indicates, however, that the drop of the mean temperature in the tropical belt during the glaciation cannot have amounted to more than 1° or 2° C., and this could not nearly have accomplished a junction of the various mountain forests (Steenis 1934–1935:391ff.). Not even the most liberal construction of land bridges could overcome this difficulty. The same is true of the spread of all animals that are in one way or another narrowly adapted to a discontinuous habitat. Stresemann (1939) calls attention to the fact that there is a considerable variability in the dispersal faculty of mountain birds. Some seem to be poorly endowed with it, and this is particularly true of the mountain species of old islands (like Celebes); others have a highly developed dispersal faculty (like most of the above-listed species).

The above-quoted evidence, in addition to the data accumulated by Gulick (1932), Darlington (1938), and Zimmerman (1942), makes it apparent that a good deal of historical zoogeography will have to be rewritten. Land bridges have been erected in the past with such utter disregard for geological, ecological, and phylogenetic data that almost any land bridge is now under suspicion. The only safe method seems to be to reject any land bridge that does not satisfy a number of basic conditions. The following are some of the criteria that have been developed by recent authors (Stresemann 1939; Simpson 1940; and Mayr 1941; see Essay 40).

1. No species can be quoted as evidence for a former land bridge which occurs also on islands known to be oceanic.

2. All land bridges are temporary, hence all the species that arrived on a certain island by the same land bridge must have the same age. Strikingly different degrees of speciation indicate colonization at different times, provided all other factors are equivalent. This is well demonstrated by the bird faunas of the Hawaiian Islands, of the Galápagos, of Biak, of New Caledonia, and of other islands.

3. A land bridge must have been utilized by a high percentage of the fauna to which it was available. In cases where only a small part of a fauna has used a possible connection, it seems more logical to assume chance dispersal across incomplete barriers rather than to postulate a highly selective land bridge.

4. There are facile trans-oceanic colonizers in nearly every group of land animals, and others that need true land bridges for dispersal. Each case must be treated on its own merits.

The validity of every land bridge must not only be tested by these four criteria, but also be subjected to the additional considerations outlined by Simpson (1940).

The evidence on land bridges can be summarized as follows. The majority of the land bridges postulated during the past 50 years are to be rejected for three reasons. First, they are in conflict with the facts of geology. Second, they cause more difficulties to the zoogeographer than they explain; the observed distribution patterns are not those that one would expect on the basis of the postulated land bridges. The third reason is that these land bridges are unnecessary, because the existing patterns of distribution can be well explained without them. This is obvious as soon as one realizes the enormous dispersal faculties of most animals, as well as the fact that extinction in intermediate areas is the reason for most of the now existing discontinuities.

In sum, it is now apparent that the patterns of distribution in island regions depend more on the dispersal faculties and ecological requirements of the involved species than on former land connections (particularly where there never have been any). This means that birds, butterflies, and lizards, three groups containing many facile colonizers, will have a different distribution pattern from snakes, mammals, or anurans, which are severely handicapped by oceanic gaps, or urodeles and true freshwater fishes, which cross ocean straits only in exceptional cases. Most plants seem to have still greater dispersal faculties than even birds or butterflies. Plants, however, have much more rigid ecological requirements than most animals. It is for this reason that phytogeographic regions seem to coincide rather closely with climatic regions, while zoogeographic regions, at least in insular districts, tend to be identical with areas of former continental connections, or at least of accessibility.

REFERENCES

Beaufort, L. F. de. 1926. *Zoogeographie van den Indischen Archipel.* Volksuniversiteits Bibliotheek, Haarlem.

Darlington, P. 1938. The origin of the fauna of the Greater Antilles, with discussion of dispersal of animals over water and through the air. *Quart. Rev. Biol.,* 13: 274–300.

Gulick, A. 1932. Biological peculiarities of oceanic islands. *Quart. Rev. Biol.,* 7: 405–427.

Matthew, W. 1915. Climate and evolution. *Ann. N.Y. Acad. Sci.,* 24:171–318.

Mayr, E. 1941. The origin and the history of the bird fauna of Polynesia. *Proc. 6th Pacific Sci. Congr.* (1939), 4:197–216.

Mayr, E., and R. M. de Schauensee. 1939. The birds of the island of Biak. *Proc. Acad. Nat. Sci.,* 91:1–37.

Myers, G. S. 1938. Freshwater fishes and West Indian zoogeography. *Smiths. Report for 1937,* pp. 339–364.

Rensch, B. 1936. *Die Geschichte des Sundabogens.* Borntraeger, Berlin.

Simpson, G. G. 1940. Mammals and landbridges. *Jour. Wash. Acad. Sci.,* 30:137–163.

Smith, M. 1943. The divisions [of the Indo-Australian Archipelago] as indicated by the vertebrata. *Proc. Linn. Soc. London,* 154th session:138–142.

Steenis, C. G. G. J. Van. 1934-1935. On the origin of the Malaysian mountain flora. *Bull. Jard. Bot. Buitenzorg,* pp. 135-262, 289-417.

Stresemann, E. 1927-1934. Geographische Verbreitung. *Handbuch der Zool.,* ed. W. Kükenthal, vol. 7, pt. 2, pp. 633-658.

—— 1939. Die Vögel von Celebes. Zoogeographie. *Jour. f. Ornith.,* 87:312-425.

Zimmerman, E. 1942. Distribution and origin of some eastern oceanic insects. *Amer. Nat.,* 76:280-307.

42
Wallace's Line in the
Light of Recent
Zoogeographic Studies

Zoogeography has had a fate very much like taxonomy. It was flourishing during the descriptive period of biological sciences. Its prestige, however, declined rapidly when experimental biology began to come to the foreground. Again as with taxonomy, a new interest in zoogeography has been noticeable in recent years. It seems to me that this revival has had two causes. One is the interest of the student of geographic speciation in the findings of the zoogeographer. A study of past and present distributions yields much information on isolation of populations and on the dispersal of species. It is in this connection that I became interested in zoogeography.

The other reason is the introduction of new methods. The intensive exploration of all corners of the globe during the past 50 years has led to an accumulation of sufficient faunistic data to permit the application of statistical methods. Furthermore, the science of ecology has reached a level of maturity at which it is beginning to affect profoundly zoogeographic methods and principles. It seemed worth while to me to study the controversial and still wide-open subject of the borderline between the Australian and Oriental regions with the help of such modern methods.

A. R. Wallace, who is generally considered the foremost representative of classical zoogeography, states in his famous essay "On the Zoological Geography of the Malay Archipelago" (1860): "The western and eastern islands of the archipelago belong to regions more distinct and contrasted than any other of the great zoological divisions of the globe. South America and Africa, separated by the Atlantic, do not differ so widely as Asia and Australia." There is much truth in this statement. Except for bats and a few rodents, the only native mammals of Australia are marsupials and monotremes. These same two groups are entirely lacking in Asia and are replaced by a wide variety of placental

Reprinted with minor revisions from "Wallace's line in the light of recent zoogeographic studies," *The Quarterly Review of Biology*, 29(1954):1–14; reproduced with the permission of *The Quarterly Review of Biology*; © Stony Brook Foundation for *The Quarterly Review of Biology*.

mammals, such as monkeys, shrews, squirrels, ungulates, and so forth. An equally pronounced faunal difference exists among birds, insects, and other groups of animals of the two regions.

Australia and Asia are connected by a belt of islands, the Malay Archipelago, and the question naturally comes up as to where in this island region the borderline is to be drawn between these two fundamentally different faunas. After reviewing the zoological evidence known to him, Wallace (1860) comes to the following conclusion: "We may consider it established that the Strait of Lombok [between Bali and Lombok] (only 15 miles wide) marks the limit and abruptly separates two of the great zoological regions of the globe." With these words he drew a zoogeographic boundary which was destined to gain fame under the name of its author: "Wallace's Line," a term first used by Huxley (1868) (see Figure 42–1). It runs between Bali and Lombok in the south, then through Makassar Strait between Borneo and Celebes, and finally turns into the open Pacific between Mindanao

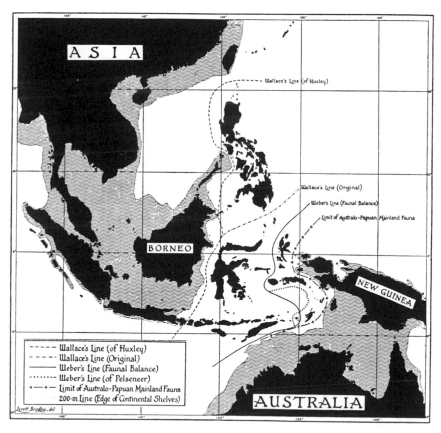

Figure 42–1. Zoogeographic borderlines in the Malay Archipelago. The shaded areas are the continental shelves.

(Philippines) and the Sanghir Islands. This convenient borderline found quick acceptance in the zoological literature and was without hesitation adopted by nearly all the zoogeographers publishing between 1860 and 1890. Sarasin (1901) and Pelseneer (1904) should be consulted for a historical survey of the earlier literature. The echo in the popular literature of this period was even more enthusiastic. A mysterious line, only 15 miles wide, that separates marsupials from tigers, and honeyeaters and cockatoes from barbets and trogons, could not fail to appeal to the imagination of the layman. E. Haeckel (1893) outdid all his contemporaries by asserting: "Crossing the narrow but deep Lombok Strait we go with a single step from the Present Era to the Mesozoicum."

Such exaggerated statements call for refutation, and shortly after 1890 doubts were expressed more and more frequently as to the validity of Wallace's Line, particularly after the distributional facts became better known. Wallace himself was much less positive in his later writings. Since then many writers have insisted that Wallace's Line was entirely imaginary (Weber 1902; Pelseneer 1904; Mertens 1930; Brongersma 1936; and others). Van Kampen (1909), for example, asserted: "Such a sharp boundary as Wallace drew it does not exist. Not only is there none where he drew it, but no such line exists anywhere in the archipelago." On the other hand, Wallace's Line has been vigorously defended by such serious authors as Dickerson et al. (1928), Raven (1935), and Rensch (1936). Curiously enough, most of the writers on this subject seem to be definitely in one or the other camp, either they are for Wallace's Line or they are against it, and they tend to present their data accordingly. Others treat one aspect only of this diversified problem. An impartial study of the situation is still lacking at the present time.

Actually, a whole complex of questions is involved, of which the following seem to be the most important ones:

1. Is Wallace's Line the borderline between the Oriental and the Australian regions, and if not, where is this borderline?

2. Does Wallace's Line represent the line of a major faunal break, and if this is true, how did such a break develop?

IS WALLACE'S LINE THE BOUNDARY BETWEEN THE ORIENTAL AND THE AUSTRALIAN REGIONS?

The fauna of the Malay Archipelago was rather poorly known in Wallace's days. Where he knew 20 species of birds, we now know 120; where he knew 5 species of reptiles, we know 40, and so forth. This lack of information caused Wallace to single out what he considered typical representatives of the respective faunas, and to use the borderline of their ranges as zoogeographic boundaries. The tiger, the squirrels and other mammals go as far east as Bali, but are absent from Lombok.

Among birds the barbets (*Capitonidae*) and many other Oriental groups are abruptly brought to a halt by Lombok Strait. The Australian honey-eaters (genera *Philemon* and *Meliphaga*) and the cockatoe (*Cacatua*) reach Lombok, but not Bali. The faunal difference on either side of Makassar Strait is even more striking: a rich Oriental fauna on Borneo and a marsupial (*Phalanger*) on Celebes. It was on the basis of such data that Wallace came to the conclusion that Lombok and Makassar Straits form the boundary between the Oriental and the Australian regions.

An analysis of the extensive faunal lists now available does not bear out Wallace's conclusion. After eliminating a few widespread species, the fauna of each of the islands of the Malay Archipelago can be divided readily into two groups of species, western ones, that is, species which are derived from the Oriental fauna, and eastern ones, that is, those which are derived from the Australian fauna. In a few species it is apparent that the genus or the family to which they belong was origi-nally of western origin, but that the particular species arrived in the island belt from the east as a descendant of a group of species that was isolated in Australia at an early date. Such secondarily eastern elements, like *Merops ornatus* among the birds, are included with the eastern group. The classification of a few species will always remain open to doubt, but a different decision in these cases would change the per-centages only slightly and would not basically affect the following figures. A specialist on a given group usually has no difficulties in deciding which species are Indo-Malayan and which Australian.

Celebes

Weber (1902), the Sarasins (1901), de Beaufort (1926), Stresemann (1939), and other recent authors agree that at least three-fourths of the Celebes animals are of western origin. According to Rensch (1936:252) the figures are: reptiles at least 88 percent, Amphibia 80 percent, and butterflies 86 percent. In birds the figure is slightly lower. Among 74 species of passerine birds 67.6 percent are western. The percentage for the old endemics (genera and good species) and for the more recent immigrants is quite similar. There is no doubt that Celebes must be included with the Oriental region.

Lesser Sunda Islands

Table 42–1 shows the ratio of the western and the eastern elements on a number of islands (the data on reptiles and amphibians are from Mertens 1930; the data on birds are original). Rensch's (1936) careful analysis shows that the Indo-Malayan element prevails numerically as far east as the islands of the Timor group. This is equally true for flying animals (birds and butterflies) and for flightless groups (mammals, land snails).

The figures in Table 42–1 permit only a single conclusion: Wallace's

Table 42-1. Percentage of western and eastern species
on Lesser Sunda Islands.

	Reptiles and amphibians	Birds		Change of percentage in birds
	Western	Western	Eastern	
Bali	94	87.0	13.0	14.5
Lombok	85	72.5	27.5	4.5
Sumbawa	87	68.0	32.0	5.0
Flores	78	63.0	37.0	5.5
Alor group	—	57.5	42.5	—

Line is not the borderline between the Australian and the Oriental regions. The first of the questions asked above is thus answered in favor of Wallace's opponents.

DOES WALLACE'S LINE INDICATE A MAJOR FAUNAL BREAK?

The fact that Wallace's Line is not the border between the Oriental and the Australo-Papuan regions is not the complete answer to our problem. A line that has been defended so vigorously by so many zoogeographers must have some significance. It is worthy of notice that its staunchest defenders were those naturalists who actually studied and collected the animal life on both sides of the line, like Dickerson and his associates in the Philippines, like Raven, who repeatedly crossed Makassar Strait in a sailboat from Borneo to Celebes and back, and like Wallace and Rensch, who crossed back and forth between Bali and Lombok. The actual impressions of these workers are vividly depicted in a quotation from one of Rensch's books. Arriving on Bali after a prolonged exploration of Lombok, Sumbawa, and Flores, he asks himself:

What about the animal life? Is it really as different from that of Lombok, as has been claimed by so many other travellers? Is the small strait between the two islands actually a sharp faunal division? A strait, which even the smallest bird could cross without any difficulties? . . . And the difference is indeed quite extraordinary! Much more conspicuous than I would have ever imagined. As soon as I entered the woods on a small native trail a whole chorus of strange bird songs greets me—in fact, among the real songsters there is not a single one with which I was familiar [from the islands east of Wallace's Line]. . . . One surprise follows the other. The very species that are most common on Bali, are absent on the islands to the east. The most characteristic bird of these woods is a green barbet . . . it belongs to the family Capitonidae which is entirely absent on Lombok! The woodpeckers also, which are represented on the islands farther east by a single species only, are found on Bali in five different species. On the other hand I missed a whole number of species of birds which are characteristic for the islands visited previously. (Rensch 1930)

An unemotional statistical analysis of the faunal data tends to support Rensch's assertions. The most striking feature of Wallace's Line is that it separates a zone with a rich animal life from a zone with a badly impoverished one. Borneo has about 420 species of breeding birds, Celebes only 220. Java has about 340 breeding species, Lombok only 120. The difference is even more obvious in freshwater fish: Borneo has 162 species of the carp family *Cyprinidae,* Celebes has none; Java has 55 species, Lombok has apparently only a single one. Raven (1935) shows that the mammalian fauna is equally impoverished. The same is true for the Philippines; their fauna is badly depleted, as compared with that of Borneo and Palawan (Dickerson et al. 1928).

THE GEOLOGY OF THE MALAY ARCHIPELAGO

Why the islands Sumatra, Java, Borneo, and Palawan should have a rich animal life, whereas the Philippines, Celebes, and the Lesser Sunda Islands have a poor one, cannot be understood without a study of the geological conditions. The British geologist Earle pointed out, as early as 1845, that geologically the Malay Archipelago consists of three parts, a western one comprising the Greater Sunda Islands and the adjoining parts of Asia, which was very stable during the Tertiary; an eastern one consisting of New Guinea and Australia, which was also stable; and an unstable island belt in between. The unstable area, comprising the Philippines, Celebes, the Moluccas, and the Lesser Sunda Islands, has a most complicated geological structure. Deep sea basins, grabens, geosynclines, and geanticlines are scrambled together in a bewildering manner. Geologists are still far from agreement in regard to the interpretation of these structures. So much, however, is clear—that this area is highly unstable and that it has seen many and violent changes in the recent past.

Three conclusions of zoogeographic significance seem to emerge from the geological data: (1) there is no evidence whatsoever for any continental connection between Borneo and Celebes; in fact, the distance between the two islands was, up to the Pleistocene, greater than it is today; (2) Java, Bali, Lombok, and the other islands of the inner Banda arc are situated on the same geanticline; and (3) there is no geological evidence for any cross connections between the inner and outer Banda arcs, except possibly between Sumba and Flores.

The first of these three conclusions shows that Makassar Strait is an ancient ocean barrier and that at least this particular part of Wallace's

Line is geologically well founded. Geologists and zoogeographers are in full agreement on this point.

Tertiary geology supplies, however, no explanation for a faunal difference between Bali and Lombok, a difference which seems to be due to events of a more recent geological past. A considerable quantity of ocean water accumulated in the polar ice caps during the Pleistocene glaciations. It has been calculated that this resulted in a lowering of the sea level of the tropical ocean by at least 70 m, but more probably by 150 m. This caused the drying up of all shallow seas and resulted in a considerable extension of land on the Sunda and Sahul shelves (see Figure 42–1). Sumatra, Java, and Borneo united with the Malay Peninsula in the formation of "Sundaland," an extension of the Asiatic mainland, and Bali became attached to this continent. Lombok, however, which is separated from Bali by a strait of a depth of 312 m, remained separated, even though it was fused temporarily with Sumbawa.

The geological background of Wallace's Line is thus as follows: In its central part, between Borneo and Celebes, it follows the edge of the continental Sunda shelf, in the south between Bali and Lombok (and the same is true in the north between Borneo-Palawan and the Philippines) it indicates the eastern edge of the Pleistocene Sundaland. The faunal break, which I have shown to exist along Wallace's Line, appears now in a new light. It is due to the fact that the line separates, on the whole, a continental from an insular fauna. This separation is clear-cut in Makassar Strait, but it is rather obscured along the Sunda arc, where the geanticline of the inner Banda arc protrudes from Sundaland like a peninsula. Faunal breaks along this chain of islands occur not only on Lombok Strait, but also on all the other interisland straits. A number of authors, among whom Mertens (1930) is foremost, have contended that some of the other straits, such as that between Java and Bali, or the one between Sumbawa and Flores, are even more efficient distribution barriers than Lombok Strait. This assertion is in conflict with the findings of Pleistocene geology, and it therefore becomes necessary to examine the relative efficiency of these water barriers in more detail.

THE EFFICIENCY OF THE WATER BARRIERS
BETWEEN THE LESSER SUNDA ISLANDS

The faunal change between Borneo and Celebes is abrupt, but it is much more gradual along the rest of Wallace's Line. The number of species of birds on this island chain is as follows: Sumatra, about 440; Java, 340; Bali, 166; Lombok, 119; Sumbawa, 123; Flores, 143; and Timor, 137. In the freshwater fish family Cyprinidae, Sumatra has 115 species; Java, 55; and Lombok only a single one. Of butterflies Sumatra has 334 species; Java, 270 species. Of reptiles Sumatra has 193 species; Java, 136 species (Rensch 1936). It is obvious from these figures that the animal life of Java is considerably impoverished as compared with that of Sumatra (or Borneo). The reasons for this are not entirely clear, but three factors seem to be most important. One is the smaller size of Java. Another is the heavy activity of the Javanese volcanoes, particularly during the Pleistocene, which covered a good part of the island with lava and ashes and may have exterminated a number of localized species. The third and most important reason is that Java is less humid and poorer in habitats than Sumatra, also more peripheral and thus less accessible to colonists from the Asiatic mainland. The climatic deterioration, which is already indicated in western Java, accelerates rapidly in the eastern part of the island, where in the lowlands true tropical rain forest seems to be largely replaced by monsoon forest. The result is that many of the most characteristic Java elements (including nearly all of the well-known endemics) are restricted to western Java. Of the 340 species of Java birds only 245 are found in the eastern half of the island, and it is reasonable to believe that some 70 of these species drop out before the eastern tip of Java is reached, leaving only about 170 species for the eastern tip of the island. No natural history survey has ever been made of this section of Java. This is unfortunate because the fauna of a small area of easternmost Java, equivalent in size to Bali, must be compared with the Bali fauna if one wants to test the significance of Bali Strait as a zoogeographic barrier. It would be entirely misleading to subtract the number of Bali species from the total number of Java species and state that the difference comprises the species that are unable to cross Bali Strait. This method was actually applied by Mertens (1930) and Brongersma (1936), who arrived thereby at the erroneous conclusion that Bali Strait was the most important barrier along the Sunda chain.

A faunal change between the western and the eastern end occurs probably not only on Java, but on all elongated islands of the Sunda chain, such as Sumbawa, Flores, and Timor. This fact invalidates to some extent the figures on the subsequent calculations, but it is fortunately of minor importance in respect to the small and rather round islands of Bali and Lombok.

This is by no means the only difficulty that is encountered in the attempt to determine the relative efficiency of the various straits in the Sunda chain. It happens that there is a gradual but steady change of climate and plant cover from west to east. Each more easterly island is somewhat more arid than its western neighbor, and one after the other of the humidity-loving species drops out because the habitat becomes unsuitable and not necessarily because it cannot cross the water barrier separating it from the next island.

The effect of six inter-island straits on the distribution of birds is illustrated in Figure 42–2. The top line of figures records the number of eastern species that find the western limit of their ranges on the inter-island straits. Lombok Strait shows the highest figure with 15 species, but, on the whole, the difference between the various straits is rather

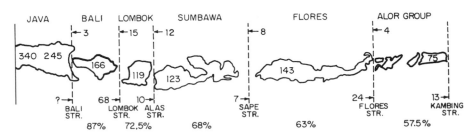

Figure 42-2. Interisland straits in the Lesser Sunda Islands and their efficiency as distributional barriers for birds (explanation of the figures is given in the text).

slight. This is not surprising, since all the eastern species have the ability to jump water barriers and it is probable that the ecological factors on the islands have as much or more to do with the limits of the ranges than age or width of the straits between them. The second row of figures gives the number of breeding species known from each island (the second figure on Java gives the number of species on the eastern half of Java). The third row of figures, and this is the most important one of all, gives the number of western species that are halted by the various straits. The significance of Lombok Strait at once becomes apparent. It prevents the passage of 68 (41 percent) of the 166 Bali species. No other strait approximates this figure. The last row of figures gives the percentage of western species on each of the islands.

The relative efficiency of Lombok, Alas, Sape, and Flores Straits can be expressed by calculating what percentage of the species occurring on either side are stopped by the straits. Lombok Strait, for example, is a barrier for 83 (= 68 + 15) species of a total of 285 (166 + 119), that is, 29.1 percent. The percentages for the other straits and for a number of other animal groups are given in Table 42-2.

The figures of Table 42-2 have, of course, only a relative value since the species totals include many species twice, once east and once west of the straits. Still they are valid as indicators of the relative efficiency of these straits and of their rank. Many of the smaller islands (Penida, Komodo, Sangeang, Rintja, etc.) are insufficiently explored and have therefore been omitted from the tabulation. Solor, Adonara, Pantar, and Alor have been united as the Alor group. In the tabulation of the borders of western species only the easternmost occurrence has been used. This explains a seeming discrepancy in the figures. Flores Strait, for example, stops only 24 of the 143 species on Flores. One would expect the Alor group to have 119 species (143 less 24), but it actually has only 75. The "missing" 44 (119 less 75) species are, however, found on Wetar, Timor, or other more easterly islands, which proves that Flores Strait is not the eastern limit of their range.

The data presented in Figure 42-2 and Table 42-2 can be sum-

Table 42–2. Relative efficiency of straits in Lesser Sunda Islands.

Strait	Birds (orig.)	Reptiles, amphibia, butterflies and land mollusks (Rensch)	Rank (for birds)
Lombok Strait (Bali-Lombok)	83 of 285 = 29%[a]	84 of 377 = 22%	1
Alas Strait (Lombok-Sumbawa)	22 of 242 = 9%	34 of 367 = 9%	3
Sape Strait (Sumbawa-Flores)	15 of 266 = 5%	52 of 364 = 15%	4
Flores Strait (Flores-Alor)	28 of 218 = 13%		2

[a]This figure indicates what percentage of the sum of the species of the two islands on either side of the strait have not crossed the strait.

marized as follows: Each of the straits in the Lesser Sunda Islands is a zoogeographic barrier. Lombok Strait, however, is more effective than any of the others. This is apparently due to the fact that this strait persisted throughout the Pleistocene, whereas Bali Strait and Alas Strait dried up at the height of the Pleistocene glaciation. Rensch's data (Table 42–2) indicate that reptiles, amphibia, butterflies, and land mollusks show conditions similar to the conditions in birds, and it is possible that a more thorough exploration of Bali, Flores, and Alor would make the two sets of data even more similar.

Freshwater fish are useful as negative zoogeographic indicators. The fact that primary freshwater fish (see Myers 1938 for a definition of this term) are absent from Seran and Kei indicates, for example, that these islands have had no continental connection with New Guinea. The presence of four species of freshwater fish in the Lesser Sunda Islands— *Rasbora elberti* on Lombok and Sumbawa; *Clarias batrachus* on Bali, Lombok, and Sumbawa; *Aplocheilus javanicus* on Lombok; and *A. celebensis* on Timor—does not necessarily prove continental connections for these islands, but it casts doubt on the means of dispersal of these species. The slight, or absent, differentiation of these species demands that these islands had a recent continental connection. However, if such had existed one would expect a much richer fish fauna. The transport of fish by water spouts is well substantiated and it is also possible that Lombok Strait had occasionally a surface sheet of fresh water while it was the outlet for the large streams of Pleistocene Sundaland. It would be dangerous to go too far in such speculations of possible chance dispersal but it is even more dangerous to base sweeping zoogeographic conclusions on the presence of a few species of so-called freshwater fish.

THE EASTERN COUNTERPART OF WALLACE'S LINE

It is obvious that there must be a line at the eastern edge of the island belt which corresponds to Wallace's Line in the west. Such a line would separate the zone of a more or less pure Australo-Papuan mainland fauna from the islands to the west with an impoverished Papuan fauna and an Indo-Malayan admixture. This line has been vaguely referred to by Lydekker and other nineteenth-century writers, but I believe de Beaufort (1913) was the first to point out its true significance. It is not difficult to trace, since it follows, except for a short stretch in the north, the 100m depth line, that is, the edge of that part of the Sahul shelf that was dry land at the height of the Pleistocene glaciation (Figure 42–1). It passes between the Aru Islands, which have a pure Papuan fauna, and the Kei Islands, with an impoverished fauna with Oriental elements. Of birds, for example, 166 species are known from the Aru Islands, including birds of paradise and many other typical Papuan types, while only 84 species are known from the Kei Islands, including some western elements. The line then passes between the mainland of New Guinea and Seran Island. There are 115 species of birds (about 30 percent western) known from Seran as against more than 300 species from the Vogelkop, the neighboring part of New Guinea. The line that separates the Papuan mainland fauna from the island fauna swings from Seran north and passes through the Gilolo passage separating the western Papuan Islands (Waigeu, Batanta, Salwati, and Misol) from the Northern Moluccas. In this section the line does not follow entirely the 100 m contour, which would exclude Koffiau, Gebe, Batanta, and Waigeu. However, all these islands are so purely Papuan and form such a well-defined faunistic unit that it seems justified to be slightly inconsistent. It might be worthwhile to emphasize that the line, as just drawn, gives a better defined delimitation of the "Papuan mainland" and "Papuan island" fauna than does Wallace's Line in the west for the Indo-Malayan fauna. Its validity is particularly apparent for all groups with a limited dispersal faculty, for example freshwater fish. De Beaufort's map (1926:103) of the range of the subfamily *Melanotaeniinae* illustrates it quite graphically. This Australian group extends westward as far as the Aru Islands and Waigeu, but is absent from the Kei Islands and from the Northern and Southern Moluccas.

The significance of this eastern line has been emphasized by a number of authors. It indicates, like Wallace's Line, a major faunal break; it separates, like its western counterpart, a continental from an island zone, as well as a zone with a more or less undiluted Papuan fauna from an area with a mixed Papuan-Oriental fauna, a contrast which is least apparent in the north. It is for all these reasons that this line must be considered a major zoogeographic boundary.

SHOULD AN INTERMEDIATE ZOOGEOGRAPHIC
REGION BE RECOGNIZED?

A gradual transition between the Oriental and the Australian faunas takes place in the island belt between Wallace's Line and its eastern counterpart. This was realized quite clearly by Salomon Müller (1846), the earliest zoogeographer of the Malay Archipelago. He lists correctly "Celebes, Flores, Timor, Gilolo and perhaps Mindanao" as islands on which a mixture of Indian and Australian elements is found. Wallace also, in his later publications, admitted the intermediate position of this region and stated of Celebes that it "hardly belongs to either [Oriental or Australian] region." Pelseneer (1904:1007) lists a whole group of workers who recognized the transitional character of this region.

There are other factors, in addition to the lack of continental connection, which contribute to the poverty of the fauna of this island belt. Müller (1846) very ably pointed out some of the reasons, such as the small size of most of the islands, their low elevation, and their aridity. There is a more or less arid corridor extending from the Philippines and Celebes to Buru and to the Sunda Islands from eastern Java to Timorlaut. This zone has acted as a barrier to many humidity-loving forms and has prevented their passage from Sundaland to the Papuan region and vice versa. Additional reasons for the faunal poverty of this zone are the young geological age of many of the islands, which limits the number of chance colonizations, and the heavy volcanic activity over part of the region. There are three lines of volcanoes in this transition zone, one extending from Sumatra through Java to the inner Banda arc, a second one following the western edge of the Northern Moluccas, and a third one reaching from north Celebes through the Sanghir Islands to the Philippines. The volcanic activity is thus strictly localized, but where it occurs it may be a very serious factor indeed. As mentioned, it seems to be one of the reasons why Java's animal life is so much poorer than that of Borneo or Sumatra (Rensch 1936). There are not only 59 young volcanoes of more than 2000 m altitude on Java, but also many extinct later Tertiary ones. This factor is even more evident on Lombok, where heavy Pleistocene eruptions of Mount Rindjani seem to have destroyed much of the mountain fauna. The same is true for the volcano on Ternate Island (Stresemann 1939:381).

All the mentioned factors combine to give the fauna of the transition zone a peculiar character. This has impressed some of the authors to such an extent that they have proposed giving formal recognition to this fauna and elevating the island belt to the rank of a separate zoogeographic region or subregion.

Dickerson et al. (1928), who coined the term *Wallacea* for this region, and Rensch (1936), who simply calls it *Zwischengebiet* (region of intermediacy), are the most recent champions of such an arrangement. This region would include four different groups of islands: (1) the Lesser Sunda Islands from Lombok eastward; (2) the Moluccas and other outliers of the Papuan region (Tenimber, Kei), (3) the Celebes group (with Sula and Talaut), and (4) the Philippines. Two reasons are usually quoted in favor of recognizing such a transition region. One is that many endemic species and genera are confined to it. The other reason is that all of the islands included in this transition zone are populated by a mixture of Indo-Malayan and Australo-Papuan elements. Against these points favoring the recognition of a transition region there are some very strong objections. Pelseneer (1904) has stated them clearly. He points out that it is only natural that

> a zoogeographic border is not a line without width and that by necessity there is a mixture of faunal elements along the border of two zoogeographic regions, caused by a reciprocal penetration.
>
> But if one would admit for this reason a special "transition region" or a "region of intermediacy," one would obviously double the difficulties of delimitation. For now it would be necessary to trace both of the border lines which separate the transition region from either of the two adjoining zoogeographic regions.

These difficulties of delimitation are fully confirmed by the two most recently proposed transition regions. Dickerson et al. (1928:297) define theirs as follows: "Wallacea is outlined sharply by Wallace's Line (as modified) on the west and Weber's Line upon the east." It thus includes the Philippines, but it excludes the Moluccas, Timorlaut, and Kei Islands. Rensch (1936:265), however, includes in his *Zwischengebiet* "Celebes, the Lesser Sunda Islands, Timorlaut (perhaps also Kei) and the Moluccas (at least the southern Moluccas)." He definitely excludes the Philippines. Celebes and the Lesser Sunda Islands are, thus, the only two districts that the two transition regions have in common.

The "degree of intermediacy" of the various sections of the transition region is very uneven. It seems, for example, that the percentage of Australo-Papuan species in the Philippines (which are included in Wallacea by Dickerson and Merrill) is smaller than the percentage of Oriental species in New Guinea or Australia. Still, nobody would want to suggest including Australo-Papua in the transition zone.

Stresemann (1939:403) adds another weighty objection. He points out that the transition zone comprises four separate districts which have much less in common with one another than each one has with some outside region: The Moluccas are faunistically closest to New Guinea, and Celebes to the Philippines, but the Philippines are closer to Malaysia than to Celebes. The Lesser Sunda Islands, finally, have a close faunal relationship with Java and Australo-Papua, but only a very slight and recent one with Celebes. To unite four such heterogeneous districts in a single "region" violates all principles of regional zoogeography.

After all, if a zoogeographic region means anything, it means the home of a more or less homogeneous characteristic fauna. "Wallacea," however, is the home of four different faunas. It is self-evident that the formal recognition of a zoogeographic region of such heterogeneity is neither practical nor scientifically defensible. The term "transition zone" is justified only if applied informally as a descriptive attribute.

WEBER'S LINE

It is apparent from the preceding discussion that neither Wallace's Line nor the formal acceptance of a transition zone is a satisfactory way of delimiting the Oriental against the Australian region. This leaves, to my mind, only one other alternative solution, namely, the recognition of a line east of Wallace's Line. Before attempting to draft the best possible course of such a line, a few words must be said about the validity of any zoogeographic borderline.

A zoogeographic region is usually defined as a geographic subdivision of the earth that is the home of a peculiar fauna. Such a region is characterized by the presence of many endemic genera and families and by the absence of the characteristic genera and families of other zoogeographic regions. Its border should be drawn along the line where this specific fauna is replaced by a different fauna. This procedure is logical and presents no difficulties in all the cases where an efficient barrier separates the two regions, such as is formed by the South Atlantic between Africa and South America. However, an intermingling of the two faunas takes place in a border zone whenever two such regions come into direct contact. This is exactly what has happened in the island belt between Asia and Australia. Both the Indo-Malayan and the Australo-Papuan mainland faunas have spilled over into the intermediate island belt, and it might seem impossible in such a mixed region to delimit one fauna from the other one. However, as Pelseneer (1904) says correctly, "it is evident that there must be a line . . . within the region of mixture, on one side of which the faunal elements of one region prevail and on the other side those of the second region. This line can serve usefully to mark the borderline between the two biogeographic regions."

On the basis of these considerations, Pelseneer established a borderline between the Oriental and the Australian regions, which he called "Weber's Line." Pelseneer drew the course of this line on the basis of nonzoological data, primarily the soundings and other oceanographic results of the Siboga Expedition, many of which are no longer valid today. However, Weber's Line actually separates the islands with a more than 50 percent Indo-Malayan fauna from the islands with a more than 50 percent Papuan fauna, as is evident from Rensch's (1936) careful data and from all the other zoogeographic work of the region. With insignificant modification the line suggested by Pelseneer is still acceptable as the best possible borderline between the Oriental and the Australo-Papuan regions.

The course of Weber's Line (Figure 42–1) is as follows: In the north it begins between Talaut and Celebes in the west and the northern Moluccas in the east. In this section the line is extremely well defined, since the fauna of the northern Moluccas consists of about 80–90 percent and that of Celebes of about 20–40 percent Papuan elements. The line continues from here between the Sula Islands in the west and Obi in the east and then swings around Buru. The fauna of the Sula Islands is insufficiently known, but it is close to that of Celebes except much poorer and with a stronger Moluccan element. Still the Papuan component is probably less than 40 percent, while it is about 63 percent on Buru and even higher on Obi. It is difficult to trace Weber's Line from Buru on. Pelseneer attempted to follow the contour of the ocean bottom and this caused him to run the line between Banda ("Indo-Malayan") and Seran ("Papuan") and between Sermatta ("Indo-Malayan") and Babber ("Papuan"). The much more detailed information on the fauna of these islands that is now available indicates that a different course might be preferable. The young volcanic Banda Islands have a fauna which almost completely lacks endemic elements, and which is very close to that of Ambon, Seran, Seranlaut, etc. There is no doubt that the Banda Islands must be included in the southern Moluccas. Babber, on the other hand, has a fauna which is closer to that of Dammer and Sermatta than to that of Timorlaut. It is, therefore, preferable to place the line between Babber and Timorlaut. Rensch (1936:206) has already pointed out the impossibility of separating Babber from the closely related Sermatta and Dammer. The fauna of Timorlaut is about 62.5 percent Australo-Papuan. The South West Islands, from Roma and Kisar to Dammer and Babber, are a faunistic unit, but the progressive decrease of Indo-Malayan elements which started on Java and Bali continues on these islands. It is possible that a future analysis may show that the eastern element on Babber and Dammer is already slightly more than 50 percent of the total fauna of these islands. Even then I would be inclined to retain them in the Oriental region rather than to draw a line through the middle of the South West Islands.

One glance at the map shows that Weber's Line is situated much closer to the Australo-Papuan than to the Asiatic shelf. The reason for this is twofold, faunal pressure and accessibility. The faunal pressure of the Indo-Malayan fauna is greater than that of the Papuan fauna because it is much richer in species and families. The sphere of influence of this rich fauna will, therefore, extend farther into the island belt than that of the poorer Papuan fauna. The second reason is that the chain of the Lesser Sunda Islands, forming practically a peninsula of Sundaland, was infinitely more easily accessible to colonists from the west than to those from the east, which had to jump the wide gap either from Australia to Timor or from New Guinea (and Aru) to the islands of the Banda Sea. The preponderance of Oriental species in the Lesser Sunda Islands would be even more pronounced if ecological factors (aridity) had not favored colonization by Australian elements. These various factors explain the present course of the line of faunal balance, Weber's Line. Wallace's argument that Celebes should be included in the Australian Region because it had so few Oriental species as compared with Borneo, is beside the point. Every true island has, of course, a much impoverished fauna, but its zoogeographic position is determined by an analysis of its existing fauna and not by the elements it lacks. With an 80 percent Oriental fauna Celebes cannot be included in the Australian region!

Weber's Line has found curiously few adherents among zoogeographers; Boden Kloss (1929) is one of the exceptions. There is nothing spectacular about this line and by crossing it one encounters a smaller faunal change than is found between Borneo and Celebes, or between New Guinea and Seran, or in general between the "mainland" and the "island" faunas (Figure 42–1). The difference between the faunas of Sula and Buru and of Babber and Timorlaut is, indeed, rather small. Weber's Line is not acceptable to those who look for a strikingly conspicuous borderline between the Oriental and Australian regions (Rensch 1936:265).

Other objections have been raised against Weber's Line. Some authors, for example, have objected to Weber's Line because it separates islands which lie on the same submarine ridges. Thus it cuts between Babber and Timorlaut, between Dammer and Banda, and between Sula and Obi, each of these three pairs of islands lying on the same submarine ridge. It seems to me that this argument is another instance of confusing zoogeographic and geological interpretations, exactly as in the case of continental versus oceanic islands (Essay 40). The geology of an island, and particularly of an oceanic island, is of no concern whatsoever when we are attempting to classify its fauna. If the fauna of Seran and Kei is prevailingly Papuan, I shall classify these islands with the Papuan region. The fact that Timor and Sumba with a prevailingly Indo-Malayan fauna lie on the same tectonic arc has absolutely no bearing on this decision. In fact there is no evidence that any of these arcs were ever raised to the extent that they were exposed for their full length, and it is obvious that undersea geology can have no influence over the distribution of forms that are dispersed across the water.

I know of only a single valid argument against the adoption of Weber's Line as the boundary between the Australian and the Oriental regions. It is the objection against dividing arbitrarily any continuous series of values at the halfway point between the extremes. In the case of Weber's Line the situation is aggravated by the fact that the 50-50 balance between the Indo-Malayan and the Australian elements is not always the same in the various taxonomic groups. The bird fauna of Wetar Island, for example, is more than 50 percent Australo-Papuan, while in other groups the Oriental element seems to prevail. On Celebes about 67 percent of the birds are of western origin, while among mammals, butterflies, reptiles, amphibians, and land snails the figure is more than 80 percent. On the whole it seems as if among reptiles and butterflies the western element pushes farther eastward than among birds and snails. However, taking the fauna as a whole, Weber's Line seems to separate rather neatly the islands with a prevailingly Oriental fauna in the west from the islands with a prevailingly Australo-Papuan fauna in the east. Different faunal regions are generally indicated on zoogeographic maps by different colors. It is obvious that the 50:50 line is the most convenient place to change from one color to another. It is in this sense that Weber's Line (as modified above) may be accepted as the boundary between the region with a prevailingly Oriental and the region with a prevailingly Australo-Papuan fauna.

UNSOLVED PROBLEMS OF
INDO-AUSTRALIAN ZOOGEOGRAPHY

The conclusions at which I arrived in the present analysis are not final. Many of the islands are insufficiently explored and it is certain that future exploration will add a good deal to our knowledge. A further refinement in zoogeographic methods is also expected to yield increased results. Salomon Müller, P. L. Sclater, A. R. Wallace, and other early representatives of the classical school of zoogeography selected arbitrarily a number of indicator species and based the outlines of the zoogeographic regions and subregions on the distribution of these species. The preferred technique of the present paper is to calculate in percent the proportion of faunal element in the total number of species of certain localities. All the percentages in Figure 42–2 and Table 42–2 are derived by this method.

In the matter of faunal composition an even superior method might be to determine the faunal relationship of the dominant species of each habitat. It seems, for example, to judge by Rensch's description (see above) that the differences between the dominant species of birds of Bali and Lombok is even more striking than is apparent from a statistical analysis of the total faunas. Such a comparison of the dominant types of local faunas must be based on accurate census data gathered in the field, and such data are not yet available. To gather them would be a worthwhile task for future explorers of the Malay Archipelago.

The combination of ecological and zoogeographic methods promises to yield data of considerable interest. It seems, for example, that the faunal composition of each habitat is different. Of the 11 species of birds that are restricted to the mountain forest of Timor (above 4000 feet) only a single one is Papuan; the other 10 are Indo-Malayan. The ratio is even, if not reversed, among the birds of the tree savanna of Timor. Lack of exact ecological data prevents a more accurate analysis at the present time. Steenis and other botanists have shown that a similar difference of floristic composition exists between different plant associations. Here is a practically untouched field for future investigators.

The delimitation of biogeographic regions depends to a considerable extent on the dispersal faculties and on the nature of the speciation processes of the organisms whose distribution is studied. It has become evident in recent years that there is much difference between phytogeographic and zoogeographic classifications. The major floristic regions coincide fairly well with the major climatic regions. The major zoogeographic regions, however, indicate primarily the extent of formerly (or currently) isolated land areas. The biogeographic classification of New Guinea is a good illustration for this. New Guinea is, for the phytogeographer, a part of the Malayan region, but faunistically it is at least as close or even closer to Australia. A comparison of phytogeographic and zoogeographic maps indicates that it is impractical at the present time to construct biogeographic maps, that is, maps that intend to illustrate simultaneously the distribution of plants and of animals.

This is also true for animal groups with different dispersal faculties. I have already mentioned above the differences between birds and reptiles in regard to the faunal composition of some of the islands. Much more accurate data are needed. It is possible that some of the invertebrates show a distributional pattern that is much more similar to that of plants than to that of mammals or birds. Progress in this field depends largely on a more thorough faunistic exploration of the Indo-Australian Region.

SUMMARY

1. Wallace's Line is not the boundary between the Indo-Malayan and the Australian regions; rather it indicates the edge of the area (Sunda shelf) that was dry at the height of the Pleistocene glaciations.

2. The equivalent line along the edge of the Sahul Shelf separates New Guinea and the Aru Islands from the Moluccas and Kei Islands.

3. Weber's Line separates the islands in the west on which the Indo-Malayan element is predominant from the islands in the east on which the Australo-Papuan element has a numerical superiority.

REFERENCES

Beaufort, L. F. de. 1913. Fishes of the eastern part of the Indo-Australian Archipelago, with remarks on its zoogeography. *Bijdr. tot de Dierk.*, 19:95–164.
———1926. *Zoogeographie van den Indischen Archipel.* Volksuniversiteits Bibliotheek, Haarlem.

Brongersma, L. D. 1936. Some comments upon H. C. Raven's paper: "Wallace's Line and the distribution of Indo-Australian mammals." *Arch. Néerl. Zool.*, 2: 240–256.

Dickerson, R. E., et al. 1928 *Distribution of life in the Philippines.* Bur. Sci. Monogr. (Manila), no. 21.

Haeckel, E. 1893. Zur Phylogenie der australischen Fauna. *Denkschr. Medic. Nat. Ges. Jena,* 4:5.

Huxley, T. H. 1868. On the classification and distribution of the Alectoromorphae and Heteromorphae. *Proc. Zool. Soc. London,* pp. 294–319.

Kampen, P. N. van. 1909. *De zoogeographie van den Indische Archipel.* Bijbl. 3 en 4, Natuurk. Tijdschr. Ned. Indie.

Kloss, C. B. 1929. The zoo-geographical boundaries between Asia and Australia and some oriental subregions. *Bull. Raffl. Mus.*, 2:1–10.

Kuenen, P. H. 1935. Geological interpretation of the bathymetrical results. In *The Snellius Expedition in the eastern part of the Netherlands East-Indies, 1929–1930,* vol. 5:1–124.

Mayr, E. 1941. The origin and the history of the bird fauna of Polynesia. *Proc. 6th Pacific Sci. Congr.* (1939), 4:197–216.

———1944. Notes on the zoogeography of Timor and Sumba. *Bull. Amer. Mus. Nat. Hist.,* 83:171–194.

Mertens, R. 1930. Die Amphibien und Reptilien der Inseln Bali, Lambok, Sumbawa und Flores. *Abh. Senckenb. Naturf. Ges.,* 42:115–344.

Molengraaff, G. A. F. 1922. Geologie. In *De Zeen van Nederlandsch Oost-Indie,* pp. 272–357. Leiden.

Müller, Salomon. 1846. Uber den Charakter der Thierwelt auf den Inseln des indischen Archipels, etc. *Arch. Naturg.,* 12:109–128.

Myers, G. S. 1938. Fresh-water fishes and West Indian zoogeography. *Smiths. Rept. for 1937,* pp. 339–364.

Pelseneer, P. 1904. La "Ligne de Weber," limite zoologique de l'Asie et de l'Australie. *Bull. Classe. Sci. Acad. R. Belgique,* 1904:1001–1022.

Raven, H. C. 1935. Wallace's Line and the distribution of Indo-Australian mammals. *Bull. Amer. Mus. Nat. Hist.,* 68:177–293.

Rensch, B. 1930. *Eine biologische Reise nach den Kleinen Sunda Inseln.* Berlin.

———1936. *Die Geschichte des Sundabogens.* Borntraeger, Berlin.

Sarasin, P., and F. Sarasin. 1901. *Über die geologische Geschichte der Insel Celebes auf Grund der Tierverbreitung.* Wiesbaden.

Scrivenor, J. B., T. H. Burkill, Malcolm A. Smith, A. St. Corbet, H. K. Airy Shaw, P. W. Richards, and F. E. Zeuner. 1943. A discussion of the biogeographic division of the Indo-Australian Archipelago, with criticism of the Wallace and Weber lines and of any other dividing lines and with an attempt to obtain uniformity in the names used for the divisions. *Proc. Linn. Soc. London,* 154th session: 120–165.

Stresemann, E. 1939. Die Vögel von Celebes. Zoogeographie. *Jour. f. Ornith.,* 87: 312–425.

Umbgrove, J. H. F. 1932. Het Neogen in den Indischen Archipel. *Tijdsch. Kon. Nederl. Aaardrijksk. Genootsch.,* 49(2):769–833.

———1938. Geological History of the East Indies. *Bull. Amer. Assoc. Petr. Geol.,* 22:1–70.

Umbgrove, J. H. F., et al. 1934. Gravity, geology and morphology of the East

Indian archipelago. In *Gravity Expeditions at Sea, 1923–1932,* vol. 2. Netherl. Geod. Commission.

Wallace, A. R. 1860. On the zoological geography of the Malay Archipelago. *Journ. Linn. Soc. London,* 4:172–184.

Weber, M. 1902. Der Indo-australische Archipel und die Geschichte seiner Tierwelt. Jena.

43
Fragments of a Papuan Ornithogeography

The bird fauna of New Guinea was, a generation ago, the least known bird fauna of any part of the world. The bird life of the interior and particularly of the mountains was largely unknown. However, during the past 30 years expeditions have gone to New Guinea at the rate of about one every 2 years and have brought back an enormous amount of information. This permitted the preparation of a systematic and distributional *List of New Guinea Birds* (Mayr 1941a), which may serve as a convenient basis for a zoogeographical analysis. A complete analysis of the 300 genera, 650 species, and over 1500 subspecies of New Guinea birds would require much more space than is available here.

THE ZOOGEOGRAPHIC POSITION OF THE NEW GUINEA BIRD FAUNA

New Guinea is classified as part of the Australian region in nearly every zoogeographical treatise. The lack of placental mammals (except for bats and some rodents) and the wealth of marsupials seem to confirm this classification to the fullest extent. The phytogeographers, however, are by no means happy about this delimitation of biogeographic regions because the flora of the tropical belt from Malaya to the Solomons seems to form a unit that is strikingly different from that of temperate and arid Australia (for example, Burkill 1943). Among the students of invertebrates, also, much doubt has been expressed as to the justification for including the Papuan area with the Australian region, for example, Michaelsen (1922) for the Oligochaeta and Karny (1929) for Gryllacridians. Whether the bird fauna of New Guinea agrees better with the pattern indicated by the mammals or that indicated by the plants will be investigated in the present section.

An analysis of the New Guinea bird fauna on the basis of non-

Adapted from "Fragments of a Papuan ornithogeography," *Proceedings of the Seventh Pacific Science Congress,* 4 (1954):11–19.

passerine families is not very revealing. If we omit sea birds, representatives of 36 families of nonpasserines breed on New Guinea. Of these families no less than two-thirds (24) are worldwide, and additional ones are widespread in the Old World (6) or in the Old World tropics (1). This leaves precisely 5 small families. Three of these (Casuariidae, Megapodiidae, and Aegothelidae) are Australo-Papuan, one is Australo-Papuan and Oriental (Podargidae), and one Oriental and Papuan (Hemiprocnidae).

The passerine birds are more revealing. There are 29 families and (within the Muscicapidae *sensu lato*) subfamilies. Of these only 4 are worldwide (Hirundinidae, Motacillidae, Turdinae, and Corvidae). Five are Old World (Alaudidae [1 part-American species], Timaliinae [1 American species], Muscicapinae, Oriolidae, and Ploceidae): 4 are Old World tropical with an occasional temperate zone species or subspecies (Pittidae, Campephagidae, Dicruridae, Zosteropidae); 4 are Old World, excepting Australia (Sylviinae, Laniidae, Sturnidae [1 part-Australian species], Nectariniidae). This leaves 12 families and subfamilies that more clearly reveal their origin and zoogeographic relation. Ten of these are Australo-Papuan (Malurinae, Pachycephalinae, Falcunculinae, Artamidae, Grallinidae, Cracticidae, Philonorhynchidae, Neosittidae, Climacterinae [Certhiidae, part], and Meliphagidae). One is almost exclusively Papuan (Paradisaeidae, with 4 of its forms spilled over into north Australia), and one is Oriental and Australo-Papuan (Dicaeidae). The families that are Papuan or Australo-Papuan include most of those with the highest number of New Guinea species, like the Meliphagidae (61), Malurinae (24), and Pachycephalinae (24). On the other hand, there is not a single family whose chief area of distribution is the Oriental and the Papuan region (Australia excluded).

An analysis of the genera provides an even clearer picture. In order to avoid any bias, I used with only slight changes the genera admitted in my New Guinea list (1941a). I omitted all oceanic and freshwater birds, including Ciconiiformes and Anseriformes. [In the original paper, I listed the genera by source in a detailed table, here omitted.] Two obvious conclusions can be drawn from these data.

1. The Australo-Papuan element by far outweighs the Oriental element. Even if we include all the widespread elements with the Oriental group and omit the Papuan endemics from the calculations there are 45.7 percent Australian or Australo-Papuan genera as against 35 percent Oriental.

2. There is a high degree of striking endemism in the Papuan region, in addition to other typically Papuan elements that are not endemic. This amounts to about 45 percent of all the genera. This element includes entire subendemic families or subfamilies (like the Paradisaeidae, Casuariidae), as well as such striking genera as *Harpyopsis, Megacrex, Trugon, Otidiphaps, Goura, Psittacella, Clytoceyx, Ifrita, Melampitta, Eugerygone, Peltops, Psittrichas, Androphobus, Eulaces-*

toma, Amblyornis, Daphoenositta, Timeliopsis, Melipotes, Oreocharis, and *Paramythia.*

DIFFERENCES BETWEEN BIRDS AND PLANTS

These first results of the analysis indicate that the bird fauna of New Guinea resembles the mammalian fauna in the strong prevalence of Australian types, or at least of types it has in common with Australia. There is an apparent discrepancy between these findings and those of the students of plants and invertebrates, referred to above. In an analysis of the flora of New Guinea the botanist Lam (1934) summarizes his conclusions in four points. These points are herewith quoted, together with my own comments based on the distribution of birds.

"(a) There are no relic-endemics known at present."

What "relic endemics" are is more or less a matter of definition. There are no endemic families in New Guinea, although the Paradisaeidae are almost endemic. However, there is a large number of endemic or subendemic genera, including many that are quite isolated. I would therefore be inclined to consider that this statement does not apply to birds.

"(b) There is a strong neo-endemism, especially at higher altitudes."

This is also true for birds. This endemism concerns genera of widespread families as well as endemic species of nonendemic genera.

"(c) Asiatic floral elements prevail, even in the mountains; Polynesian ones come next, and Australian-Antarctic ones apparently last, Australian ones being best represented in the mountains and savannahs."

In birds the opposite is true. The Australian and Australo-Papuan elements make up the majority of the bird fauna, and the Asiatic element is a poor second. There is no Polynesian element. Much of this biota was first discovered in Polynesia and has therefore been listed in biogeographical papers as "Polynesian." The heartland of this biota, however, is New Guinea, as I have emphasized previously (Essay 40), and it is misleading to label this genuinely Papuan fauna as Polynesian. Merrill (1945:212) comes to a similar conclusion as far as plants are concerned. It seems that there is no "Polynesian" biota in the strict sense of the word (but see Copeland 1948). Even those genera that reached the height of development in the Polynesian area are originally of Papuan or Australian descent. Only a few of the Papuan elements have sufficiently great dispersal facilities to reach Polynesia, hence the Polynesian fauna is composed of only a fraction of the Papuan one. Ripley and Birckhead (1942) have analyzed in detail the colonization of Polynesia by doves of the *Ptilinopus* group as a typical illustration of the Papuan origin of a Polynesian element.

"(d) Western relations are particularly strong with the Moluccas, the Philippines and Celebes."

Yes and no. The Moluccas are the western outlier of the Papuan Region and contain an impoverished Papuan bird fauna (with a considerable admixture of Asiatic elements). The Papuan element in Celebes is no larger than one would expect on the basis of the distance involved and the Australo-Papuan element in the Philippine bird fauna is negligible. It consists of *Cacatua* (1), *Trichoglossus* (1), *Gallicolumba* (1) *Irediparra* (1), *Megapodius* (1), *Ptilinopus* (6), *Poliolimnas* (1), *Eudynamis* (1), *Ninox* (1), *Gerygone* (1), *Rhipidura* (2), *Artamus* (1), *Pachycephala* (2), and possibly also *Apoia* (1) (the number of Philippine species is given in parentheses).

To this are to be added some widespread species of eastern relationship: *Rallus philippensis*, *Porzana tabuensis*, *Amaurornis olivaceus*, *Chalcophaps indica*, *Cacomantis variolosus*, *Hirundo tahitica*, *Edolisoma morio*, and *Megalurus timoriensis*.

So much about Lam's theses and my own comments. I might mention that not all phytogeographers are equally impressed with the strictly Asiatic nature of the New Guinea flora. Merrill (1945), although following Wallace in applying the term Malaysia to the entire region from the Malay Peninsula to the Papuan region, recognizes that "two sub-areas seem to be clearly distinguishable, one centred around the Sunda Islands—Sumatra, Java, Borneo, and of course the Malay Peninsula—and the other about New Guinea and its adjacent islands." But Merrill agrees with the other phytogeographers in considering the New Guinea flora to belong to the Malaysian one and to differ strikingly from the Australian flora.

THE CAUSES FOR THE CONTRAST
BETWEEN PLANTS AND BIRDS

What is the reason for this discrepancy between the findings of the ornithologist and the botanist? Since both discuss the same area, it is obvious that geographical–geological controversies can be eliminated from the discussion. This leaves two possible causes:

1. The dispersal or colonization facilities (or both) of plants and of birds are very different (ecological differences).

2. The rates of evolution of the two groups are very different.

The facts on hand seem to indicate that both causes are responsible, but there are too many unknowns involved to be certain.

Ecological Differences

As Rensch (1936:22) and others (see Essay 40) have pointed out, birds in spite of their excellent dispersal potentialities are more easily stopped by water gaps than most other groups of animals. At the same time, there is much evidence that plants disperse more easily than is admitted by botanists such as Steenis (1934–1935) and Skottsberg (1925). This is the view held by Guppy (1906), Gibbs (1917), Setchell (1935), and others. As far as colonization is concerned, conditions are reversed for birds and plants. Birds are easy colonizers since as mobile warm-blooded animals they carry their own environment with them, while the germinating seed must find a sympathetic environment if it is to succeed.

Can these ecological differences between birds and plants account for their distributional difference? It is now generally agreed that during the Tertiary there were no permanent land connections in the unstable area between the Sunda and the Sahul shelves. Distribution had to go on by island hopping. This was done by plants, birds, insects, and by other animals. The earliest birds that reached Australo-Papua became the progenitors of a vigorous fauna, the Australo-Papuan fauna, but frequent repetition of these colonizations was prevented by the respect most birds have for water gaps. By contrast, the repeated and some-times long-lasting connections between New Guinea and Australia per-mitted such a free faunal interchange that it is impossible to say whether such families and subfamilies as the honeyeaters (Meliphagi-dae), the brush-tongued lories (Loriinae), the wren-warblers (Maluri-nae), and the whistlers (Pachycephalinae) are more typically Papuan or Australian. The independence of birds from habitat restrictions leads to the remarkable phenomenon that the nearest relatives of many birds of the central Australian brush savannahs and semideserts are found in the steaming lowland or the misty mountain forest of New Guinea, such as relatives of *Malurus (Todopsis, Clytomyias), Acanthiza, Gerygone, Pachycephala, Cinclosoma, Coracina,* and so on. A shift of habitat seems not nearly the problem to a bird that it is to a plant. The Malayan plants with their comparatively great dispersal facilities found congenial conditions throughout the equatorial belt as far east as Fiji and Samoa, except in the few places (Lesser Sunda Islands, parts of Celebes, Moluccas, and of southern New Guinea) where rainfall is insuffi-cient. The result has been the development of a Malayo-Papuan flora that extends through this entire humid equatorial belt. A similar distri-butional picture has been found for certain groups of animals, for example, by Karny (1929) for the Gryllacridians. The reason this biota has been unable to prosper in Australia, except to a very slight extent in humid eastern Australia, appears to be twofold, the principal cause being these forms' inability to undertake the drastic ecological changes of which some birds are capable.

This point is further emphasized by the behavior of the Australian plant forms that have penetrated the Malay Archipelago. On Borneo, for example, there are five habitats in which edaphic conditions have permitted colonization by a not inconsiderable number of Australian species (Richards 1943). These include the high altitudes in the mountains, sandy beach forests, heaths, and some swampy habitats. There is no evidence that a similar infiltration of Australian animal types has occurred in these habitats, unless they are directly dependent on some of the plant species. It is evident from these data that the distribution of plants is much more immediately dependent on climate and edaphic factors than is that of animals.

A second reason for the difference between the distribution of plants and that of birds seems to be that plants (and some groups of animals) seem to be older than birds and that, as far as plants are concerned, the Australian habitat was already filled to capacity by well-adapted indigenous species when the later colonists from Malaya appeared.

Different Rates of Evolution

It has recently been shown that many plants, particularly trees, have had an exceedingly slow evolution. Others, like some herbaceous plants, have had evolutionary rates that compare well with those of birds and mammals. If, on the average, the evolutionary rate of plants is considerably slower than that of birds, one would expect a higher degree of endemism in the birds. This would mean that the following evolutionary levels might have been reached by those forms that became isolated in the New Guinea region in early Tertiary: birds into subfamilies and isolated genera, plants into species and less well-differentiated genera. This degree of differentiation, of course, is what is reported in the literature for the two groups.

It seems to me that the facts stated above help to dissolve most of the contradictions that seem to exist between the origin of plants and birds in the Papuan area.

REFERENCES

Burkill, I. H. 1942. The biogeographic division of the Indo-Australian Archipelago. 2. A history of the divisions which have been proposed. *Proc. Linn. Soc. London,* 154th session:127–138.

Copeland, E. G. 1948. The origin of the native flora of Polynesia. *Pacific Sci.,* 2:293–296.

Gibbs, L. S. 1917. *A contribution to the phytogeography and flora of Dutch N. W. New Guinea.* Taylor and Francis, London.

Guppy, H. B. 1906. Observations of a naturalist in the Pacific between 1896–1899. *Plant Dispersal,* vol. 2.

Karny, H. H. 1929. On the geographical distribution of the Pacific Gryllacrids. *Proc. 4th Pacific Sci. Congr.,* 3:157–172.

Lam, H. J. 1934. Materials towards a study of the flora of the islands of New Guinea. *Blumea,* 1:115–159.

Mayr, E. 1941a. *List of New Guinea birds.* Amer. Mus. Nat. Hist., New York.

———1941b. The origin and the history of the bird fauna of Polynesia. *Proc. 6th Pacific Sci. Congr.,* 4:197–216.

——1944. The birds of Timor and Sumba. *Bull. Amer. Mus. Nat. Hist.*, 83:127–194.

Merrill, E.D. 1945. *Plant life of the Pacific world*. Macmillan, New York.

Michaelsen, W. 1922. Die Verbreitung der Oligochaeten im Lichte der Wegenerschen Theorie der Kontinentalverschiebungen. *Verhandl. Nat. Ver. Hamburg*, 3:45–79.

Rensch, B. 1936. *Die Geschichte des Sundabogens*. Borntraeger, Berlin.

Richards, P. W. 1942. The biogeographic division of the Indo-Australian Archipelago: 6. The ecological segregation of the Indo-Malayan and Australian elements in the vegetation of Borneo. *Proc. Linn. Soc. London*, 154th session: 154–156.

Ripley, S. D., and H. Birckhead. 1942. Birds collected during the Whitney South Sea expedition. 51. On the fruit pigeons of the *Ptilinopus purpuratus* group. *Amer. Mus. Nov.*, no. 1192:1–14.

Setchell, W. A. 1935. Pacific insular floras and Pacific paleogeography. *Amer. Nat.*, 69:289–310.

Simpson, G. G. 1943. Mammals and the nature of continents. *Amer. Jour. Sci.*, 241:1–31.

Skottsberg, C. 1925. Juan Fernandez and Hawaii: a phytogeographical discussion. *B. P. Bishop Mus. Bull.*, 16:3–47.

Steenis, D. G. G. J. van. 1934–1935. On the origin of the Malaysian mountain flora. *Bull. Jard. Bot. Buitenzorg*, 13:135–262, 289–417.

44

The Ornithogeography of the Hawaiian Islands

Islands offer a special problem to the zoogeographer. As long as they are small and not too far offshore, they can safely be included in the same zoogeographic region with the nearest mainland. Doubts, however, arise in regard to the larger and more isolated islands. Most of the birds of New Zealand, for example, apparently arrived there from Australia. But some of the endemics are so unique, and the unchanged Australian element so small, that it seems hardly justified to include New Zealand in the Australian region. The same is true for Madagascar. In addition to an unquestionable African element the bird fauna of this island has a large endemic element (including 8 families) and a surprisingly large Oriental element (Rand 1936). An island with such a faunal composition obviously cannot be included with the Ethiopian region, as strictly defined.

Another island group that causes difficulties, at least to the ornithologist, is the Hawaiian Archipelago. For about a hundred years zoogeographers have associated these islands with Polynesia, apparently for reasons of geographical position and because the native humans are Polynesians. The fauna was practically unknown at the time when this classification was first proposed. In more formal zoogeographic studies the archipelago either was included in the "Polynesian subregion of the Australian region" or was accorded the rank of a "Hawaiian subregion," a course which I followed in my recent study of the borders of the Polynesian subregion (1941).

Since then I have made a more thorough analysis of the Hawaiian bird fauna and have found that its relationship with the Polynesian fauna is slight indeed. The total number of species of native Hawaiian land birds is open to doubt, since many of them are geographic representatives of each other and are considered full species by some authors, subspecies by others. There is, however, little doubt that these birds owe their origin to 14 separate invasions. These invasions are the

Revised from "The zoogeographic position of the Hawaiian Islands," *The Condor*, 45, no. 2(1943):45–48.

following, here listed according to their probable age with indication of relationships and sources.

Endemic Family
1. Drepanididae; related to cardueline finches or tanagers (American or Palearctic)

Endemic Genera
2. *Pennula*, rail; probably related to Porzana (Holarctic)
3. *Moho, Chaetoptila*—honeyeaters; related to *Amoromyza* and other honeyeaters (Australasian)
4a. *Phaeornis*, thrush; related to *Myadestes* (American)
4b. *Chasiempis*, flycatcher; related to the *Monarcha* group (*Pomarea, Mayrornis*, etc.) (Polynesian)
4c. *Nesochen*, goose; related to *Branta* (American)

Endemic Species
5a. *Corvus tropicus*, crow; related to continental *Corvus* (Holarctic)
5b. *Buteo solitarius*, hawk; related to *Buteo swainsoni* (American)
5c. *Anas wyvilliana*, duck; related to *Anas platyrhynchos* (Holarctic)

Endemic Subspecies
6a. *Asio flammeus sandwichensis*, short-eared owl (Holarctic)
6b. *Himantopus himantopus knudseni*, stilt (American)
6c. *Gallinula chloropus sandwichensis*, waterhen or gallinule (American)
6d. *Fulica americana alai*, coot (American)

Not Endemic
7. *Nycticorax nycticorax*, black-crowned night heron (American)

THE TAXONOMIC POSITION OF SOME OF THE HAWAIIAN ENDEMICS

The reliability of zoogeographic conclusions depends to a large extent on the soundness of the taxonomic work carried out on the groups studied. It is therefore of primary importance to determine the taxonomic position and nearest relatives of the Hawaiian endemics. There is no difficulty in regard to 7, 6d, 5b, and 4c of the preceding list; they are unquestionable American elements. The species to which 6b and 6c belong are widespread, but the endemic Hawaiian subspecies are closer to the American than to the Eurasian forms. One glance at the map is sufficient to convince one that the Holarctic immigrants (Drepanididae, *Corvus, Anas, Asio*) probably also came from North America, since it is considerably closer to Hawaii than is Asia. The

Hawaiian thrushes (*Phaeornis*) are descendants of the American *Myadestes* (Amadon 1942).

The taxonomic position of the Hawaiian flightless rail (*Pennula*) is and will probably remain doubtful. This genus has lost all distinctive characters and is now merely a nondescript-looking, small, brownish rail with a reduced wing. It is possibly related to the Polynesian *Aphanolimnas-Porzanoidea-Nesophylax* stock, but it seems more probable that it is a descendant of one of the genera of Holarctic rails (*Porzana*), as argued by Olson (1973).

The Hawaiian crow has certain peculiar characters that have appeared repeatedly in island forms of the genus *Corvus*. (Compare, for example, *C. jamaicensis* with *C. ossifragus*, and *C. fuscicapillus* with *C. validus*.) The plumage has lost its gloss and the individual feathers tend to be looser, more decomposed. The deep bill and the graduated tail suggest that the raven might be its nearest relative on the American mainland. However, according to Peale, the voice resembles that of the fish crow (*C. ossifragus*). North America is almost certainly the home of the ancestor of the Hawaiian Crow, even though the exact ancestral species may be in doubt.

All the genera and families mentioned up to now are represented in Hawaii by only one or two species. All of them together do not add up to the number of species by which the single family Drepanididae is represented on the Hawaiian Islands. The exact determination of the relationship of this diversified family is obviously of paramount importance. This group has had a tortuous taxonomic history. The genera belonging to it were originally scattered among the Fringillidae, Dicaeidae, Nectariniidae and Meliphagidae. Gadow (in Wilson and Evans 1891–1899) finally united them in the family Drepanididae, on the insistence of the field naturalist Perkins, whose observations of live birds had convinced him that the thick-billed and long-billed forms were closely related. Gadow concluded that they were more closely related to the Coerebidae than to the Tanagridae, but he did not make a very good case for this assumption since he did not compare them with typical tanagers. Sushkin (1929), on the other hand, presented some seemingly strong arguments in favor of a cardueline relationship. Birds like the goldfinch, purple finches, pine grosbeaks, and crossbills would, according to him, be the nearest relatives of the Drepanididae. Sushkin, however, fails to answer some of Gadow's objections against a cardueline relationship (nasal apertures, crop, etc.) and it seems, therefore, as if the last word has not yet been said. Actually the argument whether the Drepanididae are derived from the cardueline finches or from the tanagers is rather unimportant as far as the zoogeographic position of the islands is concerned since both groups are entirely absent from Polynesia and from the Australo-Papuan region. Hence, America (or northern Asia) must have been the home of the ancestors of the Drepanididae.

TIME OF COLONIZATION

The Hawaiian fauna has all the earmarks of that of an oceanic island. There is not a single serious modern student (I use the term *serious* advisedly) who believes in the former existence of land bridges between America and Hawaii, or between Polynesia and Hawaii. Oceanic islands are colonized at different times by accidental stragglers. The different degree of distinctness which the descendants of the bird settlers on Hawaii have attained can be regarded as irrefutable proof for the fact that they did not reach the islands simultaneously.

FAUNAL ORIGIN OF THE HAWAIIAN BIRDS

The faunal origin of the Hawaiian birds can be summarized as follows (the numbers on the left refer to the preceding list):

American: 4a, 4c, 5b, 6b, 6c, 6d, 7	7
Holarctic: 1, 5a, 5c, 6a	4
Polynesian: 3, 4b	2
Unknown: 2	1
	—
Total	14

Eliminating the single doubtful element (*Pennula*), we find that 11 of the 13 colonizations of Hawaiian birds probably came from North America and only two from Polynesia (honeyeaters, *Chasiempis*). All nonpasserine immigrants came from America, whereas the 5 colonizations of song birds are rather evenly divided (2 from Polynesia, 3 from America). Both Polynesian colonizations must be of considerable antiquity, since one of them produced an endemic genus and the other evolved even into two endemic genera (*Moho, Chaetoptila*). The oldest and most diversified group of Hawaiian birds, the Drepanididae, has branched out into about 12 genera, 22 species, and a total of 42 recognizable species and subspecies.

THE ZOOGEOGRAPHIC POSITION OF HAWAII

It is self-evident from the preceding remarks that the Hawaiian avifauna shows an overwhelming preponderance of North American, that is, Holarctic elements. The Polynesian element consists merely of one monotypic species of honeyeaters (*Chaetoptila angustipluma*), of one superspecies of honeyeaters (*Moho nobilis*), of one polytypic species of Old World flycatchers (*Chasiempis sandvicensis*), and possibly of a single polytypic species of rails (*Pennula sandwichensis*). The American or Holarctic element consists of at least 32 species or superspecies. On this basis the Hawaiian Islands should be included with the Nearctic region, in fact a case might even be made for including them in the next "A.O.U. Check-list"! However, it seems that the case of the birds is unique. All the workers on plants, insects, arachnids, and mollusks agree that the Hawaiian fauna of these groups is overwhelmingly Polynesian. It will, therefore, be wisest to evaluate the ornithological data in conjunction with the evidence from all the other groups, and to associate the Hawaiian Islands with the Australian Region, provided one believes at all in the principle of zoogeographic regions. There is, of course, a growing school of students who deny the validity of zoogeographic regions (Essay 38). They claim that there are faunas, but not regions. They say that one can speak of Nearctic and Palearctic faunas, but not of Palearctic or Nearctic regions. This is true for continents and even truer for islands.

I would now like to revert to the discussion at the beginning of this paper. Should the West Indies be included with the Neotropical or with the Holarctic region, should Madagascar be associated with the Ethiopian or with the Oriental region, should Celebes or New Guinea be included with the Oriental or with the Australian region, or should perhaps all of these islands be raised to the rank of separate regions, in addition to New Zealand and perhaps Hawaii? In each case, no decisive answer can be given. One can prepare a faunal analysis of each of these islands, but it shows in each case that the fauna is very heterogeneous. The faunas of these islands consist of a strong endemic element, as well as of immigrant components of various derivation. The same is true for all continental regions that are geographically intermediate between other major continents, such as North America or the East Indies. A faunal analysis will permit in such cases a much more accurate representation of facts than a regional analysis. Most of the "regions" of the regional zoogeographer coincide anyhow more or less with the major geographic subdivisions of the earth. To say that the bird fauna of North America consists of 23 percent Neotropical, 46 percent Nearctic, and 31 percent Palearctic elements gives a much more accurate picture of the composition of its fauna than to say that the Nearctic subregion is part of the Holarctic region. The time seems to have come to revise our zoogeographic classifications on the basis of this new concept. Its application by Stegmann (1938) to the birds of the Palearctic Region and by Stresemann (1939) to the birds of Celebes has been extremely fruitful.

REFERENCES

Amadon, D. 1942. Relationships of the Hawaiian avifauna. *Condor,* 44:280–281.

Baldwin, P. 1953. Annual cycle, environment and evolution in the Hawaiian honeycreepers (Aves: Drepanididae). *Univ. Calif. Publ. Zool.,* 52:285–398.

Bock, W. 1970. Microevolutionary sequences as a fundamental concept in macroevolutionary models. *Evolution,* 24:704–722.

Gadow, H. 1891–1899. Vögel. In *Klassen und Ordnungen des Thierreich's,* ed. Brown. C. F. Winter, Leipzig.

Mayr, E. 1939. The origin and the history of the bird fauna of Polynesia. *Proc. 6th Pacific Sci. Congr.* (1939), 4:197–217.

Olson, S. L., 1973. Evolution of the rails of the South Atlantic islands. *Smiths. Contr. Zool.,* no. 152:1–43.

Rand, A. L. 1936. The distribution and habits of Madagascar birds. *Bull. Amer. Mus. Nat. Hist.,* 72:294–299.

Stegmann, B. 1938. *Oiseaux.* In *Faune de l'URSS,* vol. 1, no. 2.

Stresemann, E. 1939. Die Vögel von Celebes. Zoogeographie. *Jour. f. Ornith.,* 87:312–425.

Sushkin, P. 1929. On the systematic position of the Drepanididae. *Verh. 6th Internat. Ornith. Kongr.* (1926), pp. 379–381.

Wilson, S. B., and A. H. Evans. *Aves Hawaiienses.* R. H. Porter, London.

45

The Nature of
Colonizations in Birds

A student of birds and other terrestrial vertebrates can make only an indirect contribution to the problem of the genetics of colonizing species. Not a single species of birds, whether colonizing or not, is genetically well known. The ornithologist can say nothing about the genotype of colonists or about genetic differences between successful and unsuccessful colonists. He can say nothing about the rate of genetic change in successful colonists. And yet ornithologists can make a contribution to this subject, because they have perhaps better information on historical changes in the biota than have students of any other component of the fauna and flora. The history of colonization is extremely well known for birds of many areas. The ornithologist thus may not be able to make a contribution on the genetics of colonization, but he can say a great deal about the phenotypes of colonizing species and from this draw certain inferences on the underlying genotypes.

There is such an abundance of phenomena of colonization described in the ornithological literature that it requires a theme to organize the raw data. The profound genetic difference between, on the one hand, the rich and diversified gene pools of central populations in widespread species and, on the other hand, the impoverished gene pools of isolated populations, particularly those that originated as founder populations, may serve as an ordering principle. As we shall presently see, there is indeed a considerable difference between the two kinds of populations in their ability to launch successful colonizations and with respect to their resistance to invading colonists.

COLONIZATIONS AND MAINLAND FAUNAS

Birds are not "weedy" like so many of the plants and small arthropods (insects, spiders, millipedes, etc.) that have followed man into disturbed habitats and are now essentially cosmopolitan. In nearly all cases in which birds seem to have followed man, the colonizations were the re-

Reprinted from "The nature of colonization in birds," pp. 30–47 in *The genetics of colonizing species,* ed. H. G. Baker and G. Ledyard Stebbins (New York: Academic Press, 1965).

sult of deliberate releases. Most frequently, so-called game birds are involved, such as ducks and gallinaceous birds, which were considered by local "sportsmen" suitable to be added to the local game. Almost equally frequent are releases of songbirds transported to new places in order to enrich the local bird chorus. Finally, in a few cases, owls were transported, as to Lord Howe Island and Hawaii, in the hope that they would help to control local rodent plagues. I have not been able to find a single definite record of an inadvertent passive dispersal of a bird by man, even though this is not entirely impossible. There are several reports in the literature of steamers carrying small groups of birds for several days that did not leave until land was reached after a lengthy ocean crossing (for instance, MacArthur and Klopfer 1958).

The impact of deliberate introductions on the bird faunas of fully occupied mainlands has been, on the whole, rather slight. In North America only 4 species have become thoroughly common among more than 50 species that were released, the starling (*Sturnus vulgaris*), the house sparrow (*Passer domesticus*), the pheasant (*Phasianus colchicus*), and the rock dove (*Columba livia*). About 6 other species have become locally established, including the Eurasian partridge (*Perdix perdix*), the Chinese starling (*Acridotheres cristatellus*), the tree sparrow (*Passer montanus*), and the mute swan (*Cygnus olor*) (Phillips 1928). There is no evidence that these introductions have led to the extinction of a single native species, although it is believed that the meadowlark (*Sturnella magna*) suffers in winter from food competition by starlings and some of the woodpeckers, particularly the red-headed woodpecker, from competition for nesting sites, also from the starling. The house sparrow (*Passer domesticus*) is believed to compete with the bluebird (*Sialia sialis*) and with some of the swallows for nesting sites.

The introduction of mammals and birds into Europe has been well told by Niethammer (1963). At least 7 species of mammals and 85 species of birds were introduced into parts of Europe where they had been previously absent. In the case of mammals, about two-thirds were successful, whereas only 13 successful establishments of birds are recorded. Most of these led to purely local success (mostly in the British Isles) and the ring-necked pheasant is the only truly successful introduction of a bird into Europe.

The isolated continent of Australia would, on first sight, appear to have a somewhat impoverished fauna. One might, therefore, expect that releases of alien bird species would be more successful than they were on the saturated northern continents. This is, however, not the case. Releases were only slightly more successful there than in Europe or North America. For instance, in the county in New South Wales in which Sydney is located, there are 204 breeding species of birds. Of the 50 or more species that were introduced only 15 became established and only 8 became common or reasonably common. There is no evidence that the introductions have caused a decrease in or the extinction of any native species (Hindwood and McGill 1958). In the state of Victoria an equally large number of species were introduced, but again only 10–15 became established. Considering how drastically the face of many parts of Australia has been changed by the destruction of the native vegetation, the impact of the importations is remarkably small. It is largely limited to the cities and towns.

Not all colonizations on mainlands are due to artificial releases. Some of them, at least, are the result of spontaneous range expansions. Every faunal history records numerous cases of species that have more or less dramatically crossed the borders of their range and expanded into vast areas from which they were previously absent.

The first bird species of which the colonization was studied in detail is the serin finch (*Serinus serinus*), which until the eighteenth century was largely restricted to the Mediterranean region, extending northward to the upper Rhine Valley in the west and to Carinthia in the east. During the nineteenth and the first half of the twentieth century, the species expanded rapidly through Germany, Austria, Hungary, and France, so that it is now found throughout western and central Europe north to the Baltic, occasionally even as far as Scandinavia (Mayr 1926). Even more rapid was the colonization of Europe by the collared dove (*Streptopelia decaocto*), excellently described by Stresemann and Nowak (1958). This species, which prior to 1925 was essentially limited to Asia and to the Balkan Peninsula south of the Danube and of Belgrade, began in the 1920s to expand into Rumania, Hungary, and Dalmatia, reaching Austria, Czechoslovakia, Poland, Italy, and Germany in the 1940s. The species is now extraordinarily common in many of the freshly occupied areas, with advance colonies existing in Scotland, Scandinavia, and Russia.

In both cases it seems that the beginning of the northward movement was favored by an amelioration of the climate and a spread of the favorite habitat of these birds. Once the species had made a certain amount of ecological shift, presumably favored by selection in favor of gene complexes suitable for colonization of the new climatic zones, the advance could proceed like an avalanche.

It must be emphasized, however, that the rapidity of the advance and the total area occupied is extreme in both cases, compared with numerous other northward range expansions of birds in Europe and North America that accompanied the amelioration of the climate. These have been studied particularly by Scandinavian and Finnish authors (for example, Kalela 1944). The rapid colonization of temperate Europe by the serin finch and collared dove thus seems to have had genetic reasons in addition to the external ones. Both species feed on seeds, particularly weed seeds, and the rapid spread of weeds and of areas favorable to weeds has surely been one reason for their success. However, the long period during which they stood, so to speak, poised at the threshold of the territory to be occupied suggests that some sort of genetic change triggered the final expansion. The two species are presumably another illustration of the importance of genetic changes in peripheral or peripherally isolated populations.

COLONIZATIONS AND ISLAND FAUNAS

The history of faunal changes on islands proves that island faunas offer far less resistance to immigrants than mainland faunas. There are now more introduced birds on the Hawaiian Islands as far as both species (44:28) and individuals are concerned than native ones (Table 45-1). This cannot be entirely ascribed to the replacement of extinct species because some 10 extinctions are compensated by 44 successful introductions. Obviously the native fauna was "unsaturated," although this is only one of a number of factors. Man introduced a great deal of new ecological diversity on Hawaii which has permitted colonizations that presumably would not have been successful in primeval Hawaii.

On New Zealand, the indigenous fauna of about 90 species of land birds was "enriched" by numerous introductions, of which about 19 species of non-Passeres and 15 species of Passeres became at least locally established, many of them now being widespread on both North Island and South Island. A similar story is true for most oceanic islands. The 19 species of native (or "self-introduced") land birds on Lord Howe Island, 300 miles east of the coast of Australia, were supplemented by 8 introductions, of which 6 were at least temporarily successful.

Some 40 years after all life on Krakatau had been completely destroyed by a volcanic eruption in August 1883, a fauna of 27 species of breeding land and freshwater birds had again become established. Considering that smaller, lower, and less fertile islands of the Sunda shelf have "so many more resident birds than Krakatau, we may safely conclude that the ornis of Krakatau will become twice as rich as it is at present" (Dammerman 1948:88). A figure of 54 species is perhaps a little optimistic, but 40-45 species would surely describe the capacity of Krakatau far better than 27 species.

Table 45-1. Success of introductions in certain islands and Australia.

Location	Native species		Introduced species	
	Living	Extinct	Successful	Unsuccessful
New Zealand	79	38	34	25-90
Hawaii	28	10	44	47
Lord Howe	10	7[a]	6	2+
Bermuda	6	0	7	1
Sydney County, Australia	204	0	15	35+

[a]Owing to invasion of rats (1918).

The 6 native land bird species of Bermuda have apparently not become rarer owing to introductions. Of 8 species known to have been introduced on this island, 3 have become tolerably common, 4 are only partially successful, and 1 has definitely become extinct.

The degree to which the fauna of an island is unsaturated and susceptible to enrichment through importation is different for different groups of animals. Newfoundland, a continental island that has not yet fully recovered from the effects of glaciation, is a good illustration of this statement. We have an excellent record of successful importations for Newfoundland, thanks to the researches of Palmen and Lindroth (Lindroth 1963). Among well-studied families and orders of arthropods and land mollusks, 146 species in a total of 1080 are regarded as introduced from Europe (=14 percent). Lindroth's tabulation shows how greatly the various kinds of animals differ from each other in their suitability for passive transport (Table 45-2). Among the 106 species of Newfoundland land and freshwater birds, only two species are of European derivation and both of these were probably derived from the United States.

NATURAL COLONIZATIONS

Records on the distribution of birds go back several hundred years for many places, and this provides an opportunity for exceptionally accurate records of new natural colonizations. For instance, 5–7 of the 19 native land birds of Lord Howe Island immigrated there during the last 100 years without assistance by man. At least 16 additional species have been found on this island as occasional stragglers or regular migrants, but have not yet taken up residence (Hindwood 1940). The story for New Zealand is similar. Seven of the 91 species of land birds established themselves during historic times. The most spectacular of these is the Tasmanian white-eye (*Zosterops lateralis*) which was first recorded on New Zealand around 1856, but soon spread like wildfire and is now the most common songbird of New Zealand. This aggressive

Table 45–2. Proportions of introduced species
in the Newfoundland fauna.

Classification	Total species	Introduced from Europe
Terrestrial isopods	12	12 = 100%
Millipedes	18	16 = 89%
Centipedes	12	9 = 75%
Carabid beetles	166	23 = 14%
Hemiptera (bugs)	72	9 = 13%
Birds (land and freshwater)	106	2 = 2%
Butterflies	280	4 = 1%

little colonist has since invaded the Chatham Islands, various of the sub-Antarctic islands (Snares, Auckland, Campbell, and Macquarie Islands), Kermadec Island, Norfolk Island, and apparently also Lord Howe Island. The original colonization required a jump of more than 1000 miles, and many of the subsequent jumps covered distances of several hundred miles in a region of high winds and inclement weather.

We are fortunate in having an almost blow by blow description of another transoceanic colonization by a bird, that of south Greenland by the fieldfare (*Turdus pilaris*) (Salomonsen 1951). Presumably on January 19, 1937, a flock of these thrushes left southwest Norway, some birds being recorded on January 20 at Jan Mayen and Ymer Island (off Greenland). In the period January 27–31, numerous records were made along the southwest coast of Greenland from Godthab south to Namortalik. Although the species is normally migratory in its more northerly populations, the Greenland population apparently became sedentary and produced a flourishing colony in the Julianehab district. As Salomonsen points out correctly, migratory individuals probably were eliminated from this population by natural selection, and this has strengthened its inherent sedentary tendencies. The recent amelioration of the climate in south Greenland, as well as the fact that there is no other large passerine bird in the area that might compete with the fieldfare, has contributed to the success of this colonization. Many similar accounts of successful transoceanic colonizations by birds can be found in the literature. Williams (1953) has followed, for instance, the fate of passerine birds that had been introduced originally on New Zealand, but have succeeded in colonizing outlying islands by spontaneous dispersal flights. Eleven species have participated in this colonizing expansion and the following numbers of species have reached these islands: Chatham (10–11), Campbell (9), Snares (8), Auckland (8), Sunday Island (Kermadec) (6), Norfolk (5), Lord Howe (4), Antipodes (2), and Macquarie (2). Most of these colonizations took place between 10 and 25 years after the species became common on New Zealand. Lindroth (1963 and earlier) and Elton (1958) have described many similar colonizations in other areas. The cattle egret (*Bubulcus ibis*) crossed the Atlantic early in the twentieth century (from Southern Europe or Africa) and established itself in northern South America. Within a few decades the species became widespread in the Caribbean and the southern United States and now breeds regularly as far north as Rhode Island. In this area there are now numerous colonies with many hundreds of breeding birds.

COLONIZATIONS AND FAUNAL TURNOVER

In the case of the Hawaiian Islands there is much circumstantial evidence to indicate that the introduced alien element had a considerable impact on the native fauna. The extinction of many native species may have been the direct result of the importations. The newcomers may affect the members of the indigenous fauna in two very different ways, as competitors and as vectors of disease organisms. There are reasons to believe that, in the case of birds, the effect of disease is far more drastic and hence more important than that of competition. The rapidity with which Hawaiian honeycreepers became extinct after the introduction of various Asiatic birds has focused attention on the possible importance of the disease factor (Amadon 1950; Hardy 1964; Warner 1968).

The vulnerability of island birds is best illustrated by the fact that more than 80 percent of all extinct birds were island birds, even though the number of island species is less than 10 percent of all bird species. The introduction of an alien species was in many cases the apparent cause of extinction. Predators, like the mongoose (*Herpestes burmanicus*) on Fiji and in the Antilles, and like certain aggressive species of ants, have, at least locally, produced equally devastating effects. An invasion of rats on Lord Howe Island, in the wake of a shipwreck, caused within a few years the extermination of 7 of the 12 endemic species or subspecies of birds on that island. A similar devastation of native faunas by rats has been described for many islands (Greenway 1958).

Man's impact on local faunas is thus undeniable, particularly on islands such as the Seychelles, Mauritius, the Hawaiian Islands, and New Zealand. Man has been the cause of drastic changes in the faunal composition of these islands either by the introduction of alien elements or by a complete change of the landscape. What is, however, often overlooked by zoogeographers is the fact that a comparatively rapid faunal turnover has been characteristic for islands long before man arrived on them. This is a phenomenon of such great evolutionary interest that it deserves far more attention than it has received until now.

I pointed out 25 years ago (Essay 40) that it is remarkable for such an old island as New Caledonia to have so recent an avifauna. Although the size of this island fluctuated throughout the Tertiary, there is no doubt that there was always room for an endemic land bird fauna of considerable size. Nevertheless, New Caledonia has only a single bird, the kagu (*Rhynochetos jubatus*), that can be called a truly old element. The island has 23 other endemics (3-4 weak monotypic genera and 19 species). About two-thirds of the 68 species in the land bird fauna are only subspecifically different from their Australian or Papuan relatives or even entirely identical with them. We thus find that on an island which is presumably more than 60 million years old most of the species seem to have arrived within the last 50,000 years. There is no evidence that would suggest that the island was not always able to accommodate at least 30 or 40 species of land birds, and we must therefore conclude that there was a continuous cycle of colonization and rapid extinction (Table 45-3).

A recent analysis of the birds of New Zealand by Fleming (1962) leads to the same conclusion (Figure 45-1). Among 93 species of New

Table 45-3. Faunal turnover as indicated by ratio of endemic and
nonendemic elements.

Location	Endemic genera	Endemic species	Nonendemic species with endemic subspecies	Nonendemic species with nonendemic subspecies
New Caledonia	5	19	23	19
Pantepui	2	30	55	11
New Zealand	31	15	17	19

Zealand land and freshwater birds, 19 are still identical with populations in the country of origin (mostly Australia), 17 have differentiated subspecifically, 15 have reached the status of endemic species, and only in the case of 29 colonizations has the rate of divergence reached the categorical level of genus or higher. Again assuming that there is no reason why New Zealand should not always have been occupied by at least 50 species (where there are now more than 90), we must assume a continuous process of rapid extermination and replacement.

Exactly the same is indicated by a continental insular fauna, the fauna of the Venezuelan highlands ("Pantepui"). The birds isolated on these tablelands owe their existence to about 96 colonizations. Of the geographical isolates of Pantepui, 11 are still subspecifically identical with the populations from the presumed place of origin. Fifty-five isolates belong to nonendemic species that have evolved endemic subspecies on Pantepui (sometimes several endemics per species), while 30 other isolates have reached the level of endemic species or superspecies. Only 2 endemics are classed as endemic genera, both of them being monotypic and rather slight. No endemic avian family occurs in the area (Mayr and Phelps 1965). The picture presented by Pantepui is thus exactly like those of New Caledonia and New Zealand: an isolated fauna of considerable size, but with more than two-thirds of the species evidently recent immigrants, and with only a rather small number of species old endemics, none of them very old (Table 45-3).

What is the conclusion we must draw from these figures? Quite obviously, the situation described by ornithologists differs rather drastically from the findings of the phytogeographers who have analyzed some of the same areas. Plants have a far higher autochthonous element and a relatively much smaller recent element. Speciation has been far more extensive among the plants of old insular areas, and there is much less evidence for a rapid replacement of floras.

It would be interesting to speculate on the reasons for the far more rapid turnover in birds than in plants. Two reasons can be suggested. First, there is nothing in birds that corresponds to the tightly knit "plant association," the entrance into which apparently causes such

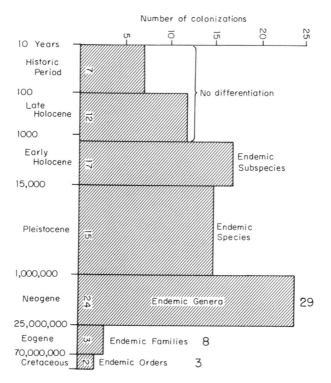

Figure 45-1. Histogram showing numbers of bird colonizations on New Zealand plotted against inferred date of colonization. Absolute age is shown on a logarithmic scale of years before the present. (From Fleming 1962.)

difficulties to a newcomer. In birds, each species is essentially on its own and more directly in competition with other species that occupy the same or similar niches. The second reason, perhaps only a different aspect of the first, is that many species of birds, such as the serin finch and the collared dove, seem to go through active expansion stages and that during such a stage they have a greatly increased ability to colonize new areas, provided they succeed in overcoming the geographical barrier separating them from the new area. This brings us to our next question.

WHAT DETERMINES SUCCESS IN COLONIZATION?

There is no simple answer to this question. The problem is perhaps most graphically demonstrated by a comparison of the distribution of floras with that of bird or mammal faunas. Let us take, for instance, the vast insular area between southeast Asia and Polynesia. On a phytogeographic map (see Burkill 1943) the greater portion of this region from Malaysia to at least the Solomons is included in a single vegetational region, the "Eastern Equatorial Region." Zoogeographically the area is sharply divided by Wallace's and Weber's Lines into a western portion, assigned to the Oriental region, and an eastern portion, assigned to the Australo-Papuan region. What is the reason for this dramatic difference between the distribution of higher vertebrates and plants? It may be postulated that three factors are responsible (Essay 43):

1. Plants have greater dispersal powers than higher vertebrates and therefore find it less difficult to bridge water gaps.

2. Whenever mammals or birds are successful in crossing a gap, however, they find it easier to establish themselves in a new area than do plants. In other words mammals and birds are better colonizers than plants. Plants are superior during dispersal within a climatic zone, whereas mammals and birds are superior in dispersal from one climatic zone to another.

3. Since the turnover of colonization and extinction in insular areas is more rapid in mammals and birds than in plants, the nature of the biota on the nearest larger land mass (source area) is of greater importance for mammals and birds than for plants. The contrast is between plants and higher vertebrates, not between plants and animals. Indeed, many groups of insects and other land invertebrates have distribution patterns that far more closely resemble those of plants than those of birds.

Faunal turnover, the balance between immigration and extinction, has been the subject of a special study by MacArthur and Wilson (1963). Their model allows for the fact that larger islands permit the establishment of richer faunas and that the number of species present is determined by the equilibrium between immigration and extinction. This work is a significant beginning in the analysis of faunal turnover, even though it does not endeavor to explain the causation of extinction. I venture to guess that the reasons for this are often as much genetic as ecological.

CHARACTERISTICS OF COLONIZING SPECIES OF BIRDS

There seems to be no single diagnostic characteristic by which one can distinguish those birds that are successful colonizers from those that are not. One cannot predict on the basis of morphological, ecological, or physiological characteristics whether or not a given species will be a successful colonizer. The Tasmanian white-eye (*Zosterops lateralis*), one of the most successful of all colonizers among birds, is in most respects exceedingly similar to many other species of *Zosterops* that are prevented by water gaps as narrow as 2 miles from undertaking colonizing flights. Nevertheless, when we look at the total assemblage of colonizing species of birds we find that on the whole they have one or several of the following characteristics.

1. A tendency to be social, that is to travel in small flocks. Groups of birds that show such a tendency, like the white-eyes, starlings, sparrows,

and certain thrushes, are among the most successful colonizers of highly isolated islands, whereas birds like woodpeckers that have a strong tendency to be solitary are easily stopped by geographical barriers.

2. Feeding characteristics. Among colonists in manmade habitats seed eaters are conspicuously more successful than insect eaters. The 2 species of birds that had such a spectacular expansion in western Europe in recent decades are both seed eaters. Of the 15 most successful species of birds introduced into Australia no less than 9 are seed eaters. This is a far higher percentage of seed eaters than found in the total bird fauna.

3. Habitat preference. Most of the widespread, more or less cosmopolitan species of birds (excluding sea birds) are associated with fresh water. Fresh water always has a scattered distribution and is often temporary. This places a premium on mobility and a willingness to undertake long colonizing flights. All 4 African species of birds that have invaded South America within recent times (2 species of ducks and 2 species of herons) are freshwater birds. Occupants of several other habitat types are also superior colonizers.

4. Considerable ecological flexibility. Successful colonizers must have the ability to fly across large stretches of ecologically unsuitable habitat. This is true not only for the freshwater birds just mentioned but also for the mountain birds, such as, for instance, the colonists on the Venezuelan tabletop mountains (Pantepui), which had to cross many hundreds of miles of lowland forest and savannahs. Several occupants of the mountain forest in the Indo-Australian archipelago have demonstrated the same capacity. The pigmy parrot (*Micropsitta bruijni*), the mountain leaf warbler (*Phylloscopus trivirgatus*), and the island thrush (*Turdus poliocephalus*) are typical examples. Stresemann (1939:379) and Mayr (Essay 41) have discussed the unexpected ability of mountain birds to cross from island to island.

5. A great ability to discover unoccupied habitats. This ability has often been described for "weedy" species, particularly those that can live successfully in manmade habitats. It is well known for all species that live in temporary habitats, second growth formations, and along the sea coast. It is virtually never pointed out that, at least in birds, entirely nonweedy species also have this ability. On the island of Timor, for instance, 14 of the 137 species of breeding birds are restricted to the monsoon forest in the mountains. Only one of these species (*Ducula cineracea*) is endemic on Timor (and Wetar). All the others are clearly recent Malaysian immigrants, more or less widespread in the mountain forests of the Sunda Islands and southeastern Malaysia. Not one of them has the characteristics of a weedy species. Several of them have evolved endemic subspecies on Timor, yet on the whole this mountain fauna seems more recent than much of the lowland bird fauna. This is in striking contrast to the usual situation where the mountains are occupied by an older element (including relics) than the lowlands. It is possible that the monsoon forest in the mountains of

Timor had become completely extinct during some phase of the Pleistocene and was reconstituted only recently.

6. An ability to shift habitat preference. Some of the most successful colonizers have a remarkable capacity for shifting into new kinds of habitats. The island thrush (*Turdus poliocephalus*), for instance, which is only found in the high mountain forest, near the upper tree line, in New Guinea and the islands west of New Guinea (as far as Sumatra and the Philippines), is limited to this same habitat on the main islands of the Solomons. It descends, however, to sea level on Rennell Island and some of the islands of Melanesia and western Polynesia.

All the more successful colonizers among birds have one or several of the six characteristics listed in the preceding paragraphs. Dispersal, then, is not a more or less random phenomenon. If we take a rich mainland fauna, certain families of birds will make a greater contribution to the colonization of surrounding islands than others. Of the 62 families of birds that occur in New Guinea, 36 (=58 percent) reached the Solomons and 23 (=37 percent) Fiji. While New Guinea has 535 species of native land and freshwater birds, the Solomons have 126 (23.5 percent of the New Guinea total) and Fiji 55 (10 percent). However, if we look at the bird fauna of the Solomon Islands and Fiji it at once becomes apparent that certain among the families of New Guinea birds, 16 families to be exact, have contributed the major share of the colonizers of the islands; 315 (=59 percent) of the New Guinea birds, 98 (=78 percent) of the Solomon Island species, and 46 (=84 percent) of the Fijian species belong to these 16 families (Table 45-4). The further away from New Guinea one gets, the higher the percentage of species that belong to characteristically colonizing genera.

THE GENETICS OF FOUNDER POPULATIONS

I have discussed in several recent publications (1954, 1963) what seem to be the genetic characteristics of founder populations. Two aspects are important for this discussion. The first is that such populations are definitely in a precarious position. It is quite typical for spontaneous

Table 45-4. Land and freshwater birds of Papua and Melanesia
(in percent).

Classification	New Guinea	Solomons	Fiji
Families	100	58	37
Species	100	23.5	10
Species in 16 most successful families	100	31	14.5
Percentage of species of colonizing families in total species	59	78	84

immigrants as well as for artificial importations to become extinct again after a few years or decades in spite of a temporary initial success. This has been described for colonists on New Zealand, Lord Howe Island, and Hawaii. Just what the sudden resistance of the environment is that converts apparent success into failure is not known.

The second point to be stressed is that many of these founder populations began with a rather uniform gene pool. There is much direct and even more indirect evidence that many successful colonizations were started by a couple of pairs or, in the case of insects, perhaps even by a single fertilized female. As soon as such founders have succeeded in establishing a founder population, they will be able to accumulate new genetic variation and, in due time, acquire the level of genetic diversity that seems to be necessary to compensate for the deleterious effects of inbreeding. I have recently listed (1963) a number of cases where founder populations were astonishingly successful in spite of having passed through the bottleneck of very small population size or perhaps even because of it.

It appears from these observations that a particular structure of the genotype in the founder population may well be less important for the success of a colonization than certain phenotypic characters of the founders, such as the ability to undertake social dispersal flights, to convert after arrival from a migratory to a sedentary habit, and to make a necessary shift in the niche after arrival in the new home.

REFERENCES

Amadon, D. 1950. The Hawaiian honeycreepers (Aves, Drepanididae). *Bull. Amer. Mus. Nat. Hist.,* 95:151–262.

Burkill, I. H. 1943. The biogeographic division of the Indo-Australian Archipelago. 2. A history of the divisions which have been proposed. *Proc. Linn. Soc. London,* 154:127–138.

Dammerman, K. W. 1948. The fauna of Krakatau. *Verhandel. Koninkl. Ned. Akad. Wetenschap. Afdel. Naturk. Sect. II,* 44:1–594.

Elton, C. S. 1958. *The ecology of invasions.* Methuen, London.

Fleming, C. A. 1962. History of the New Zealand land bird fauna. *Notornis,* 9: 270–274.

Greenway, J. C., Jr. 1958. *Extinct and vanishing birds of the world.* Amer. Comm. Intern. Wildlife Protection Ser. Publ. 13, New York.

Hardy, D. E. 1964. *Insects of Hawaii.* University of Hawaii Press, Honolulu.

Hindwood, K. A. 1940. The birds of Lord Howe Island. *Emu,* 40:1–86.

Hindwood, K. A., and A. R. McGill. *The Birds of Sydney.* Roy. Zool. Soc. N.S.W., Sydney.

Kalela, O. 1944. Zur Frage der Ausbreitungstendenz der Tiere. *Ann Zool. Soc. Zool. Botan. Fennicae Vanamo,* 10:1–23.

Lindroth, C. H. 1963. The faunal history of Newfoundland. *Opuscula Entomol., Suppl.,* 23:1–112.

MacArthur, R. H., and P. Klopfer. 1958. North American birds staying on board ship during Atlantic crossing. *Brit. Birds,* 51:358.

MacArthur, R. H., and E. O. Wilson. 1963. An equilibrium theory of insular zoogeography. *Evolution,* 17:373–387.

Mayr, E. 1926. Die Ausbreitung des Girlitz (*Serinus canaria serinus* L.). *Jour. f.
Ornith.,* 74:571–671.

——— 1954. Change of genetic environment and evolution. In *Evolution as a Process,* ed. J. Huxley, A. C. Hardy, and E. B. Ford. Allen & Unwin, London.

——— 1963. *Animal species and evolution.* Belknap Press of Harvard University
Press, Cambridge, Mass.

Mayr, E., and W. H. Phelps, Jr. 1967. The origin of the bird fauna of the southern
Venezuelan highlands. *Bull. Amer. Mus. Nat. Hist.,* 136:269–328.

Niethammer, G. 1963. Die Einbürgerung von Säugetieren und Vögeln in Europa.
Parey, Berlin.

Phillips, J. C. 1928. Wild birds introduced or transplanted in North America. *U.S.
Dept. Agr. Tech. Bull.,* 61:1–63.

Salomonsen, F. 1951. The immigration and breeding of the fieldfare (*Turdus pilaris*
L.) in Greenland. *Proc. 10th Int. Ornith. Congr.* (Uppsala, 1950), pp. 515–526.

Stresemann, E. 1939–1941. Die Vögel von Celebes. *Jour. f. Ornith.,* 87:299–425
(1939); 88:1–135, 389–487 (1940); 89:1–102 (1941).

Stresemann, E., and E. Nowak. 1958. Die Ausbreitung der Türkentaube in Asien
und Europa. *Jour. f. Ornith.,* 99:243–296.

Warner, R. E. 1968. The role of introduced diseases in the extinction of the endemic
Hawaiian avifauna. *Condor,* 70:101–120.

Williams, G. R. 1953. The dispersal from New Zealand and Australia of some introduced European passerines. *Ibis,* 95:676–692.

IX
BEHAVIOR

Introduction

Much of my youth was devoted to the intimate study of bird behavior, and a thorough knowledge of bird behavior and bird song was a key to my success as collector in New Guinea and the Solomon Islands. Professional commitments prevented me from undertaking a major monographic study of a single species of birds, but I reported some observations in short papers, listed below. Like all students of bird behavior, I was impressed by species-specific differences in behavior as well as by the selective significance of behavior. My territory study (1935) is in this tradition, as are the two following essays on the relation between behavior and evolution. For a long time I was particularly interested, as part of my concern with speciation, in behavioral isolating mechanisms between closely related species. A number of experimental studies on two sibling species of Drosophila, D. pseudoobscura *and* D. persimilis *(in part jointly with T. Dobzhansky), were the outcome of this interest (see the list below). It has been a source of particular satisfaction to me that behavior has become the deep interest of so many of my students (John Alcock, Robert Barth, George Hunt, Robert E. Jenkins, Ross Lein, Andrew J. Meyerriecks, W. John Smith, Susan T. Smith, and Roderick Suthers).*

Considerations of space preclude the reprinting of all my publications on behavior. Papers not reprinted here are given in the list below.

OBSERVATIONS ON BIRD BEHAVIOR

1928. Weidenmeisen-Beobachtungen (*Parus atricapillus salicarius* Brehm). *Jour. f. Ornith.*, 76:462–470.

1930. Beobachtungen über die Brutbiologie der Grossfusshühner von Neuguinea (*Megapodius, Talegallus* und *Aepypodius*). *Ornith. Monatsb.*, 38:101–106.

1935. Bernard Altum and the territory theory. *Proc. Linn. Soc. N.Y.*, nos. 45, 46:24–38.

1941. Red-wing observations of 1940. *Proc. Linn. Soc. N.Y.*, nos. 52–53:75–83.

1948. Gulls feeding on flying ants. *Auk*, 65:600.

1948. Repeated anting by a song sparrow. *Auk*, 65:600.

EXPERIMENTAL STUDIES ON *DROSOPHILA* BEHAVIOR

1944. Experiments on sexual isolation in *Drosophila*. I. Geographic Strains of *Drosophila willistoni*. *Proc. Nat. Acad. Sci.*, 30(9):238–244. (With T. Dobzhansky.)

1945. Experiments on sexual isolation in *Drosophila*. IV. Modification of the degree of isolation between *Drosophila pseudoobscura* and *Drosophila persimilis* and of sexual preferences in *Drosophila prosaltans*. *Proc. Nat. Acad. Sci.*, 31(2):75–82. (With T. Dobzhansky.)

1946a. Experiments on sexual isolation in Drosophila. VI. Isolation between *Drosophila pseudoobscura* and *Drosophila persimilis* and their hybrids. *Proc. Nat. Acad. Sci.*, 32(3):57–59.

1946b. Experiments on sexual isolation in Drosophila. VII. The nature of the isolating mechanisms between *Drosophila pseudoobscura* and *Drosophila persimilis*. *Proc. Nat. Acad. Sci.*, 32(5):128–137.

1950. The role of the antennae in the mating behavior of female *Drosophila*. *Evolution*, 4(2):149–154.

46

Behavior and Systematics

Ideally, a comparative study of behavior should include every single behavior element in all the species of an entire higher taxon of animals. A systematic effort to achieve this is at the present time impossible for two reasons. Unfortunately, no consistent classification of behavior elements is available, nor is any group of animals sufficiently well known to permit tracing individual behavior elements or behavior patterns throughout the entire group.

Work in mammals, for instance, deals almost exclusively with single species in a genus, or at best a few species in an entire family. Information on birds is more complete owing to systematic work such as that of Tinbergen and his associates on various species of the gull family, of Heinroth and Lorenz on ducks (Anatidae), and of Hinde and others on finches. Work in fishes has concentrated on sticklebacks and cichlids, where good comparative work is available. A great deal of information on insects, particularly social insects, has been published, and our information is building up rapidly. The same is true for grasshoppers as a result of the work of Faber and Jacobs. The most complete study of the behavior of any group of related species is that of Spieth on *Drosophila.* Much work has also been done on spiders (Crane), fiddler crabs (Crane), slugs (Gerhardt), and other isolated species and genera. Still, not a complete or even nearly complete inventory of the behavior patterns in a single family of animals is as yet available.

The backward state of the field of animal behavior is, to a considerable extent, due to the former absence of working hypotheses and heuristic schemes. As a consequence, most earlier work amounted to little more than the accumulation of raw data. It is the particular merit of Lorenz to have provided a set of hypotheses and theories which have tremendously stimulated research in this area. This is a merit which is not decreased even if some of these hypotheses were oversimplifications, or even entirely wrong. The main body of his concepts has been accepted by most students of animal behavior. There is now hope for a

Revised from "Behavior and systematics," pp. 341–362 in *Behavior and evolution,* ed. A. Roe and G. G. Simpson (New Haven: Yale University Press, 1958).

synthesis in this field, such as has been achieved recently in the field of evolution.

The comparative method is even more important for the study of the evolution of behavior than for that of morphological features. The conclusions of comparative anatomy can be tested by the paleontologist, who either substantiates or disproves them with the help of fossil material. There is no such recourse for the ethologist, since behavior does not fossilize.

CLASSIFICATION AND THE PHYLOGENETIC METHOD

Evolutionists have frequently expressed their surprise at how little the classification of better-known groups of animals was affected by the establishment and acceptance of the theory of evolution. Actually, there should have been no surprise, in view of the taxonomic method and the premises of the theory of evolution. Good classifications are based on a multitude of characters, and the only probable way to account for two organisms agreeing in the majority of their characteristics is to ascribe this agreement to descent from a common ancestor. The phylogenetic method is largely based on this reasoning.

The extension of the phylogenetic method from purely morphological to behavior characters is based on the same consideration. Students of behavior have found again and again that species or genera that had been placed next to each other on the basis of morphological characteristics also agreed or were similar in their behavior patterns. Again this should not have been a surprise, since these forms share a common heritage and since much behavior, particularly species-specific behavior, has a genetic basis. Following the pioneer efforts of Whitman and Heinroth, increasing attention is being paid to systematists to study of the behavior element.

Similarity in behavior between two species does not, however, necessarily mean common descent. The student of behavior, just like the morphologist, must make the distinction between homology and analogy. The decision between these two alternatives is even more difficult for the ethologist than for the morphologist. When in doubt, the morphologist can always fall back on Owen's criterion of homology, that of position. The impossibility of a strict application of this criterion to behavior has induced at least one morphologist to deny the propriety of extending the concept "homology" to behavior elements. This would seem an unnecessarily restricted position in view of the modern biological and evolutionary meaning of the term "homology." But what criterion can the behavior student use to establish a homology? His method will be the same as that of the pre-Darwinian taxonomist. He will base his conclusions on the sum total of behavior characters. The more behavior elements are consistent with a postulated phylogeny, the greater the probability that the phylogeny has validity.

If the postulate of the equivalence of morphological and behavior characters is correct, then we should find the same phenomena among behavior characters as among morphological ones. There should be nongenetic variation, intrapopulation variation, geographic variation, species-specific characters, group characters, and polyphyletic characters; there should be primitive and advanced characters, and parallel evolution. To be sure, the available evidence is as yet somewhat scanty, but it indicates that all this variation of morphological characters is indeed paralleled by behavior characters. Some of the evidence for this statement will be presented in subsequent sections of this essay.

From the point of view of usefulness, taxonomic characters range between two extremes: those that are so invariant in a large taxonomic group that they are useless for classification, like the two eyes of vertebrates; and those that are so variable among different individuals (or so easily affected by the environment) that they permit no discrimination whatsoever between closely related taxa. Between these extremes are those characters which are constant within a given taxonomic group (species, genus, family) but vary between it and another taxonomic group. There are many behavior characters that fall into this category.

THE UTILIZATION OF BEHAVIOR
CHARACTERISTICS IN CLASSIFICATION

The evidence supplied by morphological characters is sometimes ambiguous, and valuable supplementary information is in some cases provided by behavior. Some structural characters are very superficial and lead to evidently artificial groupings. If there is a conflict between the evidence provided by morphological characters and that of behavior, the taxonomist is increasingly inclined to give greater weight to the ethological evidence. This procedure has led to a number of recent improvements in the classification of certain groups of animals. The following may be selected from a large number of cases.

1. Three European species of grasshoppers, *Parapleurus alliaceus*, *Mecostethus grossus*, and *Ailopus thalassinus*, used to be classified with the subfamily Acridinae on the basis of morphological characters. However, Jacobs (1953) found that these three species agree in various behavior characteristics much better with members of the subfamily Oedipodinae, and subsequent analysis has revealed some morphological characteristics that support this shift.

2. The crag martins (*Ptyonoprogne*) have been customarily placed near the bank swallows (*Riparia*) or even been united with this genus. Mayr and Bond (1943) pointed out that the nesting habits of the two kinds of birds differ drastically and suggested that the crag martins be placed near the barn swallow group (*Hirundo*) because, like the latter, they build a nest from pellets of mud, while the bank swallow digs

tunnels into sand banks. The new taxonomic placement is supported by
the voice of these swallows, and by concealed white spots on the tail
feathers of crag martins, a sign stimulus also found in *Hirundo* but not
in *Riparia.* This case illustrates how even the slightest clue must be
utilized to determine relationships in a group as morphologically
uniform as are these swallows. Nest architecture has been shown by
Lack (1956) to provide valuable clues in the family of swifts.

3. Studies by Heinroth (1911) on the behavior of ducks yielded
many results at variance with the accepted classification, which was
based on morphological characters. Using this ethological information
as well as additional characters (including the color of the downy
young), Delacour and Mayr (1945) proposed a radically new classification. In this the diving ducks were split into two groups, freshwater
divers (pochards) and sea ducks; the mergansers (*Mergus*) were
associated with the golden-eyes (*Glaucionetta*); the wood duck (*Aix*)
removed from the river ducks (Anatini), and so on. These findings
have been largely confirmed by subsequent investigations. Where
modifications were proposed (for example, separation of the eider
group and the merganser group), the reasons again were in part the
result of behavior studies.

4. In a recent reclassification of finchlike birds, Tordoff (1954)
associated the genus *Fringilla* (chaffinch and relatives) with the New
World finches. Andrews and Hinde (1956) were able to show, however,
that *Fringilla* agrees in its behavior much better with the Old World
finches (Carduelinae), and Mayr (1956) came independently to the
same conclusion on the basis of morphological criteria.

In all the stated cases a species, a genus, or a group of genera was
shifted from its traditional place in the zoological system to a new
position as a result of the study of behavioral criteria. The new arrangement was subsequently confirmed in all these cases by new or
reevaluated morphological evidence.

BEHAVIOR ELEMENTS AS TAXONOMIC CHARACTERS

Taxonomic characters have a dual function (Mayr, Linsley, and
Usinger 1953). They have a diagnostic value (permitting discrimination
between similar taxa) and they have an associative value (permitting the
grouping together of related taxa). This is as true for behavior characters as it is for other taxonomic characters.

Behavior Characters in Taxonomic Discrimination

In a number of recently described cases, a study of behavioral attributes permitted much finer taxonomic discrimination than was possible
with the use of morphological criteria. A few examples of this may be
mentioned.

The conventionally recognized family of titmice ("Paridae") of the ornithologists has no morphological characters that would permit further subdivision. Yet the nest-building habits indicate that the so-called family is a somewhat artificial assemblage of four, perhaps not closely related, groups of birds. The first comprises the true titmice (*Parus* and relatives), which apparently always nest in hollow trees or other cavities. The second group consists of the long-tailed tit (*Aegithalos*), the bush tit (*Psaltriparus*) and related forms which build an oval nest with lateral entrance in bushes and trees and which are very social birds; all members of this group have essentially the same habits and call notes. The third group consists of the penduline titmouse (*Remiz*) and its relatives, which build a peculiar retort-shaped nest of plant down worked into feltlike consistency (the nest being similar to that of some flowerpeckers, Dicaeidae). The fourth group consists of the bearded titmouse (*Panurus*), which builds a stick nest with a lateral entrance and which by this and other habits is unmasked as belonging to the babbler family (Timaliidae). The study of behavior thus not only indicates that these four groups of genera are not very closely related but also provides clues for the proper placing of some of the forms.

The weaver finches (Estrildidae) have always been considered one of the groups of weaver birds (Ploceidae), and some of the older authors did not even separate them in a distinct subfamily. Steiner (1955), however, presents evidence, most of it derived from the study of behavior, to show that these two groups of seed-eating birds of the Old World tropics are not as closely related as previously believed (Table 46-1).

A study of the method employed by various groups of grasshoppers in cleaning their antennae has confirmed the justification of giving them family status. Gryllidae, Tettigoniidae, and other groups with long antennae clean them with the help of their maxillae. The Acrididae place a leg on one antenna and clean the antenna by pulling it through between leg and substrate. The Tetrigidae (and this is one reason why this group of genera is placed in a separate family) clean the antennae by stroking them with the legs, and the latter in turn by pulling them through the mouth (Jacobs 1953).

Behavior characters have proven particularly useful in distinguishing morphologically very similar species, the so-called sibling species. This is true for sibling species in many genera of animals (Essay 34). For example, in a study of the wasp *Ammophila campestris*, Adriaanse (1947) noticed that some individuals had a behavior pattern that agreed with previous descriptions, while others had an aberrant pattern (Table 46-2). The latter turned out to be a new species (*A. adriaansei*). In several other cases behavior gave the first clue to the discovery of sibling species. In the North American fireflies of the genus *Photuris*, Barber (1951) discovered several sibling species on the basis of the number, timing, and coloring of the flash signals. For further examples see Evans 1953.

Table 46–1. Differences between Estrildidae and Ploceidae.

Character	Estrildidae	Ploceidae
1. Nest	A globular structure of small twigs or grass stems with lateral entrance	Usually a finely woven structure, often hanging from twigs
2. Pair bond	Tightly knit, often lasting through years	Either no pair formation, or polygamy, or a pair bond of short duration
3. Parental care	Nest building, incubation, feeding of young jointly done by both parents	Most parental duties performed by the female alone
4. Tail movements	Lateral	Vertical
5. Courtship posture	Stiff upright, with wings pressed against the body; stereotyped repetition of song strophe	Excited courtship dances with wing fluttering and occasional display flights; noisy chatter
6. Incubation	From first egg on; young hatch with daily intervals	After completion of clutch; young hatch simultaneously
7. Tongue, gape, and throat of nestlings	With peculiar species- or genus-specific pigment spots and papillae	Without markings or papillae
8. Feeding of nestlings	Take regurgitated food from the crop of the parent	Normal food begging; parents feed with the bill
9. Begging of young	Without wing flutter	With wing flutter
10. Nest hygiene	Droppings of young not removed	Droppings of young removed by parents

Behavior Characteristics in Taxonomic Grouping

Behavior characters are taxonomically useful not only for the separation of taxa but also in giving clues as to the relationship of taxa of uncertain taxonomic position. This is well illustrated by the family of bowerbirds (Ptilinorhynchidae). All the 8 genera of this fascinating family are rather different from each other in color, and they were variously separated into 3 to 5 subfamilies on the basis of morphological criteria. However, Stresemann (1953) and Marshall (1954) showed that previous classifications were artificial and that on the basis of bower construction the following three subdivisions could be recognized: (1) stagemakers and catbirds (*Ailuroedus*, *Scenopœëtes*, and perhaps the poorly known *Archboldia*), (2) maypole bower builders (*Prionodura*, *Amblyornis*), and (3) avenue bower builders (*Sericulus*,

Table 46-2. Behavioral differences between *Ammophila campestris*
and *A. adriaansei.*

Ethological character	campestris	adriaansei
Nest hole filled with material from	A quarry	Flown in
Choice of food	Sawflies	Caterpillars
Sequence of egg laying and provisioning	First egg, then prey	First prey, then egg
Breeding season	Earlier, ending in August	Later, until middle of September

Ptilonorhynchus, Chlamydera). Subdivisions 1 and 2 have white eggs; 3
has colored eggs with a highly characteristic pattern of streaking. The
absence of bower building in *Cnemophilus,* always considered a bower-
bird, resulted in anatomical investigation which showed that it is
actually a bird of paradise.

The relationship of the spider wasps (*Pompilidae*) is another problem
solved with the help of ethology. On morphological grounds they are so
distinctly set off from other wasps that they are not placed in any of
the major families. Yet a study of the behavior traits of this family
leaves no doubt that they were derived from bethyloid-scolioid stock
and have evolved independently of other stocks of wasps (Evans 1953).
The following combination of behavioral characters defines the family,
according to Evans, as clearly as any comparable set of morphological
characters: (1) All utilize spiders as larval food. (This occurs, of course,
also in some other families, as in the mud daubers). (2) All stock the
nest cell with a single paralyzed prey (a habit otherwise exhibited
principally by certain Scolioidea and Bethyloidea). (3) In transportation to the nest, the spider is seized in the wasp's mandibles and
dragged backward over the ground (some Bethylidae and Ampulicidae
share this behavior, some Pompilidae lack it). (4) To close the nest the
female wasp pounds down the earth with the apex of the abdomen, or
in mud users uses it as a trowel for manipulating the mud (other mud
daubers use the mandibles and legs for this purpose). (5) The nest is
often prepared after the prey has been taken (as with many Bethylidae).

In all the cited instances, behavior characters have established previously unknown relationships or have confirmed a relationship that
had previously been only tentatively established. It is important to
stress that there are many behavior traits, such as the method of drink-
ing of pigeons and sand grouse, which are characteristic for entire
higher categories (genera, families, orders, or classes). All the available
evidence indicates that in their genetic basis as well as in their phylo-

genetic history such characters are completely equivalent to morphological characters which have a similar taxonomic distribution.

POLYPHYLETIC AND ANALOGOUS CHARACTERS

The systematist must be aware at all times of the shortcomings of the phylogenetic method. Similarities between two kinds of organisms do not always prove common descent. The same behavior may occasionally be acquired independently in unrelated or distantly related forms. A "rattle flight" has evolved not only in some grasshoppers of the subfamily Oedipodinae but also in unrelated genera of Acrididinae (Jacobs 1953). Small species of herons (Meyerriecks 1960) tend to utilize aerial displays much more commonly than large species, regardless of relationship.

Parallelisms and analogies are particularly common in all types of behavior that are strictly functional, such as food getting or locomotion. Birds that have similar food habits, such as those that crack seeds (Emberizidae, Fringillidae, Ploceidae, Estrildidae, Psittacidae, etc.), that prey on mice and other small vertebrates (hawks, owls), that feed like flycatchers (Tyrannidae, Muscicapidae, Todidae, even some Alcedinidae, etc.), that feed on nectar (Trochilidae, Nectariniidae, Meliphagidae), and that have otherwise similar feeding habits, show many similarities mistakenly regarded as indicating relationship. The same is true for birds with similar locomotion, such as diving (grebes, loons, *Hesperornis*), running (ostrich, emu, rhea, moa), wading (storks, cranes, flamingos, shore birds), hawking (swallows, Artamidae, swifts, nightjars), and so forth.

Perhaps one reason for the slowness of getting behavior characteristics used in systematics is that early authors, from Aristotle to about 1700, did actually use behavior (and associated adaptations) largely as the basis of their systems of classifications. Unfortunately, however, they emphasized food getting and locomotor behavior, which among birds and mammals are, on the whole, of rather low taxonomic value. This disappointment had to be thoroughly forgotten before behavior could again assert its rightful place in systematics.

THE VARIABILITY OF BEHAVIOR

Genetic variability in animals is universal, a fact that is significant not only for the student of morphology but also for the student of behavior. It is not only wrong to speak of *the* monkey but even of the behavior of *the* rhesus monkey. The variability of behavior is evident in the study not only of such a genetically plastic species as man but even of forms with very rigid, stereotyped behaviors, such as hunting wasps. The Peckhams (1898) give a delightful description of such behavior differences between individuals of the wasp *Ammophila urnaria*: "While one [individual] was beguiled from her hunting by every sorrel blossom she passed, another stuck to her work with indefatigable perseverance. While one stung her caterpillar so carelessly and made her nest in so shiftless a way that her young could only survive through some lucky chance, another devoted herself to these duties not only with conscientious thoroughness, but with an apparent craving after artistic perfection."

Whitman emphasized as early as 1899 that "the clocklike regularity and inflexibility of instinct . . . have been greatly exaggerated. They imply nothing more than a low degree of variability under normal conditions . . . close study and experiment with the most machine-like instincts always reveal some degree of adaptability to new conditions." One would like to have more precise information on the nature of this variation: How variable is a given behavior within a single population? How much variation is there from population to population within a species? Finally, one would like to know what portion of the variability is due to nongenetic modification.

"Innate" is of course only the reaction norm, which has a more or less wide range of phenotypic expression. The term "innate" is meaningful only if it is interpreted epigenetically (rather than preformistically!). This is fully understood by the geneticist, who states that a certain flower color or the presence of wing veins is "inherited." The fact that the tendency to hoard is "innate" in the Norway rat is not negated by the fact that certain treatments or experiences may reduce this tendency or obliterate it altogether. Most mammals cannot be induced to hoard no matter what treatment they get. The time has come to stress the existence of genetic differences in behavior, in view of the enormous amount of material the students of various forms of learning have accumulated on nongenetic variation in behavior. Striking individual differences have been described for predator-prey relations, for the reaction of birds to mimicking or to warning colorations, for child care among primates, and for maternal behavior in rats. It is generally agreed by observers that much of this individual difference is not affected by experience but remains essentially constant throughout the entire lifetime of the individual. Such variability is of the greatest interest to the student of evolution, and it is to be hoped that it will receive more attention from the experimental psychologist than it has in the past.

From the evolutionary point of view, there is an interesting conflict between two opposing selective forces. One selects that behavior which is optimal for the species and places a premium on uniformity of behavior within the species. The other favors variability as a means of preserving evolutionary plasticity. The most important mechanisms maintaining genetic variability of populations are perhaps geographic variation and gene flow.

The study of morphological characters has shown that the differences between species are often foreshadowed by minor or incipient differences between geographic races. The study of such geographic variation has shed a great deal of light on the origin of taxonomic characters and of new species. Studies of this sort with respect to behavior are at the very beginning. However, there is evidence for geographic variation of song in birds and in grasshoppers. In some species of the bee genus *Halictus* some geographic races are solitary while others are colonial (Michener 1953a). The begging posture of young mockingbirds of the genus *Nesomimus* in the Galápagos Islands is supposed to vary between islands. The difficulty of this field is that most behavior studies are made at a single locality, while geographic variation in behavior can be demonstrated only if the behavior of different, geographically segregated, populations of the same species is investigated.

THE PHYLOGENETIC ORIGIN OF NEW BEHAVIOR

There are at least two different possibilities for the acquisition of a new behavior pattern by a species.

1. The new behavior may have a genetic basis right from the beginning. Since much behavioral variability is correlated with the genetic variability of the species, any factor affecting the gene content of the species may also affect behavior. Some of these effects may occur as an incidental by-product of genes selected for very different properties. Some of the behavioral variability described above may have this source.

2. A new behavior is at first a nongenetic modification of an existing behavior, as a result of learning, conditioning, or habituation, and is gradually replaced by genetically controlled behavior.

The study of a new behavior "fashion" might be very revealing. When titmice in England acquired the habit of opening milk bottles, it was observed that the technique was highly variable (Fisher and Hinde 1949).

Daanje, Hinde, and other ethologists have pointed out that in birds much instinctive behavior seems to have started as intention movements or as displacement activities. Indeed, what is an intention movement or a displacement activity in one species may be incorporated into the courtship repertoire in a related species. The genetic basis of this process is still completely obscure. This is one of the few evolutionary phenomena where the "Baldwin effect" might have played a role, although the behavior, after its incorporation in the courtship, is in a different neural "environment" than it was previously. How such a change of neural tie-up may be achieved is still quite puzzling.

A thorough study of individual variation of behavior is perhaps the most promising approach to the problem of the origin of new behavior. Typological concepts have, in the past, retarded the analysis. There was a search for specific mutations producing specific new behavior patterns. It is much more likely that most kinds of behavior have a multiple genetic basis. This hypothesis is supported by the observation that much of the behavior of interspecific and intergeneric hybrids is somewhat intermediate between that of the two parental species.

Even more important for an understanding of the origin of behavior than a balanced concept of mutation is the proper consideration of natural selection. There is nearly always a dynamic balance between opposing selective forces, as is clearly evident, for instance, in the case of the so-called superoptimal stimuli. The study of these various selective forces, even though they were already considered by Darwin, is still in its infancy. Indeed, the whole subject of the evolutionary origin of new behavior and the remodeling of behavior by natural selection is a much neglected field. The study of entire populations would seem to be the most promising approach to a solution.

NEW BEHAVIOR AND MORPHOLOGICAL STRUCTURES

In the days of mutationism a heated controversy raged as to whether behavior precedes structure or vice versa. The mutationists postulated that mutation caused structural changes and that the organism then developed the behavior which best permitted it to cope with the new structure. This purely typological approach was, of course, wrong. We now know that matters are much more complex. When it comes to structure we have to specify whether we are dealing with the structure of the nervous system or the structure of peripheral organs that facilitate or emphasize behavior. On the whole, it seems correct to state, as Lorenz has emphasized, that behavior movements often precede phylogenetically the special structures that make these movements particularly conspicuous. Many birds raise the feathers of the crown or hind neck and bow their heads toward females or competitors. However, only in a limited number of species have long crests developed that emphasize these movements. Likewise, in grasshoppers there is a widespread intention movement of flexing the legs prior to jumping. However, only in a limited number of higher groups has this movement been incorporated into courtship and led to a conspicuous coloration of those parts of the legs that are shown or to the development of sound organs which produce sound during the repeated flexing of the legs. Yet it is obvious that without legs this movement could not have been made at all, just as wing displays in birds would be impossible without the prior presence of wings. Likewise, all behavior depends on certain structural components in the nervous system. It is now obvious that there is no general answer to the question "Structure first or behavior first?" Each case must be analyzed separately to determine all of its components.

TRENDS IN THE EVOLUTION OF BEHAVIOR

The student of evolution is not satisfied merely to prove that evolution has occurred and to reconstruct phylogenies; he also wants to know whether it is possible to make generalizations concerning the course of evolution and to express observed regularities in the form of "rules" or "laws" with wide applicability. It would be very satisfying if this could also be done for the evolution of behavior. "Instincts, like corporeal structures, may be said to have a phylogeny . . . The main reliance in getting at the phyletic history must be a comparative study" (Whitman 1899). Such a study is now being conducted by Tinbergen and his associates for various species of gulls. They find that hostile and courtship movements are essentially homologous in all species of the family but may differ in relative frequency, intensity, and specific form. A movement that is hardly noticeable in one species may be a conspicuous component of courtship in another species. Similar comparative studies on birds, with similar results, have been carried out by Lorenz (1941, 1952) for ducks (Anatidae), by Meyerriecks (1960) for herons (Ardeidae), by Hinde and associates for finches, by Morris and associates for weaver finches (Estrildidae), and by Whitman (1919) for pigeons. Along the same line is much work cited in the references, such as that on cichlids (Baerends and Baerends 1950), on grasshoppers (Jacobs 1953), on *Drosophila* (Spieth 1952), and on various hymenopterans (Evans 1953; Michener 1953; Iwata 1942; Lindauer 1956), to mention merely a few.

Students of evolution have long indulged in constructing phylogenetic trees on the basis of morphological characters. Where such constructions are based on fossil material, they tend to have a degree of reality. Where they are merely based on degrees of morphological complexity, they are usually pure speculation and often demonstrably wrong. The wishful hope that no secondary simplification has occurred is not always fulfilled. Attempts to reconstruct behavior phylogenies face the same difficulties as similar attempts of the morphologist. Yet their value is not to be minimized and such reconstructions have a considerable heuristic value. Evans (1953), for instance, has reconstructed the development of behavior in Pompilidae from the simplest to the most complex and derived pattern. Desneux (1952) has traced the probable development of "architectural styles" in the nests of termites of the African *Apicotermes* group. In the hive bees of the genus *Apis* a number of living species permit a reconstruction of the evolution of the dancing behavior (von Frisch 1955). One can discern certain evolutionary trends in the courtship behavior of *Drosophila* (Spieth 1952) and in the bower building of bowerbirds (Marshall 1954).

Students of vertebrates and particularly of mammals are aware of a broad phylogenetic trend toward an increased role of the higher centers of the brain at the expense of purely instinctive behavior. No other type of behavior has been studied as intensively as cortex-centered behavior in the primates and in man. It must not be forgotten, however, that, taking the animal kingdom as a whole, this is a very exceptional type of behavior. Similar trends are found in some groups of birds

(Corvidae, Sturnidae, Psittacidae), but the enlarged central nervous system of even the most advanced insects is not in this category.

"An instinct may sometimes run through a whole group of organisms with little or no modification" (Whitman 1899). An analysis of behavior differences among species, genera, and higher categories, such as was presented in the preceding sections, shows that in addition to very conservative types of behavior there are some that change rapidly in evolution. It would be intriguing to find out whether or not different components of behavior have different rates of evolution. This indeed seems to be the case.

Among the many components of behavior two stand out: (1) the ability to react selectively to specific objects and (2) the specific actions of which the behavior consists. This may be illustrated for the case of predatory wasps: (1) the object is the spider, the caterpillar, or the sawfly larva; (2) the action consists of grasping and stinging. Both 1 and 2 may and do change in the course of evolution, but it seems that in the majority of cases the action (the pattern of locomotion) is more stereotyped, less plastic than the choice of the object. Wasps of many different genera may have a rather stereotyped stinging movement, but the prey that is stung may vary from species to species or from genus to genus. In pollen-collecting bees there may be great similarity in pollen-collecting equipment and method, yet each species or genus may have a decided preference for the pollen of specific flowers. This species specificity of food preference has an obvious selective value because it will reduce competition for food among species and perhaps permit a denser occupation of the habitat.

In the courtship of birds likewise, the courtship movements seem much more conservative than the "releasers" that elicit them. Many responses to special components of the environment either are learned during the lifetime of the individual or have the earmarks of recent evolutionary acquisition. For instance, the "following response" of young ducks consists of very stereotyped movements; the object of the response may be learned by imprinting.

Since the choice of an object, the reaction to a stimulus, is determined by a perception pattern, it can be suggested that the locomotory components of behavior patterns tend to be more conservative in evolution than the perceptual. This is a working hypothesis that seems worth further testing.

REFERENCES

Adriaanse, A. 1947. *Ammophila campestris* Latr. und *Ammophila adriaansei* Wilcke. Ein Beitrag zur vergleichenden Verhaltenforschung. *Behaviour*, 1:1–34.

Armstrong, E. A. 1947. *Bird display and behaviour.* Oxford University Press, New York.

———1950. The nature and function of displacement activities. In *Physiological mechanisms in animal behaviour.* Symp. Soc. Exp. Biol., no. 4. Cambridge University Press, Cambridge.

Baerends, G. P. 1950. Specializations in organs and movements with a releasing function. In *Physiological mechanisms in animal behaviour*, Symp. Soc. Exp. Biol., no. 4. Cambridge University Press, Cambridge.

Baerends, G. P., and J. M. Baerends. 1950. *An introduction to the study of the ethology of cichlid fishes. Behaviour*, suppl. 1.

Barber, H. S. 1951. North American fireflies of the genus *Photuris. Smiths. Misc. Coll.*, 117(1):58.

Bastock, M., D. Morris, and M. Moynihan. 1954. Some comments on conflict and thwarting in animals. *Behaviour*, 6:66–84.

Berland, L. 1943. Les classifications des naturalistes confirmées par l'instinct. *Rev. Sci.*, 81:59–64.

Crane, J. 1949. Comparative biology of salticid spiders at Rancho Grande, Venezuela. Pt. IV. An analysis of display. *Zoologica*, 34:159–214.

———1952. A comparative study of innate defensive behavior in Trinidad mantids (Orthoptera: Mantoidea). *Zoologica*, 37:259–293.

Daanje, A. 1951. On locomotory movements in birds and the intention movements derived from them. *Behaviour*, 3:48–98.

Davis, D. E. 1942. The phylogeny of social nesting habits in the Crotophaginae. *Quart. Rev. Biol.*, 17:115–134.

Delacour, J., and E. Mayr. 1945. The family Anatidae. *Wils. Bull.*, 57:3–55.

Desneux, J. 1952. Les constructions hypogées des *Apicotermes* termites de l'Afrique tropical. *Ann. Mus. Roy. Congo Belge*, 17:9–120.

Drees, O. 1952. Untersuchungen über die angeborenen Verhaltensweisen bei Spring-Spinnen (Salticidae). *Z. Tierpsychol.*, 9:169–207.

Evans, H. E. 1953. Comparative ethology and the systematics of spider wasps. *Syst. Zool.*, 2:155–172.

———1955. An ethological study of the digger wasp *Bembecinus neglectus*, with a review of the ethology of the genus. *Behaviour*, 7:287–303.

Evans, H. E., C. S. Lin, and C. M. Yoshimoto. 1953. A biological study of *Anoplius spiculatus autumnalis* (Banks) and its parasite, *Evagetes mohave* (Banks) (Hymenoptera, Pompilidae). *Jour. N.Y. Entomol. Soc.*, 61:61–78.

Faber, A. 1929. Die Lautäusserungen der Orthopteren. I. *Z. Morphol. u. Ökol.*, 13:745–803.

———1937. Die Laut– und Bewegungäusserungen der Oedipodinen. *Z. wiss. Zool.*, 149:1–85.

———1953. *Laut– und Gebärdensprache bei Insekten. Orthoptera I*. Ges. Freunde Mus. Naturkunde, Stuttgart.

Fabricius, E. 1954. Aquarium observations on the spawning behaviour of the burbot, *Lota vulgaris* L. *Rep. Inst. Freshwater Res.* (Drottningholm), 35:51–57.

Fisher, J., and R. A. Hinde. 1949. The opening of milk bottles by birds. *British Birds*, 42:347–357.

Friedmann, H. 1949. The breeding habits of the weaverbirds. A study in the biology of behavior patterns. *Ann. Rept. Smiths. Inst. for 1949*, pp. 293–316.

——— 1955. *The honey-guides.* Bull. U.S. Nat. Mus., no. 208.

Frisch, K. von. 1955. Beobachtungen und Versuche M. Lindauers an indischen Bienen. Sitzungsber. Bayer. *Akad. Wiss., Math.–Nat. Klasse.*, pp. 209–216.

Gerhardt, U. 1929. Zur vergleichenden Sexualbiologie primitiver Spinnen, insbesondere der Tetrapneumonen. *Z. Morphol. u Ökol.*, 14:699–764.

—— 1934–1941. [Studies on Limacidae.] *Z. Morphol. u. Ökol.* vols. 27, 28, 30, 31, 32, 34, 35, 36, 37.

Gruhl. K. 1924. Paarungsgewohnheiten der Dipteren. *Z. wiss. Zool.*, 122:205–280.

Hall, C. 1951. Psychogenetics. In *Handbook of experimental psychology*, ed. S. S. Stevens. Wiley, New York.

Heinroth, O. 1911. Beiträge zur Biologie namentlich Ethologie und Psychologie der Anatiden, *Verh. 5th Int. Ornith. Kongr.* (Berlin, 1910), pp. 589–702.

——1930. Über bestimmte Bewegungsweisen der Wirbeltiere. *Sitzber. Ges. Naturforsch. Freunde* (Berlin), pp. 333–342.

Hinde, R. A. 1952. The behaviour of the great tit (*Parus major*) and some other related species. *Behaviour*, suppl. 2.

——1953. The conflict between drives in the courtship and copulation of the chaffinch. *Behaviour*, 5:1–31.

——1955. A comparative study of the courtship of certain finches (Fringillidae). *Ibis*, 97:706–745.

——1956. A comparative study of the courtship of certain finches (Fringillidae). *Ibis*, 98:1–23.

Iwata, K. 1942. Comparative studies on the habits of solitary wasps. *Tenthredo*, 4:1–146.

Jacobs, W. 1953. *Verhaltensbiologische Studien an Feldheuschrecken. Z. Tierpsychol*, suppl. 1.

Jameson, D. L. 1955. Evolutionary trends in the courtship and mating behavior of Salientia. *Syst. Zool.*, 4:105–119.

Kessel, E. L. 1955. The mating activities of balloon flies. *Syst. Zool.*, 4:97–104.

Krumbiegel, I. 1938. Physiologisches Verhalten als Ausdruck der Phylogenese. *Zool. Anz.*, 123:225–240.

Lack, D. 1956. A review of the genera and nesting habits of swifts. *Auk*, 73:1–32.

Lindauer, M. 1956. Orientierung bei indischen Bienen. *Z. f. vergl. Physiol.*, 38:521–557.

Lorenz, K. 1941. Vergleichende Bewegungsstudien an Anatinen. *Jour. f. Ornith.*, 89(suppl. 3): 194–294.

—— 1951–1952. Comparative studies on the behaviour of the *Anatinae* (trans. of Lorenz 1941), trans. C. H. D. Clarke. *Avicult. Mag.*, 57:157–182; 58:8–17, 61–72, 86–93, 172–183.

——1952. Die Entwicklung der vergleichenden Verhaltensforschung in den letzten 12 Jahren. *Verh. Deutsch. Zool. Gesell.*, suppl. 17:36–58.

Maidl, F. 1934. *Die Lebensgewohnheiten und Instinkte der staatenbildenden Insekten*. Vienna.

Makkink, G. F. 1936. An attempt at an ethogram of the European avocet (*Recurvirostra avosetta* L.) with ethological and psychological remarks. *Ardea*, 25:1–62.

Marshall, A. J. 1954. *Bower-birds: their displays and breeding cycles*. Clarendon Press, Oxford.

Mayr, E. 1948. *The bearing of the new systematics on genetical problems. The nature of species. Advances in genetics*, 2, 205–231.

Mayr, E., R. J. Andrews, and R. A. Hinde. 1956. Die systematische Stellung der Gattung *Fringilla. Jour. f. Ornith.* 97:258–273.

Mayr, E., and J. Bond. 1943. Notes on the classification of the swallows, Hirundinidae. *Ibis*, 85:334–341.

Mayr, E., E. G. Linsley, and R. L. Usinger. 1953. *Methods and principles of systematic zoology*. McGraw-Hill, New York.

Meyerriecks, A. J. 1960. *Comparative breeding behavior of North American herons.* Nuttall Ornith. Club, Cambridge, Mass.

Michener, C. D. 1953a. Problems in the development of social behavior and communication among insects. *Trans. Kansas Acad. Sci.,* 56:1–15.

——1953b. Life-history studies in insect systematics. *Syst. Zool.,* 2:112–118.

Morris, D. 1954a. The reproductive behavior of the zebra finch (*Poephila guttata*), with special reference to pseudofemale behaviour and displacement activities. *Behaviour,* 6:271–322.

——1954b. The courtship behaviour of the cutthroat finch. *Avicult. Mag.,* 60:169–177.

Moynihan, M. 1954. *Some aspects of reproductive behavior in the black-headed gull* (Larus ridibundus L.) *and related species. Behaviour,* suppl. 4.

—— 1962. *Hostile and sexual patterns of South American and Pacific Laridae.* E. J. Brill, Leiden.

Moynihan, M., and F. Hall, 1956. Hostile, sexual, and other social behavior patterns of the spice finch (*Lonchura punctulata*), in captivity. *Behaviour,* 7:33–76.

Peckham, G. W., and E. G. Peckham. 1898. *On the instincts and habits of the solitary wasps.* Wisc. Geol. Nat. Hist. Surv. Bull., no. 2.

Petrunkevitch, A. 1926. The value of instinct as a taxonomic character in spiders. *Biol. Bull.,* 50:427–432.

Rau, P. 1942. The nesting habits of *Polistes* wasps as a factor in taxonomy. *Ann. Ent. Soc. Amer.,* 35:335–338.

Reed, C. A. 1946. The copulatory behavior of small mammals. *Jour. Comp. Psychol.,* 39:185–206.

Regen, J. 1913. Über die Anlockung des Weibchens von *Gryllus campestris* L. durch telephonisch übertragene Stridulationslaute des Männchens. *Pflügers Arch. ges. Physiol.,* 155:193–200.

——1926. Über die Beeinflussung der Stridulation von *Thamnotrizon apterus* Fabr. Männchen durch künstlich erzeugte Töne und verschiedenartige Geräusche. *Sitz. Ber. Akad. Wiss. Wien,* 135:329–368.

Richards, O. W. 1927. Sexual selection and allied problems in insects. *Biol. Rev.,* 2:298–364.

Schmidt, R. S. 1955. The evolution of nestbuilding behavior in Apicotermes. *Evolution,* 9:157–181.

Scott, J. P., and E. Frederisson. 1951. The causes of fighting in mice and rats. *Physiol. Zool.,* 24:273–309.

Seitz, A. 1940–1941. Die Paarbildung bei einigen Cichliden. *Z. Tierpsychol.,* 4:40–84.

Spieth, H. T. 1947. Sexual behavior and isolation in *Drosophila.* I. The mating behavior of species of the *willistoni* group. *Evolution,* 1:17–31.

——1949. Sexual behavior and isolation in *Drosophila.* II. The interspecific mating behavior of species of the *willistoni* group. *Evolution,* 3:67–81.

——1951. Mating behavior and sexual isolation in the *Drosophila virilis* species group. *Behaviour,* 3:105–145.

——1952. Mating behavior within the genus *Drosophila* (Diptera). *Bull. Amer. Mus. Nat. Hist.,* 99:395–474.

Steiner, H. 1955. Das Brutverhalten der Prachtfinken, Spermestidae, als Ausdruck ihres selbständigen Familiencharakters. *Acta 11th Congr. Int. Ornith.* (Basel), pp. 350–355.

Stresemann, E. 1953. Spielplätze und Balz der Laubenvögel. *Jour. f. Ornith.*, 94:367–368.

Thorpe, W. H. 1954. Some concepts of ethology. *Nature*, 174:101–105.

Tinbergen, N. 1948. Social releasers and the experimental method required for their study. *Wils. Bull.*, 60:6–51.

——1951. *The study of instinct.* Oxford University Press, Oxford.

——1952. "Derived" activities: their causation, biological significance, origin, and emancipation during evolution. *Quart. Rev. Biol.*, 27:1–32.

——1953a. Fighting and threat in animals. *New Biol.*, 14:9–24.

——1953b. *The herring gull's world.* Collins, London.

——1954. The origin and evolution of courtship and threat display. In *Evolution as a process,* ed. J. Huxley, A. C. Hardy, and E. B. Ford. Allen and Unwin, London.

Tordoff, H. B. 1954. *A systematic study of the avian family Fringillidae based on the structure of the skull.* Misc. Publ. Mus. Zool. Univ. Mich., no. 81.

Weidmann, U. 1951. Über den systematischen Wert von Balzhandlungen bei *Drosophila. Rev. suisse zool.,* 58:502–511.

Weyrauch, W. 1939. Zur Systematik der paläarktischen Polistinen auf biologischer Grundlage. *Arch. Naturg.,* n.s., 8:145–197.

Whitman, C. O. 1899. Animal behavior. In *Biological lectures delivered at the Marine Biological Laboratory at Woods Hole in 1898.*

——1919. The behavior of pigeons. In *Posthumous works of Charles Otis Whitman.* Carnegie Inst. Wash. Publ. no. 257, vol. 3.

Zimmerman, K. 1956. Gattungstypische Verhaltensformen von Gelbhals-, Wald-, und Brandmaus. *Zool. Garten,* n.s., 22:162–171.

Zippelius, H. M. 1949. Die Paarungsbiologie einiger Orthopteren-Arten. *Z. Tierpsychol.,* 6:372–390.

47
Behavior Programs and Evolutionary Strategies

When Descartes proclaimed in the seventeenth century that an animal was nothing but a machine, he started a controversy that, in ever-changing form, has continued to the present day. As far as behavior is concerned, Descartes saw only the stimulus and the response, as if the response was a necessary and automatic reaction to the stimulus. The contribution the animal itself, particularly its central nervous system, makes to the response was entirely ignored. This omission became the object of a great controversy in the eighteenth century. Condillac (1755), attacking Buffon's Cartesian interpretation, stressed the ability of animals to learn from experience and to act intelligently. This, in turn, was objected to by Reimarus (1760), who pointed out, with well-chosen examples, how much of the adaptive behavior of animals is already present at birth, being part of the heritage of the species. He presented the first comprehensive treatment of instinctive behavior.

More recently, Jacques Loeb revived in part the Cartesian tradition, the behaviorists that of Condillac, and the ethologists that of Reimarus. However, it would be misleading to interpret these historical connections too literally. We are rapidly approaching a synthesis in which valid elements of all three interpretations are combined.

No matter how different their explanations were, all these schools agreed that behavior is remarkably adaptive. Descartes's animal machines responded to their environment just as appropriately as the instinct animals of Reimarus or the conditioned-reflex animals of Pavlov. No one questioned the economic efficiency of the hexagonal cells built by the honeybee or the benefits of homing and migration. For the devout of past centuries such perfection of adaptation seemed to provide irrefutable proof of the wisdom of the Creator. For the modern biologist it is evidence for the remarkable effectiveness of natural selection.

Darwin's *Origin of Species* brought about a fundamental reorientation in the study of behavior. Previously behavior was studied either as

Reprinted from "Behavior programs and evolutionary strategies," *American Scientist,* 62, no. 6 (1974):650–659; reprinted by permission of *American Scientist;* copyright © 1973 by Sigma Xi, The Scientific Research Society of North America.

a physiological process (without a time dimension), or (by the natural theologians) as a manifestation of the wisdom of the Creator, or finally (by the naturalists) as an entertaining component of the diversity of nature. With Darwin entirely new questions came to the fore, such as: Which selective forces cause the adaptive modification of a given behavior? What do comparative studies reveal concerning the phylogeny of particular behavior patterns? And finally: What components of behavior are genetically controlled and to what extent? These questions are of obvious concern to an evolutionist, since selection would not be able to modify behavior during evolution if it were not for the genetic components.

Biologists in the immediate post-Darwinian period were so preoccupied with other problems that they displayed only slight interest in behavior. To make matters worse, those who were interested were Lamarckians, who postulated the ready conversion of "habit" into "instinct." It was not until Weismann established in 1883 his thesis of the noninheritance of acquired characters that a new wind began to blow. Building on this new insight, Lloyd Morgan (1891) and McDougall (1912) made a good beginning of a theory of behavior based on evolution. Unfortunately, their explanations had a number of fatal weaknesses, and the new development soon dried up as the result of the attacks of the reflexologists and behaviorists.

It was the ethologists who revived an interest in the evolutionary aspects of behavior, and this emphasis of theirs was considerably more important in some respects than the details of ethological theory. It has led in recent years to a growing synthesis of behavioral biology and evolutionary biology—to the posing of a whole battery of new questions on the evolution of behavior patterns and on the impact of behavior on the course of evolution.

The history of behavioral biology is a history of controversies, many of which were caused by misconceptions or misunderstandings. Let me therefore try to make a few things clear before beginning the discussion of our main topic. Much of the argument concerning animal behavior dealt with the question: How much of it is *innate* and how much *acquired* through experience? The trouble with this terminological dichotomy is that "innate" refers to the genotype and "acquired" to the phenotype, and consequently neither term is the exact opposite of the other one. It occurred to me that this difficulty might be overcome by relating behavior to the concept of a genetic program. This concept resulted from an interaction of molecular biology and information theory, and has proved very illuminating in many branches of biology.

CLOSED AND OPEN PROGRAMS

We can ask what differences exist between genetic programs responsible for behaviors formerly called innate and those regarded as experientially acquired. A genetic program that does not allow appreciable modifications during the process of translation into the phenotype I call a *closed program*—closed because nothing can be inserted in it through experience. Such closed programs are widespread among the so-called lower animals. A genetic program that allows for additional input during the lifespan of its owner I call an *open program* (Essay 2).

Even this improvement in terminological precision does not remove all our difficulties. A particular instinctive behavior act is, of course, never controlled directly by the genotype but rather is controlled by a behavior program in the nervous system that resulted from the translation of the original genetic program. It is particularly important to make this distinction for the open program. The new information acquired through experience is inserted into the translated program in the nervous system rather than into the genetic program because, as we know, there is no inheritance of acquired characters. Neurobiology is still a long way from being able to study such behavior programs directly. All we can do at the present time is to infer their existence and nature by indirect methods.

Ethologists and evolutionary biologists have been concerned primarily with the evolutionary history of genetic programs and with the selective advantage of certain types of genetic programs. In other words, they have concentrated on "ultimate causes"—that is, on the explanation of the adaptive significance of various behaviors and the pathways of their origin. Physiologists and most experimental psychologists have been far more, if not entirely, concerned with the physiological explanation of neural programs and with the pathway of their translation into observed behavior. This has led them, on the whole, to ask questions entirely different from those asked by the evolutionary behavior student. In terms of causation, physiologists and experimental psychologists have concentrated on the study of "proximate causes"—that is, the explanation of phenomena during or subsequent to the decoding of the genetic program. They share this interest with the students of learning and of the ontogeny of behavior. The psychologists' approach was originally very one-sided and rather intolerant of any other approach. But drastic changes took place in the 1950s and 1960s, and the comparative psychology of today is a very different field from classical comparative psychology (Lockard 1971).

A good deal of uncertainty has also afflicted some of the more specific concepts of behavioral biology. The instinct concept is a typical example of one of the vague concepts that ethology has clarified, in this case by helping to break it up into three components: the cognitive, the conative, and the executive. Just how much will ultimately be left of the original formulations of the ethologists is still undetermined. For instance, as far as the cognitive component is concerned, one can ask: Do all animals have *innate releasing mechanisms* that respond to specific sign stimuli (releasers), resulting in appropriate responses (displays, etc.)? Is a given releasing mechanism controlled by the genetic program of the species (innate), or is it individually acquired either by gradual learning or by a specially rapid learning process (imprinting)? Is the releasing mechanism located in the central nervous system or is it part of the screening mechanisms in the peripheral system (including the sense organs) (Konishi 1971)? Many answers have been given to these questions through the researches of the last forty years. The most important, but for a biologist by no means unexpected, finding is that these answers may differ from species to species. No one will be able to develop a universally valid theory of behavior by studying only a single species of animal. What is needed are comparative studies—that is, the comparison of equivalent behavior manifestations in many different kinds of animals. Let me confirm this need by an example.

Male animals preferably display to females of their own species, while females normally respond only to the displays of their own males. For instance, if one keeps a freshly hatched *Drosophila* female in complete isolation until she is ready to mate and then offers her simultaneously a choice of males of several species, including her own, all these males may display to her, but she almost unerringly accepts the courtship only of the male of her own species. She acts as if she has an "innate" knowledge of the diagnostic characters of her own species. The same has been found for the display behavior of most animal species (Essay 46). Actually in most cases only a few key stimuli are crucially involved, the so-called releasers of the ethologists. Indeed, as evolutionists have now clearly established, most of the mechanisms of animals that guarantee reproductive isolation between related species consist of acts or structures that provide "species recognition" information (Wickler 1971; Brown 1965).

Let me give another example. There are several groups of birds—for instance Old World cuckoos and some New World cowbirds—in which the female lays her eggs in the nest of a foster species, let us say, in the case of the cowbird, in the nest of a song sparrow or yellow warbler. The foster parents raise the young cowbird after it hatches and continue to feed it for a period of two to three weeks after it leaves the nest until the fledgling becomes independent. It then, so to speak, says goodbye to its foster parents and searches out the company of other cowbirds, which form flocks usually composed of both immatures and adults. These flocks stay together for the entire fall, winter, and the beginning of the next spring. When the mating season arrives, they break up. The young cowbird mates with another member of his own species, and the female starts searching the nest of a foster species in which to place her eggs. Quite clearly, in this case the program for species recognition—that is, the program for recognizing the appropriateness of the future mate—was contained completely in the original fertilized zygote. This is what I call a *closed* genetic program. With few exceptions, species recognition in animals is controlled by a closed genetic program.

Let me now report some exceptional cases in which there is an *open* program for species recognition. As was first discovered by Oskar Heinroth and later confirmed by Konrad Lorenz, in some species of birds, for instance the greylag goose, the freshly hatched chick will follow the first moving object making sounds and adopt it as parent and sometimes even as potential mate. This very rapid and largely irreversible form of learning is called imprinting (Hess 1973). Imprinting is defined as a process of learning in which the object—parent, mate, companion— of an otherwise instinctive (but previously object-less) action is acquired. The best-known cases of imprinting are the "following reaction" of recently hatched precocial birds and the sexual response in many species of animals. Imprinting of a somewhat less drastic and less irreversible form may sometimes occur with respect to choice of food, nest material, song, habitat selection, sociability, roosting places, and other components of behavior. This method of enriching the behavioral program has the great advantage of being able to store far more complex objects of behavior than the genetic program (Immelmann (1969).

A particularly extreme form of imprinting was discovered by K. Immelmann (1972) in the African and Australian estrildid finches. If an egg of species A is transferred to the nest of species B, and the chick is raised by species B parents, it becomes completely imprinted on species B. Later in life, when an imprinted adult male is put in a cage with a female of his own species (A), they will finally mate, after considerable reluctance. The female will lay eggs, the eggs will be incubated and hatch, and the male will participate in feeding the young. But if, at that moment, a female of species B is introduced into the cage, male A will at once abandon his own female and brood, pair up with the female of the species of his foster parents, help in the nest building, and raise the young if this happens not to be a sterile pair. Species recognition in these finches is clearly controlled by a largely open program, which is filled by imprinting and which governs the behavior from that point on. In many species of birds the target of highly stereotyped displays is "learned" by the imprinting process. In most lower animals, however, the sign stimuli characterizing the appropriate display partner are laid down in the closed genetic program.

An open program is by no means a tabula rasa; certain types of information are more easily incorporated than others. This is as true for imprinting as for ordinary learning. Two geographic races of the honey bee, *Apis mellifica ligustica* and *A. m. carnica,* agree in their capacity for sun orientation, but *A. m. carnica* has a much greater ability to learn the position of landmarks near the source of food than the other race. There is remarkably little variation among individuals of the same race whether or not they come from the same hive (Lauer and Lindauer 1971).

Whether natural selection will favor the evolution of an open or a closed program for a given behavior depends on the circumstances. For instance, the shorter the life span of an individual, the smaller the opportunity to learn from experience. In many species, particularly short-lived ones, mating may occur only a single time in the whole life of an individual, often within hours after metamorphosis or emergence from the pupa (particularly in the case of the female). There is no opportunity whatsoever for learning through experience; indeed the correct display—in the case of the female, the correct response to the display of the male—must be produced at once. This is possible only if the pattern of the display is laid down in a closed genetic program.

The situation is radically different in species with a more or less extended period of parental care. Here, the fixed responses of the newborn can be quite few in number, being limited primarily to adequate responses to the parents. More and more behavior patterns are added through maturation and will be appropriately invariant, provided the maturation takes place in the customary environment of individuals of the species. However, if, for example, the newly hatched gosling encounters as the first living being a human rather than the mother goose, it will become imprinted to this inappropriate parent object. Similar shifts are possible for choice of food, social companions, "recognized" enemies, and so forth. Obviously, numerous open slots exist in the behavior repertory of such species.

The longer the period of parental care, the more time will be available for learning, hence, the greater the opportunity to replace the closed genetic program by an open program. The great selective advantage of a capacity for learning is, of course, that it permits storing far more experiences, far more detailed information about the environment, than can be transmitted in the DNA of the fertilized zygote. Considering this great advantage of learning, it is rather curious in how relatively few phyletic lines genetically fixed behavior patterns have been replaced by the capacity for the storage of individually acquired information. Such a replacement is most conspicuous in the higher mammals, but it occurs also, to a more limited extent, in other groups of animals, for instance in certain types of birds, particularly geese, cranes, and parrots. It is possible that the mouthbreeding habit in the cichlid fishes contributes to "teaching" the young to become habituated to their neighborhood, which results in a strict localization of populations and has been one of the causes of the enormously rapid speciation in the cichlids.

I sometimes refer rather casually to behavior as largely controlled by a closed or an open genetic program. Actually deciding to what extent a particular behavior has genetic determinants is by no means easy, as indicated by the drawn-out controversies of the last 40 years. There are quite a few clues that aid in making the decision, but it would not be correct to say that any one of them is entirely conclusive. It might help, however, to list six kinds of evidence that are suggestive of genetic influences. A combination of several of these alternate kinds of evidence is usually available in any given case to help decide to what extent the genetic program is closed or open.

1. Parallelism between behavioral and structural phylogeny. As was first shown by Heinroth (1911) for ducks and by Whitman (1919) for pigeons, if one constructs the phylogeny of a group of animals on the basis of behavior traits, it is exceedingly similar to a phylogeny of the same group independently designed on the basis of strictly morphological characters. The simplest interpretation of this parallelism is that it is due to the fact that both sets of characters are the product of the same genotype.

2. Occurrence of behavior in naive individuals prior to any opportunity for acquiring experience. As far back as Reimarus (1760) students of instinct have listed innumerable cases where individuals, right after birth or hatching, or when completely isolated from all other members of the species, have performed the appropriate species-typical behavior. This includes such diverse behaviors as response to the characteristic sign stimulus of their parents, food-taking movements, courtship displays (often including vocalization), nest construction, migratory orientation, and often even choice of food.

3. Stereotypy of behavior. Many behaviors are extremely stereotyped. It is this, in part, which led to the concept of ritualization. The Mandarin drake (*Aix galericulata*), in one of his displays, performs a pseudo-preening movement across a greatly enlarged ornamental wing covert. When this feather is cut short, the bill is moved above this covert at exactly the place where the intact feather would be. The weaving movements of an orb-weaving spider or the nest-building movement of birds, particularly weaver birds (Figure 47–1), are other examples. I do not know of a single instance of a change in the form of such a behavior through conditioning (but see Barlow 1968).

4. Segregation of the behavior patterns of the parent species in species hybrids. The behavior of hybrids may be intermediate between those of the parent, or the pattern of one of the two species may dominate (von de Wall 1963; Frank 1970; Dilger 1962; Davies 1969). Genetic recombination may occasionally result in the emergence of entirely novel behavior in the hybrid.

5. Susceptibility to selection. Work, particularly in the genus *Drosophila*, has shown that certain components of behavior can be modified by selection (Sherwin and Spiess 1973). It has often been pointed out that tameness and other behavior characters of domestic animals are the result of selection, as are many of the behavioral differences between breeds, particularly in the dog (Fuller and Thompson 1960; Hirsch 1967). Where different races of the same species live in different climates, selected adaptation may result in racial behavior differences that are correlated with the climatic differences, as found for the honeybee (Lauer and Lindauer 1971).

6. The persistence of a behavior pattern after a change of conditions. Many "atavistic" behaviors have been described in the literature. In the

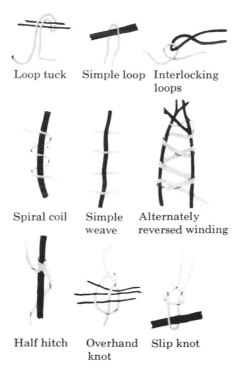

Loop tuck   Simple loop   Interlocking
                              loops

Spiral coil   Simple   Alternately
              weave   reversed winding

Half hitch   Overhand   Slip knot
             knot

Figure 47-1. Various common stitches and fastenings are used by the true weaver birds (Ploceinae). (From Collias and Collias 1964; reprinted by permission of The Regents of the University of California.)

three-spined stickleback (*Gasterosteus aculeatus*), a fish of circumpolar distribution, the male has a clear red abdomen during aggressive and courtship behavior. In one stream system on the Olympic Peninsula, however, courting males are jet black. As a defense against a black predatory fish, the black stickleback is estimated to have evolved about 8,000 years ago in a glacial lake, and since that time the females in this stream have been courted by black males. Is there now anything left of the old genetic program that made females respond to a red-bellied male? In an experiment made by McPhail (1969), the black female, when given a free choice between a black and a red-bellied male, chose the red male 5 out of 6 times after 8,000 years of "habituation" to black males! This strikes me as an extraordinarily long retention of an innate releasing mechanism in the absence of any reinforcement by natural selection.

CLASSIFICATION OF BEHAVIOR

The evolutionary biologist is concerned with the origin of genetic programs, with the elaboration of new genotypes, and with the selective value of adaptations. Applied to the study of animal behavior, these concerns lead to the posing of questions such as: How do behavior patterns change in the course of evolution and how do new behaviors arise? When does a change in behavior have a major impact on evolutionary trends? Or, more generally: What role does behavior play as a selection pressure (Tinbergen 1965)?

It is not easy to answer these questions. Since there is no direct fossil record of ancestral behavior, it is necessary to make use of indirect approaches. If evolution of behavior proceeds like the evolution of structural or molecular characters, then, according to the Darwinian interpretation, it must have two characteristics. First, in order to be able to respond to selection pressures, such behavior must at least in part have a genetic basis, and, second, the genetic basis must be somewhat variable, that is, it must be able to supply the material on which natural selection can act. Behavioral characteristics thus would share, whenever they evolve, the two most important aspects of evolving structural characters: variability and a genetic basis. These postulates permit us to apply to behavior characters the same methods of evolutionary analysis that we normally apply in the study of structural characters, such as the search for homologies and the determination of adaptive significance.

Studies of structural evolution are based primarily on two methodologies: comparison and classification. If we want to apply them to the evolutionary study of behavior, we are confronted by a major difficulty. Comparisons are biologically meaningful only if appropriate items are compared. In other words, comparisons must be preceded by sound classification, a guiding principle of comparative anatomy since the days of Cuvier. But it is difficult to apply this principle to the behavior field because no one quite knows how to classify different behaviors.

Neurologists and psychologists have tried for generations to classify types of behavior largely on a physiological basis—for example, reflexes, conditioned reflexes, tropisms, taxes, instinctive behavior, and learned behavior. But, as we have seen, even such simple abilities as species recognition or parent recognition in closely related families of birds are sometimes based on a closed program and sometimes on an open program. Heritability thus is obviously not a good classifying criterion. Similar objections can be raised against other previously proposed classifications because the criteria on which they are based either intergrade with each other or are unsuitable for other reasons. I am not aware of the availability of any other reasonably well-tested classification of types of behavior. When we adopt a purely functional approach we get such behavior categories as display, feeding, grooming, migration, exploration, communication, and other similar classes. These are disjunct groups for which no reasonable classification suggests itself. It would seem that we must experiment with an entirely new criterion of classification, and I have chosen as the discriminating criterion the potential response of the recipient of the behavior. Either the recipient is strictly passive, as in much of food-getting behavior or in habitat selection, or else the recipient is potentially capable of responding by behavior of its own. Behavior resulting, at least potentially, in an active response can be designated quite broadly as communicative behavior. For the sake of convenience it may be subdivided into behavior directed toward a member of one's own species and behavior toward an individual of another species. [A further category should probably be recognized for all behaviors of an individual toward itself, which would include grooming, bathing, all comfort movements, and similar activities. The validity of such a category remains to be studied (Heinroth 1930; Jander 1966).]

I will pose and attempt to answer two questions with respect to these three types of behavior, which I will call intraspecific, interspecific, and noncommunicative. How does selection affect the nature of the behavior program (favoring either a closed or an open genetic program)? and What role do the three kinds of behavior play in macroevolution? Do they differ in their effect on major trends in evolution and on the origin of major evolutionary inventions?

Intraspecific Behavior

Courtship displays between males and females have several functions, as ethologists have pointed out repeatedly, including the stimulation of the sex partner and the synchronization of the courtship of the two participants. Yet in recent years it has become quite evident that one of the most important functions of courtship displays is that they serve as isolating mechanisms. They prevent the female of one species from responding to the courting of the male of another species and thereby wasting his and her gametes in the production of inviable or inferior offspring. This explains why there is such a high selection pressure in favor of the perfecting of isolating mechanisms.

When we analyze courtship more closely we recognize that it consists invariably of an exchange of signals between the potential sex partners. Incorrect or imprecise signals may fail to evoke the desired response from the other sex partner. As a consequence, those signals will be favored among various competing ones that are most unmistakable, and stabilizing selection will constantly weed out all the more deviant, aberrant behaviors until a rigid and stereotyped ritual is achieved. This is one of the aspects of the process generally referred to as "ritualization," and is equally true for visual, acoustic, and chemical signals (Huxley 1966; Wiley 1973).

Let me now turn to another kind of intraspecific behavior—parent-offspring interactions in species with parental care. Both parents and offspring have various behavior components that are strictly species-specific and indispensable for the successful raising of the brood. In several species of wild pheasants, for instance, the downy young have uniquely characteristic color patterns on the crown and the back. If one places the eggs of the wrong species into the nest of a hen pheasant, the mother treats the chicks like alien intruders as soon as they hatch and kills them. In the turkey, the chicks have to utter a characteristic call. If they fail to do so, they are killed by the mother hen. If one deafens the hen, she kills her own young (W. Schleidt and M. Schleidt 1960). In many other species, and particularly in the domestic fowl, there is no instantaneous recognition of one's own young. Everyone is familiar with the picture of the mother hen leading a brood of little ducklings.

In the case of colonial birds and herd-forming mammals, a capacity for species recognition is not enough. There is a selective premium for the parents to be able to recognize their own young individually and for the young to recognize their parents. Such individual recognition is achieved in mammals soon after birth and in colonial birds within 2 to 5 days after hatching. Indeed, the young may learn the individuality of the calls of their parents before they have hatched from the egg (Gottlieb 1968). It is astonishing to us how unerringly the parents recognize their own young in big seabird colonies and in the large penguin colonies in the Antarctic.

Obviously the capacity for storing the individual characteristics of one's young or parents requires an elaborate genetic program which is sufficiently open to permit the insertion of this special kind of information. Owing to its uniqueness and to the unpredictable components of genetic recombination, it would be impossible to code such information in a closed genetic program. It must be acquired after conception.

A third type of intraspecific behavior is the behavior between males of the same species; many manifestations of hostility and various threat behaviors during territory defense may differ strikingly or subtly in closely related species. Much of this signaling and fighting seems to be controlled by closed genetic programs. Students of African antelopes have observed that members of each species have a somewhat different way of using their heads and horns in intraspecific hostilities. The fighting often becomes a sham ritual, and various kinds of bluffing often replace actual combat.

Displays that tend to weaken the aggressive spirit of the opponent are widespread in the animal world. This is particularly true of carnivores with their powerful teeth. When an individual (primarily within a group) loses a fight, he tends to adopt a genetically fixed, or largely fixed, highly species-specific type of submission behavior or appeasement ceremony. In canids, for instance, throwing oneself on the back and showing the abdomen sometimes completely immobilizes the aggressor.[1] As Moynihan (1962) and Tinbergen (1959) have shown,

1. Lorenz (1943) originally assumed that submission behavior functioned to terminate serious fights, particularly in carnivores. This has not been confirmed by subsequent studies. Fights among male carnivores, primarily when they do not belong to the same social group (pack), can end with the killing of the vanquished, both in felids (e.g., lions) and canids. However, submission behavior within social groups helps to strengthen social integration (Schenkel 1967; Mech 1970).

black-headed gulls turn away the front part of the head with its sharp bill and intimidating brown mask, and show the back, a display which also remarkably reduces hostility (in this case, particularly between the sexes). The important point to realize is that hostile behavior among members of the same species contains numerous ritualized signaling displays, most of which are controlled by the closed genetic program of these species.

Interspecific Behavior

This type of behavior comprises the vast class of behavioral interactions between different species. It includes such heterogeneous phenomena as mixed-species social aggregations (Moynihan 1968), symbiosis, parasitism, and all kinds of predator-prey relationships that result in adaptations in the searching behavior of the predator and in predator-thwarting behavior of the prey. The behavior of the prey toward the predator clearly is interspecific behavior, while the predator very often treats the prey as if it were simply part of the inanimate substrate. Interspecific behavior consists to a very large extent of specialized signals exchanged between the two species, whether the interaction is beneficial (e.g., cleaning fishes) or inimical (e.g., predator-prey interaction). These signals must be unequivocally understood by the recipients, or else they must clearly "fool" the predator, as do the "broken-wing display" of many birds with young, various warning displays, and mimicry. In other words, they must be able to evoke a predictable reaction in the recipient (Blest 1957; Wickler 1968). They are therefore usually highly stereotyped—that is, the programs controlling them are largely closed. The young individual in species without parental care has little opportunity to learn, but must have a ready answer available for the most important encounters with other sympatric species. One special class of potential interspecific behavior interactions—courting by individuals of closely related species—has already been mentioned. To prevent interspecific courtship is the function of behavioral isolating mechanisms, and these almost invariably are controlled by a largely closed genetic program.

Noncommunicative Behavior

The number of components in the environment toward which an individual may react is very large. Among the behaviors that are particularly important in higher organisms, the two which students of behavior have studied most often are food selection and habitat selection. Yet, the study of these two kinds of behavior has, on the whole, been rather neglected by ethologists, perhaps because they are so different from intraspecific behavior. For instance, such behavior does not consist of signals, because there is no responding partner, no recipient of the behavior who would answer. For this reason there is no selection pressure in favor of well-defined, clear-cut displays. Furthermore, such behavior is usually controlled by a largely open program and is therefore rather variable. For many species it is of considerable selective advantage to retain extensive flexibility toward components of the environment. Food sources come and go and so do competitors. Habitats change and an individual will encounter different substrates. Phenotypic flexibility rather than genetic precision is at a selective premium under these circumstances. Recent research has, indeed, substantiated that much of the behavior toward the environment, particularly among higher organisms, is modifiable by learning and habituation.

Salmon fry, for instance, are imprinted to the stream to which they will eventually return for spawning (Hasler 1966). It has been demonstrated for many species of insects, turtles, and birds that as adults they may prefer the food on which they were raised or which they first encountered when young (Newton 1967). Young birds generally try out a much wider variety of foods than those to which they eventually confine themselves. Such a trial-and-error period may be a typical stage in the growing up of the young in many species of animals (de Ruiter 1967). Whenever a novel habit was acquired by troops of Japanese macaques, it was invariably a young individual that took the lead (Kawai 1965).

One may generalize these observations by stating that in no other group of behavior phenomena does the genetic program seem so often wide open as in noncommunicative behavior. However, there are genetic components even in habitat and food selection. Indeed, among many species of host-specific insects, parasites, and many other lower invertebrates, food choice seems to be rigidly determined by a closed genetic program. This is even true for certain species of birds and mammals. One need only think of the panda (bamboo) and the koala (a few species of eucalyptus); these species have surely abandoned almost all flexibility in food choice. Likewise, in the case of habitat selection, observations and experiments have revealed the presence of a considerable amount of genetic determination.

One must make another restriction. The object—e.g., food, habitat—of the behavior is often acquired through experience, but the executive locomotion (*fixed action pattern* of the ethologists) which the object elicits may be rigidly determined genetically. This is particularly conspicuous in the case of construction behavior (e.g., hexagonal cells of the beehive, silkworm chrysalis, spider webs, bird nests), where extremely stereotyped and precise motions are evidently controlled by a closed program (Collias 1964). Equally invariant may be—but is not always—the choice of the material used for the construction.

MACROEVOLUTIONARY CONSEQUENCES

The first studies of ethologists that dealt with macroevolutionary aspects of behavior were limited almost exclusively to a reconstruction of the phylogeny of behavior patterns. More recently, however, there has been increasing interest in the evolutionary consequences of shifts in behavior. It is quite evident that a change in behavior that results in the choice of a new habitat or a new kind of food will set up new selection pressures and may even permit the invasion of a new niche or adaptive zone (Essay 9). But what is the relative frequency of behavioral shifts with macroevolutionary consequences in the stated behavior categories?

Intraspecific communicative behavior, at least so far as courtship and male aggression behavior is concerned, has, on the whole, little impact on major evolutionary developments. Of greater importance is parental behavior, which, if developed into long-continued parental care, permits the replacement and supplementation of closed by open genetic programs. The origin of man, for instance, is intimately related to this development. Another exception is sociality. Degrees and kind of social behavior have been of major evolutionary significance, as for instance in the case of social insects and higher primates. Rather little thought has been given, up to now, to the nature of the changes in the genetic program which have accompanied the acquisition of such sociality. Communicative interspecific behavior, such as in predation, in cleaning relationships (fishes), and in mixed herds and flocks, seems to be of relatively low macroevolutionary significance.

Noncommunicative behavioral shifts in the utilization of the animate and inanimate environment are by far the most important factors in macroevolution. They are involved in all major adaptive radiations and in the development of all major evolutionary novelties (Essay 9). New structures that evolve under the new selection pressures may, in turn, permit the development of new behaviors. Virtually all orders of mammals, such as the ungulates (all herbivores) and carnivores, owe their origin to the invasion of, or restriction to, a new food niche. Bats are normally insect eaters, but the megachiropterans feed on fruit, and even among the microchiropterans there are a few specialists, such as the vampire bats, which suck blood, and certain fish-eating bats. It is quite probable that shifts in habitat and food niche have played a decisive role in the evolution of our own hominid ancestors. Here the shift from a prevailingly vegetarian diet to the hunting of large ungulates surely exerted entirely new selection pressures.

Shifts in locomotory habits were key events in the origin of the land-living vertebrates and in the conquest of the air space by insects, pterodactyls, birds, and bats. The habit of certain fishes living in stagnant waters of rising to the surface to gulp air probably produced the selection pressure which eventually led to the evolution of lungs. A capacity for behavioral shifts was a prerequisite for the acquisition of these evolutionary novelties. Those individuals had the greatest evolutionary potential who were able to undergo the most rapid adjustment to changes in the environment or to adopt a new way of life.

In the beginning I asked the question: Under what circumstances is a closed genetic program favored and under what others an open one?

The answer is now quite clear. Since much of the behavior directed toward other conspecific individuals consists of formal signals and of appropriate responses to signals, and since there is a high selective premium for these signals to be unmistakable, the essential components of the phenotype of such signals must show low variability and must be largely controlled genetically. In other words, the genetic program for formal signaling must be essentially closed; to state it more generally, selection should favor the evolution of a closed program when there is a reliable relationship between a stimulus and only one correct response. This conclusion is, on the whole, confirmed by the observed facts.

By contrast, noncommunicative behavior leading to an exploitation of natural resources should be flexible, permitting an opportunistic adjustment to rapid changes in the environment and also permitting an enlargement of the niche as well as a shift into a new niche. Such flexibility would be impossible if such behavior were too rigidly determined genetically. Again, this is largely confirmed by our observations.

The longer the life span of an individual, the greater will be the selective premium on replacing or supplementing closed genetic programs by open ones. In the most primitive organisms we find that most behavior is genetically fixed and largely predictable. The direction of many evolutionary pathways, thus, is clear. It often leads to a gradual opening up of the genetic program, permitting the incorporation of personally acquired information to an ever-greater extent. There are two prerequisites for this to happen. Storage of personally acquired information requires a far greater storage capacity than is needed for the carefully selected information of a closed genetic program; in other words, it requires a larger central nervous system. Indeed it has long been known that brain size and "intelligence"—defined as the ability to learn from experience—are closely correlated. A subsidiary factor favoring the development of an open program is prolonged parental care. When the young of a species grow up under the guidance of their parents, they have a long period of opportunity to learn from them—to fill their open programs with useful information on enemies, food, shelter, and other important components of their immediate environment.

Much of the recent controversy in the literature on animal behavior can be better understood now that we are aware of the important differences between behaviors controlled by closed and by open genetic programs. Ethologists have been primarily interested in species-specific signals and in their evolution. Comparison of different species has been of great concern to them. The classical experimental psychologists, who were principally interested in the neurophysiological and developmental aspects of behavior, almost invariably worked only with a single species (Beach 1950). Their primary interest was in learning, conditioning, and other modifications of behavior. They approached behavior with the interests of the physiologist, and the phenomena they studied were, to a large extent, aspects of noncommunicative behavior, such as maze running or food selection.

No wonder the conclusions of the two groups of investigators were so different. The descriptions of animal behavior by the two schools were like the proverbial blind men's description of the elephant. Such one-sidedness is inevitable as long as one treats behavior typologically— that is, as long as one considers it a unitary phenomenon. My analysis has shown, however, how different various kinds of behavior can be, if they are classified according to the target of the behavior and to the nature of the selection pressure to which they are exposed.

One could classify behavior also according to the role of the individual in the behavior act: the individual can be either the target of the behavior (cognitive aspects) or the actor of the behavior (executive aspects). The work of ethologists has shown that displays (signals) are much more often (and more rigidly) genetically determined than the cognitive aspects. The imprinting studies in particular have shown that the "object" of a behavior act (parent, young, enemy, food) is often contained only very vaguely in an open genetic program and that additional detail—often a great deal of it—is added to the open program through subsequent experience. In these cases the recipient *learns* what the stimulus is to which he is to respond. Behavior is also strongly affected by many components of the life cycle, such as life span, kind of pair formation, migratory habits, and reproductive strategies. Each of these is apt to exert a different selection pressure on the various components of behavior.

It should be obvious from this analysis how manifold the relations are between behavior and evolution. Not only does behavior set up diverse selection pressures and thus become an important initiating force in evolution, but also, in turn, it is itself molded by evolution. This evolutionary change of behavior provides material for the reconstruction of phylogenies of species-specific behavior patterns and shows what a significant portion of the phenotype behavior is. Behavior constantly interacts with both the living and the inanimate environment and is thus constantly the target of natural selection. In order to provide the optimal response to these pressures, it is sometimes advantageous for the genetic program governing the behavior to be largely closed, while in other behavior interactions and in other types of organisms an open behavior program is favored by selection. There is a wide and largely unexplored field of research to determine the selective advantages of various possible options in different organisms and under different conditions.

REFERENCES

Alcock, J. 1973. The feeding response of hand-reared red-winged blackbirds (*Agelaius phoeniceus*) to a stinkbug (*Euschistus conspersus*). *Amer. Midl. Nat.*, 89:307–313.

Barlow, G. 1968. Ethological units of behavior. In *The central nervous system and fish behavior*, ed. D. Ingle, pp. 217–232. University of Chicago Press, Chicago.

Beach, F. 1950. The snark was a boojum. *Amer. Psych.*, 5:115–124.

Blest, D. 1957. The evolution of protective displays in the Saturnioidea and Sphingidae. *Behaviour*, 11:257–309.

Brown, R. G. B. 1965. Courtship behavior in the *Drosophila obscura* group II. *Behaviour*, 25:281–323.

Caspari, E., ed. 1964. Refresher course on behavior genetics. *Amer. Zool.*, 4:97–173.

Collias, N. E., et al. 1964. The evolution of external construction in animals. *Amer. Zool.*, 4:175–243.

Collias, N. E., and E. C. Collias. 1964. Evolution of nest-building in the weaverbirds (*Ploceinae*). *Univ. Cal. Publ. Zool.*, vol. 73.

Condillac, E. B. de. 1755. *Traité des animaux*. De Bure, Amsterdam.

Cooper, J. C., and A. D. Hasler. 1974. Electroencephalographic evidence for retention of olfactory cues in homing coho salmon. *Science*, 183:336–338.

Cullen, J. J. 1959. Behavior as a help in taxonomy. *Syst. Assoc. Publ.*, no. 3:131–140.

Davies, S. J. J. F. 1969. Patterns of inheritance in the bowing display and associated behaviour of some hybrid Streptopelia doves. *Behaviour*, 26:187–214.

Delacour, J., and E. Mayr. 1945. The family Anatidae. *Wils. Bull.*, 57:3–54.

Dilger, W. C. 1962. Behavior and genetics. In *Roots of behavior*, ed. E. I. Bliss. Harper & Bros., New York.

Frank, D. 1970. Verhaltensgenetische Untersuchungen an Arbastarden der Gattung Xiphophorus. *Z. Tierpsychol.*, 27:1–34.

Fuller, J., and W. R. Thompson. 1960. *Behavior genetics*. Wiley, New York.

Gottlieb, G. 1968. Prenatal behavior of birds. *Quart. Rev. Biol.*, 43:148–174.

Hasler, A. D. 1966. *Underwater guideposts: homing of salmon*. University of Wisconsin Press, Madison.

Heinroth, O. 1911. Beiträge zur Biologie, namentlich Ethologie und Psychologie der Anatiden. *Verh. 5th Int. Ornith. Kongr.* (Berlin 1910), pp. 589–702.

—— 1930. Über bestimmte Bewegungsweisen bei Wirbeltieren. *Sitz.-Ber. Ges. Naturf. Freunde* (1929), pp. 333–342.

Hess, E. H. 1973. *Imprinting: early experience and the developmental psychobiology of attachment*. Van Nostrand Reinhold, New York.

Hirsch, J., ed. 1967. *Behavior: genetic analysis*. McGraw-Hill, New York.

Huxley, J. S., et al. 1966. A discussion on ritualization of behavior in animals and man. *Phil. Trans. Roy. Soc. London*, ser. B, 251:247–256.

Immelmann, K. 1969. Ökologische und stammesgeschichtliche Betrachtungen zum Prägungsphänomen. *Zool. Anz.*, 183:1–12.

—— 1972. The influence of early experience upon the development of social behaviour in estrildine finches. *Proc. 15th Int. Ornith. Congr.*, pp. 316–338.

Itani, J. 1958. On the acquisition and propagation of a new food habit in the troop of Japanese monkeys at Takasakiyama. *Primates*, 1:131–148.

Jander, U. 1966. Untersuchungen zur Stammesgeschichte von Putzbewegungen von Tracheaten. *Z. Tierpsychol.*, 23:799–844.

Kahl, P. 1971. Social behavior and taxonomic relationships of the storks. *Living Bird*, 10:151–170.

Kawai, M. 1965. Newly acquired precultural behavior of the natural troop of Japanese monkeys on Koshima island. *Primates*, 8:35–74.

Kawamura, S. 1963. The process of subcultural propagation among Japanese macaques. In *Primate social behavior*, ed. C. H. Southwick. Van Nostrand, Toronto.

Konishi, M. 1971. Ethology and neurobiology. *Amer. Sci.,* 59:56-63.

Lauer, J., and M. Lindauer. 1971. Genetisch fixierte Lerndisposition bei der Honig-biene. *Akad. Wiss. Lit. Mainz. Math. nat. Kl.,* 1:1-87.

Lockard, R. B. 1971. Reflections on the fall of comparative psychology: is there a message for us all? *Amer. Psych.,* 26:168-179.

Lorenz, K. 1941. Vergleichende Bewegungsstudien an Anatiden. *Jour. f. Ornith.,* 3:194-293.

—— 1943. Die angeborenen Formen der möglichen Erfahrung. *Z. Tierpsychol.,* 5:235-409.

McDougall, W. 1912. *Psychology: the study of behavior.* Methuen, London.

McPhail, J. D. 1969. Predation and the evolution of a stickleback (*Gasterosteus*). *J. Fish. Res. Board Canada,* 26:183-208.

Mech, L. D. 1970. *The wolf.* Natural History Press, Garden City, N.Y.

Morgan, L. 1891. *Animal life and intelligence.* Edward Arnold, London.

Moynihan, M. 1962. *Hostile and sexual behavior patterns of South American and Pacific Laridae. Behaviour,* suppl. 8.

—— 1968. Social mimicry: character convergence versus character displacement. *Evolution,* 22:315-331.

Newton, I. 1967. The adaptive radiation and feeding ecology of some British finches. *Ibis,* 109:33-98.

Reimarus, H. S. 1760. *Allgemeine Betrachtungen über die Triebe der Thiere, hauptsächlich über ihre Kunsttriebe* . . . Bohn, Hamburg.

Ruiter, L. de. 1967. Feeding behavior of vertebrates in the natural environment. In *Handbook of Physiol. Alim. Canal,* chap. 7.

Schenkel, R. 1967. Submission: its features and function in the wolf and dog. *Amer. Zool.,* 7:319-329.

Schleidt, W., and M. Schleidt, 1960. Störung der Mutter-Kind Bezichung bei Trut-hühnern durch Gehörverlust. *Behaviour,* 16:3-4.

Scholz, A. T., et al. 1973. Olfactory imprinting in coho salmon. *Proc. 16th Conf. Great Lakes Res.,* pp. 143-153.

Sherwin, R. N., and E. B. Spiess. 1973. Chromosomal control of mating activity in *Drosophila pseudoobscura. Proc. Nat. Acad. Sci.,* 70:459-461.

Spieth, H. T. 1952. Mating behavior within the genus *Drosophila. Bull. Amer. Mus. Nat. Hist.,* 99:395-474.

Tinbergen, N. 1959. Comparative studies of the behaviour of gulls (Laridae): a progress report. *Behaviour,* 15:1-70.

—— 1965. Behavior and natural selection. In *Ideas in modern biology,* ed. J. A. Moore, pp. 521-545. Natural History Press, Garden City, N.Y.

Wall, W. von de. 1963. Bewegungsstudien an Anatiden. *Jour. f. Ornith.,* 104:1-15.

Whitman, C. O. 1919. *The behavior of pigeons.* Carnegie Inst. Washington, publ. 257.

Wickler, W. 1968. *Mimicry in plants and animals.* World University Library (McGraw-Hill), New York.

—— 1971. Ökologie und Stammesgeschichte von Verhaltensweisen. *Fortschritte Zool.,* 13:303-365.

Wiley, R. H. 1973. The strut display of male sage grouse: a "fixed" action pattern. *Behaviour,* 47:129-152.

Index